D0706766

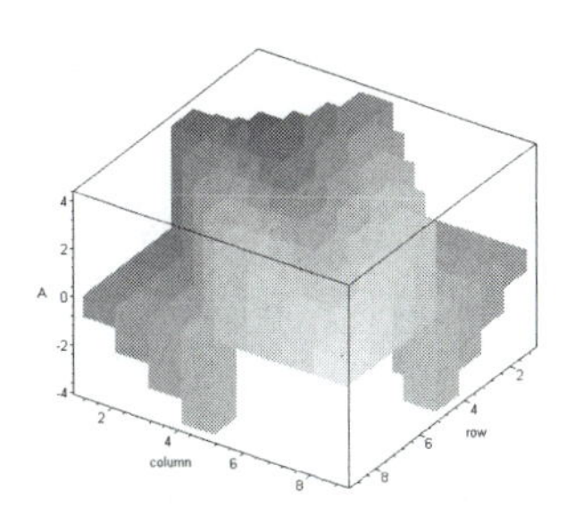

LINEAR ALGEBRA

An Introduction Using *Maple*®

Fred Szabo

Department of Mathematics and Statistics
Concordia University

A Harcourt Science and Technology Company

San Diego San Francisco New York Boston
London Toronto Sydney Tokyo

Sponsoring Editor: Barbara Holland
Production Editor: Angela Dooley
Cover Design: Dick Hannus/Hannus Design Assoc.
Copyeditor: Elliot Simon
Composition: Integre Technical Publishing Co., Inc.
Printer: Sheridan Books

This book is printed on acid-free paper. ∞

Academic Press
A Harcourt Science and Technology Company
525 B Street, Suite 1900, San Diego, CA 92101-4495, USA
http://www.academicpress.com

Academic Press
Harcourt Place, 32 Jamestown Road, London NW1 7BY, UK
http://www.academicpress.com

Harcourt/Academic Press
A Harcourt Science and Technology Company
200 Wheeler Road, Burlington, Massachusetts 01803, USA
http://www.harcourt-ap.com

Library of Congress Catalog Card Number: 2001090602
International Standard Book Number: 0-12-680140-1

Printed in the United States of America
01 02 03 04 05 06 SB 9 8 7 6 5 4 3 2 1

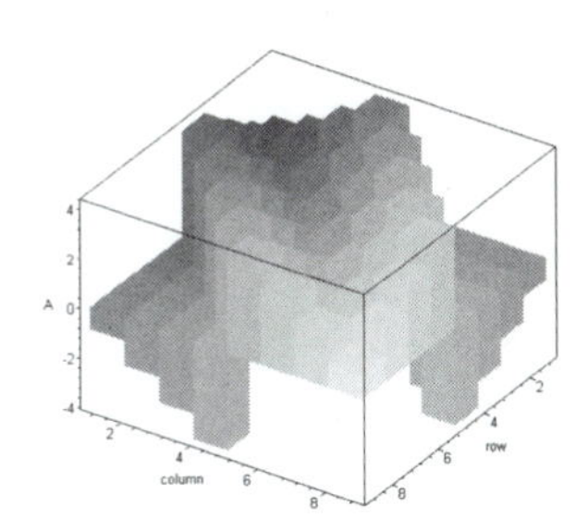

About the Cover

The cover image is a histogram representing the sum of a Hilbert and a Toeplitz matrix. Both types of matrices arise in theoretical and practical aspects of linear algebra. The image illustrates the three-dimensional graphic features of *Maple* and is generated with the following sequence of commands:

```
[> with(plots):
[> with(linalg):
[> A:=hilbert(8):
[> B:=toeplitz([1,2,3,4,-4,-3,-2,-1]):
[> matrixplot(A+B,heights=histogram,axes=frame,gap=0.25,
    style=patchnogrid);
```

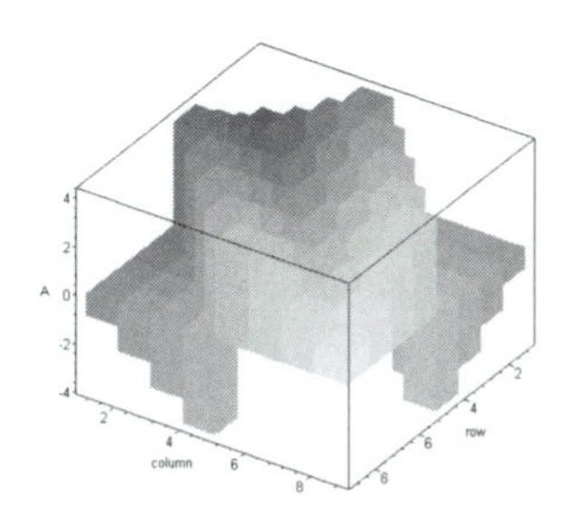

Contents

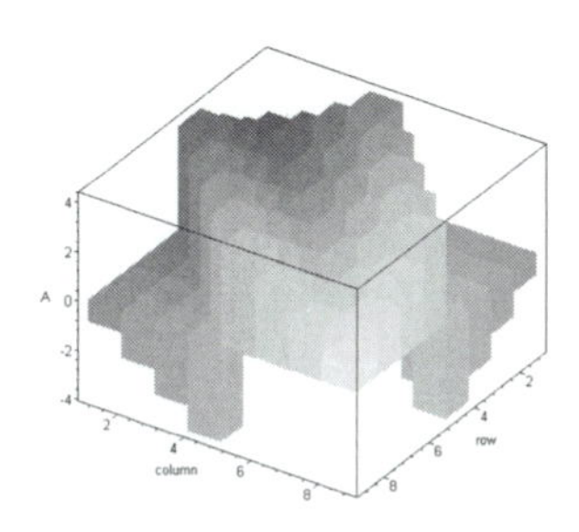

PREFACE

This text is meant for a first undergraduate course in linear algebra. It is written for college and university students who have completed two semesters of college-level mathematics. Experience with calculus is helpful for some of the applications but is not essential for most of the topics covered. The text is designed for the learner. Its primary goal is for students to have a positive and fruitful experience in their first encounter with linear algebra.

Students majoring in mathematics and client disciplines, such as computer science, engineering, physics, chemistry, economics, statistics, and actuarial mathematics, should be able to take this course. The presentation is matrix-based, and the material includes the standard topics for a first course recommended by the Linear Algebra Curriculum Study Group.

The aim of the book is to make linear algebra accessible to college majors through a narrative presentation of the material, enriched by opportunities for interactive teaching and learning using the *Maple* computer algebra system. The text provides a gentle, step-by-step introduction to technology. It contains routine *Maple* examples and exercises to build up the students' confidence and experience with the technology. This gradual approach has been effective among my students, who often have varying degrees of proficiency with technology when they begin the course.

We present linear algebra as a story. The theme is the solution of linear systems, and the bookends for the course are Gaussian elimination and the singular value decomposition. The text begins with the problem of finding the exact solutions of a linear system $A\mathbf{x} = \mathbf{b}$, using elementary row operations. It ends with the solution $\widehat{\mathbf{x}} = A^{+}\mathbf{b}$, where A^{+} is the pseudoinverse of A, derived from the singular value decomposition of the matrix A. If A is invertible, then $\widehat{\mathbf{x}} = A^{+}\mathbf{b}$ is the exact solution $A^{-1}\mathbf{b}$. If A is not invertible, then $\widehat{\mathbf{x}}$ is a best approximation. Most of the material in between leads up to and is unified in the singular value decomposition.

TO THE TEACHER

This text can be used in at least two different ways. It can serve as the main text for a traditional course in linear algebra, enriched and facilitated using *Maple V* or *Maple 6*. It can also be used as the textbook for a linear algebra course in which technology is fully integrated from the very beginning. In both cases, the *Maple computer algebra system* can serve as a conceptual motivator and a computational facilitator. Since matrices are central to the course, for example, array routines are frequently used. However, these routines not only serve to create matrices. They also motivate students to think about functions and functional expressions. Furthermore, the dual computational power of *Maple*, with its distinction between symbolic and numerical calculations, helps lay the foundation for an appreciation of the difference between exact and approximate solutions of mathematical problems. As such, it gives concrete meaning to the leading problem around which this course is built: the problem of solving linear systems.

One of the advantages of using technology in the teaching of linear algebra is that it eliminates the arithmetical tedium associated with certain aspects of the subject. Students are able to tackle realistic problems and can be sure of the accuracy of their solutions. Furthermore, technology provides the opportunity to engage in modest forms of research so that individuals as well as groups can work on meaningful projects. This book is not, however, about programming or computational complexity. Most basic exercises can be solved by hand, and all exercises can be solved with the functions built into *Maple*, supplemented with the additional routines supplied in the packages specified in Appendix E.

In many ways it can be argued that computers play an essential role not only in using linear algebra, but also in learning it. Many algorithms used in linear algebra, such as Gaussian elimination by elementary row operations, calculation of determinants by Laplace expansions, and matrix inversion using elementary matrices, are arithmetically simple but space and time intensive. Once the algorithms have been mastered, there is little gain in spending extended periods of time on manual manipulations at the expense of learning other material. Moreover, certain key objects of linear algebra, such as eigenvalues, are algorithmically inaccessible in all but the simplest cases.

In this text, we are aiming to strike a balance between the time spent on the conceptual aspects of the subject and that spent on the computational aspects. At every step along the way, *Maple* serves to remove arithmetical obstacles, to build confidence, to illustrate concepts and constructions, and to facilitate teaching and learning. Experience in the classroom suggests that this approach makes linear algebra more accessible and the learning and teaching of it more enjoyable and rewarding. *Maple* is an easy tool for exploring the consequences of changes in assumptions and data, and is ideal for practicing precision and rigor. Many topics of current interest, such as large matrices and singular value decompositions, are impossible to learn and teach without technology.

Maple is an integrated electronic environment in which mathematics can be studied, discovered, and presented. It provides a wonderful framework for computer-assisted symbolic and numerical computation. While this textbook takes advantage of this technology, most of the material is not dependent on it. *Maple* routines can usually be converted into manual steps. Moreover, the details of the presented calculations are sufficient to allow students to rework *Maple* examples by hand, if such practice is needed. In most cases, it is left to the instructor to decide where exercises should be done with pencil and paper and where with a computer. At times, both methods are appropriate. Here are some examples.

Mathematics with *Maple*

The mathematics in this text is *Maple*-enhanced but not *Maple*-dependent. Usually, *Maple* is called upon to do the donkey work when no pedagogical value is derived from spending time on routine calculations.

On other occasions, *Maple* is actually used to carry out computations that are impossible to do by hand. Good examples are the calculation of eigenvalues for matrices 5×5 and larger, determinants, characteristic polynomials and inverses of large matrices, LU and QR decompositions, the orthogonalization of large bases, orthogonal projections in spaces of high dimension, and so on.

Sometimes *Maple* also helps with the organization of a piece of mathematics that involves precise but complex information. The use of quaternions to calculate three-dimensional rotations is conceptually simple but computationally complex. In *Maple*, it easy to set up the basic ingredients required for the calculations. In most of these uses of *Maple*, students must know what they are doing before they are able to let *Maple* carry out the calculations. *Maple* assists but does not justify. This is so since definitions and theorems in this text are totally independent of the software in which they are illustrated, studied, or explored.

There are occasions, of course, when *Maple* can be made to lead the way. *Maple* routines can usually be executed using implemented functions and examples in the Help file, without any understanding of what is going on. However, a judicious use of this feature also creates an environment for discovery through pattern recognition and reflection. One of the topics of importance in linear algebra is the distinction between exact and approximate computation. Overdetermined linear systems without exact solutions are among many linear algebra topics that have assumed critical importance because of the advent of technology. Since the roots of polynomials of degree five or higher cannot in general be found exactly, the eigenvalues of large matrices cannot be found exactly either. Hence an understanding of what is involved in approximating these values is crucial to a reliable use of linear algebra techniques in practice.

If we compare the details in traditional examples in linear algebra texts with those presented here with *Maple*, we can safely say that in many cases, the *Maple*-based solutions are as informative for students as the traditional examples. Here is an illustration. Let us consider how a *Maple* calculation of the eigenvalues of a 3×3 real matrix may differ from a traditional pencil-and-paper calculation provided in a standard textbook.

Suppose we are asked to find the eigenvalues and some corresponding eigenvectors of the matrix

$$A = \begin{bmatrix} 14 & 0 & -3 \\ 54 & 5 & -18 \\ 45 & 0 & -10 \end{bmatrix}$$

in three ways: using the **linalg** package of *Maple V*, using the **LinearAlgebra** package of *Maple 6*, and using the traditional methods of pencil and paper. What is the difference between the three methods?

Case 1. (*Maple V*) We load the **linalg** package, define the matrix A as an object of type **matrix**, and apply the **eigenvectors** function to the matrix A.

```
[> with(linalg):
[> A:=matrix([[14,0,-3],[54,5,-18],[45,0,-10]]):
```

```
[> evA:=eigenvectors(A);
```

$$evA := [-1, 1, \{[1, 6, 5]\}], [5, 2, \{[1, 0, 3], [0, 1, 0]\}]$$

From the *Maple* output we can tell that the eigenvalues of the matrix A are -1 and 5, that the vector $(1, 6, 5)$ is an eigenvector of A belonging to the eigenvalue -1, and that the vectors $(1, 0, 3)$ and $(0, 1, 0)$ are eigenvalues of A belonging to the eigenvalue 5.

Case 2. (*Maple 6*) We load the **LinearAlgebra** package, define the matrix A as an object of type **Matrix**, and apply the **Eigenvectors** function to the matrix A.

The dialog

```
[> with(LinearAlgebra):
[> A:=Matrix([[14,0,-3],[54,5,-18],[45,0,-10]]);
```

$$A := \begin{bmatrix} 14 & 0 & -3 \\ 54 & 5 & -18 \\ 45 & 0 & -10 \end{bmatrix}$$

```
[> evA:=Eigenvectors(A);
```

$$evA := \begin{bmatrix} -1 \\ 5 \\ 5 \end{bmatrix}, \begin{bmatrix} 1 & 1 & 0 \\ 6 & 0 & 1 \\ 5 & 3 & 0 \end{bmatrix}$$

produces a comparable answer, in different notation. The column vector on the left lists the eigenvalues of A, and the columns of the matrix on the right are the corresponding eigenvectors.

The matrix A in *Maple V* is an object of a different type from the matrix A in *Maple 6*. The objects `matrix(A)` and `Matrix(A)` cannot, in general, be used as arguments of the same functions. Since linear algebra is a subject that, by its very nature, is built on two types of objects, scalars and vectors, there is some pedagogical value in being forced to pay attention to the type of an object.

Case 3. (Pencil and Paper) How do we solve this problem with pencil and paper? We calculate the eigenvalues of A by finding the roots of the polynomial $(-1)^3 \det(A - \lambda I)$. We then calculate eigenvectors $\mathbf{x}$ of A belonging to an eigenvalue λ by solving the homogeneous linear system $(A - \lambda I)\mathbf{x} = \mathbf{0}$.

Step 1. We form the characteristic matrix

$$A - \lambda I = \begin{bmatrix} 14 & 0 & -3 \\ 54 & 5 & -18 \\ 45 & 0 & -10 \end{bmatrix} - \lambda \begin{bmatrix} 1 & 0 & 0 \\ 0 & 1 & 0 \\ 0 & 0 & 1 \end{bmatrix} = \begin{bmatrix} 14-\lambda & 0 & -3 \\ 54 & 5-\lambda & -18 \\ 45 & 0 & -10-\lambda \end{bmatrix}$$

Step 2. We calculate the characteristic polynomial.

$$\begin{aligned} (-1)^3 \det(A - \lambda I) &= (-1)^3 \det \begin{bmatrix} 14-\lambda & 0 & -3 \\ 54 & 5-\lambda & -18 \\ 45 & 0 & -10-\lambda \end{bmatrix} \\ &= \lambda^3 - 9\lambda^2 + 15\lambda + 25 \end{aligned}$$

Step 3. We factor the characteristic polynomial, if possible.

$$\lambda^3 - 9\lambda^2 + 15\lambda + 25 = (\lambda + 1)(\lambda - 5)^2$$

Step 4. We read off the eigenvalues -1 and 5.

Step 5. We solve the homogeneous linear system $(A - 5I)\mathbf{x} = \mathbf{0}$ and $(A + I)\mathbf{x} = \mathbf{0}$. For $\lambda = 5$, we solve

$$\left\{ \begin{aligned} 14x - 3z - 5x &= 0 \\ 54x + 5y - 18z - 5y &= 0 \\ 45x - 10z - 5z &= 0 \end{aligned} \right.$$

and find that all nonzero vectors of the form $(x, y, 3x)$ are eigenvectors belonging to $\lambda = 5$. Similarly we find that all nonzero vectors of the form $(x, 6x, 5x)$ are eigenvectors belonging to $\lambda = -1$.

Thus students need to know the definitions of eigenvalues and eigenvectors and need to be familiar with the process of setting up and solving equations of the form $\det(A - \lambda I) = 0$ and $(A - \lambda I)\mathbf{x} = \mathbf{0}$. A traditional textbook solution of this problem usually provides the solutions of these equations and refers to other sources for explanations of how the solutions can be found. In many applications the next step consists of finding enough linearly independent eigenvectors to diagonalize the matrix A, if possible. Finding such vectors by pencil and paper is often laborious and not necessarily instructive. *Maple* provides them immediately.

It is certainly true that when using *Maple*, students do not need to know this procedure. They do not have to be familiar with the definitions of eigenvalues and eigenvectors to be able to compute them. When using the traditional method, they must know how determinants, characteristic matrices, and homogeneous linear systems are involved. However, these ideas can only be rehearsed with very specific examples where the arithmetic is kept simple and the matrices are small. The time involved in producing solutions for examples of even relatively small dimension can be horrendous. By using *Maple* for the calculations, many hours of study time can be freed up and used to explore some of the properties of eigenvalues and eigenvectors. The time saved can be spent on such instructive activities as exploring of the geometric meaning of eigenvalues and eigenvectors. For example, the graphing power of *Maple* in two and three dimensions can be used to acquire a deeper geometric insight into the role of eigenvalues and eigenvectors in mathematics. The principal axis theorem, for example, can be graphically illustrated, the connection between eigenvalues and maximum stretch can be geometrically discussed, and so on. In this way, *Maple* facilitates a pedagogically fruitful reallocation of teaching and learning resources.

The given *Maple* solutions of many examples in this text provide a complete list of steps required for the solution of a problem. They therefore encourage students to develop structural thinking. What are we talking about? What concepts are involved, what definitions? What mathematical algorithms and procedures are required to solve a problem? After all, the names of built-in functions may suggest the right tool, but their syntax usually requires a deeper understanding of the properties and limitations of these functions. Here again, analytical skills come into play.

Experience shows that learning linear algebra with the help of technology is intellectually rewarding and pedagogically sound, as long as we keep focused on the fact that technology is a means to an end, not the end itself. As students work their way through this thematically organized text, they will acquire both a solid, theoretical, *Maple*-independent knowledge of linear algebra as well as a collection of powerful computational techniques for applying these skills in practice.

What follows are some of the many questions that can be addressed by integrating *Maple* into the day-to-day learning and teaching process.

EXPERIMENTAL QUESTIONS

How would a change in numerical content affect the solvability of a linear system, the invertibility of a matrix, the eigenvalues of a matrix, the linear independence of the columns of a matrix, and so on? *Maple* usually produces quick answers to these questions. This what-if approach to learning is enjoyable and rewarding and often leads to the discovery of mathematical facts. What does a three-dimensional surface determined by a quadratic form look like? How would the graph of a quadratic form change if we changed the content of a matrix? The three-dimensional graphing tools of *Maple* are particularly helpful in answering such questions.

ANALYTICAL QUESTIONS

What steps are required to prove a certain theorem? By rewriting a proof of the theorem in *Maple*, using text and math cells, students can cut-and-paste and explore the steps of the proof easily and effectively. Math cells can be used to illustrate a general calculation with simple numerical examples, and text cells can be used to rewrite a proof as a list of steps. Cross-references can be replaced electronically by the mathematical statements to which they refer.

What are the components of a certain definition? Again *Maple* can be used to make an electronic list of the components of the definition, with a specific example or counterexample for each concept or qualifying clause.

What is the sequence of ideas needed to prove a certain theorem? *Maple* can be used to create an electronic sequence of the concepts used in a proof, enriched with simple numerical examples, calculations, and graphical illustrations.

COMPUTATIONAL QUESTIONS

What are the solutions of a large linear system? What is the inverse of a given large matrix? What are the roots of a given real polynomial? What are the eigenvalues of a given matrix? Are the columns of a matrix linearly independent? In most cases, *Maple* provides quick and instructive answers to these and similar questions.

PEDAGOGICAL QUESTIONS

How can we best motivate a given topic? How can students best learn linear algebra? How can we free up time to discuss mathematical ideas and results? How can we teach students the conceptual benefits of notational rigor? Over the last five years, I have discovered plausible answers to many of these questions using technology. The presentation of linear algebra and the choice and arrangement of topics in this text is my answer to such questions.

TO THE STUDENT

You will find that this book presents linear algebra as a unified subject, leading to a complete solution of a well-defined and easily understood problem: the problem of solving linear systems. You will discover that Gaussian elimination can be used to find exact solutions whenever they exist. The singular value decomposition, on the other hand, yields best approximations when exact solutions are impossible. The text begins with Gaussian elimination and ends with singular value decomposition. Most of the material in between serves to build a bridge between these two methods for solving linear systems.

The goals of this book are to stimulate you intellectually, to expose you to the beauty and utility of linear algebra, and to develop your analytical skills. The systematic interplay between theory and practice throughout the text is designed for this purpose. In this connection, technology acts as both a catalyst and a facilitator. You will find it easier to discuss mathematical ideas with your classmates since you will develop confidence in the computational correctness of your work. You will also find that learning and retention can be reinforced through easy electronic experimentation. Theoretical material therefore becomes more accessible. By using *Maple* to replicate the worked examples and illustrations in this book, you can intersperse your studies with periods of playful exploration. As you study difficult concepts, theorems, and definitions, you can use *Maple* to generate examples and counterexamples to test and deepen your understanding of challenging material, even when some of the subject matter is not yet within your grasp.

You will find that the exercises are grouped into two levels: exercises to be done using pencil and paper, and exercises to be done with *Maple.* At the beginning it may be faster for you to learn a particular technique by doing the *Maple* exercises first. You can then test your understanding by doing the remaining exercises by hand. Once you have a thorough understanding of a topic, you will usually find it more rewarding to use *Maple* to solve difficult problems than to spend time on solving easy problems by pencil and paper. However, you usually need enough manual practice to be able to succeed in tests and examinations.

The examples in the text are not intended to provide optimal *Maple* code for the solution of numerical problems. They are written in a way that is intended to suggest different programming approaches using different syntax. In particular, we often make interchangeable use of the two linear algebra packages **linalg,** available in both *Maple V* and *Maple 6,* and **LinearAlgebra**, first introduced in *Maple 6*. One standard exercise for users of *Maple 6* would be to rewrite the solutions in one package using the other package, whenever possible. The required changes in syntax are often obvious and are usually documented in the *Maple* Help file.

ORGANIZATION

Chapter 1 introduces systems of linear equations and their matrix forms. We cover Gaussian elimination applied to augmented matrices and define the matrix form of linear systems. We discuss basic and free variables, pivots, the row echelon and reduced row echelon forms, and solving linear systems by back substitution. We also cover the definition of matrices and their *Maple* representations. We then discuss homogeneous linear systems and use them to study dependence relations among the columns of a matrix. Applications at the end of the chapter include the balancing of chemical equations, the analysis of resistor circuits, the rewriting of higher-order differential equations as systems of first-order equations, heat transfer problems, and population dynamics.

Chapter 2 explores the algebra of matrices. We cover addition, scalar multiplication, matrix products, matrix transposition, matrix inversion, and the trace. In addition, we present a lexicon of the special matrices encountered in the course and establish some of their distinguishing properties. We also discuss the LU and PLU decompositions of matrices and use them to solve linear systems. Furthermore, we present several tests for matrix invertibility. One of the key results proved in this chapter is that invertible matrices are products of elementary matrices. The chapter ends with applications in geometry, computer animation, coding theory, economics, and graph theory.

Chapter 3 deals with determinants. We use the Laplace expansion to define them. However, we also show that this definition is equivalent to the definition based on permutations. We use this second approach to prove elementary properties of determinants. This part of the course also encourages students to reflect on the usually unstated importance of order in linear algebra. The evaluation of determinants is based on ordered lists of scalars. The row echelon reduction of matrices involves the order of rows. Matrix representations of linear transformations are built on ordered sets of vectors, and so on. The chapter ends with a variety of applications, including the use of determinants in geometry and calculus.

Chapter 4 introduces real vector spaces. It deals with linear combinations and the linear dependence and independence of sets of vectors. To provide a tangible conceptual framework, we classify vector spaces into coordinate spaces, polynomial spaces, matrix spaces, and others. We define bases, standard bases, and dimension. We then discuss in detail the idea of coordinate vectors and coordinate conversion. Moreover, we spend considerable time on the four fundamental subspaces of a rectangular matrix and illustrate their importance in linear algebra. The chapter ends with a discussion of complex vector spaces and an application of quaternions to three-dimensional rotations.

Chapter 5 deals with linear transformations and their basic properties. We discuss their images and kernels, as well as the singularity and invertibility of linear transformations. We prove that every finite-dimensional real vector space is isomorphic to a coordinate space. We use this fact to link matrices and linear transformations and to explain why linear transformations can be regarded as matrix transformations. We then show that matrix multiplication corresponds to functional composition. We conclude the chapter with a discussion of similar matrices and show that similar matrices are connected by coordinate conversion matrices.

Chapter 6 deals with eigenvalues and eigenvectors. We discuss the mapping properties of diagonal matrices and study diagonalizable linear transformations. In addition, we analyze the connection between the eigenvalues of a linear transformation and invertibility. We also discuss the use of Hessenberg matrices and Householder transformations for finding characteristic polynomials. Two key results, proved in this chapter, are that eigenvectors belonging to distinct eigenvalues are linearly independent and that a linear transformation is diagonalizable if and only if it has enough linearly independent eigenvectors. The chapter includes a discussion of the problem of finding approximate eigenvalues of large matrices. Applications at the end of the chapter deal with the eigenvalues of differential operators, third-order differential equations, biological rates of change, discrete dynamical systems, and Markov chains.

Chapter 7 introduces the basic geometric concepts of vector geometry. The chapter begins with a discussion of Euclidean vector and matrix norms and their relation to length, distance, and angles. We then discuss non-Euclidean norms, real inner products, angles, and quadratic forms. This material also prepares students for the applications of the singular value decomposition in Chapter 9. Applications integrated throughout the chapter include angles in statistics, the principal axis theorem for quadratic forms applied to conic sections, and Sylvester's theorem for quadratic forms applied to quadric surfaces. This application shows how the diagonalization of matrices can be used to classify geometric objects.

Chapter 8 is devoted to orthogonality. The basic construction around which this chapter is organized is that of an orthogonal projection. We introduce the concept of an orthogonal basis and show how the Gram–Schmidt process can be used to construct orthogonal bases. Applications throughout the chapter include the following: proof that real inner products are dot products relative to orthonormal bases; the QR decomposition of matrices; proof that the fundamental subspaces determined by a matrix are orthogonal in pairs; and the method of least squares. A key result established in this chapter is that inner-product-preserving linear transformations are self-adjoint linear transformations, represented by symmetric matrices. We use this fact to prove the spectral theorem for symmetric matrices, the basis for the singular value decomposition.

Chapter 9 consolidates the ideas and techniques studied in this course. We bring them together in the definition and characterization of the singular values and singular vectors of real rectangular matrices. We show that this topic uses most of the properties of finite-dimensional linear algebra covered in the preceding chapters: linear systems, eigenvalues and eigenvectors, diagonalization, inner products, norms, orthogonal vectors and orthonormal bases, and the orthogonal diagonalization of real symmetric matrices. We discuss the computational properties of invertible matrices as well as the two-norm and Frobenius norm of a square matrix in terms of singular values. In the Applications section, we use singular value decomposition to construct orthonormal bases for fundamental subspaces. We also use the singular value decomposition to define the pseudoinverse of a matrix and use it to show that all linear systems have solutions, either exact or best possible approximations. Other applications deal with the effective rank of a matrix, digital image compression, the orthogonal approximation of invertible matrices, the inversion of large matrices, and the condition numbers of a matrix.

APPENDICES

The book ends with short appendices on several topics.

Appendix A consists of a brief discussion of complex numbers and a statement of the fundamental theorem of algebra. This theorem is required in Chapter 8 to prove that real symmetric matrices have real eigenvalues.

Appendix B discusses numerical calculations and illustrates round-off errors.

Appendix C introduces the principle of mathematical induction. Mathematical induction is used to prove that square row echelon matrices are upper triangular, to define the determinant of a square matrix, to prove that eigenvectors belonging to distinct eigenvalues are linearly independent, to define the Gram–Schmidt orthogonalization algorithm, and to prove that a self-adjoint linear transformation on a finite-dimensional real inner product space can be represented by an orthogonal matrix.

Appendix D explains the use of the sigma notation and ellipses.

Appendix E itemizes the *Maple* packages used in this book.

Appendix F provides answers to selected exercises.

LANGUAGE

By its very nature, the language of linear algebra is heavily *typed*. It speaks of matrices, row vectors, column vectors, and scalars, for example. These objects interact in subtle ways, depending on their ***type***. Scalars and vectors cannot be added, but they can be multiplied. Matrices and vectors cannot usually be added, but they can be multiplied if they have the right shape and type. Vector spaces, the basic structures in which linear operations take place, have two sets of rules, those for adding vectors and those for multiplying vectors and scalars. Vector spaces with the same sets of vectors may differ because they have different sets of scalars. The language of *Maple* assigns to each of these objects a type. Whether or not a specific *Maple* function can be applied to an object often depends on the type assigned to the object. For example, in *Maple 6,* rectangular arrays of objects can have three different types: ***array***, ***matrix***, and ***Matrix***. Many functions built into the kernel of *Maple* can be applied to an object of *array* type. Objects of *matrix* type can usually be processed only by functions in the **linalg** package, and objects of *Matrix* type can be processed only by functions in the **LinearAlgebra** package. The **whattype** function of *Maple* makes is easy to verify the type of an object in case of a conflict of types.

When learning linear algebra it is often important to be aware of both the name and the type of an object. In this text, this is achieved by usually referring to an object both by its type and its name. We always speak of *the matrix* A, *the eigenvalues* λ, *the vector* $\mathbf{x}$, and so on, instead of simply referring to these objects as A, λ, and $\mathbf{x}$. In this way students always know what they are talking about. We use italic capital letters $A, B, C, \ldots$ for matrices, bold lowercase letters $\mathbf{u}, \mathbf{v}, \mathbf{w}, \mathbf{x}, \mathbf{y}, \mathbf{z}, \ldots$ for vectors, and italic lowercase letters $a, b, c, \ldots$ for scalars.

Every effort has been made to provide students with a clean, descriptive terminology. For example, we have introduced the suggestive term ***coordinate conversion matrix*** for the matrices relating coordinate vectors in different bases. We have also made an effort to minimize the use of special symbols and have used descriptive phrases in their place. Thus we usually write A *is approximately equal to* A^+, for example, instead of writing $A \approx A^+$.

NOTATION

1. The expression $\{x_1, \ldots, x_n\}$ denotes a ***set*** containing the elements $x_1, \ldots, x_n$.

2. The expression $[x_1, \ldots, x_n]$ denotes a ***list*** containing the elements $x_1, \ldots, x_n$. It also denotes a ***row vector*** with components $x_1, \ldots, x_n$.

3. The expression $\langle x_1, \ldots, x_n \rangle$ denotes a *Maple* ***input sequence*** for a column vector containing the components $x_1, \ldots, x_n$.

4. The expression $\langle x_1 | \cdots | x_n \rangle$ denotes a *Maple* ***input sequence*** for a row vector containing the components $x_1, \ldots, x_n$.

5. The expression $(x_1, \ldots, x_n)$ denotes a ***point*** with the ***coordinates*** $x_1, \ldots, x_n$.

6. The expression $[x_1 \ \cdots \ x_n]$ denotes both a ***row vector*** with the ***components*** $x_1, \ldots, x_n$, and a $1 \times n$ ***matrix***. When the intended meaning is clear from the context, we sometimes write a row vector $[x_1 \ \cdots \ x_n]$ in point form as $(x_1, \ldots, x_n)$.

7. The expression

$$\begin{bmatrix} x_1 \\ \vdots \\ x_n \end{bmatrix}$$

denotes both a ***column vector*** with ***components*** $x_1, \ldots, x_n$ and an $n \times 1$ ***matrix***. To save space, we often write column vectors in point form as $(x_1, \ldots, x_n)$.

8. The array

$$\begin{bmatrix} a_{11} & \cdots & a_{1j} & \cdots & a_{1n} \\ \vdots & \ddots & \vdots & \ddots & \vdots \\ a_{i1} & \cdots & a_{ij} & \cdots & a_{in} \\ \vdots & \ddots & \vdots & \ddots & \vdots \\ a_{m1} & \cdots & a_{mj} & \cdots & a_{mn} \end{bmatrix}$$

denotes a ***matrix*** with m rows and n columns and ***entry*** a_{ij} in row i and column j.

9. For each natural number n, the boldface symbol **n** denotes the list $[1, \ldots, n]$ of natural numbers from 1 to n, in their usual order. We call **n** an ***indexing set.***

10. For any indexing set **n**, and one–one onto function $\pi : \mathbf{n} \to \mathbf{n}$, the list $[\pi(1), \ldots, \pi(n)]$ denotes the sequence of values of π. We call $[\pi(1), \ldots, \pi(n)]$ a ***permutation*** of **n**.

ACKNOWLEDGMENTS

Teaching has never been as rewarding or as much fun as it is now. In response to the enthusiasm of my students and colleagues, I have written this book as the background material for an interactive first course in linear algebra. The encouragement and positive feedback that I have received during the design and development of the book have given me the energy required to complete the project. I am grateful to several of my colleagues, especially Dr. Jean Turgeon, for reading earlier versions of the manuscript and for providing encouraging comments. I would also like to thank all of my students who have joyfully suffered through the completion of this project and have contributed to its success. Many of them are now using linear algebra in their daily work.

I am grateful to Dr. Todd Arbogast of the University of Texas at Austin, Dr. William Bauldry of Appalachian State University, Dr. Eric Carlen of Georgia Tech, Dr. Thomas W. Cusick of SUNY at Buffalo, Dr. Karin Reinhold of SUNY at Albany, and Dr. Saleem Watson of California State University at Long Beach for their careful reading of the manuscript and their incisive comments and constructive criticism. The book is much the better for their input.

I would also like to thank Dr. Mirek Majewski of the Inter-University Institute of Macau for helping with the preparation of some of the answers to the exercises. In addition, my son, Stuart, has worked through the examples and many of the exercises. He has critiqued them with the keen eye of a Harvard undergraduate and has helped make this book more accessible.

I have learned a great deal from conversations with my former students Andreas Soupliotis of Microsoft Research and of Peter Wulfraat of Nuance about computer applications of linear algebra and I value their continued interest in the completion of this project.

I am especially grateful to my editor, Barbara Holland, for helping shape the nature of this text and for sharing my goal to make the teaching and learning of linear algebra with *Maple* accessible, enjoyable, and rewarding. Additional thanks go to my publisher, Mike Sugarman, and to Victor Curran, Diane Grossman, and Angela Dooley of Harcourt/Academic Press as well as to Elliot Simon, the copyeditor, for their enthusiastic support and skillful technical and production assistance.

Finally, I am profoundly grateful to my family, Isabel, Julie, and Stuart, for their patience, encouragement, and understanding throughout this project.

The manuscript was written and typeset using Scientific WorkPlace, the TeX-based document preparation system, produced by MacKichan Software Inc. I would like to thank Roger Hunter, George Pearson, Steve Swanson, and Barry MacKichan and his team for having developed this wonderful writing tool for mathematics. The final version of the book was produced in LaTeX by Integre Technical Publishing.

Fred Szabo

Using *Maple*

After loading the *Maple* program, you are presented with an empty screen and the *Maple* prompt:

```
[>
```

You are now ready to do linear algebra with *Maple.* You have three choices.

1. You can initiate an input-output dialog.

```
[>A:=array([[1,2],[3,4]]);
```

$$A := \begin{bmatrix} 1 & 2 \\ 3 & 4 \end{bmatrix}$$

The command **A:=array([[1,2],[3,4]])** instructs *Maple* to construct a 2×2 array with the entries 1, 2, 3, 4, arranged in two rows and columns. If you terminate your command with a semicolon (;) and press Enter, the result of a defined action is displayed in the center of the screen. If you had terminated the command with a colon (:), the letter A would have become a name for the defined matrix, but the result of the definition would not have been displayed.

2. You can load the **linalg** package and increase the number of basic linear algebra functions available to you. The full list of functions in this package is shown in Appendix E. The commands

```
[>with(linalg):
[>A:=matrix([[1,2],[3,4]]):
[>d:=det(A);
```

$$d := -2$$

load the package, define the matrix A, and output the determinant -2 of the matrix A.

3. Instead of loading the **linalg** package, you can load the **LinearAlgebra** package and increase the number of available linear algebra functions. The commands

```
[> with(LinearAlgebra):
[> A:=Matrix([[1,2],[3,4]]):
[> d:=Determinant(A);
```

$$d := -2$$

load the package, define the matrix A, and output the determinant -2 of the matrix A.

You can also load both linear algebra packages at the same time. In that case, a *Maple* message appears explaining any changes in the active functions caused by this dual loading.

```
[> with(linalg):
[> with(LinearAlgebra):
Warning, the assigned name GramSchmidt now has a global
binding
```

A different message appears if the loading order is reversed.

```
[> with(LinearAlgebra):
[> with(linalg):
Warning, the previous binding of the name GramSchmidt has
been removed and it now has an assigned value
Warning, the protected names norm and trace have been
redefined and unprotected
```

The messages show that at a fundamental level, the conflicts in the two loaded function sets are minimal.

The basic structures used in this text are two-dimensional arrays of numbers called ***matrices***. The three preceding simple input-output dialogs illustrate that there are at least three ways of forming matrices in *Maple.* We can use the **array** function, built into the kernel of *Maple*, the **matrix** function in the **linalg** package, and the **Matrix** function in the **LinearAlgebra** package. In some situations, the three types of objects are interchangeable, but usually they are not.

The **LinearAlgebra** package is available only in *Maple 6.* If you are using *Maple V*, you will have to translate some of the *Maple* examples and exercises in this text into the language of **linalg**. This will work in most cases. However, the syntax, the commands, and the options in **linalg** and **LinearAlgebra** are often different. For this reason, we have included

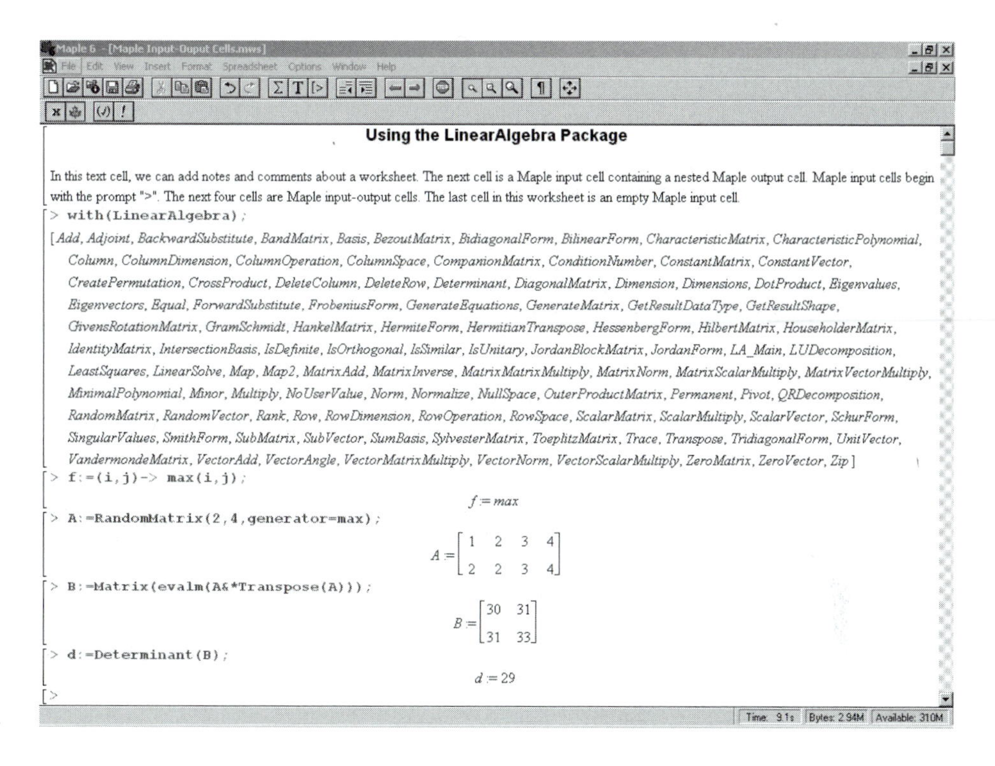

FIGURE 1 Input-Output Cells in a *Maple* Worksheet.

a variety of similar calculations, using the two packages. By studying these examples you will learn how to use both sets of commands and functions. The commands and functions in the **LinearAlgebra** package are also shown in Appendix E.

In Figure 1, we show a *Maple* worksheet. As you can see, the page is divided into a sequence of ***cells***. Each cell has a certain property and is bounded on the left by a square bracket. A cell may be a ***text cell*** or a math ***input–output cell***. By default, worksheet cells are input-output cells unless this option is changed from the toolbar. This means that an input is interpreted as an instruction to carry out a mathematical operation.

The first cell in Figure 1 is a text cell. It contains the title **Using the LinearAlgebra Package**. The next cells are input-output cells.

RANDOM MATRICES

Throughout the text, we need numerical matrices to illustrate definitions, theorems, and constructions. Most of the time, the specific entries of the matrices are irrelevant. In such cases, we use randomly generated matrices.

At the basic level of *Maple*, we can combine the **rand()**, **seq**, and **evalm** functions to generate a random array.

```
[> A:=evalm([seq([rand(),rand(),rand()],i=1..2)]);
```

$$A := \begin{bmatrix} 951053530086 & 146486307198 & 155590763466 \\ 429392673709 & 525428510973 & 272600608981 \end{bmatrix}$$

Although not explicitly defined as a matrix, the array A can be used as an argument in some of the functions in the **linalg** package.

```
[> linalg[transpose](A);
```

$$\begin{bmatrix} 951053530086 & 429392673709 \\ 146486307198 & 525428510973 \\ 155590763466 & 272600608981 \end{bmatrix}$$

However, most functions in the **LinearAlgebra** package will not accept a randomly generated array as an argument, unless the array has been declared to be of **Matrix** type.

```
[> LinearAlgebra[Transpose](A);
Error, LinearAlgebra:- Transpose expects its 1st argument,
MV, to be of type {Matrix, Vector, scalar}, but received A
```

We can remedy this situation by declaring A to be of **Matrix** type.

```
[> LinearAlgebra[Transpose](Matrix(A));
```

$$\begin{bmatrix} 951053530086 & 429392673709 \\ 146486307198 & 525428510973 \\ 155590763466 & 272600608981 \end{bmatrix}$$

These syntactic difficulties can usually be avoided if we use the appropriate random matrix generators built into the **linalg** and **LinearAlgebra** packages.

The following commands use the **linalg** package to generate two random matrices with integer entries and calculate their sum:

```
[> with(linalg):
[> A:=randmatrix(2,2);
```

$$A := \begin{bmatrix} 79 & 56 \\ 49 & 63 \end{bmatrix}$$

```
[> B:=randmatrix(2,2);
```

$$B := \begin{bmatrix} 57 & -59 \\ 45 & -8 \end{bmatrix}$$

```
[> C:=evalm(A+B);
```

$$C := \begin{bmatrix} 136 & -3 \\ 94 & 55 \end{bmatrix}$$

The next commands use the **LinearAlgebra** package to generate two random matrices with integer entries and calculate their product:

```
[> with(LinearAlgebra):
[> A:=RandomMatrix(2,2);
```

$$A := \begin{bmatrix} 62 & -71 \\ -79 & 28 \end{bmatrix}$$

```
[> B:=RandomMatrix(2,2);
```

$$B := \begin{bmatrix} -56 & -50 \\ -8 & 30 \end{bmatrix}$$

```
[> C:=A.B;
```

$$C := \begin{bmatrix} -2904 & -5230 \\ 4200 & 4790 \end{bmatrix}$$

Occasionally, we will also need matrices with real numbers as entries. We have several options for defining them. We can combine the **rand()** function with a function such as the sine function and generate a matrix whose entries are random values of the sine function. A simpler method is to evaluate a random matrix with integer entries as a floating-point object.

In *Maple*, real numbers are defined using floating-point numbers. In Appendix B we briefly discuss such numbers. A *Maple* matrix with integer entries can be converted to a floating-point matrix by applying the function **evalf** to it.

```
[>A:=evalf(linalg[randmatrix](2,3));
```

$$A := \begin{bmatrix} -18. & 31. & -26. \\ -62. & 1. & -47. \end{bmatrix}$$

The decimal point behind each of the entries of A indicates that *Maple* considers the entries to be real numbers, rather than integers. The important difference between the real number $(-18.)$ and the integer (-18) is that in calculations, *Maple* treats $(-18.)$ as an approximate number, whereas it treats (-18) as an exact number. In various places in the text we discuss the need for approximate calculations.

ROW AND COLUMN VECTORS

The rows and columns of a matrix are examples of objects called ***vectors***. In the **LinearAlgebra** package of *Maple 6*, a distinction is made between ***row vectors*** and ***column vectors.*** In the **linalg** package of *Maple V*, on the other hand, the distinction is built into the contexts in which vectors are used. The following example illustrates the difference.

```
[>with(LinearAlgebra):
[>A:=RandomMatrix(2,3);
```

$$A := \begin{bmatrix} -50 & 62 & -71 \\ 30 & -79 & 28 \end{bmatrix}$$

```
[>c1A:=Column(A,1);
```

$$c1A := \begin{bmatrix} -50 \\ 30 \end{bmatrix}$$

The **whattype** function tells us the ***type*** of the object that *Maple* has produced.

```
[>whattype(c1A);
```

$$Vector_{column}$$

As we can see, *Maple 6* treats **c1A** as a column vector.

```
[> r1A:=Row(A,1);
```

$$r1A := [-50, 62, -71]$$

```
[> whattype(r1A);
```

$$Vector_{row}$$

This shows that *Maple 6* treats **r1A** as a row vector.

Now consider the corresponding operations carried out using *Maple V* functions.

```
[> with(linalg):
[> B:=randmatrix(2,3);
```

$$B := \begin{bmatrix} 79 & 56 & 49 \\ 63 & 57 & -59 \end{bmatrix}$$

```
[> c1B:=col(B,1);
```

$$c1B := [79, 63]$$

```
[> whattype(c1B);
```

$$symbol$$

```
[> r1B:=row(B,1);
```

$$r1B := [79, 56, 49]$$

```
[> whattype(r1B);
```

$$symbol$$

As we can see, the distinction between row and column vectors is lost. The difference in type between the vectors **r1A** and **c1A** and **r1B** and **c1B** is algebraically significant. We can, for example, form the sum **c1A** + **c1A** and obtain directly the output

$$\begin{bmatrix} -100 \\ 60 \end{bmatrix}$$

On the other hand, forming the sum **c1B** + **c1B** produces the output

$$2\,c1B$$

We must apply the **evalm** function if we want to obtain a numerical output.

```
[> evalm(c1B+c1B);
```

$$[158, 126]$$

In this text, we highlight the differences between the properties of vectors in *Maple V* and *Maple 6* by varying the functions used to solve problems. Both versions of *Maple* provide a variety of syntactic options for defining vectors. They are conveniently summarized in the *Maple* Help file and are illustrated in examples and exercises. Globally, it can be said that vectors used in *Maple V* functions must usually be of type **vector**, whereas vectors in *Maple 6* functions must be of type **Vector**.

DOT PRODUCTS AND MATRIX MULTIPLICATION

The multiplication of matrices is one of the most important operations in linear algebra. It is based on the ***dot product***. The entries of the matrix

$$C := \begin{bmatrix} -2904 & -5230 \\ 4200 & 4790 \end{bmatrix}$$

for example, are the dot products formed with the rows of the matrix

$$A := \begin{bmatrix} 62 & -71 \\ -79 & 28 \end{bmatrix}$$

and the columns of the matrix

$$B := \begin{bmatrix} -56 & -50 \\ -8 & 30 \end{bmatrix}$$

Given the prominence of this operation in matrix multiplication, it is not surprising that *Maple* has a dot product function built explicitly into each of the two linear algebra packages.

```
[>with(linalg):
[>A:=matrix([[62,-71],[-79,28]]):
[>B:=matrix([[-56,-50],[-8,30]]):
[>p:=(i,j)->dotprod(row(A,i),col(B,j)):
```

```
[>P:=matrix([[p(1,1),p(1,2)],[p(2,1),p(2,2)]]);
```

$$P := \begin{bmatrix} -2904 & -5230 \\ 4200 & 4790 \end{bmatrix}$$

It is easy to verify that $p(1,1) = 62 \times (-56) + (-71) \times (-8)$ and that the other entries of P are obtained a similar way. We could also have calculated the matrix product using the **DotProduct** function in the **LinearAlgebra** package. We would have had to load the package, define A and B with the commands

```
[>A:=Matrix([[62,-71],[-79,28]]):
[>B:=Matrix([[-56,-50],[-8,30]]):
```

and would have had to define the entries of P by

```
[>p:=(1,1)->DotProduct(Row(A,i),Column(B,j)):
```

Maple provides two other options for calculating the dot product of a row and a column vector.

```
[>x:=<62,-71>;
```

$$x := \begin{bmatrix} 62 \\ -71 \end{bmatrix}$$

```
[>y:=<-56,-8>:
```

$$y := \begin{bmatrix} -56 \\ -8 \end{bmatrix}$$

```
[>Transpose(x).y;
```

$$-2904$$

In this case, **x** and **y** are defined as column vectors, and the dot product is evaluated using the dot function. We will use this form of the dot product in most of the text.

We could also have used the **Row** and **Column** functions in the **LinearAlgebra** package to simplify the data entry.

```
[> with(LinearAlgebra):
[> A:=Matrix([[62,-71],[-79,28]]):
[> B:=Matrix([[-56,-50],[-8,30]]):
[> rA:=i->Row(A,i):
[> cB:=i->Column(B,i):
```

```
[> P:=Matrix([[rA(1).cB(1),rA(1).cB(2)],
      [rA(2).cB(1),rA(2).cB(2)]]);
```

$$P := \begin{bmatrix} -2904 & -5230 \\ 4200 & 4790 \end{bmatrix}$$

As we can see, the matrix P is also the product of the matrices A and B.

ASSIGNMENTS AND PROCEDURES

The commands `rA:=i->Row(A,i)` and `cB:=j->Column(B,j)` are examples of *Maple* assignments. The first command assigns the name rA to the function that selects the ith row from the matrix A, and the second command assigns the name cB to the function that selects the jth column from the matrix B. The command `A:=Matrix([[1,2],[3,4]])` assigns the letter A to the matrix

$$\begin{bmatrix} 1 & 2 \\ 3 & 4 \end{bmatrix}$$

Until these names are unassigned, *Maple* replaces every occurrence of the expression rA(1) with the first row of A, every occurrence of cB(2) with the second column of B, and every occurrence of A with the matrix

$$\begin{bmatrix} 1 & 2 \\ 3 & 4 \end{bmatrix}$$

In the last case, however, A is often treated as a symbol and must be evaluated as a matrix for the assignment to take place. The command `evalm(A)` has the desired effect.

As we saw earlier, functions in more than one variable can also be defined by assignments. The definition

```
[> f:=(x,y)->3*x+5*y:
```

for example, defines a function f whose value at (x, y) is $3x + 5y$.

Unassigning the Values of Variables

When *Maple* is first loaded, we are free to use letters and other combinations of symbols as names for objects to be processed. We refer to these names as ***variables***. It sometimes happens that we would like to use a particular letter repeatedly in a lengthy calculation, but with more than one meaning. We can do so by overriding an assignment and making a new one, or by first restoring a variable to an unassigned state. The command

```
[>unassign('x'):
```

for example, removes any meaning we might have assigned to the letter x. The unassigning of variables is particularly important in nested calculations, where *Maple* may be using an unintended value in a hidden context.

SOLVING LINEAR SYSTEMS

Maple has powerful tools for solving linear systems. We can use augmented matrices and various forms of Gaussian elimination. For example, we can solve the linear system

$$\begin{cases} 2x - y = 18 \\ x + 3y = 2 \end{cases}$$

by using the basic **solve** function.

```
[> solve({2*x-y=18,x+3*y=2});
```

$$\{y = -2, x = 8\}$$

We can also form the augmented matrix of the given system and use Gaussian elimination.

```
[> with(linalg):
[> A:=genmatrix([2*x-y=18,x+3*y=2],[x,y],'b'):
[> Ab:=augment(A,b);
```

$$Ab := \begin{bmatrix} 2 & -1 & 18 \\ 1 & 3 & 2 \end{bmatrix}$$

```
[> B:=rref(Ab);
```

$$B := \begin{bmatrix} 1 & 0 & 8 \\ 0 & 1 & -2 \end{bmatrix}$$

From the matrix B we can now read off the solution of the given linear system: $x = 8$ and $y = -2$, as before. In addition, we can convert the given system to a matrix equation of the form $A\mathbf{x} = \mathbf{b}$ and use the **linsolve** function in the **linalg** package.

```
[> linsolve(A,b);
```

$$[8, -2]$$

The **linsolve** function outputs the values of the variables x and y in the order in which the variables were specified in the **genmatrix** function. In this special case, we could also have found the solution $\mathbf{x} = [8, -2]$ by forming the inverse A^{-1} of the matrix A and multiplying it with **b**.

```
[> evalm(inverse(A)&*b);
```

$$[8, -2]$$

Other techniques for solving linear system discussed in this text include the LU decomposition. This method involves decomposing the coefficient matrix A of a linear system $A\mathbf{x} = \mathbf{b}$ into a product LU of two triangular matrices. The solution $\mathbf{x}$ can then be found by back and forward substitution.

MATRIX DECOMPOSITION

In addition to the LU decomposition, *Maple* provides us with other tools for decomposing matrices. Each method has its uses. We can form the QR decomposition, for example, where in some cases, A is the product of an orthogonal matrix Q and an invertible triangular matrix R. We can also form the singular value decomposition of A by decomposing A into a product UDV^T, where U and V are orthogonal matrices and D is a generalized diagonal matrix. The singular value decomposition always exists for real matrices. It generalizes the eigenvector-eigenvalue decomposition PDP^{-1}, which only exists in certain cases.

Using functions from the **LinearAlgebra** package, we can calculate these decompositions of a real matrix A as follows:

```
[> with(LinearAlgebra):
[> QRDecomposition(A):
[> SingularValues(A,output=['U','S','Vt']):
```

The uses of the QR decomposition are discussed in Chapter 8. Its applications include the solution of the normal equations of overdetermined linear systems and the calculation of approximate eigenvalues. The singular value decomposition and some of its applications form the unifying climax of this book and are presented in Chapter 9.

Most matrix decompositions are computationally complex. Without the use of *Maple* or other such systems, these topics would therefore be difficult to teach and learn in a first course in linear algebra. The advent of computers and the development of powerful mathematical software have changed all that.

DETERMINANTS AND NORMS

The linear algebra packages of *Maple* provide an extensive range of tools for a quantitative analysis of the properties of matrices. For each square matrix A, we can, for example, calculate the determinant of A. Chapter 3 deals with determinants and some of their uses in geometry and calculus. We have already illustrated the basic syntax of determinants earlier in this chapter.

The packages also contain functions for calculating various vector and matrix norms. Chapter 7 deals with these measures in depth. The commands

```
[> with(LinearAlgebra):
[> A:=Matrix([[1,2],[3,4]]):
[> fnA:=Norm(A,Frobenius);
```

$$fnA := \sqrt{30}$$

calculate the square root of the sum of the squares of 1, 2, 3, and 4. This norm exists for all rectangular real matrices and generalizes the Euclidean length of vectors in $\mathbb{R}^n$. For example, the length of the vector (3, 4) is $\sqrt{3^2 + 4^2} = 5$. We can use *Maple* to calculate this length.

```
[> with(LinearAlgebra):
[> v:=Vector([3,4]):
[> Norm(v,Euclidean);
                                    5
```

The Frobenius norm of the corresponding column matrix has the same value.

```
[> V:=Matrix([[3],[4]]):
[> Norm(V,Frobenius);
                                    5
```

This example explains in which sense the Frobenius norm of matrices in $\mathbb{R}^{m \times n}$ generalize the Euclidean norm of column vectors in $\mathbb{R}^{mn}$.

GEOMETRY

The dot product is a special case of a more general function called an *inner product.* The **BilinearForm** function in the **LinearAlgebra** package and the **innerprod** function in the **linalg** package calculate inner products. Inner products are used in Chapter 7 to study geometry in vector spaces.

For any two compatible vectors $\mathbf{x}$ and $\mathbf{y}$ and suitable matrix A, we can use expressions of the form $\mathbf{x}^T A\mathbf{y}$ to define length, distance, and angles. In its simplest form, the **BilinearForm** function calculates such dot products. The calculation

```
[>x:=<2,3>:  y:=<4,5>:
[> LinearAlgebra[BilinearForm](x,y);
                                    23
```

produces the values $2 \times 4 + 3 \times 5 = 23$. However, if we let A be the diagonal matrix

$$\begin{bmatrix} 9 & 0 \\ 0 & 11 \end{bmatrix}$$

for example, then the calculation

```
[> with(LinearAlgebra):
[> x:=<2,3>:  y:=<4,5>:
[> A:=DiagonalMatrix([9,11]):
[> BilinearForm(x,y,A);
```

$$237$$

yields the value $9 \times 2 \times 4 + 11 \times 3 \times 5 = 237$. In Chapter 7, we discuss the meaning of this result. The same value can be produced with the **innerprod** function.

```
[> innerprod(x,A,y);
```

$$237$$

By suppressing the reference to the matrix A, or by using an identity matrix for A, we can also use this function to calculate the dot product of $\mathbf{x}$ and $\mathbf{y}$:

```
[> innerprod(x,y);
```

$$23$$

If we use a nonidentity matrix A as an argument in the **innerprod** function, we obtain different values.

```
[> x:=vector(2):
[> y:=vector(2):
[> A:=diag(3,5):
```

```
[> innerprod(x,A,y);
```

$$3x_1y_1 + 5x_2y_2$$

If the matrix A is symmetric and $\mathbf{x}^T A\mathbf{x}$ is positive for all nonzero vectors $\mathbf{x}$, then **innerprod(x,A,y)** is a function that can be used to define non-Euclidean measures for distance, length, and angles. The matrices in these functions are called ***positive definite.*** They are discussed in detail in Chapter 7. It is often difficult to decide by hand whether a matrix is positive definite. Fortunately, *Maple* provides a test function facilitating the task. The function **IsDefinite** in the **LinearAlgebra** package decides whether a given symmetric matrix is positive definite. Using the assignment

```
[> d:=x->sqrt(innerprod(x,A,x)):
```

we can generalize the idea of Euclidean length. Whereas the commands

```
[> x:=[3,4]:
[> A:=matrix([[1,0],[0,1]]):
[> d(x);
                              5
```

calculate the Euclidean length $\sqrt{3^2 + 4^2} = \sqrt{25} = 5$ of the vector (3, 4), the commands

```
[> A:=diag(5,7):
[> d(x);
                              √157
```

produce a different value. It, too, can be interpreted as the length of a vector.

For any general vectors $\mathbf{x} = (x, y)$, and diagonal matrix

$$A := \begin{bmatrix} a & 0 \\ 0 & b \end{bmatrix}$$

the commands

```
[> X:=[x,y]:
[> A:=diag(a,b):
[> q:=X->innerprod(X,A,X):
[> q(X,A);
                              x^2 a + y^2 b
```

define a quadratic form $q(x, y) = ax^2 + by^2$ in the variables x and y. For a fixed constant k, we get the corresponding quadratic equation $q(x, y) = k$. Using *Maple* graphing tools, we can easily graph quadratic forms and their associated quadratic equations. The commands

```
[> A:=diag(5,7):
[> q(X,A):
                              5x^2 + 7y^2
```

```
[> plot3d(5*x^2+7*y^2,x=-10..10,y=-10..10);
```

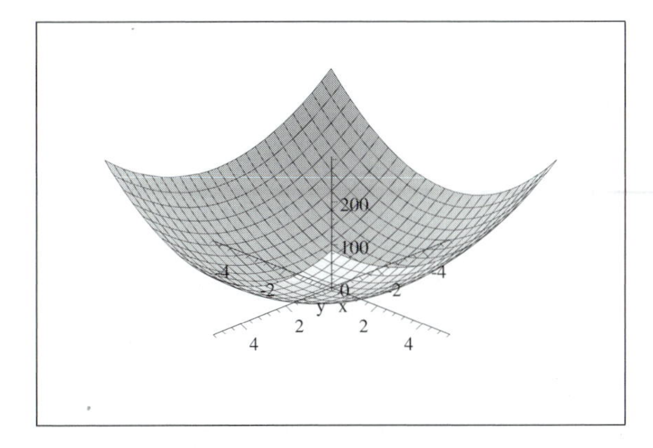

produce a three-dimensional graph of the quadratic form $q(x, y) = 5x^2 + 7y^2$. To graph a particular quadratic equation determined by this quadratic form, we can use the **implicitplot** function in the **plots** package.

```
[> with(plots);
[> implicitplot(5*x^2+7*y^2=35,x=-10..10,y=-10..10);
```

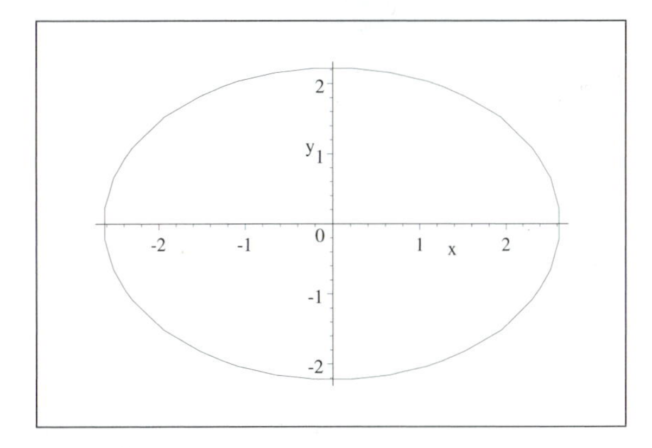

The available functions in the package are listed in Appendix E.

SPECIAL SYMBOLS

In *Maple* input statements, the symbols **exp(1)**, **I**, and **Pi** must be used for the constants e, i, and π. However, we can also enter these constants from the **Symbol** palette by clicking on the appropriate buttons. For easier readability, we usually write these constants in the traditional notation as e, i, and π when not defining *Maple* commands.

MAPLE HELP

Whenever *Maple* is launched, the kernel of the program is loaded, and all features implemented in the kernel can be used. In addition, the entire interactive *Maple* manual is available online through the ***Help*** menu. Also online is a ***New User's Tour*** section. Beginning users should familiarize themselves with these features of *Maple*.

Help Topics

The Introduction to the Help file provides global guidelines on using Help. The Using Help section is a comprehensive tutorial on *Maple* Help. It is divided into eighteen sections:

Close All Help, Close Help Topic, Collapse Sections, Context-sensitive Help, Display a Known Help Topic, Display Balloon Help, Expand Sections, Find Out What's New, Glossary, Help Toolbar, New User's Tour, Overview of Help Menu, Overview of Help Page Menu Bar,

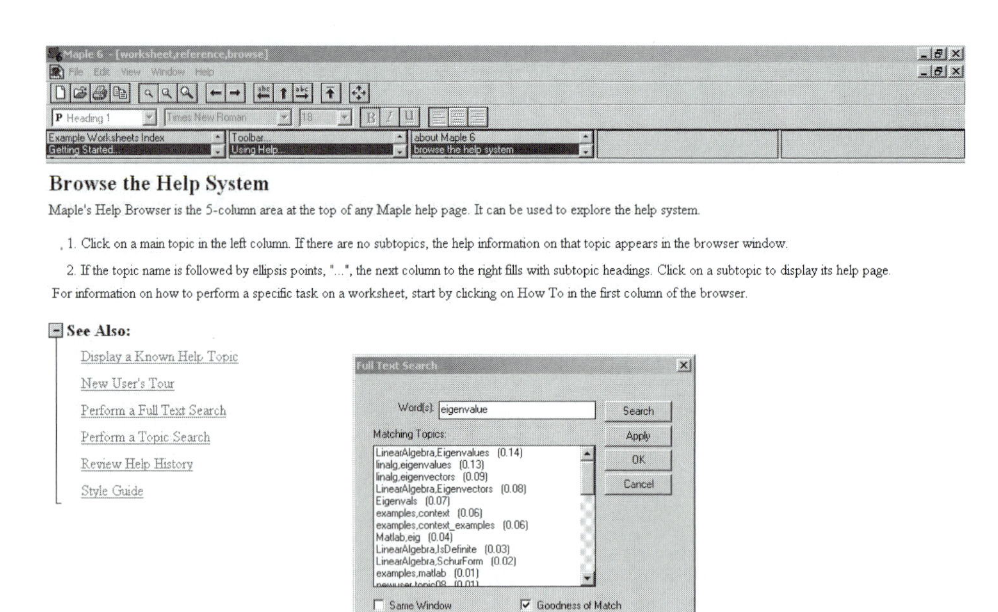

FIGURE 2 A Full Text Search.

Perform a Full Text Search, Perform a Topic Search, Review Help History, Status Bar, and Style Guide.

The Full Text Search facility is particularly helpful for less well-defined searches and comparisons between comparable functions in different packages. Searching for eigenvalues, for example, produces the result in Figure 2. By clicking on the appropriate topics in the output box, we can explore the different aspects of eigenvalues implemented in *Maple 6*.

INTEGRATED TUTORIAL

In this text, *Maple* is integrated into both the elementary and the more advanced aspects of the subject. Students can therefore develop simultaneously the manipulative and conceptual experience required to use *Maple* with confidence for more difficult problems later in the text. This approach eliminates the need for a separate *Maple* tutorial for linear algebra. All step-by-step solutions are complete and can easily be replicated by students new to *Maple*. Whenever manual practice is also required, many of the given examples and exercises can usually be solved independently with pencil and paper. The details in the worked examples suggest the necessary steps. The methods used to solve the examples throughout the text have been varied whenever possible to provide students with different *Maple* programming options.

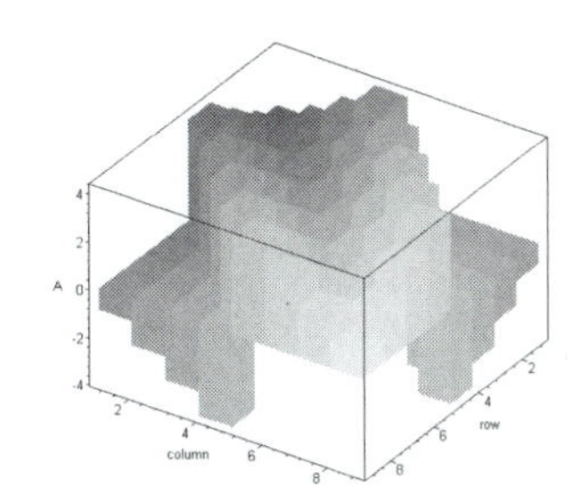

1

LINEAR SYSTEMS

Systems of linear equations occur in such diverse fields as image processing, computer animation, electrical engineering, aerospace engineering, chemistry, physics, economics, population biology, geometry, statistics, and many others. This book is about the mathematics required to solve such systems. The subject is called linear algebra. It is based on an exciting and harmonious blend of ideas from algebra and geometry.

LINEAR EQUATIONS

Linear algebra begins with the study of linear equations. Consider, for example, the equation

$$ax + by = c$$

in the variables x and y and the real numbers a, b, c. This equation is linear in an algebraic sense since the variables x and y are synonymous with their first powers x^1 and y^1. It is also linear in a geometric sense. If $b \neq 0$, we can solve the equation for y and obtain

$$y = -\frac{a}{b}x + \frac{c}{b}$$

This equation determines a straight line in the plane, with slope $-a/b$ and y-intercept c/b.

Two real numbers r_1 and r_2 are considered to be a solution of $ax + by = c$ if $ar_1 + br_2 = c$. In plane geometry, the solutions of this equation correspond to the points of the line determined by the equation.

In order to write linear equations, we need variables and constants. We will use the letters from the end of the alphabet as ***variables*** and letters from the beginning of the alphabet as ***constants***. Unless we specify the contrary, we assume constants to be real numbers and variables to range over the set $\mathbb{R}$ of real numbers. We will also use subscripts to enlarge our supply of variables and constants. Thus

$$x, y, z, x_1, y_1, z_1, x_2, y_2, z_2, x_3, y_3, z_3, x_4, y_4, z_4, \ldots$$

denote variables, and

$$a, b, c, d, a_1, a_2, a_3, b_1, b_2, b_3, c_1, c_2, c_3, d_1, d_2, d_3, \ldots$$

denote constants. Occasionally, we also use the letter t as a variable and the letters m and r as constants. The context always makes it clear whether a letter is used as a variable or a constant. Additional symbols such as 3, 5, and π of course also denote constants. The difference is that $a, b, c, \ldots$ stand for fixed but arbitrary numbers, whereas $3, 5, \pi, \sqrt{2}$, and others stand for specific numbers. Depending on the context, we will sometimes refer to variables as ***unknowns***. Linear equations can be formed in any number of variables.

DEFINITION 1.1 *A **linear equation** in the variables $x_1, \ldots, x_n$ is an equation of the form*

$$a_1 x_1 + \cdots + a_n x_n = b$$

*where $a_1, \ldots, a_n$ and b are real or complex numbers. The numbers $a_1, \ldots, a_n$ are the **coefficients** of the equation and b is the **constant term**.*

In working with linear equations, it will sometimes be necessary to identify the first variable in the equation whose coefficient is nonzero. It is convenient to give a name to this variable.

DEFINITION 1.2 *The **leading variable** of a linear equation $a_1 x_1 + \cdots + a_n x_n = b$ is the first variable x_i in the list $(x_1, \ldots, x_i, \ldots, x_n)$ for which $a_i \neq 0$.*

Since the coefficients a_i can be any real numbers, an expression of the form

$$0x_1 + \cdots + 0x_n = b$$

is a legitimately formed linear equation. We refer to such an equation as a ***degenerate linear equation***.

EXAMPLE 1.1 A Linear Equation

Use the coefficients $a_1 = 3$, $a_2 = 5$, $a_3 = -4$, the corresponding variables x_1, x_2, x_3, and the constant term 25 to construct a linear equation.

Solution. The desired equation is $3x_1 + 5x_2 - 4x_3 = 25$. ▲

A more interesting way of notating linear equations is to write them in *Maple,* using the function **dotprod** in the **linalg** package or the function **DotProduct** in the **LinearAlgebra** package.

```
[>with(linalg):
[>dotprod([3,5,-4],[x[1],x[2],x[3]],orthogonal)=25;
```

$$3x_1 + 5x_2 - 4x_3 = 25$$

The function **dotprod** allows us to separate the list of coefficients $[3, 5, -4]$ from the list of variables $[x_1, x_2, x_3]$ in the construction process. The optional argument *orthogonal* tells *Maple* to interpret the variables over the real and not the complex numbers. In most applications, we will be able to omit this argument.

The **DotProduct** is another *Maple* function that can be used to produce a linear equation.

```
[> with(LinearAlgebra):
[> DotProduct(Vector([3,5,-4]),Vector([x[1],x[2],x[3]]),
     conjugate=false)=25;
```

$$3x_1 + 5x_2 - 4x_3 = 25$$

The optional argument *conjugate=false* is required to instruct *Maple* to let the variables x_1, x_2, and x_3 range over the real and not the complex numbers. This argument can also usually be omitted.

The functions **dotprod** and **DotProduct** produce the same equations, but for different inputs. The inputs of **dotprod** are simple ***lists***, whereas the function **DotProduct** requires as inputs special objects called ***vectors***. If we use the **DotProduct** function with a list, *Maple* replies with an error message.

```
[> with(LinearAlgebra):
[> DotProduct([3,5,-4],[x[1],x[2],x[3]],conjugate=false)=25;
Error, LinearAlgebra:-DotProduct expects its 1st argument,
V1, to be of type Vector, but received [3, 5, -4]
```

One of the pedagogical virtues of *Maple* is its requirement to pay close attention to the types of objects used in specific contexts. This is particularly helpful in linear algebra, which deals with so many different types of objects.

One of the basic units of information in linear algebra is that of a list of items. We use boldface letters as names for lists, and we write

$$\mathbf{r} = (r_1, \ldots, r_n)$$

for the list named $\mathbf{r}$, consisting of the items $r_1, \ldots, r_n$.

As mentioned in the Using Maple introduction, *Maple* allows us to interpret the list $\mathbf{r}$ in at least four additional ways: as a row vector, a column vector, a row matrix, or a column matrix. For this reason, we often refer to lists as *vectors* or even *matrices*, depending on the context.

The commands

```
[> with(linalg):
[> cons:=vector([3,5,-4]):
[> vars:=vector([x[1],x[2],x[3]]):
[> dotprod(cons,vars,orthogonal)=25;
```

$$3x_1 + 5x_2 - 4x_3 = 25$$

produce the same linear equation as the commands involving the corresponding lists. In the case of the **dotprod** function, the designation of [3, 5, −4] as a vector was tolerated but not required. The **DotProduct** function, however, is defined only for vector inputs.

```
[> with(LinearAlgebra):
[> cons:=Vector([3,5,-4]):
[> vars:=Vector([x[1],x[2],x[3]]):
[> DotProduct(cons,vars,conjugate=false)=25;
```

$$3x_1 + 5x_2 - 4x_3 = 25$$

The designation of the list [3, 5, −4] as the Vector([3,5,-4]), written with a capital V, is essential for the **DotProduct** function. For its own internal reasons, *Maple* often requires that we distinguish vector and matrix inputs for functions in the **linalg** package from those for functions in the **LinearAlgebra** package. The command

```
[> DotProduct(vector([3,5,-4]),vars,conjugate=false)=25;
```

for example, produces the error message

```
Error, LinearAlgebra:  - DotProduct expects its 1st
argument, V1, to be of type Vector, but received
array(1 ..  3,[(1)=3,(2)=5,(3)=-4])
```

We will discuss vectors in the next section, where we define them and explain their notation. Let us anticipate these definitions by pointing out that for reasons that will become clear later in the course, we will often write lists of data vertically as well as horizontally. Both the **linalg** and the **LinearAlgebra** packages contain an operation for converting horizontal lists and vectors to vertical ones and vertical ones to horizontal ones. The operation is referred to as the ***transpose*** operation.

```
[> with(LinearAlgebra):
[> x:=Vector([a,b]):   y:=Vector([3,4]):
[> DotProduct(x,y,conjugate=false);
```

$$3a + 4b$$

```
[> DotProduct(Transpose(x),y,conjugate=false);
```

$$3a + 4b$$

```
[> DotProduct(x,Transpose(y),conjugate=false);
```

$$3a + 4b$$

```
[> DotProduct(Transpose(x),Transpose(y),conjugate=false);
```

$$3a + 4b$$

As we can see, the **DotProduct** function is insensitive to the horizontal or vertical orientation of its inputs. The same remark applies to the **dotprod** function, although it interacts less smoothly with the **transpose** function in **linalg**. In this text, we use the *Maple* output notation for vectors and their transposes. *Maple 6* also provides both an input and an output notation for the two orientations of vectors.

```
[> with(LinearAlgebra):
[> x:=<a|b>;
```

$$x := \begin{bmatrix} a & b \end{bmatrix}$$

```
[> y:=<3,4>;
```

$$y := \begin{bmatrix} 3 \\ 4 \end{bmatrix}$$

```
[> transx:=Transpose(x);
```

$$transx := \begin{bmatrix} a \\ b \end{bmatrix}$$

```
[> transy:=Transpose(y);
```

$$transy := \begin{bmatrix} 3 & 4 \end{bmatrix}$$

The outputs **x** and **transy** are called ***row vectors***, and the vectors **y** and **transx** are called ***column vectors***. The term ***vector*** refers both to a row and to a column vector. The elements a, b, 3, and 4 are called the ***components*** of the vectors. We write

$$\mathbf{x} = (x_1, \ldots, x_n)$$

for an arbitrary row or column vector. The context will make it clear which orientation of the vector is intended.

Using the language of vectors, we now define their dot product. In the definition, the expressions $x_1 y_1, \ldots, x_n y_n$ denote the results of multiplication. In most instances in this text, the components of vectors will be constants, variables, or polynomials.

DEFINITION 1.3 *The dot product* $\mathbf{x} \cdot \mathbf{y}$ *of the vectors* $\mathbf{x} = (x_1, \ldots, x_n)$ *and* $\mathbf{y} = (y_1, \ldots, y_n)$ *is the sum* $x_1 y_1 + \cdots + x_n y_n$.

As we see, the dot product associates with a pair of vectors the sum of the products of the corresponding components of the vectors.

If **x** and **y** are both given as column vectors, we usually write $\mathbf{x}^T \mathbf{y}$ for $\mathbf{x} \cdot \mathbf{y}$. The reasons for this notation will become clear when we discuss matrix multiplication. In such cases, the vector $\mathbf{x}^T$ will be a row of a matrix, and the vector **y** will be a column of another matrix. Although a row vector is essentially the same as a matrix consisting of a single row, and a column vector is essentially the same as a matrix consisting of a single column, *Maple* often requires us to treat them as different types of objects. The function **whattype** is often useful for verifying the type of *Maple* object.

Since matrices will be enclosed in square brackets, we use this notation to write dot product of the form $\mathbf{x}^T \mathbf{y}$. Thus, we will usually use the matrix notion

$$\mathbf{x}^T \mathbf{y} = \begin{bmatrix} x_1 \\ \vdots \\ x_n \end{bmatrix}^T \begin{bmatrix} y_1 \\ \vdots \\ y_n \end{bmatrix} = \begin{bmatrix} x_1 & \cdots & x_n \end{bmatrix} \begin{bmatrix} y_1 \\ \vdots \\ y_n \end{bmatrix}$$

for dot products.

EXAMPLE 1.2 Dot Products

Form the dot products $\mathbf{x} \cdot \mathbf{y}$ of **x** and **y** for the following pairs of vectors:

a. $\mathbf{x} = (5, 9), \mathbf{y} = (3, 7)$ b. $\mathbf{x} = (a, b), \mathbf{y} = (x, y)$ c. $\mathbf{x} = (a, b), \mathbf{y} = (a, b)$

Solution. By definition,

a. $\mathbf{x} \cdot \mathbf{y} = (5, 9) \cdot (3, 7) = 5 \times 3 + 9 \times 7 = 78$

b. $\mathbf{x} \cdot \mathbf{y} = (a, b) \cdot (x, y) = ax + by$

c. $\mathbf{x} \cdot \mathbf{y} = (a, b) \cdot (a, b) = a^2 + b^2$

If we were to assign numerical values to a, b, x, and y in cases (b) and (c), then the dot product would produce numerical values in all three cases. ▲

EXAMPLE 1.3 A Dot Product in Matrix Notation

Write the product $\mathbf{x} \cdot \mathbf{y}$ of the vectors $\mathbf{x} = (5, 9)$ and $\mathbf{y} = (3, 7)$ in matrix notation.

Solution. We write $\mathbf{x} \cdot \mathbf{y}$ as the matrix product $\mathbf{x}^T \mathbf{y}$.

$$\mathbf{x}^T \mathbf{y} = \begin{bmatrix} 5 \\ 9 \end{bmatrix}^T \begin{bmatrix} 3 \\ 7 \end{bmatrix} = \begin{bmatrix} 5 & 9 \end{bmatrix} \begin{bmatrix} 3 \\ 7 \end{bmatrix}$$

The value of $\mathbf{x}^T \mathbf{y}$ coincides with that of the dot product $\mathbf{x} \cdot \mathbf{y}$ when $\mathbf{x}^T \mathbf{y} = 5 \times 3 + 9 \times 7$. ▲

We will see later that the idea of multiplying two more general matrices is built on the dot product operation as illustrated in Example 1.3.

EXAMPLE 1.4 Using dotprod to Define a Linear Equation

Use the **dotprod** function to define the linear equation whose coefficients are the real numbers 3, 5, $-\pi$, 45, -27, with corresponding variables are u, v, w, x and y, and constant term 18.

Solution. We begin by loading the **linalg** package.

```
[> with(linalg):
```

Next we define the constant and variable vectors. We have a choice, but opt for column vectors.

```
[> cons:=<3,5,-Pi,45,-27>:
[> vars:=<u,v,w,x,y>:
[> dotprod(cons,vars,orthogonal)=18;
```

$$3u + 5v - \pi w + 45x - 27y = 18$$

As we can see, the input `-Pi` produced the constant π as output. ▲

The option "orthogonal" of the **dotprod** function is connected with the fact that the dot product is used to generalize the idea of perpendicularity. Until we study this idea in depth in Chapter 8, we will use the dot product to define the orthogonality of two vectors.

DEFINITION 1.4 *Two vectors* $\mathbf{x} = (x_1, \ldots, x_n)$ *and* $\mathbf{y} = (y_1, \ldots, y_n)$ *are **orthogonal** if their dot product* $\mathbf{x} \cdot \mathbf{y}$ *is zero.*

The orthonormal basis theorem in Chapter 8 explains why this definition of orthogonality plays such a prominent role in linear algebra.

EXAMPLE 1.5 Orthogonal Vectors

Show that the vectors $\mathbf{x} = (1, 2, -5, 0, 9)$ and $\mathbf{y} = (-2, 1, 9, 20, 5)$ are orthogonal.

Solution. By definition,

$$\begin{aligned} \mathbf{x} \cdot \mathbf{y} &= (1, 2, -5, 0, 9) \cdot (-2, 1, 9, 20, 5) \\ &= 1 \times (-2) + 2 \times 1 + (-5) \times 9 + 0 \times 20 + 9 \times 5 \\ &= 0 \end{aligned}$$

Hence $\mathbf{x}$ and $\mathbf{y}$ are orthogonal. ▲

In addition to the dot product functions, the **linalg** package of *Maple* has another built-in function that can be used to generate a linear equation.

EXAMPLE 1.6 Using geneqns to Generate a Linear Equation

Use the **geneqns** function in the **linalg** package to generate a linear equation with the coefficient vector $(2, -4, 3)$, the variable vector (x, y, z), and the constant term 8. Its input requirements, however, are stricter. The coefficients must be specified as a matrix, the variables must be specified as a list, and the constant term must be specified as a vector.

Solution. We begin by loading the **linalg** package.

```
[> with(linalg):
```

Next we use the **geneqns** function to build the require linear equation.

```
[> coefficients:=matrix([[2,-4,3]]):
[> variables:=[x,y,z]:
[> constants:=vector([8]):
[> geneqns(coefficients, variables, constants);
```

$$\{2x - 4y + 3z = 8\}$$

As we can see, the result is the expected linear equation. ▲

Instead of producing the equation $2x - 4y + 3z = 8$, *Maple* has produced a set whose only element is the desired equation. The reason for this is that the **geneqns** function is actually designed to generate several equations for more general input matrices. In the next section, we will discuss such matrices.

Linear equations are used to express quantitative relationships between variables. We refer to the process of finding real numbers r_1, r_2, r_3, r_4, r_5 for which

$$3r_1 + 5r_2 - \pi r_3 + 45r_4 - 27r_5 = 18$$

as ***solving the equation*** $3x_1 + 5x_2 - \pi x_3 + 45x_4 - 27x_5 = 18$. This process may or may not lead to a solution.

DEFINITION 1.5 *A vector* $(r_1, \ldots, r_n)$ *of real numbers is a* ***solution*** *of the linear equation* $a_1x_1 + \cdots + a_nx_n = b$ *if* $a_1r_1 + \cdots + a_nr_n = b$.

For example, the vectors $(1, 3, 0, 0, 0)$ and $(0, 0, 0, 1, 1)$ are solutions of the equation

$$3u + 5v - \pi w + 45x - 27y = 18$$

The set of all solutions of an equation is called the ***solution set*** of the equation. Instead of speaking of a solution vector, we also say that the values

$$x_1 = r_1, \ldots, x_n = r_n$$

are a solution of an equation, or that the set

$$\{x_1 = r_1, \ldots, x_n = r_n\}$$

is a solution.

Two steps are required to express and solve linear equations in *Maple:* First we must write the equation in *Maple* form. Then we can use the **solve** function to find the solutions. If an equation contains more than one variable, we must specify the variables for which the equation is to be solved.

EXAMPLE 1.7 Using *Maple* to Solve a Linear Equation

Write the following linear equations in *Maple* notation and find their solutions.

a. $5x_1 = 16$ b. $5x_1 - x_2 = 16$

c. $x_1 + 45x_2 + 7x_3 = 0$ d. $x_1 + 5x_2 + 5x_3 - \pi x_4 = 1$

Solution. Equation (a) is written as `5*x[1]=16`. We then use the **solve** function to solve for the variable x_1.

```
[> solve(5*x[1]=16);
```

$$\frac{16}{5}$$

Equation (b) contains two variables. We use the **solve** function to solve for the variable x_1 in terms of the variable x_2.

```
[> solve(5*x[1]-x[2]=16,{x[1]});
```

$$\{x_1 = \frac{1}{5}x_2 + \frac{16}{5}\}$$

Equation (c) contains three variables. We use the **solve** function to solve for x_1 in terms of x_2 and x_3.

```
[> solve(x[1]+45*x[2]+7*x[3]=0,{x[1]});
```

$$\{x_1 = -45x_2 - 7x_3\}$$

Equation (d) contains four variables, and we use the **solve** function again to solve for the variable x_1 in terms of the variables x_2, x_3, and x_4.

```
[> solve(x[1]+5*x[2]+5*x[3]-Pi*x[4]=1,{x[1]});
```

$$\{x_1 = -5x_2 - 5x_3 + \pi\, x_4 + 1\}$$

As we can see, only equation (a) has an explicit solution. In the other cases, we must assign values to the variable for which we have solved the equations in order to obtain a specific numerical solution. ▲

We saw that in the case of the **geneqns** function, the coefficients of the variables had to be entered as a matrix and the variables as a list. Essentially this means that they had to be entered as vectors. This prompts us to consider the idea of writing linear equations as ***matrix equations***. We conclude this section with a discussion of how *Maple* can be used to solve a linear equation written in matrix form.

Let us begin by defining a matrix equation. We will see later that this definition is a special case of the definition of a system of linear equations written in matrix form.

DEFINITION 1.6 *For any row vector* $\mathbf{a} = (a_1, \ldots, a_n)$ *of constants, column vector*

$$\mathbf{x} = \begin{bmatrix} x_1 \\ \vdots \\ x_n \end{bmatrix}$$

of variables, and constant b*, the equation* $\mathbf{a}^T\mathbf{x} = b$ *is the* ***matrix equation*** *associated with the linear equation* $a_1x_1 + \cdots + a_nx_n = b$.

EXAMPLE 1.8 A Linear Equation as a Matrix Equation

Write the linear equation $5x_1 - x_2 = 16$ as a matrix equation.

Solution. If we let $\mathbf{a} = (5, -1)$ be a coefficient vector and let $\mathbf{x} = (x_1, x_2)$ be a variable vector, then

$$\mathbf{a}^T\mathbf{x} = \begin{bmatrix} 5 \\ -1 \end{bmatrix}^T \begin{bmatrix} x_1 \\ x_2 \end{bmatrix} = \begin{bmatrix} 5 & -1 \end{bmatrix} \begin{bmatrix} x_1 \\ x_2 \end{bmatrix} = 16$$

is the required matrix equation. ▲

EXAMPLE 1.9 Using *Maple* to Solve a Matrix Equation

Use *Maple* to solve the matrix equation

$$\begin{bmatrix} 5 & -1 \end{bmatrix} \begin{bmatrix} x_1 \\ x_2 \end{bmatrix} = 16$$

Solution. The given matrix contains the variables x_1 and x_2. We use *Maple* to solve for x_1 in terms of x_2.

```
[> A:=<5|-1>;
```

$$A := \begin{bmatrix} 5 & -1 \end{bmatrix}$$

```
[> X:=<[x[1],x[2]>;
```

$$X := \begin{bmatrix} x_1 \\ x_2 \end{bmatrix}$$

```
[> A.X;
```

$$5x_1 - x_2$$

```
[> solve(A.X=16,{x[1]});
```

$$\{x_1 = \tfrac{1}{5}x_2 + \tfrac{16}{5}\}$$

As we can see, the solutions of the matrix equation agree with those found for the corresponding linear equation in Example 1.7. ▲

EXERCISES 1.1

1. Form all possible linear equations $a_1x_1 + a_2x_2 = 5$ using the constants 0, 1, and 2 as coefficients.

2. Rewrite the linear equations constructed in Exercise 1 as matrix equations.

3. Use the variables x and y to construct a linear equation for which $x = 9$ and $y = -7$ is a solution.

4. Find coefficients a and b for which the equation $ax + by = 12$ has the solution $x = 1$ and $y = 1$.

5. Find coefficients a and b for which the equation $ax + by = 12$ has the solution $x = 1$ and $y = 0$.

6. Why is it impossible to find coefficients a and b for which the equation $ax + by = 12$ has the solution $x = 0$ and $y = 0$?

7. Calculate the dot products of the following pairs of vectors.

   a. $\mathbf{x} = (1, 2, 3)$ and $\mathbf{y} = (2, 2, 1)$ b. $\mathbf{x} = (6, 9, -1)$ and $\mathbf{y} = (6, 0, 7)$
   c. $\mathbf{x} = (5, 0, 1, 6)$ and $\mathbf{y} = (a, b, c, d)$ d. $\mathbf{x} = (6, 9, -1, 4)$ and $\mathbf{y} = (w, x, y, z)$

8. Assign nonzero values to a, b, c, d and w, x, y, z in Exercise 7b so that the vectors $\mathbf{x}$ are orthogonal to the vectors $\mathbf{y}$.

9. (*Maple V*) Use the **genmatrix** function to convert the equation $3x_1 + 7x_2 - 9x_3 = 12$ to a matrix equation.

10. (*Maple V*) Set up the **solve** command for solving the equation $3x_1 + 7x_2 - 9x_3 = 12$ in terms of x_2 and x_3, but do not solve.

11. (*Maple V*) Rewrite the *Maple* command in Exercise 10 in vector form.

12. (*Maple V*) Solve the equation in Exercise 9 for x_1 in terms of x_2 and x_3.

13. (*Maple V*) Solve the equation in Exercise 9 for x_2 in terms of x_1 and x_3.

14. (*Maple V*) Explore the properties of the **transpose** function in the **linalg** package.

15. (*Maple V*) Use the **dotprod** function in the **linalg** package to solve the equation in Exercise 9 for x_1 in terms of x_2 and x_3.

16. (*Maple 6*) Use the **DotProduct** function in the **LinearAlgebra** package to solve the equation in Exercise 9 for x_2 in terms of x_1 and x_3.

17. (*Maple 6*) Use the **GenerateMatrix** and **GenerateEquations** functions in the **LinearAlgebra** package to create a linear equation and its vector form from the following data:

Coefficient Vector:	$(3, -8, 9, 0, 11, 12, 1)$
Variable Vector:	(t, u, v, w, x, y, z)
Constant:	-7

18. (*Maple 6*) Explore the properties of the **Transpose** function in the **LinearAlgebra** package.

Points, Arrows, and Vectors

Most methods for visualizing fundamental concepts in linear algebra are based on our understanding of Euclidean analytic geometry. We are all familiar with the idea of placing dots on a piece of paper and calling them ***points***. It was Rene Descartes who thought of labeling these points with numerical coordinates that identify the locations of the points in a unique way, relative to a fixed set of coordinate axes. This link between geometry and algebra makes it possible for us to think of points as sets of numbers and of other geometric objects as equations.

A point in the plane requires a list (x, y) consisting of two numbers, and a point in space requires a list (x, y, z) consisting of three numbers. For algebraic reasons, we will think of such lists as column vectors. We treat the points (x, y) in $\mathbb{R}^2$ and the points (x, y, z) in $\mathbb{R}^3$ as the column vectors

$$\begin{bmatrix} x \\ y \end{bmatrix} \quad \text{and} \quad \begin{bmatrix} x \\ y \\ z \end{bmatrix}$$

respectively. To save space, we often write these vectors in their point form as (x, y) and (x, y, z). When thinking geometrically, we refer to the numbers x, y, and z as the ***coordinates*** of the points (x, y) and (x, y, z). When thinking algebraically, we refer to them as the ***components*** of the corresponding column vectors. We will use boldface letters $\mathbf{u}, \mathbf{v}, \mathbf{w}, \mathbf{x}, \mathbf{y}, \mathbf{z}, \ldots$ for column vectors and italic letters $u, v, w, x, y, z, \ldots$ for their components.

Algebraically, we can of course form column vectors of any height. We therefore define $\mathbb{R}^n$ to be the set of column vectors of height n, with real numbers as their components. Here n can be any natural number. In particular, $\mathbb{R}^1$ is interpreted as the real line $\mathbb{R}$. We can scale the vectors in $\mathbb{R}^n$ by multiplying all components of a vector by the constant, and we can add such vectors by adding their corresponding components. With these operations, vector algebra is born. The operations are called ***scalar multiplication*** and ***vector addition***. In $\mathbb{R}^n$, we define them as follows.

DEFINITION 1.7 *For any column vectors $(x_1, \ldots, x_n)$ and $(y_1, \ldots, y_n)$ in $\mathbb{R}^n$, and any real number a, the scalar multiple $a\mathbf{x}$ and the vector sum $(\mathbf{x} + \mathbf{y})$ are defined by*

$$a \begin{bmatrix} x_1 \\ \vdots \\ x_n \end{bmatrix} = \begin{bmatrix} ax_1 \\ \vdots \\ ax_n \end{bmatrix} \quad \text{and} \quad \begin{bmatrix} x_1 \\ \vdots \\ x_n \end{bmatrix} + \begin{bmatrix} y_1 \\ \vdots \\ y_n \end{bmatrix} = \begin{bmatrix} x_1 + y_1 \\ \vdots \\ x_n + y_n \end{bmatrix}$$

In $\mathbb{R}^2$, these operations have elegant geometric interpretations. Given any vector $\mathbf{u} \in \mathbb{R}^2$, the vector $a\mathbf{u}$ is parallel to $\mathbf{u}$ and is scaled by a factor a. If a is negative, then the arrow $a\mathbf{u}$ points in the direction opposite to that of $\mathbf{u}$. However, the vectors $\mathbf{u}$ and $a\mathbf{u}$ always lie on the same straight line through the origin (0, 0).

We can use parallel arrows to depict the quantitative relationships between these vectors. The **plots** and **plottools** packages of *Maple* make it easy to depict parallel arrows and to illustrate scaling.

```
[> with(plottools):
[> arrow1:=arrow([0,0],[4,3],.1,.4,.1):
[> arrow2:=arrow([2,4], [6,7],.1,.4,.1):
[> arrow3:=arrow(3*[2,4],3*[6,7],.1,.4,.1):
[> arrow4:=arrow((-2)*[6,2],(-2)*[10,5],.1,.4,.1):
```

```
[> plots[display](arrow1,arrow2,arrow,arrow4);
```

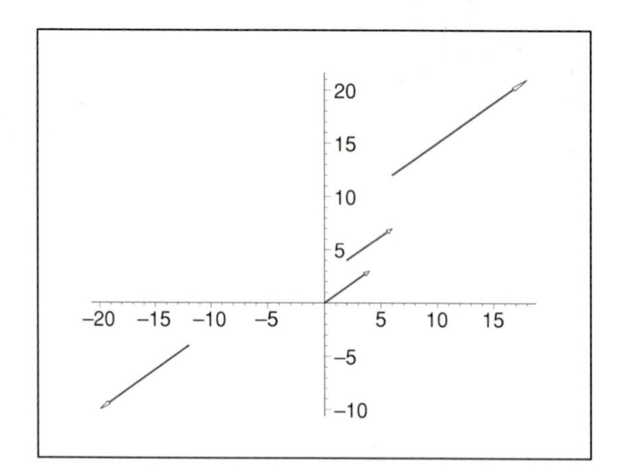

This graph shows that the four arrows are parallel and the three of them point in the same direction. We can use the same packages to illustrate vector addition in $\mathbb{R}^2$.

```
[> with(plottools):
[> arrow1:=arrow([0,0],[4,3],.05,.4,.1):
[> arrow2:=arrow([0,0],[2,5],.05,.4,.1):
[> arrow3:=arrow([4,3],[6,8],.05,.4,.1):
[> arrow4:=arrow([2,5],[6,8],.05,.4,.1):
[> arrow5:=arrow([0,0],[6,8],.05,.4,.05):
```

```
[> plots[display](arrow1,arrow2,arrow3,arrow4,arrow5);
```

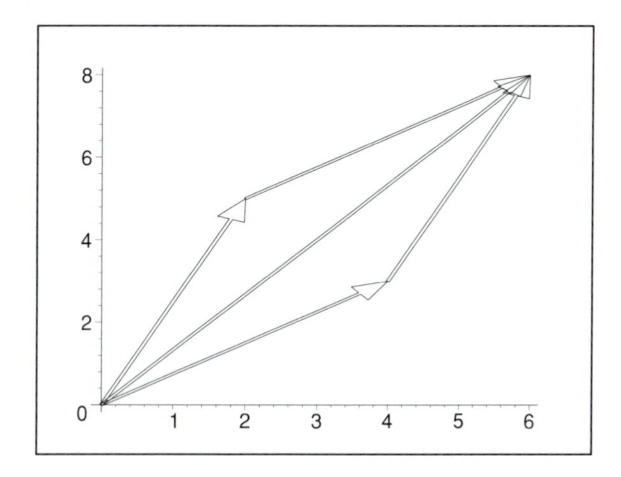

The graph shows that the sum of the vectors $(4, 3)$ and $(2, 5)$, for example, is the vector $(6, 8)$. It can be built from the vectors $(4, 3)$ and $(2, 5)$ using the ***parallelogram law***. If we copy the position vector $(2, 5)$ and displace this copy parallel to itself so that its initial point coincides with the terminal point of $(4, 3)$, then the sum vector $(6, 8)$ is a diagonal of the parallelogram determined by the vectors $(4, 3)$ and $(2, 5)$. In Chapter 4, we show that the operations of scalar multiplication and vector addition as defined satisfy the laws of a structure called a ***real vector space***.

Geometry and Linear Equations

The graphic capabilities of *Maple* also allow us to illustrate the geometric properties of the solutions of linear equations in $\mathbb{R}^2$ and $\mathbb{R}^3$. Here are some examples.

EXAMPLE 1.10 Graphing a Linear Equation in $\mathbb{R}^2$

Use *Maple* to graph the solutions of the equation $x + 2y - 7 = 0$.

Solution. We use the **implicitplot** function to plot the graph. First we load the **plots** package.

```
[> with(plots):
```

We now instruct *Maple* to plot the equation. We assume that the variables x and y range over the interval $-10 \leq x, y \leq 10$.

```
[> implicitplot(x+2*y-7=0,x=-10..10,y=-10..10);
```

The result is the following graph.

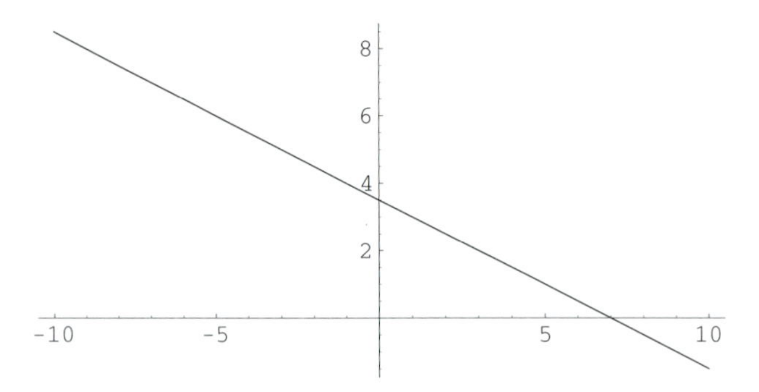

As we can see, the solution $x = 1$ and $y = 3$ corresponds to a point on the graph. ▲

EXAMPLE 1.11 Finding Points on the Graph of a Linear Equation in $\mathbb{R}^2$

Use *Maple* to find two points on the graph of the equation $x + 2y - 7 = 0$.

Solution. We decide, arbitrarily, to find the points $(7, y)$ and $(-15, y)$.

```
[> x:=7:
[> solve(x+2*y-7=0,y);
                                0
```

This shows that the point $(7, 0)$ lies on the graph of the given equation. Next we find the point $(-15, y)$.

```
[> x:=-15:
[> y=(7-x)/2;
                              y = 11
```

This shows that the point $(-15, 11)$ also lies on the graph of the given equation. ▲

EXAMPLE 1.12 Finding a Line in $\mathbb{R}^2$ Determined by Two Points

Use *Maple* to find the line in $\mathbb{R}^2$ determined by the points (1, 1) and (3, 5).

Solution. Let $y = mx + b$ be the equation of the line determined by (1, 1) and (3, 5). We need to find the slope m and the y-intercept b of the line. The calculation

```
[> solve({1=m+b,5=3*m+b});
```

$$\{m = 2, b = -1\}$$

tells us that $y = 2x - 1$ is the equation of the required line. ▲

EXAMPLE 1.13 Graphing a Linear Equation in $\mathbb{R}^3$

Consider the equation $x + 2y - 3z - 7 = 0$ in the variables x, y, and z and the real constants 1, 2, -3, and -7. This equation is algebraically and geometrically linear. Three real numbers r_1, r_2, and r_3 are a solution of $x + 2y - 3z - 7 = 0$ if $r_1 + 2r_2 - 3r_3 - 7 = 0$.

In geometry, the solution $x = r_1$, $y = r_2$, and $z = r_3$ can be viewed as the point (r_1, r_2, r_3) in the plane $x + 2y - 3z - 7 = 0$. The following graph displays the solutions of the equation.

```
[> with(plots):
[> implicitplot3d(x+2*y-3*z-7=0,x=-10..10,y=-10..10,
      z=-10..10);
```

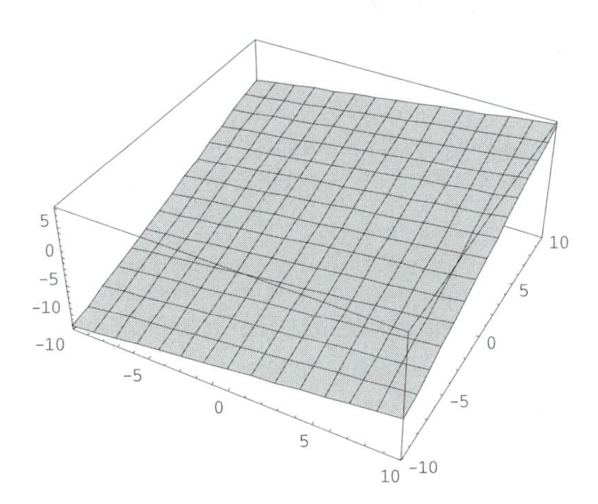

As we can see, the solutions of $x + 2y - 3z - 7 = 0$ form a plane in space. ▲

EXAMPLE 1.14 Verifying That a Point Lies on a Plane

Use *Maple* to determine which of the points (1, 2, 3) and (2, 4, 1) lie in the plane $x + 2y - 3z - 7 = 0$.

Solution. We assign the values 1, 2, and 3 to x, y, and z and evaluate $x + 2y - 3z - 7$.

```
[> x:=1:  y:=2:  z:=3:
[> x+2*y-3*z-7;
                              -11
```

This tells us that the point $(1, 2, 3)$ does not lie in the plane $x + 2y - 3z - 7 = 0$ since the calculated value of $x + 2y - 3z - 7$ is not 0. We therefore investigate the second point.

```
[> x:=2:  y:=4:  z:=1:
[> x+2*y-3*z-7;
                              0
```

Hence the point $(2, 4, 1)$ lies in the given plane. ▲

Vector Equations

We can use the vectors in $\mathbb{R}^2$ or $\mathbb{R}^3$ to define new kinds of equations known as ***vector equations.*** They are different from the matrix equations discussed earlier. For example, we can form the equation

$$\begin{bmatrix} x \\ y \end{bmatrix} = \begin{bmatrix} 1 \\ 2 \end{bmatrix} + t \begin{bmatrix} 5 \\ 8 \end{bmatrix}$$

in which x, y, and t are variables. The associated linear equations

$$\begin{cases} x = 1 + 5t \\ y = 2 + 8t \end{cases}$$

are known as ***parametric equations***. The variable t is called the ***parameter*** of the equations. The general notation for a vector equation is

$$\mathbf{x} = \mathbf{a} + t\mathbf{d}$$

with $\mathbf{x}$, $\mathbf{a}$, and $\mathbf{d}$ denoting vectors and t denoting a parameter. The solutions of a vector equation are found by letting t take on all real numbers as values. The vector $\mathbf{a}$ is viewed as a ***position vector*** and is called a ***point***, whereas the vector $\mathbf{d}$ is referred to a ***direction arrow***. The components of $\mathbf{d}$ are called ***direction numbers.*** It is easy to see that in $\mathbb{R}^2$ and $\mathbb{R}^3$, the sets of solutions of vector equations can be viewed as straight lines. The equation $\mathbf{x} = \mathbf{a} + t\mathbf{d}$ is then called the ***parametric representation*** of the line determined by $\mathbf{a}$ and $\mathbf{d}$.

EXAMPLE 1.15 Parametric Representation of a Straight Line in $\mathbb{R}^2$

Show that the vector equation $(x, y) = (2, 3) + t(5, 7)$ represents the straight line through the point $(2, 3)$ and in the direction of the arrow determined by the points $(0, 0)$ and $(5, 7)$.

Solution. We first use the point–slope formula

$$y - y_1 = m(x - x_1)$$

to find the equation of the straight line determined by the given point and arrow. Since the arrow starts at $(0, 0)$ and ends at $(5, 7)$, we know that it determines a line whose slope is $7/5$. Therefore

$$y - 3 = \tfrac{7}{5}(x - 2)$$

is the required equation. We can rewrite this equation as $7x - 5y + 1 = 0$. On the other hand, the vector equation

$$(x, y) = (2, 3) + t(5, 7)$$

corresponds to the two equations $x = 2 + 5t$ and $y = 3 + 7t$. If we solve both equations for t and equate the results, we get

$$\frac{x - 2}{5} = t = \frac{y - 3}{7}$$

This equation converts to $7x - 5y + 1 = 0$. ▲

In the case of $\mathbb{R}^2$, we have a choice of defining a line in parametric or nonparametric form. In the case of $\mathbb{R}^3$, however, we have no such choice. A linear equation of the form $ax + by + cz + d = 0$ always determines a plane rather than a straight line. We must therefore use the parametric form to define a line in $\mathbb{R}^3$.

EXAMPLE 1.16 Parametric Representation of a Straight Line in $\mathbb{R}^3$

Find the parametric equations of the straight line determined by the point $\mathbf{a} = (1, 2, 3)$ and the direction arrow $\mathbf{d} = (5, -2, 6)$.

Solution. In the parameter t, the required vector equation is $(x, y, z) = (1, 2, 3) + t(5, -2, 6)$. We can rewrite this equation by separating out its components:

$$\begin{cases} x = 1 + 5t \\ y = 2 - 2t \\ z = 3 + 6t \end{cases}$$

In its component form, the straight line is represented by three linear equations. ▲

EXERCISES 1.2

1. Write the position vectors $(2, 3)$, $(-2, 3)$, $(2, -3)$, and $(-2, -3)$ in column form.

2. Write the column vectors

$$\text{a.}\begin{bmatrix} 8 \\ -2 \end{bmatrix} \quad \text{b.}\begin{bmatrix} 7 \\ 3 \end{bmatrix} \quad \text{c.}\begin{bmatrix} -15 \\ -1 \end{bmatrix} \quad \text{d.}\begin{bmatrix} -8 \\ 9 \end{bmatrix}$$

in point form.

3. Use the parallelogram law to find the points determined by the following sum vectors.

$$\text{a.}\begin{bmatrix} 1 \\ 2 \end{bmatrix} + \begin{bmatrix} 3 \\ 4 \end{bmatrix} \quad \text{b.}\begin{bmatrix} 1 \\ 2 \end{bmatrix} + \begin{bmatrix} -3 \\ 4 \end{bmatrix}$$

$$\text{c.}\begin{bmatrix} 1 \\ 2 \end{bmatrix} + \begin{bmatrix} 3 \\ -4 \end{bmatrix} \quad \text{d.}\begin{bmatrix} 1 \\ 2 \end{bmatrix} + \begin{bmatrix} -3 \\ -4 \end{bmatrix}$$

4. Find the points corresponding to the column vectors

$$\text{a. } 3\begin{bmatrix} 1 \\ 2 \end{bmatrix} + 0\begin{bmatrix} 4 \\ 5 \end{bmatrix} \quad \text{b. } 3\begin{bmatrix} 1 \\ 2 \end{bmatrix} - 2\begin{bmatrix} 4 \\ 5 \end{bmatrix}$$

5. Find the equations of the straight lines determined by the following pairs of points. Write each equation in the form $ax + by = c$.

a. (3, 4) and (6, 2) b. (3, 4) and (−6, 2)
c. (3, 4) and (6, −2) d. (3, 4) and (0, 0)

6. Determine which of the following sets of points satisfy the same linear equation. Give a geometric explanation of your answer.

a. (1, 10), (−2, 1), (3, 16) b. $(0, -\frac{100}{3}, -16)$, (−2, 0, 2), (−1, −7, −5)
c. (1, −2), (−5, 4), (1, 2) d. (1, 1, −9), (3, −1, −7), (0, −4, −16)

7. Find the vector equation of the straight line determined by the point $a = (4, -2)$ and the direction arrow (7, 12).

8. Find the vector equation of the straight line determined by the point $a = (4, -2, 6)$ and the direction arrow (7, 12, −5).

9. (*Maple V*) Write the column vectors in Exercise 1 in *Maple V* notation.

10. (*Maple V*) Use the **arrow** function in the **plottools** package to graph the position vectors in Exercise 1.

11. (*Maple V*) Use the **vector** function to represent the vectors in *Maple*.

a. $\begin{bmatrix} 1 \\ 2 \end{bmatrix}$ b. $\begin{bmatrix} 4 \\ 5 \end{bmatrix}$ c. $\begin{bmatrix} -6 \\ 3 \end{bmatrix}$ d. $\begin{bmatrix} 1 \\ 2 \end{bmatrix} + \begin{bmatrix} 4 \\ 5 \end{bmatrix}$ e. $\begin{bmatrix} 1 \\ 2 \end{bmatrix} + \begin{bmatrix} -6 \\ 3 \end{bmatrix}$

12. (*Maple V*) Use the **implicitplot** function in the **plots** package to graph the following lines in $\mathbb{R}^2$, using $-5 \leq x, y \leq 5$.

a. $x + y = 0$ b. $x - y = 0$
c. $x + y = 1$ d. $3x - 4y = 2$

13. (*Maple V*) Use the **plot3d** function to graph the following planes in $\mathbb{R}^3$, using $-5 \leq x \leq 5$ and $-10 \leq y \leq 10$.

a. $x + y + z = 0$ b. $x - y + z = 0$
c. $x + y - z = 1$ d. $3x - 4y + 5z = 2$

14. (*Maple V*) Find the lines in $\mathbb{R}^2$ determined by the following pairs of points.

a. $(2, 2)$ and $(1, -1)$ b. $(2, 2)$ and $(-1, 1)$
c. $(3, 7)$ and $(2, 2)$ d. $(0, 5)$ and $(8, 1)$

15. (*Maple V*) Determine which of the points $(0, 5)$, $(8, 3)$, and $(6, 5)$ lie on the line $-28x + 32y + 128 = 0$.

16. (*Maple V*) Graph the planes in $\mathbb{R}^3$ determined by the following triples of points.

a. $(2, 2, 3)$, $(7, 6, 5)$ and $(1, -1, 1)$ b. $(2, 2, -2)$, $(7, 6, 5)$, and $(1, -1, 1)$
c. $(2, 2, 6)$, $(7, 6, 5)$, and $(-1, 1, 0)$ d. $(0, 5, 0)$, $(8, 1, 3)$, and $(7, 6, 5)$

17. (*Maple V*) Determine which of the three points $(8, 3, 1)$, $(0, 5, -72)$, and $(6, 5, -30)$ lie in the plane $-28x + 32y + 4z + 128 = 0$.

18. (*Maple V*) Graph the following linear equations and describe how they differ.

a. $4x + 3y = 7$ b. $-4x + 3y = 7$ c. $4x - 3y = 7$

19. (*Maple V*) Graph the following linear equations by treating z as a function of x and y. Describe the graphs.

a. $x + y + z = 7$ b. $3x + y + z = 7$
c. $x + 3y + z = 7$ d. $x + y + 3z = 7$

20. (*Maple 6*) Use the **Vector** function to represent the vectors

a. $\begin{bmatrix} 1 \\ 2 \end{bmatrix}$ b. $\begin{bmatrix} 4 \\ 5 \end{bmatrix}$ c. $\begin{bmatrix} -6 \\ 3 \end{bmatrix}$ d. $\begin{bmatrix} 1 \\ 2 \end{bmatrix} + \begin{bmatrix} 4 \\ 5 \end{bmatrix}$ e. $\begin{bmatrix} 1 \\ 2 \end{bmatrix} + \begin{bmatrix} -6 \\ 3 \end{bmatrix}$

LINEAR SYSTEMS

Linear algebra provides a unifying language, a simplified notation, and comprehensive techniques for solving systems of linear equations.

DEFINITION 1.8 *A **linear system** in the variables $x_1, \ldots, x_n$ is a sequence*

$$\begin{cases} a_{11}x_1 + \cdots + a_{1n}x_n &= b_1 \\ &\vdots \\ a_{m1}x_1 + \cdots + a_{mn}x_n &= b_m \end{cases}$$

*of equations. The real numbers $a_{11}, \ldots, a_{mn}$ are the **coefficients** of the system and the real numbers $b_1, \ldots, b_m$ are its **constant terms**.*

To save space, we sometimes write a linear system horizontally in set form as

$$\{a_{11}x_1 + \cdots + a_{1n}x_n = b_1, \ldots, a_{m1}x_1 + \cdots + a_{mn}x_n = b_m\}$$

We refer to a linear system in m equations and n variables as an $m \times n$ linear system. If $m = n$, we call the system ***determined***. If $m < n$, we call it ***underdetermined***. If $m > n$, we call it ***overdetermined***. The coefficients and constant terms of a linear system will be assumed to be real numbers, unless otherwise specified. Since it may happen that all coefficients of a linear system are zero, we give a name to such a system. We call a linear system ***degenerate*** if all of its equations are degenerate.

One of our goals is to develop systematic ways for finding the solutions of linear systems. A list $(r_1, \ldots, r_n)$ of real numbers is a ***solution*** of an $m \times n$ linear system if it is a solution of each equation in the linear system. We also often say that the values $r_1, \ldots, r_n$ of the variables $x_1, \ldots, x_n$ are a solution of the system.

We will see later that all linear systems either have 0, 1, or infinitely many solutions. Table 1 lists the three possibilities.

TABLE 1 SOLUTIONS OF LINEAR SYSTEMS.

LINEAR SYSTEM	EQUATIONS/VARIABLES	POSSIBLE SOLUTIONS
Determined	$m = n$	$0, 1, \infty$
Overdetermined	$m > n$	$0, 1, \infty$
Underdetermined	$m < n$	$0, \infty$

If a linear system has at least one solution, we call the system ***consistent***. If it has no solution, we call it ***inconsistent***. If the constant terms $b_1, \ldots, b_m$ of an $m \times n$ linear system are all zero, we refer to the system as ***homogeneous***. All $m \times n$ homogeneous linear systems are consistent since they always have the solution $x_1 = 0, \ldots, x_n = 0$.

Determined Linear Systems

The linear system

$$\begin{cases} x + y = 1 \\ x + y = 2 \end{cases}$$

is a 2×2 determined linear system. It has no solution and is therefore inconsistent.

The 2×2 determined linear system

$$\begin{cases} x + 2y = 5 \\ 2x + y = 7 \end{cases}$$

on the other hand, has the unique solution $(3, 1)$ since $3 + 2 \times 1 = 5$ and $2 \times 3 + 1 = 7$. It is therefore consistent.

The 2×2 determined linear system

$$\begin{cases} x + y = 1 \\ -x - y = -1 \end{cases}$$

is consistent since it has infinitely many solutions. For every real number s, the point $(s, 1-s)$ is a solution.

We can use *Maple* to illustrate these three linear systems graphically using the **implicitplot** function in the **plots** package, which allows us to plot linear systems in two variables.

```
[>plots[implicitplot]({x+y=1,x+y=2},x=-5..10,y=-4..4);
```

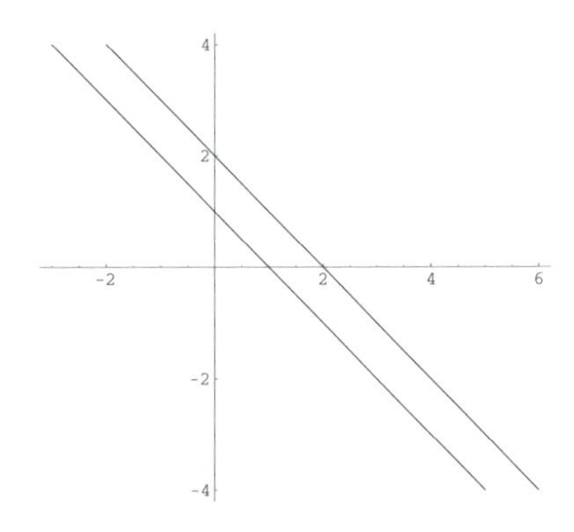

```
[>plots[implicitplot]({x+y=7,x-y=3},x=-5..10,y=-4..10);
```

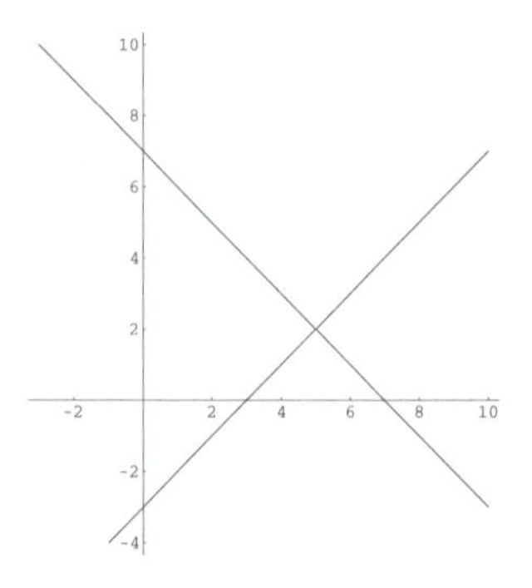


```
[>plots[implicitplot]({x+y=1,-x-y=-1},x=-5..10,y=-4..10);
```

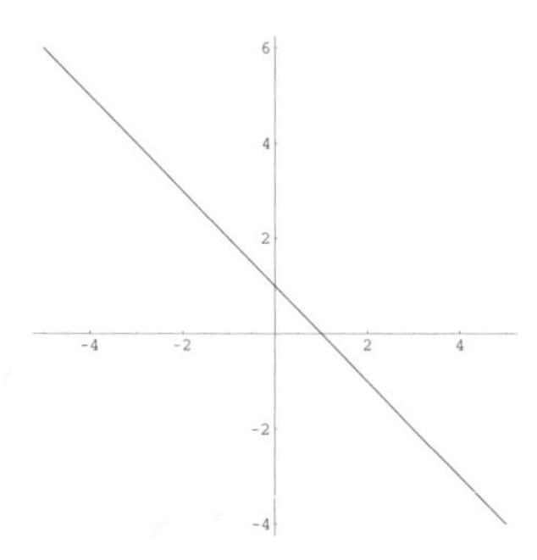

As we can see, an inconsistent 2×2 linear system corresponds to two parallel lines. A 2×2 linear system with a unique solution corresponds to two lines intersecting at a single point. A 2×2 linear system with infinitely many solutions corresponds to two overlapping lines.

These examples show that there are determined linear systems with no solution, with one solution, and with infinitely many solutions. We will see later on that there are no other possibilities. There are no linear systems with exactly two solutions, for example.

Overdetermined Linear Systems

Let us consider the case of 3×2 linear systems. The system

$$\begin{cases} x + 2y = 5 \\ x - 2y = 1 \\ 2x + y = 6 \end{cases}$$

for example, is inconsistent. In any solution satisfying the first two equations, x would have to take the value 3, and in any solution satisfying the last two equations, x would have to take the value $13/5$. Hence there is no common value for x that satisfies all three equations.

The consistent system

$$\begin{cases} x + 2y = 5 \\ x - 2y = 1 \\ 2x + y = 7 \end{cases}$$

on the other hand, has the unique solution $x = 3$ and $y = 1$.

The consistent system

$$\begin{cases} x + 2y = 5 \\ 3x + 6y = 15 \\ -x - 2y = -5 \end{cases}$$

has infinitely many solutions. For any real number s, the list $(-2s + 5, s)$ is a solution.

Underdetermined Linear Systems

Now let us consider the case of 2×3 linear systems. The system

$$\begin{cases} x + y + z = 25 \\ x + y + z = 2 \end{cases}$$

for example, is inconsistent. It implies that $x + y + z = 5 = x + y + z = 20$. This is obviously false. Graphically, this inconsistency is illustrated by the two parallel planes in Figure 1.

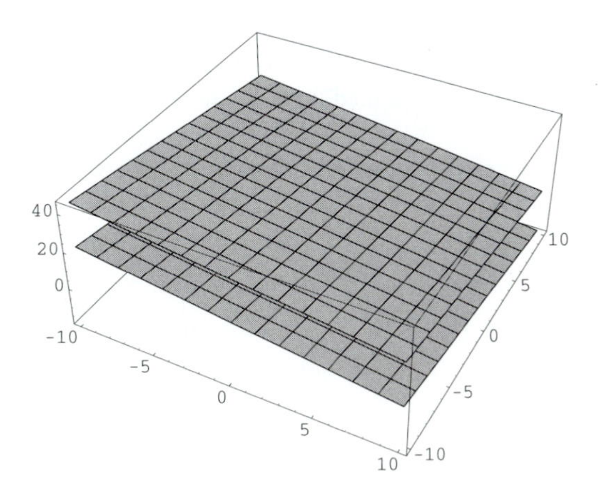

FIGURE 1 Two Parallel Planes.

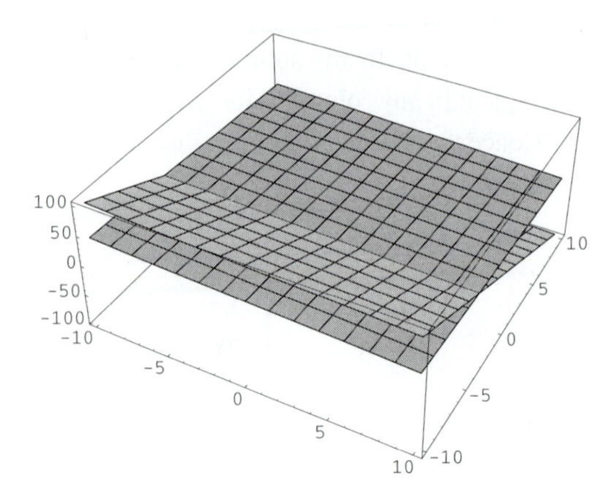

FIGURE 2 Two Intersecting Planes.

The system

$$\begin{cases} x + y + z = 5 \\ x + y + 2z = 2 \end{cases}$$

has infinitely many solutions. For all real numbers s, the list $(-s - 10, s, 15)$ is a solution. Figure 2 shows that the set of solutions forms a straight line, determined by the intersection of the planes $x + y + z = 5$ and $x + y + 2z = 20$.

EXERCISES 1.3

1. Find a linear system determined by four numbers whose sum is 100, with the first three numbers adding up to 50 and the last three numbers adding up to 75. Explain why the system is an underdetermined linear system with infinitely many solutions.

2. Construct linear systems with the following properties and explain why the systems have these properties.

   a. Consistent and 2×2
   b. Inconsistent and 2×2
   c. Consistent and 2×3
   d. Inconsistent and 2×3
   e. Unique solution and 3×3
   f. Infinitely many solutions and 3×3
   g. Inconsistent and 4×3
   h. Infinitely many solutions and 4×4

3. Show that all $m \times n$ linear systems whose equations are of the form $a_1x_1 + \cdots + a_nx_n = 0$ are consistent.

4. Determine by inspection which of the following linear systems are consistent.

$$\text{a. } \begin{cases} x + 2y = 2 \\ x - y = 4 \end{cases} \quad \text{b. } \begin{cases} x + 2y - z = 2 \\ x - y + z = 4 \\ 2x + y - z = 5 \end{cases} \quad \text{c. } \begin{cases} x + 2y - z = 2 \\ x - y + z = 4 \end{cases}$$

5. Classify the following linear systems as determined, overdetermined, or underdetermined. Explain why the systems are inconsistent.

$$\text{a. } \begin{cases} 2x + y - z = 2 \\ 2x + y - z = 6 \end{cases} \quad \text{b. } \begin{cases} 2x + y - z = 2 \\ x - 3y + z = 1 \\ 2x + y - z = 6 \end{cases} \quad \text{c. } \begin{cases} 2x + 3y = 4 \\ x - y = 1 \\ x + y = 3 \end{cases}$$

6. Classify the following linear systems as determined, overdetermined, or underdetermined. Determine by inspection which of the systems are consistent and have one solution and which are consistent and have infinitely many solutions.

$$\text{a. } \begin{cases} 3x + y = 6 \\ x - 2y = 3 \end{cases} \quad \text{b. } \begin{cases} 3x + y = 6 \\ 6x + 2y = 12 \end{cases} \quad \text{c. } \begin{cases} 3x + y = 6 \\ 6x + 2y = 1 \\ x + y = 7 \end{cases}$$

$$\text{d. } \begin{cases} 3x + y - z = 6 \\ x - 2y + z = 3 \\ 4x - 8y + 2z = 12 \end{cases} \quad \text{e. } \begin{cases} 3x + y - z + w = 6 \\ x - 2y + z - w = 3 \\ 4x - 8y + 2z = 12 \end{cases} \quad \text{f. } \begin{cases} x - 2y + z = 3 \\ 4x - 8y + 2z = 12 \end{cases}$$

7. (*Maple V*) Use the **implicitplot** function in the **plots** package to describe the geometric properties of the following inconsistent overdetermined 3×2 linear systems.

$$\text{a. } \begin{cases} x + 2y = 5 \\ x - 2y = 1 \\ 2x + y = 6 \end{cases} \quad \text{b. } \begin{cases} x + 2y = 5 \\ x - 2y = 1 \\ 2x + y = 7 \end{cases} \quad \text{c. } \begin{cases} x + 2y = 5 \\ 3x + 6y = 15 \\ -x - 2y = -5 \end{cases}$$

Geometry and Linear Systems

We conclude this section with some additional examples from three-dimensional geometry.

EXAMPLE 1.17 Three Intersecting Planes

Use *Maple* to verify that $(11, 8, -18)$ is the point of intersection of the three planes given by the equations $a + b + c = 1$, $3a + 5b + 4c = 1$, and $-a + 6b + 2c = 1$.

Solution. We first assign the values 11, 8, and -18 to the variables a, b, and c, respectively.

```
[> a:=11:  b:=8:  c:=-18:
```

Next we verify that with this assignment, the three given equations hold.

```
[> [evalb(a+b+c=1),evalb(3*a+5*b+4*c=1),
     evalb(-a+6*b+2*c=1)];
```

$$[true, true, true]$$

Our calculations show that $a = 11$, $b = 8$, and $c = -18$ is a solution of the linear system

$$\begin{cases} a + b + c = 1 \\ 3a + 5b + 4c = 1 \\ -a + 6b + 2c = 1 \end{cases}$$

and that $(11, 8, -18)$ is therefore the point of intersection of the three given planes. ▲

EXAMPLE 1.18 Four Points and a Plane

Show that there is no plane containing the points $(1, 1, 1)$, $(3, 5, 4)$, $(-1, 6, 2)$, and $(1, 2, 3)$.

Solution. Suppose that the given points lie on the plane $ax + by + cz + d = 0$. Then the linear system

$$\begin{cases} a + b + c + d = 0 \\ 3a + 5b + 4c + d = 0 \\ -a + 6b + 2c + d = 0 \\ a + 2b + 3c + d = 0 \end{cases}$$

must have a common solution. If we examine these equations more closely, we discover that the only common solution is $a = b = c = d = 0$. Thus the plane containing the four given points would have to be $0x + 0y + 0z + 0 = 0$. However, this is not a plane. ▲

EXAMPLE 1.19 Two Parallel Planes

Show that the two planes $x + 2y + 3z = 1$ and $x + 2y + 3z = 2$ have no point of intersection.

Solution. Suppose that the point (a, b, c) lies on the given planes. Algebraically this means that the linear system

$$\begin{cases} a + 2b + 3c = 1 \\ a + 2b + 3c = 2 \end{cases}$$

has a solution. This implies that $a + 2b + 3c = 1$ and that $a + 2b + 3c = 2$. In other words, it implies that $1 = 2$. This is a contradiction. ▲

EXAMPLE 1.20 Four Points and a Line

Show that there is no line in $\mathbb{R}^2$ containing the points $(1, 1)$, $(3, 5)$, $(-1, 6)$, and $(7, 2)$.

Solution. Suppose that there is a line $y = mx + b$ containing the four points. Since each point must satisfy this equation, the given points determine a linear system

$$\begin{cases} 1 = m + b \\ 5 = 3m + b \\ 6 = -m + b \\ 2 = 7m + b \end{cases}$$

consisting of four equations in two unknowns. If we add the first equation to the third, we get $7 = 2b$. If we subtract the first equation from the third, we get $5 = -2m$. This means that b must be $7/2$ and m must be $-5/2$. If we substitute these values in the fourth equation, we get $2 = 7(-5/2) + 7/2 = -14$. This is a contradiction. ▲

EXAMPLE 1.21 Two Nonintersecting Nonparallel Lines in $\mathbb{R}^3$

Show that the two lines L_1 and L_2 given by the vector equations

$$L_1 : (x, y, z) = (-2, 3, 1) + s(1, 1, 1)$$
$$L_2 : (u, v, w) = (5, 1, 4) + t(-3, 2, 1)$$

are nonparallel and do not intersect.

Solution. We begin by noting that the two lines are not parallel since the direction arrows $\mathbf{d}_1 = (1, 1, 1)$ and $\mathbf{d}_2 = (-3, 2, 1)$ are not parallel. Next use *Maple* to show that the overdetermined linear system

$$\begin{cases} -2 + s = 5 - 3t \\ 3 + s = 1 + 2t \\ 1 + s = 4 + t \end{cases}$$

is inconsistent. It follows that there is no point $(a, b, c) \in \mathbb{R}^3$ that lies on both straight lines.

```
[> solve({-2+s=5-3*t,3+s=1+2*t,1+s=4+t},{s,t});
[>
```

Maple has produced the empty output since the given system has no solution. ▲

EXERCISES 1.4

1. Use the techniques discussed in this section to find the plane determined by the three points $(1, 1, 1)$, $(3, 5, 2)$, and $(7, 2, 5)$.

2. Find a plane that is parallel to the plane $x - y + 2z = 5$ and contains the point $(3, 4, 5)$. Explain why the linear system determined by the two planes is inconsistent.

3. Use **solve** function to graph the solutions of the following linear systems.

$$\text{a.} \begin{cases} x + 2y = 2 \\ x - y = 4 \end{cases} \quad \text{b.} \begin{cases} x + y = \ \ 2 \\ -2x - 2y = -4 \end{cases} \quad \text{c.} \begin{cases} x + 2y = \ \ 2 \\ 6x + 12y = 18 \end{cases}$$

4. Rewrite the linear system

$$\begin{cases} x = 4 + 3t \\ y = 8 + 2t \\ z = 3 - 7t \end{cases}$$

as a vector equation of a straight line in $\mathbb{R}^3$.

5. Convert the vector equation of the straight line in $\mathbb{R}^5$, defined by the arrow $\mathbf{a} = (1, 2, 3, 4, 5)$ and the direction arrow $\mathbf{d} = (3, -1, 2, -6, 9)$, to a system of parametric equations.

6. Show that the vectors $(1, 1)$ and $(1, -1)$ are orthogonal and interpret the result geometrically.

7. Let L be a line $(x, y, z) = (a, b, c) + t(p, q, r)$ in $\mathbb{R}^3$. Then $(x - a, y - b, z - c) = t(p, q, r)$. Hence L consists of all points (x, y, z) for which the vectors $(x - a, y - b, z - c)$ are parallel to the direction arrow (p, q, r). Show that all vectors $(x - a, y - b, z - c) \in \mathbb{R}^3$ orthogonal to a fixed vectors $\mathbf{d} = (p, q, r)$ lie in a plane through (a, b, c). The vector $\mathbf{d}$ is called a ***normal*** to the plane. Find the equation of the plane in $\mathbb{R}^3$, which passes through $(1, 1, 1)$ and has the normal $\mathbf{d} = (5, -2, 3)$.

SOLVING LINEAR SYSTEMS

We now develop a language and general techniques for describing and finding the solutions of consistent linear systems. In addition to using *Maple* to solve linear systems, we can use *Maple* to verify that the solutions found by other methods are correct. For example, the calculation

```
[> solve({x+y=5,x-y=3},{x,y});
```

$$\{x = 4, y = 1\}$$

produces the solution $(x, y) = (4, 1)$ for the linear system

$$\begin{cases} x + y = 5 \\ x - y = 3 \end{cases}$$

If a manual attempt to solve the system had produced the assignments $x = 5$ and $y = 1$, we could use *Maple* to show that this is incorrect.

```
[> x:=5:  y:=1:
[> evalb(x+y=5);
```

$$false$$

Maple is telling us that the claim that $x + y$ equals 5 is false.

EXAMPLE 1.22 A Consistent System with One Solution

Use *Maple* to show that the linear system

$$\begin{cases} x + y - z = a \\ 2x - y - 3z = b \\ -x + y + 6z = c \end{cases}$$

has a unique solution for all a, b, c.

Solution. We write the given system as a list of equations and apply the **solve** function.

```
[> solve({x+y-z=a,2*x-y-3*z=b,-x+y+6*z=c},{x,y,z});
```

$$\left\{y = \frac{9}{13}a - \frac{5}{13}b - \frac{1}{13}c, z = -\frac{1}{13}a + \frac{2}{13}b + \frac{3}{13}c, x = \frac{3}{13}a + \frac{7}{13}b + \frac{4}{13}c\right\}$$

The *Maple* output shows that the given system has a unique solution for each assignment of values to a, b, and c. ▲

EXAMPLE 1.23 A Consistent System with Infinitely Many Solutions

Use *Maple* to show that the linear system

$$\begin{cases} x + y - z = 2 \\ x - y + z = 4 \\ 2x + y - z = 5 \end{cases}$$

has infinitely many solutions.

Solution. Again we use the **solve** function.

```
[> solve({x+y-z=2,x-y+z=4,2*x+y-z=5});
```

$$\{x = 3, z = z, y = -1 + z\}$$

Maple is telling us that although the system consists of three equations in three unknowns, two of the variables depend on the third. If we replace z by 10, for example, the replacements of x by 3 and of y by $-1 + 10$ produce a solution of the system. Since we have infinitely many choices for z, the system has infinitely many solutions. If we were to instruct *Maple* to solve the system for two of the three variables, say, for x and y, the statement $z = z$ would disappear:

```
[> solve({x+y-z=2,x-y+z=4,2*x+y-z=5},{x,y});
```

$$\{y = -1 + z, x = 3\}$$

As expected, the solution $x = 3$ and $y = -1 + z$ is a function of z. ▲

EXAMPLE 1.24 An Underdetermined Linear System

Use *Maple* to solve the system

$$\begin{cases} x + y - z = 2 \\ x - y + z = 4 \end{cases}$$

obtained from the previous example by deleting the third equation.

Solution. Again we use the **solve** function. Since the system has two equations, we solve for two of the three variables.

```
[> solve({x+y-z=2,x-y+z=4},{x,y});
```

$$\{y = z - 1, x = 3\}$$

As we can see, the system has the same solutions as those in the previous example. ▲

The next example explains the behavior of *Maple* when it encounters an inconsistent system.

EXAMPLE 1.25 An Inconsistent Linear System

Use *Maple* to show that the system

$$\begin{cases} x + y - z = 2 \\ x - y + z = 4 \\ 2x + y - z = 6 \end{cases}$$

is inconsistent.

Solution. We try to use **solve** function.

```
[> solve({x+y-z=2,x-y+z=4,2*x+y-z=6},{x,y});
[>
```

As we can see, the *Maple* output is empty since the system has no solution. ▲

We conclude this section by considering an inconsistent overdetermined system.

EXAMPLE 1.26 An Overdetermined Linear System

Use *Maple* to show that the system

$$\begin{cases} x + y = 2 \\ x - y = 4 \\ 2x + y = 6 \end{cases}$$

has no solution.

Solution. *Maple* treats an overdetermined system like any other system. Since the system has at least as many equations as variables, we can use the **solve** function without having to specify any variables to solve for.

```
[> solve({x+y=2,x-y=4,2*x+y=6});
[>
```

The empty output tells us that the system is inconsistent. ▲

Gaussian Elimination

The solutions of a linear system of the form

$$\begin{cases} a_{11}x_1 + \cdots + a_{1n}x_n = b_1 \\ \vdots \\ a_{m1}x_1 + \cdots + a_{mn}x_n = b_m \end{cases}$$

can be found by a method known as ***Gaussian elimination.*** Gauss discovered that three basic operations could be used to devise a systematic procedure for solving linear systems.

1. ***Multiply all terms of an equation of the system by a nonzero constant.***
2. ***Add a multiple of one equation of the system to another equation of the system.***
3. ***Change the order in which the equations of a system are listed.***

Instead of the addition of two equations, we sometime speak of the ***subtraction*** of one equation from another. By the ***subtraction*** of an equation $p = q$ from an equation $r = s$ we mean the addition of the equation $-p = -q$ to the equation $r = s$.

In general, Gaussian elimination steps can be applied to a linear system in many different ways. One of the objectives of this text is to develop a systematic approach to the choice and order of application of these rules. Our first step toward systematizing the process is to use the steps to convert a given system to ***row echelon form***.

DEFINITION 1.9 *A linear system in the variables $x_1, \ldots, x_n$ is in **row echelon form** if none of its equations is degenerate and the first equation containing the leading variable x_i occurs above the first equation containing the leading variable x_j if $i < j$.*

In the case of $n \times n$ linear systems, it is also customary to refer to the row echelon form of the system as a ***triangular form***.

In Table 2 we describe an algorithm for converting a nondegenerate linear system in the variables $x_1, \ldots, x_n$ to row echelon form.

TABLE 2 CONVERTING A LINEAR SYSTEM TO ROW ECHELON FORM.

Step 1. Take a linear system in the variables $x_1, \ldots, x_n$.

Step 2. Interchange the equations so that x_1 is the leading variable of the first equation.

Step 3. Add appropriate multiples of this equation to every equation below it until the variable x_1 has been eliminated from these equations.

Step 4. Delete all degenerate equations of the form $0x_1 + \cdots + 0x_n = 0$, if there are any.

Step 5. If one of the new equations is of the form $0x_1 + \cdots + 0x_n = b$, with $b \neq 0$, then stop, since the system has no solution.

Step 6. Omit the first equation and repeat Steps 1–4 for the new subsystem.

Step 7. Continue this process until the final system is in row echelon form.

A system of three equations in the variables x, y, and z, is in row echelon form, for example, if the second equation does not contain the variable x and the third equation contains neither the variable x nor the variable y. If a consistent linear system is in row echelon form, we can easily find a solution for it using a method called ***back substitution***.

EXAMPLE 1.27 Gaussian Elimination and Back Substitution

Use Gaussian elimination and back substitution to solve the system

$$\begin{cases} x + 2y - z = 2 \\ x - y + z = 4 \\ 2x + y - z = 5 \end{cases}$$

Solution. Here is a possible sequence of operations that converts the given system to row echelon form from which a solution can easily be found by back substitution.

1. We begin by subtracting the first equation from the second:

$$\begin{cases} x + 2y - z = 2 \\ x - y + z = 4 \\ 2x + y - z = 5 \end{cases} \to \begin{cases} x + 2y - z = 2 \\ -3y + 2z = 2 \\ 2x + y - z = 5 \end{cases}$$

2. We then subtract twice the first equation from the third:

$$\begin{cases} x + 2y - z = 2 \\ -3y + 2z = 2 \\ 2x + y - z = 5 \end{cases} \to \begin{cases} x + 2y - z = 2 \\ -3y + 2z = 2 \\ -3y + z = 1 \end{cases}$$

3. Next we subtract the third equation from the second:

$$\begin{cases} x + 2y - z = 2 \\ -3y + 2z = 2 \\ -3y + z = 1 \end{cases} \to \begin{cases} x + 2y - z = 2 \\ z = 1 \\ -3y + z = 1 \end{cases}$$

4. Now we interchange the second and third equations.

$$\begin{cases} x + 2y - z = 2 \\ z = 1 \\ -3y + z = 1 \end{cases} \to \begin{cases} x + 2y - z = 2 \\ -3y + z = 1 \\ z = 1 \end{cases}$$

5. At this stage, we can see that in any solution, z must be 1.

6. We then subtract the third equation from the second and obtain a linear system in row echelon form.

$$\begin{cases} x + 2y - z = 2 \\ -3y + z = 1 \\ z = 1 \end{cases} \to \begin{cases} x + 2y - z = 2 \\ -3y = 0 \\ z = 1 \end{cases}$$

7. Next we divide the second equation by -3:

$$\begin{cases} x + 2y - z = 2 \\ -3y = 0 \\ z = 1 \end{cases} \to \begin{cases} x + 2y - z = 2 \\ y = 0 \\ z = 1 \end{cases}$$

8. This tells us that y must be 0.

9. We therefore replace y by 0 and z by 1 in the first equation:

$$\begin{cases} x + 2y - z = 2 \\ \qquad\quad y = 0 \\ \qquad\quad z = 1 \end{cases} \rightarrow \begin{cases} x - 1 = 2 \\ \quad y = 0 \\ \quad z = 1 \end{cases}$$

   This step is called ***back substitution***.

10. By adding 1 to both sides of the first equation, we get the solution of the system: $x = 3$, $y = 0$, and $z = 1$. ▲

If the linear system has infinitely many solutions, back substitution may require us to assign a value to one or more variables in the last equation of a row echelon form of the system before being able to find a specific solution. Suppose, for example, that the last line of the row echelon form of a linear system is

$$x + y = 5$$

Then we can assign a value such as 4 to x and force the corresponding value of y to be 1. We can then substitute the values $x = 4$ and $y = 1$ back into the equations higher up and find values for all other variables in terms of the assigned value for x.

The Gaussian elimination process also detects the case of a linear system without a solution. Suppose, for example, that the linear system is

$$\begin{cases} x + y = 2 \\ x + y = 1 \end{cases}$$

Then subtracting the second equation form the first yields the contradiction $0 = 1$. This tells us that the given system has no solution.

EXERCISES 1.5

1. Find the points of intersection of the following pairs of lines.

   a. The lines $x + y = 5$ and $x - y = 3$
   b. The lines $2x - 7y = 1$ and $x + 3y = 4$
   c. The lines $x = 5$ and $x + y = 7$
   d. The lines $x - y = 0$ and $y = 9$
   e. The lines $x = 5$ and $y = 21$

2. Use Gaussian elimination and back substitution to solve the following linear systems.

   a. $\begin{cases} x + 2y = 2 \\ x - y = 4 \end{cases}$ b. $\begin{cases} x + 2y - z = 2 \\ x - y + z = 4 \\ 2x + y - z = 5 \end{cases}$ c. $\begin{cases} x + 2y - z = 2 \\ x - y + z = 4 \end{cases}$

3. (*Maple V*) Use the **solve** function to solve the following linear systems.

a. $\begin{cases} 2x + y - z = 2 \\ x - 3y + z = 1 \\ 2x + y - 2z = 6 \end{cases}$ b. $\begin{cases} 2x + y - z = 2 \\ x - 3y + z = 1 \\ 2x + y - z = 6 \end{cases}$ c. $\begin{cases} 2x + y - z = 2 \\ x - 3y + z = 1 \end{cases}$

4. (*Maple V*) Use the function **gausselim** and **backsub** in the **linalg** package to find the points of intersection of the following sets of planes.

   a. The planes $x + y + z = 5$, $x - y + 2z = 3$, and $x + 3y - z = 4$
   b. The planes $2x - 7y = 1$, $x - z = 0$, and $x + 3y - z = 4$
   c. The planes $14x + 7y + z = 5$, $y = 3$, and $x + y + z = 7$
   d. The planes $x - y + 4z = 12$, $y = 7$, and $x + y = 9$
   e. The planes $x - y - z = 5$, $x - 2y + 7z = 21$, and $y - z = 1$

MATRICES AND LINEAR SYSTEMS

We can associate matrices with linear systems in two basic ways. We can form the ***augmented matrix*** of a system, or we can rewrite the system as a ***matrix equation***.

In this section, we first discuss the idea of converting a linear system to an augmented matrix. We define the row echelon and reduced row echelon forms of a matrix and show how to adapt the Gaussian elimination process to a procedure for converting augmented matrices to row echelon and reduced row echelon form. We show that the solutions of the linear system can be obtained by ***back substitution*** using echelon matrices. We then discuss the conversion of a linear system to a matrix equation and show how matrix algebra can be used to solve the system. Let us begin with a general introduction to matrices and their *Maple* form.

DEFINITION 1.10 *A **matrix** is a rectangular array of mathematical objects.*

Most matrices studied in this text are rectangular arrays of real numbers. The array

$$A = \begin{bmatrix} 1 & 2 & -1 & \pi \\ 1 & -1 & 1.5 & 4 \\ 2 & 1 & -1 & 5 \end{bmatrix}$$

for example, is a matrix of real numbers. The objects in the cells of the array are called the ***entries*** of the matrix.

If the entries of a matrix are all zero, we call the matrix a ***zero matrix*** and denote it by **0**. We use the same symbol no matter how many rows and columns a zero matrix may have. The meaning of the symbol **0** will always be clear from the context.

Maple provides several options for defining matrices. We can write them as a sequence of rows of entries; we can use matrix palettes as entry devices; or we can use functions and expressions, in combination with the built-in **array**, **Array,** and **seq** functions, to create them.

Arrays and Functions

Suppose that f is a function in two variables, either built into *Maple* or userdefined. Then we can use the **array** function to generate a matrix whose ijth entry is $f(i, j)$.

EXAMPLE 1.28 Using array to Generate a Matrix

1. The **max** function is a built-in *Maple* function whose value at (i, j) is the maximum of i and j. We can combine it with the **array** function to generate a matrix.

```
[> array([
    [seq(max(1,i),i=1..4)],
    [seq(max(2,i),i=1..4)],
    [seq(max(3,i),i=1..4)]]);
```

$$\begin{bmatrix} 1 & 2 & 3 & 4 \\ 2 & 2 & 3 & 4 \\ 3 & 3 & 3 & 4 \end{bmatrix}$$

The output is an array with the three rows [1, 2, 3, 4], [2, 2, 3, 4], and [3, 3, 3, 4].

2. The **rand** function is a built-in *Maple* function whose values are pseudorandom integers lying in a specified range $min..max$. The default outputs are 12-digit nonnegative integers. We can combine the **array** function with the **rand** function to generate a ***random matrix***. Here is an example of a random matrix with integer entries between -8 and 8.

```
[> Null;
[> r:=rand(-8..8):
[> array([[r(),r(),r()],[r(),r(),r()]]);
```

$$\begin{bmatrix} 2 & 8 & -8 \\ 6 & 5 & -5 \end{bmatrix}$$

3. The next example is a random matrix with real entries between 0 and 1. Since rand() produces 12-digit integers r, we multiply the integers r by 10^{-12} to produce real numbers less than 1. We use the function **evalf** to calculate numerical approximations.

```
[> a:=x->evalf(rand())*10^(-12):
[> A:=array([[a(1),a(1),a(1)],[a(1),a(1),a(1)],
     [a(1),a(1),a(1)]]);
```

$$A := \begin{bmatrix} .2197600994 & .6759829338 & .8454735095 \\ .6764707883 & .2813387792 & .7924959004 \\ .7512095393 & .6283634430 & .3137460865 \end{bmatrix}$$

4. We can generate a matrix by combining the **array** function with a user-defined function.

```
[> f:=(i,j)->3*i+5*j:
[> array([
    [seq(f(1,i),i=1..4)],
    [seq(f(2,i),i=1..4)],
    [seq(f(3,i),i=1..4)]]);
```

$$\begin{bmatrix} 8 & 13 & 18 & 23 \\ 11 & 16 & 21 & 26 \\ 14 & 19 & 24 & 29 \end{bmatrix}$$

The output is a matrix with the three rows $[8, 13, 18, 23]$, $[11, 16, 21, 26]$, and $[14, 19, 24, 29]$.

5. We can define a function by specifying its values, one by one, and by then using its name to generate a matrix.

```
[> g:=(a,b)->piecewise(
    a=1 and b=1, 8,
    a=1 and b=2, 9,
    a=2 and b=1, 7,
    a=2 and b=2, 11):
```

```
[> array([[seq(g(1,i),i=1..2)],[seq(g(2,i),i=1..2)]]);
```

$$\begin{bmatrix} 8 & 9 \\ 7 & 11 \end{bmatrix}$$

The output is a matrix whose two rows are $[8, 9]$ and $[7, 11]$.

6. The entries of a matrix can also be nonnumeric mathematical objects. The following matrix, for example, has polynomials as entries.

```
[> g:=(a,b)->piecewise(
    a=1 and b=1, 1-t,
    a=1 and b=2, t^2,
    a=2 and b=1, 9,
    a=2 and b=2,11+t-t^3):
```

```
[> array([[seq(g(1,i),i=1..2)],[seq(g(2,i),i=1..2)]]);
```

$$\begin{bmatrix} 1-t & t^2 \\ 9 & 11+t-t^3 \end{bmatrix}$$

The most appropriate method for generating a matrix depends on the context. ▲

EXERCISES 1.6

1. (*Maple V*) Use the **array** function, together with a built-in *Maple* function in two variables, to generate a matrix with four rows and five columns.

2. (*Maple V*) Use the **array** function, together with a user-defined function in two variables, to generate a matrix with four rows and five columns.

3. (*Maple V*) Use the **array** function, together with afunction in two variables defined by listing its values, to generate a matrix with four rows and five columns.

4. (*Maple 6*) Use the functions and arrays methods in Example 1.28 to generate the 3×3 Hilbert matrix defined in the **LinearAlgebra** package.

5. (*Maple 6*) Use the **Array** function, together with the **rand** function, to generate a matrix with four rows and five columns.

6. (*Maple 6*) Use the **Array** function, together with a user-defined function in two variables, to generate a matrix with four rows and five columns whose entries are polynomials.

7. (*Maple V, 6*) Make a table that lists and compares the results of using the following *Maple* functions to construct vectors and matrices: **array**, **Array**, **vector**, **Vector**, **matrix**, and **Matrix.**

Lists and Expressions

Suppose that *expr* is an expression in two variables such as $3i + 5j$. Then we can use the **seq** function to generate a matrix whose ijth entry is $3i + 5j$.

EXAMPLE 1.29 Using seq to Generate an Exact Matrix

Use the **seq** function to generate a matrix with three rows and four columns whose ijth entry is $3i + 5j$.

Solution. We first define the expression $3i + 5j$ and then use it as an argument in the **seq** function.

```
[> expr:=(i,j)->3*i+5*j:
[> array([
    [seq(expr(1,i),i=1..4)],
    [seq(expr(2,i),i=1..4)],
    [seq(expr(3,i),i=1..4)]]);
```

$$\begin{bmatrix} 8 & 13 & 18 & 23 \\ 11 & 16 & 21 & 26 \\ 14 & 19 & 24 & 29 \end{bmatrix}$$

The output is a matrix consisting of the three rows [8, 13, 18, 23], [11, 16, 21, 26], and [14, 19, 24, 29]. ▲

Suppose that *expr* is an expression in two variables such as **N[Exp[i]+Log[j]].** Then we can use the **seq** function again to generate a matrix whose ijth entry is **N[Exp[i]+Log[j]]**.

EXAMPLE 1.30 Using seq to Generate an Approximate Matrix

```
[> expr:=(i,j)->evalf(exp(i)+log(j)):
[> array([
    [seq(expr(1,i),i=1..4)],
    [seq(expr(2,i),i=1..4)],
    [seq(expr(3,i),i=1..4)]]);
```

$$\begin{bmatrix} 2.718281828 & 3.411429009 & 3.816894117 & 4.104576189 \\ 7.389056099 & 8.082203280 & 8.487668388 & 8.775350460 \\ 20.08553692 & 20.77868410 & 21.18414921 & 21.47183128 \end{bmatrix}$$

The output is a 3×4 matrix whose entries are approximations of sums of values of the exponential and logarithmic functions. ▲

We can also use the **seq** function in combination with random expressions.

EXAMPLE 1.31 Using seq to Generate a Random Matrix

```
[> A:=evalf(evalm(
    [seq([sin(rand()),sin(rand()),sin(rand()),sin(rand())],
    i=1..3)]),5);
```

$$A := \begin{bmatrix} .85131 & .99740 & -.87017 & -.99702 \\ .58712 & -.71663 & .52331 & .79995 \\ -.75725 & .27703 & -.96087 & -.39769 \end{bmatrix}$$

The output is a matrix whose three rows consist of truncated random numbers between -1 and 1. ▲

EXERCISES 1.7

1. (*Maple V*) Write three expressions in i and j, and use them in conjunction with the **seq** function to generate three matrices with four rows and five columns.
2. (*Maple V*) Use three expressions based on built-in *Maple* functions, together with the **seq** function, to generate three matrices with four rows and five columns.
3. (*Maple V*) Use the **seq** function, together with expressions built with the **rand** function, to generate three matrices with four rows and five columns.

Matrices as Functions on Indexing Sets

If we take a closer look at a system such as

$$\begin{cases} x + 2y - z = 2 \\ x - y + z = 4 \\ 2x + y - z = 5 \end{cases}$$

we realize that it is uniquely determined by the coefficients of its variables and the constants on the right-hand side of the equations. We can record this information in a matrix

$$A = \begin{bmatrix} 1 & 2 & -1 & 2 \\ 1 & -1 & 1 & 4 \\ 2 & 1 & -1 & 5 \end{bmatrix}$$

The matrix consists of three rows and four columns. We know that the coefficients of x are listed in the first column, the coefficients of y in the second column, and the coefficients of z in the third column. The last column contains the constants occurring on the right-hand side of the equations.

If we write $A(i, j)$ for the information in the cell located in the ith row and jth column of the matrix A, then

$$A = \begin{bmatrix} A(1,1) & A(1,2) & A(1,3) & A(1,4) \\ A(2,1) & A(2,2) & A(2,3) & A(2,4) \\ A(3,1) & A(3,2) & A(3,3) & A(3,4) \end{bmatrix}$$

with $A(1,1)= 1$, $A(1,2)= 2$, $A(1,3)= -1$, and so forth. Since $A(i, j)$ occurs in the ith row and jth column of A, we call it the ijth ***entry*** of A. In *Maple* it is denoted by $\mathbf{A}_{i,j}$. If we use the list [1, 2, 3] to label the rows of A and the list [1, 2, 3, 4] to label the columns, we have an address system for A where each cell is labeled by a pair of integers taken from the set

$$[1,2,3] \times [1,2,3,4]$$

consisting of the ordered pairs of integers (1, 1), (1, 2), (1, 3), (1, 4), (2, 1), (2, 2), (2, 3), (2, 4), (3, 1), (3, 2), (3, 3), and (3, 4). We refer to lists such as [1], [1, 2], [1, 2, 3], [1, 2, 3, 4] as ***indexing sets***. We write $\mathbf{n}$ for the indexing set $[1, 2, \ldots, n]$ consisting of the integers from 1 and n. (We use a boldface $\mathbf{n}$ to distinguish the list $[1, 2, \ldots, n]$ from its largest element n.) As we can see, the location of a cell in a matrix with m rows and n columns can be specified by a pair of integers taken from the set

$$\mathbf{m} \times \mathbf{n} = [1, 2, \ldots, m] \times [1, 2, \ldots, n]$$

We call $m \times n$ the ***dimension*** of A.

Since the entry $A(i, j)$ of a matrix A is uniquely identified by the pair (i, j), we can think of a matrix with m rows and n columns as a ***function*** from $\mathbf{m} \times \mathbf{n}$ to the mathematical system $\mathbb{X}$ from which the entries $A(i, j)$ are taken. In this text, typical systems in which matrices take their values are the set $\mathbb{R}$ of real numbers and the set $\mathbb{R}[t]$ of polynomials in t with coefficients in $\mathbb{R}$. In some of the calculus applications, the system $\mathbb{X}$ is more general.

DEFINITION 1.11 *An **$\mathbb{X}$-valued matrix** of dimension $m \times n$ is a function $A : \mathbf{m} \times \mathbf{n} \to \mathbb{X}$.*

We call this the ***indexed form*** of A. We assume that $\mathbb{X} = \mathbb{R}$, unless otherwise specified, and refer to such matrices as ***real matrices***. We often simplify the terminology by speaking simply of *matrices* if it is clear from the context that their entries are intended to be real numbers. If $m = n = 1$, we usually identify the matrix $[A(1,1)]$ with its entry $A(1,1)$.

EXAMPLE 1.32 An $\mathbb{R}$-valued 2 x 3 Matrix

Suppose that $A : \mathbf{2} \times \mathbf{3} \to \mathbb{R}$ is the function defined by $A(i, j)= \max(i, j)$. Then

$$A = \begin{bmatrix} A(1,1) & A(1,2) & A(1,3) \\ A(2,1) & A(2,2) & A(2,3) \end{bmatrix} = \begin{bmatrix} 1 & 2 & 3 \\ 2 & 2 & 3 \end{bmatrix}$$

As we can see, the matrix A is a rectangular array with real entries, arranged in two rows and three columns. ▲

EXAMPLE 1.33 An $\mathbb{R}[t]$-Valued 2 x 2 Matrix

Suppose that $A : \mathbf{2} \times \mathbf{2} \to \mathbb{R}[t]$ is the function defined by t if $i + j$ is even and by t^3 otherwise. Then

$$A = \begin{bmatrix} A(1,1) & A(1,2) \\ A(2,1) & A(2,2) \end{bmatrix} = \begin{bmatrix} t & t^3 \\ t^3 & t \end{bmatrix}$$

Here the matrix A is a rectangular array with polynomial entries, arranged in two rows and two columns. ▲

Maple has a built-in routine for creating the indexing sets.

EXAMPLE 1.34 Indexing Sets

Use *Maple* to create the indexing sets **3** and **4**.

Solution. The **seq** function can be used to create indexing sets.

```
[> [seq(i,i=1..3)];
                    [1, 2, 3]
[> [seq(i,i=1..4)];
                    [1, 2, 3, 4]
```

As we can see, the outputs are the indexing sets **3** and **4**. ▲

In the next example, we show that *Maple* also has built-in functions for calculating the dimensions of a matrix.

EXAMPLE 1.35 The Dimension of a Matrix

Use the **dim, rowdim,** and **coldim** functions to calculate the dimension of the matrix

$$A = \begin{bmatrix} 1 & 2 & 3 & 4 & 5 \\ 6 & 7 & 8 & 9 & 10 \\ 11 & 12 & 13 & 14 & 15 \end{bmatrix}$$

in two different ways.

Solution. First we define the matrix A, and then we apply the dimensions function to A.

```
[> A:=matrix([[1,2,3,4,5],[6,7,8,9,10],[11,12,13,14,15]]):
[> dim:=A->[linalg[rowdim](A),linalg[coldim](A)];
```

$$dim := A- > [linalg_{rowdim}(A), linalg_{coldim}(A)]$$

```
[> [dim(A),dim(A)[1],dim(A)[2]];
```

$$[[3, 5], 3, 5]$$

The first number in the list [3, 5] counts the number of rows of A, and the second number counts the number of columns. ▲

We could also have calculated the dimension of the matrix A in Example 1.35 using the **nops** function, which counts the number of elements of a list. Since each element [1, 2, 3, 4, 5], [6, 7, 8, 9, 10], [11, 12, 13, 14, 15] of A is a row of A, the **nops** function, applied to one of these lists, computes the number of columns of A.

```
[> nops([1,2,3,4,5]);
```

$$5$$

To find the number of rows of A, we apply the **nops** function to the list of lists

$$[[1, 2, 3, 4, 5], [6, 7, 8, 9, 10], [11, 12, 13, 14, 15]]$$

```
[> nops([[1,2,3,4,5],[6,7,8,9,10],[11,12,13,14,15]]);
```

$$3$$

Thus, *Maple* provides us with several ways of calculating the dimension of a matrix. We note that, as expected, the command

```
[> nops(matrix([[1,2,3,4,5],[6,7,8,9,10],[11,12,13,14,15]]));
```

$$1$$

outputs the value 1, since the argument of the **matrix** function is the single list

[[1,2,3,4,5],[6,7,8,9,10],[11,12,13,14,15]]

which, itself, is made up of three other lists. ▲

EXERCISES 1.8

1. Write the following matrices in rectangular form.

   a. The matrix $A : \mathbf{2} \times \mathbf{3} \to \mathbb{R}$ defined by

   $$\begin{array}{lll} A(1,1) = 3, & A(1,2) = -7, & A(1,3) = \pi, \\ A(2,1) = 2.8, & A(2,2) = 21, & A(2,3) = -100 \end{array}$$

   b. The matrix $A : \mathbf{3} \times \mathbf{2} \to \mathbb{R}$ defined by

   $$\begin{array}{lll} A(1,1) = 3, & A(1,2) = 2.8, & A(2,1) = -7, \\ A(2,2) = 21, & A(3,1) = \pi, & A(3,2) = -100 \end{array}$$

   c. The matrix $A : \mathbf{1} \times \mathbf{2} \to \mathbb{R}$ defined by $A(1,1) = 3$, $A(1,2) = 2.8$

   d. The matrix $A : \mathbf{2} \times \mathbf{1} \to \mathbb{R}$ defined by $A(1,1) = 3$, $A(2,1) = 2.8$

   e. The matrix $A : \mathbf{4} \times \mathbf{4} \to \mathbb{R}$ defined by $A(i, j) = 1$ if $i = j$ and 0 otherwise

   f. The matrix $A : \mathbf{4} \times \mathbf{4} \to \mathbb{R}$ defined by $A(i, j) = i$ if $i = j$ and 0 otherwise

   g. The matrix $A : \mathbf{4} \times \mathbf{4} \to \mathbb{R}$ defined by $A(i, j) = \sqrt{j}$ if $i = j$ and 0 otherwise

   h. The matrix $A : \mathbf{3} \times \mathbf{3} \to \mathbb{R}$ defined by $A(i, j) = i + j$ for $1 \leq i, j \leq 3$

   i. The matrix $A : \mathbf{1} \times \mathbf{1} \to \mathbb{R}$ defined by $A(1,1) = 99$

   j. The matrix $A : \mathbf{3} \times \mathbf{3} \to \mathbb{R}$ defined by $A(i, j) = i^2$ if $i + j$ is even and 0 otherwise

2. Write the following matrices in indexed form.

   $$\text{a.} \begin{bmatrix} -85 & -55 & -37 & -35 \\ 97 & 50 & 79 & 56 \\ 49 & 63 & 57 & -59 \end{bmatrix} \quad \text{b.} \begin{bmatrix} 1 & 0 & 0 \\ 0 & 1 & 0 \\ 0 & 0 & 1 \end{bmatrix} \quad \text{c.} \begin{bmatrix} 3 & 4 \\ 2 & 3 \\ 1 & 2 \end{bmatrix}$$

3. Explain why the set $\{A(1,1) = 7, A(2,2) = 5, A(2,3) = 0\}$ is not a matrix.

4. (*Maple V*) Convert the following matrices to lists of lists.

   $$\text{a.} \begin{bmatrix} -85 & 0 & -37 \\ 97 & 50 & 4 \end{bmatrix} \quad \text{b.} \begin{bmatrix} t & 0 \\ 0 & t^2 - 1 \\ 3 & 0 \end{bmatrix} \quad \text{c.} \begin{bmatrix} 2 & -3 & 4 \\ 3 & 4 & 7 \\ 4 & 5 & 6 \end{bmatrix}$$

5. (*Maple V*) Use the list of lists method to create the following matrices.

   $$\text{a.} \begin{bmatrix} 1 & 0 \\ 0 & 1 \end{bmatrix} \quad \text{b.} \begin{bmatrix} -85 \\ 97 \\ 49 \end{bmatrix} \quad \text{c.} \begin{bmatrix} -85 & 97 & 49 \end{bmatrix} \quad \text{d.} \begin{bmatrix} 0 & 1 & 0 \\ 1 & 0 & 0 \\ 0 & 0 & 1 \end{bmatrix}$$

6. (*Maple V*) Use the **Matrix** palette to create the following matrices.

$$\text{a.}\begin{bmatrix} 4 \\ 5 \\ 6 \end{bmatrix} \quad \text{b.}\begin{bmatrix} 1 & 2 & 3 & 4 & 5 \\ -9 & 3 & 7 & 7 & 8 \\ 0 & 12 & 45 & \pi & 27 \end{bmatrix} \quad \text{c.}\begin{bmatrix} 1 & 0 & 0 \\ 6 & 0 & 0 \\ 2 & 0 & 0 \end{bmatrix}$$

7. (*Maple V*) Use the **seq** function to generate the following indexing sets.

a. $\mathbf{1} = \{1\}$ b. $\mathbf{2} = \{1, 2\}$ c. $\mathbf{5} = \{1, 2, 3, 4, 5\}$ d. $\mathbf{6} = \{1, 2, 3, 4, 5, 6\}$

8. (*Maple V*) Use the **Matrix** palette of to create the matrices defined in Exercise 1.

Vectors as Matrices

Now that we have the language of matrices available to us, we can redefine vectors as special matrices. We will use this opportunity to define two types of vectors, column vectors and row vectors.

DEFINITION 1.12 *A* ***column vector*** *is a matrix of the form* $A : \mathbf{m} \times \mathbf{1} \to \mathbb{X}$.

We call m the ***height*** of A. Usually we drop any reference to the set **1** and write the matrix A as

$$\begin{bmatrix} a_1 \\ \vdots \\ a_m \end{bmatrix}$$

with $a_1 = A(1, 1), a_2 = A(2, 1), \ldots, a_m = A(m, 1)$. We refer to the entries a_i as the ***components*** of the vector A. We can use the **Matrix** function in the **LinearAlgebra** package to represent column vectors. For example, if $m = 3$ and $a_1 = A(1, 1) = 1$, $a_2 = A(2, 1) = 2$, and $a_3 = A(3, 1) = 3$, then A can be represented as follows:

```
[> A:=Matrix([[1],[2],[3]]);
```

$$A := \begin{bmatrix} 1 \\ 2 \\ 3 \end{bmatrix}$$

We can also use the **Vector[column]** function in the **LinearAlgebra** package to achieve the same result:

```
[> A:=Vector[column]([1,2,3]);
```

$$A := \begin{bmatrix} 1 \\ 2 \\ 3 \end{bmatrix}$$

We write vectors as **x**, **y**, **z**, **u**, **v**, **w**, ... to distinguish them from more general matrices A, B, C, D, and so on.

EXAMPLE 1.36 A Column Vector

Write the column vector $A : \mathbf{3} \times \mathbf{1} \to \mathbb{R}$, defined by $A(1, 1) = 1$, $A(2, 1) = 5$, $A(3, 1) = 9$, as a *Maple* expression and output the vector in matrix form.

Solution. We use the **matrix** function in the **linalg** package.

```
[> A:=linalg[matrix]([[1],[5],[9]]);
```

$$A := \begin{bmatrix} 1 \\ 5 \\ 9 \end{bmatrix}$$

As we can see, *Maple* displays A in column format. ▲

DEFINITION 1.13 *A **row vector** is a matrix* $A : \mathbf{1} \times \mathbf{n} \to \mathbb{X}$.

We call n the ***length*** of A. Usually we drop the reference to the set **1** and write A as $[a_1 \ \cdots \ a_n]$, with $a_1 = A(1, 1), a_2 = A(1, 2), \ldots, a_n = A(1, n)$. As in the case of column vectors, we refer to the entries a_i of A as the ***components*** of A. We can use the **Matrix** function in the **LinearAlgebra** package to represent row vectors as $1 \times n$ matrices. For example, if $n = 3$ and $a_1 = A(1, 1) = 1$, $a_2 = A(1, 2) = 2$, and $a_3 = A(1, 3) = 3$, then A can be represented as follows:

```
[> A:=Matrix([[1,2,3]]);
```

$$A := \begin{bmatrix} 1 & 2 & 3 \end{bmatrix}$$

We can also use the **Vector[row]** function in the **LinearAlgebra** package to represent A. However, the notational results are not the same. The command

```
[> A:=Vector[row]([1,2,3]);
```

$$A := [1, 2, 3]$$

produces a vector in list form.

EXAMPLE 1.37 A Row Vector

Write the row vector $A : \mathbf{1} \times \mathbf{3} \to \mathbb{R}$, defined by $A(1, 1) = 1$, $A(1, 2) = 5$, $A(1, 3) = 9$, as a *Maple* expression, and output the vector in list form.

Solution. We use the **Vector[row]** function in the **LinearAlgebra** package.

```
[> with(LinearAlgebra):
[> A:=Vector[row]([1,5,9]);
```

$$A := [1, 5, 9]$$

As expected, *Maple* produces a row vector in list form. ▲

We use the term ***vector*** ambiguously for both row and column vectors. However, in most cases, vectors will be column vectors.

In *Maple*, the interpretation of vectors as special matrices is not always appropriate. In *Maple V*, for example, vectors are objects whose orientation as row or column objects is determined by the context. This is not the case for matrices. The orientation of a matrix is always fixed. In *Maple 6,* on the other hand, vectors must usually be specified as either row or column objects. Nevertheless, both *Maple V* and *Maple 6* usually distinguish algebraically between vectors and matrices. Some operations can be carried out only if the objects involved are designated as vectors or matrices, as required.

EXERCISES 1.9

1. Rewrite the row vectors

$$\text{a. } [1, 2, -1] \quad \text{b. } [-8, 2, 99] \quad \text{c. } [x_1, x_2, x_3]$$

as matrices in indexed form.

2. Rewrite the column vectors

$$\text{a. } \begin{bmatrix} 1 \\ 2 \\ -1 \end{bmatrix} \quad \text{b. } \begin{bmatrix} -8 \\ 2 \\ 99 \end{bmatrix} \quad \text{c. } \begin{bmatrix} x_1 \\ x_2 \\ x_3 \end{bmatrix}$$

matrices in indexed form.

3. (*Maple V*) Use the **linalg[vector]** function to express the row vectors in Exercise 1.

4. (*Maple V*) Use the **linalg[vector]** function to express the column vectors in Exercise 2.

5. (*Maple V*) Evaluate the row vector

$$3[1, 2, -1] - 7[-8, 2, 99] + [1, 0, -100]$$

6. (*Maple V*) Evaluate the column vector

$$3\begin{bmatrix} 1 \\ 2 \\ -1 \end{bmatrix} - 7\begin{bmatrix} -8 \\ 2 \\ 99 \end{bmatrix} + \begin{bmatrix} 1 \\ 0 \\ -100 \end{bmatrix}$$

7. (*Maple 6*) Use the **Vector[row]** function to express the row vectors in Exercise 1.

8. (*Maple 6*) Use the **Vector[column]** function to represent the following column vectors in Exercise 2.

AUGMENTED MATRICES

As we have seen, a linear system is determined by the coefficients of its variables and its equational constants. We can arrange these data in matrix form and develop a procedure for finding the solutions of a linear system by manipulating the associated matrices.

DEFINITION 1.14 *The* ***augmented matrix*** *of the linear system*

$$\begin{cases} a_{11}x_1 + \cdots + a_{1n}x_n & = & b_1 \\ \vdots & \vdots & \vdots \\ a_{m1}x_1 + \cdots + a_{mn}x_n & = & b_m \end{cases}$$

is the $m \times (n + 1)$ *matrix*

$$[A \mid \mathbf{b}] = \begin{bmatrix} a_{11} & \cdots & a_{1n} & b_1 \\ \vdots & \vdots & \vdots & \vdots \\ a_{m1} & \cdots & a_{mn} & b_m \end{bmatrix}$$

EXAMPLE 1.38 An Augmented Matrix

Find the augmented matrix $[A \mid \mathbf{b}]$ of the linear system

$$\begin{cases} x + 2y - z = 2 \\ x - y + z = 4 \\ 2x + y - z = 5 \end{cases}$$

Solution. Since the given system consists of three equations in three unknowns, the augmented matrix $[A \mid \mathbf{b}]$ will be a 3×4 matrix. Its first column will contain the coefficients of x, its second column the coefficients of y, and its third column the coefficients of z. The last column will contain the constant terms. The matrix

$$\begin{bmatrix} 1 & 2 & -1 & 2 \\ 1 & -1 & 1 & 4 \\ 2 & 1 & -1 & 5 \end{bmatrix}$$

clearly satisfies these conditions. ▲

The rectangular array made up of the first three columns of A is the ***coefficient matrix*** of the given system and the fourth column is the ***vector of constants*** of the system. The **augment** function of the **linalg** package make it easy to create augmented matrices. They produce identical results.

```
[> A:=matrix([[1,2,-1],[1,-1,1],[2,1,-1]]):
[> b:=matrix([[2],[4],[5]]);
[> linalg[augment](A,b);
```

$$\begin{bmatrix} 1 & 2 & -1 & 2 \\ 1 & -1 & 1 & 4 \\ 2 & 1 & -1 & 5 \end{bmatrix}$$

We can also use the **GenerateMatrix** function in the **LinearAlgebra** package of *Maple 6* to calculate the augmented matrix of a linear system.

```
[> sys:=[x+2*y-z=2,x-y+z=4,2*x+y-z=5]:
[> vars:=[x,y,z]:
[> LinearAlgebra[GenerateMatrix](sys,vars,augmented=true);
```

$$\begin{bmatrix} 1 & 2 & -1 & 2 \\ 1 & -1 & 1 & 4 \\ 2 & 1 & -1 & 5 \end{bmatrix}$$

As we can see, both methods yield the augmented matrix of the given linear system. ▲

Gaussian elimination steps can be translated into operations on augmented matrices:

1. The multiplication of an equation of a system by a nonzero constant translates into the multiplication of every entry in a row of a matrix by a constant.
2. The addition of a multiple of one equation of a system to another equation of the system translates into multiplying each entry in a row by the same constant and adding the new entries to the corresponding entries of another row.
3. The permutation of the order of two equations of a system translates into the permutation of two rows of an augmented matrix.

Later in this chapter, we will introduce appropriate notation for these operations, known as ***elementary row operations***.

We now illustrate how these operations can be used to solve a linear system.

EXAMPLE 1.39 Gaussian Elimination and Augmented Matrices

Use elementary row operations on augmented matrices to convert the system

$$\begin{cases} x + 2y - z = 2 \\ x - y + z = 4 \\ 2x + y - z = 5 \end{cases}$$

to the system

$$\begin{cases} x = 3 \\ y = 0 \\ z = 1 \end{cases}$$

Solution. We saw earlier that the augmented matrix $[A \mid \mathbf{b}]$ of the given system is

$$\begin{bmatrix} 1 & 2 & -1 & 2 \\ 1 & -1 & 1 & 4 \\ 2 & 1 & -1 & 5 \end{bmatrix}$$

We use elementary row operations to convert $[A \mid \mathbf{b}]$ to a matrix $[B \mid \mathbf{c}]$ and then rewrite this matrix in the form of a linear system.

1. First we subtract the first row from the second:

$$\begin{bmatrix} 1 & 2 & -1 & 2 \\ 1 & -1 & 1 & 4 \\ 2 & 1 & -1 & 5 \end{bmatrix} \rightarrow \begin{bmatrix} 1 & 2 & -1 & 2 \\ 0 & -3 & 2 & 2 \\ 2 & 1 & -1 & 5 \end{bmatrix}$$

2. Then we subtract twice the first row from the third:

$$\begin{bmatrix} 1 & 2 & -1 & 2 \\ 0 & -3 & 2 & 2 \\ 2 & 1 & -1 & 5 \end{bmatrix} \rightarrow \begin{bmatrix} 1 & 2 & -1 & 2 \\ 0 & -3 & 2 & 2 \\ 0 & -3 & 1 & 1 \end{bmatrix}$$

3. Next we subtract the third row from the second:

$$\begin{bmatrix} 1 & 2 & -1 & 2 \\ 0 & -3 & 2 & 2 \\ 0 & -3 & 1 & 1 \end{bmatrix} \rightarrow \begin{bmatrix} 1 & 2 & -1 & 2 \\ 0 & 0 & 1 & 1 \\ 0 & -3 & 1 & 1 \end{bmatrix}$$

4. Then we subtract the second row from the third:

$$\begin{bmatrix} 1 & 2 & -1 & 2 \\ 0 & 0 & 1 & 1 \\ 0 & -3 & 1 & 1 \end{bmatrix} \rightarrow \begin{bmatrix} 1 & 2 & -1 & 2 \\ 0 & 0 & 1 & 1 \\ 0 & -3 & 0 & 0 \end{bmatrix}$$

5. Next we multiply the third row by $-1/3$:

$$\begin{bmatrix} 1 & 2 & -1 & 2 \\ 0 & 0 & 1 & 1 \\ 0 & -3 & 0 & 0 \end{bmatrix} \rightarrow \begin{bmatrix} 1 & 2 & -1 & 2 \\ 0 & 0 & 1 & 1 \\ 0 & 1 & 0 & 0 \end{bmatrix}$$

6. Then we subtract twice the third row from the first.

$$\begin{bmatrix} 1 & 2 & -1 & 2 \\ 0 & 0 & 1 & 1 \\ 0 & 1 & 0 & 0 \end{bmatrix} \rightarrow \begin{bmatrix} 1 & 0 & -1 & 2 \\ 0 & 0 & 1 & 1 \\ 0 & 1 & 0 & 0 \end{bmatrix}$$

7. Next we add the second row to the first:

$$\begin{bmatrix} 1 & 2 & -1 & 2 \\ 0 & 0 & 1 & 1 \\ 0 & 1 & 0 & 0 \end{bmatrix} \rightarrow \begin{bmatrix} 1 & 0 & 0 & 3 \\ 0 & 0 & 1 & 1 \\ 0 & 1 & 0 & 0 \end{bmatrix}$$

8. We conclude by exchanging the second row and the third:

$$\begin{bmatrix} 1 & 0 & 0 & 3 \\ 0 & 0 & 1 & 1 \\ 0 & 1 & 0 & 0 \end{bmatrix} \rightarrow \begin{bmatrix} 1 & 0 & 0 & 3 \\ 0 & 1 & 0 & 0 \\ 0 & 0 & 1 & 1 \end{bmatrix}$$

The last matrix is the matrix $[B \mid \mathbf{c}]$. It is the augmented matrix of

$$\begin{cases} x = 3 \\ y = 0 \\ z = 1 \end{cases}$$

This completes the conversion. ▲

In the next example, we show how a linear system can be solved by simplifying its augmented matrix to the point where we can read off the values of some of the variables. By back substituting these values into the remaining equations, we can find a solution for the entire system.

EXAMPLE 1.40 Augmented Matrices and Back Substitution

Use back substitution to solve the linear system determined by the augmented matrix

$$[A \mid \mathbf{b}] = \begin{bmatrix} -5 & 1 & 7 & 3 \\ 9 & 0 & 9 & 2 \\ 4 & 3 & 5 & 1 \end{bmatrix}$$

Solution. We first use the described row operations to convert $[A \mid \mathbf{b}]$ to

$$B = \begin{bmatrix} -5 & 1 & 7 & 3 \\ 0 & \frac{9}{5} & \frac{108}{5} & \frac{37}{5} \\ 0 & 0 & -35 & -\frac{110}{9} \end{bmatrix}$$

From the third row of B we know that $-35z = -\frac{110}{9}$. Hence $z = 22/63$. By substituting this value of z back into the equation $(9/5)\,y + (108/5)\,z = 37/5$ we get $(9/5)\,y + (108/5)\,(22/63) = 37/5$. Hence $y = -5/63$. We then substitute the values of z and y back into the equation

$$-5x + y + 7z = 3$$

and obtain $x = -8/63$. ▲

EXAMPLE 1.41 Gaussian Elimination in *Maple*

Use *Maple* to reduce the augmented matrix

$$[A \mid \mathbf{b}] = \begin{bmatrix} 3 & 2 & 2 & 3 \\ 1 & 0 & 3 & 2 \\ 3 & 5 & 0 & 4 \end{bmatrix}$$

to the augmented matrix of the system

$$\begin{cases} 3x + 2y + 2z = 3 \\ \quad -2y + 7z = 3 \\ \qquad 17z = 11 \end{cases}$$

Solution. The required matrix is

$$[B \mid \mathbf{c}] = \begin{bmatrix} 3 & 2 & 2 & 3 \\ 0 & -2 & 7 & 3 \\ 0 & 0 & 17 & 11 \end{bmatrix}$$

We use the **ffgausselim** function in the **linalg** package to find it.

```
[> with(linalg):
[> Ab:=matrix([[3,2,2,3],[1,0,3,2],[3,5,0,4]]):
[> Bc:=ffgausselim(Ab);
```

$$\begin{bmatrix} 3 & 2 & 2 & 3 \\ 0 & -2 & 7 & 3 \\ 0 & 0 & 17 & 11 \end{bmatrix}$$

As we can see, the final output is the matrix $[B \mid \mathbf{c}]$. ▲

In the next example, we show that augmented matrices and Gaussian elimination can be used to calculate the coefficients of interpolating polynomials. A polynomial $p(t) = a_0 + a_1 t + \cdots + a_{n-1}t^{n-1}$ is an ***interpolating polynomial*** for the points $(x_1, y_1), (x_2, y_2), \ldots,$ (x_n, y_n) if $p(x_i) = y_i$ for all $1 \leq i \leq n$, since then the graph of $p(t)$ passes through all of the given points. We assume that $x_i < x_j$ for $i < j$.

EXAMPLE 1.42 Augmented Matrices and Interpolating Polynomials

Use augmented matrices and Gaussian elimination to find the interpolating polynomial passing through the points $(-1, 5)$, $(2, 7)$, and $(3, 4)$.

Solution. Let $p(t) = a_0 + a_1 t + a_2 t^2$ be the required polynomial. If the given points lie on the graph of $p(t)$, then $p(-1) = 5$, $p(2) = 7$, and $p(3) = 4$. This information determines the linear system

$$\begin{cases} a_0 + a_1(-1) + a_2(-1)^2 = 5 \\ a_0 + a_1(2) + a_2(2)^2 = 7 \\ a_0 + a_1(3) + a_2(3)^2 = 4 \end{cases}$$

whose augmented matrix $[A \mid \mathbf{b}]$ is

$$\begin{bmatrix} 1 & -1 & 1 & 5 \\ 1 & 2 & 4 & 7 \\ 1 & 3 & 9 & 4 \end{bmatrix}$$

We can use elementary row operations to convert $[A \mid \mathbf{b}]$ to the augmented matrix

$$[B \mid \mathbf{c}] = \begin{bmatrix} 1 & -1 & 1 & 5 \\ 0 & 3 & 3 & 2 \\ 0 & 0 & 12 & -11 \end{bmatrix}$$

of the linear system

$$\begin{cases} a_0 - a_1 + a_2 = 5 \\ 3a_1 + 3a_2 = 2 \\ 12a_2 = -11 \end{cases}$$

By back substitution we can determine that the coefficients of $p(t)$ are $a_0 = 15/2$, $a_1 = 19/12$, and $a_2 = -11/12$. Hence

$$p(t) = \tfrac{15}{2} + \tfrac{19}{12}t - \tfrac{11}{12}t^2$$

is the required polynomial. ▲

EXAMPLE 1.43 Using Gaussian Elimination to Solve a Large System

Let $[A \mid \mathbf{b}]$ be a randomly generated 20×21 matrix, considered as the augmented matrix of a linear system $A\mathbf{x} = \mathbf{b}$ in the variables $x_1, \ldots, x_{20}$. Use the **gausselim** function in the **linalg** package to find the value of x_{20}, if it exists.

Solution. The *Maple* Help file explains that if we apply the **gausselim** function to a 20×21 matrix A, the result is a 20×21 upper triangular matrix $B = \left[b_{ij}\right]$. As explained in Chapter 2, this means that $b_{ij} = 0$ for all $i > j$. In other words, the last row of B is of the form

$$\mathbf{r}_{20}[B \mid \mathbf{c}] = \begin{bmatrix} b_{20,1} & b_{20,2} & \cdots & b_{20,19} & b_{20,20} & b_{20,21} \end{bmatrix}$$

$$= \begin{bmatrix} 0 & 0 & \cdots & 0 & b_{20,20} & b_{20,21} \end{bmatrix}$$

with the linear system $B\mathbf{x} = \mathbf{c}$ connected to $A\mathbf{x} = \mathbf{b}$ by Gaussian elimination. The last row $\mathbf{r}_{20}B$ of B corresponds to the equation

$$\left(b_{20,20}\right) x_{20} = b_{20,21}$$

The value of x_{20} is therefore $b_{20,21}/b_{20,20}$, provided that $b_{20,20}$ is nonzero. Because of the way *Maple* calculates random matrices, this condition almost certainly holds if we precede our calculation with the **restart** command.

```
[> restart;
[> with(linalg):
[> Ab:=randmatrix(20,21):
[> Bc:=gausselim(Ab):
[> is(Bc[20,20]=0);
```

$$false$$

```
[> x20:=evalf(Bc[20,21]/Bc[20,20],5);
```

$$x20 := -39.384$$

This shows that -39.384 is the approximate values of x_{20} in the linear system $A\mathbf{x} = \mathbf{b}$. ▲

EXAMPLE 1.44 Using Maple to Generate a Large Homogeneous Linear System

Use the **geneqns** function in the **linalg** package to generate a homogeneous linear system in 20 variables and display the first equation of the system.

Solution. We generate a random 20×20 matrix A and apply the **geneqns** function to A.

```
[> A:=randmatrix(20,20):
```

Next we use A as the coefficient matrix of a homogeneous linear system, generate the associated equations, and extract the first equation.

```
[> with(linalg):
[> vars:=[seq(x[i],i=1..20)]:
[> S:=geneqns(A,vars):
```

The command `S[1]` produces the output

$$-22x_1 + 90x_2 + 20x_3 - 26x_4 + 27x_5 - 32x_6 + 85x_7 + 74x_8 + 46x_9 - 50x_{10} - 2x_{11}$$
$$-45x_{12} - 74x_{13} - 60x_{14} - x_{15} - 31x_{16} - 43x_{17} + 73x_{18} - 82x_{19} + 75x_{20} = 0$$

and 19 other linear equations. ▲

EXERCISES 1.10

1. Find the augmented matrices of the following linear systems.

a. $\begin{cases} 2x + y - z = 2 \\ x - 3y + z = 1 \\ 2x + y - 2z = 6 \end{cases}$ b. $\begin{cases} 2x + y - z = 2 \\ x - 3y + z = 1 \\ 2x + y - z = 6 \end{cases}$ c. $\begin{cases} 2x + y - z = 2 \\ x - 3y + z = 1 \end{cases}$

d. $\begin{cases} x - z + w = 2 \\ -y + z - 7w = 4 \\ 2x + y = 5 \end{cases}$ e. $\begin{cases} x + 2y - z = 2 \\ -y + z = 4 \\ y + 2x = 5 \end{cases}$ f. $\begin{cases} x = 2 \\ y = 4 \\ y + 2x = 5 \end{cases}$

2. Construct linear systems with the following augmented matrices.

a. $\begin{bmatrix} 3 & 2 & 3 \\ 5 & 4 & 2 \\ 5 & 2 & 1 \\ 1 & 1 & 3 \end{bmatrix}$ b. $\begin{bmatrix} 3 & 5 & 5 & 1 \\ 2 & 4 & 2 & 1 \\ 3 & 2 & 1 & 3 \end{bmatrix}$ c. $\begin{bmatrix} 3 & 5 & 5 & 1 \\ 2 & 4 & 2 & 1 \end{bmatrix}$

3. Use augmented matrices and Gaussian elimination to find the following interpolating polynomials.

   a. The polynomial $p(t) = a_0 + a_1 t + a_2 t^2$ passing through the points $(-1, 3)$, $(0, 5)$, $(2, 1)$

   b. The polynomial $p(t) = a_0 + a_1 t + a_2 t^2 + a_3 t^3$ passing through the points $(-1, 3)$, $(0, 5)$, $(2, 1)$, $(7, 4)$

   c. The polynomial $p(t) = a_0 + a_1 t + a_2 t^2 + a_3 t^3 + a_4 t^4$ passing through the points $(-1, 3)$, $(0, 5)$, $(2, 1)$, $(7, 4)$, $(8, 1)$

4. Suppose that $p(t) = a_0 + a_1 t + a_2 t^2$ and $q(t) = b_0 + b_1 t + b_2 t^2$ are two polynomials passing through the points $(1, 2)$, $(2, 3)$, and $(3, 4)$. Show that $p(t) = q(t)$, using the fact that the homogeneous linear system

$$\begin{cases} x + 2y + 4z = 0 \\ x + 3y + 9z = 0 \\ x + 4y + 16z = 0 \end{cases}$$

has only the zero solution.

5. (*Maple V*) Use augmented matrices to solve the following linear systems.

a. $\begin{cases} x + 2y - z = 2 \\ x - y + z = 4 \\ 2x + y - z = 5 \end{cases}$ b. $\begin{cases} x - y + z = 2 \\ x - y + z = 4 \\ 2x + y - z = 5 \end{cases}$ c. $\begin{cases} x + 2y - z + w = 2 \\ x - y + z = 4 \\ 2x + y - z = 5 \end{cases}$

6. (*Maple V*) Use the **matrix** and **augment** functions in the **linalg** package to create the augmented matrices of the following linear systems.

a. $\begin{cases} x + y = 3 \\ y = \frac{37}{5} \end{cases}$ b. $\begin{cases} x + y + z = 3 \\ y + z = \frac{37}{5} \end{cases}$ c. $\begin{cases} x + 2y - z = 2 \\ x - y + z = 4 \\ 2x + y - z = 5 \end{cases}$

7. (*Maple V*) Use the **linalg** package to generate a homogeneous linear system in 15 variables, and display the third equation.

8. (*Maple 6*) Use the **linalg** package to generate a nonhomogeneous linear system in 15 variables, and display the seventh equation.

9. (*Maple 6*) Use the **GenerateMatrix** function and its **augmented** option in the **LinearAlgebra** package to create the augmented matrices of the linear systems in Exercise 6.

10. (*Maple 6*) Use the **LinearAlgebra** package to generate the augmented matrix of a 30×30 linear system, and find the values of two of the variables.

ROW ECHELON MATRICES

The Gaussian elimination steps can be adapted to matrices and used to convert augmented matrices to ***row echelon form***. The row echelon form of an augmented matrix is the matrix analog of the row echelon form of the underlying linear system. The basic strategy in the reduction is to convert the augmented matrices to new matrices containing as many zeros as possible, where these zeros occur in a prescribed order.

The definition of a row echelon matrix uses the ideas of a zero row and the leading entry of a row. Consider any $m \times n$ matrix A of the form

$$\begin{bmatrix} a_{11} & a_{12} & \cdots & a_{1n} \\ & & \vdots & \\ a_{i1} & a_{i2} & \cdots & a_{in} \\ & & \vdots & \\ a_{m1} & a_{m2} & \cdots & a_{mn} \end{bmatrix} = \begin{bmatrix} \mathbf{r}_1 A \\ \vdots \\ \mathbf{r}_i A \\ \vdots \\ \mathbf{r}_m A \end{bmatrix}$$

with $\mathbf{r}_i A = [a_{i1} \ \cdots \ a_{in}]$ denoting the ith row of A. We call $\mathbf{r}_i(A)$ a ***zero row*** of A if $a_{i1} = \cdots = a_{in} = 0$. All rows that are not zero rows are called ***nonzero rows***. A ***leading entry*** of a nonzero row is the leftmost nonzero entry of the row.

DEFINITION 1.15 *An $m \times n$ matrix is in* **row echelon form** *if it has the following three properties: (1) All zero rows occur below all nonzero rows. (2) All entries below a leading entry are zero. (3) The leading entry of a nonzero row occurs in a column to the right of the column containing the leading entry of the row above it.*

Elementary Row Operations

The operations required to reduce a matrix to row echelon form are the ***elementary row operations***. They correspond to the operations on linear systems and augmented matrices described earlier. Suppose that A is an $m \times n$ matrix with rows $\mathbf{r}_1 A, \ldots, \mathbf{r}_m A$.

The operation $(R_i \to sR_i)$, for $s \neq 0$, replaces the row $\mathbf{r}_i A$ with the row $[b_{i1} \cdots b_{in}]$, in which $b_{ik} = sa_{ik}$ for $1 \leq k \leq n$. We write $A\ (R_i \to sR_i)\ B$ to express the fact that the matrix B has been obtained from the matrix A by the operation $(R_i \to sR_i)$.

The operation $(R_i \rightleftharpoons R_j)$ interchanges the rows $\mathbf{r}_i A$ and $\mathbf{r}_j A$ of A. We write $A\ (R_i \rightleftharpoons R_j)\ B$ to indicate that the matrix B has been obtained from the matrix A by the operation $(R_i \rightleftharpoons R_j)$.

The operation $(R_i \to R_i + sR_j)$ replaces the row $\mathbf{r}_i A$ with the row $[b_{i1} \cdots b_{in}]$, in which $b_{ik} = a_{ik} + sa_{jk}$ for $1 \leq k \leq n$ and $i \neq j$. We write $A\ (R_i \to R_i + sR_j)\ B$ to express the fact that the matrix B has been obtained from the matrix A by the operation $(R_i \to R_i + sR_j)$.

By insisting that $i \neq j$ in $(R_i \to R_i + sR_j)$, we are disallowing the operation of adding a multiple of a row to the same row. Such an operation would not be helpful for solving linear systems and could even destroy the nature of the solutions of a system. If $s = -1$, for example, then the operation

$$(R_i \to R_i + sR_i) = (R_i \to R_i + (-1)\, R_i) = (R_i \to (1-1)\, R_i) = (R_i \to 0R_i)$$

corresponds to the multiplication of the ith row of A by 0. However, we have disallowed this operation by insisting that $s \neq 0$ in $(R_i \to sR_i)$. Here is what could happen if we allowed this operation. The conversion

$$\begin{cases} x + y = 4 \\ x - y = 2 \end{cases} \to \begin{cases} 0x + 0y = 0 \times 4 \\ x - y = 2 \end{cases}$$

would destroy the solutions of the original system. The first system has the unique solution $(x, y) = (3, 1)$, whereas the second system has the infinitely many solutions $(x, x - 2)$.

EXAMPLE 1.45 Reducing a Matrix to Row Echelon Form

Use elementary row operations to reduce the matrix

$$A = \begin{bmatrix} 0 & 2 & 3 & 4 \\ 0 & 0 & 0 & 0 \\ 6 & 7 & 0 & 8 \\ 0 & 4 & 1 & 8 \end{bmatrix}$$

to row echelon form.

Solution. The following steps convert A to row echelon form.

1. First we move all zero rows to the bottom of the matrix.

2. Then we modify the nonzero rows using interchanges and row additions until we have the desired form. The following are three possible steps achieving the conversion.

   a. We interchange the second and fourth rows:

$$\begin{bmatrix} 0 & 2 & 3 & 4 \\ 0 & 0 & 0 & 0 \\ 6 & 7 & 0 & 8 \\ 0 & 4 & 1 & 8 \end{bmatrix} (R_2 \rightleftharpoons R_4) \begin{bmatrix} 0 & 2 & 3 & 4 \\ 0 & 4 & 1 & 8 \\ 6 & 7 & 0 & 8 \\ 0 & 0 & 0 & 0 \end{bmatrix}$$

   b. We interchange the first and third rows:

$$\begin{bmatrix} 0 & 2 & 3 & 4 \\ 0 & 4 & 1 & 8 \\ 6 & 7 & 0 & 8 \\ 0 & 0 & 0 & 0 \end{bmatrix} (R_3 \rightleftharpoons R_1) \begin{bmatrix} 6 & 7 & 0 & 8 \\ 0 & 4 & 1 & 8 \\ 0 & 2 & 3 & 4 \\ 0 & 0 & 0 & 0 \end{bmatrix}$$

   c. We subtract a multiple of the second row from the third row:

$$\begin{bmatrix} 6 & 7 & 0 & 8 \\ 0 & 4 & 1 & 8 \\ 0 & 2 & 3 & 4 \\ 0 & 0 & 0 & 0 \end{bmatrix} \left(R_3 \to R_3 - \frac{1}{2}R_2\right) \begin{bmatrix} 6 & 7 & 0 & 8 \\ 0 & 4 & 1 & 8 \\ 0 & 0 & \frac{5}{2} & 0 \\ 0 & 0 & 0 & 0 \end{bmatrix}$$

As we can see, the last matrix is in row echelon form. ▲

If we wanted a fraction-free row echelon form of the matrix A in Example 1.45, we would add a fourth step to the reduction:

$$\begin{bmatrix} 6 & 7 & 0 & 8 \\ 0 & 4 & 1 & 8 \\ 0 & 0 & \frac{5}{2} & 0 \\ 0 & 0 & 0 & 0 \end{bmatrix} (R_3 \to 2R_3) \begin{bmatrix} 6 & 7 & 0 & 8 \\ 0 & 4 & 1 & 8 \\ 0 & 0 & 5 & 0 \\ 0 & 0 & 0 & 0 \end{bmatrix}$$

The elementary row operations $(R_i \to sR_i)$, $(R_i \rightleftharpoons R_j)$, and $(R_i \to R_i + sR_j)$ are implemented in the **linalg** package by the functions **linalg[mulrow]**, **linalg[swaprow]**, and **linalg[addrow].** Using these functions, we can reproduce the previous example in *Maple* as follows:

```
[> A:=matrix([[0,2,3,4],[0,0,0,0],[6,7,0,8],[0,4,1,8]]):
[> A1:=swaprow(A,2,4):
[> A2:=swaprow(A1,1,3):
[> A3:=addrow(A2,2,3,-1/2):
[> A4:=mulrow(A3,3,2);
```

$$A4 := \begin{bmatrix} 6 & 7 & 0 & 8 \\ 0 & 4 & 1 & 8 \\ 0 & 0 & 5 & 0 \\ 0 & 0 & 0 & 0 \end{bmatrix}$$

We note that new zero rows may arise at various stages in a reduction to row echelon form. Moreover, no interchanges may be required to place the zero rows below all nonzero rows. The reduction

$$\begin{bmatrix} 1 & 1 & 1 \\ 1 & 1 & 1 \\ 1 & 1 & 1 \end{bmatrix} (R_2 \to R_2 - R_1) \begin{bmatrix} 1 & 1 & 1 \\ 0 & 0 & 0 \\ 1 & 1 & 1 \end{bmatrix} (R_3 \to R_3 - R_1) \begin{bmatrix} 1 & 1 & 1 \\ 0 & 0 & 0 \\ 0 & 0 & 0 \end{bmatrix}$$

for example, produces a new zero row at each stage of the reduction. Moreover, the reduction involves new interchanges of rows.

EXERCISES 1.11

1. Extract the rows of the following matrices. Label each row as $\mathbf{r}_i A$ or $\mathbf{r}_i B$, as appropriate.

$$A = \begin{bmatrix} 1 & 0 & 0 & 0 & 3 \\ 1 & 2 & 0 & 5 & 3 \\ 0 & 0 & 0 & 4 & -1 \end{bmatrix} \quad B = \begin{bmatrix} 1 & 0 & 0 & 0 & 3 \\ 0 & 2 & 0 & 5 & 3 \\ 0 & 0 & 0 & 0 & 0 \end{bmatrix}$$

2. Identify the zero rows of each of the matrices in Exercise 1.

3. For each matrix C in Exercise 1, carry out the following operations on the rows of the matrix.

   a. $\mathbf{r}_i C \to 4\mathbf{r}_i C$
   b. $\mathbf{r}_2 C \to \mathbf{r}_2 C + 7\mathbf{r}_1 C$
   c. $\mathbf{r}_2 C \rightleftharpoons \mathbf{r}_3 C$

4. Determine which of the following are row echelon matrices. Explain your answers.

$$\text{a.}\begin{bmatrix} 1 & 0 & 0 & 0 & 3 \\ 1 & 2 & 0 & 5 & 3 \\ 0 & 0 & 0 & 4 & -1 \end{bmatrix} \quad \text{b.}\begin{bmatrix} 1 & 0 & 0 & 0 & 3 \\ 0 & 2 & 0 & 5 & 3 \\ 0 & 0 & 0 & 4 & -1 \end{bmatrix} \quad \text{c.}\begin{bmatrix} 0 & 2 & 0 & 5 & 3 \\ 1 & 0 & 0 & 0 & 3 \\ 0 & 0 & 0 & 4 & -1 \end{bmatrix}$$

5. Use elementary row operations to convert the following matrices to row echelon matrices.

$$\text{a.}\begin{bmatrix} -1 & 1 & 4 & 1 \\ -3 & 3 & 5 & 5 \\ -2 & -3 & 3 & -3 \end{bmatrix} \quad \text{b.}\begin{bmatrix} 2 & -3 & 0 \\ 0 & 4 & 3 \\ 1 & 0 & 5 \\ 2 & 5 & 2 \end{bmatrix} \quad \text{c.}\begin{bmatrix} 0 & 20 & -40 \\ 0 & -405 & -499 \\ 0 & 5840 & 4522 \end{bmatrix}$$

Explain why the resulting matrices are in row echelon form.

6. Use elementary row operations to convert matrix A to matrix B.

$$A = \begin{bmatrix} 4 & -13 & 5 & 0 & 5 \\ 4 & 4 & 2 & 5 & -5 \\ 1 & 0 & -3 & -5 & -8 \\ -5 & 4 & 7 & -9 & 5 \end{bmatrix} \quad B = \begin{bmatrix} 4 & -13 & 5 & 0 & 5 \\ 0 & 68 & -12 & 20 & -40 \\ 0 & 0 & -250 & -405 & -499 \\ 0 & 0 & 0 & 5840 & 4522 \end{bmatrix}$$

7. (*Maple V*) Repeat Exercise 5 using the functions **addrow**, **mulrow**, and **swaprow** in the **linalg** package.

8. (*Maple V*) Repeat Exercise 6 using the functions **addrow**, **mulrow**, and **swaprow** in the **linalg** package.

9. (*Maple V*) Apply the **linalg[rref]** function to the two matrices in Exercise 6. Explain the result.

10. (*Maple V*) Use the **randmatrix** function in the **linalg** package to generate two 3×3 random matrices. Apply the **linalg[rref]** functions to these matrices. There is a high probability that the resulting matrices will all be of the form

$$\begin{bmatrix} 1 & 0 & 0 \\ 0 & 1 & 0 \\ 0 & 0 & 1 \end{bmatrix}$$

Try to explain this result. Modify the entries of the two matrices so that the function **linalg[rref]** produces the matrix

$$\begin{bmatrix} 1 & 0 & 0 \\ 0 & 0 & 1 \\ 0 & 0 & 0 \end{bmatrix}$$

11. (*Maple 6*) Repeat Exercise 5 using the function **RowOperation** in the **LinearAlgebra** package.

12. (*Maple 6*) Repeat Exercise 6 using the function **RowOperation** in the **LinearAlgebra** package.

Inverse of Elementary Row Operations

Our later work involving elementary row operations depends on the fact that these operations are invertible.

1. If $s \neq 0$ and B was obtained from A by $(R_i \to sR_i)$, then A can be recovered from B by applying $(R_i \to \frac{1}{s}R_i)$ to B.

2. If B was obtained from A by $(R_i \to R_i + sR_j)$, then A can be recovered from B by applying $(R_i \to R_i - sR_j)$ to B.

3. If B was obtained from A by $(R_i \rightleftharpoons R_j)$, then A can be recovered from B by applying the operation $(R_i \rightleftharpoons R_j)$ to B.

The invertibility of these operations provides the rationale for introducing the following relationship among matrices.

DEFINITION 1.16 *A matrix A is **row equivalent** to a matrix B if we can convert A to B by a finite number of elementary row operations.*

In the exercises, we prove that row equivalence is an equivalence in the sense that it is a reflexive, symmetric, and transitive relation. What makes row equivalence useful is the fact that the linear systems determined by row equivalent augmented matrices have the same solutions. This follows immediately from the fact that Gaussian elimination preserves the solutions of linear systems because the operations involved are invertible.

EXAMPLE 1.46 Row Equivalent Matrices

Show that the matrices

$$A = \begin{bmatrix} 1 & 2 & 3 \\ 4 & 5 & 6 \\ 7 & 8 & 9 \end{bmatrix} \quad \text{and} \quad B = \begin{bmatrix} 39 & 48 & 57 \\ -8 & -10 & -12 \\ 1 & 2 & 3 \end{bmatrix}$$

are row equivalent.

Solution. We use elementary row operations to convert A to B. The following are three possible steps converting A to B.

1. We interchange the first and third rows:

$$\begin{bmatrix} 1 & 2 & 3 \\ 4 & 5 & 6 \\ 7 & 8 & 9 \end{bmatrix} (R_1 \rightleftharpoons R_3) \begin{bmatrix} 7 & 8 & 9 \\ 4 & 5 & 6 \\ 1 & 2 & 3 \end{bmatrix}$$

2. We multiply the second row by a nonzero real number:

$$\begin{bmatrix} 7 & 8 & 9 \\ 4 & 5 & 6 \\ 1 & 2 & 3 \end{bmatrix} (R_2 \to -2R_2) \begin{bmatrix} 7 & 8 & 9 \\ -8 & -10 & -12 \\ 1 & 2 & 3 \end{bmatrix}$$

3. We subtract a multiple of the second row from the first row:

$$\begin{bmatrix} 7 & 8 & 9 \\ -8 & -10 & -12 \\ 1 & 2 & 3 \end{bmatrix} (R_1 \to R_1 - 4R_2) \begin{bmatrix} 39 & 48 & 57 \\ -8 & -10 & -12 \\ 1 & 2 & 3 \end{bmatrix}$$

Since B can be obtained from A by three elementary row operations, A and B are row equivalent. ▲

Every matrix can be reduced to row echelon form. The proof of the next theorem provides an algorithm for doing so.

THEOREM 1.1 (Gaussian elimination theorem) *Every matrix is row equivalent to a row echelon matrix.*

Proof. Let A be a given $m \times n$ matrix. Locate the leftmost nonzero column $\mathbf{c}_j A$ of A. If the first entry of $\mathbf{c}_j A$ is zero, use a row interchange $(\mathbf{r}_1 A \rightleftharpoons \mathbf{r}_i A)$ to convert A to a matrix A' with a nonzero entry in the first row of column $\mathbf{c}_i A'$. If the first entry of $\mathbf{c}_j A$ is nonzero, omit this step. Next, use appropriate operations $\left(\mathbf{r}_s A' \to \mathbf{r}_s A' + s\mathbf{r}_r A'\right)$ to convert the matrix A' to a matrix A'' whose jth column is of the form

$$\mathbf{c}_j A'' = \begin{bmatrix} c_{1j} \\ 0 \\ \vdots \\ 0 \end{bmatrix}$$

Apply this construction to the $(m-1) \times n$ submatrix A''' obtained from A'' by neglecting the first row of A''. Continue this process for successive $(m-k) \times n$ submatrices until the resulting matrix is in echelon form. ■

EXERCISES 1.12

1. Use elementary row operations to show that the following matrices are row equivalent.

a. $\begin{bmatrix} 4 & -13 & 5 & 0 & 5 \\ 4 & 4 & 2 & 5 & -5 \\ 1 & 0 & -3 & -5 & -8 \\ -5 & 4 & 7 & -9 & 5 \end{bmatrix}$ b. $\begin{bmatrix} 4 & -13 & 5 & 0 & 5 \\ 0 & 68 & -12 & 20 & -40 \\ 0 & 0 & -250 & -405 & -499 \\ 0 & 0 & 0 & 5840 & 4522 \end{bmatrix}$

c. $\begin{bmatrix} 8 & -9 & 7 & 5 & 0 \\ 4 & 4 & 2 & 5 & -5 \\ 9 & -9 & 4 & 0 & -8 \\ -5 & 4 & 7 & -9 & 5 \end{bmatrix}$ d. $\begin{bmatrix} 8 & -9 & 7 & 5 & 0 \\ 0 & 9 & -31 & -45 & -64 \\ 0 & 0 & 52 & -126 & -59 \\ 0 & 0 & 0 & 5840 & 4522 \end{bmatrix}$

2. Show that if A and B are row equivalent augmented matrices, then the linear systems determined by A and B have the same solutions.

3. Show that row equivalence is a reflexive, symmetric, and transitive relation on matrices. It is ***reflexive*** in the sense that every matrix is row equivalent to itself. It is ***symmetric*** in the sense that if A is row equivalent to B, then B is row equivalent to A. It is also ***transitive*** in the sense that if A is row equivalent to B and B is row equivalent to C, then A is row equivalent to C.

4. (*Maple V*) Repeat Exercise 1 using the functions **addrow**, **mulrow**, and **swaprow** in the **linalg** package.

5. (*Maple V*) Apply the **linalg[rref]** function to the matrices in Exercise 1 and explain the result.

6. (*Maple V*) Use the function **gausselim** in the **linalg** package to reduce the following matrices to row echelon form. Describe what you think is special about the resulting matrices.

a. $\begin{bmatrix} 2 & -3 & 0 \\ 0 & 4 & 3 \\ 1 & 0 & 5 \\ 2 & 5 & 2 \end{bmatrix}$ b. $\begin{bmatrix} 0 & 0 & 0 & 0 & 3 \\ 0 & 2 & 0 & 5 & 3 \\ 1 & 0 & 0 & 4 & -1 \end{bmatrix}$ c. $\begin{bmatrix} 0 & 2 & 0 & 5 & 3 \\ 0 & 0 & 0 & 0 & 0 \\ 0 & 0 & 0 & 4 & -1 \end{bmatrix}$

d. $\begin{bmatrix} 2 & 2 & 2 & 2 & 2 \\ 2 & 2 & 2 & 2 & 2 \\ 2 & 2 & 2 & 2 & 2 \end{bmatrix}$ e. $\begin{bmatrix} 1 & 0 & 0 & 0 & 3 \\ 1 & 2 & 0 & 5 & 3 \end{bmatrix}$ f. $\begin{bmatrix} 0 & 0 & 3 & 0 & 2 \\ 0 & 2 & 2 & 2 & 4 \\ 2 & 0 & 0 & 8 & 0 \end{bmatrix}$

7. (*Maple 6*) Repeat Exercise 1 using the function **RowOperation** in the **LinearAlgebra** package.

REDUCED ROW ECHELON MATRICES

It is often helpful to use elementary row operations to reduce a row echelon matrix further to a unique, still simpler form in which all diagonal entries are 1 or 0. By the ***diagonal entries*** of a matrix $A = [a_{ij}]$, we mean all entries a_{ij} for which $i = j$. The following example illustrates a row reduction that produces 1s and 0s on the diagonal of a matrix.

EXAMPLE 1.47 Reducing a Matrix to a Matrix with 1s and 0s on the Diagonal

Use elementary row operations to reduce the matrix

$$A = \begin{bmatrix} 1 & -3 & 0 & 2 \\ 2 & 6 & 0 & 7 \\ 0 & 3 & 0 & 5 \end{bmatrix}$$

to the row echelon form

$$B = \begin{bmatrix} 1 & -3 & 0 & 2 \\ 0 & 1 & 0 & \frac{5}{3} \\ 0 & 0 & 0 & -17 \end{bmatrix}$$

Solution. The following steps achieve the desired reduction.

1. We subtract a multiple of the first row from the second row:

$$\begin{bmatrix} 1 & -3 & 0 & 2 \\ 2 & 6 & 0 & 7 \\ 0 & 3 & 0 & 5 \end{bmatrix} (R_2 \to R_2 - 2R_1) \begin{bmatrix} 1 & -3 & 0 & 2 \\ 0 & 12 & 0 & 3 \\ 0 & 3 & 0 & 5 \end{bmatrix}$$

2. We add a multiple of the third row to the second row:

$$\begin{bmatrix} 1 & -3 & 0 & 2 \\ 0 & 12 & 0 & 3 \\ 0 & 3 & 0 & 5 \end{bmatrix} (R_2 \to R_2 - 4R_3) \begin{bmatrix} 1 & -3 & 0 & 2 \\ 0 & 0 & 0 & -17 \\ 0 & 3 & 0 & 5 \end{bmatrix}$$

3. We interchange the second and third rows:

$$\begin{bmatrix} 1 & -3 & 0 & 2 \\ 0 & 0 & 0 & -17 \\ 0 & 3 & 0 & 5 \end{bmatrix} (R_2 \rightleftharpoons R_3) \begin{bmatrix} 1 & -3 & 0 & 2 \\ 0 & 3 & 0 & 5 \\ 0 & 0 & 0 & -17 \end{bmatrix}$$

4. We multiply the second row by a nonzero real number:

$$\begin{bmatrix} 1 & -3 & 0 & 2 \\ 0 & 3 & 0 & 5 \\ 0 & 0 & 0 & -17 \end{bmatrix} \left(R_2 \to \frac{1}{3}R_2\right) \begin{bmatrix} 1 & -3 & 0 & 2 \\ 0 & 1 & 0 & \frac{5}{3} \\ 0 & 0 & 0 & -17 \end{bmatrix}$$

As we can see, the diagonal entries of B are 1. ▲

Pivots

If we take a closer look at matrix B in Example 1.47, we see that B can be reduced further. We can eliminate the -3 in the first row by adding 3 times the second row to the first:

$$\begin{bmatrix} 1 & -3 & 0 & 2 \\ 0 & 1 & 0 & \frac{5}{3} \\ 0 & 0 & 0 & -17 \end{bmatrix} (R_1 \to R_1 + 3R_2) \begin{bmatrix} 1 & 0 & 0 & 7 \\ 0 & 1 & 0 & \frac{5}{3} \\ 0 & 0 & 0 & -17 \end{bmatrix} = C$$

The matrix C that results has certain special features. We describe them in terms of **pivots**.

DEFINITION 1.17 *The leading entry of a nonzero row in a row echelon matrix is a* ***pivot*** *of the matrix.*

DEFINITION 1.18 *A row echelon matrix is in* ***reduced row echelon form*** *either if it is a zero matrix or if all of its pivots are* 1 *and all entries above its pivots are* 0.

The reduced row echelon form of a nonzero row echelon matrix $A = [a_{ij}]$ is obtained by applying the following reduction steps to the matrix.

1. If a_{ij} is a leading entry of A, apply the operation $\left(R_i \to \frac{1}{a_{ij}} R_i\right)$ to A. Repeat this operation until all leading entries are 1.

2. Let $B = [b_{ij}]$ be the matrix resulting from Step 1. If b_{ij} is a leading entry of B and b_{kj} is a nonzero entry above b_{ij}, apply the operation $(R_k \to R_k - b_{kj} R_i)$ to B. Repeat this operation until all entries above a leading entry of B are zero.

This conversion is known as ***Gauss–Jordan elimination.*** We formulate it as a theorem.

THEOREM 1.2 (Row echelon form theorem) *Every matrix is row equivalent to a matrix in reduced row echelon form.*

Proof. Suppose that A is a given matrix. We first use Gaussian elimination to convert A to a matrix B in row echelon form. We then use the Steps 1 and 2 in Definition 1.18 to convert B to a matrix C in reduced row echelon form. ■

We summarize the steps discussed in the proof of Theorem 1.2 in Table 3 on page 89.

We will show later that matrix C in the proof of Theorem 1.2 is unique. This implies that if $A = [a_{ij}]$ and $B = [b_{ij}]$ are two row equivalent matrices in row echelon form, then a_{ij} is a pivot of A if and only if b_{ij} is a pivot of B, since Steps 1 and 2 in the reduction to reduced row

TABLE 3 FINDING A REDUCED ROW ECHELON FORM.

Step 1. Use row reduction to convert A to a matrix B in row echelon form.
Step 2. Multiply appropriate rows of B by nonzero constants to convert all leading entries to 1.
Step 3. Locate the leftmost column containing a leading entry that has nonzero entries above it.
Step 4. Add suitable multiples of the row with the leading entry to the rows above it to change all nonleading entries in that column to 0 .
Step 5. Repeat Steps 3–4 until all entries above leading entries are 0.

echelon form preserve the positions of leading entries. We can therefore consider the position of the pivots of the leading entries of the row echelon forms of a matrix to be a characteristic of the original matrix. This leads to the following definition.

DEFINITION 1.19 *A **pivot position** of a matrix A is the position of a pivot in a row echelon form of A.*

We can now use these unique pivot positions to identify the specific columns of a matrix in which the pivot positions occur.

DEFINITION 1.20 *A **pivot column** of a matrix is a column with a **pivot position**.*

EXAMPLE 1.48 Reduction of a Matrix to Reduced Row Echelon Form

Use elementary row operations to reduce the matrix

$$A = \begin{bmatrix} 4 & 0 & 8 \\ -9 & 0 & 5 \\ 0 & 0 & 4 \end{bmatrix}$$

to reduced row echelon form.

Solution. The following is a possible reduction.

1. $$\begin{bmatrix} 4 & 0 & 8 \\ -9 & 0 & 5 \\ 0 & 0 & 4 \end{bmatrix} (R_1 \to \tfrac{1}{4}R_1) \begin{bmatrix} \left(\frac{1}{4}\right)4 & 0 & \left(\frac{1}{4}\right)8 \\ -9 & 0 & 5 \\ 0 & 0 & 4 \end{bmatrix} = \begin{bmatrix} 1 & 0 & 2 \\ -9 & 0 & 5 \\ 0 & 0 & 4 \end{bmatrix}$$

2. $$\begin{bmatrix} 1 & 0 & 2 \\ -9 & 0 & 5 \\ 0 & 0 & 4 \end{bmatrix} (R_2 \to R_2 + 9R_1) \begin{bmatrix} 1 & 0 & 2 \\ -9+(9)\,1 & 0 & 5+(9)\,2 \\ 0 & 0 & 4 \end{bmatrix} = \begin{bmatrix} 1 & 0 & 2 \\ 0 & 0 & 23 \\ 0 & 0 & 4 \end{bmatrix}$$

3. $\begin{bmatrix} 1 & 0 & 2 \\ 0 & 0 & 23 \\ 0 & 0 & 4 \end{bmatrix} (R_3 \to \frac{1}{4} R_3) \begin{bmatrix} 1 & 0 & 2 \\ 0 & 0 & 23 \\ 0 & 0 & \left(\frac{1}{4}\right) 4 \end{bmatrix} = \begin{bmatrix} 1 & 0 & 2 \\ 0 & 0 & 23 \\ 0 & 0 & 1 \end{bmatrix}$

4. $\begin{bmatrix} 1 & 0 & 2 \\ 0 & 0 & 23 \\ 0 & 0 & 1 \end{bmatrix} (R_2 \to R_2 - (23)\, R_3) \begin{bmatrix} 1 & 0 & 2 \\ 0 & 0 & 23 - (23)\, 1 \\ 0 & 0 & 1 \end{bmatrix} = \begin{bmatrix} 1 & 0 & 2 \\ 0 & 0 & 0 \\ 0 & 0 & 1 \end{bmatrix}$

5. $\begin{bmatrix} 1 & 0 & 2 \\ 0 & 0 & 0 \\ 0 & 0 & 1 \end{bmatrix} (R_1 \to R_1 - (2)\, R_3) \begin{bmatrix} 1 & 0 & 2 - (2)\, 1 \\ 0 & 0 & 0 \\ 0 & 0 & 1 \end{bmatrix} = \begin{bmatrix} 1 & 0 & 0 \\ 0 & 0 & 0 \\ 0 & 0 & 1 \end{bmatrix}$

6. $\begin{bmatrix} 1 & 0 & 0 \\ 0 & 0 & 0 \\ 0 & 0 & 1 \end{bmatrix} (R_2 \rightleftharpoons R_3) \begin{bmatrix} 1 & 0 & 0 \\ 0 & 0 & 1 \\ 0 & 0 & 0 \end{bmatrix} = B$

As we can see, the matrix has been reduced to the reduced row echelon B whose first and last columns are pivot columns. ▲

From Diagonal Entries to Pivots

In light of Example 1.48, we revise our strategy for simplifying the row echelon form of a matrix. We replace the goal of converting the diagonal entries of a matrix to 1s and 0s by the goal of converting the matrix to reduced row echelon form. The result will be a matrix in which all nonzero pivots are 1, all entries above a pivot are 0, and all zero rows occur at the bottom of the matrix.

EXAMPLE 1.49 The Pivots of a Matrix

Use *Maple* to generate a random 4×5 matrix A, and find the pivot positions and pivot columns of A.

Solution. First we generate a random matrix A.

```
[> A:=linalg[randmatrix](4,5);
```

$$A := \begin{bmatrix} 54 & -5 & 99 & -61 & -50 \\ -12 & -18 & 31 & -26 & -62 \\ 1 & -47 & -91 & -47 & -61 \\ 41 & -58 & -90 & 53 & -1 \end{bmatrix}$$

The five $\mathbf{c}_1 A, \ldots, \mathbf{c}_5 A$ columns of A are

$$\begin{bmatrix} 54 \\ -12 \\ 1 \\ 41 \end{bmatrix}, \begin{bmatrix} -5 \\ -18 \\ -47 \\ -58 \end{bmatrix}, \begin{bmatrix} 99 \\ 31 \\ -91 \\ -90 \end{bmatrix}, \begin{bmatrix} -61 \\ -26 \\ -47 \\ 53 \end{bmatrix}, \begin{bmatrix} -50 \\ -62 \\ -61 \\ -1 \end{bmatrix}$$

We use *Maple* to find the reduced row echelon form of A.

```
[> linalg[rref](A);
```

$$\begin{bmatrix} 1 & 0 & 0 & 0 & \frac{18700781}{31981699} \\ 0 & 1 & 0 & 0 & \frac{50010042}{31981699} \\ 0 & 0 & 1 & 0 & \frac{-13142038}{31981699} \\ 0 & 0 & 0 & 1 & \frac{17341232}{31981699} \end{bmatrix}$$

This shows that the pivot positions of A are a_{11}, a_{22}, a_{33}, and a_{44} and that the columns $\mathbf{c}_1 A$, $\mathbf{c}_2 A$, $\mathbf{c}_3 A$, and $\mathbf{c}_4 A$ are therefore pivot columns. ▲

Because of the fact that the pivot columns of a matrix are uniquely determined by the matrix, we cannot of course expect that the pivot columns of a matrix are necessarily contiguous. Here is an example.

EXAMPLE 1.50 Noncontiguous Pivot Columns

Construct a 4×5 matrix A whose pivot columns are $\mathbf{c}_1 A$, $\mathbf{c}_2 A$, and $\mathbf{c}_4 A$.

Solution. First we create two columns $\mathbf{c}_1 A$ and $\mathbf{c}_2 A$ with the property that $a\mathbf{c}_1 A + b\mathbf{c}_2 A = \mathbf{0}$ if and only if $a = b = 0$. For example, if

$$\mathbf{c}_1 A = \begin{bmatrix} 0 \\ 1 \\ 2 \\ 3 \end{bmatrix} \quad \text{and} \quad \mathbf{c}_2 A = \begin{bmatrix} -5 \\ -2 \\ 0 \\ 4 \end{bmatrix}$$

then

$$a \begin{bmatrix} 0 \\ 1 \\ 2 \\ 3 \end{bmatrix} + b \begin{bmatrix} -5 \\ -2 \\ 0 \\ 4 \end{bmatrix} = \begin{bmatrix} -5b \\ a - 2b \\ 2a \\ 3a + 4b \end{bmatrix} = \begin{bmatrix} 0 \\ 0 \\ 0 \\ 0 \end{bmatrix}$$

if and only if $a = b = 0$. We therefore choose two nonzero values for a and b, the values $a = 2$ and $b = 3$, for example, and let $\mathbf{c}_3 A = 2\mathbf{c}_1 A + 3\mathbf{c}_2 A$. We continue to experiment with the columns $\mathbf{c}_1 A$ and $\mathbf{c}_2 A$, as well as a possible candidate for $\mathbf{c}_4 A$, until we have found a column $\mathbf{c}_4 A$ with the property that $a\mathbf{c}_1 A + b\mathbf{c}_2 A + c\mathbf{c}_4 A = \mathbf{0}$ if and only if $a = b = c = 0$. Column $\mathbf{c}_5 A$ can be arbitrary. Here is the result of one such choice. Let A be the matrix

$$\begin{bmatrix} 0 & -5 & -15 & 4 & 7 \\ 1 & -2 & -4 & 3 & 6 \\ 2 & 0 & 4 & 2 & 1 \\ 3 & 4 & 18 & 1 & 4 \end{bmatrix}$$

Then its reduced row echelon form is

$$\begin{bmatrix} 1 & 0 & 2 & 0 & -\frac{25}{4} \\ 0 & 1 & 3 & 0 & 4 \\ 0 & 0 & 0 & 1 & \frac{27}{4} \\ 0 & 0 & 0 & 0 & 0 \end{bmatrix}$$

The pivot positions of A are a_{11}, a_{22}, and a_{34} and its pivot columns are the columns $\mathbf{c}_1 A$, $\mathbf{c}_2 A$, and $\mathbf{c}_4 A$. ▲

We can use existence and uniqueness of the reduced row echelon form of a matrix to characterize consistent linear systems.

THEOREM 1.3 *An $m \times n$ linear system $A\mathbf{x} = \mathbf{b}$ in the variables $x_1, \ldots, x_n$ is consistent for all $\mathbf{b} \in \mathbb{R}^m$ if and only if the reduced row echelon form of A has m pivots.*

Proof. Let $[A \mid \mathbf{b}] = \left[r_{ij}\right]$ be the augmented matrix of the linear system $A\mathbf{x} = \mathbf{b}$, and convert $[A \mid \mathbf{b}]$ to its reduced row echelon form B. If the last column $\mathbf{c}_{n+1} B$ of B is a pivot column, then the system is inconsistent since B contains a row $\mathbf{r}_i B$ of the form

$$[b_{in} \ \cdots \ b_{in} \ b_{i(n+1)}] = [0 \ \cdots \ 0 \ 1]$$

that corresponds to the contradictory equation

$$0x_1 + \cdots + 0x_n = 1$$

Hence the number of pivot columns of $[A \mid \mathbf{b}]$ cannot exceed the number of pivot columns of A, if the system is to be consistent. Moreover, the number of pivot columns of A cannot exceed the number of rows of A, since no two pivots of a matrix can occur in the same row.

Suppose, therefore, that the reduced row echelon form B of $[A \mid \mathbf{b}]$ has m pivots. Then the mth row of B is of the form

$$\begin{aligned} \mathbf{r}_m B &= [b_{m1} \ \cdots \ b_{m(j-1)} \ 1 \ b_{m(j+1)} \ \cdots \ b_{mn} \ b_{m(n+1)}] \\ &= [0 \ \cdots \ 0 \ 1 \ b_{m(j+1)} \ \cdots \ b_{mn} \ b_{m(n+1)}] \end{aligned}$$

This row corresponds to the equation

$$x_j + b_{m(j+1)}x_{j+1} + \cdots + b_{mn}x_n = b_{m(n+1)}$$

In that case, we can assign values to the variables $x_{j+1}, \ldots, x_n$, solve for x_m, and use back substitution to build a solution $\mathbf{b} \in \mathbb{R}^m$ for the system. This clearly works for any assignment of values to the variables x_i. On the other hand, if the number of pivots of B is less than m, then B contains a row of the form

$$\mathbf{r}_i B = [b_{in} \ \cdots \ b_{in} \ b_{i(n+1)}] = [0 \ \cdots \ 0 \ b_{i(n+1)}]$$

If we let $\mathbf{b} = (b_1, \ldots, b_i, \ldots, b_m)$ be any vector in $\mathbb{R}^m$, in which $b_i = b_{i(n+1)} \neq 0$, the systems $A\mathbf{x} = \mathbf{b}$ is inconsistent. Hence there exists a vector $\mathbf{b} \in \mathbb{R}^m$ that is not a solution of $A\mathbf{x} = \mathbf{b}$. This proves the theorem. ■

Basic and Free Variables

We can use the pivot columns of the augmented matrix of a linear system to classify the variables of the system.

DEFINITION 1.21 *The variables of a linear system $A\mathbf{x} = \mathbf{b}$ corresponding to the pivot columns of the coefficient matrix A are called* ***basic variables*** *of the system. All other variables of the system are called* ***free variables.***

Here is an example. The linear system

$$\begin{cases} -5x_2 - 15x_3 + 4x_4 = 7 \\ x_1 - 2x_2 - 4x_3 + 3x_4 = 6 \\ 2x_1 + 4x_3 + 2x_4 = 1 \\ 3x_1 + 4x_2 + 18x_3 + x_4 = 4 \end{cases}$$

in Example 1.50 is a linear system in the four variables x_1, x_2, x_3, and x_4. The coefficient matrix of the system is the matrix

$$A = \begin{bmatrix} \mathbf{c}_1 A & \mathbf{c}_2 A & \mathbf{c}_3 A & \mathbf{c}_4 A \end{bmatrix} = \begin{bmatrix} 0 & -5 & 15 & 4 \\ 1 & -2 & -4 & 3 \\ 2 & 0 & 4 & 3 \\ 3 & 4 & 18 & 2 \end{bmatrix}$$

Since the reduced row echelon form of A is

$$B = \begin{bmatrix} 1 & 0 & 2 & 0 & -\frac{25}{4} \\ 0 & 1 & 3 & 0 & 4 \\ 0 & 0 & 0 & 1 & \frac{27}{4} \\ 0 & 0 & 0 & 0 & 0 \end{bmatrix}$$

we can tell that the pivot columns of A are $\mathbf{c}_1 A$, $\mathbf{c}_2 A$, and $\mathbf{c}_4 A$. This means that the basic variables of the system are x_1, x_2, and x_4. The variable x_3 is therefore free. In other words, we are free to choose any value for x_3. Once this choice is made, the values of x_1, x_2, and x_4 required for a solution of $A\mathbf{x} = \mathbf{b}$ are uniquely determined. The linear system

$$\begin{cases} x_1 = -2x_3 - \frac{25}{4} \\ x_2 = -3x_3 + 4 \\ x_4 = \frac{27}{4} \end{cases}$$

determined by the reduced row echelon form of the augmented matrix of the system $A\mathbf{x} = \mathbf{b}$ shows why this is the case. It is called the ***general solution*** of the original system since it provides explicit formulas in the free variable x_3 for finding all solutions of the system.

EXERCISES 1.13

1. Determine which of the following matrices are in reduced row echelon form. Explain your answers.

a. $\begin{bmatrix} 1 & 0 & 0 & 0 & 3 \\ 1 & 2 & 0 & 5 & 3 \\ 0 & 0 & 0 & 4 & -1 \end{bmatrix}$ b. $\begin{bmatrix} 1 & 0 & 0 & 0 & 3 \\ 0 & 1 & 0 & 5 & 3 \\ 0 & 0 & 0 & 1 & -1 \end{bmatrix}$ c. $\begin{bmatrix} 0 & 1 & 0 & 5 & 3 \\ 1 & 0 & 0 & 0 & 3 \\ 0 & 0 & 0 & 1 & -1 \end{bmatrix}$

d. $\begin{bmatrix} 1 & 2 & 2 & 2 & 2 \\ 0 & 1 & 2 & 2 & 2 \\ 0 & 0 & 1 & 2 & 2 \end{bmatrix}$ e. $\begin{bmatrix} 1 & 0 & 0 & 0 & 3 \\ 0 & 1 & 0 & 5 & 3 \end{bmatrix}$ f. $\begin{bmatrix} 1 & 0 & 0 & 0 & 3 \end{bmatrix}$

2. Use elementary row operations to reduce the matrix

$$\begin{bmatrix} -5 & -1 & 9 & 2 & 1 \\ 5 & 1 & 5 & 9 & 2 \\ 7 & -2 & 4 & -8 & 3 \end{bmatrix}$$

to reduced row echelon form.

3. Find the pivot columns of the following matrices. Explain your answers.

a. $\begin{bmatrix} 1 & 0 & 0 & 0 & 3 \\ 1 & 2 & 0 & 5 & 3 \\ 0 & 0 & 0 & 4 & -1 \end{bmatrix}$ b. $\begin{bmatrix} 0 & 0 & 0 & 0 & 3 \\ 0 & 2 & 0 & 5 & 3 \\ 0 & 0 & 0 & 4 & -1 \end{bmatrix}$ c. $\begin{bmatrix} 0 & 2 & 0 & 5 & 3 \\ 1 & 0 & 0 & 0 & 3 \\ 0 & 0 & 0 & 4 & -1 \end{bmatrix}$

d. $\begin{bmatrix} 2 & 2 & 2 & 2 & 2 \\ 2 & 2 & 2 & 2 & 2 \\ 2 & 2 & 2 & 2 & 2 \end{bmatrix}$ e. $\begin{bmatrix} 1 & 0 & 0 & 0 & 3 \\ 0 & 2 & 0 & 5 & 3 \end{bmatrix}$ f. $\begin{bmatrix} 1 & 0 & 0 & 0 & 3 \end{bmatrix}$

4. (*Maple V*) Apply the **rref** function in the **linalg** package to the following matrices, and explain why the resulting matrices are in reduced row echelon form.

$$\text{a.}\begin{bmatrix} -1 & 1 & 4 & 1 \\ -3 & 3 & 5 & 5 \\ -2 & -3 & 3 & -3 \end{bmatrix} \quad \text{b.}\begin{bmatrix} 2 & -3 & 0 \\ 0 & 4 & 3 \\ 1 & 0 & 5 \\ 2 & 5 & 2 \end{bmatrix} \quad \text{c.}\begin{bmatrix} 5 & -1 & 3 \\ -1 & 0 & -4 \\ 3 & -4 & -5 \end{bmatrix}$$

5. (*Maple V*) Construct two 3×3 matrices A and B with nonzero entries whose reduced row echelon form is

$$\begin{bmatrix} 1 & 2 & 0 \\ 0 & 0 & 1 \\ 0 & 0 & 0 \end{bmatrix}$$

and explain your result.

6. (*Maple V*) Restart *Maple,* load the **linalg** package, and use the command

```
A:=randmatrix(25,25)
```

to generate a random 25×25 matrix A. Let B be the transpose of A, and explain why the command `equal(rref(A),rref(B))` is likely to return the value *true*.

7. (*Maple V*) Apply the **rref** function in the **linalg** package to the following matrices, and explain why the resulting matrices are in reduced row echelon form.

$$\text{a.}\begin{bmatrix} -1 & 1 & 4 & 1 \\ -3 & 3 & 5 & 5 \\ -2 & -3 & 3 & -3 \end{bmatrix} \quad \text{b.}\begin{bmatrix} 2 & -3 & 0 \\ 0 & 4 & 3 \\ 1 & 0 & 5 \\ 2 & 5 & 2 \end{bmatrix} \quad \text{c.}\begin{bmatrix} 5 & -1 & 3 \\ -1 & 0 & -4 \\ 3 & -4 & -5 \end{bmatrix}$$

8. (*Maple V*) Use *Maple* to write out all 3×3 reduced row echelon matrices whose entries are either 0 or 1, and use the **gausselim** function in the **linalg** package to verify that for each constructed matrix A, the command `equal(A,gausselim(A))` returns the value *true*.

MATRIX EQUATIONS

We now introduce a second matrix form for linear systems in which we can calculate the solutions of the systems.

The Systems $A\mathbf{x} = \mathbf{b}$

We recall from our earlier discussion that in addition to being able to represent the system

$$\begin{cases} 3x + 4y - 7z = 1 \\ 2x + y = 2 \\ 8x - 6y - z = 15 \end{cases}$$

by the augmented matrix

$$[A \mid \mathbf{b}] = \begin{bmatrix} 3 & 4 & -7 & 1 \\ 2 & 1 & 0 & 2 \\ 8 & -6 & -1 & 15 \end{bmatrix}$$

we can also write it as

$$\begin{bmatrix} 3 & 4 & -7 \\ 2 & 1 & 0 \\ 8 & -6 & -1 \end{bmatrix} \begin{bmatrix} x \\ y \\ z \end{bmatrix} = \begin{bmatrix} 1 \\ 2 \\ 15 \end{bmatrix}$$

We call this equation the ***matrix equation*** of the system and write it as $A\mathbf{x} = \mathbf{b}$. The operation that produces the column vector

$$A\mathbf{x} = \begin{bmatrix} 3x + 4y - 7z \\ 2x + y \\ 8x - 6y - z \end{bmatrix}$$

from

$$A = \begin{bmatrix} 3 & 4 & -7 \\ 2 & 1 & 0 \\ 8 & -6 & -1 \end{bmatrix} \quad \text{and} \quad \mathbf{x} = \begin{bmatrix} x \\ y \\ z \end{bmatrix}$$

will later be called ***matrix multiplication***. The general form of this operation is discussed in Chapter 2.

Although the equation $A\mathbf{x} = \mathbf{b}$ links the matrix A and the vectors $\mathbf{x}$ and $\mathbf{b}$, we nevertheless call it a matrix equation to distinguish it from a ***vector equation*** such as

$$a_1\mathbf{x}_1 + \cdots + a_n\mathbf{x}_n = \mathbf{b}$$

in which a combination of column vectors $\mathbf{x}_1, \ldots, \mathbf{x}_n$ is linked to a column vector $\mathbf{b}$.

DEFINITION 1.22 *The* ***matrix equation*** $A\mathbf{x} = \mathbf{b}$ *determined by the* $m \times n$ *linear system*

$$\begin{cases} a_{11}x_1 + \cdots + a_{1n}x_n = b_1 \\ \vdots \\ a_{m1}x_1 + \cdots + a_{mn}x_n = b_m \end{cases}$$

is the equation

$$\mathbf{Ax} = \begin{bmatrix} a_{11} & \cdots & a_{1n} \\ \vdots & \ddots & \vdots \\ a_{m1} & \cdots & a_{mn} \end{bmatrix} \begin{bmatrix} x_1 \\ \vdots \\ x_n \end{bmatrix} = \begin{bmatrix} b_1 \\ \vdots \\ b_m \end{bmatrix} = \mathbf{b}$$

in which $a_{i1}x_1 + \cdots + a_{in}x_n = b_i$.

If we assign the constants $c_1, \ldots, c_n$ to the variables $x_1, \ldots, x_n$ and let $\widehat{\mathbf{x}} = (c_1, \ldots, c_n)$, then

$$A\widehat{\mathbf{x}} = \begin{bmatrix} a_{11}c_1 + \cdots + a_{1n}c_n \\ \vdots \\ a_{m1}c_1 + \cdots + a_{mn}c_n \end{bmatrix}$$

is a constant vector.

DEFINITION 1.23 *A vector* $\widehat{\mathbf{x}}$ *is a solution of the matrix equation* $A\mathbf{x} = \mathbf{b}$ *if the vectors* $A\widehat{\mathbf{x}}$ *and* $\mathbf{b}$ *are equal.*

EXAMPLE 1.51 A Linear System in Matrix Form

Use *Maple* to convert the linear system

$$\begin{cases} x + 3y - 7z + w - 45 = 0 \\ 2x - z + 8w - 1 = 0 \\ x + y + z + 9 = 0 \\ x - 8y - 8z + 12w + 12 = 0 \end{cases}$$

to a system in matrix form.

Solution. The **GenerateMatrix** function in the **LinearAlgebra** package makes it easy to convert linear systems to their matrix form.

```
[> with(LinearAlgebra):
[> sys:=[
    x+3*y-7*z+w-45=0,
    2*x-z+8*w-1=0,
    x+y+z+9=0,
    x-8*y-8*z+12*w+12=0]:
[> vars:=[x,y,z,w]:
```

```
[> A,b:=GenerateMatrix(sys,vars);
```

$$A, b := \begin{bmatrix} 1 & 3 & -7 & 1 \\ 2 & 0 & -1 & 8 \\ 1 & 1 & 1 & 0 \\ 1 & -8 & -8 & 12 \end{bmatrix}, \begin{bmatrix} 45 \\ 1 \\ -9 \\ -12 \end{bmatrix}$$

The output is a pair $(A, \mathbf{b})$ consisting of the coefficient matrix A and the vector of constants $\mathbf{b}$. We use A and $\mathbf{b}$ to construct the required matrix equation $A\mathbf{x} = \mathbf{b}$. By writing $A\mathbf{x}$ as **A&*X** and not evaluating this expression with **evalm**, we can display the structure of the equation.

```
[> X:=Vector(vars);
```

$$X := \begin{bmatrix} x \\ y \\ z \\ w \end{bmatrix}$$

```
[> A&*X=b;
```

$$\begin{bmatrix} 1 & 3 & -7 & 1 \\ 2 & 0 & -1 & 8 \\ 1 & 1 & 1 & 0 \\ 1 & -8 & -8 & 12 \end{bmatrix} \&* \begin{bmatrix} x \\ y \\ z \\ w \end{bmatrix} = \begin{bmatrix} 45 \\ 1 \\ -9 \\ -12 \end{bmatrix}$$

The result is a linear system in matrix form. ▲

If we had written the equation **A&*X=b** as **A.X=b**, then *Maple 6* would have evaluated the product $A\mathbf{x}$ and returned the vector equation

$$\begin{bmatrix} x + 3y - 7z + w \\ 2x - z + 8w \\ x + y + z \\ x - 8y - 8z + 12w \end{bmatrix} = \begin{bmatrix} 45 \\ 1 \\ -9 \\ -12 \end{bmatrix}$$

We saw in Example 1.42 that in certain applications, the coefficient matrix A of a linear system $A\mathbf{x} = \mathbf{b}$ has the special form

$$\begin{bmatrix} 1 & x_1 & x_1^2 & \cdots & x_1^{n-1} \\ 1 & x_2 & x_2^2 & \cdots & x_2^{n-1} \\ & & & \vdots & \\ 1 & x_{n-1} & x_{n-1}^2 & \cdots & x_{n-1}^{n-1} \\ 1 & x_n & x_n^2 & \cdots & x_n^{n-1} \end{bmatrix}$$

This type of matrix is known as a ***Vandermonde matrix***. Among other things, it provides a template for generating interpolating polynomials. Suppose, for example, that two parallel scientific experiments involve taking five readings and that we would like to use the graphs of the associated interpolating polynomials to compare the results of the two experiments. The next example illustrates how Vandermonde matrices can be used for this purpose.

EXAMPLE 1.52 A Vandermonde Matrix

Suppose that the observation times of two parallel experiments are 1, 2, 3, 4, and 5, with the results of the experiments shown in the following chart.

EXPERIMENT 1	$\{(1,3), (2,4), (3,2), (4,6), (5,7)\}$
EXPERIMENT 2	$\{(1,2), (2,3), (3,2), (4,6), (5,8)\}$

Use a Vandermonde matrix to find the interpolating polynomials for the experiments, and overlap the graphs of the polynomials.

Solution. First we define the matrix A and the vector $\mathbf{x}$.

```
[> with(LinearAlgebra):
[> A:=VandermondeMatrix([1,2,3,4,5],5,5);
```

$$A := \begin{bmatrix} 1 & 1 & 1 & 1 & 1 \\ 1 & 2 & 4 & 8 & 16 \\ 1 & 3 & 9 & 27 & 81 \\ 1 & 4 & 16 & 64 & 256 \\ 1 & 5 & 25 & 125 & 625 \end{bmatrix}$$

```
[> x:=<seq(a[i],i=1..5)>:
```

Next we solve the equations $A\mathbf{x} = \mathbf{b}$ and $A\mathbf{x} = \mathbf{c}$, where $\mathbf{b}$ and $\mathbf{c}$ are the vectors containing the data of two experiments.

```
[>b:=<3,4,2,6,7>:
[>eqnsb:={seq((A.x)[i]=b[i],i=1..5)};
```

$$eqnsb := \left\{ \begin{array}{r} a_1 + a_2 + a_3 + a_4 + a_5 = 3, \\ a_1 + 2a_2 + 4a_3 + 8a_4 + 16a_5 = 4, \\ a_1 + 3a_2 + 9a_3 + 27a_4 + 81a_5 = 2, \\ a_1 + 4a_2 + 16a_3 + 64a_4 + 256a_5 = 6, \\ a_1 + 5a_2 + 25a_3 + 125a_4 + 625a_5 = 7 \end{array} \right\}$$

We note that the linear system *eqnsb* consists precisely of the equations required to find the coefficients of the interpolating polynomial.

```
[>solve(eqnsb,{seq(a[i],i=1..5)});
```

$$\left\{a_1 = -28, a_3 = \tfrac{-147}{4}, a_5 = \tfrac{-3}{4}, a_2 = \tfrac{119}{2}, a_4 = 9\right\}$$

Therefore, the interpolating polynomial for Experiment 1 is

$$p(x) = -28 + \tfrac{119}{2}x - \tfrac{147}{4}x^2 + 9x^3 - \tfrac{3}{4}x^4$$

We now find the interpolating polynomial for Experiment 2.

```
[>c:=<2,3,2,6,8>:
[>eqnsc:={seq((A.x)[i]=c[i],i=1..5)}:
[>solve(eqnsb,{seq(a[i],i=1..5)});
```

$$\{a_1 = -22, a_3 = \tfrac{-341}{12}, a_5 = \tfrac{-7}{12}, a_2 = 46, a_4 = 7\}$$

Therefore, the interpolating polynomial for Experiment 2 is

$$q(x) = -22 + 46x - \tfrac{341}{12}x^2 + 7x^3 - \tfrac{7}{12}x^4$$

Finally, we graph the two polynomials.

```
[>p := -28 + 119/2*x - 147/4*x^2 + 9*x^3 - 3/4*x^4:
[>q := -22 + 46*x - 341/12*x^2 + 7*x^3 - 7/12*x^4:
```

```
[>plot({p,q},x=0..5);
```

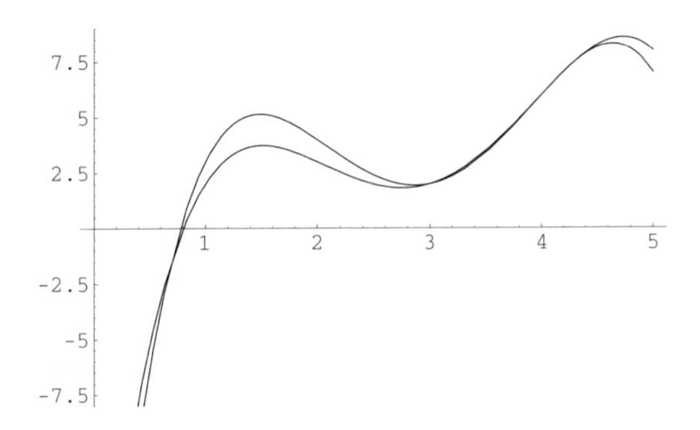

The two overlapping graphs illustrate the comparative results of Experiments 1 and 2. ▲

EXERCISES 1.14

1. Rewrite the following linear systems as matrix equations.

a. $\begin{cases} 2x + y - z = 2 \\ x - 3y + z = 1 \\ 2x + y - 2z = 9 \end{cases}$ b. $\begin{cases} 2x + y - z - 5w = 0 \\ x - 3y + z = 0 \\ 2x + y - z + w = 0 \end{cases}$ c. $\begin{cases} 2x + y - z = 1 \\ x - 3y + z = 2 \\ 6x - 3y + 7z = 3 \\ x - 3y + z = 4 \end{cases}$

and

d. $\begin{cases} 2x + y - z = 2 \\ z - y + x = 1 \\ 2x + z - 2y = 9 \end{cases}$ e. $\begin{cases} 2x + y - z = 1 \\ x - 3y + x = 2 \\ 6z - 3y + 7w = 3 \\ w - 3y + z = 4 \end{cases}$ f. $\begin{cases} 6z - 3y + 7w = 3 \\ w - 3y + z = 4 \end{cases}$

2. Rewrite the vector equation $(x, y) = (4, 8) + t(3, 2)$ of a straight line in $\mathbb{R}^2$ as a matrix equation of the form $A\mathbf{x} = \mathbf{b}$, where $\mathbf{x} = (x, y, t)$.

3. Rewrite the vector equation $(x, y, z) = (4, 8, 3) + t(3, 2, 7)$ of a straight line in $\mathbb{R}^3$ as a matrix equation of the form $A\mathbf{x} = \mathbf{b}$, where $\mathbf{x} = (x, y, z, t)$.

4. (*Maple V*) Use a Vandermonde matrix to find the interpolating polynomial of degree 5 for the following experimental data.

EXPERIMENT	$\{(1, 1), (2, -1), (3, 2), (4, 3), (5, -2), (6, 5)\}$

Construct the necessary matrix using the **linalg[vandermonde]** function.

5. (*Maple 6*) Use the **VandermondeMatrix** function to find a polynomial curve passing through the points $\{(1, 3), (2, -3), (3, 3), (4, -3), (5, 3), (6, -3), (7, 3)\}$.

6. (*Maple 6*) Explore the properties of the **VandermondeMatrix** function in the **LinearAlgebra** package.

7. (*Maple V, 6*) Explore the properties of the **genmatrix** function in the **linalg** package and the **GenerateMatrix** function in the **LinearAlgebra** package.

The Systems $A\mathbf{x} = \mathbf{0}$

We conclude this chapter with a discussion of the linear systems of the form $A\mathbf{x} = \mathbf{0}$. As mentioned earlier, these systems are called ***homogeneous*** and arise in many contexts. In Chapter 4, for example, we will use homogeneous systems to show that a set $\{\mathbf{x}_1, \ldots, \mathbf{x}_n\}$ of column vectors is ***linearly independent*** by forming the matrix $[\mathbf{x}_1 \ \cdots \ \mathbf{x}_n]$ and showing that the equation

$$\begin{bmatrix} \mathbf{x}_1 & \cdots & \mathbf{x}_n \end{bmatrix} \begin{bmatrix} a_1 \\ \vdots \\ a_n \end{bmatrix} = \mathbf{0}$$

has only the solution $a_1 = \cdots = a_n = 0$.

We begin our analysis of homogeneous linear systems by showing that they are consistent since they always have the solution $\mathbf{x} = \mathbf{0}$. This solution is called the ***trivial solution.***

THEOREM 1.4 *The zero vector is a solution of the homogeneous linear system* $A\mathbf{x} = \mathbf{0}$.

Proof. It follows from the definition of $A\mathbf{0}$ that

$$A\mathbf{0} = A \begin{bmatrix} 0 \\ \vdots \\ 0 \end{bmatrix} = \begin{bmatrix} a_{11}0 + \cdots + a_{1n}0 \\ \vdots \\ a_{m1}0 + \cdots + a_{mn}0 \end{bmatrix} = \begin{bmatrix} 0 \\ \vdots \\ 0 \end{bmatrix} = \mathbf{0}$$

Hence the zero vector is a solution of the system. ■

All nonzero solutions of a homogeneous linear system are called ***nontrivial***. Systems that are not homogeneous are called ***nonhomogeneous.***

Homogeneous linear systems are consistent since they always have the trivial solution. For underdetermined homogeneous systems, more is true.

THEOREM 1.5 *Every underdetermined homogeneous linear system* $A\mathbf{x} = \mathbf{0}$ *has a nontrivial solution.*

Proof. A linear system is underdetermined if it contains more variables than equations. Consider, therefore, an $m \times n$ linear system $A\mathbf{x} = \mathbf{0}$ in the variables $x_1, \ldots, x_n$, and suppose that $m < n$. If A is a zero matrix, then any vector $\mathbf{x} \in \mathbb{R}^n$ is a solution. If A contains a nonzero entry, then the associated augmented matrix $[A \mid \mathbf{0}]$ contains a nonzero entry. According to

Theorem 1.2, the matrix $[A \mid \mathbf{0}]$ can be reduced to a matrix $[B \mid \mathbf{b}]$ in reduced row echelon form. Since elementary row operations preserve zero columns, it is clear that $\mathbf{b} = \mathbf{0}$. Since A is nonzero and $m < n$, matrix B must have $1 \leq p \leq m$ pivots associated with p basic variables in $\mathbf{x}$. Since the number of rows of B is less than the number of its columns, p must be less than n. Hence the system must have $n - p > 0$ free variables. Each nonzero assignment of values to these variables yields a nontrivial solution of the given homogeneous system. ■

EXAMPLE 1.53 An Underdetermined Homogeneous System

Use *Maple* to show that the underdetermined homogeneous system

$$\begin{cases} 3x - y + z = 0 \\ x + 2y - 4z = 0 \end{cases}$$

has infinitely many solutions.

Solution. Since the given system consists of two equations in three variables, we elect to solve it for two of the three variables.

```
[> solve({3*x-y+z=0,x+2*y-4*z=0},{x,y});
```

$$\{x = \tfrac{2}{7}z, y = \tfrac{13}{7}z\}$$

This shows that for any assignment of a value to z we obtain a solution. Since we have infinitely many choices of values for z, the given system has infinitely many solutions.▲

The next two examples show that the number of equations does not determine the number of solutions of a homogeneous linear system. A system with the same number of equations as variables may have one or infinitely many solutions.

EXAMPLE 1.54 A Homogeneous System with One Solution

Use *Maple* to show that the homogeneous system

$$\begin{cases} 2x + y - z = 0 \\ x - y + z = 0 \\ x + 2y - z = 0 \end{cases}$$

consisting of three equations in three variables has only the trivial solution.

Solution. We use the **solve** function.

```
[> solve({2*x+y-z=0,x-y+z=0,x+2*y-z=0},{x,y,z});
```

$$\{z = 0, y = 0, x = 0\}$$

This shows that the system $A\mathbf{x} = \mathbf{0}$ has only the trivial solution. ▲

EXAMPLE 1.55 A Homogeneous System with Infinitely Many Solutions

Use *Maple* to show that the homogeneous system

$$\begin{cases} x + y - 2z = 0 \\ x - 2y + z = 0 \\ 2x - y - z = 0 \end{cases}$$

of three equations in three unknowns has infinitely many solutions.

Solution. We begin by calculating the number of basic variables of the system by finding the number of pivot columns of the coefficient matrix.

```
[> A:=matrix([[1,1,-2],[1,-2,1],[2,-1,-1]]);
```

$$A := \begin{bmatrix} 1 & 1 & -2 \\ 1 & -2 & 1 \\ 2 & -1 & -1 \end{bmatrix}$$

```
[> B:= linalg[rref](A);
```

$$B := \begin{bmatrix} 1 & 0 & -1 \\ 0 & 1 & -1 \\ 0 & 0 & 0 \end{bmatrix}$$

This shows that the columns $\mathbf{c}_1 B$ and $\mathbf{c}_2 B$ are the pivot columns of B. Therefore, the variables x and y are the basic variables of the system $A\mathbf{x} = \mathbf{0}$. We therefore solve for x and y in terms of z.

```
[> solve({x+y-2*z=0,x-2*y+z=0,2*x-y-z=0},{x,y});
```

$$\{y = z, x = z\}$$

Since we are free to assign any value to z, the system has infinitely many solutions of the form $x = y = z$. ▲

One of the characteristic features of homogeneous linear systems is that their sets of solutions contain the trivial solution and are closed under addition and multiplication by constants. In Chapter 4, we will interpret this property as saying that the set of solutions of a homogeneous linear system forms a ***vector space***.

EXAMPLE 1.56 A Sum of Solutions

Use *Maple* to show that if **u** and **v** are solutions of

$$\begin{cases} 3x - y + z - w = 0 \\ x + 2y - 4z + w = 0 \end{cases}$$

and a and b are two arbitrary real numbers, then $a\mathbf{u} + b\mathbf{v}$ is a solution of the given system.

Solution. First we find the solutions of the given system. Since we have two equations in four variables, we solve for two of them.

```
[> solve({3*x-y+z-w=0,x+2*y-4*z+w=0},{x,y});
```

$$\{x = \tfrac{2}{7}z + \tfrac{1}{7}w, y = \tfrac{13}{7}z - \tfrac{4}{7}w\}$$

Let $\mathbf{u} = [1/7, -4/7, 0, 1]$ and $\mathbf{v} = [11/7, 40/7, 4, 3]$ be the solutions of the system determined by the assignment $w = 1$ and $z = 0$ and the assignment $w = 3$ and $z = 4$. We show that for any scalars a and b, the vector $a\mathbf{u} + b\mathbf{v}$ is a solution of the given system. We begin by inputting A, $a\mathbf{u}$, and $b\mathbf{v}$.

```
[> A:=matrix([[3,-1,1,-1],[1,2,-4,1]]):
[> au:=a*matrix([[1/7],[-4/7],[0],[1]]):
[> bv:=b*matrix([[11/7],[40/7],[4],[3]]):
```

```
[> evalm(A&*(au+bv));
```

$$\begin{bmatrix} 0 \\ 0 \end{bmatrix}$$

This tells us that

$$\begin{bmatrix} 3 & -1 & 1 & -1 \\ 1 & 2 & -4 & 1 \end{bmatrix} \left[a \begin{bmatrix} \frac{1}{7} \\ -\frac{4}{7} \\ 0 \\ 1 \end{bmatrix} + b \begin{bmatrix} \frac{11}{7} \\ \frac{40}{7} \\ 4 \\ 3 \end{bmatrix} \right] = \begin{bmatrix} 0 \\ 0 \end{bmatrix}$$

This shows that if **u** and **v** are solutions of the given linear system, then $a\mathbf{u} + b\mathbf{v}$ is also a solution of the system, for any two constants a and b. ▲

Linear Dependence Relations and Homogeneous Systems

The left-hand side of $A\mathbf{x} = \mathbf{0}$ can be written as a sum of column vectors. Thus the system

$$\begin{cases} a_{11}x_1 + \cdots + a_{1n}x_n = 0 \\ a_{21}x_1 + \cdots + a_{2n}x_n = 0 \\ \vdots \\ a_{m1}x_1 + \cdots + a_{mn}x_n = 0 \end{cases}$$

can be rewritten as $x_1\mathbf{c}_1 A + \cdots + x_n\mathbf{c}_n A = \mathbf{0}$, where

$$\mathbf{c}_i A = \begin{bmatrix} a_{1i} \\ \vdots \\ a_{mi} \end{bmatrix}$$

and $1 \leq i \leq n$. The equation $x_1\mathbf{c}_1 A + \cdots + x_n\mathbf{c}_n A = \mathbf{0}$ is another example of a ***vector equation*** as defined earlier in this chapter.

The trivial solution and the nontrivial solutions of a vector equation provide examples of two of the most fundamental relationships among vectors. The general importance of these relationships in linear algebra will become obvious in Chapter 4.

DEFINITION 1.24 *The column vectors $\mathbf{c}_1, \ldots, \mathbf{c}_n$ are **linearly independent** if the vector equation $x_1\mathbf{c}_1 + \cdots + x_n\mathbf{c}_n = \mathbf{0}$ has only the trivial solution. They are **linearly dependent** if they are not linearly independent.*

DEFINITION 1.25 *A **linear dependence relation** among the columns vectors $\mathbf{c}_1, \ldots, \mathbf{c}_n$ is any nontrivial solution of the equation $x_1\mathbf{c}_1 + \cdots + x_n\mathbf{c}_n = \mathbf{0}$.*

EXAMPLE 1.57 Linear Dependence Relations

Use *Maple* to show that the linear dependence relations of the system

$$\begin{bmatrix} -5 & 5 & 1 & 8 \\ 6 & -1 & 2 & 3 \\ 3 & 0 & 1 & 2 \end{bmatrix} \begin{bmatrix} a \\ b \\ c \\ d \end{bmatrix} = \begin{bmatrix} 0 \\ 0 \\ 0 \end{bmatrix}$$

are the same as those of the reduced row echelon form of the system.

Solution. We show that the two systems have the same nontrivial solutions. We begin by solving the given system $A\mathbf{x} = \mathbf{0}$ using the **linsolve** function of the **linalg** package.

```
[> with(linalg):
[> A:=matrix([[-5,5,1,8],[6,-1,2,3],[3,0,1,2]]):
[> linsolve(A,[0,0,0]);
```

$$[\_t[1], -8\_t[1], -19\_t[1], 8\_t[1]]$$

This tells us that for every nonzero value assigned to t_1, the vector $(t_1, -8t_1, -19t_1, 8t_1)$ is a nontrivial solution of the system $A\mathbf{x} = \mathbf{0}$. We now show that the homogeneous linear system $B\mathbf{x} = \mathbf{0}$ has precisely the same solutions, where B is the reduced row echelon form of A.

```
[> B:=rref(A):linsolve(B,[0,0,0]);
```

$$[\_t[1], -8\_t[1], -19\_t[1], 8\_t[1]]$$

This shows that the two systems have the same solutions. They therefore have the same linear dependence relations. ▲

Since any homogeneous system $A\mathbf{x} = \mathbf{0}$ can be written in the form

$$x_1\mathbf{c}_1 A + \cdots + x_n\mathbf{c}_n A = \mathbf{0}$$

it is clear that the linear dependence relations among the columns of A correspond precisely to the nontrivial solutions of the equation $A\mathbf{x} = \mathbf{0}$. We simplify this statement by saying that a nontrivial solution of a homogeneous system is a ***linear dependence relation*** of the system.

THEOREM 1.6 *If a homogeneous linear system results from another system by elementary row operations, then the two systems have the same linear dependence relations.*

Proof. Since elementary row operations preserve the solutions of the linear systems involved, a system $B\mathbf{x} = \mathbf{0}$ that is row equivalent to a system $A\mathbf{x} = \mathbf{0}$ has the same linear dependence relations as $A\mathbf{x} = \mathbf{0}$. ■

Theorem 1.6 implies the following important properties of row equivalent matrices.

COROLLARY 1.7 *Row equivalent matrices satisfy the same linear dependence relations.*

Proof. Let A and B be two row equivalent $m \times n$ matrices with columns $\mathbf{c}_1 A, \ldots, \mathbf{c}_n A$ and $\mathbf{c}_1 B, \ldots, \mathbf{c}_n B$. Then by Theorem 1.6, the homogeneous linear systems $A\mathbf{x} = \mathbf{0}$ and $B\mathbf{x} = \mathbf{0}$ have the same linear dependence relations. Therefore, $x_1\mathbf{c}_1 A + \cdots + x_n\mathbf{c}_n A = \mathbf{0}$ if and only if $x_1\mathbf{c}_1 B + \cdots + x_n\mathbf{c}_n B = \mathbf{0}$. This proves the corollary. ■

We can also say that the columns of A are linearly independent if and only if those of B are. A basic property of real matrices is that their pivot columns are linearly independent and that their nonpivot columns are linearly dependent on their pivot columns.

EXAMPLE 1.58 Linear Independence of Pivot Columns

Show that the pivot columns of the reduced row echelon matrix

$$A = \begin{bmatrix} 1 & 0 & 4 & -2 \\ 0 & 1 & 5 & 0 \end{bmatrix}$$

are linearly independent.

Solution. It is clear from the definition that the pivot columns of A are

$$\mathbf{c}_1 A = \begin{bmatrix} 1 \\ 0 \end{bmatrix} \quad \text{and} \quad \mathbf{c}_2 A = \begin{bmatrix} 0 \\ 1 \end{bmatrix}$$

They are linearly independent since

$$a \begin{bmatrix} 1 \\ 0 \end{bmatrix} + b \begin{bmatrix} 0 \\ 1 \end{bmatrix} = \begin{bmatrix} a \\ b \end{bmatrix} = \begin{bmatrix} 0 \\ 0 \end{bmatrix}$$

if and only if $a = b = 0$. The nonpivot columns of A are

$$\mathbf{c}_3 A = \begin{bmatrix} 4 \\ 5 \end{bmatrix} \quad \text{and} \quad \mathbf{c}_4 A = \begin{bmatrix} -2 \\ 0 \end{bmatrix}$$

As we can see, $\mathbf{c}_3 A$ and $\mathbf{c}_4 A$ are the unique sums $4\mathbf{c}_1 A + 5\mathbf{c}_2 A$ and $-2\mathbf{c}_1 A + 0\mathbf{c}_2 A$ of the pivot columns $\mathbf{c}_1 A$ and $\mathbf{c}_2 A$. ▲

Uniqueness of the Reduced Row Echelon Form

The observant reader may have noticed that on several occasions we have implicitly used the fact that the reduced row echelon form of a matrix is unique. We now prove this fact.

THEOREM 1.8 (Reduced row echelon form theorem) *The reduced row echelon form of a matrix is unique.*

Proof. Suppose that the matrices B and C are reduced row echelon forms of a matrix A. We must show that $B = C$.

If A is an $m \times n$ zero matrix, then it follows from the nature of elementary row operations that B and C are also $m \times n$ zero matrices. Hence $B = C$.

Suppose now that A is not a zero matrix and that the reduced row echelon matrices B and C are row equivalent to A. We must again show that $B = C$.

We first note that B and C are row equivalent, since both are row equivalent to A. It therefore follows from Corollary 1.7 that they have the same linear dependence relations. This means that if $\mathbf{c}_i B$ is the ith column of B and $\mathbf{c}_i C$ is the ith column of C, then $\mathbf{c}_i B = x_1 \mathbf{c}_1 B + \cdots + x_k \mathbf{c}_k B$ if and only if $\mathbf{c}_i C = x_1 \mathbf{c}_1 C + \cdots + x_k \mathbf{c}_k C$.

Since A is not zero, the matrices B and C are not zero either and must both have a pivot column. Moreover, by Corollary 1.7, B and C must have the same number of pivot columns. Since B and C are in reduced row echelon form, these columns must be of the form

$\mathbf{e}_1, \ldots, \mathbf{e}_p \in \mathbb{R}^m$, where $\mathbf{e}_i = (\delta_1, \ldots, \delta_j, \ldots, \delta_m)$, with $\delta_j = 1$ if $i = j$ and $\delta_j = 0$ otherwise.

Suppose now that $\mathbf{c}_q B$ is a pivot column of B. Then $\mathbf{c}_q B = \mathbf{e}_i$ for some $1 \leq i \leq p$. Therefore $\mathbf{c}_q C = \mathbf{e}_i$ since B and C have the same pivot columns. On the other hand, if $\mathbf{c}_q B$ is not a pivot column of B, then $\mathbf{c}_q C$ is not a pivot column of C. Therefore $\mathbf{c}_q B$ is a linear combination $x_1\mathbf{e}_1 + \cdots + x_j\mathbf{e}_j$ of the pivot columns to its left in B, and $\mathbf{c}_q C$ is a linear combination $\mathbf{c}_q C = x_1\mathbf{e}_1 + \cdots + x_j\mathbf{e}_j$ of precisely the same pivot columns to its left in C. It follows that B and C have identical columns. Hence $B = C$. ■

EXERCISES 1.15

1. Find the pivot columns of the following matrices. If a column is not a pivot column, explain why.

$$\text{a.} \begin{bmatrix} 1 & 3 & 3 & 1 \\ 5 & 0 & 0 & 3 \\ 3 & 1 & 3 & 0 \\ 3 & 4 & 2 & 3 \end{bmatrix} \quad \text{b.} \begin{bmatrix} 2 & 3 & 3 & 2 \\ 5 & 5 & 4 & 5 \\ 0 & 2 & 1 & 1 \end{bmatrix} \quad \text{c.} \begin{bmatrix} 4 & 0 & 4 & 1 & 4 \\ 5 & 4 & 3 & 3 & 5 \\ 3 & 5 & 1 & 2 & 1 \end{bmatrix}$$

2. Find linear systems whose augmented matrices are the matrices in Exercise 1. Explain the different possible choices of basic and free variables in each case.

3. Use the following matrices to create homogeneous linear systems $A\mathbf{x} = \mathbf{0}$. Determine in each case whether the set of solutions is a point, a line, a plane, or all of $\mathbb{R}^3$.

$$\text{a. } A = \begin{bmatrix} -8 & -5 & 7 \\ 9 & 0 & 9 \\ 1 & 2 & 3 \end{bmatrix} \quad \text{b. } A = \begin{bmatrix} 0 & 0 & 0 \\ 0 & 0 & 0 \\ 0 & 0 & 0 \end{bmatrix} \quad \text{c. } A = \begin{bmatrix} 1 & 0 & 0 \\ 0 & 1 & 0 \\ 0 & 0 & 1 \end{bmatrix}$$

$$\text{d. } A = \begin{bmatrix} 1 & 7 & -3 \\ 9 & 9 & 6 \\ 8 & 2 & 9 \end{bmatrix} \quad \text{e. } A = \begin{bmatrix} 3 & 2 & 1 \\ 4 & 3 & 2 \end{bmatrix} \quad \text{f. } A = \begin{bmatrix} 1 & -5 & 7 \\ 2 & -10 & 14 \\ 3 & -15 & 21 \end{bmatrix}$$

4. Use Theorem 1.5 to show that for the vectors $\mathbf{x} = (1, 2)$, $\mathbf{y} = (3, 4)$, and $\mathbf{z} = (-2, 7)$ there exist real numbers a, b, c, not all zero, for which $a\mathbf{x} + b\mathbf{y} + c\mathbf{z} = \mathbf{0}$.

5. Find a real number λ for which the homogeneous system

$$\begin{bmatrix} -\lambda & -6 \\ 1 & 5-\lambda \end{bmatrix} \begin{bmatrix} x \\ y \end{bmatrix} = \begin{bmatrix} 0 \\ 0 \end{bmatrix}$$

has a nontrivial solution.

6. Use Theorem 1.6 to show that the homogeneous systems

$$\begin{bmatrix} 2 & -1 & 1 & 4 \\ 1 & -3 & 3 & 5 \\ 5 & -2 & -3 & 3 \end{bmatrix} \begin{bmatrix} w \\ x \\ y \\ z \end{bmatrix} = \begin{bmatrix} 0 \\ 0 \\ 0 \end{bmatrix}$$

and

$$\begin{bmatrix} 2 & -1 & 1 & 4 \\ 0 & -5 & 5 & 6 \\ 0 & 0 & 25 & 32 \end{bmatrix} \begin{bmatrix} w \\ x \\ y \\ z \end{bmatrix} = \begin{bmatrix} 0 \\ 0 \\ 0 \end{bmatrix}$$

have the same dependence relations.

7. Show that there exist no linear dependence relations among the columns of the matrix

$$\begin{bmatrix} 1 & 0 & 0 \\ 0 & 1 & 0 \\ 0 & 0 & 1 \end{bmatrix}$$

8. (*Maple V*) Show that the column vectors determined by the following matrices are linearly dependent.

a. $\begin{bmatrix} 1 & 3 & 4 & 1 \\ 5 & 0 & 5 & 3 \\ 3 & 1 & 4 & 0 \\ 3 & 4 & 7 & 3 \end{bmatrix}$ b. $\begin{bmatrix} 2 & 3 & 3 & 1 & 0 \\ 5 & 5 & 4 & 3 & 2 \\ 0 & 2 & 1 & 6 & 5 \end{bmatrix}$ c. $\begin{bmatrix} 4 & 0 & 8 \\ 5 & 4 & 2 \\ 3 & 5 & -4 \end{bmatrix}$

9. (*Maple V*) Show that the column vectors determined by the following matrices are linearly independent.

a. $\begin{bmatrix} 1 & 3 & 3 & 1 \\ 5 & 0 & 0 & 3 \\ 3 & 1 & 3 & 0 \\ 3 & 4 & 2 & 3 \end{bmatrix}$ b. $\begin{bmatrix} 2 & 3 & 3 \\ 5 & 5 & 4 \\ 0 & 2 & 1 \end{bmatrix}$ c. $\begin{bmatrix} 1 & 3 & 3 & 1 \\ 5 & 0 & 0 & 3 \\ 3 & 1 & 3 & 0 \\ 3 & 4 & 2 & 3 \\ 2 & 1 & 3 & 5 \end{bmatrix}$

10. (*Maple V*) Show that the homogeneous systems

$$\begin{cases} 5x + 4y + 2z = 0 \\ 5x + 2y + z = 0 \end{cases} \quad \text{and} \quad \begin{cases} 5x + 4y + 2z + s\,(5x + 2y + z) = 0 \\ 5x + 2y + z = 0 \end{cases}$$

have the same solutions for all scalars s.

11. (*Maple V*) Use *Maple* Help to discover how you can apply the function **geneqns** in the **linalg** package to solve matrix equations.

12. (*Maple 6*) Use the **RowOperation** function in the **LinearAlgebra** package to show that that the homogeneous systems

$$\begin{cases} 3x - y + z = 0 \\ x + 2y - 4z = 0 \end{cases} \quad \text{and} \quad \begin{cases} x - \frac{2}{7}z = 0 \\ y - \frac{13}{7}z = 0 \end{cases}$$

have the same solution.

13. (*Maple 6*) Use *Maple* Help to discover how you can employ the function **Generate-Equations** in the **Linear Algebra** package to solve matrix equations.

Matrices as Linear Functions

Earlier we discussed matrices as functions on indexing sets. It allowed us to discuss the structure of a matrix and provided us with a convenient vocabulary for talking about matrices as two-dimensional objects, made up of rows and columns. From a mathematical point of view, matrices are functions in a much deeper sense. The equations $A\mathbf{x} = \mathbf{b}$ provide us with a clue. Given any vector $\mathbf{x}$, the matrix A associates with $\mathbf{x}$ a unique vector $\mathbf{b}$ satisfying the equation $A\mathbf{x} = \mathbf{b}$. Thus the matrix A defines a function T_A determined by the rule $T_A(\mathbf{x}) = A\mathbf{x}$.

Chapter 5 deals in detail with functions defined by matrices. It will follow from the properties of matrix-vector products that if a and b are two constants and $\mathbf{x}$ and $\mathbf{y}$ are two suitable vectors, then

$$A\,(a\mathbf{x} + b\mathbf{y}) = aA\mathbf{x} + bA\mathbf{y}$$

This property is known as ***linearity***. In functional notation, this equation expresses the fact that

$$T_A\,(a\mathbf{x} + b\mathbf{y}) = aT_A\,(\mathbf{x}) + bT_A\,(\mathbf{y})$$

The domains and codomains of linear functions are usually structures known as ***vector spaces***. Chapter 4 deals with such spaces. Although this text deals almost exclusively with linear functions defined by matrices, it is important to note that there are many well-known linear functions that are not based on matrices. We know from calculus, for example, that

$$\frac{d}{dt}(af(x) + bg(x)) = a\frac{d}{dt}f(x) + b\frac{d}{dt}g(x)$$

for any two constants a and b and any two differentiable function f and g, the derivatives $\frac{d}{dt}f(x)$ and $\frac{d}{dt}g(x)$ are rarely matrix-vector products. For geometric reasons, we refer to the linear functions studied in this text as ***linear transformations*** and to linear transformations determined by matrices as ***matrix transformations***.

EXAMPLE 1.59 A Matrix Transformation from $\mathbb{R}^3$ to $\mathbb{R}^2$

Explain in what sense the 2×3 matrix

$$A = \begin{bmatrix} 1 & 2 & 3 \\ 4 & 5 & 6 \end{bmatrix}$$

defines a linear function T_A from $\mathbb{R}^3$ to $\mathbb{R}^2$.

Solution. Let $\mathbf{x} = (x, y, z)$ be any vector in $\mathbb{R}^3$. We can use the operation involved in converting linear systems to matrix form and define

$$A\mathbf{x} = \begin{bmatrix} 1 & 2 & 3 \\ 4 & 5 & 6 \end{bmatrix} \begin{bmatrix} x \\ y \\ z \end{bmatrix} = \begin{bmatrix} x + 2y + 3z \\ 4x + 5y + 6z \end{bmatrix}$$

The result is a unique vector $A\mathbf{x}$ in $\mathbb{R}^2$. We let T_A be the linear transformation defined by the matrix A in this way. The domain of T_A is $\mathbb{R}^3$ and the codomain is $\mathbb{R}^2$. In the language of functions, a vector $\mathbf{x} \in \mathbb{R}^3$ is a solution of the linear system $A\mathbf{x} = \mathbf{b}$ if and only if the function T_A maps the vector $\mathbf{x} \in \mathbb{R}^3$ to the vector $\mathbf{b} \in \mathbb{R}^2$. ▲

EXAMPLE 1.60 A Matrix Transformation from $\mathbb{R}^2$ to $\mathbb{R}^3$

Explain in what sense the 3×2 matrix

$$A = \begin{bmatrix} 1 & 4 \\ 2 & 5 \\ 3 & 6 \end{bmatrix}$$

defines a linear function T_A from $\mathbb{R}^2$ to $\mathbb{R}^3$, and use linear systems to show that not all vectors in $\mathbb{R}^3$ are values of T_A.

Solution. Let $\mathbf{x} = (x, y)$ be any vector in $\mathbb{R}^2$. If we define

$$A\mathbf{x} = \begin{bmatrix} 1 & 4 \\ 2 & 5 \\ 3 & 6 \end{bmatrix} \begin{bmatrix} x \\ y \end{bmatrix} = \begin{bmatrix} x + 4y \\ 2x + 5y \\ 3x + 6y \end{bmatrix}$$

then $A\mathbf{x}$ is a unique vector in $\mathbb{R}^3$. We let T_A be the linear transformation defined by the matrix A in this way. The domain of T_A is $\mathbb{R}^2$ and the codomain is $\mathbb{R}^3$. By trial and error, we find a vector $\mathbf{b} \in \mathbb{R}^3$ that is not a solution of the linear system

$$\begin{cases} x + 4y = a \\ 2x + 5y = b \\ 3x + 6y = c \end{cases}$$

Let $a = 1$, $b = 2$, and $c = 1$. We can verify easily that the linear system

$$\begin{cases} x + 4y = 1 \\ 2x + 5y = 2 \\ 3x + 6y = 1 \end{cases}$$

is inconsistent. Therefore the vector $(1, 2, 1) \in \mathbb{R}^3$ is not a value of T_A. ▲

EXAMPLE 1.61 A Matrix Transformation from $\mathbb{R}^3$ to $\mathbb{R}^3$

Explain in what sense the matrix 3×3 matrix

$$A = \begin{bmatrix} 1 & 2 & 3 \\ 4 & 5 & 6 \\ 7 & 8 & 9 \end{bmatrix}$$

defines a linear function T_A from $\mathbb{R}^3$ to $\mathbb{R}^3$, and explain in terms of functions the statement that the linear system $A\mathbf{x} = \mathbf{b}$ is inconsistent for some $\mathbf{b} \in \mathbb{R}^3$.

Solution. Let $\mathbf{x} = (x, y, z)$ be any vector in $\mathbb{R}^3$. If we define

$$A\mathbf{x} = \begin{bmatrix} 1 & 2 & 3 \\ 4 & 5 & 6 \\ 7 & 8 & 9 \end{bmatrix} \begin{bmatrix} x \\ y \\ z \end{bmatrix} = \begin{bmatrix} x + 2y + 3z \\ 4x + 5y + 6z \\ 7x + 8y + 9z \end{bmatrix}$$

then $A\mathbf{x}$ is a unique vector in $\mathbb{R}^3$. We let T_A be the linear transformation defined by the matrix A in this way. The domain and codomain of T_A are both $\mathbb{R}^3$. Every vector $\mathbf{x} \in \mathbb{R}^3$ determines a unique vector $A\mathbf{x} \in \mathbb{R}^3$, but a vector $\mathbf{b} \in \mathbb{R}^3$ is of the form $A\mathbf{x}$ if and only if the linear system $A\mathbf{x} = \mathbf{b}$ is consistent. However, it is easy to verify that the linear system

$$\begin{cases} x + 2y + 3z = 1 \\ 4x + 5y + 6z = 2 \\ 7x + 8y + 9z = 1 \end{cases}$$

is inconsistent. Hence the vector $\mathbf{b} = (1, 2, 1)$ is not a value of T_A. ▲

EXERCISES 1.16

1. Use the matrix transformation T_A defined in Example 1.61 to map the vectors $(1, 3, 3)$, $(-3, 2, 1)$, $(0, 0, 0)$, $(1, 1, 1)$ in $\mathbb{R}^3$ to other vectors in $\mathbb{R}^3$.

2. Use the matrix transformation T_A defined in Example 1.59 to map the vectors $A\mathbf{x}$ in $\mathbb{R}^3$ found in Exercise 1 to vectors in $\mathbb{R}^2$.

3. A function $f : A \to B$ is ***one–one*** if $f(x) = f(y)$ implies that $x = y$ for all $x, y \in A$ and is ***onto*** if for every $y \in B$ there exists an $x \in A$ such that $f(x) = y$. Let A be the 2×2 coefficient matrix

$$\begin{bmatrix} 2 & 0 \\ 0 & 3 \end{bmatrix}$$

Use the language of linear systems to explain why the matrix transformation T_A determined by A is a one–one onto function from $\mathbb{R}^2$ to $\mathbb{R}^2$.

4. Let A be the 3×3 coefficient matrix

$$\begin{bmatrix} -2 & -2 & 0 \\ 0 & -1 & 3 \\ 0 & 0 & 0 \end{bmatrix}$$

Use the language of linear systems to explain why the matrix transformation T_A from $\mathbb{R}^3$ to $\mathbb{R}^3$ determined by A is neither one–one nor onto.

5. For any given matrix transformation T_A, we refer to the set of all values $A\mathbf{x}$ as the ***range*** of the function. Discuss the range of a matrix transformations T_A determined by 2×3 homogeneous linear systems $A\mathbf{x} = \mathbf{0}$. Explain why the domain of T_A is all of $\mathbb{R}^3$.

6. Show by direct calculation that if $a = 3, 4$, $b = -1, 2$, $\mathbf{x} = (3, 3)$, and $\mathbf{y} = (2, 4)$, then $A(a\mathbf{x} + b\mathbf{y}) = aA\mathbf{x} + bA\mathbf{y}$, where A is the 3×2 matrix [3 4].

7. Show by direct calculation that if $a = 3, 4$, $b = -1, 2$, $\mathbf{x} = (3, 3, 3)$, and $\mathbf{y} = (0, 2, 4)$, then $A(a\mathbf{x} + b\mathbf{y}) = aA\mathbf{x} + bA\mathbf{y}$, where A is the 3×3 matrix in Exercise 4.

8. (*Maple V*) Use *Maple* to find a 3×3 matrix A without zero entries for which the function T_A is one–one for all $\mathbf{x} \in \mathbb{R}^3$. Explain your reasoning.

9. (*Maple V*) Find two distinct vectors $\mathbf{x} = (x_1, x_2, x_3)$ and $\mathbf{y} = (y_1, y_2, y_3)$ for which

$$\begin{bmatrix} 1 & 2 & 3 \\ 0 & 0 & 0 \\ 0 & 0 & 0 \end{bmatrix}\begin{bmatrix} x_1 \\ x_2 \\ x_3 \end{bmatrix} = \begin{bmatrix} 1 & 2 & 3 \\ 0 & 0 & 0 \\ 0 & 0 & 0 \end{bmatrix}\begin{bmatrix} y_1 \\ y_2 \\ y_3 \end{bmatrix}$$

Use this example to explain why a matrix transformation $T_A : \mathbb{R}^3 \to \mathbb{R}^3$ determined by a 3×3 matrix with two identical rows can never be one–one.

10. (*Maple V*) A matrix transformation $T_A : \mathbb{R}^n \to \mathbb{R}^m$ defined by an $m \times n$ matrix A is ***onto*** if for every $\mathbf{b} \in \mathbb{R}^m$ there exists an $\mathbf{x} \in \mathbb{R}^n$ for which $T_A(\mathbf{x}) = \mathbf{b}$. Find constants a, b, and

c for which the overdetermined linear system

$$\begin{bmatrix} 1 & 4 \\ 2 & 5 \\ 3 & 6 \end{bmatrix} \begin{bmatrix} x \\ y \end{bmatrix} = \begin{bmatrix} a \\ b \\ c \end{bmatrix}$$

has no solution. Use this example to explain why a matrix transformation $T_A : \mathbb{R}^2 \to \mathbb{R}^3$ can never be onto.

APPLICATIONS

Chemical Equations

Stoichiometry is the study of quantitative relationships in chemical reactions. One of the steps in the standard procedure for solving problems in stoichiometry is to write balanced chemical equations. Suppose that a chemical reaction is described by the equation

$$H_2S + O_2 \to H_2O + SO_2$$

expressing the fact that if oxygen (O) is added to a compound made up of hydrogen (H) and sulfur (S), then water (H_2O) and sulfur dioxide (SO_2) result. We would like to solve the mass-mass problem of finding positive integers a, b, c, and d that balance the equation. The chemicals H_2S and O_2 are called the ***reactants,*** and H_2O and SO_2 are called the ***products***. The balancing integers a, b, c, and d are called ***stoichiometric coefficients***.

In the language of linear algebra, this problem consists of finding constants a, b, c, and d for which the following two vector sums are equal:

$$a \begin{bmatrix} 2\text{H} \\ 1\text{S} \\ 0\text{O} \end{bmatrix} + b \begin{bmatrix} 0\text{H} \\ 0\text{S} \\ 2\text{O} \end{bmatrix} = c \begin{bmatrix} 2\text{H} \\ 0\text{S} \\ 1\text{O} \end{bmatrix} + d \begin{bmatrix} 0\text{H} \\ 1\text{S} \\ 2\text{O} \end{bmatrix}$$

This is equivalent to solving the linear system

$$\begin{cases} 2a - 2c = 0 \\ a - d = 0 \\ 2b - c - 2d = 0 \end{cases}$$

A solution is easily read off the equations. If we let $a = c = d = 2$ and $b = 3$, then we get a balanced chemical equation. Hence

$$2H_2S + 3O_2 \to 2H_2O + 2SO_2$$

is a feasible mass–mass relationship between the chemical substances involved in the process.

Electric Circuits

Systems of linear equations can be used to determine the properties of simple steady-state circuits involving elements such as resistors, capacitors, and batteries. Such circuits can be analyzed through the use of two sets of principles:

1. **Ohm's law** $V = IR$, which describes a voltage V, measured in ***volts***, as the amount of a current I, measured in ***amps***, passing through a resistor R, whose capacity is measured in ***ohms***.

2. **Kirchhoff's laws** for parallel and series combinations of the circuit elements. Kirchhoff's laws simplify the analysis of complicated circuits and make it possible to calculate the current flowing through the circuit by representing the circuit with a linear system according to the following two principles.

   a. Currents entering any junction of a circuit must equal the sum of the currents leaving the junction. This principle expresses the law of the conservation of charge. Consider the junction in Figure 3. By Kirchhoff's current law, $i_1 = i_2 + i_3$.

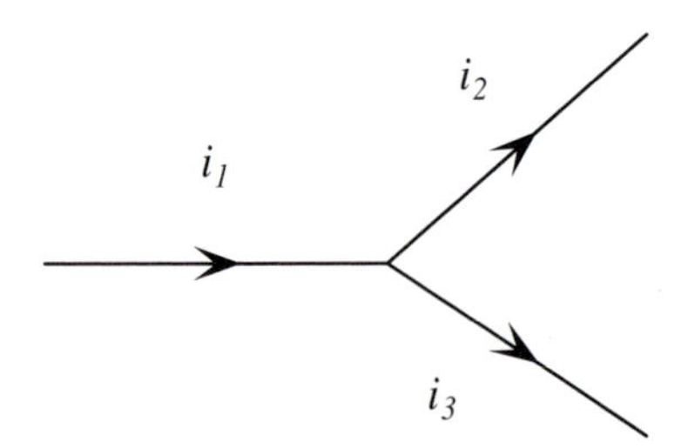

FIGURE 3 Electric Circuit Junction.

   b. The sum of the changes in potential voltage across the elements around any closed circuit loop must be zero. This is a statement of the law of conservation of energy.

In a circuit containing resistors and electromotive forces, certain conventions determine the signs of the potential differences.

1. If a resistor is traversed in the direction of the current, as shown in Figure 4(a), then the potential difference across the resistor is IR.

2. If a resistor is traversed in the direction opposite to the current, as shown in Figure 4(b), then the potential difference across the resistor is $-IR$.

3. If a source of an electromotive force is traversed in the direction of the force, as shown in Figure 4(c), then the potential difference is $+E$.

4. If a source of an electromotive force is traversed in the opposite direction of the force, as shown in Figure 4(d), then the potential difference is $-E$.

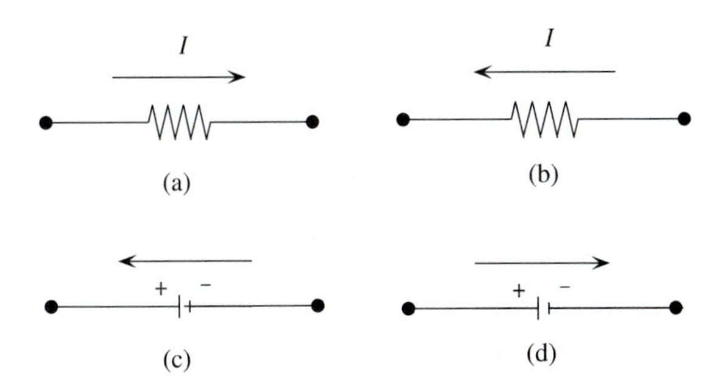

FIGURE 4 Electrical Current Flow.

EXAMPLE 1.62 A Resistor Circuit

Use *MAPLE* to calculate the currents in the circuit shown in Figure 5 for $R_1 = R_3 = R_5 = 1$, $R_2 = R_4 = R_6 = 2$, and $E_1 = 2$ and $E_2 = 3$.

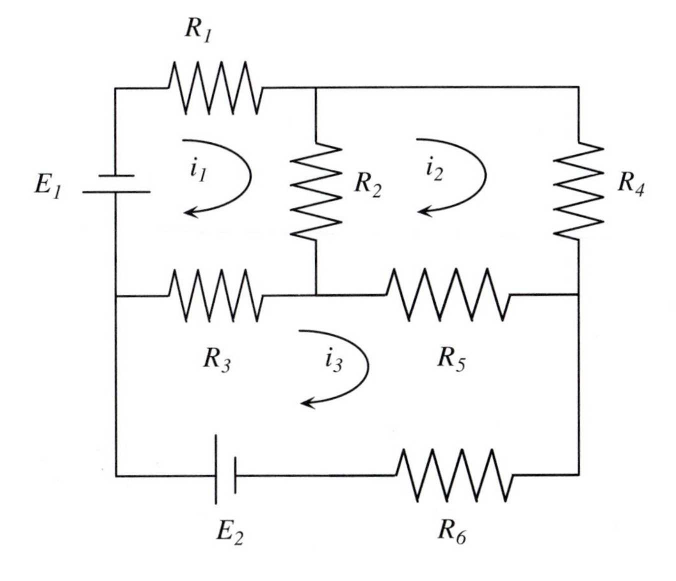

FIGURE 5 Electrical Resistor Circuit.

Solution. By Kirchhoff's voltage law, the loop currents i_1, i_2, and i_3 are determined by the linear system

$$\begin{cases} R_1 i_1 + R_2 (i_1 - i_2) + R_3 (i_1 - i_3) = -E_1 \\ R_2 (i_2 - i_1) + R_4 i_2 + R_5 (i_2 - i_3) = 0 \\ R_3 (i_3 - i_1) + R_5 (i_3 - i_2) + R_6 i_3 = E_2 \end{cases}$$

We use the *Maple* to solve this system for i_1, i_2, and i_3 by applying Gaussian elimination to the augmented matrix of the system. For the given values for R_i and E_j, the augmented matrix of the system is

$$\begin{bmatrix} R_1 + R_2 + R_3 & -R_2 & -R_3 & -E_1 \\ -R_2 & R_2 + R_4 + R_5 & -R_5 & 0 \\ -R_3 & -R_5 & R_3 + R_5 + R_6 & E_2 \end{bmatrix} = \begin{bmatrix} 4 & -2 & -1 & -2 \\ -2 & 5 & -1 & 0 \\ -1 & -1 & 4 & 3 \end{bmatrix}$$

We begin by defining the matrix A.

```
[> with(linalg):
[> A:=matrix([[4,-2,-1,-2],[-2,5,-1,0],[-1,-1,4,3]]):
```

Next we row-reduce A.

```
[> rref(A);
```

$$\begin{bmatrix} 1 & 0 & 0 & -\frac{1}{3} \\ 0 & 1 & 0 & 0 \\ 0 & 0 & 1 & \frac{2}{3} \end{bmatrix}$$

It follows from the definition of augmented matrices that $i_1 = -1/3$ amps, $i_2 = 0$ amps, and $i_3 = 2/3$ amps. ▲

Higher-Order Differential Equations

Suppose that $y^{(1)}(t), y^{(2)}(t), \ldots, y^{(n)}(t)$ are the first n derivatives of a function $y(t)$. Our goal is to rewrite any nth-order differential equation involving these functions as a linear system of first-order differential equations. Here is the method, described for $n = 3$. Consider the differential equation

$$y^{(3)}(t) + a_3 y(t) + a_2 y^{(1)}(t) + a_1 y^{(2)}(t) = 0$$

We rewrite this equation as

$$\begin{aligned} y^{(3)}(t) &= -a_3 y(t) - a_2 y^{(1)}(t) - a_1 y^{(2)}(t) \\ &= -a_3 y_1(t) - a_2 y_2(t) - a_1 y_3(t) \end{aligned}$$

$$= \begin{bmatrix} -a_3 & -a_2 & -a_1 \end{bmatrix} \begin{bmatrix} y_1(t) \\ y_2(t) \\ y_3(t) \end{bmatrix}$$

where $y_1(t) = y(t)$, $y_2(t) = y^{(1)}(t)$, and $y_3(t) = y^{(2)}(t)$. In this notation, $y_3^{(1)}(t) = y^{(3)}(t)$.

Using these equations, we form the matrix equation

$$\mathbf{y}' = \begin{bmatrix} y_1^{(1)}(t) \\ y_2^{(1)}(t) \\ y_3^{(1)}(t) \end{bmatrix} = \begin{bmatrix} 0 & 1 & 0 \\ 0 & 0 & 1 \\ -a_3 & -a_2 & -a_1 \end{bmatrix} \begin{bmatrix} y_1(t) \\ y_2(t) \\ y_3(t) \end{bmatrix} = A\mathbf{y}$$

The result is a linear system of first-order equations. Suppose now that there exists an invertible matrix P and a diagonal matrix D for which $A = PDP^{-1}$. Then the linear system $\mathbf{y}' = A\mathbf{y}$ can be converted to a linear system $\mathbf{x}' = D\mathbf{x}$ by a change of variables whose general solution is of the form

$$\mathbf{x}(s) = c_1 e^{\lambda_1 s}\mathbf{x}_1 + c_2 e^{\lambda_2 s}\mathbf{x}_2 + c_3 e^{\lambda_3 s}\mathbf{x}_3$$

where λ_1, λ_2, and λ_3 are the diagonal entries of D, and the vectors $\mathbf{x}_1, \mathbf{x}_2$, and $\mathbf{x}_3$ are the columns of A. Chapter 6 deals with methods for finding the scalars λ_i—called ***eigenvalues*** of A—and associated vectors $\mathbf{x}_i$—called ***eigenvectors*** of A.

This method works in general. Let $y_1(t) = y(t)$, and let $y_i(t) = y_{i-1}^{(1)}(t)$ be the ith derivative of a function $y(t)$, for $i = 2, \ldots, n$, and then suppose that

$$y^{(n)}(t) + a_1 y^{(n-1)}(t) + \cdots + a_{n-1} y^{(1)}(t) + a_n y(t) = 0$$

is a homogeneous nth–order linear differential equation. We use the functions $y_i(t)$ to form the matrix equation

$$\begin{bmatrix} y_1'(t) \\ y_2'(t) \\ \vdots \\ y_{n-1}'(t) \\ y_n'(t) \end{bmatrix} = \begin{bmatrix} 0 & 1 & 0 & \cdots & 0 \\ 0 & 0 & 1 & \cdots & 0 \\ \vdots & \vdots & \vdots & \vdots & \vdots \\ 0 & 0 & 0 & \cdots & 1 \\ -a_n & -a_{n-1} & -a_{n-2} & \cdots & -a_1 \end{bmatrix} \begin{bmatrix} y_1(t) \\ y_2(t) \\ \vdots \\ y_{n-1}(t) \\ y_n(t) \end{bmatrix}$$

The result is a linear system of first-order differential equations corresponding to the given nth-order equation. In Chapter 5, we will show that this reduction can be used to solve certain higher-order differential equations.

EXAMPLE 1.63 Reducing a Differential Equation to a First-Order System

Find a first-order linear system for the third-order differential equation

$$y^{(3)}(t) + 5y^{(2)}(t) - 2y(t) = 0$$

and write the result in matrix form.

Solution. We let $y_1(t) = y(t)$, $y_2(t) = y_1^{(1)}(t)$, and $y_3(t) = y_2^{(1)}(t)$. Using these definitions, we rearrange the given equation as

$$\begin{aligned} y^{(3)}(t) &= 2y(t) - 5y^{(2)}(t) \\ &= 2y(t) + 0y^{(1)}(t) - 5y^{(2)}(t) \\ &= 2y_1(t) + 0y_2(t) - 5y_3(t) \end{aligned}$$

We then form linear system

$$\begin{cases} y_1^{(1)}(t) = y_2(t) \\ y_2^{(1)}(t) = y_3(t) \\ y_3^{(1)}(t) = 2y_1(t) + 0y_2(t) - 5y_3(t) \end{cases}$$

and write it as the matrix equation

$$\begin{bmatrix} y_1^{(1)}(t) \\ y_2^{(1)}(t) \\ y_3^{(1)}(t) \end{bmatrix} = \begin{bmatrix} 0 & 1 & 0 \\ 0 & 0 & 1 \\ 2 & 0 & -5 \end{bmatrix} \begin{bmatrix} y_1(t) \\ y_2(t) \\ y_3(t) \end{bmatrix}$$

The result is the required first-order linear system, written in matrix form. ▲

Heat Transfer

In physics and engineering, linear systems are used to solve heat transfer problems. The method is based on the assumption that the temperature at a point in a thin metal plate, for example, is approximately equal to the average of the temperatures of surrounding points. The approximation improves with the number of surrounding points chosen.

EXAMPLE 1.64 Heat Transfer

Consider a thin rectangular metal plate as shown in Figure 6, with temperatures measured in degrees Celsius. Calculate the temperature of the plate at points $P_1, \ldots, P_6$.

Solution. It is clear from the discussion preceding this example that if t_i is the temperature of the plate at point P_i, then

$$\begin{cases} t_1 = \frac{1}{4}(80 + t_2 + t_6 + 120), & t_2 = \frac{1}{4}(80 + t_3 + t_5 + t_1) \\ t_3 = \frac{1}{4}(80 + 60 + t_4 + t_2), & t_4 = \frac{1}{4}(t_3 + 60 + 80 + t_5) \\ t_5 = \frac{1}{4}(t_2 + t_4 + 80 + t_6), & t_6 = \frac{1}{4}(t_1 + t_5 + 80 + 120) \end{cases}$$

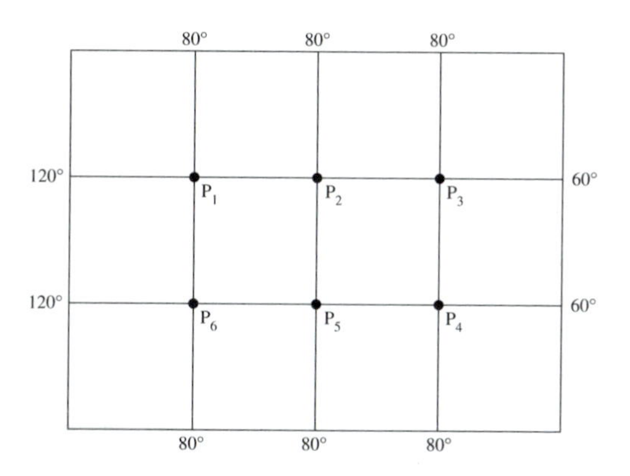

FIGURE 6 Cross Section of a Metal Plate.

We use *Maple* to solve this system.

```
[> solve({
   4*t1=80+t2+t6+120,4*t2=80+t3+t5+t1,
   4*t3=80+60+t4+t2,4*t4=t3+60+80+t5,
   4*t5=t2+t4+80+t6,4*t6=t1+t5+80+120},
   {t1,t2,t3,t4,t5,t6}):
```

```
[> evalf(%);
```

$$\{t3 = 73.55263158, t2 = 82.43421053, t6 = 94.04605263,$$
$$t1 = 94.12006579, t5 = 82.06414474, t4 = 71.77631579\}$$

Thus the temperature of the plate at points P_1 and P_6 is approximately 94 °C, at P_2 and P_5 approximately 82°C, and at P_3 and P_4 approximately 72°C. ▲

Population Dynamics

Suppose that the females in a family of animals are divided into n age groups, with y_i the number of females in the ith group. The length of the equal time intervals will depend on the nature of the animals. Suppose also that the probability that a female in the ith age group will be still be alive in the $(i+1)$th age group is p_i, and let a_i be the average number of females born to a female in the ith group. Then

$$\mathbf{y} = \begin{bmatrix} y_1 \\ y_2 \\ \vdots \\ y_{n-1} \\ y_n \end{bmatrix} = \begin{bmatrix} a_1 & a_2 & \cdots & a_{n-1} & a_n \\ p_1 & 0 & \cdots & \cdots & 0 \\ 0 & p_2 & \cdots & \cdots & 0 \\ \vdots & \vdots & \vdots & \vdots & \vdots \\ 0 & 0 & \cdots & p_{n-1} & 0 \end{bmatrix} \begin{bmatrix} x_1 \\ x_2 \\ \vdots \\ x_{n-1} \\ x_n \end{bmatrix} = A\mathbf{x}$$

where the vector $\mathbf{x} = \mathbf{x}(0)$ represents the population distribution of the family at time $t = 0$ and $\mathbf{y} = \mathbf{y}(1)$ represents the distribution at time $t = 1$. After k time intervals, the population distribution is given by the vector $\mathbf{y}(k) = A^k\mathbf{x}(0)$. The matrix A is called the ***Leslie matrix*** for the given family.

EXAMPLE 1.65 Population Dynamics

Use a Leslie matrix to calculate the female population after five time intervals of a family divided into three age groups, with the first family consisting of 300 females, the second family of 700 females, and the third family of 500 females. The probability that a female in the first group will be alive in the second group is $1/2$, and the probability that a female in the second group will be alive in the third group is $3/4$. Suppose that no female in the first age group gives birth to an offspring, that the average number of female offspring of a female in the second group is 7, and that in the third group is 3.

Solution. We know from the foregoing discussion that the population vector

$$\mathbf{y}(5) = \begin{bmatrix} 0 & 7 & 3 \\ 0.5 & 0 & 0 \\ 0 & 0.75 & 0 \end{bmatrix}^5 \begin{bmatrix} 300 \\ 700 \\ 500 \end{bmatrix}$$

yields the required data. We use *Maple* to calculate $\mathbf{y}(5)$.

```
[> with(linalg):
[> A:=matrix([[0,7,3],[0.5,0,0],[0,0.75,0]]):
[> A^5&*[[300],[700],[500]]:
[> evalm(%);
```

$$\begin{bmatrix} 82534.3750 \\ 8193.7500 \\ 8526.5625 \end{bmatrix}$$

After five time intervals, the family consists of approximately 82,534 females in the first age group, 8,194 females in the second group, and 8,527 females in the third group. ▲

EXERCISES 1.17

1. Balance the following chemical equations.

   a. $N_2 + H_2 \to NH_3$

   b. $C_3H_8 + O_2 \to CO_2 + H_2O$

   c. $CH_4 + NH_3 + O_2 \to HCN + H_2O$

2. Consult a textbook on chemistry and use linear algebra to solve additional stoichiometric problems found there.

3. Use Ohm's and Kirchhoff's laws to verify the equation for calculating the current shown in Figure 7.

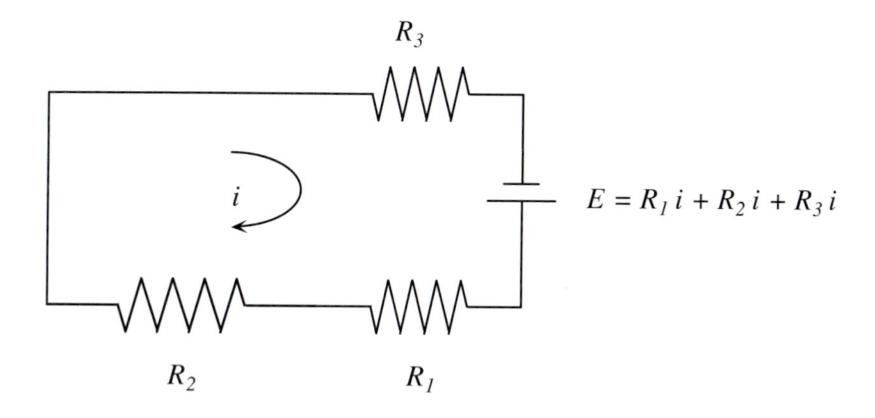

FIGURE 7 A Three-Resistor Circuit.

4. Use Ohm's and Kirchhoff's laws, in conjunction with Gaussian elimination, to calculate the currents in the loops in Figure 8.

5. Consult a textbook on physics and electrical engineering and use linear algebra to solve additional electrical circuit problems found there.

6. Reduce the following higher-order differential equations to systems of first-order equations.

   a. $y^{(4)}(t) + y^{(3)}(t) - 2y(t) = 0$ b. $y^{(2)}(t) - 8y^{(1)}(t) + y(t) = 0$

   c. $y^{(3)}(t) + y^{(2)}(t) + y(t) = 0$ d. $y^{(5)}(t) + 5y^{(3)}(t) - 6y(t) = 0$

7. Consult a textbook on differential equations and use linear systems to reduce higher-order homogeneous linear differential equations found there to systems of first-order differential equations.

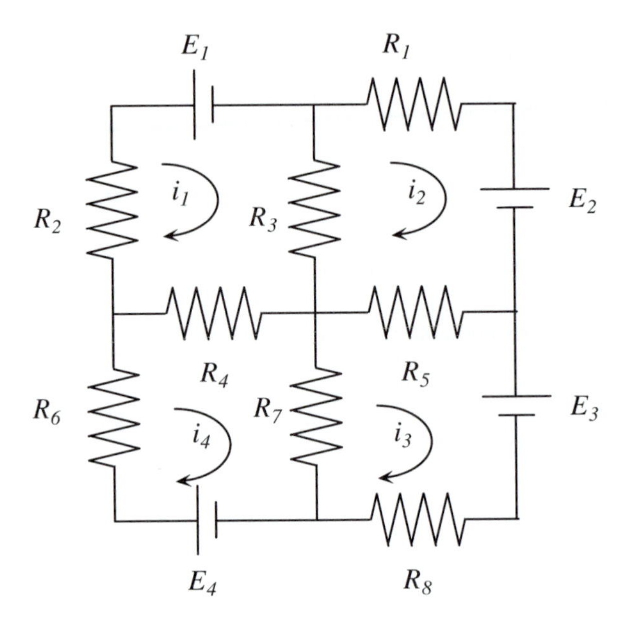

FIGURE 8 An Eight-Resistor Circuit.

8. Use a linear system to calculate the average temperature at 16 equidistant interior points of a thin square metal plate whose boundary temperatures are 100°C on the top and bottom of the plate, 60°C on the left side of the plate, and 120°C on the right side of the plate.

9. Suppose that the females of a certain family of animals are divided into five age groups and that the Leslie matrix for this family is

$$A = \begin{bmatrix} 0.0 & 3.0 & 2.0 & 1.5 & 0.1 \\ 0.9 & 0.0 & 0.0 & 0.0 & 0.0 \\ 0.0 & 0.6 & 0.0 & 0.0 & 0.0 \\ 0.0 & 0.0 & 0.5 & 0.0 & 0.0 \\ 0.0 & 0.0 & 0.0 & 0.5 & 0.0 \end{bmatrix}$$

Calculate the number of females of the different age groups after eight years if at time $t = 0$ each age group contains 1,000 females.

10. Consult a textbook on population biology and use Leslie matrices to solve additional population problems.

REVIEW

KEY CONCEPTS ▶ Define and discuss each of the following.

Equivalence

Row equivalence.

Geometry

Arrow, line, plane, point.

Linear Systems

Basic variable, free variable. Consistent system, determined system, inconsistent system, overdetermined system, underdetermined system. Back substitution, Gaussian elimination, Gauss-Jordan elimination. Homogeneous linear system, linear dependence relation, nontrivial solution, trivial solution.

Lists

Indexing list, list of lists.

Matrices

Augmented matrix, coefficient matrix, row echelon matrix, reduced row echelon matrix, Vandermonde matrix. Dimension. Entry. Pivot, pivot column, pivot position.

Matrix Operations

Elementary row operation.

Vectors

Column vector, row vector, zero vector.

KEY FACTS ▶ Explain and illustrate each of the following.

1. Every matrix is row equivalent to a row echelon matrix.
2. Every matrix is row equivalent to a matrix in reduced row echelon form.
3. The zero vector is a solution of the homogenous linear system $A\mathbf{x} = \mathbf{0}$.
4. An $m \times n$ linear system in the variables $x_1, \ldots, x_n$ with augmented $m \times (n+1)$ matrix $[A \mid \mathbf{b}]$ is consistent for all $\mathbf{b} \in \mathbb{R}^m$ if and only if the matrix A has m pivots.
5. Every underdetermined homogeneous linear system has a nontrivial solution.
6. If a homogeneous linear system results from another system by elementary row operations, then the two systems have the same linear dependence relations.
7. The reduced row echelon form of a matrix is unique.

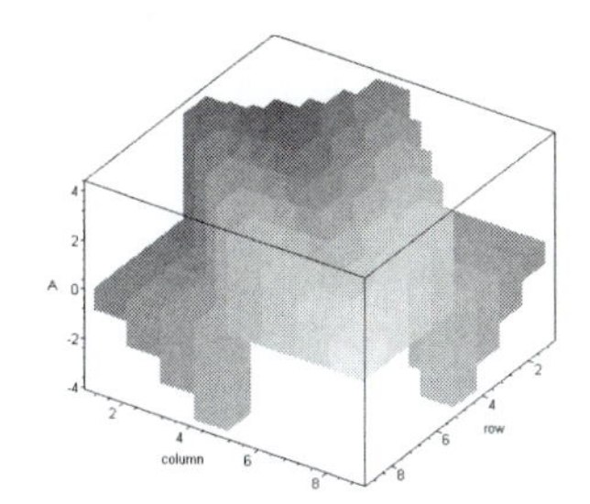

2

MATRIX ALGEBRA

INTRODUCTION

As we saw in Chapter 1, it is natural to represent linear systems in matrix form. The following examples illustrate additional situations where matrix algebra is natural and useful.

EXAMPLE 2.1 A Cost Analysis

A furniture manufacturer has received orders to furnish several houses in a new housing development, 5 in an Alpine style, 7 in a Colonial style, and 14 in a Spanish style. To produce this furniture, the manufacturer needs different quantities of wood, metal, plastic, glass, and leather. The unit costs of the raw materials are \$120, \$150, \$40, \$80, and \$200, respectively. Table 1 lists the quantities of raw materials required for the three types of house.

TABLE 1 REQUIRED RAW MATERIALS.

	Wood	Metal	Plastic	Glass	Leather
Alpine	20	8	2	18	4
Colonial	18	7	4	15	2
Spanish	15	9	3	21	18

Use matrix algebra to find the total cost of the raw material required for this project.

Solution. First we convert the given table into the matrix

$$A = \begin{bmatrix} 20 & 8 & 2 & 18 & 4 \\ 18 & 7 & 4 & 15 & 2 \\ 15 & 9 & 3 & 21 & 18 \end{bmatrix}$$

If each type of house requires 10 pieces of comparable furniture, then the manufacturer must produce 50 pieces of Alpine furniture, 70 pieces of Colonial furniture, and 140 pieces of Spanish furniture. The vector $\mathbf{x} = (50, 70, 140)$ lists the number of different pieces of furniture of each style to be produced. The matrix

$$\mathbf{x}A = \begin{bmatrix} 50 & 70 & 140 \end{bmatrix} \begin{bmatrix} 20 & 8 & 2 & 18 & 4 \\ 18 & 7 & 4 & 15 & 2 \\ 15 & 9 & 3 & 21 & 18 \end{bmatrix} = \begin{bmatrix} 4360 & 2150 & 800 & 4890 & 2860 \end{bmatrix}$$

tabulates the total number of units of wood, metal, plastic, glass, and leather required for the entire project. Moreover, the vector $\mathbf{y} = (120, 150, 40, 80, 200)$ lists the unit costs of these raw materials. The product

$$\mathbf{x}A\mathbf{y} = \begin{bmatrix} 4360 & 2150 & 800 & 4890 & 2860 \end{bmatrix} \begin{bmatrix} 120 \\ 150 \\ 40 \\ 80 \\ 200 \end{bmatrix} = 1840900$$

shows that the total cost of the raw materials for the entire project is $\$1,840,900$. ▲

EXAMPLE 2.2 Mortgage Payments

Use matrix algebra to calculate the monthly payments required for a $\$100,000$ mortgage if the annual interest rate is 9% per year and the mortgage is to be paid off in 30 years.

Solution. Suppose that M is the required monthly payment, that $\mathbf{x}$ is the vector

$$[100000, 100000, -M]$$

whose first two components are the amount of the mortgage and whose third component is $-M$, and that A is the 3×3 matrix

$$\begin{bmatrix} 1 & 1 & 0 \\ \left(\frac{1}{100}\right)\left(\frac{9}{12}\right) & \left(\frac{1}{100}\right)\left(\frac{9}{12}\right) & 0 \\ 1 & 1 & 1 \end{bmatrix}$$

containing the monthly rate interest. The product

$$\mathbf{x}A = \begin{bmatrix} 100000 & 100000 & -M \end{bmatrix} \begin{bmatrix} 1 & 1 & 0 \\ \left(\frac{1}{100}\right)\left(\frac{9}{12}\right) & \left(\frac{1}{100}\right)\left(\frac{9}{12}\right) & 0 \\ 1 & 1 & 1 \end{bmatrix}$$

of $\mathbf{x}$ and A produces the vector $\mathbf{y} = [100750 - M, 100750 - M, -M]$, whose first component

$$100750 - M = 100000 + \left[100000 \times \left(\frac{1}{100}\right)\left(\frac{9}{12}\right)\right] - M$$

is the amount of the mortgage still owed after one month. If we now use the vector $\mathbf{y}$ in place of $\mathbf{x}$ and calculate the vector-matrix product

$$\begin{bmatrix} 100750 - M & 100750 - M & -M \end{bmatrix} \begin{bmatrix} 1 & 1 & 0 \\ \left(\frac{1}{100}\right)\left(\frac{9}{12}\right) & \left(\frac{1}{100}\right)\left(\frac{9}{12}\right) & 0 \\ 1 & 1 & 1 \end{bmatrix}$$

we obtain the vector

$$\mathbf{z} = \begin{bmatrix} 1.0151 \times 10^5 - 2.0075M & 1.0151 \times 10^5 - 2.0075M & -1.0M \end{bmatrix}$$

whose first component, $1.0151 \times 10^5 - 2.0075M$, represents the amount of the mortgage still owed after two months. The vector $\mathbf{z}$ can also be calculated directly from $\mathbf{x}$ and A by using $\mathbf{x}$ and the *square* of A. The first component of the vector-matrix product

$$\begin{bmatrix} 100000 & 100000 & -M \end{bmatrix} \begin{bmatrix} 1 & 1 & 0 \\ \left(\frac{1}{100}\right)\left(\frac{9}{12}\right) & \left(\frac{1}{100}\right)\left(\frac{9}{12}\right) & 0 \\ 1 & 1 & 1 \end{bmatrix}^2$$

is also $\mathbf{z}$. Using these ideas, we can express the amount of the mortgage owed after $12 \times 30 = 360$ months as first component

$$\begin{bmatrix} 1.4731 \times 10^6 - 1830.7M & 1.4731 \times 10^6 - 1830.7M & -1.0M \end{bmatrix}$$

of the vector-matrix product

$$\begin{bmatrix} 100000 & 100000 & -M \end{bmatrix} \begin{bmatrix} 1 & 1 & 0 \\ \left(\frac{1}{100}\right)\left(\frac{9}{12}\right) & \left(\frac{1}{100}\right)\left(\frac{9}{12}\right) & 0 \\ 1 & 1 & 1 \end{bmatrix}^{360}$$

Since the mortgage is paid off at that time,

$$1.4731 \times 10^6 - 1830.7M = 0$$

If we solve this equation for M, we get $M = 804.66$. Our monthly payments are therefore approximately \$805.00. We will see later that the particular forms of the vectors $\mathbf{x}$, $\mathbf{y}$, $\mathbf{z}$ and of the matrix A are required because of the nature of the rules for matrix multiplication. ▲

Real Matrices

We usually simplify the exposition by writing the entries of matrices as integers. Occasionally, however, *Maple* needs to be told explicitly that the entries of a matrix are ***real numbers*** since the algorithms used for calculating with integers and real numbers often are not the same.

Telling *Maple* that the entries of a matrix are real numbers is very simple: We place a decimal point behind one of the entries. As far as *Maple* is concerned, the matrix

$$A = \begin{bmatrix} 20 & 8 & 2 & 18 & 4 \\ 18. & 7 & 4 & 15 & 2 \\ 15 & 9 & 3 & 21 & 18 \end{bmatrix}$$

is a matrix with real number entries since one of its entries is written as "18." We could have written "18.0" in place of "18." and achieved the same result. At times, we will require random real matrices as illustrative examples.

EXAMPLE 2.3 A Random Matrix with Real Entries

Use *Maple* to generate a random 3×4 matrix with real entries.

Solution. We use the **RandomMatrix** function in the **LinearAlgebra** package, in combination with the function **evalf.**

```
[> evalf(LinearAlgebra[RandomMatrix](3,4));
```

$$\begin{bmatrix} -34. & -21. & -50. & -79. \\ -62. & -56. & 30. & -71. \\ -90. & -8. & 62. & 28. \end{bmatrix}$$

The result is a real 3×4 matrix with entries between -100 and 100. ▲

EXERCISES 2.1

1. Set up a total cost matrix for the basic ingredients of three meals at a major resort hotel. One night the hotel serves filet mignon with vegetables. The meal requires 80 lb of meat, 100 lb of vegetables, 2 lb of spices, and 3 lb of other ingredients. The next night the hotel serves Irish stew. The meal requires 100 lb of meat, 300 lb of vegetables, 4 lb of spices, and 5 lb of other ingredients. The third night the hotel serves Hungarian goulash. The meal requires 150 lb of meat, 350 lb of vegetables, 5 lb of spices, and 7 lb of other ingredients. The average cost per pound of the meat is \$8.00, of the vegetables \$3.00, of the spices \$9.00, and of the other ingredients \$4.00. Calculate the total cost of the three meals.

2. A businessman returned from a European business trip with the following currencies: 100 German marks, 7000 French francs, 100, 000 Italian lire, and 120 British pounds. The next week he had to visit Canada and decided to convert these currencies to Canadian dollars. At the time, one Canadian dollar was worth 1.0776 German marks, 3.61496 French francs, 1066.51257 Italian lira, and 0.38288 British pounds.

   a. Write a row vector $[q_1, q_2, q_3, q_4]$ for the quantity of each currency the businessman brought back from Europe.

b. Express the values c_1, c_2, c_3, and c_4 of one unit of each currency in Canadian dollars as a column vector.

c. Show that the amount of Canadian dollars the businessman obtained for the European currencies is given by the product

$$\begin{bmatrix} q_1 & q_2 & q_3 & q_4 \end{bmatrix} \begin{bmatrix} c_1 \\ c_2 \\ c_3 \\ c_4 \end{bmatrix} = q_1c_1 + q_2c_2 + q_3c_3 + q_4c_4$$

3. Use the ideas discussed in Example 2.2 to calculate the monthly payments required for a $95,000 mortgage to be paid off in 25 years if the annual interest rate is 6%.

4. (*Maple V*) Define a function that generates random integers between 50 and 100.

5. (*Maple V*) Define a function that generates random real numbers between 98 and 100.

6. (*Maple V*) Use the **randmatrix** function in the **linalg** package to generate a 3×4, a 2×5, and a 4×4 random matrix with integer entries.

7. (*Maple 6*) Use the **RandomMatrix** function in the **LinearAlgebra** package to generate a 3×4, a 2×5, and a 4×4 random matrix with real entries.

8. (*Maple 6*) Explore the properties of the **RandomMatrix** function in the **LinearAlgebra** package.

BASIC MATRIX OPERATIONS

In this section, we discuss the *sum* $A + B$ of two matrices, the *scalar multiples* sA of a scalar and a matrix, the *product* AB of two matrices, and the *transpose* A^T of a matrix.

Matrix Addition

One of most basic operation on matrices of the same dimension is their addition.

DEFINITION 2.1 *If* $A = [a_{ij}]$ *and* $B = [b_{ij}]$ *are two* $m \times n$ *matrices, the* **sum** $A + B$ *of* A *and* B *is the matrix* $[c_{ij}] = [a_{ij} + b_{ij}]$.

The operation of forming the sum of two matrices is called ***matrix addition***. This definition also covers the ***subtraction*** of matrices, since we can think of $A - B$ as $A + (-B)$, where $-B = [-b_{ij}]$. The matrix $A - B$ is called the ***difference*** of A and B.

EXAMPLE 2.4 A Sum of Matrices

Consider the two 3×4 matrices

$$A = \begin{bmatrix} 1 & 2 & 0 & 3 \\ 5 & 3 & 0 & 1 \\ 9 & 2 & 1 & 6 \end{bmatrix} \quad \text{and} \quad B = \begin{bmatrix} 1 & 2 & 3 & 4 \\ 2 & 6 & 7 & 8 \\ 9 & 10 & 11 & 12 \end{bmatrix}$$

The matrix

$$A + B = \begin{bmatrix} 2 & 4 & 3 & 7 \\ 7 & 9 & 7 & 9 \\ 18 & 12 & 12 & 18 \end{bmatrix}$$

is the sum of the matrices A and B. ▲

Scalar Multiplication

Another basic operation on matrices is the multiplication of each entry of a matrix by the same scalar.

DEFINITION 2.2 *If $A = [a_{ij}]$ is an $m \times n$ matrix and s is a scalar, then the* ***scalar multiple*** *sA of A is the matrix $[sa_{ij}]$.*

The operation of forming a scalar multiple of a matrix is called ***scalar multiplication.***

EXAMPLE 2.5 A Scalar Multiple of a Matrix

Let $s = 7$ be a given scalar, and let

$$A = \begin{bmatrix} 1 & 2 & 0 & 3 \\ 5 & 3 & 0 & 1 \\ 9 & 2 & 1 & 6 \end{bmatrix}$$

be a 3×4 numerical matrix. Then

$$sA = 7A = \begin{bmatrix} 7\times1 & 7\times2 & 7\times0 & 7\times3 \\ 7\times5 & 7\times3 & 7\times0 & 7\times1 \\ 7\times9 & 7\times2 & 7\times1 & 7\times6 \end{bmatrix} = \begin{bmatrix} 7 & 14 & 0 & 21 \\ 35 & 21 & 0 & 7 \\ 63 & 14 & 7 & 42 \end{bmatrix}$$

The matrix $7A$ is the scalar multiple of A by 7. ▲

Matrix Multiplication

In Chapter 1, we introduced the matrix equations $A\mathbf{x} = \mathbf{b}$ and mentioned the fact that the vector $A\mathbf{x}$ results from the matrix A and the vector $\mathbf{x}$ by "matrix multiplication." We will now define this operation in general. We denote the result of multiplying two matrices A and B

by AB and call it the ***product*** of A and B. The matrix AB is built from the *dot products* of the rows of A and the columns of B. We emphasize the role of these dot products in matrix multiplication by calling them ***row-column products.***

DEFINITION 2.3 *If $\mathbf{r}_i A$ is the ith row an $m \times n$ matrix A and $\mathbf{c}_j B$ is the jth columns of an $n \times p$ matrix B, then the **row-column product** of $\mathbf{r}_i A$ and $\mathbf{c}_j B$ is the dot product*

$$\mathbf{r}_i A\, \mathbf{c}_j B = \begin{bmatrix} a_{i1} & \cdots & a_{in} \end{bmatrix} \begin{bmatrix} b_{1j} \\ \vdots \\ b_{nj} \end{bmatrix} = a_{i1}b_{1j} + \cdots + a_{in}b_{nj}$$

EXAMPLE 2.6 A Row-Column Product

Suppose that $\mathbf{x} = [a, b, c]$ is a row vector of length 3 and that $\mathbf{y} = (x, y, z)$ is a column vector of height 3. Then the product of $\mathbf{x}$ and $\mathbf{y}$ is the expression

$$\mathbf{xy} = \begin{bmatrix} a & b & c \end{bmatrix} \begin{bmatrix} x \\ y \\ z \end{bmatrix} = ax + by + cz$$

For any given assignment of constants to a, b, c, x, y, z, the product $\mathbf{xy}$ is a constant. ▲

DEFINITION 2.4 *The **product** AB of an $m \times n$ matrix A and $n \times p$ matrix B is the $m \times p$ matrix*

$$\begin{bmatrix} \mathbf{r}_1 A\, \mathbf{c}_1 B & \cdots & \mathbf{r}_1 A\, \mathbf{c}_p B \\ & \vdots & \\ \mathbf{r}_m A\, \mathbf{c}_1 B & \cdots & \mathbf{r}_m A\, \mathbf{c}_p B \end{bmatrix}$$

whose ijth entry is row-column product $\mathbf{r}_i A\, \mathbf{c}_j B$.

The operation of forming the product of two matrices is called ***matrix multiplication***.

EXAMPLE 2.7 A Product of Matrices

Calculate the matrix products AB and BA for

$$A = \begin{bmatrix} 1 & 2 \\ 3 & 4 \end{bmatrix} \quad \text{and} \quad B = \begin{bmatrix} a & b \\ c & d \end{bmatrix}$$

Solution. We begin by calculating AB. By definition,

$$AB = \begin{bmatrix} \mathbf{r}_1 A \\ \mathbf{r}_2 A \end{bmatrix} \begin{bmatrix} \mathbf{c}_1 B & \mathbf{c}_2 B \end{bmatrix} = \begin{bmatrix} \mathbf{r}_1 A \mathbf{c}_1 B & \mathbf{r}_1 A \mathbf{c}_2 B \\ \mathbf{r}_2 A \mathbf{c}_1 B & \mathbf{r}_2 A \mathbf{c}_2 B \end{bmatrix}$$

Therefore

$$AB = \begin{bmatrix} \begin{bmatrix} 1 & 2 \end{bmatrix}\begin{bmatrix} a \\ c \end{bmatrix} & \begin{bmatrix} 1 & 2 \end{bmatrix}\begin{bmatrix} b \\ d \end{bmatrix} \\ \begin{bmatrix} 3 & 4 \end{bmatrix}\begin{bmatrix} a \\ c \end{bmatrix} & \begin{bmatrix} 3 & 4 \end{bmatrix}\begin{bmatrix} b \\ d \end{bmatrix} \end{bmatrix} = \begin{bmatrix} a+2c & b+2d \\ 3a+4c & 3b+4d \end{bmatrix}$$

Next we calculate BA. By definition,

$$BA = \begin{bmatrix} \mathbf{r}_1 B \\ \mathbf{r}_2 B \end{bmatrix}\begin{bmatrix} \mathbf{c}_1 A & \mathbf{c}_2 A \end{bmatrix} == \begin{bmatrix} \mathbf{r}_1 B \mathbf{c}_1 A & \mathbf{r}_1 B \mathbf{c}_2 A \\ \mathbf{r}_2 B \mathbf{c}_1 A & \mathbf{r}_2 B \mathbf{c}_2 A \end{bmatrix}$$

Therefore

$$BA = \begin{bmatrix} \begin{bmatrix} a & b \end{bmatrix}\begin{bmatrix} 1 \\ 3 \end{bmatrix} & \begin{bmatrix} a & b \end{bmatrix}\begin{bmatrix} 2 \\ 4 \end{bmatrix} \\ \begin{bmatrix} c & d \end{bmatrix}\begin{bmatrix} 1 \\ 3 \end{bmatrix} & \begin{bmatrix} c & d \end{bmatrix}\begin{bmatrix} 2 \\ 4 \end{bmatrix} \end{bmatrix} = \begin{bmatrix} a+3b & 2a+4b \\ c+3d & 2c+4d \end{bmatrix}$$

The matrices AB and BA are equal if and only if

$$\begin{cases} a+2c &= a+3b \\ b+2d &= 2a+4b \\ 3a+4c &= c+3d \\ 3b+4d &= 2c+4d \end{cases}$$

It is easy to assign values to a, b, c, d for which $AB \neq BA$. ▲

In the next two examples, we illustrate briefly how matrices are multiplied in *Maple.*

EXAMPLE 2.8 A *Maple* Product of Arrays

Use *Maple* to calculate the product of two arrays

$$A = \begin{bmatrix} a & b \\ c & d \end{bmatrix} \quad \text{and} \quad B = \begin{bmatrix} u & v \\ x & y \end{bmatrix}$$

Solution. We first define the arrays A and B and then apply the function **evalm** to $(A\&*B)$.

```
[>A:=array([[a,b],[c,d]]):
[>B:=array([[u,v],[x,y]]):
```

```
[> AB:=evalm(A&*B);
```

$$AB := \begin{bmatrix} au + bx & av + by \\ cu + dx & cv + dy \end{bmatrix}$$

As we can see, the entries of AB are the dot products of the rows of A with the columns of B. ▲

EXAMPLE 2.9 A Matrix Product in *Maple 6*

Assign the type *Matrix* to the arrays in Example 2.8 and use the *Maple* operation $A.B$ to calculate the matrix AB.

Solution. First we define the A and B as arrays.

```
[> restart;
[> A:=array([[a,b],[c,d]]):
[> B:=array([[u,v],[x,y]]):
```

Next we assign the type *Matrix* to A and B and calculate their product.

```
[> Matrix(A).Matrix(B);
```

$$\begin{bmatrix} au + bx & av + by \\ cu + dx & cv + dy \end{bmatrix}$$

As we can see, the output is the same as that in Example 2.8. Instead of defining arrays and then assigning them the type *Matrix*, we can of course define the matrices A and B from scratch and obtain the same result.

```
[> A:=Matrix([[a,b],[c,d]]):
[> B:=Matrix([[u,v],[x,y]]):
[> A.B;
```

$$\begin{bmatrix} au + bx & av + by \\ cu + dx & cv + dy \end{bmatrix}$$

Note that we did not have to load the **LinearAlgebra** package to evaluate the matrix products. On the other hand, if we had assigned the type *matrix* to A and B, the calculation would have failed. In combination with the **evalm** function, it would have produced an unexpected result. ▲

EXAMPLE 2.10 Dot Products and Matrix Products

Use the **DotProduct** function in the **LinearAlgebra** package to calculate the product of two general 2×2 matrices.

Solution. With first load the **LinearAlgebra** package.

```
[> with(LinearAlgebra):
```

Next we define the general matrices A and B.

```
[> A:=Matrix(2,2,a);
```

$$A := \begin{bmatrix} a(1,1) & a(1,2) \\ a(2,1) & a(2,2) \end{bmatrix}$$

```
[> B:=Matrix(2,2,b);
```

$$B := \begin{bmatrix} b(1,1) & b(1,2) \\ b(2,1) & b(2,2) \end{bmatrix}$$

We now form the matrix whose entries are the dot products of the rows of A and the columns of B.

```
[> rA:=i->Row(A,i):  cB:=i->Column(B,i):
```

```
[> AdotB:=Matrix([[rA(1).cB(1),rA(1).cB(2)],
    [rA(2).cB(1),rA(2).cB(2)]]);
```

$$AdotB := \begin{bmatrix} a(1,1)b(1,1) + a(1,2)b(2,1) & a(1,1)b(1,2) + a(1,2)b(2,2) \\ a(2,1)b(1,1) + a(2,2)b(2,1) & a(2,1)b(1,2) + a(2,2)b(2,2) \end{bmatrix}$$

We conclude by asking *Maple* to compare the matrix $AdotB$ with the matrix product AB.

```
[> Equal(AdotB,A.B);
```

true

As expected, the two matrices are equal. ▲

EXAMPLE 2.11 Two Matrices for Which $AB = \mathbf{0}$ and $BA \neq \mathbf{0}$

Show that if

$$A = \begin{bmatrix} 0 & 1 \\ 0 & 2 \end{bmatrix} \quad \text{and} \quad B = \begin{bmatrix} 3 & 4 \\ 0 & 0 \end{bmatrix}$$

then $AB = \mathbf{0}$ and $BA \neq \mathbf{0}$.

Solution. By the definition of matrix products,

$$AB = \begin{bmatrix} 0 & 1 \\ 0 & 2 \end{bmatrix} \begin{bmatrix} 3 & 4 \\ 0 & 0 \end{bmatrix} = \begin{bmatrix} 0 & 0 \\ 0 & 0 \end{bmatrix}$$

and

$$BA = \begin{bmatrix} 3 & 4 \\ 0 & 0 \end{bmatrix} \begin{bmatrix} 0 & 1 \\ 0 & 2 \end{bmatrix} = \begin{bmatrix} 0 & 11 \\ 0 & 0 \end{bmatrix}$$

As we can see, $AB = \mathbf{0}$, but $BA \neq \mathbf{0}$. ▲

EXAMPLE 2.12 The Product of a 2 x 4 and a 4 x 3 Matrix

Use *Maple* to calculate the product AB of the matrices

$$A = \begin{bmatrix} 1 & 2 & 0 & 3 \\ 5 & 3 & 0 & 1 \end{bmatrix} \quad \text{and} \quad B = \begin{bmatrix} 1 & 2 & 1 \\ 2 & 6 & 1 \\ 9 & 10 & 1 \\ 3 & 8 & 1 \end{bmatrix}$$

and attempt to calculate the product BA.

Solution. We use the **Matrix** palette to input A and B.

First we create the template for a 2×4 matrix.

```
[>A:=Matrix([[%?,%?,%?,%?],[%?,%?,%?,%?]]);
```

Then we fill in the blanks.

```
[>A:=Matrix([[1,2,0,3],[5,3,0,1]]);
```

$$\begin{bmatrix} 1 & 2 & 0 & 3 \\ 5 & 3 & 0 & 1 \end{bmatrix}$$

Next we create a template for a 4×3 matrix.

```
[> B:=Matrix([[%?,%?,%?],[%?,%?,%?],[%?,%?,%?],[%?,%?,%?]]);
```

Then we again fill in the blanks.

```
[> B:=Matrix([[1,2,1],[2,6,1],[9,10,1],[3,8,1]]);
```

$$\begin{bmatrix} 1 & 2 & 1 \\ 2 & 6 & 1 \\ 9 & 10 & 1 \\ 3 & 8 & 1 \end{bmatrix}$$

We then calculate the product matrices AB.

```
[> A.B;
```

$$\begin{bmatrix} 14 & 38 & 6 \\ 14 & 36 & 9 \end{bmatrix}$$

Next we ask *Maple* to calculate the product $B.A$.

```
[> B.A;
Error, (in LinearAlgebra:-MatrixMatrixMultiply) first
matrix column dimension (3) <> second matrix row
dimension (2)
```

As expected, *Maple* returns an error message, since the product BA is not defined. ▲

It is often helpful to express a product AB in terms of A and the columns of B.

THEOREM 2.1 (Matrix-column product theorem) *If A is an $m \times n$ matrix and B an $n \times p$ matrix with columns $\mathbf{c}_1 B, \ldots, \mathbf{c}_p B$, then AB is the $m \times p$ matrix with columns $A\mathbf{c}_1 B, \ldots, A\mathbf{c}_p B$..*

Proof. This result follows from the fact that

$$A\,\mathbf{c}_i B = \begin{bmatrix} \mathbf{r}_1 A \\ \vdots \\ \mathbf{r}_m A \end{bmatrix} (\mathbf{c}_i B) = \begin{bmatrix} \mathbf{r}_1 A\ \mathbf{c}_i B \\ \vdots \\ \mathbf{r}_m A\ \mathbf{c}_i B \end{bmatrix}$$

for all $1 \leq i \leq p$. ■

We note that strictly speaking, the columns of the matrix AB are not the column vectors $A\mathbf{c}_1 B, \ldots, A\mathbf{c}_p B$. The matrix AB is obtained from the column vectors $A\mathbf{c}_1 B, \ldots, A\mathbf{c}_p B$ by rewriting them as the columns of a new matrix. The **concat** function in the **linalg** package of *Maple* serves precisely this purpose. Here is a simple example illustrating the concatenation of three 2×1 column vectors into a 2×3 matrix. Let

```
[> x:=<x1,x2>;
```

$$x := \begin{bmatrix} x1 \\ x2 \end{bmatrix}$$

```
[> y:=<y1,y2>;
```

$$y := \begin{bmatrix} y1 \\ y2 \end{bmatrix}$$

```
[> z:=<z1,z2>;
```

$$z := \begin{bmatrix} z1 \\ z2 \end{bmatrix}$$

be three 2×1 column vectors. The concatenation

```
[> A:=linalg[concat](x,y,z);
```

$$A := \begin{bmatrix} x1 & y1 & z1 \\ x2 & y2 & z2 \end{bmatrix}$$

produces a 2×3 matrix whose columns correspond to the given vectors. However, if we form the list [x,y,z], we get a different object.

```
[> [x,y,z];
```

$$\left[\begin{bmatrix} x1 \\ x2 \end{bmatrix}, \begin{bmatrix} y1 \\ y2 \end{bmatrix}, \begin{bmatrix} z1 \\ z2 \end{bmatrix} \right]$$

We will nevertheless continue our harmless abuse of language and speak of the list $[\mathbf{x}_1, \ldots, \mathbf{x}_n]$ of $m \times 1$ vectors or matrices as *the matrix whose columns are* $\mathbf{x}_1, \ldots, \mathbf{x}_n$.

COROLLARY 2.2 *If A is an $m \times n$ matrix and $\mathbf{b} = (b_1, \ldots, b_n)$ is an $n \times 1$ column vector, then $A\mathbf{b} = b_1\mathbf{c}_1 A + \cdots + b_n\mathbf{c}_n A$.*

Proof. Let $A = \left[a_{ij}\right]$. Then it follows from Theorem 2.1 that $A\mathbf{b}$ is the vector

$$\begin{bmatrix} \mathbf{r}_1 A \\ \vdots \\ \mathbf{r}_m A \end{bmatrix} \begin{bmatrix} b_1 \\ \vdots \\ b_n \end{bmatrix} = \begin{bmatrix} a_{11}b_1 + \cdots + a_{1n}b_n \\ \vdots \\ a_{m1}b_1 + \cdots + a_{mn}b_n \end{bmatrix}$$

and the definition of vector addition implies that $b_1\mathbf{c}_1 A + \cdots + b_n\mathbf{c}_n A$ is the vector

$$\begin{bmatrix} a_{11}b_1 \\ \vdots \\ a_{m1}b_1 \end{bmatrix} + \cdots + \begin{bmatrix} a_{1n}b_n \\ \vdots \\ a_{mn}b_n \end{bmatrix} = \begin{bmatrix} a_{11}b_1 + \cdots + a_{1n}b_n \\ \vdots \\ a_{m1}b_1 + \cdots + a_{mn}b_n \end{bmatrix}$$

Hence $A\mathbf{b} = b_1\mathbf{c}_1 A + \cdots + b_n\mathbf{c}_n A$. ■

COROLLARY 2.3 *If $\mathbf{a} = (a_1, \ldots, a_n)$ is a $1 \times n$ row vector and $B = \left[b_{ij}\right]$ is an $n \times p$ matrix, then $\mathbf{a}B = a_1\mathbf{r}_1 B + \cdots + a_n\mathbf{r}_n B$.*

Proof. It follows immediately from the definition of matrix products that

$$\mathbf{a}B = \begin{bmatrix} a_1 & \cdots & a_n \end{bmatrix} \begin{bmatrix} \mathbf{r}_1 B \\ \vdots \\ \mathbf{r}_n B \end{bmatrix} = a_1\mathbf{r}_1 B + \cdots + a_n\mathbf{r}_n B$$

This proves the corollary. ■

COROLLARY 2.4 *If A is an $m \times n$ matrix and $a\mathbf{x} + b\mathbf{y}$ an $n \times 1$ vector, then $A(a\mathbf{x} + b\mathbf{y}) = aA\mathbf{x} + bA\mathbf{y}$.*

Proof. Suppose that $A = [\mathbf{c}_1 A \ \cdots \ \mathbf{c}_n A]$ is an $m \times n$ matrix with columns $\mathbf{c}_i A$ and that

$$a\mathbf{x} + b\mathbf{y} = a \begin{bmatrix} x_1 \\ \vdots \\ x_n \end{bmatrix} + b \begin{bmatrix} y_1 \\ \vdots \\ y_n \end{bmatrix} = \begin{bmatrix} ax_1 + by_1 \\ \vdots \\ ax_n + by_n \end{bmatrix}$$

Then Theorem 2.1 and Corollary 2.2 imply that

$$A\,(a\mathbf{x} + b\mathbf{y}) = \begin{bmatrix} \mathbf{c}_1 A & \cdots & \mathbf{c}_n A \end{bmatrix} \begin{bmatrix} ax_1 + by_1 \\ \vdots \\ ax_n + by_n \end{bmatrix}$$

$$
\begin{aligned}
&= (ax_1 + by_1)\,\mathbf{c}_1 A + \cdots + (ax_n + by_n)\,\mathbf{c}_n A \\
&= (ax_1\,\mathbf{c}_1 A + \cdots + ax_n\,\mathbf{c}_n A) + (by_1\,\mathbf{c}_1 A + \cdots + by_n\,\mathbf{c}_n A) \\
&= a\,(x_1\,\mathbf{c}_1 A + \cdots + x_n\,\mathbf{c}_n A) + b\,(y_1\,\mathbf{c}_1 A + \cdots + y_n\,\mathbf{c}_n A) \\
&= aA\mathbf{x} + bA\mathbf{y}
\end{aligned}
$$

This proves the corollary. ■

The following theorems describe some of the basic properties of matrix products.

THEOREM 2.5 (Composition theorem) *For all matrices A and B and column vectors $\mathbf{x}$ of compatible dimension, $A(B\mathbf{x}) = (AB)\mathbf{x}$.*

Proof. Suppose that A is an $m \times n$ matrix, that $B = \begin{bmatrix}\mathbf{c}_1 B & \cdots & \mathbf{c}_p B\end{bmatrix}$ is an $n \times p$ matrix with columns $\mathbf{c}_i B$, and that $\mathbf{x} = (x_1, \ldots, x_p)$ is a $p \times 1$ column vector. Then it follows from Theorem 2.1 and Corollary 2.2 that $B\mathbf{x} = x_1\mathbf{c}_1 B + \cdots + x_p\mathbf{c}_p B$ and

$$
\begin{aligned}
A\,(B\mathbf{x}) &= A\left(x_1\mathbf{c}_1 B + \cdots + x_p\mathbf{c}_p B\right) = x_1 A\mathbf{c}_1 B + \cdots + x_p A\mathbf{c}_p B \\
&= \begin{bmatrix} A\mathbf{c}_1 B & \cdots & A\mathbf{c}_p B \end{bmatrix} \begin{bmatrix} x_1 \\ \vdots \\ x_p \end{bmatrix} = (AB)\,\mathbf{x}
\end{aligned}
$$

where

$$
\begin{bmatrix} A\mathbf{c}_1 B & \cdots & A\mathbf{c}_p B \end{bmatrix}
$$

is the matrix whose columns are the vectors $A\,\mathbf{c}_i B$. This proves the theorem. ■

THEOREM 2.6 *For all matrices A, B, C of compatible dimension and all scalars s,*

1. $(AB)\,C = A\,(BC)$
2. $A\,(B + C) = AB + AC$
3. $(A + B)\,C = AC + BC$
4. $s\,(AB) = (sA)\,B = A\,(sB)$
5. $0A = \mathbf{0}$
6. $A\mathbf{0} = \mathbf{0}$

Proof. We show that $(AB)C = A(BC)$. Suppose that $C = \begin{bmatrix}\mathbf{c}_1 C & \cdots & \mathbf{c}_p C\end{bmatrix}$. Then it follows from Theorem 2.1 and Theorem 2.5 that

$$
\begin{aligned}
A\,(BC) &= A\left(B\begin{bmatrix}\mathbf{c}_1 C & \cdots & \mathbf{c}_p C\end{bmatrix}\right) \\
&= A\begin{bmatrix} B\,(\mathbf{c}_1 C) & \cdots & B\left(\mathbf{c}_p C\right)\end{bmatrix} \\
&= \begin{bmatrix} A\,(B\,(\mathbf{c}_1 C)) & \cdots & A\left(B\left(\mathbf{c}_p C\right)\right)\end{bmatrix}
\end{aligned}
$$

$$= \left[(AB)(\mathbf{c}_1 C) \ \cdots \ (AB)(\mathbf{c}_p C)\right]$$
$$= (AB)\,C$$

We leave the proofs of statements 2 through 4 as exercises. ■

Although the intended meaning of the equation $A\mathbf{0} = \mathbf{0}$ in Theorem 2.6 is clear, the notation is ambiguous. The two zero matrices in the equation may in fact be of different dimension. Here is an example:

$$\begin{bmatrix} 1 & 0 & 7 & 3 \\ 6 & 8 & 5 & 8 \\ 1 & 9 & 5 & 3 \end{bmatrix} \begin{bmatrix} 0 & 0 \\ 0 & 0 \\ 0 & 0 \\ 0 & 0 \end{bmatrix} = \begin{bmatrix} 0 & 0 \\ 0 & 0 \\ 0 & 0 \end{bmatrix}$$

We use Theorem 2.6 to show that the set of solutions of a homogeneous linear system has the following basic closure property.

THEOREM 2.7 (Homogeneous systems theorem) *If* $\mathbf{y}$ *and* $\mathbf{z}$ *are solutions of a homogeneous system* $A\mathbf{x} = \mathbf{0}$ *and* a *and* b *are scalars, then* $a\mathbf{y} + b\mathbf{z}$ *is also a solution of* $A\mathbf{x} = \mathbf{0}$.

Proof. Suppose that $A\mathbf{y} = A\mathbf{z} = \mathbf{0}$, and let a and b be two arbitrary scalars. Then it follows from Theorem 2.6 that

$$A\,(a\mathbf{y} + b\mathbf{z}) = A\,(a\mathbf{y}) + A\,(b\mathbf{z}) = a\,(A\mathbf{y}) + b\,(A\mathbf{z}) = a\mathbf{0} + b\mathbf{0} = \mathbf{0}$$

Hence $a\mathbf{y} + b\mathbf{z}$ is a solution of $A\mathbf{x} = \mathbf{0}$. ■

Matrix Transpose

With every matrix we can associate another matrix obtained by interchanging the rows and columns of the given matrix.

DEFINITION 2.5 *The* ***transpose*** *of an* $m \times n$ *matrix* $A = \left[a_{ij}\right]$ *is the* $n \times m$ *matrix* $A^T = \left[b_{ji}\right]$ *whose entries are linked to those of* A *by the equations* $b_{ji} = a_{ij}$ *of* A.

EXAMPLE 2.13 The Transpose of a Matrix

Use *Maple* to find the transpose of the matrix

$$A = \begin{bmatrix} 1 & 2 & 9 & 3 \\ 2 & 6 & 7 & 8 \end{bmatrix}$$

Solution. We use the **transpose** function in the **linalg** package.

```
[> B:=linalg[transpose](matrix([[1,2,9,3],[2,6,7,8]]));
```

$$B := \begin{bmatrix} 1 & 2 \\ 2 & 6 \\ 9 & 7 \\ 3 & 8 \end{bmatrix}$$

The result is the transpose A^T of the matrix A. ▲

Example 2.13 illustrates why it is useful to link the entries of A^T to those of A in terms of the location indices i and j. If

$$A = \begin{bmatrix} 1 & 2 & 9 & 3 \\ 2 & 6 & 7 & 8 \end{bmatrix} = \begin{bmatrix} a_{11} & a_{12} & a_{13} & a_{14} \\ a_{21} & a_{22} & a_{23} & a_{24} \end{bmatrix}$$

then

$$A^T = \begin{bmatrix} 1 & 2 \\ 2 & 6 \\ 9 & 7 \\ 3 & 8 \end{bmatrix} = \begin{bmatrix} b_{11} & b_{12} \\ b_{21} & b_{22} \\ b_{31} & b_{32} \\ b_{41} & b_{42} \end{bmatrix} = \begin{bmatrix} a_{11} & a_{21} \\ a_{12} & a_{22} \\ a_{13} & a_{23} \\ a_{14} & a_{24} \end{bmatrix}$$

The claim that $b_{ij} = a_{ji}$ for all $1 \leq i \leq 4$ and $1 \leq j \leq 2$ establishes the necessary connection between A and A^T.

Matrix Multiplication and the Transpose

We frequently need to calculate matrix products involving transposed matrices. The following lemma will be helpful.

LEMMA 2.8 *For any $m \times n$ matrix $A = [a_{ij}]$ and any $n \times p$ matrix $B = [b_{jk}]$, the row-column products $\mathbf{r}_i A\, \mathbf{c}_j B$ and $\mathbf{r}_j \left(B^T\right) \mathbf{c}_i (A^T)$ are equal.*

Proof. By the definition of row-column products,

$$\mathbf{r}_i A\, \mathbf{c}_j B = \begin{bmatrix} a_{i1} & \cdots & a_{in} \end{bmatrix} \begin{bmatrix} b_{1j} \\ \vdots \\ b_{nj} \end{bmatrix} = a_{i1}b_{1j} + \cdots + a_{in}b_{nj}$$

and

$$\mathbf{r}_j B^T \, \mathbf{c}_i A^T = \begin{bmatrix} b_{1j} & \cdots & b_{nj} \end{bmatrix} \begin{bmatrix} a_{i1} \\ \vdots \\ a_{in} \end{bmatrix} = b_{1j}a_{i1} + \cdots + b_{nj}a_{in}$$

Therefore $\mathbf{r}_i A \, \mathbf{c}_j B = \mathbf{r}_j B^T \, \mathbf{c}_i A^T$ since $a_{ik}b_{kj} = b_{kj}a_{ik}$ for all i, j, and k. ■

We illustrate Lemma 2.8 with an example.

EXAMPLE 2.14 $\mathbf{r}_i A \, \mathbf{c}_j B = \mathbf{r}_j \left(B^T\right) \mathbf{c}_i(A^T)$

Show that $\mathbf{r}_1 A \, \mathbf{c}_2 B = \mathbf{r}_2 \left(B^T\right) \mathbf{c}_1(A^T)$, where

$$A = \begin{bmatrix} a_{11} & a_{12} & a_{13} \\ a_{21} & a_{22} & a_{23} \end{bmatrix} \quad \text{and} \quad B = \begin{bmatrix} b_{11} & b_{12} \\ b_{21} & b_{22} \\ b_{31} & b_{32} \end{bmatrix}$$

Solution. By the definition of the transpose,

$$A^T = \begin{bmatrix} a_{11} & a_{21} \\ a_{12} & a_{22} \\ a_{13} & a_{23} \end{bmatrix} \quad \text{and} \quad B^T = \begin{bmatrix} b_{11} & b_{21} & b_{31} \\ b_{12} & b_{22} & b_{32} \end{bmatrix}$$

Therefore,

$$\mathbf{r}_1 A \, \mathbf{c}_2 B = \begin{bmatrix} a_{11} & a_{12} & a_{13} \end{bmatrix} \begin{bmatrix} b_{12} \\ b_{22} \\ b_{32} \end{bmatrix} = b_{12}a_{11} + b_{22}a_{12} + b_{32}a_{13}$$

and

$$\mathbf{r}_2 B^T \, \mathbf{c}_1 A^T = \begin{bmatrix} b_{12} & b_{22} & b_{32} \end{bmatrix} \begin{bmatrix} a_{11} \\ a_{12} \\ a_{13} \end{bmatrix} = b_{12}a_{11} + b_{22}a_{12} + b_{32}a_{13}$$

This shows that $\mathbf{r}_1 A \, \mathbf{c}_2 B = \mathbf{r}_2 \left(B^T\right) \mathbf{c}_1(A^T)$. ▲

The following theorem summarizes the interactions between sums, scalar multiples, products, and transposes.

THEOREM 2.9 (Transpose theorem) *For any two matrices A and B of compatible dimension,*

1. $(A^T)^T = A$
2. $(sA)^T = sA^T$
3. $(A + B)^T = A^T + B^T$
4. $(AB)^T = B^T A^T$

Proof. We prove the theorem in four steps.

1. Suppose $A = [a_{ij}]$ is an $m \times n$ matrix. Then A^T is the $n \times m$ matrix $B = [b_{ij}]$ whose entries are related to those of A by the equations $b_{ij} = a_{ji}$. Moreover, A^{TT} is the $m \times n$ matrix $C = [c_{ij}]$ whose entries satisfy the equations $c_{ij} = b_{ji} = a_{ij}$. Hence $c_{ij} = a_{ij}$ for all i and j. Therefore, $A^{TT} = A$.

2. For any scalar s, $(sA)^T = (sa_{ij})^T = [sb_{ji}] = s[b_{ji}] = s(A^T)$. Hence $(sA)^T = s(A^T)$.

3. Now suppose that B is another $m \times n$ matrix $[b_{ij}]$ and that $A^T = \left[a'_{ji}\right]$ and $B^T = \left[b'_{ji}\right]$. Then

$$(A+B)^T = [a_{ij} + b_{ij}]^T = \left[a'_{ji} + b'_{ji}\right] = \left[a'_{ji}\right] + \left[b'_{ji}\right] = A^T + B^T$$

4. It remains to prove that $(AB)^T = B^T A^T$. We simplify the notation by assuming that A has two rows and B has three columns.

$$A = \begin{bmatrix} \mathbf{r}_1 A \\ \mathbf{r}_2 A \end{bmatrix} \quad \text{and} \quad B = \begin{bmatrix} \mathbf{c}_1 B & \mathbf{c}_2 B & \mathbf{c}_3 B \end{bmatrix}$$

By the definition of matrix products and transposes,

$$AB = \begin{bmatrix} \mathbf{r}_1 A\, \mathbf{c}_1 B & \mathbf{r}_1 A\, \mathbf{c}_2 B & \mathbf{r}_1 A\, \mathbf{c}_3 B \\ \mathbf{r}_2 A\, \mathbf{c}_1 B & \mathbf{r}_2 A\, \mathbf{c}_2 B & \mathbf{r}_2 A\, \mathbf{c}_3 B \end{bmatrix}$$

and

$$(AB)^T = \begin{bmatrix} \mathbf{r}_1 A\, \mathbf{c}_1 B & \mathbf{r}_2 A\, \mathbf{c}_1 B \\ \mathbf{r}_1 A\, \mathbf{c}_2 B & \mathbf{r}_2 A\, \mathbf{c}_2 B \\ \mathbf{r}_1 A\, \mathbf{c}_3 B & \mathbf{r}_2 A\, \mathbf{c}_3 B \end{bmatrix}$$

It therefore follows from Lemma 2.8 that

$$\begin{aligned}(AB)^T &= \begin{bmatrix} \mathbf{r}_1 (B^T)\, \mathbf{c}_1(A^T) & \mathbf{r}_1 (B^T)\, \mathbf{c}_2(A^T) \\ \mathbf{r}_2 (B^T)\, \mathbf{c}_1(A^T) & \mathbf{r}_2 (B^T)\, \mathbf{c}_2(A^T) \\ \mathbf{r}_3 (B^T)\, \mathbf{c}_1(A^T) & \mathbf{r}_3 (B^T)\, \mathbf{c}_2(A^T) \end{bmatrix} \\ &= \begin{bmatrix} \mathbf{r}_1 (B^T) \\ \mathbf{r}_2 (B^T) \\ \mathbf{r}_3 (B^T) \end{bmatrix} \begin{bmatrix} \mathbf{c}_1(A^T) & \mathbf{c}_2(A^T) \end{bmatrix} \\ &= B^T A^T \end{aligned}$$

This proves the theorem. ■

EXAMPLE 2.15 Matrix Operations and the Transpose

Use *Maple* to generate two random square matrices, and employ them to illustrate the previous theorem.

Solution. Since the theorem holds for square matrices of any dimension, it is instructive to illustrate the capacity of *Maple* by calculating with matrices too large to print on a single page.

```
[>with(linalg):
[>A:=randmatrix(100,100):
[>B:=randmatrix(100,100):
```

```
[>equal(transpose(transpose(A)),A);
                              true
```

```
[>equal(s*transpose(A),transpose(s*A));
                              true
```

```
[>equal(transpose(A+B),transpose(A)+transpose(B));
                              true
```

```
[>equal(transpose(evalm(A&*B)),
   evalm(transpose(B)&*transpose(A)));
                              true
```

As we can see, all identities hold. ▲

The Trace

The sum of the diagonal entries of a square matrix often gives us useful information about the properties the matrix. For any $n \times n$ matrix $A = [a_{ij}]$ with entries in a mathematical system $\mathbb{X}$ equipped with addition, the function trace $A = a_{11} + \cdots + a_{nn}$ produces an element in the system $\mathbb{X}$.

DEFINITION 2.6 *The **trace** of an $n \times n$ matrix $A = [a_{ij}]$ is the sum $a_{11} + \cdots + a_{nn}$ of the diagonal entries of A.*

EXAMPLE 2.16 The Trace of a Matrix

Suppose that

$$A = \begin{bmatrix} a_{11} & a_{12} & a_{13} \\ a_{21} & a_{22} & a_{23} \\ a_{31} & a_{32} & a_{33} \end{bmatrix} = \begin{bmatrix} 1 & 2 & 3 \\ 5 & 6 & 7 \\ 9 & 10 & 11 \end{bmatrix}$$

Then trace $A = a_{11} + a_{22} + a_{33} = 1 + 6 + 11 = 18$. ▲

Next we describe the interaction between trace and basic matrix operations.

THEOREM 2.10 (Trace theorem) *If A and B are two $n \times n$ matrices, then*

1. $\text{trace } A^T = \text{trace } A$
2. $\text{trace } sA = s \text{ trace } A$
3. $\text{trace}(A + B) = \text{trace } A + \text{trace } B$
4. $\text{trace } AB = \text{trace } BA$

Proof. Suppose that $A = [a_{ij}]$ and $B = [b_{ij}]$ are two $n \times n$ matrices. Since A and A^T have identical diagonals, it is clear that they have the same trace. Moreover, the trace of sA is s times the trace of A since

$$\begin{aligned} \text{trace } sA &= sa_{11} + \cdots + sa_{nn} \\ &= s\,(a_{11} + \cdots + a_{nn}) \\ &= s \text{ trace } A \end{aligned}$$

The fact that $\text{trace}(A + B) = \text{trace } A + \text{trace } B$ is also immediate since

$$\begin{aligned} \text{trace } A + \text{trace } B &= (a_{11} + \cdots + a_{nn}) + (b_{11} + \cdots + b_{nn}) \\ &= (a_{11} + b_{11}) + \cdots + (a_{nn} + b_{nn}) \\ &= \text{trace } (A + B) \end{aligned}$$

It remains for us to consider the case of matrix products. By definition,

$$\text{trace } AB = r_1 A \times c_1 B + \cdots + r_n A \times c_n B$$

Since trace AB is a scalar, we can consider it to be a 1×1 matrix and conclude that

$$\text{trace } AB = (\text{trace } AB)^T$$

Hence

$$\begin{aligned} \text{trace } AB &= \mathbf{r}_1 A\, \mathbf{c}_1 B + \cdots + \mathbf{r}_n A\, \mathbf{c}_n B \\ &= \sum_{i \in \mathbf{n}} \sum_{k \in \mathbf{n}} a_{ik} b_{ki} \end{aligned}$$

$$= \sum_{k \in \mathbf{n}} \sum_{i \in \mathbf{n}} b_{ki} a_{ik}$$
$$= \mathbf{r}_1 B\, \mathbf{c}_1 A + \cdots + \mathbf{r}_n B\, \mathbf{c}_n A$$
$$= \text{trace } BA$$

This proves the theorem. ■

EXAMPLE 2.17 Calculating the Trace of a Matrix

Use *Maple* to calculate the trace of a general 9×9 matrix $A = [a_{ij}]$.

Solution. We use the **linalg** package to generate A and apply the **trace** function.

```
[> A:=matrix(9,9);
```

$$A := array(1..9, 1..9, [])$$

```
[> linalg[trace](A);
```

$$A_{1,1} + A_{2,2} + A_{3,3} + A_{4,4} + A_{5,5} + A_{6,6} + A_{7,7} + A_{8,8} + A_{9,9}$$

As expected, *Maple* outputs the trace of A. ▲

EXERCISES 2.2

1. Calculate the sums $A + B$ and $B + A$ of the matrices

$$A = \begin{bmatrix} 1 & 2 & 3 & 4 & 5 \\ 2 & -1 & 0 & 5 & 6 \\ -6 & 4 & 3 & 1 & 2 \end{bmatrix}, \quad B = \begin{bmatrix} 2 & 5 & 6 & -2 & 0 \\ 2 & 2 & 7 & 4 & 3 \\ 2 & 1 & 8 & -8 & 6 \end{bmatrix}$$

and explain why $A + B = B + A$.

2. Calculate the matrices $3A - 5B$ and $-(5B - 3A)$ determined by the matrices A and B in Exercise 1 and explain why the two matrices are equal.

3. Write out the rows $\mathbf{r}_i A$ and the columns $\mathbf{c}_j B$ of the matrices

$$A = \begin{bmatrix} 1 & 2 & 3 & 4 & 5 \\ 2 & 1 & 0 & 5 & 6 \\ 6 & 4 & 3 & 1 & 2 \end{bmatrix}, \quad B = \begin{bmatrix} 2 & 5 & 6 \\ 2 & 2 & 7 \\ 2 & 1 & 8 \\ 3 & 8 & 2 \\ 3 & 9 & 4 \end{bmatrix}$$

Use the row-column products $\mathbf{r}_i \mathbf{c}_j = \mathbf{r}_i A\, \mathbf{c}_j B$ to calculate AB in the form of

$$\begin{bmatrix} \mathbf{r}_1\,\mathbf{c}_1 & \mathbf{r}_1\,\mathbf{c}_2 & \mathbf{r}_1\,\mathbf{c}_3 \\ \mathbf{r}_2\,\mathbf{c}_1 & \mathbf{r}_2\,\mathbf{c}_2 & \mathbf{r}_2\,\mathbf{c}_3 \\ \mathbf{r}_3\,\mathbf{c}_1 & \mathbf{r}_3\,\mathbf{c}_2 & \mathbf{r}_3\,\mathbf{c}_3 \end{bmatrix}$$

4. Try to form the matrix product AB using the matrices in Example 1. Explain the result.

5. Calculate the products AB and BA of the matrices A and B in Exercise 3. Explain why $AB \neq BA$.

6. Find two nonzero 3×3 matrices A and B for which $AB = BA$ and two other nonzero matrices A and B for which $AB \neq BA$.

7. Suppose that

$$A = \begin{bmatrix} 1 & 2 & 3 \\ 2 & 1 & 0 \\ 6 & 4 & 3 \end{bmatrix}, \quad B = \begin{bmatrix} 3 & 4 & 5 \\ 0 & 5 & 6 \\ 3 & 1 & 2 \end{bmatrix}, \quad \mathbf{x} = \begin{bmatrix} x \\ y \\ y \end{bmatrix}$$

Calculate $B\mathbf{x}$, $A(B\mathbf{x})$, and $(AB)\mathbf{x}$. Explain why $A(B\mathbf{x}) = (AB)\mathbf{x}$.

8. Use the matrices A and B in Exercise 3 and calculate $(AB)^T$, A^T, and B^T. Explain why $B^T A^T = (AB)^T$.

9. Suppose that

$$A = \begin{bmatrix} 1 & 2 & 3 \\ 2 & 1 & 0 \\ 6 & 4 & 3 \end{bmatrix}, \quad \mathbf{x} = \begin{bmatrix} x \\ y \\ z \end{bmatrix}$$

Show that

$$A\mathbf{x} = x \begin{bmatrix} 1 \\ 2 \\ 6 \end{bmatrix} + y \begin{bmatrix} 2 \\ 1 \\ 4 \end{bmatrix} + z \begin{bmatrix} 3 \\ 0 \\ 3 \end{bmatrix}$$

10. Show that if $\mathbf{0}$ is the $m \times n$ zero matrix, then $A + \mathbf{0} = A$ for all $m \times n$ matrices A.

11. Show that for all matrices A and B of compatible dimension, $A + B = B + A$.

12. Show that for all matrices A and B of compatible dimension and all scalars s, $s\,(A + B) = sA + sB$.

13. Show that for all matrices A and all scalars r and s, $(r + s)\,A = rA + sA$.

14. Show that for all matrices A and all scalars r and s, $(rs)\,A = r\,(sA)$.

15. Show that for all matrices A, B, C of compatible dimension, $(A + B) + C = A + (B + C)$.

16. Show that if A, B, C are matrices of compatible dimension, then $A(B + C) = AB + AC$.

17. Show that if A, B, C are matrices of compatible dimension, then $(A + B)C = AC + BC$.

18. Show that for all matrices A and B of compatible dimension and all scalars s, $s(AB) = (sA)B = A(sB)$.

19. Show that for any $m \times n$ matrix A, the matrix $0A$ is the $m \times n$ zero matrix.

20. Show that for any $m \times n$ matrix A and any $n \times p$ zero matrix $\mathbf{0}$, the matrix $A\mathbf{0}$ is the $m \times p$ zero matrix.

21. Show that, in general, trace $(AB) \neq$ (trace A) (trace B).

22. Let $A = (a_{ij})$ be a real $m \times p$ matrix and B be a real $n \times q$ matrix. Then the $(mn) \times (pq)$ matrix

$$A \otimes B = \begin{bmatrix} a_{11}B & a_{12}B & \cdots & a_{1p}B \\ a_{21}B & a_{22}B & \cdots & a_{2p}B \\ \vdots & \vdots & \ddots & \vdots \\ a_{m1}B & a_{m2}B & \cdots & a_{mp}B \end{bmatrix}$$

built by concatenation from the matrix blocks $a_{ij}B$, is called the *Kronecker product* of A and B. Calculate the Kronecker product of the matrices

$$A = \begin{bmatrix} -91 & -47 \\ 41 & -58 \end{bmatrix} \quad B = \begin{bmatrix} 83 & -86 & 23 \\ -84 & 19 & -50 \end{bmatrix}$$

23. (*Maple V*) Try to use the *Maple* command **`evalm(A&*B)`** to find the matrix product AB of a 3×5 matrix A and a 4×3 matrix B. Explain the resulting message "Error, (in linalg[multiply]) nonmatching dimensions for vector/matrix product."

24. (*Maple V*) Suppose that

$$A = \begin{bmatrix} 1 & 2 & 3 \\ 2 & 1 & 0 \end{bmatrix} \quad \mathbf{x} = \begin{bmatrix} 3 \\ 0 \\ 3 \end{bmatrix} \quad \mathbf{y} = \begin{bmatrix} 5 \\ 6 \\ 2 \end{bmatrix}$$

Show that $A(a\mathbf{x} + b\mathbf{y}) = aA\mathbf{x} + bA\mathbf{y}$ for all a, b.

25. (*Maple 6*) Use the trace function **LinearAlgebra[Trace]** to calculate trace A, trace B, trace$(A + B)$, and trace AB, where

$$A = \begin{bmatrix} -91 & -47 & -61 \\ 41 & -58 & -90 \\ 53 & -1 & 94 \end{bmatrix}, \quad B = \begin{bmatrix} 83 & -86 & 23 \\ -84 & 19 & -50 \\ 88 & -53 & 85 \end{bmatrix}$$

26. (*Maple 6*) Generate a random 3×5 matrix A. Copy the output matrix to a new input locations and form the matrices $A^T A$ and AA^T. What is special about these matrices? Repeat your experiment for several other dimensions.

27. (*Maple 6*) Restart *Maple,* load the **LinearAlgebra** package, and execute the following sequence of commands:

    a. `A:=matrix([[a,b],[c,d]])`

    b. `B:=matrix([[u,v],[x,y]])`

    c. `evalm(A.B)`

    Explain the result and compare it a similar result produced in Exercise 22. The assignment of the type **matrix** instead of **Matrix**to A and B is deliberate.

28. (*Maple 6*) Use the **RandomMatrix** and **Trace** functions to calculate the trace of a random real square matrix of dimensions 12×12.

29. (*Maple 6*) Use the **LinearAlgebra[RandomMatrix]** function to generate a 3×5 matrix A and a 5×3 matrix B. Then use *Maple* to calculate the products AB and BA.

30. (*Maple V, 6*) Compare the **trace** function in the **linalg** package with the **Trace** function in the **LinearAlgebra** package.

A LEXICON OF MATRICES

In this section we define and illustrate the basic types of matrices encountered in this text.

Square Matrices

A matrix $A : \mathbf{m} \times \mathbf{n} \to \mathbb{X}$ is ***square*** if $\mathbf{m} = \mathbf{n}$. All other matrices will be referred to as ***rectangular*** matrices.

EXAMPLE 2.18 A Square Matrix with Entries in $\mathbb{R}$

The real matrix

$$A = \begin{bmatrix} 1 & 2 \\ 5 & \pi \end{bmatrix}$$

is the function $A : \mathbf{2} \times \mathbf{2} \to \mathbb{R}$ defined by $A(1,1) = 1$, $A(1,2) = 2$, $A(2,1) = 5$, and $A(2,2) = \pi$. ▲

EXAMPLE 2.19 A Square Matrix with Entries in $\mathbb{R}[t]$

The polynomial matrix

$$A = \begin{bmatrix} 1 & t^2 & t+3 \\ 5t^2 - t + 1 & -6 & 7t \\ 9t^3 & 10 & t - \pi \end{bmatrix}$$

is the function $A : \mathbf{3}\times\mathbf{3} \to \mathbb{R}[t]$ be the function defined by $A(1,1) = 1$, $A(1,2) = t^2$, $A(1,3) = t+3$, $A(2,1) = 5t^2 - t + 1$, $A(2,2) = -6$, $A(2,3) = 7t$, $A(3,1) = 9t^3$, $A(3,2) = 10$, and $A(3,3) = t - \pi$. ▲

Lower-Triangular Matrices

A square matrix $A : \mathbf{n} \times \mathbf{n} \to \mathbb{X}$ is ***lower triangular*** if $A(i,j) = 0$ for all $i < j$.

EXAMPLE 2.20 A Lower-Triangular Matrix with Entries in $\mathbb{R}$

The real lower-triangular matrix

$$A = \begin{bmatrix} 1 & 0 & 0 \\ 5 & 6 & 0 \\ 9 & 10 & 11 \end{bmatrix}$$

is the function $A : \mathbf{3}\times\mathbf{3} \to \mathbb{R}$ defined by $A(1,1) = 1$, $A(1,2) = 0$, $A(1,3) = 0$, $A(2,1) = 5$, $A(2,2) = 6$, $A(2,3) = 0$, $A(3,1) = 9$, $A(3,2) = 10$, and $A(3,3) = 11$. ▲

EXAMPLE 2.21 A Lower-Triangular Matrix with Entries in $\mathbb{R}[t]$

The polynomial lower-triangular matrix

$$A = \begin{bmatrix} t^2 & 0 \\ t & t^3 - 6 \end{bmatrix}$$

is the function $A : \mathbf{2} \times \mathbf{2} \to \mathbb{R}[t]$ defined by $A(1,1) = t^2$, $A(1,2) = 0$, $A(2,1) = t$, and $A(2,2) = t^3 - 6$. ▲

Upper-Triangular Matrices

A square matrix $A : \mathbf{n} \times \mathbf{n} \to \mathbb{X}$ is ***upper triangular*** if $A(i,j) = 0$ for all $i > j$.

EXAMPLE 2.22 An Upper-Triangular Matrix with Entries in $\mathbb{R}$

The real upper-triangular matrix

$$A = \begin{bmatrix} 1 & 5 \\ 0 & 6 \end{bmatrix}$$

is a function $A : \mathbf{2} \times \mathbf{2} \to \mathbb{R}$ defined by $A(1,1) = 1$, $A(1,2) = 5$, $A(2,1) = 0$, and $A(2,2) = 6$. ▲

EXAMPLE 2.23 An Upper-Triangular Matrix with Entries in $\mathbb{R}[t]$

The matrix

$$A = \begin{bmatrix} 3t^5 - 77 & 5t^3 + t^2 - 1 & 9t^{33} \\ 0 & -t^3 + 6t + 44 & 10 \\ 0 & 0 & t - 11 \end{bmatrix}$$

is the function $A : \mathbf{3} \times \mathbf{3} \to \mathbb{R}[t]$ defined by $A(1,1) = 3t^5 - 77$, $A(1,2) = 5t^3 + t^2 - 1$, $A(1,3) = 9t^{33}$, $A(2,1) = 0$, $A(2,2) = -t^3 + 6t + 44$, $A(2,3) = 10$, $A(3,1) = 0$, $A(3,2) = 0$, and $A(3,3) = t - 11$. ▲

Diagonal Matrices

The list $[a_{11}, a_{22}, \ldots, a_{nn}]$ of entries of an $n \times n$ matrix $A = \left[a_{ij}\right]$ is the ***diagonal*** of A. An entry on the diagonal is called a ***diagonal entry*** of A. An $n \times n$ matrix $\left[a_{ij}\right]$ is ***diagonal*** if $a_{ij} = 0$ for all $i \neq j$. We sometimes write a diagonal matrix $A = \left[a_{ij}\right]$ as $\operatorname{diag}(a_{11}, \ldots, a_{nn})$.

EXAMPLE 2.24 A Diagonal Matrix with Entries in $\mathbb{R}$

The matrix

$$\begin{bmatrix} a_{11} & a_{12} & a_{13} & a_{14} \\ a_{21} & a_{22} & a_{23} & a_{24} \\ a_{31} & a_{32} & a_{33} & a_{34} \\ a_{41} & a_{42} & a_{43} & a_{44} \end{bmatrix} = \begin{bmatrix} 6 & 0 & 0 & 0 \\ 0 & 8 & 0 & 0 \\ 0 & 0 & 3 & 0 \\ 0 & 0 & 0 & -12 \end{bmatrix}$$

is a diagonal matrix with diagonal $[a_{11}, a_{22}, a_{33}, a_{44}] = [6, 8, 3, -12]$. ▲

Symmetric Matrices

A matrix A is ***symmetric*** if $A = A^T$.

EXAMPLE 2.25 A Symmetric Matrix

The matrix

$$\begin{bmatrix} a_{11} & a_{12} & a_{13} \\ a_{21} & a_{22} & a_{23} \\ a_{31} & a_{32} & a_{33} \end{bmatrix} = \begin{bmatrix} 6 & 5 & -1 \\ 5 & 8 & \pi \\ -1 & \pi & 3 \end{bmatrix}$$

is a symmetric matrix with entries in $\mathbb{R}$. ▲

Identity Matrices

A diagonal matrix diag(1, ... , 1) whose diagonal entries are all 1 is called an ***identity matrix.*** We write I_n for the $n \times n$ identity matrix. If n is clear from the context, we use I in place of I_n.

EXAMPLE 2.26 The 4 x 4 Identity Matrix

Use *Maple* to generate the 4×4 identity matrix I_4.

Solution. The **linalg** package provides us with several options for defining the matrix I_4. Here are three of them. We use the general 4×4 matrix

```
[> A:=matrix(4,4,a);
```

$$A := \begin{bmatrix} a(1,1) & a(1,2) & a(1,3) & a(1,4) \\ a(2,1) & a(2,2) & a(2,3) & a(2,4) \\ a(3,1) & a(3,2) & a(3,3) & a(3,4) \\ a(4,1) & a(4,2) & a(4,3) & a(4,4) \end{bmatrix}$$

to show that multiplication of A by each of the defined matrices leaves the matrix A unchanged.

1. We can use the **diag** function.

```
[> with(linalg):
[> firstId4:=diag(1,1,1,1);
```

$$firstId4 := \begin{bmatrix} 1 & 0 & 0 & 0 \\ 0 & 1 & 0 & 0 \\ 0 & 0 & 1 & 0 \\ 0 & 0 & 0 & 1 \end{bmatrix}$$

```
[> equal(A,evalm(A&*firstId4));
```

true

Hence the matrix $firstId4$ is an identity matrix for the set of 4×4 matrices.

2. We can use the **matrix** function. The calculation

```
[> secondId4:=evalm(matrix(4,4,0)+1);
```

$$secondId4 := \begin{bmatrix} 1 & 0 & 0 & 0 \\ 0 & 1 & 0 & 0 \\ 0 & 0 & 1 & 0 \\ 0 & 0 & 0 & 1 \end{bmatrix}$$

```
[> equal(A,evalm(A&*secondId4));
```

true

shows that the matrix $secondId4$ is also a 4×4 identity matrix. *Maple* treats the input

```
(matrix(4,4,0)+1)
```

as an abbreviation for the input

```
(matrix(4,4,0)+diag(1,1,1,1))
```

as the following calculation shows:

```
[> equal(evalm(matrix(4,4,0)+diag(1,1,1,1)),
   evalm(matrix(4,4,0)+1));
```

true

3. We can use the **LinearAlgebra[IdentityMatrix]** function instead.

```
[> thirdId4:=LinearAlgebra[IdentityMatrix](4,4);
```

$$thirdId4 := \begin{bmatrix} 1 & 0 & 0 & 0 \\ 0 & 1 & 0 & 0 \\ 0 & 0 & 1 & 0 \\ 0 & 0 & 0 & 1 \end{bmatrix}$$

```
[> equal(Matrix(A),Matrix(A).thirdId4);
```

true

Of course, we could also have defined the required matrix by specifying its 16 entries, one by one. ▲

The key property of an $n \times n$ identity matrix I_n is that if A is an $m \times n$ matrix and B is an $n \times p$ matrix, then $AI_n = A$ and $I_n B = B$. We leave the verification of this property as an exercise.

It is clear from the definition that the $n \times n$ identity matrix I_n is in reduced row echelon form and that each of its columns is a pivot column. It therefore has n pivots since each of its diagonal entries is a pivot.

Elementary Matrices

An $n \times n$ ***elementary matrix*** is a matrix E obtained from the identity matrix I_n by one elementary row operation. It is clear that identity matrices are elementary since the operations $R_i \to 1R_i$ preserve identity matrices. Gaussian and Gauss-Jordan elimination can be described in terms of left-multiplication by elementary matrices.

EXAMPLE 2.27 Elementary Matrices

Use *Maple* to calculate the elementary matrices corresponding to the elementary row operation $(R_1 \rightleftharpoons R_3)$, $(R_2 \to -6R_2)$, $(R_1 \to R_1 + (-3)\, R_2)$.

Solution. We begin by loading the **linalg** package.

```
[> with(linalg):
```

Next we define the identity matrix I_4.

```
[> Id4:=LinearAlgebra[IdentityMatrix](4,4);
```

$$Id4 := \begin{bmatrix} 1 & 0 & 0 & 0 \\ 0 & 1 & 0 & 0 \\ 0 & 0 & 1 & 0 \\ 0 & 0 & 0 & 1 \end{bmatrix}$$

Now we create the matrix E_1 by applying the operation $(R_1 \rightleftharpoons R_3)$ to I_4.

```
[> E1:=swaprow(Id4,1,3);
```

$$\begin{bmatrix} 0 & 0 & 1 & 0 \\ 0 & 1 & 0 & 0 \\ 1 & 0 & 0 & 0 \\ 0 & 0 & 0 & 1 \end{bmatrix}$$

Next we create the matrix E_2 corresponding to the operation $(R_2 \to (-6)\, R_2)$.

```
[> E2:=mulrow(Id4,2,-6);
```

$$\begin{bmatrix} 1 & 0 & 0 & 0 \\ 0 & -6 & 0 & 0 \\ 0 & 0 & 1 & 0 \\ 0 & 0 & 0 & 1 \end{bmatrix}$$

We conclude by constructing the matrix E_3 obtained from I_4 by applying the operation $(R_1 \rightarrow R_1 + (-3)\, R_2)$.

```
[> E3:=addrow(Id4,2,1,-3);
```

$$\begin{bmatrix} 1 & -3 & 0 & 0 \\ 0 & 1 & 0 & 0 \\ 0 & 0 & 1 & 0 \\ 0 & 0 & 0 & 1 \end{bmatrix}$$

It is easy to check that the matrices E_1, E_2, and E_3 have the specified properties. ▲

The calculations in Example 2.27 show that there are situations where operations defined in **linalg** can be applied to objects of type *Matrix*. However, the *Maple* Help file does not provide a complete table of compatibilities.

Elementary Permutation Matrices

Among the elementary row operations are the row interchanges. An ***elementary permutation matrix*** is an elementary matrix obtained from an identity matrix by a single operation of the form $(R_i \rightleftharpoons R_j)$.

EXAMPLE 2.28 Elementary Permutation Matrices

Use *Maple 6* to create three 3×3 elementary permutation matrices.

Solution. We choose to create the matrices E_1, E_2, and E_3 corresponding to the elementary row operations $(R_1 \rightleftharpoons R_2)$, $(R_3 \rightleftharpoons R_1)$, and $(R_2 \rightleftharpoons R_3)$.

```
[> with(LinearAlgebra):
[> E1:=RowOperation(IdentityMatrix(3,3),[1,2]);
```

$$\begin{bmatrix} 0 & 1 & 0 \\ 1 & 0 & 0 \\ 0 & 0 & 1 \end{bmatrix}$$

```
[> E2:=RowOperation(IdentityMatrix(3,3),[1,3]);
```

$$\begin{bmatrix} 0 & 0 & 1 \\ 0 & 1 & 0 \\ 1 & 0 & 0 \end{bmatrix}$$

```
[> E3:=RowOperation(IdentityMatrix(3,3),[2,3]);
```

$$\begin{bmatrix} 1 & 0 & 0 \\ 0 & 0 & 1 \\ 0 & 1 & 0 \end{bmatrix}$$

It is easy to see that for any given 3×3 matrix A, the matrices $E_1 A$, $E_2 A$, and $E_3 A$ are the same as the matrices obtained from A by the given permutations of rows. ▲

The next example shows that ***left-multiplication*** by an elementary permutation matrix corresponds to the interchange of the corresponding rows of the matrix.

EXAMPLE 2.29 Left-Multiplication by an Elementary Permutation Matrix

Use *Maple* to illustrate the effect of a left-multiplication by an elementary permutation matrix.

Solution. The following calculation shows that left-multiplication by the elementary matrix

$$E = \begin{bmatrix} 0 & 1 & 0 \\ 1 & 0 & 0 \\ 0 & 0 & 1 \end{bmatrix}$$

permutes the first two rows of the matrix A.

```
[> A:=array([[1,2,3],[4,5,6],[7,8,9]]):
[> E:=array([[0,1,0],[1,0,0],0,0,1]]):
```

```
[> E.A;
```

$$\begin{bmatrix} 4 & 5 & 6 \\ 1 & 2 & 3 \\ 7 & 8 & 9 \end{bmatrix}$$

As we can see, applying the elementary row operation ($R_1 \rightleftharpoons R_2$) to the matrix A has the same effect as forming the EA. ▲

Right-multiplication by an elementary permutation matrix corresponds to an ***elementary column operation.***

EXAMPLE 2.30 Right-Multiplication by an Elementary Permutation Matrix

Use *Maple* to illustrate the effect of right-multiplication by an elementary matrix.

Solution. We let A and E be the matrices in Example 2.29 and form the matrix product AE.

```
[> A:=array([[1,2,3],[4,5,6],[7,8,9]]):
[> E:=array([[0,1,0],[1,0,0],[0,0,1]]):
[> evalm(A&*E1);
```

$$\begin{bmatrix} 2 & 1 & 3 \\ 5 & 4 & 6 \\ 8 & 7 & 9 \end{bmatrix}$$

As we can see, the matrix AE results from A by the interchange of the first two columns. ▲

We summarize the connection between elementary row operations and elementary matrices as a theorem.

THEOREM 2.11 *If E is an $m \times m$ elementary matrix obtained from I_m by an elementary row operation R, then every $m \times n$ matrix EA is the matrix obtained from A by performing the same operation R on A.*

Proof. Let $\mathbf{r}_i E$ be the ith row of E and $\mathbf{c}_j A$ be the jth column of A. Then ijth entry of the matrix EA is the dot product $\mathbf{r}_i E\, \mathbf{c}_j A$. Suppose that $m = 3$ and that

$$A = \begin{bmatrix} a_{11} & a_{12} & a_{13} & a_{14} \\ a_{21} & a_{22} & a_{23} & a_{24} \\ a_{31} & a_{32} & a_{33} & a_{34} \end{bmatrix}$$

is a 3×4 matrix. We must consider three cases.

Case 1. Let E be the matrix obtained by I $(R_1 \rightarrow sR_1)$ E. Then

$$EA = \begin{bmatrix} s & 0 & 0 \\ 0 & 1 & 0 \\ 0 & 0 & 1 \end{bmatrix} \begin{bmatrix} a_{11} & a_{12} & a_{13} & a_{14} \\ a_{21} & a_{22} & a_{23} & a_{24} \\ a_{31} & a_{32} & a_{33} & a_{34} \end{bmatrix}$$

$$= \begin{bmatrix} \mathbf{r}_1 E\, \mathbf{c}_1 A & \mathbf{r}_1 E\, \mathbf{c}_2 A & \mathbf{r}_1 E\, \mathbf{c}_3 A & \mathbf{r}_1 E\, \mathbf{c}_4 A \\ a_{21} & a_{22} & a_{23} & a_{24} \\ a_{31} & a_{32} & a_{33} & a_{34} \end{bmatrix}$$

$$= \begin{bmatrix} sa_{11} & sa_{12} & sa_{13} & sa_{14} \\ a_{21} & a_{22} & a_{23} & a_{24} \\ a_{31} & a_{32} & a_{33} & a_{34} \end{bmatrix}$$

Case 2. Let E be the matrix obtained by $I\ (R_1 \to R_1 + sR_2)\ E$. Then

$$EA = \begin{bmatrix} 1 & s & 0 \\ 0 & 1 & 0 \\ 0 & 0 & 1 \end{bmatrix} \begin{bmatrix} a_{11} & a_{12} & a_{13} & a_{14} \\ a_{21} & a_{22} & a_{23} & a_{24} \\ a_{31} & a_{32} & a_{33} & a_{34} \end{bmatrix}$$

$$= \begin{bmatrix} \mathbf{r}_1 E\, \mathbf{c}_1 A & \mathbf{r}_1 E\, \mathbf{c}_2 A & \mathbf{r}_1 E\, \mathbf{c}_3 A & \mathbf{r}_1 E\, \mathbf{c}_4 A \\ a_{21} & a_{22} & a_{23} & a_{24} \\ a_{31} & a_{32} & a_{33} & a_{34} \end{bmatrix}$$

$$= \begin{bmatrix} a_{11} + sa_{21} & a_{12} + sa_{22} & a_{13} + sa_{23} & a_{14} + sa_{24} \\ a_{21} & a_{22} & a_{23} & a_{24} \\ a_{31} & a_{32} & a_{33} & a_{34} \end{bmatrix}$$

Case 3. Let E be the matrix obtained by $I\ (R_1 \rightleftharpoons R_2)\ E$. Then

$$EA = \begin{bmatrix} 0 & 1 & 0 \\ 1 & 0 & 0 \\ 0 & 0 & 1 \end{bmatrix} \begin{bmatrix} a_{11} & a_{12} & a_{13} & a_{14} \\ a_{21} & a_{22} & a_{23} & a_{24} \\ a_{31} & a_{32} & a_{33} & a_{34} \end{bmatrix}$$

$$= \begin{bmatrix} \mathbf{r}_1 E\, \mathbf{c}_1 A & \mathbf{r}_1 E\, \mathbf{c}_2 A & \mathbf{r}_1 E\, \mathbf{c}_3 A & \mathbf{r}_1 E\, \mathbf{c}_4 A \\ \mathbf{r}_2 E\, \mathbf{c}_1 A & \mathbf{r}_2 E\, \mathbf{c}_2 A & \mathbf{r}_2 E\, \mathbf{c}_3 A & \mathbf{r}_2 E\, \mathbf{c}_4 A \\ a_{31} & a_{32} & a_{33} & a_{34} \end{bmatrix}$$

$$= \begin{bmatrix} a_{21} & a_{22} & a_{23} & a_{24} \\ a_{11} & a_{12} & a_{13} & a_{14} \\ a_{31} & a_{32} & a_{33} & a_{34} \end{bmatrix}$$

In each case, we can see that the matrix E results from I_3 by the same operation as that required to convert the matrix A to EA. The general proof is based on the same analysis. ■

COROLLARY 2.12 *A matrix A is row equivalent to a matrix B if and only if there exist elementary matrices $E_1, \ldots, E_p$ such that $A = E_1 \cdots E_p B$.*

Proof. This proof is left as an exercise. ■

Permutation Matrices

A ***permutation matrix*** P is a product $E_1 \cdots E_n$ of elementary permutation matrices.

EXAMPLE 2.31 A Permutation Matrix

Find two elementary permutation matrices E_1 and E_2 whose product $P = E_2E_1$ converts the matrix

$$A = \begin{bmatrix} a & b & c & d \\ e & f & g & h \\ i & j & k & l \end{bmatrix}$$

to the matrix

$$PA = \begin{bmatrix} i & j & k & l \\ a & b & c & d \\ e & f & g & h \end{bmatrix}$$

Solution. We can convert A to PA by first applying the operation $(R_1 \rightleftharpoons R_3)$ and then the operation $(R_2 \rightleftharpoons R_3)$. We know from Example 2.28 that the elementary permutation matrices corresponding to these operations are

$$E_1 = \begin{bmatrix} 0 & 0 & 1 \\ 0 & 1 & 0 \\ 1 & 0 & 0 \end{bmatrix} \quad \text{and} \quad E_2 = \begin{bmatrix} 1 & 0 & 0 \\ 0 & 0 & 1 \\ 0 & 1 & 0 \end{bmatrix}$$

We therefore form the product

$$P = E_2E_1 = \begin{bmatrix} 1 & 0 & 0 \\ 0 & 0 & 1 \\ 0 & 1 & 0 \end{bmatrix} \begin{bmatrix} 0 & 0 & 1 \\ 0 & 1 & 0 \\ 1 & 0 & 0 \end{bmatrix} = \begin{bmatrix} 0 & 0 & 1 \\ 1 & 0 & 0 \\ 0 & 1 & 0 \end{bmatrix}$$

and use *Maple* to confirm that PA is the intended matrix.

```
[> A:=array([[a,b,c,d],[e,f,g,h],[i,j,k,l]]):
[> P:=array([[0,0,1],[1,0,0],[0,1,0]]):
[> evalm(P&*A);
```

$$\begin{bmatrix} i & j & k & l \\ a & b & c & d \\ e & f & g & h \end{bmatrix}$$

As we can see, the matrix P represents the appropriate permutation. ▲

Generalized Diagonal Matrices

In Chapter 9, we will study the decomposition of rectangular matrices into products of simpler matrices. To be able to describe this decomposition elegantly, we extend the idea of a diagonal matrix to the rectangular case. We say that an $m \times n$ matrix $A = \left[a_{ij}\right]$ is ***generalized diagonal*** if $a_{ij} = 0$ for all $i \neq j$. If no confusion arises, we refer to a generalized diagonal matrix as a diagonal matrix.

EXAMPLE 2.32 A 2 x 3 Generalized Diagonal Matrix

The matrix

$$A = \begin{bmatrix} a_{11} & a_{12} & a_{13} \\ a_{21} & a_{22} & a_{23} \end{bmatrix} = \begin{bmatrix} -1 & 0 & 0 \\ 0 & 4 & 0 \end{bmatrix}$$

is diagonal since $a_{12} = a_{13} = a_{21} = a_{23} = 0$. ▲

EXAMPLE 2.33 A 3 x 2 Generalized Diagonal Matrix

The matrix

$$A = \begin{bmatrix} a_{11} & a_{12} \\ a_{21} & a_{22} \\ a_{31} & a_{32} \end{bmatrix} = \begin{bmatrix} -1 & 0 \\ 0 & 4 \\ 0 & 0 \end{bmatrix}$$

is diagonal since $a_{12} = a_{21} = a_{31} = a_{32} = 0$. ▲

A square diagonal matrix is obviously a generalized diagonal matrix, and so is every rectangular zero matrix.

EXAMPLE 2.34 A 2 x 4 Diagonal Zero Matrix

The matrix

$$A = \begin{bmatrix} a_{11} & a_{12} & a_{13} & a_{14} \\ a_{21} & a_{22} & a_{23} & a_{24} \end{bmatrix} = \begin{bmatrix} 0 & 0 & 0 & 0 \\ 0 & 0 & 0 & 0 \end{bmatrix}$$

is diagonal since $a_{12} = a_{13} = a_{14} = a_{21} = a_{23} = a_{24} = 0$. ▲

Submatrices

A matrix obtained from a matrix A by deleting some of its rows and columns is called a ***submatrix*** of A. In this text, submatrices are required, for example, to define and calculate determinants, to define and calculate eigenvalues, and to define and calculate pseudoinverses of rectangular matrices.

Maple has two built-in operations for extracting submatrices. The function **submatrix** in the **linalg** package extracts a specified submatrix from a given matrix. The command

```
submatrix(A, p..q, r..s)
```

extracts the submatrix determined by rows p to q and columns r to s. The function **SubMatrix** in the **LinearAlgebra** package can be used for the same purpose, but has additional features. In addition, we can use the **delrows** and **delcols** functions in the **linalg** package to extract submatrices step by step.

EXAMPLE 2.35 A Submatrix Determined by Two Entries

Use the **delrows** and **delcols** functions in the **linalg** package to extract from the matrix

$$A = \begin{bmatrix} 0 & 0 & 1 & 1 & 5 & 3 & 8 \\ 5 & 7 & 6 & 5 & 7 & 1 & 9 \\ 3 & 9 & 2 & 0 & 7 & 8 & 0 \\ 8 & 5 & 3 & 8 & 2 & 2 & 3 \end{bmatrix}$$

the submatrix B determined by a_{23} and a_{44}.

Solution. We define the matrix A and delete the required rows and columns.

```
[> with(linalg):
[> A:=matrix([
    [0,0,1,1,5,3,8],
    [5,7,6,5,7,1,9],
    [3,9,2,0,7,8,0],
    [8,5,3,8,2,2,3]]):
```

```
[> A1:=delrows(A,1..1):  A2:=delcols(A1,1..2):
   B:=delcols(A2,3..5);
```

$$B := \begin{bmatrix} 6 & 5 \\ 2 & 0 \\ 3 & 8 \end{bmatrix}$$

The matrix B is the 2×3 matrix consisting of all entries between a_{23} and a_{44}. ▲

It is often helpful that the **delrows** and **delcols** functions can be nested. The single command `delcols(delcols(delrows(A,1..1),1..2),3..5)` is equivalent to the three separate commands in Example 2.35.

EXAMPLE 2.36 A Permuted Submatrix Determined by a Range of Entries

Use the **SubMatrix** function in the **LinearAlgebra** package to illustrate the extraction of a permuted submatrix from a matrix.

Solution. We define a 4×5 matrix A, extract the last three columns from A, and permute the rows of the resulting submatrix using the permutation [4, 1, 3, 2].

```
[> with(LinearAlgebra):
[> A:=Matrix([
    [1,2,3,4,5],
    [6,7,8,9,10],
    [11,12,13,14,15],
    [16,17,18,19,20]]):
```

```
[> S:=SubMatrix(A,[4,1,3,2],[3..5]);
```

$$S := \begin{bmatrix} 18 & 19 & 20 \\ 3 & 4 & 5 \\ 13 & 14 & 15 \\ 8 & 9 & 10 \end{bmatrix}$$

As we can see, the matrix S consists of the last three columns of A, suitably permuted. ▲

EXERCISES 2.3

1. Determine which of the following matrices are upper triangular, lower triangular, diagonal, or symmetric. Justify your answers by appealing to the definitions of the matrix properties involved. Keep in mind that a matrix may have several or none of these properties.

$$\text{a.} \begin{bmatrix} 1 & 0 & -2 & 6 \\ 0 & 5 & 7 & 1 \\ 0 & 0 & 0 & 1 \\ 0 & 0 & 0 & 1 \end{bmatrix} \quad \text{b.} \begin{bmatrix} 0 & 0 & 0 & 0 \\ 0 & 0 & 0 & 0 \\ 0 & 0 & 0 & 0 \\ 0 & 0 & 0 & 0 \end{bmatrix} \quad \text{c.} \begin{bmatrix} 45 & -8 & -93 \\ 8 & -62 & 77 \\ 93 & -77 & 99 \end{bmatrix}$$

and

$$\text{d.} \begin{bmatrix} 1 & 0 & 0 & 0 \\ 0 & 1 & 0 & 0 \\ 0 & 0 & 1 & 0 \\ 0 & 0 & 0 & 1 \end{bmatrix} \quad \text{e.} \begin{bmatrix} 1 & 0 & 0 & 0 \\ 0 & 3 & 0 & 0 \\ 0 & 0 & 1 & 0 \\ 0 & 0 & 0 & -11 \end{bmatrix} \quad \text{f.} \begin{bmatrix} 1 & 0 & 0 & 9 \\ 0 & 1 & 7 & 0 \\ 0 & 7 & 1 & 0 \\ 9 & 0 & 0 & 1 \end{bmatrix}$$

2. Determine which of the following matrices are elementary matrices and explain why.

$$\text{a.} \begin{bmatrix} 1 & 0 & 0 & 0 \\ 0 & 1 & 0 & 0 \\ 0 & 0 & 1 & 0 \\ 0 & 0 & 0 & 1 \end{bmatrix} \quad \text{b.} \begin{bmatrix} 0 & 0 & 1 & 0 \\ 0 & 1 & 0 & 0 \\ 1 & 0 & 0 & 0 \\ 0 & 0 & 0 & 1 \end{bmatrix} \quad \text{c.} \begin{bmatrix} 0 & 0 & 0 & 1 \\ 0 & 0 & 1 & 0 \\ 0 & 1 & 0 & 0 \\ 1 & 0 & 0 & 0 \end{bmatrix}$$

3. Describe the effect of multiplying the matrix

$$A = \begin{bmatrix} 45 & -8 & -93 & 92 \\ 43 & -62 & 77 & 66 \\ 54 & -5 & 99 & -61 \\ -50 & -12 & -18 & 31 \end{bmatrix}$$

on the right by the matrices in Exercise 2. Repeat this exercise by multiplying A on the left by these matrices.

4. Show that if I_n is the $n \times n$ identity matrix, then $AI_n = A$ and $I_nB = B$ for any $m \times n$ matrix A and any $n \times p$ matrix B.

5. Define a real generalized diagonal 7×4 matrix with integer entries.

6. Prove that if A and B are two $n \times n$ diagonal matrices, then the matrix AB is diagonal.

7. Find two 3×3 matrices A and B for which $AB = 0$ and $BA \neq 0$.

8. Find two nonzero 10×10 matrices A and B for which $A + B = 0$ and $AB \neq 0$.

9. Show that for any elementary 3×3 permutation matrix E and any 3×3 matrix A, $(EA)^T = A^T E$.

10. Prove that every square matrix is the sum of a symmetric and a skew-symmetric matrix.

11. Prove that a matrix A is row equivalent to a matrix B if and only if there exist elementary matrices $E_1, \ldots, E_p$ such that $A = E_1 \cdots E_p B$.

12. Show that the right-multiplication of a matrix by an elementary matrix corresponds to carrying out the corresponding elementary column operation on the matrix.

13. Explain why every 3×3 row echelon matrix is upper triangular.

14. Prove that if A and B are two diagonal $n \times n$ matrices, then $AB = BA$.

15. Find all 2×2 submatrices

$$\begin{bmatrix} a_{ij} & a_{i(j+1)} \\ a_{(i+1)j} & a_{(i+1)(j+1)} \end{bmatrix}$$

of the matrix

$$A = \begin{bmatrix} 1 & 2 & 3 \\ 4 & 5 & 6 \\ 7 & 8 & 9 \end{bmatrix}$$

16. Find all 3×3 submatrices of the matrix

$$A = \begin{bmatrix} 19 & 12 & 17 & 13 \\ 25 & 12 & 20 & 12 \\ 17 & 20 & 12 & 18 \\ 24 & 24 & 22 & 13 \end{bmatrix}$$

17. (*Maple V*) Use the **diag** function in the **linalg** package and additional user-defined functions, if necessary, to create 3×3 diagonal matrices with integer entries, with real entries, and with polynomial entries.

18. (*Maple V*) Use the function **randmatrix** in the **linalg** package to generate a 2×4 matrix with integer entries between 200 and 300.

19. (*Maple V*) Use the **randmatrix** function in the **linalg** package to generate a symmetric 25×25 matrix of the form $A^T A = [a_{ij}]$ and display the entries $a_{18,18}$ and $a_{25,20}$.

20. (*Maple V*) Use the **submatrix** function of the **linalg** package to produce the following four submatrices of the matrix

$$A = [a_{ij}] = \begin{bmatrix} 3 & 0 & 1 & 7 & 1 \\ -1 & 2 & 0 & -1 & 1 \\ 0 & -2 & 21 & 1 & 1 \end{bmatrix}$$

a. The submatrix A_1 consists of the first four columns of A

b. The submatrix A_2 consists of the first two and last two columns of A

c. The submatrix A_3 consists of all of the entries of A between a_{22} and a_{34}

d. The submatrix A_4 consist of the first and third rows of A

21. (*Maple V*) Find two 2×2 matrices A and B for which $AB \neq BA$.

22. (*Maple V*) Find two distinct nonzero 2×2 matrices A and B for which $AB = BA$.

23. (*Maple V*) Find two distinct 3×3 polynomial matrices A and B for which $AB \neq BA$.

24. (*Maple V*) Find two nonzero 3×3 matrices A and B for which $AB = BA$.

25. (*Maple V*) Generate a random 3×7 real matrix A, form the matrix $B = AA^T$, and show that $B - B^T$ is the real 3×3 zero matrix.

26. (*Maple V*) Show that $AA^T = (AA^T)^T$ for any 3×4 matrix

$$A = \begin{bmatrix} a_{11} & a_{12} & a_{13} & a_{14} \\ a_{21} & a_{22} & a_{23} & a_{24} \\ a_{31} & a_{32} & a_{33} & a_{34} \end{bmatrix}$$

27. (*Maple V*) Generate a random 6×6 matrix A and show that $A^T A$ and AA^T are symmetric.

28. (*Maple V*) A square matrix A is ***skew-symmetric*** if $A = -A^T$. Use the **evalb** function to verify that the matrix

$$A = \begin{bmatrix} a_{11} & a_{12} & a_{13} & a_{14} \\ a_{21} & a_{22} & a_{23} & a_{24} \\ a_{31} & a_{31} & a_{33} & a_{34} \\ a_{41} & a_{42} & a_{43} & a_{44} \end{bmatrix} = \begin{bmatrix} 0 & -2 & 9 & -3 \\ 2 & 0 & -7 & -8 \\ -9 & 7 & 0 & 5 \\ 3 & 8 & -5 & 0 \end{bmatrix}$$

is skew-symmetric.

29. (*Maple V*) Use function **KroneckerDelta** defined by the procedure

```
KroneckerDelta := proc(i, j) if i = j then 1 else 0 fi; end:
```

to generate the 5×5 identity matrix.

30. (*Maple 6*) Generate three lower-triangular 5×5 matrices with integer entries, with real entries, and with polynomial entries.

31. (*Maple 6*) Generate three upper-triangular 5×5 matrices with integer entries, with real entries, and with polynomial entries.

32. (*Maple 6*) Generate symmetric 5×5 matrices with integer entries, with real entries, and with polynomial entries.

33. (*Maple 6*) Construct the elementary matrices in Example 2.27.

34. (*Maple 6*) Explore the properties of the **SubMatrix** function of the **LinearAlgebra** package.

INVERTIBLE MATRICES

Often we can solve a linear system $A\mathbf{x} = \mathbf{b}$ by finding a unique matrix, denoted by A^{-1}, that has the property that $\mathbf{x} = A^{-1}\mathbf{b}$. We now study this possibility.

DEFINITION 2.7 *An $n \times n$ matrix A is **invertible** if there exists a matrix B such that $AB = BA = I_n$.*

If A and B are two $n \times n$ matrices and AB is the $n \times n$ identity matrix I_n, then A is called a ***left inverse*** of B and B is called a ***right inverse*** of A. We will show that left and right inverses are unique and that a square matrix has a left inverse if and only if it has a right inverse. We will also show that the left inverse, if it exists, equals the right inverse. Hence we call the matrix B for which $AB = BA = I_n$ the ***inverse*** of A. For algebraic reasons, we denote the inverse of a matrix A by A^{-1}. The operation of forming the inverse of a matrix is called ***matrix inversion***. A square matrix that is not invertible is called ***noninvertible*** or ***singular.*** In this terminology, an invertible matrix is also called ***nonsingular.***

At various stages in our work, we will give criteria for invertibility. For example, we will show that an $n \times n$ matrix is invertible if and only if it has n pivots. This is equivalent to showing that A is row equivalent to I_n. In Chapter 7, we will also show that a matrix transformation $T_A : \mathbb{R}^n \to \mathbb{R}^n$ preserves Euclidean lengths and angles if and only if A is invertible and $A^{-1} = A^T$. Several other characterizations are given when they are needed. In Chapter 4, for example, we show that A is invertible if and only if the columns of A are linearly independent. In this section, we limit our discussion to showing that matrix inverses are unique and that the product of two invertible matrices is again an invertible matrix. We begin by proving two uniqueness properties of inverses. We recall from Exercise 2.3.4 that for any $n \times n$ matrix A, the $n \times n$ identity matrix I acts as an identity for matrix multiplication: $AI = IA = A$.

THEOREM 2.13 *If B is a right inverse of A and C is a left inverse of A, then $B = C$.*

Proof. Suppose that $AB = CA = I$. Then $B = IB = (CA)B = C(AB) = CI = C$. ■

THEOREM 2.14 *If B and C are inverses of A, then $B = C$.*

Proof. Since B and C are inverses, B is certainly a right inverse and C is a left inverse. Therefore $B = C$ by Theorem 2.13. ■

One of the important properties of invertible matrices is their closure under multiplication.

THEOREM 2.15 *If A and B are invertible matrices, then so is AB.*

Proof. Suppose that A^{-1} and B^{-1} are the inverses of A and B. We show that $B^{-1}A^{-1}$ is both a left inverse and a right inverse of AB. Since $B^{-1}A^{-1}AB = B^{-1}B = I$, the matrix $B^{-1}A^{-1}$ is a left inverse. Moreover, since $ABB^{-1}A^{-1} = A^{-1}A = I$, it is also a right inverse. Hence it is an inverse. By Theorem 2.14, $B^{-1}A^{-1}$ is the only matrix with this property. ■

Invertibility and Gaussian Elimination

Elementary row operations can be used to find the inverse A^{-1} of an invertible $n \times n$ matrix A. The procedure will be justified in Theorem 2.17. It consists of writing the matrix A and the identity matrix I_n side by side and applying the same sequence of elementary row operations to A and to I_n, until A has been reduced to the identity matrix. This process transforms I_n to A^{-1}. The following example illustrates the procedure.

EXAMPLE 2.37 Elementary Row Operations and Matrix Inversion

Use elementary row operations to find the inverse of the matrix

$$A = \begin{bmatrix} 1 & 2 \\ 3 & 4 \end{bmatrix}$$

Solution. We apply elementary row operations to the following matrix:

$$[A \mid I_2] = \begin{bmatrix} 1 & 2 & 1 & 0 \\ 3 & 4 & 0 & 1 \end{bmatrix} \rightarrow \begin{bmatrix} 1 & 2 & 1 & 0 \\ 0 & -2 & -3 & 1 \end{bmatrix}$$

$$\rightarrow \begin{bmatrix} 1 & 0 & -2 & 1 \\ 0 & -2 & -3 & 1 \end{bmatrix} \rightarrow \begin{bmatrix} 1 & 0 & -2 & 1 \\ 0 & 1 & \frac{3}{2} & -\frac{1}{2} \end{bmatrix} = [I_2 \mid B]$$

We use *Maple* to confirm that the matrix B is the inverse of A.

```
[>with(linalg):
[>A:=array([[1,2],[3,4]]):
[>B:=evalm(inverse(A));
```

$$B := \begin{bmatrix} -2 & 1 \\ \frac{3}{2} & -\frac{1}{2} \end{bmatrix}$$

As we can see, *Maple* confirms that the matrix B is the inverse of the matrix A. ▲

Invertibility and Elementary Matrices

In this section, we show that elementary matrices are the building blocks of invertible matrices: Every invertible matrix is a product of elementary matrices. We begin by showing that elementary matrices are invertible.

THEOREM 2.16 *Every elementary matrix is invertible.*

Proof. Let E be an elementary matrix. To prove that E is invertible we consider three cases, depending on which type of elementary row operation was used to create E.

1. If E is the result of the operation $(R_i \rightleftharpoons R_j)$, then $E^{-1} = E$.
2. If E is the result of the operation $(R_i \to sR_i)$, then E^{-1} is the matrix obtained from I by the operation $(R_i \to \frac{1}{s}R_i)$.
3. If E is the result of the operation $(R_i \to R_i + sR_j)$, then E^{-1} is the matrix obtained from I by the operation $(R_i \to R_i - sR_j)$.

It is easy to see that the matrices defined in this way have the required properties. ■

EXAMPLE 2.38 Inverses of Elementary Matrices

Illustrate the steps involved in the previous proof by examining the case of 3×3 elementary matrices.

Solution. The following three cases illustrate how elementary matrices are inverted.

```
[> with(linalg):
[> I3:=matrix([[1,0,0],[0,1,0],[0,0,1]]):
[> E1:=matrix([[0,1,0],[1,0,0],[0,0,1]]):
[> E2:=matrix([[1,0,0],[0,1/3,0],[0,0,1]]):
[> E3:=matrix([[1,0,0],[0,3,0],[0,0,1]]):
[> E4:=matrix([[1,0,0],[0,1,0],[-5,0,1]]):
[> E5:=matrix([[1,0,0],[0,1,0],[5,0,1]]):
[> [equal(E1&*E1,I3),equal(E2&*E3,I3),equal(E4&*E5,I3)];
```

$$[true, true, true]$$

We can see that in each case, the second matrix is the inverse of the first. ▲

Example 2.38 suggests the following useful test for matrix invertibility.

THEOREM 2.17 *An $n \times n$ matrix A is invertible if and only if it is row equivalent to I_n.*

Proof. Suppose that A is invertible. Theorem 1.2 tells us that A is row equivalent to a matrix B in reduced row echelon form. By Corollary 2.12, there exist elementary matrices $E_1, \ldots, E_p$ for which $E_1 \cdots E_p A = B$. Since all of the matrices $E_1, \ldots, E_p$ and A are invertible and since, by Theorem 2.15, the product of invertible matrices is invertible, the matrix B is invertible. If B is in reduced row echelon form and is not I_n, then B must have a zero row. This contradicts the fact that B is invertible, since the matrices BC must then have a zero row for all C. Hence $B = I_n$. Conversely, suppose that A is row equivalent to I_n. Then we know from Corollary 2.12 that A is a product $E_1 \cdots E_p$ of elementary matrices. Since elementary matrices are invertible, it follows that A is a product of invertible matrices and is therefore invertible. ■

COROLLARY 2.18 *An $n \times n$ matrix A is invertible if and only if its reduced row echelon form is I_n.*

Proof. By Theorem 2.17, A is invertible if and only if it can be reduced to I_n by elementary row operations. On the other hand, I_n is a matrix in reduced row echelon form. By Theorem 1.8, the reduced row echelon form of A is unique. Hence A is invertible if and only if its reduced row echelon form is I_n. ■

Theorem 2.17 also characterizes invertible matrices in terms of elementary matrices.

COROLLARY 2.19 *A matrix A is invertible if and only if there exist elementary matrices $E_1, \ldots, E_p$ for which $A = E_1 \cdots E_p$.*

Proof. This corollary follows immediately from the proof of Theorem 2.17 and the fact that the inverse of an elementary matrix is elementary. ■

COROLLARY 2.20 *An $n \times n$ matrix A is invertible if and only if it has n pivots.*

Proof. This corollary follows from Theorem 2.17 since the identity matrix I_n is the only $n \times n$ matrix in reduced row echelon form with n pivots. ■

The next theorem establishes a connection between invertibility and homogeneous linear systems.

THEOREM 2.21 *An $n \times n$ matrix A is row equivalent to I_n if and only if the system $A\mathbf{x} = \mathbf{0}$ has only the trivial solution.*

Proof. Suppose that $A\mathbf{x} = \mathbf{0}$ and that A is row equivalent to I_n. Then there exist elementary matrices $E_1, \ldots, E_p$ for which $E_1 \cdots E_p A = I_n$. Suppose further that $A\mathbf{x} = \mathbf{0}$. Then

$$E_1 \cdots E_p A\mathbf{x} = I_n\mathbf{x} = \mathbf{x} = E_1 \cdots E_p\mathbf{0} = \mathbf{0}$$

Therefore $\mathbf{x} = \mathbf{0}$. Conversely, if $A\mathbf{x} = \mathbf{0}$ has only the trivial solution, then A must be row equivalent to a reduced row echelon matrix B with n leading 1s. Hence $B = I_n$. ■

COROLLARY 2.22 *An $n \times n$ matrix A is invertible if and only if the equation $A\mathbf{x} = \mathbf{0}$ has only the trivial solution.*

Proof. This corollary follows immediately from the last two theorems. ■

Next we show that a square matrix is invertible if and only if it has either a left or a right inverse. The existence of one of these inverses is therefore sufficient for invertibility.

COROLLARY 2.23 (Invertible product theorem) *If A and B are square matrices and $AB = I$, then both A and B are invertible.*

Proof. Suppose that $AB = I$. Then A is a left inverse of B. We show that A is also a right inverse of B. If $B\mathbf{x} = \mathbf{0}$, then $AB\mathbf{x} = A\mathbf{0} = \mathbf{0}$. Therefore $\mathbf{x} = I\mathbf{x} = \mathbf{0}$ since $AB = I$. Thus Corollary 2.22 tells us that B is invertible, with inverse B^{-1}. Since A is a left inverse and B^{-1} is a right inverse, Theorems 2.13 and 2.14 guarantee that $B^{-1} = A$. Hence, $B = A^{-1}$. ■

Let us now take a closer look at the connection between the invertibility of a coefficient matrix A and the entries of the reduced row echelon form of the augmented matrix $[A \mid \mathbf{b}]$.

Suppose, for example, that

$$\begin{bmatrix} a_{11} & a_{12} & a_{13} & b_1 \\ a_{21} & a_{22} & a_{23} & b_2 \\ a_{31} & a_{32} & a_{33} & b_3 \end{bmatrix} \to \begin{bmatrix} 1 & 0 & 0 & c_1 \\ 0 & 1 & 0 & c_2 \\ 0 & 0 & 1 & c_3 \end{bmatrix}$$

is the result of a reduction of $[A \mid \mathbf{b}]$ to reduced row echelon form. Then the vector $\widehat{\mathbf{x}} = (c_1, c_2, c_3)$ is clearly a solution of the system $A\mathbf{x} = \mathbf{b}$. Conversely, if

$$\begin{bmatrix} a_{11} & a_{12} & a_{13} & c_1 \\ a_{21} & a_{22} & a_{23} & c_2 \\ a_{31} & a_{32} & 1 & c_3 \end{bmatrix}$$

is the reduced row echelon form of $[A \mid \mathbf{b}]$, then the definition of reduced row echelon matrices implies that

$$\begin{bmatrix} a_{11} & a_{12} & a_{13} & c_1 \\ a_{21} & a_{22} & a_{23} & c_2 \\ a_{31} & a_{32} & 1 & c_3 \end{bmatrix} = \begin{bmatrix} 1 & 0 & 0 & c_1 \\ 0 & 1 & 0 & c_2 \\ 0 & 0 & 1 & c_3 \end{bmatrix}$$

On the other hand if $a_{33} = 0$, then no vector of the form $\mathbf{b} = (b_1, b_2, b_3)$ can be a solution of the system $A\mathbf{x} = \mathbf{b}$ if $b_3 \neq 0$. Similar considerations yield a proof of the following general fact.

LEMMA 2.24 *Suppose that A is an $n \times n$ real matrix. Then the linear system $A\mathbf{x} = \mathbf{b}$ has a solution for every $\mathbf{b} \in \mathbb{R}^n$ if and only if A is row equivalent to I_n.*

Proof. Let $\mathbf{b}$ be any vector in $\mathbb{R}^n$ and convert the augmented matrix $[A \mid \mathbf{b}]$ to its unique reduced row echelon form $[B \mid \mathbf{c}]$. If B is the $n \times n$ identity matrix, then $\widehat{\mathbf{x}} = \mathbf{c}$ is a solution of $A\mathbf{x} = \mathbf{b}$.

Conversely, suppose that the matrix B is not the identity. Then the last row of B must be a zero row. Since the matrix $[B \mid \mathbf{c}]$ was obtained from $[A \mid \mathbf{b}]$ by row reduction, there exist elementary matrices $E_1, \ldots, E_p$ for which $E_1 \cdots E_p A = B$. Let $\mathbf{e}_n = [\delta_i] \in \mathbb{R}^n$ in which $\delta_i = 0$ if $i \neq n$ and $\delta_i = 1$ if $i = n$. Using the fact that elementary matrices are invertible, we form the vector $\mathbf{b}_0 = \left(E_1 \cdots E_p\right)^{-1} \mathbf{e}_n \in \mathbb{R}^n$. Since $E_1 \cdots E_p A = B$, it follows from the definition of $\mathbf{e}_n$ that

$$\left(E_1 \cdots E_p\right) [A \mid \mathbf{b}_0] = [B \mid \mathbf{e}_n]$$

The resulting system is inconsistent since the last row of the matrix $[B \mid \mathbf{e}_n]$ is of the form $[0, 0, \ldots, 0, 1]$. Hence B is not the identity. The system $A\mathbf{x} = \mathbf{b}_0$ therefore has no solution for $\mathbf{b}_0$. ■

We can rephrase this lemma as Fredholm's alternative theorem for square linear systems.

THEOREM 2.25 (Fredholm's alternative theorem for linear systems) *For any $n \times n$ real matrix A, precisely one of the following statements holds. 1. The linear system $A\mathbf{x} = \mathbf{b}$ has a solution for all $\mathbf{b} \in \mathbb{R}^n$. 2. The homogeneous linear system $A\mathbf{x} = \mathbf{0}$ has a nontrivial solution.*

Proof. The proof is identical to proof of Lemma 2.24. ■

In Chapter 8, we will discuss a more general version of this theorem, obtained as a corollary of Theorem 8.21.

The Matrix Inversion Algorithm

Example 2.37 illustrates how Corollaries 2.18 and 2.20 can be used to develop a procedure for finding the inverse of a matrix. The following theorem is the basis for this method. Let $[A \mid I_n]$ be the matrix obtained from A by adding the columns of I_n to A, and let $[I_n \mid B]$ be the matrix obtained from I_n by adding to columns of A to I_n.

THEOREM 2.26 (Matrix inversion theorem) *A matrix B is the inverse of a matrix A if and only if the matrices $[A \mid I_n]$ and $[I_n \mid B]$ are row equivalent.*

Proof. Suppose that A is an $n \times n$ invertible matrix. By Corollary 2.18, the reduced row echelon form of A is I_n. Since each step in the row reduction of A to I_n corresponds to left-multiplication by an elementary matrix, the reduction of A to I_n is of the form

$$A \to E_1 A \to \cdots \to \left(E_p \cdots E_1\right) A = I_n$$

for some elementary matrices $E_1, \ldots, E_p$. Therefore $E_p \cdots E_1 = A^{-1}$. This tell us that the reduction

$$I_n \to E_1 I_n \to \cdots \to \left(E_p \cdots E_1\right) I_n = A^{-1}$$

yields A^{-1}. Therefore, the elementary row operations corresponding to $E_1, \ldots, E_p$, applied to the matrix $[A \mid I_n]$, result in the matrix $[I_n \mid B]$. By retracing our steps, we get the converse. ■

The steps involved in finding the inverse of a matrix are summarized in Table 2.

TABLE 2 FINDING THE INVERSE OF A MATRIX.

Step 1. Take an invertible $n \times n$ matrix.
Step 2. Form the $n \times (2n)$ matrix $[A \mid I_n]$ by concatenating the $n \times n$ identity matrix with the $n \times n$ invertible matrix A.
Step 3. Use the algorithm for reducing a matrix to reduced row echelon form to convert the matrix $[A \mid I_n]$ to the matrix $[I_n \mid B]$.
Step 4. The $n \times n$ submatrix B of $[I_n \mid B]$ is the inverse of A.

EXAMPLE 2.39 Inverting a Matrix

Use *Maple* and elementary row reduction to calculate the inverse of the matrix

$$A = \begin{bmatrix} 1 & 2 & 3 \\ 0 & 1 & 0 \\ 2 & 0 & 12 \end{bmatrix}$$

Solution. We form the matrix $[A \mid I_n]$ and reduce it to $[I_n \mid B]$. By Theorem 2.26, the matrix B will be A^{-1}.

```
[> with(linalg):
[> A1:=matrix([[1,2,3],[0,1,0],[2,0,12]]):
[> A2:= matrix([[1,0,0],[0,1,0],[0,0,1]]):
```

```
[> A3:= augment(A1,A2);
```

$$A3 := \begin{bmatrix} 1 & 2 & 3 & 1 & 0 & 0 \\ 0 & 1 & 0 & 0 & 1 & 0 \\ 2 & 0 & 12 & 0 & 0 & 1 \end{bmatrix}$$

```
[> addrow(%,1,3,-2):   addrow(%,2,1,-2):   addrow(%,2,3,4):
[> addrow(%,3,1,-1/2):   mulrow(%,3,1/6):
[> A4:=%;
```

$$A4 := \begin{bmatrix} 1 & 0 & 0 & 2 & -4 & \frac{-1}{2} \\ 0 & 1 & 0 & 0 & 1 & 0 \\ 0 & 0 & 1 & \frac{-1}{3} & \frac{2}{3} & \frac{1}{6} \end{bmatrix}$$

```
[> A5:=submatrix(A4,1..3,4..6);
```

$$\begin{bmatrix} 2 & -4 & \frac{-1}{2} \\ 0 & 1 & 0 \\ \frac{-1}{3} & \frac{2}{3} & \frac{1}{6} \end{bmatrix}$$

```
[> equal(A1&*A5,A2);
```

true

This shows that $A4$ is the *Maple* form of the matrix $[I_n \mid B]$ and that $A5$ is A^{-1}. ▲

If the matrix A is invertible, we can use matrix inversion to solve a linear system $A\mathbf{x} = \mathbf{b}$. This approach is helpful if we need to solve the system for several constant vectors $\mathbf{b}_1, \ldots, \mathbf{b}_n$. We calculate A^{-1} once and then find the solutions of the systems by multiplying each of the vectors $\mathbf{b}_1, \ldots, \mathbf{b}_n$ by A^{-1}.

EXAMPLE 2.40 Using Matrix Inversion to Solve $A\mathbf{x} = \mathbf{b}$

Use *Maple* and matrix inversion to solve the matrix equations $A\mathbf{x} = \mathbf{b}$ and $A\mathbf{x} = \mathbf{c}$, where

$$A = \begin{bmatrix} 3 & 2 & 3 \\ 5 & 4 & 2 \\ 5 & 2 & 1 \end{bmatrix}, \quad \mathbf{x} = \begin{bmatrix} x \\ y \\ z \end{bmatrix}, \quad \mathbf{b} = \begin{bmatrix} 4 \\ 5 \\ 6 \end{bmatrix}, \quad \mathbf{c} = \begin{bmatrix} -2 \\ -8 \\ 7 \end{bmatrix}$$

Solution. We begin by loading the **linalg** package.

```
[> with(linalg):
```

Next we define the matrix A and the vectors **b** and **c**.

```
[> A:=matrix([[3,2,3],[5,4,2],[5,2,1]]):
[> b:=vector([4,5,6]):
[> c:=vector([-2,-8,7]):
```

We now solve the linear systems by calculating the vectors $A^{-1}\mathbf{b}$ and $A^{-1}\mathbf{c}$.

```
[> {x1=evalm(inverse(A)&*b),x2=evalm(inverse(A)&*c)};
```

$$\left\{x2 = \left[\frac{22}{5}, \frac{-149}{20}, \frac{-1}{10}\right], x1 = \left[\frac{7}{5}, \frac{-7}{10}, \frac{2}{5}\right]\right\}$$

This solves the two matrix equations. ▲

EXERCISES 2.4

1. Show that each of the following matrices is row equivalent to the identity matrix I_2.

$$\text{a.} \begin{bmatrix} 6 & 0 \\ 0 & 6 \end{bmatrix} \quad \text{b.} \begin{bmatrix} 6 & 0 \\ 4 & 6 \end{bmatrix} \quad \text{c.} \begin{bmatrix} 6 & 3 \\ 4 & 6 \end{bmatrix} \quad \text{d.} \begin{bmatrix} 0 & -1 \\ 1 & 0 \end{bmatrix}$$

2. Use the matrix inversion algorithm to find the inverses of the matrices in Exercise 1.

3. Show that each matrix in Exercise 1 is a product of elementary matrices.

4. Show that each matrix in Exercise 1 has two pivots.

5. Show that for each matrix A in Exercise 1, the homogeneous linear system $A\mathbf{x} = \mathbf{0}$ has only the trivial solution.

6. Find the inverse of the product of the matrices in Exercise 1.

7. Show that each of the following matrices is row equivalent to the identity matrix I_3.

$$\text{a.} \begin{bmatrix} 4 & 0 & 0 \\ 0 & 5 & 0 \\ 0 & 0 & 4 \end{bmatrix} \quad \text{b.} \begin{bmatrix} 4 & 0 & 0 \\ 2 & 5 & 0 \\ 3 & 2 & 4 \end{bmatrix} \quad \text{c.} \begin{bmatrix} \frac{1}{\sqrt{2}} & \frac{1}{\sqrt{2}} & 0 \\ -\frac{1}{\sqrt{2}} & \frac{1}{\sqrt{2}} & 0 \\ 0 & 0 & 1 \end{bmatrix}$$

8. Use the matrix inversion algorithm to find the inverses of the matrices in Exercise 7.

9. Show that each matrix in Exercise 7 is a product of elementary matrices.

10. Show that each matrix in Exercise 7 has three pivots.

11. Show that for each matrix A in Exercise 7, the homogeneous linear system $A\mathbf{x} = \mathbf{0}$ has only the trivial solution.

12. Find the inverse of the product of the matrices in Exercise 7.

13. Construct a 3×2 matrix A and a 2×3 matrix B for which $BA = I_2$ and $AB \neq I_3$. Explain why the existence of A and B does not contradict Theorem 2.23.

14. Suppose that A is an invertible $n \times n$ matrix and that B is A^{-1}. Show that if $i \neq j$ and $\mathbf{r}_i\,(A)$ is the ith row of A and $\mathbf{c}_j\,(B)$ is the jth column of B, then the dot products $\mathbf{r}_i\,(A)\,\mathbf{c}_j\,(B)$ are zero.

15. (*Maple V*) Use reduced row echelon matrices to show that if A is the matrix

$$\begin{bmatrix} 4 & -1 & -4 & -1 \\ -1 & -4 & -2 & 3 \\ 4 & 4 & -3 & -3 \\ 4 & -1 & 0 & 5 \end{bmatrix}$$

then the linear system $A\mathbf{x} = \mathbf{b}$ has a solution for all $\mathbf{b} \in \mathbb{R}^4$.

16. (*Maple V*) Use matrix inversion to solve the matrix equation

$$\begin{bmatrix} 0 & 3 & -4 & -5 \\ -2 & -1 & -9 & 5 \\ 8 & 4 & -3 & 8 \\ -9 & 7 & 5 & 0 \end{bmatrix} \begin{bmatrix} w \\ x \\ y \\ z \end{bmatrix} = \begin{bmatrix} 1 \\ 2 \\ 3 \\ 4 \end{bmatrix}$$

17. (*Maple 6*) Generate a random 4×4 matrix A with four pivots. Verify that A is invertible.

18. (*Maple 6*) Let $A\mathbf{x} = \mathbf{b}$ be a linear system with a 6×6 invertible coefficient matrix A. Use *Maple* to explore the solvability of the linear systems $B\mathbf{x} = \mathbf{b}$, obtained by applying several elementary row operations to A.

19. (*Maple 6*) Use 3×3 real matrices to illustrate Fredholm's alternative theorem.

20. (*Maple 6*) Generate a 4×4 matrix A with three linear independent columns and show that Fredholm's alternative theorem holds for A.

21. (*Maple V, 6*) Explore the properties of the **pivot** function in the **linalg** package and the **Pivot** function in the **LinearAlgebra** package.

Matrix Inverses, Transposes, and Products

We conclude this section by establishing a link between matrix inversion, matrix transposes, and matrix products.

THEOREM 2.27 *If A is an invertible matrix, then so is A^T.*

Proof. Suppose that $B = (A^{-1})^T$. Then Theorem 2.9 implies that

$$A^T B = A^T (A^{-1})^T = (A^{-1}A)^T = I^T = I$$

Hence A^T has a right inverse. It follows from Theorem 2.23 that A^T is invertible. By Theorem 2.14, the matrix B is the inverse of A^T. ■

EXERCISES 2.5

1. Show that in each of the following examples, the matrices A and B are inverses.

a. $A = \begin{bmatrix} 0 & 1 \\ -1 & 0 \end{bmatrix}, B = \begin{bmatrix} 0 & -1 \\ 1 & 0 \end{bmatrix}$ b. $A = \begin{bmatrix} 2 & -3 \\ 0 & -3 \end{bmatrix}, B = \begin{bmatrix} \frac{1}{2} & -\frac{1}{2} \\ 0 & -\frac{1}{3} \end{bmatrix}$

2. Show that if a and b are any two real numbers and $b \neq 0$, then the matrix

$$A = \begin{bmatrix} a & \dfrac{1-a^2}{b} \\ b & -a \end{bmatrix}$$

is its own inverse.

3. Show that the matrix

$$A = \begin{bmatrix} 4 & -6 \\ 2 & -3 \end{bmatrix}$$

is not invertible by finding a nontrivial solution of the equation $A\mathbf{x} = \mathbf{0}$.

4. Construct the inverse of the matrix

$$A = \begin{bmatrix} 1 & -3 & -3 \\ -3 & 1 & 3 \\ -3 & 3 & -2 \end{bmatrix}$$

from the calculations that show that A is row equivalent to I_3.

5. Suppose that A, B, and C are real $n \times n$ matrices with the property that $AB = CA$. Does it follow that $A^{-1} = B = C$? Explain your answer.

6. Show that $(A^{-1})^{-1} = A$ and that if $s \neq 0$, then $(sA)^{-1} = (1/s)\,A^{-1}$.

7. (*Maple 6*) Calculate the inverses of the matrices in the following examples and show that in each case, $(AB)^{-1} = B^{-1}A^{-1}$.

a. $$A = \begin{bmatrix} 2 & -3 \\ -3 & -3 \end{bmatrix}, \; B = \begin{bmatrix} 0 & -3 \\ -3 & 4 \end{bmatrix}$$

b. $$A = \begin{bmatrix} -3 & 3 & 1 \\ 3 & 3 & 2 \\ 1 & 2 & 3 \end{bmatrix}, \; B = \begin{bmatrix} 5 & 3 & 3 \\ -1 & 5 & -1 \\ -2 & -1 & 3 \end{bmatrix}$$

8. (*Maple 6*) Show that for each of the following matrices, $(A^T)^{-1} = (A^{-1})^T$.

a. $A = \begin{bmatrix} 2 & -3 \\ -3 & -3 \end{bmatrix}$ b. $A = \begin{bmatrix} 2 & 1 \\ -3 & -2 \end{bmatrix}$ c. $A = \begin{bmatrix} -3 & 3 & 1 \\ 3 & 3 & 2 \\ 1 & 2 & 3 \end{bmatrix}$

9. (*Maple 6*) Find the inverse of the matrix

$$A = \begin{bmatrix} 1 & a & a^2 \\ 1 & b & b^2 \\ 1 & c & c^2 \end{bmatrix}$$

ORTHOGONAL MATRICES

Much of the work in Chapter 7 and all of the work in Chapter 9 are based on matrices with the remarkable property that their inverse is simply their transpose.

DEFINITION 2.8 *An invertible real matrix A is **orthogonal** if $A^{-1} = A^T$.*

The reason for this terminology will become clear in Chapter 8, where we will show that the rows and columns of orthogonal matrices are actually ***orthogonal vectors*** in a precise geometric sense.

EXAMPLE 2.41 An Orthogonal 2 x 2 Matrix

Use *Maple* to verify that the matrix

$$A = \begin{bmatrix} \frac{1}{\sqrt{2}} & -\frac{1}{\sqrt{2}} \\ \frac{1}{\sqrt{2}} & \frac{1}{\sqrt{2}} \end{bmatrix}$$

is orthogonal.

Solution. We calculate A^{-1} and A^T and compare the results.

```
[> A:=matrix([[1/sqrt(2),-1/sqrt(2)],[1/sqrt(2),1/sqrt(2)]]):
[> inverse(A);
```

$$\begin{bmatrix} \frac{1}{2}\sqrt{2} & \frac{1}{2}\sqrt{2} \\ -\frac{1}{2}\sqrt{2} & \frac{1}{2}\sqrt{2} \end{bmatrix}$$

```
[> transpose(A);
```

$$\begin{bmatrix} \frac{1}{2}\sqrt{2} & \frac{1}{2}\sqrt{2} \\ -\frac{1}{2}\sqrt{2} & \frac{1}{2}\sqrt{2} \end{bmatrix}$$

This shows that $A^{-1} = A^T$. Hence the matrix A is orthogonal. ▲

We use *Maple* to find out whether a matrix A is orthogonal without actually displaying A^{-1} and A^T.

EXAMPLE 2.42 An Orthogonal 3 x 3 Matrix

Use *Maple* to verify that the matrix

$$A = \begin{bmatrix} \cos\theta & -\sin\theta & 0 \\ \sin\theta & \cos\theta & 0 \\ 0 & 0 & 1 \end{bmatrix}$$

is orthogonal.

Solution. We calculate A^{-1} and A^T and compare the results.

```
[> A:=matrix([
    [cos(theta),-sin(theta),0],
    [sin(theta),cos(theta),0],
    [0,0,1]]):
[> equal(simplify(inverse(A)),transpose(A));
                                   true
```

This shows that $A^{-1} = A^T$. The matrix A is therefore orthogonal. ▲

Maple has a built-in function that often makes it easy to test for orthogonality in the case of exact calculations. It tells us whether a specified equation is true or false. Some caution is needed in cases where *Maple* claims that orthogonality does not hold.

EXAMPLE 2.43 A Boolean Test for Orthogonality

Use the Boolean test definable in *Maple* to verify that the matrix

$$\begin{bmatrix} \frac{1}{2}\sqrt{2} & \frac{1}{2}\sqrt{2} \\ -\frac{1}{2}\sqrt{2} & \frac{1}{2}\sqrt{2} \end{bmatrix}$$

is orthogonal.

Solution. We define A and use the **IsOrthogonal** function.

```
[> with(LinearAlgebra):
[> A:=Matrix([
    [(1/2)*sqrt(2),(1/2)*sqrt(2)],
    [-(1/2)*sqrt(2),(1/2)*sqrt(2)]]):
[> IsOrthogonal(A);
                                   true
```

This tells us that the matrix A is orthogonal. ▲

Unfortunately, the **IsOrthogonal** function may fail to detect the orthogonality of some matrices if certain algebraic simplifications are required to establish orthogonality. Hence we cannot conclude that a matrix is not orthogonal if the command `IsOrthogonal(A);` produces the output $false$. Here is an example.

Suppose we would like to *Maple* to show that the matrix

$$A := \begin{bmatrix} \cos\theta & -\sin\theta & 0 \\ \sin\theta & \cos\theta & 0 \\ 0 & 0 & 1 \end{bmatrix}$$

is orthogonal for all angles θ. Then the calculation

```
[> A:=Matrix([
    [cos(theta),-sin(theta),0],
    [sin(theta),cos(theta),0],
    [0,0,1]]):
[> LinearAlgebra[IsOrthogonal](A);
                              false
```

fails to establish this fact. However, we can use a direct test involving inverted and transposed matrices.

```
[> equal(simplify(inverse(A)),transpose(A));
                              true
```

This shows that the matrix A is orthogonal. A corresponding test using the **Linear-Algebra** package also fails.

```
[> Equal(simplify(MatrixInverse(A)),Transpose(A));
                              false
```

We know that this *Maple* statement is incorrect. ▲

EXERCISES 2.6

1. Show that the product of two orthogonal matrices is an orthogonal matrix.

2. Show that the matrix $A = \begin{bmatrix} \cos\theta & -\sin\theta \\ \sin\theta & \cos\theta \end{bmatrix}$ is orthogonal for all θ.

3. Show that the matrix $A = \begin{bmatrix} 0 & 1 \\ 1 & 0 \end{bmatrix}$ is orthogonal.

4. Show that a 2×2 matrix $A = \begin{bmatrix} 1+s & 0 \\ 0 & 1 \end{bmatrix}$ is orthogonal if and only if $s = 0$.

5. Show that each of the following matrices

a. $\begin{bmatrix} \frac{4}{5} & -\frac{3}{5} \\ -\frac{3}{5} & -\frac{4}{5} \end{bmatrix}$ b. $\begin{bmatrix} -1 & 0 \\ 0 & 1 \end{bmatrix}$

c. $\begin{bmatrix} \frac{15}{19} & -\frac{10}{19} & \frac{6}{19} \\ -\frac{10}{19} & -\frac{6}{19} & \frac{15}{19} \\ \frac{6}{19} & \frac{15}{19} & \frac{10}{19} \end{bmatrix}$ d. $I - \dfrac{2}{\mathbf{u}^T\mathbf{u}}\mathbf{u}\mathbf{u}^T$

is orthogonal. In part (d), the vector $\mathbf{u}$ is any nonzero vector in $\mathbb{R}^4$ and I is the 4×4 identity matrix.

6. Let A be a 3×3 orthogonal matrix. Show that if $\mathbf{r}_i A$ are the rows of A and the dot products

$$(\mathbf{r}_1 A)\begin{bmatrix} x \\ y \\ z \end{bmatrix}, \quad (\mathbf{r}_2 A)\begin{bmatrix} x \\ y \\ z \end{bmatrix}, \quad \text{and} \quad (\mathbf{r}_3 A)\begin{bmatrix} x \\ y \\ z \end{bmatrix}$$

are 0, then $x = y = z = 0$.

7. (*Maple V*) Repeat Exercise 5 using the **orthog** function of the **linalg** package.

8. (*Maple V*) Construct a 5×5 matrix A whose rows are mutually orthogonal in the sense that $\mathbf{r}_i A \left(\mathbf{r}_j A\right)^T = 0$ for all $i \neq j$. Use the **dotprod** function in the **linalg** package to explore the values of the dot products $\mathbf{r}_i A\, \mathbf{x}$, where $\mathbf{x}$ is any other column vector in $\mathbb{R}^5$.

9. (*Maple 6*) Let B be the 4×5 matrix obtained from the matrix A in Exercise 8 by deleting the last row of the matrix A. Assign the **Matrix** type to B and use the **DotProduct** function in the **LinearAlgebra** package to explore the connection between any two nonzero vectors $\mathbf{x}, \mathbf{y} \in \mathbb{R}^5$ for which the dot products $\mathbf{r}_i A\, \mathbf{x} = \mathbf{r}_i A\, \mathbf{y} = \mathbf{0}$ for all rows $\mathbf{r}_i A$ of A.

THE LU DECOMPOSITION

An $m \times n$ linear system $A\mathbf{x} = \mathbf{b}$ can often be solved efficiently by LU decomposition—that is, by decomposing the coefficient matrix A into a product LU, where L is a lower-triangular matrix L with 1s on the diagonal and U is a row echelon matrix that is upper triangular if $m = n$. Solving the system $A\mathbf{x} = \mathbf{b}$ is then equivalent to solving two simpler systems $L\mathbf{y} = \mathbf{b}$ and $U\mathbf{x} = \mathbf{y}$. Since L is lower triangular, the system $L\mathbf{y} = \mathbf{b}$ can be solved by ***forward***

substitution. Moreover, since U is in row echelon form, the system $U\mathbf{x} = \mathbf{y}$ can be solved by ***back substitution.***

We will use the built-in **LUDecomposition** function to solve LU-decomposition problems. Given a real nonsingular matrix A, the command `LUDecomposition(A);` returns a list of three items. The first element is a combination of the upper- and lower-triangular matrices L and U. The second element is a vector specifying rows used for pivoting, and the third element is an estimate of the condition number of A, discussed in Chapter 9. The following example illustrates these ideas.

EXAMPLE 2.44 Forward and Back Substitution

Use LU decomposition and forward and back substitution to solve the system

$$\begin{bmatrix} 1 & -3 & 2 & -2 \\ 3 & -2 & 0 & -1 \\ 2 & 36 & -28 & 27 \\ 1 & -3 & 22 & 5 \end{bmatrix} \begin{bmatrix} x_1 \\ x_2 \\ x_3 \\ x_4 \end{bmatrix} = \begin{bmatrix} -11 \\ -4 \\ 155 \\ 10 \end{bmatrix}$$

Solution. We begin by finding the LU decomposition of the given coefficient matrix. By definition, we need to find a lower-triangular matrix with 1s on the diagonal and an upper-triangular matrix for which

$$\begin{bmatrix} 1 & -3 & 2 & -2 \\ 3 & -2 & 0 & -1 \\ 2 & 36 & -28 & 27 \\ 1 & -3 & 22 & 5 \end{bmatrix} = \begin{bmatrix} 1 & 0 & 0 & 0 \\ a & 1 & 0 & 0 \\ b & c & 1 & 0 \\ d & e & f & 1 \end{bmatrix} \begin{bmatrix} 1 & -3 & 2 & -2 \\ 0 & u & v & w \\ 0 & 0 & x & y \\ 0 & 0 & 0 & z \end{bmatrix}$$

If we multiply out the right-hand side of this equation, we obtain the matrix

$$\begin{bmatrix} 1 & -3 & 2 & -2 \\ a & -3a+u & 2a+v & -2a+w \\ b & -3b+cu & 2b+cv+x & -2b+cw+y \\ d & -3d+eu & 2d+ev+fx & -2d+ew+fy+z \end{bmatrix}$$

By equating entries and solving the resulting equations, we obtain the values $a = 3$, $b = 2$, $c = 6$, $d = 1$, $e = 0$, $f = 5$, $u = 7$, $v = -6$, $w = 5$, $x = 4$, $y = 1$, and $z = 2$. Hence

$$\begin{bmatrix} 1 & -3 & 2 & -2 \\ 3 & -2 & 0 & -1 \\ 2 & 36 & -28 & 27 \\ 1 & -3 & 22 & 5 \end{bmatrix} = LU = \begin{bmatrix} 1 & 0 & 0 & 0 \\ 3 & 1 & 0 & 0 \\ 2 & 6 & 1 & 0 \\ 1 & 0 & 5 & 1 \end{bmatrix} \begin{bmatrix} 1 & -3 & 2 & -2 \\ 0 & 7 & -6 & 5 \\ 0 & 0 & 4 & 1 \\ 0 & 0 & 0 & 2 \end{bmatrix}$$

We now use the matrix L to form the linear system $L\mathbf{y} = \mathbf{b}$ and solve for $\mathbf{y}$ by forward substitution.

$$\begin{bmatrix} 1 & 0 & 0 & 0 \\ 3 & 1 & 0 & 0 \\ 2 & 6 & 1 & 0 \\ 1 & 0 & 5 & 1 \end{bmatrix} \begin{bmatrix} y_1 \\ y_2 \\ y_3 \\ y_4 \end{bmatrix} = \begin{bmatrix} -11 \\ -4 \\ 155 \\ 10 \end{bmatrix}$$

If we multiply out the left-hand side of this equation, we get

$$\begin{bmatrix} y_1 \\ 3y_1 + y_2 \\ 2y_1 + 6y_2 + y_3 \\ y_1 + 5y_3 + y_4 \end{bmatrix} = \begin{bmatrix} -11 \\ -4 \\ 155 \\ 10 \end{bmatrix}$$

This means that $y_1 = -11$. If we substitute this value in the equation $3y_1 + y_2 = -4$ and solve for y_2, we get $y_2 = 29$. By substituting the values for y_1 and y_2 in the equation $2y_1 + 6y_2 + y_3 = 155$ and solving for y_3, we get $y_3 = 3$. By substituting $y_1 = -11$ and $y_3 = 3$ in the equation $y_1 + 5y_3 + y_4 = 10$ and solving for y_4, we get $y_4 = 6$. Thus

$$\begin{bmatrix} y_1 \\ y_2 \\ y_3 \\ y_4 \end{bmatrix} = \begin{bmatrix} -11 \\ 29 \\ 3 \\ 6 \end{bmatrix}$$

is a solution of the equation $L\mathbf{y} = \mathbf{b}$. For obvious reasons, this process is known as ***forward substitution.***

Next we form the system $U\mathbf{x} = \mathbf{y}$ and solve for $\mathbf{x}$.

$$\begin{bmatrix} 1 & -3 & 2 & -2 \\ 0 & 7 & -6 & 5 \\ 0 & 0 & 4 & 1 \\ 0 & 0 & 0 & 2 \end{bmatrix} \begin{bmatrix} x_1 \\ x_2 \\ x_3 \\ x_4 \end{bmatrix} = \begin{bmatrix} -11 \\ 29 \\ 3 \\ 6 \end{bmatrix}$$

If we multiply out the left-hand side of this equation, we get

$$\begin{bmatrix} x_1 - 3x_2 + 2x_3 - 2x_4 \\ 7x_2 - 6x_3 + 5x_4 \\ 4x_3 + x_4 \\ 2x_4 \end{bmatrix} = \begin{bmatrix} -11 \\ 29 \\ 3 \\ 6 \end{bmatrix}$$

As we can see, $x_4 = 3$. If we substitute this value back into the equation $4x_3 + x_4 = 3$, we get a value for x_3. Continuing in this way, we get the vector

$$\begin{bmatrix} x_1 \\ x_2 \\ x_3 \\ x_4 \end{bmatrix} = \begin{bmatrix} 1 \\ 2 \\ 0 \\ 3 \end{bmatrix}$$

as a solution of the equation $Ux = y$. This process is known as ***back substitution.*** ▲

Example 2.44 shows that we were able to solve the given system in three steps: First we decomposed A into a product LU of a lower-triangular matrix L and an upper-triangular matrix U. We then solved the system $L\mathbf{y} = \mathbf{b}$ for $\mathbf{y}$ by forward substitution and concluded the process by solving the system $U\mathbf{x} = \mathbf{y}$ for $\mathbf{x}$ by back substitution.

The LU decomposition of a matrix is based on a closer analysis of the process of reducing a matrix to row echelon form. First we observe that the operations $(R_i \to sR_i)$ are not required since we are initially not interested in converting the pivots of the given matrix to specific values. This means that we need to consider only two cases: the ideal case, requiring only operations of the form $(R_i \to R_i + sR_j)$, and the mixed case, requiring both $(R_i \to R_i + sR_j)$ and $(R_i \leftrightharpoons R_j)$. We will show that if no permutations of rows $(R_i \leftrightharpoons R_j)$ are required, we can decompose A into the form LU. If permutations of rows are required, we can carry them out first, using a suitable permutation matrix P, and then decompose the matrix PA into LU. In both cases, the rows R_j involved in the operations $(R_i \to R_i + sR_j)$ occur above the rows R_i.

To be able to prove the LU decomposition theorem, we first need to establish some properties of triangular matrices. The first is stated in the next theorem.

THEOREM 2.28 *If A is an $n \times n$ row echelon matrix, then A is upper triangular.*

Proof. Let $A = [a_{ij}]$ be a matrix in row echelon form. If A is a zero matrix, then it is obviously upper triangular. Suppose, therefore, that A is nonzero and that $a_{1p_1}, \ldots, a_{mp_m}$ are the pivots of A. Then $1 \leq p_1 < \cdots < p_m$. If $a_{ij} \neq 0$, then $\mathbf{r}_i A$ is a nonzero row of A. Therefore, $j \geq p_i \geq i$. Hence $a_{ij} = 0$ if $i > j$. ■

The other required properties are that lower- and upper-triangular matrices are closed under multiplication.

THEOREM 2.29 *If A and B are two lower-triangular $n \times n$ matrices, then AB is a lower-triangular matrix.*

Proof. Let $A = [a_{ij}]$ and $B = [b_{ij}]$ be two $n \times n$ lower-triangular matrices, and let $C = [c_{ij}] = AB$. Then $a_{ij} = b_{ij} = 0$ if $i < j$. Therefore,

$$c_{ij} = a_{i1}b_{1j} + \cdots + a_{in}b_{nj} = a_{ij}b_{jj} + \cdots + a_{in}b_{nj}$$

since $b_{ij} = 0$ for $i < j$. On the other hand, $a_{ij} = 0$ for $i < j$. Therefore

$$a_{ij}b_{jj} + \cdots + a_{in}b_{nj} = 0$$

Hence $c_{ij} = 0$ for $i < j$. This entails that C is lower triangular. ■

COROLLARY 2.30 *The product of two upper-triangular matrices is an upper-triangular matrix.*

Proof. We get this result by reversing the inequalities in the previous proof and adapting the equations as required. ■

We are now ready to tackle the LU decomposition problem. We begin with an example.

EXAMPLE 2.45 An LU Decomposition of a Matrix A

Use elementary matrices based on the operations to find a formula for an LU decomposition for the matrix

$$A = \begin{bmatrix} a & b & c \\ d & e & f \\ g & h & i \end{bmatrix}$$

Solution. We use elementary matrices corresponding the elementary row operations of the form $(R_i \to R_i + sR_j)$ to construct a sequence of row equivalent matrices $A = A_1, A_2, A_3$ for which A_3 is a matrix in row echelon form. In the construction, we assume that a and $(-(d/a)b + e)$ are nonzero.

Step 1. To replace the entry g in A by zero, we carry out the multiplication

$$A_1 = \begin{bmatrix} 1 & 0 & 0 \\ 0 & 1 & 0 \\ -g/a & 0 & 1 \end{bmatrix} \begin{bmatrix} a & b & c \\ d & e & f \\ g & h & i \end{bmatrix}$$
$$= \begin{bmatrix} a & b & c \\ d & e & f \\ 0 & -\frac{g}{a}b + h & -\frac{g}{a}c + i \end{bmatrix} = E_1 A$$

The elementary row operation involved is $(R_3 \to R_3 + (-g/a) R_1)$.

Step 2. To replace entry d in A_1 by zero, we carry out the multiplication

$$A_2 = \begin{bmatrix} 1 & 0 & 0 \\ -d/a & 1 & 0 \\ 0 & 0 & 1 \end{bmatrix} \begin{bmatrix} a & b & c \\ d & e & f \\ 0 & -\frac{g}{a}b + h & -\frac{g}{a}c + i \end{bmatrix}$$

$$= \begin{bmatrix} a & b & c \\ 0 & -\frac{d}{a}b + e & -\frac{d}{a}c + f \\ 0 & -\frac{g}{a}b + h & -\frac{g}{a}c + i \end{bmatrix} = E_2 A_1$$

The elementary row operation involved is $(R_2 \to R_2 + (-d/a)\, R_1)$.

Step 3. To replace entry $\left(-\frac{g}{a}b + h\right)$ in A_2 by zero, we carry out the multiplication

$$A_3 = \begin{bmatrix} 1 & 0 & 0 \\ 0 & 1 & 0 \\ 0 & -\frac{-\frac{g}{a}b+h}{-\frac{d}{a}b+e} & 1 \end{bmatrix} \begin{bmatrix} a & b & c \\ 0 & -\frac{d}{a}b + e & -\frac{d}{a}c + f \\ 0 & -\frac{g}{a}b + h & -\frac{g}{a}c + i \end{bmatrix}$$

$$= \begin{bmatrix} a & b & c \\ 0 & -\frac{d}{a}b + e & -\frac{d}{a}c + f \\ 0 & 0 & -\frac{-\frac{g}{a}b+h}{-\frac{d}{a}b+e}\left(-\frac{d}{a}c + f\right) - \frac{g}{a}c + i \end{bmatrix} = E_3 A_2$$

The elementary row operation involved is

$$(R_3 \to R_3 + (-(-(g/a)\, b + h) \,/\, (-(d/a)\, b + e))\, R_2)$$

We then let $U = A_3$ and $L = E_1^{-1} E_2^{-1} E_3^{-1}$. The reduction steps were possible because the entries a and $(-(d/a)\, b + e)$ were nonzero.

The assumption that the reduction can be carried out without the interchanging of rows guarantees that such conditions are satisfied in any given numerical matrix. As we can see, the only elementary row operations required for the reduction were operations of the form $\left(R_i \to R_i + sR_j\right)$, in which $i > j$. The general decomposition uses similar ideas.

THEOREM 2.31 (LU decomposition theorem) *Suppose that A is an $m \times n$ matrix that can be reduced to row echelon form without requiring the permutation of rows. Then there exist an $m \times m$ lower-triangular matrix L with 1s on the diagonal and an $m \times n$ matrix row echelon matrix U such that $A = LU$.*

Proof. By Corollary 2.12, there exist elementary matrices $E_1, \ldots, E_p$ such that $E_p \cdots E_1 A$ is in row echelon form. Since row echelon matrices are upper triangular, we can define U to be $E_p \cdots E_1 A$. The reduction of A to U can be achieved without the permutation of rows. We can therefore assume that the required elementary matrices E_k represent operations of the form $(R_i \to R_i + sR_j)$, designed to replace a nonzero entry a_{ij} by 0, for $i > j$.

Each E_k is a lower-triangular elementary matrix with 1s on the diagonal. Hence the inverses E_k^{-1} are also lower-triangular matrices with 1s on the diagonal. By Theorem 2.29, the matrix $E_1^{-1} \cdots E_p^{-1}$ is lower triangular and has 1s on the diagonal. If we put $L = E_1^{-1} \cdots E_p^{-1}$, then $A = LU$ is the required decomposition. ■

In Table 3 on page 188, we summarize the process for finding LU decompositions described in the proof of Theorem 2.31.

TABLE 3 FINDING AN LU DECOMPOSITION WITHOUT ROW INTERCHANGES.
Step 1. Take an $m \times n$ matrix A that can be reduced to a row echelon matrix U by elementary row operations of the form $(R_i \to R_i + sR_j)$.
Step 2. Reduce A to U using row additions $(R_i \to R_i + sR_j)$, as shown in Example 2.45.
Step 3. For each i, form the elementary matrix E_i that corresponds to an operation $(R_i \to R_i + sR_j)$ required in Step 2.
Step 4. The product $\left(E_1^{-1} \cdots E_k^{-1}\right)$ is the required lower-triangular matrix L.

The next example shows that some matrices can only be decomposed into the form LU if some of their rows are first permuted.

EXAMPLE 2.46 An LU Decomposition of a Matrix *PA*

Find an LU decomposition of the matrix

$$A = \begin{bmatrix} 1 & 2 & 3 \\ 2 & 4 & 1 \\ 2 & 5 & 7 \end{bmatrix}$$

using a permutation matrix P, if necessary.

Solution. The following steps yield an LU decomposition of A.

1. We first apply $(R_2 \to R_2 - 2R_1)$ to A and get

$$\begin{bmatrix} 1 & 0 & 0 \\ -2 & 1 & 0 \\ 0 & 0 & 1 \end{bmatrix} \begin{bmatrix} 1 & 2 & 3 \\ 2 & 4 & 1 \\ 2 & 5 & 7 \end{bmatrix} = \begin{bmatrix} 1 & 2 & 3 \\ 0 & 0 & -5 \\ 2 & 5 & 7 \end{bmatrix} = A_1$$

2. Then we apply $(R_3 \to R_3 - 2R_1)$ to A_1 and get

$$\begin{bmatrix} 1 & 0 & 0 \\ 0 & 1 & 0 \\ -2 & 0 & 1 \end{bmatrix} \begin{bmatrix} 1 & 2 & 3 \\ 0 & 0 & -5 \\ 2 & 5 & 7 \end{bmatrix} = \begin{bmatrix} 1 & 2 & 3 \\ 0 & 0 & -5 \\ 0 & 1 & 1 \end{bmatrix} = A_2$$

3. Next we apply $(R_3 \rightleftharpoons R_2)$ to A_2 and obtain

$$\begin{bmatrix} 1 & 0 & 0 \\ 0 & 0 & 1 \\ 0 & 1 & 0 \end{bmatrix} \begin{bmatrix} 1 & 2 & 3 \\ 0 & 0 & -5 \\ 0 & 1 & 1 \end{bmatrix} = \begin{bmatrix} 1 & 2 & 3 \\ 0 & 1 & 1 \\ 0 & 0 & -5 \end{bmatrix} = A_3$$

4. The interchange of rows 2 and 3 in the last step was essential to the reduction of A to row echelon form. If we carry this interchange out first, we get the matrix

$$\begin{bmatrix} 1 & 0 & 0 \\ 0 & 0 & 1 \\ 0 & 1 & 0 \end{bmatrix} \begin{bmatrix} 1 & 2 & 3 \\ 2 & 4 & 1 \\ 2 & 5 & 7 \end{bmatrix} = \begin{bmatrix} 1 & 2 & 3 \\ 2 & 5 & 7 \\ 2 & 4 & 1 \end{bmatrix} = PA$$

5. Since

$$LU = \begin{bmatrix} 1 & 0 & 0 \\ 2 & 1 & 0 \\ 2 & 0 & 1 \end{bmatrix} \begin{bmatrix} 1 & 2 & 3 \\ 0 & 1 & 1 \\ 0 & 0 & -5 \end{bmatrix} = \begin{bmatrix} 1 & 2 & 3 \\ 2 & 5 & 7 \\ 2 & 4 & 1 \end{bmatrix} = PA$$

the matrices

$$L = \begin{bmatrix} 1 & 0 & 0 \\ 0 & 1 & 0 \\ -2 & 0 & 1 \end{bmatrix}^{-1} \begin{bmatrix} 1 & 0 & 0 \\ -2 & 1 & 0 \\ 0 & 0 & 1 \end{bmatrix}^{-1} = \begin{bmatrix} 1 & 0 & 0 \\ 2 & 1 & 0 \\ 2 & 0 & 1 \end{bmatrix}$$

and

$$U = \begin{bmatrix} 1 & 2 & 3 \\ 0 & 1 & 1 \\ 0 & 0 & -5 \end{bmatrix}$$

yield an LU decomposition of PA. ▲

The use of the interchange $(R_3 \leftrightharpoons R_2)$ in this decomposition cannot be avoided. The next lemma shows that we can always carry out all required permutations of rows before all operations of the form $(R_i \to R_i + sR_j)$.

LEMMA 2.32 *For every real $m \times n$ matrix A there exists a permutation matrix P such that the matrix PA can be reduced to row echelon form without the permutation of rows.*

Proof. We must show that for every real $m \times n$ matrix A there exists an $m \times m$ permutation matrix P and lower-triangular elementary matrices $E_1, \ldots, E_p$ not involving any permutations of rows $(R_i \leftrightharpoons R_j)$ such that $E_1 \cdots E_p PA = U$ is a reduced row echelon matrix.

We convert A to U column by column, starting on the left. The underlying idea is to build up the permutation matrix P step by step as we construct the matrices $E_1 \cdots E_p$. We use the fact that whereas a left-multiplication by an elementary permutation matrix P corresponds to an interchange of rows, a right-multiplication by P corresponds to an interchange of columns. We use this fact to show that for appropriate lower-triangular matrices L, there exist permutation matrices P for which $PL = L'P$. We can therefore replace PL by $L'P$ and move all required permutation to the right of all required lower-triangular matrices L. The following examples illustrate the idea. If

$$L = \begin{bmatrix} 1 & 0 & 0 \\ a & 1 & 0 \\ b & 0 & 1 \end{bmatrix}$$

then

$$PLP^{-1} = \begin{bmatrix} 1 & 0 & 0 \\ 0 & 0 & 1 \\ 0 & 1 & 0 \end{bmatrix} \begin{bmatrix} 1 & 0 & 0 \\ a & 1 & 0 \\ b & 0 & 1 \end{bmatrix} \begin{bmatrix} 1 & 0 & 0 \\ 0 & 0 & 1 \\ 0 & 1 & 0 \end{bmatrix} = \begin{bmatrix} 1 & 0 & 0 \\ b & 1 & 0 \\ a & 0 & 1 \end{bmatrix} = L'$$

and if

$$L = \begin{bmatrix} 1 & 0 & 0 & 0 \\ a & 1 & 0 & 0 \\ b & 0 & 1 & 0 \\ c & 0 & 0 & 1 \end{bmatrix}$$

then

$$\begin{aligned} PLP^{-1} &= \begin{bmatrix} 1 & 0 & 0 & 0 \\ 0 & 0 & 1 & 0 \\ 0 & 1 & 0 & 0 \\ 0 & 0 & 0 & 1 \end{bmatrix} \begin{bmatrix} 1 & 0 & 0 & 0 \\ a & 1 & 0 & 0 \\ b & 0 & 1 & 0 \\ c & 0 & 0 & 1 \end{bmatrix} \begin{bmatrix} 1 & 0 & 0 & 0 \\ 0 & 0 & 1 & 0 \\ 0 & 1 & 0 & 0 \\ 0 & 0 & 0 & 1 \end{bmatrix} \\ &= \begin{bmatrix} 1 & 0 & 0 & 0 \\ b & 1 & 0 & 0 \\ a & 0 & 1 & 0 \\ c & 0 & 0 & 1 \end{bmatrix} = L' \end{aligned}$$

In both cases, $PLP^{-1} = L'$, and therefore $PL = L'P$ since $P = P^{-1}$. We use these an analogous equations to construct a decomposition $(E_1 \cdots E_p) PA = LPA = U$. Since L^{-1} is also lower triangular, we get $L^{-1}U = LU = PA$. ■

THEOREM 2.33 (PLU decomposition theorem) *For any $m \times n$ matrix A there exist a permutation matrix P, an $m \times m$ lower-triangular matrix L with 1s on the diagonal, and an $m \times n$ row echelon matrix U such that $PA = LU$.*

Proof. By Lemma 2.32, there exists a permutation matrix P such that PA can be reduced to row echelon form without requiring further permutations of rows. Theorem 2.31 therefore guarantees that there exist suitable matrices L and U for which $PA = LU$. ■

Let us now consider the special case of an LU decomposition of an invertible matrix. We show that in this case, the decomposition is unique.

THEOREM 2.34 *Suppose that A is an invertible matrix that can be reduced to row echelon form without requiring the permutation of rows. Then there exists a unique lower-triangular matrix L with 1s on the diagonal and a unique upper-triangular matrix U such that $A = LU$.*

Proof. By Theorem 2.31, there exist matrices L and U of the required form for which $A = LU$. It remains to prove the uniqueness of L and U. Suppose, therefore, that $A = L_1U_1 = L_2U_2$. Then

$$\begin{aligned} U_1U_2^{-1} &= \left(L_1^{-1}L_1\right)\left(U_1U_2^{-1}\right) \\ &= L_1^{-1}\left(L_1U_1\right)U_2^{-1} \\ &= L_1^{-1}\left(L_2U_2\right)U_2^{-1} \\ &= \left(L_1^{-1}L_2\right)\left(U_2U_2^{-1}\right) \\ &= L_1^{-1}L_2 \end{aligned}$$

But $U_1U_2^{-1}$ is upper triangular and $L_1^{-1}L_2$ is lower triangular with 1s on the diagonal. Hence $U_1U_2^{-1} = L_1^{-1}L_2 = I$. Therefore, $L_1 = L_2$ and $U_1 = U_2$. ■

Maple has a built-in function for calculating LU and PLU decompositions. In combination with back substitution, they provide an efficient method for solving certain matrix equations.

EXAMPLE 2.47 Using *Maple* to Find an LU Decomposition

Use *Maple* to find an LU decomposition for the matrix

$$A = \begin{bmatrix} 1 & 4 & 6 & 3 \\ 2 & 1 & 3 & 0 \\ 1 & 8 & 3 & 0 \\ 4 & 5 & 1 & 5 \end{bmatrix}$$

Solution. We begin by defining the matrix A.

```
[>with(LinearAlgebra):
[>A := Matrix([[1,4,6,3],[2,1,3,0],[1,8,3,0],[4,5,1,5]]):
```

Next we use *Maple* to find an LU decomposition of A.

```
[> P,L,U:=LUDecomposition(A);
```

$$P, L, U := \begin{bmatrix} 1 & 0 & 0 & 0 \\ 0 & 1 & 0 & 0 \\ 0 & 0 & 1 & 0 \\ 0 & 0 & 0 & 1 \end{bmatrix}, \begin{bmatrix} 1 & 0 & 0 & 0 \\ 2 & 1 & 0 & 0 \\ 1 & \frac{-4}{7} & 1 & 0 \\ 4 & \frac{11}{7} & \frac{62}{57} & 1 \end{bmatrix}, \begin{bmatrix} 1 & 4 & 6 & 3 \\ 0 & -7 & -9 & -6 \\ 0 & 0 & \frac{-57}{7} & \frac{-45}{7} \\ 0 & 0 & 0 & \frac{179}{9} \end{bmatrix}$$

```
[> P.L.U;
```

$$\begin{bmatrix} 1 & 4 & 6 & 3 \\ 2 & 1 & 3 & 0 \\ 1 & 8 & 3 & 0 \\ 4 & 5 & 1 & 5 \end{bmatrix}$$

Thus the *Maple* has produced a PLU decomposition of A. Theorem 2.31 tells us that the matrices L and U are unique, since A is invertible and since P is an identity matrix.▲

EXAMPLE 2.48 Using *Maple* to Find a PLU Decomposition

Use *Maple* to find a PLU decomposition of the matrix

$$A = \begin{bmatrix} 5 & 0 & 3 \\ 5 & 0 & 5 \\ 0 & 1 & 8 \end{bmatrix}$$

Solution. We begin by loading the **LinearAlgebra** package.

```
[> with(LinearAlgebra):
```

Next we define the matrix A and calculate its PLU decomposition.

```
[> A:=Matrix([[5,0,3],[5,0,5],[0,1,8]]):
[> P,L,U:=LUDecomposition(A);
```

$$P, L, U := \begin{bmatrix} 1 & 0 & 0 \\ 0 & 0 & 1 \\ 0 & 1 & 0 \end{bmatrix}, \begin{bmatrix} 1 & 0 & 0 \\ 0 & 1 & 0 \\ 1 & 0 & 1 \end{bmatrix}, \begin{bmatrix} 5 & 0 & 3 \\ 0 & 1 & 8 \\ 0 & 0 & 2 \end{bmatrix}$$

where P is the permutation matrix corresponding to the permutation $[1, 3, 2]$ and L and U are lower- and upper-triangular matrices. ▲

The **LUDecomposition** function is designed to go hand in hand with the **ForwardSubstitute** and **BackwardSubstitute** functions to produce efficient solutions of a set matrix equation $A\mathbf{x} = \mathbf{b}_1, \ldots, A\mathbf{x} = \mathbf{b}_n$.

EXAMPLE 2.49 LU Back Substitution and Matrix Equations

Use the **LUDecomposition** function in the **LinearAlgebra** package to solve the matrix equation

$$\begin{bmatrix} 1 & 4 & 6 & 3 \\ 2 & 1 & 3 & 0 \\ 1 & 8 & 3 & 0 \\ 4 & 5 & 1 & 5 \end{bmatrix} \begin{bmatrix} w \\ x \\ y \\ z \end{bmatrix} = \begin{bmatrix} 5 \\ 2 \\ -3 \\ 1 \end{bmatrix}$$

Solution. We begin by finding an LU decomposition of the matrix A.

```
[>with(LinearAlgebra):
[>A:=Matrix([[1,4,6,3],[2,1,3,0],[1,8,3,0],[4,5,1,5]]):
[>b:=Vector[column]([5,2,-3,1]):
```

```
[>LU:=LUDecomposition(A);
```

$$LU := \begin{bmatrix} 1 & 0 & 0 & 0 \\ 0 & 1 & 0 & 0 \\ 0 & 0 & 1 & 0 \\ 0 & 0 & 0 & 1 \end{bmatrix}, \begin{bmatrix} 1 & 0 & 0 & 0 \\ 2 & 1 & 0 & 0 \\ 1 & \frac{-4}{7} & 1 & 0 \\ 4 & \frac{11}{7} & \frac{62}{57} & 1 \end{bmatrix}, \begin{bmatrix} 1 & 4 & 6 & 3 \\ 0 & -7 & -9 & -6 \\ 0 & 0 & \frac{-57}{7} & \frac{-45}{7} \\ 0 & 0 & 0 & \frac{179}{19} \end{bmatrix}$$

Next we apply the **ForwardSubstitute** function to the equation $L\mathbf{y} = \mathbf{b}$ and the function **BackwardSubstitute** to the equations $U\mathbf{x} = \mathbf{y}$ to find $\mathbf{x}$.

```
[>y:=ForwardSubstitute(LU[2],b);
```

$$y := \begin{bmatrix} 5 \\ -8 \\ \frac{-88}{7} \\ \frac{413}{57} \end{bmatrix}$$

```
[> x:=BackwardSubstitute(LU[3],y);
```

$$x := \begin{bmatrix} -\frac{8}{179} \\ -\frac{129}{179} \\ \frac{503}{537} \\ \frac{413}{537} \end{bmatrix}$$

This shows that the vector $(w, x, y, z) = (-8/179, -129/179, 503/537, 413/537)$ is a solution of the given equation. ▲

We can use the **LUDecomposition** function to solve several linear systems at once. For example, we can solve the system

$$A\mathbf{x} = \begin{bmatrix} 1 & 4 & 6 & 3 \\ 2 & 1 & 3 & 0 \\ 1 & 8 & 3 & 0 \\ 4 & 5 & 1 & 5 \end{bmatrix} \begin{bmatrix} w \\ x \\ y \\ z \end{bmatrix} = \begin{bmatrix} 5 \\ 2 \\ -3 \\ 1 \end{bmatrix} = \mathbf{b}_1$$

and the system

$$A\mathbf{x} = \begin{bmatrix} 1 & 4 & 6 & 3 \\ 2 & 1 & 3 & 0 \\ 1 & 8 & 3 & 0 \\ 4 & 5 & 1 & 5 \end{bmatrix} \begin{bmatrix} w \\ x \\ y \\ z \end{bmatrix} = \begin{bmatrix} 1 \\ 2 \\ 3 \\ 4 \end{bmatrix} = \mathbf{b}_2$$

by using $\mathbf{b}_1$ and $\mathbf{b}_2$ as the columns of a matrix B and solving the equation $A\mathbf{x} = B$.

```
[> with(LinearAlgebra):
[> A:=Matrix([[1,4,6,3],[2,1,3,0],[1,8,3,0],[4,5,1,5]]):
[> B:=Matrix([[5,1],[2,2],[-3,3],[1,4]]):
[> LU:=LUDecomposition(A):
```

```
[> Y:=ForwardSubstitute(LU[2],B);
```

$$Y := \begin{bmatrix} 5 & 1 \\ -8 & 0 \\ \frac{-88}{7} & 2 \\ \frac{413}{57} & \frac{-124}{57} \end{bmatrix}$$

```
[> X:=BackwardSubstitute(LU[3],Y);
```

$$X := \begin{bmatrix} \frac{-8}{179} & \frac{171}{179} \\ \frac{-129}{179} & \frac{50}{179} \\ \frac{503}{537} & \frac{-34}{537} \\ \frac{413}{537} & \frac{-124}{537} \end{bmatrix}$$

As we can see, the first column of X is a solution of the equation $A\mathbf{x} = \mathbf{b}_1$. Similarly, the second column of X is a solution of the equation $A\mathbf{x} = \mathbf{b}_2$.

In addition to the various built-in techniques for solving linear systems using LU decomposition, *Maple* has a very general function for solving matrix equations. A consistent linear system of the form $A\mathbf{x} = \mathbf{b}$ can always be solved using the **LinearSolve** function.

EXAMPLE 2.50 Using LinearSolve to Solve a Matrix Equation

Use the **LinearSolve** function to find a solution of the equation

$$\begin{bmatrix} 1 & 4 & 6 & 3 \\ 2 & 1 & 3 & 0 \\ 1 & 8 & 3 & 0 \end{bmatrix} \begin{bmatrix} w \\ x \\ y \\ z \end{bmatrix} = \begin{bmatrix} 1 \\ 2 \\ 1 \end{bmatrix}$$

Solution. We begin by defining the matrix A and the column vector **b**.

```
[> A:=Matrix([[1,4,6,3],[2,1,3,0],[1,8,3,0]]):
[> b:=<1,2,1>:
```

Next we apply the **LinearSolve** function to A and **b**.

```
[> LinearSolve(A,b);
```

$$\begin{bmatrix} 1 - \frac{7}{5}\_t3_3 \\ -\frac{1}{5}\_t3_3 \\ \_t3_3 \\ -\frac{19}{15}\_t3_3 \end{bmatrix}$$

If we assign the value 0 to the variable $\_t3_3$, it follows that the vector $(1, 0, 0, 0)$ is a solution of the given equation. ▲

The function **LinearSolve** can be forced to use several solution methods, including LU decomposition. For example, the command `LinearSolve(A,b,method='LU')` may produce the output

$$\begin{bmatrix} 1+\frac{21}{19}\_t28_1 \\ \frac{3}{19}\_t28_1 \\ \frac{-15}{19}\_t28_1 \\ \_t28_1 \end{bmatrix}$$

with the free variable $\_t28_1$. If we assign the value 0 to $\_t28_1$, we get the same solution as that found in Example 2.50.

EXERCISES 2.7

1. Determine which of the following matrices are upper triangular. Explain your answers.

a. $A = \begin{bmatrix} 4 & 1 & 3 & 1 \\ 0 & 1 & 4 & 3 \\ 0 & 0 & 3 & 1 \\ 0 & 0 & 0 & 2 \end{bmatrix}$ b. $A = \begin{bmatrix} 4 & 1 & 3 & 1 \\ 0 & 1 & 4 & 3 \\ 0 & 0 & 3 & 1 \\ 0 & 0 & 0 & 2 \\ 0 & 0 & 0 & 0 \end{bmatrix}$ c. $A = \begin{bmatrix} 1 & 0 & 0 & 0 & 0 \\ 0 & 1 & 0 & 0 & 0 \\ 0 & 0 & 3 & 0 & 0 \\ 0 & 0 & 0 & 0 & 0 \\ 0 & 0 & 0 & 0 & 0 \end{bmatrix}$

2. Take the transpose of the matrices in Exercise 1 and determine which of the resulting matrices are lower triangular. Explain your answers.

3. Find elementary matrices $E_1, \ldots, E_n$ and an upper-triangular matrix U for the matrix

$$A = \begin{bmatrix} 1 & 2 & 4 & 1 \\ -4 & 1 & 4 & -4 \\ 2 & -2 & -4 & 4 \\ 0 & 0 & 2 & 3 \end{bmatrix}$$

such that $A = E_1^{-1} \cdots E_n^{-1} U$.

4. Use the LU decomposition

$$\begin{bmatrix} 7 & 5 \\ 1 & 4 \end{bmatrix} = \begin{bmatrix} 1 & 0 \\ \frac{1}{7} & 1 \end{bmatrix} \begin{bmatrix} 7 & 5 \\ 0 & \frac{23}{7} \end{bmatrix}$$

and forward and back substitution to solve the linear system

$$\begin{bmatrix} 7 & 5 \\ 1 & 4 \end{bmatrix} \begin{bmatrix} x \\ y \end{bmatrix} = \begin{bmatrix} 2 \\ 5 \end{bmatrix}$$

5. Use the LU decomposition

$$\begin{bmatrix} 5 & 0 \\ -3 & 7 \end{bmatrix} = \begin{bmatrix} 1 & 0 \\ -\frac{3}{5} & 1 \end{bmatrix} \begin{bmatrix} 5 & 0 \\ 0 & 7 \end{bmatrix}$$

and forward and back substitution to solve the linear system

$$\begin{bmatrix} 5 & 0 \\ -3 & 7 \end{bmatrix} \begin{bmatrix} x \\ y \end{bmatrix} = \begin{bmatrix} 3 \\ 4 \end{bmatrix}$$

6. Use the LU decomposition

$$\begin{bmatrix} 5 & 0 & -2 \\ -3 & 7 & 5 \\ 9 & 1 & 4 \end{bmatrix} = \begin{bmatrix} 1 & 0 & 0 \\ -\frac{3}{5} & 1 & 0 \\ \frac{9}{5} & \frac{1}{7} & 1 \end{bmatrix} \begin{bmatrix} 5 & 0 & -2 \\ 0 & 7 & \frac{19}{5} \\ 0 & 0 & \frac{247}{35} \end{bmatrix}$$

and forward and back substitution to solve the linear system

$$\begin{bmatrix} 5 & 0 & -2 \\ -3 & 7 & 5 \\ 9 & 1 & 4 \end{bmatrix} \begin{bmatrix} x \\ y \\ z \end{bmatrix} = \begin{bmatrix} 3 \\ 4 \\ 5 \end{bmatrix}$$

7. Use the LU decomposition

$$\begin{bmatrix} 1 & 0 & 2 \\ 4 & 2 & 5 \\ 9 & 1 & 3 \end{bmatrix} = \begin{bmatrix} 1 & 0 & 0 \\ 4 & 1 & 0 \\ 9 & \frac{1}{2} & 1 \end{bmatrix} \begin{bmatrix} 1 & 0 & 2 \\ 0 & 2 & -3 \\ 0 & 0 & -\frac{27}{2} \end{bmatrix}$$

and forward and back substitution to solve the linear system

$$\begin{bmatrix} 1 & 0 & 2 \\ 4 & 2 & 5 \\ 9 & 1 & 3 \end{bmatrix} \begin{bmatrix} x \\ y \\ z \end{bmatrix} = \begin{bmatrix} 2 \\ 5 \\ 3 \end{bmatrix}$$

8. Use elementary row reduction to find an LU decomposition of the following matrices.

a. $\begin{bmatrix} 2 & 2 & 1 \\ 3 & 5 & 0 \end{bmatrix}$ b. $\begin{bmatrix} 2 & 2 & 1 & 0 & 3 \\ 3 & 5 & 0 & 0 & 0 \\ 1 & 3 & 1 & 0 & 2 \end{bmatrix}$ c. $\begin{bmatrix} 5 & 2 & 5 & 3 & 2 & 5 \\ 0 & 5 & 2 & 0 & 3 & 5 \\ 3 & 2 & 1 & 4 & 0 & 4 \\ 3 & 3 & 1 & 5 & 5 & 0 \end{bmatrix}$

9. Use elementary row reduction and suitable permutation matrices P to find an LU decomposition of PA for the following matrices A.

$$\text{a.} \begin{bmatrix} 0 & 3 \\ 4 & 0 \\ 5 & 5 \end{bmatrix} \quad \text{b.} \begin{bmatrix} 0 & 0 & 3 & 5 \\ 1 & 4 & 0 & 4 \end{bmatrix} \quad \text{c.} \begin{bmatrix} 0 & 0 & 3 & 1 & 3 \\ 0 & 1 & 5 & 0 & 0 \\ 3 & 0 & 1 & 3 & 0 \\ 3 & 4 & 2 & 3 & 2 \end{bmatrix}$$

10. Use LU decomposition to solve the following matrix equations $A\mathbf{x} = \mathbf{b}_i$.

$$\text{a. } A = \begin{bmatrix} 1 & 0 & 2 \\ 5 & 2 & 5 \\ 3 & 2 & 5 \end{bmatrix} \quad \mathbf{b}_1 = \begin{bmatrix} 1 \\ 2 \\ 3 \end{bmatrix} \quad \mathbf{b}_2 = \begin{bmatrix} 0 \\ 5 \\ 4 \end{bmatrix} \quad \mathbf{b}_3 = \begin{bmatrix} 6 \\ 7 \\ 1 \end{bmatrix}$$

$$\text{b. } A = \begin{bmatrix} 1 & 5 & 2 & 0 \\ 3 & 5 & 3 & 2 \\ 1 & 4 & 0 & 4 \\ 3 & 3 & 1 & 5 \end{bmatrix} \quad \mathbf{b}_1 = \begin{bmatrix} 1 \\ 2 \\ 0 \\ 2 \end{bmatrix} \quad \mathbf{b}_2 = \begin{bmatrix} 0 \\ 1 \\ 0 \\ 4 \end{bmatrix} \quad \mathbf{b}_3 = \begin{bmatrix} 2 \\ 8 \\ 4 \\ -2 \end{bmatrix}$$

11. Find two distinct LU decompositions L_1U_1 and L_2U_2 of the zero matrix

$$\begin{bmatrix} 0 & 0 & 0 \\ 0 & 0 & 0 \\ 0 & 0 & 0 \end{bmatrix}$$

and explain why this does not contradict Theorem 2.34.

12. Construct an example that illustrates Theorem 2.31.

13. Construct an example that illustrates Theorem 2.34.

14. (*Maple V*) Generate two random 3×3 matrices A and B and replace all entries above the diagonal in A and B by zero. Show that the product of the resulting matrices is lower triangular.

15. (*Maple V*) Find an LU decomposition of the following matrices.

$$\text{a.} \begin{bmatrix} 3 & 3 & 2 \\ 5 & 5 & 4 \\ 5 & 0 & 2 \end{bmatrix} \quad \text{b.} \begin{bmatrix} 1 & 1 & 4 \\ 0 & 4 & 1 \\ 4 & 5 & 4 \end{bmatrix} \quad \text{c.} \begin{bmatrix} 0 & 1 & 3 & -2 \\ 2 & -1 & 0 & 5 \\ -2 & 5 & 2 & 1 \\ 2 & -4 & 2 & -5 \end{bmatrix}$$

16. (*Maple 6*) Use the command

```
RandomMatrix(100,100,generator=1..9,
outputoptions = [shape=triangular[lower]])
```

to generate two random matrices A and B and let $C = AB$. Explain why the command

```
evalb(C[45,46]=0 and C[77,99]=0)
```

returns the values *true*.

17. (*Maple 6*) Use LU decomposition to solve the following matrix equations $A\mathbf{x} = \mathbf{b}_i$.

$$A = \begin{bmatrix} 1 & 5 & 2 & 0 \\ 3 & 5 & 3 & 2 \\ 1 & 4 & 0 & 4 \\ 3 & 3 & 1 & 5 \end{bmatrix}, \quad \mathbf{b}_1 = \begin{bmatrix} 1 \\ 2 \\ 0 \\ 2 \end{bmatrix}, \quad \mathbf{b}_2 = \begin{bmatrix} 0 \\ 1 \\ 0 \\ 4 \end{bmatrix}, \quad \mathbf{b}_3 = \begin{bmatrix} 2 \\ 8 \\ 4 \\ -2 \end{bmatrix}$$

18. (*Maple 6*) Use the **LinearAlgebra[LinearSolve]** function to find solutions for the equations in Exercise 17.

19. (*Maple 6*) Explore the properties of the **LUDecomposition**, **BackwardSubstitute,** and **ForwardSubstitute** functions in the **LinearAlgebra** package to find solutions for the equations in Exercise 17.

20. (*Maple 6*) Generate a graphic similar to that on the cover of this book using the **RandomMatrix** and **LUDecomposition** functions in the **LinearAlgebra package**, together with the **matrixplot** function in the **plots** package.

Complex Matrices

A ***complex matrix*** is a matrix $A : \mathbf{m} \times \mathbf{n} \to \mathbb{C}$ whose entries are complex numbers. A ***complex polynomial matrix*** is a matrix $A : \mathbf{m} \times \mathbf{n} \to \mathbb{C}[t]$ whose entries are complex polynomials. Since every real number is also a complex number, we can regard all real matrices as complex. Here are some examples of complex matrices that are not real matrices.

EXAMPLE 2.51 A Complex Matrix

Find a *Maple* representation of the matrix $A : \mathbf{3} \times \mathbf{4} \to \mathbb{C}$, defined by

$$\begin{aligned} &A(1,1) = 3,\ A(1,2) = 4,\ A(1,3) = \tfrac{1}{2},\ A(1,4) = \pi, \\ &A(2,1) = 2,\ A(2,2) = \sqrt{-1} = i,\ A(2,3) = -i,\ A(2,4) = 2, \\ &A(3,1) = 8,\ A(3,2) = -\tfrac{6}{7},\ A(3,3) = -1 + 2i,\ A(3,4) = \sqrt{15} \end{aligned}$$

Solution. We use the **Matrix** palette.

```
[> A:=Matrix([[%?, %?, %?, %?],[%?, %?, %?, %?],
    [%?, %?, %?, %?]]):
[> A:=Matrix([[3,4,1/2,Pi], [2,I,-I,2],
    [8,-6/7,-1+2*I,sqrt(15)]]);
```

$$A := \begin{bmatrix} 3 & 4 & \frac{1}{2} & \pi \\ 2 & I & -I & 2 \\ 8 & \frac{-6}{7} & -1+2I & \sqrt{15} \end{bmatrix}$$

As we can see, *Maple* represents the given matrix as a list of three lists. ▲

EXAMPLE 2.52 A Complex Polynomial Matrix

Write the matrix $A : \mathbf{3} \times \mathbf{4} \to \mathbb{C}[t]$, defined by

$$A(1,1) = \sqrt{-1}t^4,\ A(1,2) = \sqrt{-1},\ A(1,3) = t^2 - 1,\ A(1,4) = 2t,$$
$$A(2,1) = 5,\ A(2,2) = t^3,\ A(2,3) = 4\pi,\ A(2,4) = -t^5 + t^3 - 7,$$
$$A(3,1) = 1 + 2\sqrt{-1},\ A(3,2) = 5,\ A(3,3) = t + 8,\ A(3,4) = \tfrac{1}{3}t + 12$$

as a list of lists and convert the *Maple* output to matrix form.

Solution. We use the **matrix** function to construct A.

```
[> A := linalg[matrix]([
    [sqrt(-1)*t^4,sqrt(-1),t^2-1,2*t],
    [5,t^3,4*Pi,-t^5+t^3-7],
    [1+2*sqrt(-1),5,t+8,1/3*t+12]]);
```

$$A := \begin{bmatrix} I\,t^4 & I & t^2-1 & 2t \\ 5 & t^3 & 4\pi & -t^5+t^3-7 \\ 1+2I & 5 & t+8 & \frac{1}{3}t+12 \end{bmatrix}$$

The output is the required polynomial matrix. ▲

Hermitian Matrices

If $A = [a_{ij}]$ is an $m \times n$ matrix with complex entries, then the ***conjugate transpose*** A^H of A is the $n \times m$ matrix $[b_{ij}]$ in which $b_{ij} = \overline{a_{ji}}$, where $\overline{a_{ji}}$ is the complex conjugate of a_{ji}. For real matrices, the transpose A^T and the conjugate transpose A^H are the same, since $\overline{a_{ji}} = a_{ij}$ for all entries of A.

EXAMPLE 2.53 The Conjugate Transpose of a Matrix

Use *Maple* to calculate the conjugate transpose B of the matrix

$$A = \begin{bmatrix} 1 & 2-4i & 9 & 3 \\ i & 6 & 7+5i & 8 \end{bmatrix}$$

Solution. By combining the **conjugate** function with the **transpose** function in the **linalg** package, we can calculate A^H in one step.

```
[> B:=linalg[transpose]([
   [1,conjugate(2-4*I),9,3],
   [conjugate(I),6,conjugate(7+5*I),8]]);
```

$$B := \begin{bmatrix} 1 & -I \\ 2+4I & 6 \\ 9 & 7-5I \\ 3 & 8 \end{bmatrix}$$

This output is the *Maple* form of A^H. ▲

A ***Hermitian matrix*** is a complex matrix A with the property that $A = A^H$.

EXAMPLE 2.54 A Hermitian Matrix

The matrix

$$A = \begin{bmatrix} a_{11} & a_{12} & a_{13} & a_{14} \\ a_{21} & a_{22} & a_{23} & a_{24} \\ a_{31} & a_{31} & a_{33} & a_{34} \\ a_{41} & a_{42} & a_{43} & a_{44} \end{bmatrix} = \begin{bmatrix} 1 & 2-i & 9 & 3 \\ 2+i & 6 & i & 8 \\ 9 & -i & 4 & 5+7i \\ 3 & 8 & 5-7i & 3 \end{bmatrix}$$

is Hermitian since $a_{ij} = \overline{a_{ji}}$ for all indices $1 \leq i, j \leq 4$. ▲

Unitary Matrices

A ***unitary matrix*** is a complex matrix A with the property that $A^{-1} = A^H$.

EXAMPLE 2.55 A Unitary Matrix

Use *Maple* to show that the matrix

$$A = \begin{bmatrix} \frac{1}{2}(1-i) & -\frac{1}{2}(1-i) \\ \frac{1}{2}(1+i) & \frac{1}{2}(1+i) \end{bmatrix}$$

is unitary.

Solution. We use *Maple* to create the matrices A^{-1} and A^H and compare the results.

```
[> A:=matrix([[(1-I)/2,-(1-I)/2],[(1+I)/2,(1+I)/2]]):
[> B:=matrix([
   [conjugate(%[1,1]),conjugate(%[1,2])],
   [conjugate(%[2,1]),conjugate(%[2,2])]]):
```

```
[> linalg[inverse](A)=linalg[transpose](B);
```

$$\begin{bmatrix} \frac{1}{2}+\frac{1}{2}I & \frac{1}{2}-\frac{1}{2}I \\ -\frac{1}{2}-\frac{1}{2}I & \frac{1}{2}-\frac{1}{2}I \end{bmatrix} = \begin{bmatrix} \frac{1}{2}+\frac{1}{2}I & \frac{1}{2}-\frac{1}{2}I \\ -\frac{1}{2}-\frac{1}{2}I & \frac{1}{2}-\frac{1}{2}I \end{bmatrix}$$

As we can see, the two matrices are identical. Hence A is unitary. ▲

The **LinearAlgebra** package has a built-in test function for determining unitary matrices. The matrix

$$\begin{bmatrix} \frac{1}{2}+\frac{1}{2}I & \frac{1}{2}-\frac{1}{2}I \\ -\frac{1}{2}-\frac{1}{2}I & \frac{1}{2}-\frac{1}{2}I \end{bmatrix}$$

for example, can be shown to be unitary using the following commands:

```
[> A:=Matrix([[(1-I)/2,-(1-I)/2],[(1+I)/2,(1+I)/2]]):
[> LinearAlgebra[IsUnitary](A);
                              true
```

EXERCISES 2.8

1. Construct two complex 2×3 matrices A and B, each with three real and three imaginary entries. Calculate the sum $A + B$.

2. Calculate the conjugate transposes of the matrices A, B, and $A + B$ in Exercise 1.

3. Construct a complex 3×4 matrix A and a 4×3 matrix B, each with six real and six imaginary entries. Calculate the matrices AB and BA.

4. Calculate the conjugate transposes of the matrices A, B, AB, and BA in Exercise 3.

5. Construct a 2×2 matrix A whose four entries are complex polynomials and calculate the matrix A^2.

6. Show that the Pauli spin matrices

$$\text{a.} \begin{bmatrix} 0 & 1 \\ 1 & 0 \end{bmatrix} \quad \text{b.} \begin{bmatrix} 0 & -i \\ i & 0 \end{bmatrix} \quad \text{c.} \begin{bmatrix} 1 & 0 \\ 0 & -1 \end{bmatrix}$$

used to calculate the electron spin in physics are both Hermitian and unitary.

7. (*Maple V*) Show that the Dirac matrices

$$\text{a.} \begin{bmatrix} 0 & 0 & 0 & 1 \\ 0 & 0 & 1 & 0 \\ 0 & 1 & 0 & 0 \\ 1 & 0 & 0 & 0 \end{bmatrix} \quad \text{b.} \begin{bmatrix} 0 & 0 & 0 & -i \\ 0 & 0 & i & 0 \\ 0 & -i & 0 & 0 \\ i & 0 & 0 & 0 \end{bmatrix}$$

and

$$\text{c.} \begin{bmatrix} 0 & 0 & 1 & 0 \\ 0 & 0 & 0 & -1 \\ 1 & 0 & 0 & 0 \\ 0 & -1 & 0 & 0 \end{bmatrix} \quad \text{d.} \begin{bmatrix} 1 & 0 & 0 & 0 \\ 0 & 1 & 0 & 0 \\ 0 & 0 & -1 & 0 \\ 0 & 0 & 0 & -1 \end{bmatrix}$$

are Hermitian and unitary.

8. (*Maple V*) Find a 2×2 and a 3×3 Hermitian matrix and use *Maple* to verify your answer.

9. (*Maple 6*) Verify that the matrix

$$A = \begin{bmatrix} \frac{1}{\sqrt{2}} & 0 & \frac{i}{\sqrt{2}} \\ 0 & 1 & 0 \\ -\frac{1}{\sqrt{2}} & 0 & \frac{i}{\sqrt{2}} \end{bmatrix}$$

is unitary. Modify A so that the result is a 2×2 unitary matrix.

APPLICATIONS

Geometry and Invertible Matrices

In this section, we illustrate the geometric properties of invertible 2×2 matrices. We show that these matrices correspond to sequences of contractions, expansions, reflections, and shears.

Let

$$A = \begin{bmatrix} a & b \\ c & d \end{bmatrix}$$

be a 2×2 real invertible matrix. By Corollary 2.19, the matrix A is equal to a product $E_1 \cdots E_n$ of elementary matrices and can therefore be obtained from the 2×2 identity matrix by a sequence of elementary row operations. We now explore the geometric effect of the transformation $\mathbf{x} \mapsto A\mathbf{x}$ on a point $\mathbf{x}$ of $\mathbb{R}^2$.

We recall from our earlier work that the following cases describe the possible types of elementary 2×2 matrices.

1. The operations $(R_1 \to sR_1)$, for $s \neq 0$, yield the elementary matrices

$$\begin{bmatrix} 1 & 0 \\ 0 & 1 \end{bmatrix} (R_1 \to sR_1) \begin{bmatrix} s & 0 \\ 0 & 1 \end{bmatrix} = E_1$$

   Left-multiplication by E_1 is called an ***expansion along the x-axis*** if $s > 1$ and a ***contraction along the x-axis*** if $0 < s < 1$. If $s = -1$, then a left-multiplication by E_1 corresponds to a ***reflection about the y-axis***.

2. The operations $(R_2 \to sR_2)$, for $s \neq 0$, yield the elementary matrices

$$\begin{bmatrix} 1 & 0 \\ 0 & 1 \end{bmatrix} (R_2 \to sR_2) \begin{bmatrix} 1 & 0 \\ 0 & s \end{bmatrix} = E_2$$

   Left-multiplication by E_2 is called an ***expansion along the y-axis*** if $s > 1$ and a ***contraction along the y-axis*** if $0 < s < 1$. If $s = -1$, then a left-multiplication by E_2 corresponds to a ***reflection about the x-axis***.

3. The operation $(R_1 \rightleftharpoons R_2)$ yields the elementary matrix

$$\begin{bmatrix} 1 & 0 \\ 0 & 1 \end{bmatrix} (R_1 \rightleftharpoons R_2) \begin{bmatrix} 0 & 1 \\ 1 & 0 \end{bmatrix} = E_3$$

   Left-multiplication by E_3 is called a ***reflection about the line*** $y = x$.

4. The operations $(R_1 \to R_1 + sR_2)$ yield the elementary matrices

$$\begin{bmatrix} 1 & 0 \\ 0 & 1 \end{bmatrix} (R_1 \to R_1 + sR_2) \begin{bmatrix} 1 & s \\ 0 & 1 \end{bmatrix} = E_4$$

   Left-multiplication by E_3 is called a ***shear along the x-axis***.

5. The operations $(R_2 \to R_2 + sR_1)$ yield the elementary matrices

$$\begin{bmatrix} 1 & 0 \\ 0 & 1 \end{bmatrix} (R_2 \to R_2 + sR_1) \begin{bmatrix} 1 & 0 \\ s & 1 \end{bmatrix} = E_5$$

   Left-multiplication by E_3 is called a ***shear along the y-axis***.

We now use the **LinearAlgebra** package to define fivefunctions that provide an efficient tool for calculating the geometric effects of left-multiplication by invertible matrices.

EXAMPLE 2.56 ***Maple*** **and 2 x 2 Invertible Matrices**

Use *Maple* to define the geometric transformations corresponding to left-multiplication by elementary 2×2 matrices.

Solution. We begin by loading the **LinearAlgebra** package.

```
[>with(LinearAlgebra):
```

Next we define the transformations $m_1 - m_5$ determined by the matrices $E_1 - E_5$.

```
[>m1:=x->Vector[column](evalm(Matrix([[s,0],[0,1]])&*x)):
[>m2:=x->Vector[column](evalm(Matrix([[1,0],[0,s]])&*x)):
[>m3:=x->Vector[column](evalm(Matrix([[0,1],[1,0]])&*x)):
[>m4:=x->Vector[column](evalm(Matrix([[1,s],[0,1]])&*x)):
[>m5:=x->Vector[column](evalm(Matrix([[1,0],[s,1]])&*x)):
```

For any invertible 2×2 matrix $A = E_n \cdots E_1$, the geometric effect of left-multiplying a point (x, y) by A corresponds to the assignment

$$\begin{bmatrix} x \\ y \end{bmatrix} \mapsto (m_n \circ \cdots \circ m_1) \begin{bmatrix} x \\ y \end{bmatrix}$$

where the m_i are the *Maple* transformations determined by the matrices E_i and where

$$m_n \circ \cdots \circ m_1$$

is the composite transformation of $m_1, \ldots, m_n$. ▲

The next two examples illustrate the use of the transformations defined in Example 2.56

EXAMPLE 2.57 Expansions of the Unit Square

Use *Maple* to find the rectangle determined by a unit square with vertices $(0, 0)$, $(1, 0)$, $(1, 1)$ and $(0, 1)$ and by an expansion along the x-axis for $s = 2$.

Solution. We first load the **LinearAlgebra** package.

```
[>with(LinearAlgebra):
```

Then we define the appropriate *Maple* function.

```
[>ExpansionX:=x->evalm(Matrix([[2,0],[0,1]])&*x):
```

Next we calculate the new vertices.

```
[> map(ExpansionX,
    {Vector([0,0]),Vector([1,0]),Vector([1,1]),
    Vector([0,1])});
```

$$\{[2, 1], [0, 1], [2, 0], [0, 0]\}$$

As we can tell, the unit square is expanded along the x-axis to a rectangle with a base of length 2 and height 1. ▲

EXAMPLE 2.58 Four Geometric Transformations

Use *Maple* to find the triangle determined by the isosceles triangle with vertices (0, 0), (4, 0), and (2, 2) and by the geometric transformation corresponding to the following operations:

a. A reflection about the line $y = x$

b. A shear along the x-axis for $s = 3$

c. A contraction along the y-axis for $s = 1/4$

d. A reflection about the x-axis

Solution. We use four of the functions defined in Example 2.56.

```
[> with(LinearAlgebra):
[> m1:=x->Vector[column](evalm(Matrix([[0,1],[1,0]])&*x)):
[> m2:=x->Vector[column](evalm(Matrix([[1,3],[0,1]])&*x)):
[> m3:=x->Vector[column](evalm(Matrix([[1,0],[0,1/4]])&*x)):
[> m4:=x->Vector[column](evalm(Matrix([[1,0],[0,-1]])&*x)):
```

Next we define the appropriate *Maple* functions and compose them.

```
[> m:= m4@m3@m2@m1:
```

We now calculate the new vertices.

```
[> G:=[Vector([0,0]),Vector([4,0]),Vector([2,2])]:
[> map[m,G];
```

$$\left\{\begin{bmatrix} 0 \\ 0 \end{bmatrix}, \begin{bmatrix} 12 \\ -1 \end{bmatrix}, \begin{bmatrix} 8 \\ \frac{-1}{2} \end{bmatrix}\right\}$$

We then graph the triangle.

```
[> with(plottools):
[> v:=i->[map(m,G)[i][1],map(m,G)[i][2]]:
[> T := polygon({v(1),v(2),v(3)},color=grey):
```

```
[> plots[display](T);
```

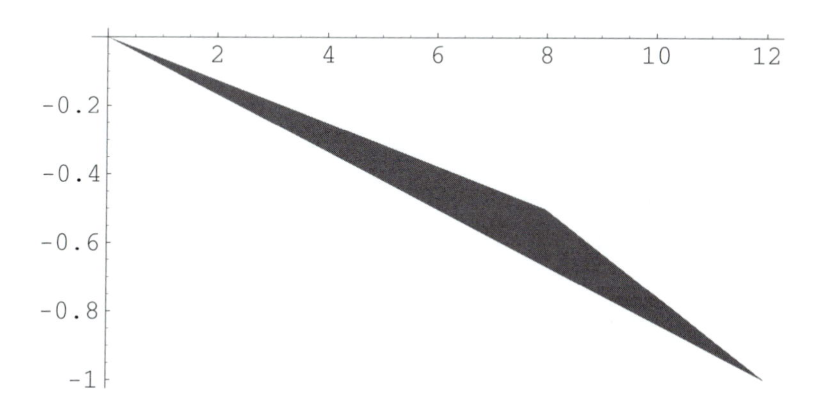

As we can see, the vertex (0, 0) remains fixed, the vertex (4, 0) is mapped to (12, −1), and the vertex (2, 2) is mapped to $(8, -\frac{1}{2})$. ▲

Reflections about the line $y = x$ are a special case of rotations about the origin. Geometrically, the matrix-vector product

$$\begin{bmatrix} \cos\theta & \sin\theta \\ -\sin\theta & \cos\theta \end{bmatrix} \begin{bmatrix} x \\ y \end{bmatrix} = \begin{bmatrix} (\cos\theta)\,x + (\sin\theta)\,y \\ -(\sin\theta)\,x + (\cos\theta)\,y \end{bmatrix}$$

converts the point (x, y) to the point $((\cos\theta)\,x + (\sin\theta)\,y, -(\sin\theta)\,x + (\cos\theta)\,y)$, obtained from (x, y) by rotating (x, y) clockwise about the origin (0, 0) through the angle θ. For example, if $\theta = \pi/4$, then

$$\begin{bmatrix} \cos\theta & \sin\theta \\ -\sin\theta & \cos\theta \end{bmatrix} \begin{bmatrix} 1 \\ 1 \end{bmatrix} = \begin{bmatrix} \sqrt{2} \\ 0 \end{bmatrix}$$

Thus the point (1, 1), which lies on a circle of radius $\sqrt{2}$ centered at the origin, is rotated to the point $(\sqrt{2}, 0)$ on the x-axis.

We now use the **plottools** and **geometry** packages to illustrate that reflections about the linear $y = x$ are a special case of rotations about the origin.

EXAMPLE 2.59 Reflection About the Line y = x

Let **x** and **y** be the vectors

$$\begin{bmatrix} \frac{1}{2}+\frac{\sqrt{3}}{2} \\ -\frac{1}{2}+\frac{\sqrt{3}}{2} \end{bmatrix} \quad \text{and} \quad \begin{bmatrix} -\frac{1}{2}+\frac{\sqrt{3}}{2} \\ \frac{1}{2}+\frac{\sqrt{3}}{2} \end{bmatrix}$$

respectively. Use *Maple* to show that **x** and **y** can be obtained from the vector $(1, 1)$ by rotations about the origin through the angles $-\pi/6$ and $\pi/6$, respectively.

Solution. We begin by loading the **plottools** and **geometry** packages.

```
[>with(plottools):  with(geometry):
```

Next we define the required rotation matrices R and S and the vector $\mathbf{v} = (1, 1)$.

```
[>R:=Matrix([[cos(-Pi/6),-sin(-Pi/6)],
   [sin(-Pi/6),cos(-Pi/6)]]):
[>S:=Matrix([[cos(Pi/6),-sin(Pi/6)],[sin(Pi/6),cos(Pi/6)]]):
[>v:=Vector([1,1]):
```

Then we rotate the vector $\mathbf{v} = (1, 1)$ through the angles $-\pi/6$ and $\pi/6$.

```
[>x:=evalm(R&*v);
```

$$x := \left[\frac{1}{2}+\frac{1}{2}\sqrt{3}, \frac{1}{2}\sqrt{3}-\frac{1}{2}\right]$$

```
[>y:=evalm(S&*v);
```

$$y := \left[\frac{1}{2}\sqrt{3}-\frac{1}{2}, \frac{1}{2}+\frac{1}{2}\sqrt{3}\right]$$

We conclude this example by graphing the three vectors.

```
[>X:=arrow(vector([0,0]),x,.001,.04,.04):
[>Y:=arrow(vector([0,0]),y,.001,.04,.04):
[>V:=arrow(vector([0,0]),vector([1.5,1.5]),.001,.04,.04):
```

```
[>plots[display]({X,Y,V},axes=normal,view=[0..2,0..2],
   color=black);
```

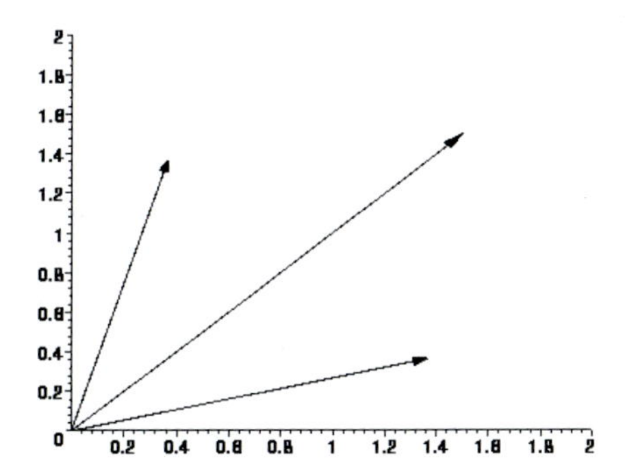

As we can see, the given vectors are symmetric about the line $y = x$. ▲

Example 2.59 shows that reflections about the line $y = x$ are special rotations. Furthermore, the rotation matrices are invertible. The matrices

$$\begin{bmatrix} \cos\theta & -\sin\theta \\ \sin\theta & \cos\theta \end{bmatrix} \text{ and } \begin{bmatrix} \cos(-\theta) & -\sin(-\theta) \\ \sin(-\theta) & \cos(-\theta) \end{bmatrix}$$

are inverses of each other. They must therefore be products of 2×2 elementary matrices. We now show how such products can be constructed.

THEOREM 2.35 *For any $\theta \in \mathbb{R}$, the rotation matrix*

$$A = \begin{bmatrix} \cos\theta & -\sin\theta \\ \sin\theta & \cos\theta \end{bmatrix}$$

is a product of elementary matrices.

Proof. Let $a = \cos\theta$ and $b = \sin\theta$. If $\theta = 2k\pi$, then $a = 1$ and $b = 0$. This means that A is the 2×2 identity matrix and is therefore elementary. Suppose now that $\theta \neq 2k\pi$. Then it is clear from trigonometry that $b \neq 0$ and $a^2 + b^2 = 1$. Therefore,

$$\begin{bmatrix} 1/b & 0 \\ 0 & 1 \end{bmatrix}\begin{bmatrix} 1 & a \\ 0 & 1 \end{bmatrix}\begin{bmatrix} -1 & 0 \\ 0 & 1 \end{bmatrix}\begin{bmatrix} 1 & 0 \\ a & 1 \end{bmatrix}\begin{bmatrix} 0 & 1 \\ 1 & 0 \end{bmatrix}\begin{bmatrix} b & 0 \\ 0 & 1 \end{bmatrix}$$

is a decomposition of A into a product of elementary matrices. ■

This shows that geometrically, any rotation can be represented by the product of an expansion, a contraction, a reflection about the y-axis, a reflection about the line $y = x$, a shear along the x-axis, and a shear along the y-axis.

EXERCISES 2.9

1. (*Maple V*) Graph the vectors $\mathbf{x}$, $A\mathbf{x}$, and $B\mathbf{x}$, where

$$A = \begin{bmatrix} 3 & 0 \\ 0 & 1 \end{bmatrix} \quad B = \begin{bmatrix} \frac{1}{3} & 0 \\ 0 & 1 \end{bmatrix} \quad \mathbf{x} = \begin{bmatrix} 9 \\ 6 \end{bmatrix}$$

   Explain the connection between the vectors.

2. (*Maple V*) Graph the vectors $\mathbf{x}$, $A\mathbf{x}$, and $B\mathbf{x}$, where

$$A = \begin{bmatrix} 1 & 0 \\ 0 & 3 \end{bmatrix} \quad B = \begin{bmatrix} 1 & 0 \\ 0 & \frac{1}{3} \end{bmatrix} \quad \mathbf{x} = \begin{bmatrix} 9 \\ 6 \end{bmatrix}$$

   Explain the connection between the vectors.

3. (*Maple V*) Graph the vectors $\mathbf{x}$, $A\mathbf{x}$, and $B\mathbf{x}$, where

$$A = \begin{bmatrix} 1 & 2 \\ 0 & 1 \end{bmatrix} \quad B = \begin{bmatrix} 1 & 0 \\ 2 & 1 \end{bmatrix} \quad \mathbf{x} = \begin{bmatrix} 4 \\ -1 \end{bmatrix}$$

   Explain the connection between the vectors.

4. (*Maple V*) Graph the vectors $\mathbf{x}$, $A\mathbf{x}$, and $B\mathbf{x}$, where

$$A = \begin{bmatrix} 0 & 1 \\ 1 & 0 \end{bmatrix} \quad B = \begin{bmatrix} 0 & -2 \\ -2 & 0 \end{bmatrix} \quad \mathbf{x} = \begin{bmatrix} 9 \\ 6 \end{bmatrix}$$

   Explain the connection between the vectors.

5. (*Maple V*) Show graphically the steps involved in transforming the point

$$\mathbf{x} = \begin{bmatrix} 2 \\ 3 \end{bmatrix}$$

   to the point $A\mathbf{x}$, where A is the matrix product

$$\begin{bmatrix} 0 & 1 \\ 1 & 0 \end{bmatrix}\begin{bmatrix} \frac{1}{3} & 0 \\ 0 & 1 \end{bmatrix}\begin{bmatrix} 1 & 0 \\ 2 & 1 \end{bmatrix}\begin{bmatrix} 0 & -3 \\ -3 & 0 \end{bmatrix}$$

6. (*Maple V*) Repeat Exercise 5 using the function

```
AffineMap:=(A,x)->
evalm(SubMatrix(A,1..2,1..2)&*x+SubMatrix(A,1..2,3))
```

   In each example, the given matrices must be augmented by a column of zeros for the application of the **AffineMap** function to make sense.

Computer Animation

Computer images are sets of points connected by lines and curves. Much of computer animation is concerned with manipulating these images using matrix multiplication. We have already seen how basic geometric transformations such as expansions, contractions, reflections, rotations, and shears can be achieved through matrix multiplication. However, moving images and three-dimensional objects in space also requires translations, which are algebraically nonlinear. The operation $T : \mathbb{R}^2 \to \mathbb{R}^2$ defined by $T(x, y) = (x + h, y + k)$ mapping the origin $(0, 0)$ to the point $(0 + h, 0 + k) = (h, k)$ is linear if and only if $h = k = 0$. It is therefore impossible to move computer images in the plane by multiplying their points by 2×2 matrices. For the same reason, three-dimensional objects cannot be moved in space by multiplying them by 3×3 matrices.

It is quite remarkable that matrix multiplication can nevertheless be used in a unifying way to reflect, rotate, shear, and translate two-dimensional computer images. The trick consists of embedding two-dimensional images in $\mathbb{R}^3$, using matrix multiplication in $\mathbb{R}^3$, and then converting the new images back into two-dimensional objects.

The embedding that makes this possible is the function $H : \mathbb{R}^2 \to \mathbb{R}^3$ defined by

$$H(x, y) = (x, y, 1)$$

The coordinates $(x, y, 1)$ are called the ***homogeneous coordinates*** of the points (x, y). It is important to note that H is not a linear transformation. However, the standard coordinates of points in $\mathbb{R}^2$ can be recovered from their homogeneous coordinates by stripping off the third coordinate. Figure 1 illustrates the relationship between the xy-plane, represented in $\mathbb{R}^3$ by the equation $z = 0$, and its image under H, represented by the equation $z = 1$.

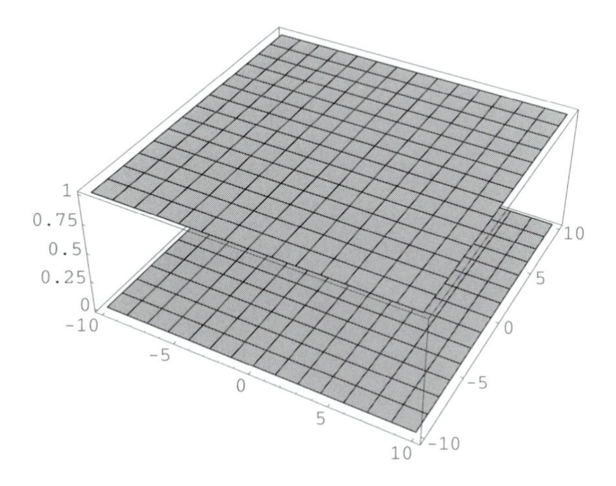

FIGURE 1 The Planes $z = 0$ and $z = 1$ in $\mathbb{R}^3$.

EXAMPLE 2.60 Translation and Matrix Multiplication

Find a 3×3 matrix that translates the homogeneous coordinates of the point (x, y) to the homogeneous coordinates of the points $(x + h, y + k)$ by matrix multiplication.

Solution. We show that the matrix

$$A = \begin{bmatrix} 1 & 0 & h \\ 0 & 1 & k \\ 0 & 0 & 1 \end{bmatrix}$$

has the required properties.

Let (x, y) be any vector in $\mathbb{R}^2$. Then

$$A\left(H\begin{bmatrix} x \\ y \end{bmatrix}\right) = \begin{bmatrix} 1 & 0 & h \\ 0 & 1 & k \\ 0 & 0 & 1 \end{bmatrix}\begin{bmatrix} x \\ y \\ 1 \end{bmatrix} = \begin{bmatrix} x+h \\ y+k \\ 1 \end{bmatrix}$$

We define the inverse of this operation by letting

$$\widehat{H}\begin{bmatrix} x+h \\ y+k \\ 1 \end{bmatrix} = \begin{bmatrix} x+h \\ y+k \end{bmatrix}$$

As we can see, the translation of (x, y) to $(x + h, y + k)$ is accomplished in $\mathbb{R}^3$ by multiplying the homogeneous coordinates of (x, y) by the matrix A. ▲

EXAMPLE 2.61 Translation of a Triangle

Use matrix multiplication to translate the triangle with vertices $(0, 0)$, $(2, 0)$, and $(2, 1)$ to the triangle with vertices $(3, 4)$, $(5, 4)$, and $(5, 5)$.

Solution. We know from Example 2.60 that the matrix

$$A = \begin{bmatrix} 1 & 0 & 3 \\ 0 & 1 & 4 \\ 0 & 0 & 1 \end{bmatrix}$$

is the appropriate translation matrix. We therefore multiply the homogeneous coordinates of the given points by A and obtain

$$\text{a. } A\begin{bmatrix} 0 \\ 0 \\ 1 \end{bmatrix} = \begin{bmatrix} 3 \\ 4 \\ 1 \end{bmatrix} \quad \text{b. } A\begin{bmatrix} 2 \\ 0 \\ 1 \end{bmatrix} = \begin{bmatrix} 5 \\ 4 \\ 1 \end{bmatrix} \quad \text{c. } A\begin{bmatrix} 2 \\ 1 \\ 1 \end{bmatrix} = \begin{bmatrix} 5 \\ 5 \\ 1 \end{bmatrix}$$

If we now apply the function $\widehat{H}$ to the vectors (a)–(c), we get the vertices $(3, 4)$, $(5, 4)$, and $(5, 5)$ of the translated triangle. ▲

We conclude this section by showing how the defined **AffineMap** function can be used in place of homogeneous coordinates to carry the calculations required for computer animation.

EXAMPLE 2.62 Homogeneous Coordinates and the AffineMap

Show that the function defined by

```
AffineMap:=(A,x)->
evalm(SubMatrix(A,1..2,1..2)&*x +SubMatrix(A,1..2,3)):
```

can be used to calculate a sequence of affine transformations.

Solution. Let $T : \mathbb{R}^2 \to \mathbb{R}^2$ be the affine transformation $T(\mathbf{x}) = A\mathbf{x} + \mathbf{b}$ consisting of a shear along the x-axis for $s = 3$, an expansion along the y-axis for $s = 2$, counterclockwise rotation by $\pi/3$, and a translation $(x, y) \mapsto (x + 3, y + 4)$. We show that the representations of T using homogeneous coordinates and using the **AffineMap** function are equivalent.

We know from our previous work that the matrix

$$A = \begin{bmatrix} 1 & 0 & 3 \\ 0 & 1 & 4 \\ 0 & 0 & 1 \end{bmatrix} \begin{bmatrix} \cos\pi/3 & -\sin\pi/3 & 0 \\ \sin\pi/3 & \cos\pi/3 & 0 \\ 0 & 0 & 1 \end{bmatrix} \begin{bmatrix} 1 & 0 & 0 \\ 0 & 2 & 0 \\ 0 & 0 & 1 \end{bmatrix} \begin{bmatrix} 1 & 3 & 0 \\ 0 & 1 & 0 \\ 0 & 0 & 1 \end{bmatrix}$$

$$= \begin{bmatrix} \frac{1}{2} & \frac{3}{2} - \sqrt{3} & 3 \\ \frac{1}{2}\sqrt{3} & \frac{3}{2}\sqrt{3} + 1 & 4 \\ 0 & 0 & 1 \end{bmatrix}$$

represents T in homogeneous coordinates. This means that

$$T\begin{bmatrix} x \\ y \end{bmatrix} = \widehat{H}\left[A\begin{bmatrix} x \\ y \\ 1 \end{bmatrix}\right] = \widehat{H}\begin{bmatrix} \frac{1}{2}x + \left(\frac{3}{2} - \sqrt{3}\right)y + 3 \\ \frac{1}{2}\sqrt{3}x + \left(\frac{3}{2}\sqrt{3} + 1\right)y + 4 \\ 1 \end{bmatrix}$$

$$= \begin{bmatrix} \frac{1}{2}x + \left(\frac{3}{2} - \sqrt{3}\right)y + 3 \\ \frac{1}{2}\sqrt{3}x + \left(\frac{3}{2}\sqrt{3} + 1\right)y + 4 \end{bmatrix}$$

We now show that we can also use the **AffineMap** function to represent T.

```
[> with(LinearAlgebra):
[> AffineMap:=(A,x)->
   evalm(SubMatrix(A,1..2,1..2)&*x+SubMatrix(A,1..2,3)):
```

Next we create the necessary 2×3 matrix. We define the matrix

$$\begin{bmatrix} \cos\pi/3 & -\sin\pi/3 \\ \sin\pi/3 & \cos\pi/3 \end{bmatrix} \begin{bmatrix} 1 & 0 \\ 0 & 2 \end{bmatrix} \begin{bmatrix} 1 & 3 \\ 0 & 1 \end{bmatrix} = \begin{bmatrix} \frac{1}{2} & \frac{3}{2} - \sqrt{3} \\ \frac{1}{2}\sqrt{3} & \frac{3}{2}\sqrt{3} + 1 \end{bmatrix}$$

and then add a third column to represent the translation $(x, y) \mapsto (x + 3, y + 4)$.

```
[> B:=Matrix([[1/2,3/2-sqrt(3),3],
   [1/2 sqrt(3),3/2*sqrt(3)+1,4]]);
```

$$\begin{bmatrix} \frac{1}{2} & \frac{3}{2} - \sqrt{3} & 3 \\ \frac{1}{2}\sqrt{3} & \frac{3}{2}\sqrt{3} + 1 & 4 \end{bmatrix}$$

Using B, we define the required affine transformation.

```
[> m:= x:->AffineMap(B,x):
```

We then evaluate m at (x, y).

```
[> m(Vector([x,y]));
```

$$\begin{bmatrix} 3 + \frac{x}{2} + \left(\frac{3}{2} - \sqrt{3}\right) y \\ 4 + \frac{\sqrt{3}x}{2} + \left(1 + \frac{3\sqrt{3}}{2}\right) y \end{bmatrix}$$

As expected, $T(x, y) = m(x, y)$. ▲

EXERCISES 2.10

1. Find the homogeneous coordinates in $\mathbb{R}^3$ of the points $(3, 5)$, $(-8, 0)$, and $(112, 115)$ in $\mathbb{R}^2$.

2. Find the points in $\mathbb{R}^2$ whose homogeneous coordinates in $\mathbb{R}^3$ are $(4, 5, 1)$, $(-8, -6, 1)$, and $(\pi, -\pi, 1)$.

3. Find the 3×3 matrix A that converts homogeneous coordinates of the image determined by the points $(0, 0)$, $(3, 0)$, $(3, 2)$, and $(0, 2)$ to the homogeneous coordinates of the new image obtained by three consecutive operations:

   a. A shear $\begin{bmatrix} 1 & \frac{1}{3} \\ 0 & 1 \end{bmatrix}$ along the x-axis

   b. A counterclockwise rotation by $\pi/4$ radians

   c. A translation $(x, y) \to (x + 2, y - 5)$

4. The homogeneous coordinates of a point (x, y, z) in $\mathbb{R}^3$ are $(x, y, z, 1)$ in $\mathbb{R}^4$. Find the homogeneous coordinates in $\mathbb{R}^4$ of the points $(3, 5, 7)$, $(-8, 0, 1)$, and $(112, 115, 0)$.

5. Find the points in $\mathbb{R}^3$ whose homogeneous coordinates in $\mathbb{R}^4$ are $(4, 5, 2, 1)$, $(\pi, -\pi, 8, 1)$, and $(-8, -6, -1, 1)$.

6. Find the 4×4 translation matrix A that corresponds to the translation of axes $(x, y, z) \to (x + h, y + k, z + j)$ in $\mathbb{R}^3$. Use homogeneous coordinates and the matrix A to construct an example of a geometric object and its translated image.

7. Show that a translation matrix

$$\begin{bmatrix} 1 & 0 & h \\ 0 & 1 & k \\ 0 & 0 & 1 \end{bmatrix}$$

is orthogonal if and only if $h = k = 0$.

8. (*Maple 6*) Repeat Exercise 3 using the function **AffineMap** defined by

```
(A,x)->evalm(SubMatrix(A,1..2,1..2)&*x+SubMatrix(A,1..2,3))
```

Message Encoding

Matrices and their inverses can be used to encode messages. Here is a simple example.

EXAMPLE 2.63 Encoding a Message

Use *Maple* to create a scheme for encoding ASCII text messages.

Solution. We assume that all messages are written in capital letters using the standard English alphabet. We therefore begin by assigning numbers to the required letters. The assignment in Table 4 will do. The symbol ♠ denotes a blank space, and ■ identifies the end of a sentence.

TABLE 4 NUMERICAL ENCODING OF THE ENGLISH ALPHABET.

$(A, 1)$	$(B, 2)$	$(C, 3)$	$(D, 4)$	$(E, 5)$	$(F, 6)$	$(G, 7)$
$(H, 8)$	$(I, 9)$	$(J, 10)$	$(K, 11)$	$(L, 12)$	$(M, 13)$	$(N, 14)$
$(O, 15)$	$(P, 16)$	$(Q, 17)$	$(R, 18)$	$(S, 19)$	$(T, 20)$	$(U, 21)$
$(V, 22)$	$(W, 23)$	$(X, 24)$	$(Y, 25)$	$(Z, 26)$	(♠, 27)	(■, 28)

Using this numerical encoding, we can replace any string of capital letters, spaces, and punctuations by a unique string of numbers. Thus the string

THE QUICK BROWN FOX JUMPS OVER THE LAZY DOG

becomes

$$\begin{bmatrix} 20 & 8 & 5 & 27 & 17 & 21 & 9 & 3 & 11 & 27 \\ 2 & 18 & 15 & 23 & 14 & 27 & 6 & 15 & 24 & 27 \\ 10 & 21 & 13 & 16 & 19 & 27 & 15 & 22 & 5 & 18 \\ 27 & 20 & 8 & 5 & 27 & 12 & 1 & 26 & 25 & 27 \\ 4 & 15 & 7 & 28 & & & & & & \end{bmatrix}$$

To be able to calculate with this matrix, we fill the empty cells with zeros. This yields the matrix

$$A = \begin{bmatrix} 20 & 8 & 5 & 27 & 17 & 21 & 9 & 3 & 11 & 27 \\ 2 & 18 & 15 & 23 & 14 & 27 & 6 & 15 & 24 & 27 \\ 10 & 21 & 13 & 16 & 19 & 27 & 15 & 22 & 5 & 18 \\ 27 & 20 & 8 & 5 & 27 & 12 & 1 & 26 & 25 & 27 \\ 4 & 15 & 7 & 28 & 0 & 0 & 0 & 0 & 0 & 0 \end{bmatrix}$$

The numerical encoding of the message determined by this matrix is achieved by premultiplying A with an invertible matrix B that is known only to the sender and the recipient of the message. Using A and B, we form the encoded message BA. Since A is a 5×10 matrix, B must be a 5×5 matrix. We use *Maple* to generate a random 5×5 matrix.

We begin by defining the matrix A.

```
[> A:=array([
    [20,8,5,27,17,21,9,3,11,27],
    [2,18,15,23,14,27,6,15,24,27],
    [10,21,13,16,19,27,15,22,5,18],
    [27,20,8,5,27,12,1,26,25,27],
    [4,15,7,28,0,0,0,0,0,0]]):
```

Then we generate a random 5×5 matrix with integer entries.

```
[> with(linalg):
[> B:=randmatrix(5,5);
```

$$B := \begin{bmatrix} 1 & 4 & 6 & 3 & 2 \\ 1 & 3 & 0 & 1 & 8 \\ 3 & 0 & 4 & 5 & 1 \\ 5 & 4 & 9 & 0 & 2 \\ 2 & 3 & 7 & 9 & 5 \end{bmatrix}$$

Next we verify that B is invertible. If not, we modify it until we have obtained an invertible matrix. We use *Maple* to verify that B is invertible by asking the program to calculate the inverse M of B without displaying the result. If B were not invertible, *Maple* would produce a message to that effect.

```
[> M:=evalm(inverse(B)):
```

Since *Maple* did not produce an error message, we know that B is invertible, and we can use it to encode A and calculate the matrix BA.

```
[> E:=evalm(B&*A);
```

$$E := \begin{bmatrix} 177 & 296 & 181 & 286 & 268 & 327 & 126 & 273 & 212 & 324 \\ 85 & 202 & 114 & 325 & 86 & 114 & 28 & 74 & 108 & 135 \\ 239 & 223 & 114 & 198 & 262 & 231 & 92 & 227 & 178 & 288 \\ 206 & 331 & 216 & 427 & 312 & 456 & 204 & 273 & 196 & 405 \\ 379 & 472 & 253 & 420 & 452 & 420 & 150 & 439 & 354 & 504 \end{bmatrix}$$

The resulting matrix E is the encoded message. To decode it, we must use the matrix M. We recover the original numerical version of the transmitted message by calculating the matrix ME.

```
[> ME:=evalm(M&*E);
```

$$ME := \begin{bmatrix} 20 & 8 & 5 & 27 & 17 & 21 & 9 & 3 & 11 & 27 \\ 2 & 18 & 15 & 23 & 14 & 27 & 6 & 15 & 24 & 27 \\ 10 & 21 & 13 & 16 & 19 & 27 & 15 & 22 & 5 & 18 \\ 27 & 20 & 8 & 5 & 27 & 12 & 1 & 26 & 25 & 27 \\ 4 & 15 & 7 & 28 & 0 & 0 & 0 & 0 & 0 & 0 \end{bmatrix}$$

The string $\mathbf{r}_1 ME\, \mathbf{r}_2 ME\, \mathbf{r}_3 ME\, \mathbf{r}_4 ME\, \mathbf{r}_5 ME$ is the original decoded message. ▲

EXERCISES 2.11

1. (*Maple V*) Use the encoding scheme defined in Example 2.63, together with the invertible matrix

$$B = \begin{bmatrix} -3 & 5 & 2 & 2 \\ -5 & -3 & 4 & 4 \\ 2 & 6 & -4 & 6 \\ 4 & 4 & 2 & 5 \end{bmatrix}$$

to encode the message

"SOME PEOPLE LIKE BASEBALL AND SOME LIKE FOOTBALL."

2. (*Maple V*) Make up your own coding scheme using a $6 \times n$ matrix for a suitable n, and encode and decode the message in Exercise 1.

3. (*Maple V*) Use the inverse of the matrix B in Exercise 1 to decode the message

$$\begin{bmatrix} 56 & 125 & 67 & 124 & 36 & 176 & 146 & 126 & 97 \\ 102 & 5 & 31 & 60 & -92 & 106 & -52 & -78 & -119 \\ 42 & 150 & 4 & 234 & 170 & 148 & 356 & 96 & 158 \\ 121 & 252 & 171 & 239 & 175 & 224 & 342 & 246 & 156 \end{bmatrix}$$

4. (*Maple 6*) Repeat Exercise 2 using a $4 \times n$ matrix for a suitable n.

Leontief Input-Output Models

Economic input-output models are well-known applications of matrix algebra. They are named after the Harvard Professor Wassily Leontief, who used a giant matrix to build a model of the U.S. economy and was awarded a Nobel prize for his work. In the following example, we illustrate the basic idea behind the construction.

EXAMPLE 2.64 Building a Leontief Economic Input-Output Model

A small island economy in a Pacific archipelago consists of four industries: agriculture, fishing, manufacturing, and tourism. Each industry can meet its own purchasing needs from these four industries. In addition, the island provides goods and services to some of the neighboring islands. Table 5 describes how the output of the four industries is consumed in a particular year.

TABLE 5 CONSUMPTION VECTORS.

Supplier/Consumer	Agriculture	Fishing	Manufacturing	Tourism
Agriculture	.10	.20	.10	.40
Fishing	.20	.10	.10	.20
Manufacturing	.30	.40	.30	.30
Tourism	.10	.10	.20	.10

Each column in Table 5 is a ***consumption vector*** $\mathbf{c} \in \mathbb{R}^4$, describing the input per unit of output from the other industries. Thus the model involves four consumption vectors:

$$\mathbf{c}_1 = \begin{bmatrix} .10 \\ .20 \\ .30 \\ .10 \end{bmatrix} \quad \mathbf{c}_2 = \begin{bmatrix} .20 \\ .10 \\ .40 \\ .10 \end{bmatrix} \quad \mathbf{c}_3 = \begin{bmatrix} .10 \\ .10 \\ .30 \\ .20 \end{bmatrix} \quad \mathbf{c}_4 = \begin{bmatrix} .40 \\ .20 \\ .30 \\ .10 \end{bmatrix}$$

The consumption of the neighboring islands of the goods and services produced by the four industries is represented by a vector $\mathbf{d} \in \mathbb{R}^4$, called the ***demand vector***. The consumption vectors and the demand vector are related by an equation of the form

$$\mathbf{x} = x_1\mathbf{c}_1 + x_2\mathbf{c}_2 + x_3\mathbf{c}_3 + x_4\mathbf{c}_4 + \mathbf{d}$$

expressing the input-output relationships between the goods and services produced by the four industries. The vector $\mathbf{x} \in \mathbb{R}^4$ is called the ***production vector*** of the entire economy. In this example, we have

$$\begin{bmatrix} x_1 \\ x_2 \\ x_3 \\ x_4 \end{bmatrix} = x_1 \begin{bmatrix} .10 \\ .20 \\ .30 \\ .10 \end{bmatrix} + x_2 \begin{bmatrix} .20 \\ .10 \\ .40 \\ .10 \end{bmatrix} + x_3 \begin{bmatrix} .10 \\ .10 \\ .30 \\ .20 \end{bmatrix} + x_4 \begin{bmatrix} .40 \\ .20 \\ .30 \\ .10 \end{bmatrix} + \begin{bmatrix} d_1 \\ d_2 \\ d_3 \\ d_4 \end{bmatrix}$$

By combining the first four vectors on the right-hand side of this equation, we get

$$\begin{bmatrix} x_1 \\ x_2 \\ x_3 \\ x_4 \end{bmatrix} = \begin{bmatrix} .1x_1 + .2x_2 + .1x_3 + .4x_4 \\ .2x_1 + .1x_2 + .1x_3 + .2x_4 \\ .3x_1 + .4x_2 + .3x_3 + .3x_4 \\ .1x_1 + .1x_2 + .2x_3 + .1x_4 \end{bmatrix} + \begin{bmatrix} d_1 \\ d_2 \\ d_3 \\ d_4 \end{bmatrix}$$

The resulting equation expresses the fact that the total production is the sum of the *internal* and *external* demand of the economy. The internal demand vector can be expressed as a matrix-vector product

$$\begin{bmatrix} .1x_1 + .2x_2 + .1x_3 + .4x_4 \\ .2x_1 + .1x_2 + .1x_3 + .2x_4 \\ .3x_1 + .4x_2 + .3x_3 + .3x_4 \\ .1x_1 + .1x_2 + .2x_3 + .1x_4 \end{bmatrix} = \begin{bmatrix} .10 & .20 & .10 & .40 \\ .20 & .10 & .10 & .20 \\ .30 & .40 & .30 & .30 \\ .10 & .10 & .20 & .10 \end{bmatrix} \begin{bmatrix} x_1 \\ x_2 \\ x_3 \\ x_4 \end{bmatrix} = C \begin{bmatrix} x_1 \\ x_2 \\ x_3 \\ x_4 \end{bmatrix}$$

The matrix C is called a ***consumption matrix***, and the equation

$$\mathbf{x} = C\mathbf{x} + \mathbf{d}$$

is known as the ***Leontief input-output model*** of the economy. We can solve for the equation $\mathbf{d}$ and express the external demand of the economy by the equation $\mathbf{x} - C\mathbf{x} = \mathbf{d}$. If $\mathbf{d}$ is known, we can therefore express the total production **x** of the economy required to meet the external demand by matrix-vector product

$$\mathbf{x} = (I - C)^{-1}\,\mathbf{d}$$

provided that the matrix $(I - C)$ is invertible. In the present example, this is the case.

We use *Maple* to verify this fact.

```
[>with(LinearAlgebra):
[>A:=IdentityMatrix(4):
[>B:=Matrix([
    [.10,.20,.10,.40],
    [.20,.10,.10,.20],
    [.30,.40,.30,.30],
    [.10,.10,.20,.10]]):
```

```
[>C:=evalf(MatrixInverse(A-B),4);
```

$$C := \begin{bmatrix} 1.632 & .7829 & .6652 & 1.121 \\ .6249 & 1.584 & .5478 & .8125 \\ 1.287 & 1.497 & 2.304 & 1.673 \\ .5367 & .5957 & .6471 & 1.698 \end{bmatrix}$$

Suppose now that the external demand for the goods and services of the economy, in units of output, is given by the vector

```
[>d:=Vector[column]([500,300,400,900]):
```

Then the calculation

```
[>evalf(C.d,6);
```

$$\begin{bmatrix} 2324.76 \\ 1736.95 \\ 3517.27 \\ 2232.91 \end{bmatrix}$$

yields the total production of goods and services required to satisfy the demand **d**. ▲

EXERCISES 2.12

1. (*Maple V*) Suppose that

$$C = \begin{bmatrix} .2 & .1 & 0 & .1 \\ .1 & .2 & .2 & .3 \\ .2 & .3 & .2 & .2 \\ 0 & .2 & .2 & .1 \end{bmatrix}$$

is the consumption matrix of a small economy. Find the consumption vectors.

2. (*Maple V*) Suppose that the vectors $\mathbf{d} = (150, 200, 100, 300)$ and $\mathbf{x} = (x_1, x_2, x_3, x_4)$ are, respectively, the demand and total production vectors for the economy whose consumption matrix is given in Exercise 1. Find the matrix equation that expresses the assumption that supply equals demand.

3. (*Maple V*) Find the internal demand vector for the economy in Exercise 2.

4. (*Maple V*) Find the production vector $\mathbf{x}$ defined in Example 2.64, if the demand vector $\mathbf{d}$ remains unchanged but the consumption matrix C is

$$C = \begin{bmatrix} .30 & .20 & .30 & .10 \\ .20 & .30 & .10 & .20 \\ .30 & .10 & .10 & .10 \\ .10 & .40 & .20 & .10 \end{bmatrix}$$

5. (*Maple V*) Find the demand vector $\mathbf{d}$ in Example 2.64 if the production vector $\mathbf{x}$ remains unchanged but the consumption matrix C is

$$C = \begin{bmatrix} .10 & .20 & .30 & .15 \\ .10 & .20 & .40 & .10 \\ .30 & .10 & .10 & .10 \\ .10 & .30 & .20 & .10 \end{bmatrix}$$

6. (*Maple V*) Suppose that an economy consists of three industries A_1, A_2, and A_3, producing quantities x_1, x_2, and x_3 of some commodities. Suppose further that these companies export 30, 20, and 40 percent of their output, respectively. Find the matrix A for which

$$A \begin{bmatrix} x_1 \\ x_2 \\ x_3 \end{bmatrix} = \begin{bmatrix} 30 \\ 20 \\ 40 \end{bmatrix}$$

when the percentages a_{ij} that company A_i consumes of the output of company A_j are given by the consumption matrix

$$\begin{bmatrix} a_{11} & a_{12} & a_{13} \\ a_{21} & a_{22} & a_{23} \\ a_{31} & a_{32} & a_{33} \end{bmatrix} = \begin{bmatrix} .30 & .15 & .4 \\ .25 & .30 & .1 \\ .20 & .10 & .3 \end{bmatrix}$$

Graph Theory

In this section, we discuss the use of matrices and matrix multiplication to describe the properties of finite graphs. First we deal with undirected graphs, then with directed ones. We motivate the definitions involved by considering the simple network of roads in Figure 2, connecting the cities of Beaconsfield (BFLD), Dorval (DVAL), Lachine (LCHN), and Montreal (MTRL). Table 6 displays the number of direct roads connecting these municipalities.

In this example, V is the set {BFLD,DVAL,LCHN,MTRL} of ***vertices***, and E is a set of ***edges*** $\{e_1, e_2, e_3, e_4, e_5, e_6, e_7\}$, connected by the function $p : E \to V \times V$ defined by

$$\begin{array}{ll} p(e_1) = (\text{LCHN,MTRL}) & p(e_2) = (\text{LCHN,MTRL}) \\ p(e_3) = (\text{BFLD,LCHN}) & p(e_4) = (\text{BFLD,DVAL}) \\ p(e_5) = (\text{DVAL,MTRL}) & p(e_6) = (\text{DVAL,LCHN}) \\ p(e_7) = (\text{DVAL,LCHN}) & \end{array}$$

An ***undirected graph*** $G = (V, E, p : E \to V \times V)$ consists of a set V of vertices, a set E of edges, and a function p that assigns to each edge e a pair $p(e) = (u, v)$ of vertices. We write $e : u \sim v$ or $u \overset{e}{\sim} v$ if $p(e) = (u, v)$.

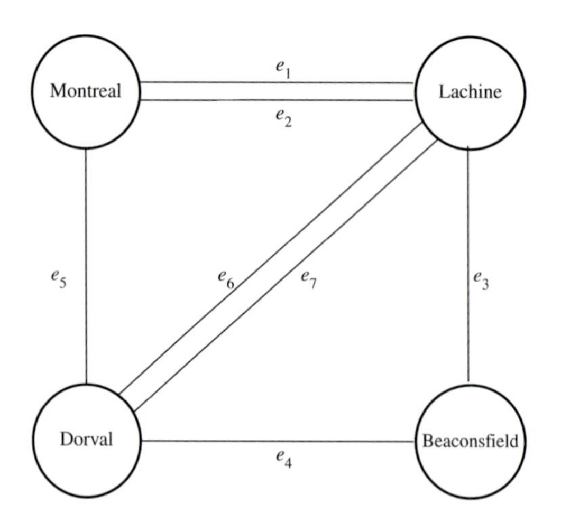

FIGURE 2 A Network of Two-Way Roads.

TABLE 6 CONNECTIONS.

	BFLD	DVAL	LCHN	MTRL
BFLD	0	1	1	0
DVAL	1	0	2	1
LCHN	1	2	0	2
MTRL	0	1	2	0

A vertex u of a graph G is ***connected*** to a vertex v of G if $(u, v) = p(e)$ for some $e \in E$. A finite sequence $u_1 \overset{e_1}{\sim} u_2 \sim \cdots \sim u_n \overset{e_n}{\sim} u_{n+1}$ is a ***path*** of length n of G.

We now continue our example. We associate two different matrices with the graph of the given network of roads. The matrix

$$M = \begin{bmatrix} 0 & 1 & 1 & 0 \\ 1 & 0 & 2 & 1 \\ 1 & 2 & 0 & 2 \\ 0 & 1 & 2 & 0 \end{bmatrix}$$

records the number of edges connecting the vertices BFLD, DVAL, LCHN, and MTRL. It is symmetric, since the edges of G are undirected. We call M the matrix of the graph G. We then form a second matrix

$$A = \begin{bmatrix} 0 & 1 & 1 & 0 \\ 1 & 0 & 1 & 1 \\ 1 & 1 & 0 & 1 \\ 0 & 1 & 1 & 0 \end{bmatrix}$$

whose entries are 1 if the graph contains at least one edge connecting two corresponding vertices and 0 otherwise. We call this the adjacency matrix of the graph G.

If $V = \{v_1, \ldots, v_n\}$ is the set of vertices of a graph G, then ***the matrix of the graph*** G is the $n \times n$ matrix $M = [m_{ij}]$ whose entries m_{ij} count the number of edges connecting the vertices v_i and v_j. The ***adjacency matrix of the graph*** G is the $n \times n$ matrix $A = [a_{ij}]$ whose entries a_{ij} are 1 if v_i and v_j are connected by an edge and 0 otherwise.

If we examine the given network of roads connecting the four municipalities, we notice that there is no direct road from Beaconsfield to Montreal. This means that the cities Beaconsfield and Montreal are not connected. But as Table 7 shows, we can get from Beaconsfield to Montreal and from Montreal to Beaconsfield via a second or a second and a third city.

Routes 1 to 4 are ***paths*** of lengths 2, 2, 2, and 3, respectively, of the graph G. Matrix multiplication allows us to calculate the number of different paths determined by a graph. For

TABLE 7 ROUTES FROM BEACONSFIELD TO MONTREAL.

Route 1	Route 2
BFLD $\overset{e_3}{\rightarrow}$ LCHN $\overset{e_1}{\rightarrow}$ MTRL	BFLD $\overset{e_3}{\rightarrow}$ LCHN $\overset{e_2}{\rightarrow}$ MTRL
Route 3	Route 4
BFLD $\overset{e_4}{\rightarrow}$ DVAL $\overset{e_5}{\rightarrow}$ MTRL	BFLD $\overset{e_4}{\rightarrow}$ DVAL $\overset{e_7}{\rightarrow}$ LCHN $\overset{e_1}{\rightarrow}$ MTRL

example, the matrix

$$M^2 = \begin{bmatrix} 0 & 1 & 1 & 0 \\ 1 & 0 & 2 & 1 \\ 1 & 2 & 0 & 2 \\ 0 & 1 & 2 & 0 \end{bmatrix} \begin{bmatrix} 0 & 1 & 1 & 0 \\ 1 & 0 & 2 & 1 \\ 1 & 2 & 0 & 2 \\ 0 & 1 & 2 & 0 \end{bmatrix} = \begin{bmatrix} 2 & 2 & 2 & 3 \\ 2 & 6 & 3 & 4 \\ 2 & 3 & 9 & 2 \\ 3 & 4 & 2 & 5 \end{bmatrix}$$

shows that there are three paths of length 2 from Beaconsfield to Montreal. If we examine Table 7, we see that these paths correspond precisely to Routes 1, 2, and 3 from Beaconsfield to Montreal. The matrix

$$M^3 = \begin{bmatrix} 2 & 2 & 2 & 3 \\ 2 & 6 & 3 & 4 \\ 2 & 3 & 9 & 2 \\ 3 & 4 & 2 & 5 \end{bmatrix} \begin{bmatrix} 0 & 1 & 1 & 0 \\ 1 & 0 & 2 & 1 \\ 1 & 2 & 0 & 2 \\ 0 & 1 & 2 & 0 \end{bmatrix} = \begin{bmatrix} 4 & 9 & 12 & 6 \\ 9 & 12 & 22 & 12 \\ 12 & 22 & 12 & 21 \\ 6 & 12 & 21 & 8 \end{bmatrix}$$

on the other hand, shows us that there are six paths of length 3 from Beaconsfield to Montreal. Route 4 is one of them. The number of routes grows quickly as we lengthen the allowable paths.

If there is a path from one vertex of a graph to another, we say that the path in question ***connects*** the two vertices. Two vertices u and v of a graph G are ***pathwise connected*** by G if G determines a path from u to v. The following theorem is proved in graph theory and describes the number of connections of a given length determined by a graph.

THEOREM 2.36 *If M_G is the matrix of a graph G and the ijth entry of M_G^p is q, then there are q paths of length p from v_i to v_j in G.* ■

A question of interest, especially if the graph is large and the number of connections between vertices is not easily gleaned from a table, is whether a given vertex can be reached from any other vertex by some path. A graph G is ***strongly connected*** if every pair of vertices of G is pathwise connected. If a graph G has n vertices, then G is strongly connected if and only if every pair of vertices of G is connected by a path of length $n - 1$. Using this fact, we have the following beautiful characterization of strongly connected graphs.

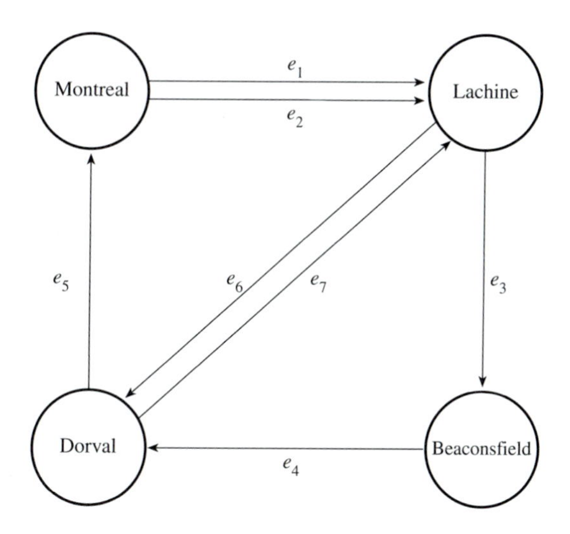

FIGURE 3 A Network of One-Way Roads.

THEOREM 2.37 *If A is the adjacency matrix of a graph G, then G is strongly connected if and only if the matrix $B = A + A^2 + \cdots + A^{n-1}$ has no zero entries.* ■

Let us now turn to directed graphs. Suppose that the four municipalities are trying to reduce the number of road accidents by converting all of their highways into one-way roads. Figure 3 describes one result of this decision.
Each edge of the diagram has a unique initial vertex and a unique terminal vertex. This leads us to the following definition.

A ***directed graph*** is a structure $G = (V, E, i : E \to V, t : E \to V)$ consisting of a set V of vertices, a set E of edges, a function i that assigns to each edge e a vertex $i(e)$ called the ***initial vertex*** of e, and a function t that assigns to each edge e a vertex $t(e)$ called the ***terminal vertex*** of e. A vertex u is ***connected*** to a vertex v in a directed graph G are ***connected*** if $i(e) = u$ and $t(e) = v$ for some $e \in E$. We write $e : u \to v$ or $u \overset{e}{\to} v$ if $i(e) = u$ and $t(e) = v$.

Table 8 shows the number of roads connecting the four municipalities.

TABLE 8 ROADS.

	BFLD	DVAL	LCHN	MTRL
BFLD	0	1	0	0
DVAL	0	0	1	1
LCHN	1	1	0	0
MTRL	0	0	2	0

The matrix of the graph G defined by this able is

$$M = \begin{bmatrix} 0 & 1 & 0 & 0 \\ 0 & 0 & 1 & 1 \\ 1 & 1 & 0 & 0 \\ 0 & 0 & 2 & 0 \end{bmatrix}$$

This matrix is no longer symmetric, as was the case for an undirected graph, since an edge from a vertex u to a vertex v is no longer also an edge from v to u. The adjacency matrix of the given graph is

$$A = \begin{bmatrix} 0 & 1 & 0 & 0 \\ 0 & 0 & 1 & 1 \\ 1 & 1 & 0 & 0 \\ 0 & 0 & 1 & 0 \end{bmatrix}$$

The definitions of a path of length q and of two vertices being pathwise connected are identical to those for undirected graphs. We have the following analogs of the previous theorems.

THEOREM 2.38 *If M is the matrix of a directed graph G and the ijth entry of M^p is q, then there are q paths of length p from the vertex v_i to the vertex v_j in G.* ■

THEOREM 2.39 *If A is the adjacency matrix of a directed graph G, then G is strongly connected if and only if the matrix $B = A + A^2 + \cdots + A^{n-1}$ has no zero entries.* ■

Here are some additional examples illustrating the preceding definitions.

EXAMPLE 2.65 The Matrix of a Finite Graph

Suppose that the set of vertices of a graph G is $V = \{v_1, v_2, v_3, v_4\}$ and the set of edges of G is $E = \{e_1 : v_1 \sim v_2, e_2 : v_1 \sim v_4, e_3 : v_2 \sim v_3, e_4 : v_2 \sim v_3\}$. Then the matrix

$$M = \begin{bmatrix} 0 & 1 & 0 & 1 \\ 1 & 0 & 2 & 0 \\ 0 & 2 & 0 & 0 \\ 1 & 0 & 0 & 0 \end{bmatrix}$$

represents the graph G. ▲

EXAMPLE 2.66 A Finite Directed Graph

Suppose that the set of vertices of a graph G is $V = \{v_1, v_2, v_3, v_4\}$ and the set of edges of G is

$$E = \{e_1 : v_1 \to v_2, e_2 : v_1 \to v_4, e_3 : v_2 \to v_3, e_4 : v_3 \to v_2, e_5 : v_3 \to v_2\}$$

Then

$$M(G) = \begin{bmatrix} 0 & 1 & 0 & 1 \\ 0 & 0 & 1 & 0 \\ 0 & 2 & 0 & 0 \\ 0 & 0 & 0 & 0 \end{bmatrix} \quad \text{and} \quad A(G) = \begin{bmatrix} 0 & 1 & 0 & 1 \\ 0 & 0 & 1 & 0 \\ 0 & 1 & 0 & 0 \\ 0 & 0 & 0 & 0 \end{bmatrix}$$

are, respectively, the matrix and the adjacency matrix of G. ▲

EXAMPLE 2.67 Paths of an Undirected Graph

Use *Maple* to find the number of paths of length 2 in the graph whose matrix is

$$A = \begin{bmatrix} 0 & 1 & 0 & 1 \\ 1 & 0 & 1 & 0 \\ 0 & 1 & 0 & 0 \\ 1 & 0 & 0 & 0 \end{bmatrix}$$

Solution. First we define the matrix A.

```
[> A:=array([[0,1,0,1],[1,0,1,0],[0,1,0,0],[1,0,0,0]]):
```

Next we calculate the powers A^2 and A^3 of A.

```
[> A2:=A^2;
```

$$A2 := \begin{bmatrix} 2 & 0 & 1 & 0 \\ 0 & 2 & 0 & 1 \\ 1 & 0 & 1 & 0 \\ 0 & 1 & 0 & 1 \end{bmatrix}$$

```
[> A3:=A^3;
```

$$A3 := \begin{bmatrix} 0 & 3 & 0 & 2 \\ 3 & 0 & 2 & 0 \\ 0 & 2 & 0 & 1 \\ 2 & 0 & 1 & 0 \end{bmatrix}$$

The matrix A^2 shows that there is one path $v_3 \sim v_1$ of length 2 from v_3 to v_1, that there are two paths $v_1 \sim v_2 \sim v_1$ and $v_1 \sim v_4 \sim v_1$ of length 2 from v_1 to v_1, and so on. The matrix also shows that there are two paths of length 2 from v_2 to v_2, namely, $v_2 \sim v_1 \sim v_2$ and $v_2 \sim v_3 \sim v_2$.

The matrix A^3 shows that there are three paths of length 3 from v_1 to v_2, namely, $v_1 \sim v_2 \sim v_1 \sim v_2$, $v_1 \sim v_2 \sim v_3 \sim v_2$ and $v_1 \sim v_4 \sim v_1 \sim v_2$, but no such paths from v_1 to v_1, and so on. ▲

EXAMPLE 2.68 Paths of a Directed Graph

Use *Maple* to find the number of paths of length 2 in the graph whose matrix is given by

$$A = \begin{bmatrix} 0 & 1 & 0 & 1 \\ 0 & 0 & 1 & 0 \\ 0 & 1 & 0 & 0 \\ 0 & 0 & 0 & 0 \end{bmatrix}$$

Solution. We begin by specifying A. Then we can calculate the powers A^2 and A^3 of A.

```
[> A:=array([[0,1,0,1],[0,0,1,0],[0,1,0,0],[0,0,0,0]]):
[> A2:=A^2;
```

$$A2 := \begin{bmatrix} 0 & 0 & 1 & 0 \\ 0 & 1 & 0 & 0 \\ 0 & 0 & 1 & 0 \\ 0 & 0 & 0 & 0 \end{bmatrix}$$

The matrix $A2$ shows that there is one path of length 2 from v_1 to v_3, one path of length 2 from v_2 to v_2, and so on. The path of length 2 from v_2 to v_2 consists of the edges $e_4 : v_3 \to v_2$ and $e_3 : v_2 \to v_3$. The two paths of length 2 from v_3 to v_3 consist of the edges $e_4 : v_3 \to v_2$ and $e_3 : v_2 \to v_3$.

```
[> A3:=A^3;
```

$$A3 := \begin{bmatrix} 0 & 1 & 0 & 0 \\ 0 & 0 & 1 & 0 \\ 0 & 1 & 0 & 0 \\ 0 & 0 & 0 & 0 \end{bmatrix}$$

The matrix $A3$ shows that there is one path of length 3 from v_2 to v_3 but no such paths from v_1 to v_1, and so on. ▲

The **networks** package makes it easy to construct and analyze both undirected and directed graphs.

EXAMPLE 2.69 Using *Maple* to Construct an Undirected Graph

Use *Maple* to construct the undirected graph defined in Table 6.

Solution. We begin by loading the **networks** package and giving a name to the graph.

```
[> with(networks):
[> new(G):
```

Next we specify the cities whose names form the edges of the graph.

```
[> cities:={Beaconsfield, Dorval, Lachine, Montreal};
```

cities := {*Beaconsfield, Montreal, Dorval, Lachine*}

We then designate the city names as vertices.

```
[> addvertex(cities,G);
```

Beaconsfield, Montreal, Dorval, Lachine

Next we define the edges labeling the roads connecting the four cities.

```
[> connect([Montreal,Montreal],[Lachine],G);
```

$e1, e2$

```
[> connect([Lachine],[Beaconsfield],G);
```

$e3$

```
[> connect([Dorval],[Beaconsfield,Montreal],G);
```

$e4, e5$

```
[> connect([Dorval,Dorval],[Lachine],G);
```

$e6, e7$

We use *Maple* to verify that the graph has the expected sets of vertices and edges.

```
[> vertices(G);
```

$$\{Beaconsfield, Montreal, Dorval, Lachine\}$$

```
[> edges(G);
```

$$\{e1, e2, e3, e4, e5, e6, e7\}$$

The command

```
M:=adjacency(G)
```

produces the adjacency matrix

$$M := \begin{bmatrix} 0 & 1 & 1 & 0 \\ 1 & 0 & 2 & 1 \\ 0 & 1 & 2 & 0 \\ 2 & 2 & 2 & 3 \end{bmatrix}$$

of the graph G and the commands

a. `TwoPaths:=evalm(M^2)`

b. `ThreePaths:=evalm(M^3)`

produce the matrices counting the number of paths of lengths 2 and 3, respectively.

We can also use *Maple* to construct directed graphs. The edges of the graph described in Table 7, for example, are specified as follows:

```
[> new(DG):
[> addvertex(cities,DG):
```

```
[> connect([Montreal],[Lachine,Lachine],'directed',DG):
[> connect([Lachine],[Beaconsfield],'directed',DG):
[> connect([Beaconsfield],[Dorval],'directed',DG):
[> connect([Dorval],[Montreal],'directed',DG):
[> connect([Lachine],[Dorval],'directed',DG):
[> connect([Dorval],[Lachine],'directed',DG):
```

where DG is the directed graph connecting the four specified cities.

EXERCISES 2.13

1. Calculate the matrix $M(G)$ and adjacency matrix $A(G)$ of the graph G with vertices $V = \{v_1, v_2, v_3, v_4\}$ and edges $E = \{e_1 : v_1 \sim v_2, e_2 : v_1 \sim v_4, e_3 : v_2 \sim v_3, e_4 : v_2 \sim v_3\}$.

2. Calculate all paths of lengths 2, 3, and 4 determined by the graph in Exercise 1.

3. Draw an undirected graph G with four vertices v_1, v_2, v_3, and v_4 in which each vertex v_i is connected to each other vertex v_j. Determine the matrix $M(G)$ and the adjacency matrix $A(G)$ of the graph G.

4. Use the powers of the adjacency matrix $A(G)$ in Exercise 3 to find the number of paths of length 3 connecting vertices v_1 and v_2.

5. Four friends, Fred, Bob, Jane, and Sarah, have decided to play a round-robin tennis tournament. In the first match, Fred plays against Bob and Jane against Sarah. In the second match, Fred plays against Jane and Bob against Sarah. In the third match, Fred plays against Sarah and Bob against Jane. Construct the three graphs depicting the order of play of each match, and show that adjacency matrices of the graphs are products of elementary matrices.

6. Draw an undirected graph G that is not pathwise connected and has four vertices. Form the adjacency matrix $A(G)$ of G. Calculate the matrix powers $A(G)^2$ and $A(G)^3$, and explain why both matrices have zero entries.

7. Draw a strongly connected directed graph G with four vertices and specify its adjacency matrix $A(G)$. Show that the matrix $A(G) + A(G)^2 + A(G)^3$ has no zero entry.

8. (*Maple V*) Find the matrix $M(G)$ and the adjacency matrix $A(G)$ of the directed graph G determined by the set of vertices $V = \{v_1, v_2, v_3, v_4\}$ and the set of edges

$$E = \{e_1 : v_1 \rightarrow v_2, e_2 : v_1 \rightarrow v_4, e_3 : v_2 \rightarrow v_3, e_4 : v_3 \rightarrow v_2, e_5 : v_3 \rightarrow v_2\}$$

9. (*Maple V*) An undirected graph is complete if every pair of vertices is connected by exactly one edge. Use the **networks** package to construct a complete graph with five vertices, and show that the graph is strongly connected.

10. (*Maple V*) Find all paths of lengths 2, 3, and 4 determined by the graph in Exercise 9.

11. (*Maple V*) A group of five friends, Stuart, Mary, Tom, Dick, and Julie, are taking the same linear algebra course. They have decided to set up a telephone chain and inform each other of important announcements. They have agreed that Stuart phones Mary, Mary phones Tom, Tom phones Dick, and Dick phones Julie. Show that the graph determined by this communication chain is not strongly connected.

12. (*Maple V*) Repeat Exercises 1–6 using the **networks** package.

REVIEW

KEY CONCEPTS ▶ Define and discuss each of the following.

Lexicon of Matrices

Diagonal matrix, elementary matrix, elementary permutation matrix, generalized diagonal matrix, Hermitian matrix, identity matrix, invertible matrix, lower-triangular matrix, orthogonal matrix, permutation matrix, rectangular matrix, square matrix, symmetric matrix, unitary matrix, upper-triangular matrix.

Linear Systems

Homogeneous linear system.

Matrix Components

Diagonal entry of a matrix, diagonal of a matrix, entry of a matrix.

Matrix Operations

Conjugate transpose of a matrix, inverse of a matrix, left inverse of a matrix, left-multiplication by a matrix, LU decomposition of a matrix, product of matrices, right inverse of a matrix, right-multiplication by a matrix, row-column product of a matrix, scalar multiple of a matrix, sum of matrices, trace of a matrix, transpose of a matrix.

KEY FACTS ▶ Explain and illustrate each of the following.

1. If A is an $m \times n$ matrix and if B is an $n \times p$ matrix whose columns are $\mathbf{c}_1 B, \ldots, \mathbf{c}_p B$, then $AB = [A\,\mathbf{c}_1 B \;\cdots\; A\,\mathbf{c}_p B]$.
2. If $\mathbf{a} = [a_1, \ldots, a_n]$ is a $1 \times n$ row vector and $B = \left[b_{ij}\right]$ is an $n \times p$ matrix, then $\mathbf{a}B = a_1\mathbf{r}_1 B + \cdots + a_n\mathbf{r}_n B$.
3. For all matrices A and B and column vectors $\mathbf{x}$ of compatible dimension, $A(B\mathbf{x}) = (AB)\mathbf{x}$.
4. For all matrices A, B, C of compatible dimension, the following equalities hold: $(AB)\,C = A\,(BC)$; $A\,(B + C) = AB + AC$; $(A + B)\,C = AC + BC$; for all s, $s\,(AB) = (sA)\,B = A\,(sB)$; $0A = \mathbf{0}$, $A\mathbf{0} = \mathbf{0}$.
5. If $\mathbf{y}$ and $\mathbf{z}$ are solutions of a homogeneous system $A\mathbf{x} = \mathbf{0}$ and a and b are scalars, then $a\mathbf{y} + b\mathbf{z}$ is also a solution of $A\mathbf{x} = \mathbf{0}$.
6. For any $m \times n$ matrix $A = \left[a_{ij}\right]$ and any $n \times p$ matrix $B = \left[b_{jk}\right]$, the row-column products $\mathbf{r}_i\,(A)\,\mathbf{c}_j\,(B)$ and $\mathbf{r}_j(B^T)\mathbf{c}_i(A^T)$ are equal.
7. For any two square matrices A and B of compatible dimension, the following equalities hold: $(A^T)^T = A$; for all s, $(sA)^T = sA^T$; $(A + B)^T = A^T + B^T$; $(AB)^T = B^T A^T$.

8. If A and B are two $n \times n$ matrices, then the following equalities hold: trace A^T = trace A; for all s, trace $(sA) = s$ trace A; trace$(A+B)$ = trace A+trace B; trace AB = trace BA.

9. If the matrix B is the result of an elementary row operation applied to a matrix A, then there exists an elementary matrix E for which $B = EA$.

10. A matrix A is row equivalent to a matrix B if and only if there exist elementary matrices $E_1, \ldots, E_p$ such that $A = E_1 \cdots E_p B$.

11. If B is a right inverse of A and C is a left inverse of A, then $B = C$.

12. If B and C are inverses of A, then $B = C$.

13. If A and B are invertible matrices, then so is AB.

14. Every elementary matrix is invertible.

15. An $n \times n$ matrix A is invertible if and only if it is row equivalent to I_n.

16. An $n \times n$ matrix A is invertible if and only if its reduced row echelon form is I_n.

17. A matrix is invertible if and only if it is a product of elementary matrices.

18. An $n \times n$ matrix A is invertible if and only if it has n pivots.

19. An $n \times n$ matrix A is row equivalent to I_n if and only if the system $A\mathbf{x} = \mathbf{0}$ has only the trivial solution.

20. An $n \times n$ matrix A is invertible if and only if the equation $A\mathbf{x} = \mathbf{0}$ has only the trivial solution.

21. Suppose that A is an $n \times n$ real matrix. Then the linear system $A\mathbf{x} = \mathbf{b}$ has a solution for every $\mathbf{b} \in \mathbb{R}^n$ if and only if A is row equivalent to I_n.

22. An $n \times n$ matrix A is invertible if and only if there exists a matrix B such that $AB = I_n$.

23. A matrix B is the inverse of a matrix A if and only if the matrices $[A \mid I_n]$ and $[I_n \mid B]$ are row equivalent.

24. If A is an invertible matrix, then $(A^{-1})^T$ is the inverse of A^T.

25. If A and B are invertible, then AB is invertible and $(AB)^{-1} = B^{-1}A^{-1}$.

26. If A and B are real $n \times n$ matrices and AB is invertible, then so are A and B.

27. If A is an $n \times n$ row echelon matrix, then A is upper triangular.

28. The product of two $n \times n$ lower-triangular matrices is a lower-triangular matrix.

29. The product of two $n \times n$ upper-triangular matrices is an upper-triangular matrix.

30. Suppose that A is an $m \times n$ matrix that can be row-reduced to an upper-triangular matrix U without interchanging any rows of A. Then there exist a lower-triangular matrix L with 1s on the diagonal such that $A = LU$.

31. For any $m \times n$ matrix A there exist a permutation matrix P, an $m \times m$ lower-triangular matrix L with 1s on the diagonal, and an $m \times n$ row echelon matrix U such that $PA = LU$.

32. Suppose that A is an invertible matrix that can be reduced to row echelon form without interchanging any rows of A. Then there exist a unique lower-triangular matrix L with 1s on the diagonal and a unique upper-triangular matrix U such that $A = LU$.

33. For any $\theta \in \mathbb{R}$, the rotation matrix

$$A = \begin{bmatrix} \cos\theta & -\sin\theta \\ \sin\theta & \cos\theta \end{bmatrix}$$

is a product of elementary matrices.

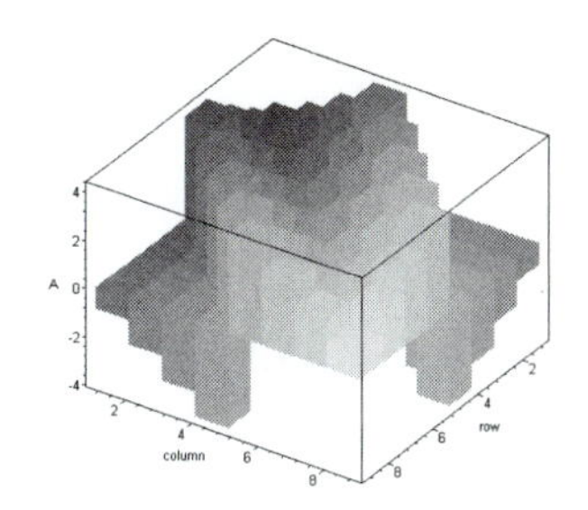

3

DETERMINANTS

With every square matrix we can associate a certain quantity called its ***determinant.*** If the matrix is a numerical matrix, the determinant is a number. If the matrix is a polynomial matrix, the determinant is a polynomial. As we study determinants, we will see that they have an amazing variety of mathematical uses. They can be used to show whether a real matrix is or is not invertible, to find matrix inverses, to show that two matrices fail to represent the same linear transformation, to calculate areas and volumes, to solve systems of linear equations, to find the partial derivatives of functions in several variables, to express relationships between multiple integrals, and so on. In Chapter 6, we will employ determinants of polynomial matrices to study the eigenvalues of square matrices.

THE LAPLACE EXPANSION

The determinant $\det A$ of an $n \times n$ matrix A, for $n \geq 2$, has a very elegant definition by an induction on n, using submatrices of A. The definition is known as the ***Laplace expansion*** of the determinant. We begin with the cases $n = 1$ and 2.

DEFINITION 3.1 *If A is the 1×1 matrix $A = [a_{11}]$, then $\det A = a_{11}$, and if A is the 2×2 matrix $A = \left[a_{ij}\right]$, then $\det A = (a_{11}a_{22} - a_{12}a_{21})$.*

To define the determinant $\det A$ for $n > 2$, we use the entries in the first row of A, together with specific submatrices. The resulting formula is called the ***Laplace expansion*** of the determinant of A along the first row.

DEFINITION 3.2 *The ijth* ***minor*** *M_{ij} of an $n \times n$ matrix A is the $(n-1) \times (n-1)$ submatrix of A obtained by deleting the ith row and the jth column of A.*

We assume that determinants $\det M_{ij}$ of the minors of A are known and use them to define the determinant of A itself. We simplify the notation by introducing the following auxiliary concept.

DEFINITION 3.3 *The ijth **cofactor** C_{ij} of an $n \times n$ matrix A is $(-1)^{i+j} \det M_{ij}$.*

The factor $(-1)^{i+j}$ keeps or reverses the sign of $\det M_{ij}$, depending on whether $i + j$ is odd or even.

DEFINITION 3.4 *If $n > 2$ and $C_{11}, \ldots, C_{1n}$ are the n cofactors of an $n \times n$ matrix $A = [a_{ij}]$ determined by the first row of A, then*

$$\det A = a_{11}C_{11} + \cdots + a_{1n}C_{1n}$$

It is easy to work with the definition of determinants in *Maple* by using the built-in **minor** function. The following calculations show, for example, that the **det** function in the **linalg** package produces the same general value for a 3×3 matrix A as the Laplace expansion of the determinant of A along the first row based on the minors of A.

EXAMPLE 3.1 Using a Laplace Expansion to Calculate a Determinant

Use the Laplace expansion along the first row to calculate the determinant of a general 33 matrix and compare the result with the value obtained by using the **det** function in the **linalg** package.

Solution. We begin by loading the **linalg** package and constructing a general 3×3 matrix.

```
[>with(linalg):
[>A:=matrix(3,3,a);
```

$$A := \begin{bmatrix} a(1,1) & a(1,2) & a(1,3) \\ a(2,1) & a(2,2) & a(2,3) \\ a(3,1) & a(3,2) & a(3,3) \end{bmatrix}$$

Next we define cofactors and combine it with a function that calculates the terms of a Laplace expansion.

```
[>cofactor:=(i,j)->(-1)^(i+j)*det(minor(A,i,j)):
[>Laplace:=(i,j)->A[i,j]*cofactor(i,j):
```

We can now calculate the determinant of A and compare it with the value obtained by **det** function.

```
[> L1:=Laplace(1,1);
```

$$L1 := a(1,1)(a(2,2)a(3,3) - a(2,3)a(3,2))$$

```
[> L2:=Laplace(1,2);
```

$$L2 := a(1,2)(-a(2,1)a(3,3) + a(2,3)a(3,1))$$

```
[> L3:=Laplace(1,3);
```

$$L3 := a(1,3)(a(2,1)a(3,2) - a(2,2)a(3,1))$$

```
[> is(det(A)=L1+L2+L3);
```

true

As we can see, the two values are identical. ▲

The determinant of a matrix can be defined along any row of A. The Laplace expansion of $\det A$ along the ith row of A, for example, is defined to be

$$\det A = a_{i1}C_{i1} + \cdots + a_{in}C_{in}$$

We could also have defined the determinant as

$$\det A = a_{1j}C_{1j} + a_{2j}C_{2j} + \cdots + a_{nj}C_{nj}$$

using the Laplace expansion of A along the jth column of A. If we were to compare the values obtained by expanding along the different columns and different rows of A in Example 3.1, for example, we would find that

$$\begin{aligned}\det A &= \text{Laplace}(1, 1) + \text{Laplace}(1, 2) + \text{Laplace}(1, 3) \\ &= \text{Laplace}(i, 1) + \text{Laplace}(i, 2) + \text{Laplace}(i, 3) \\ &= \text{Laplace}(1, j) + \text{Laplace}(2, j) + \text{Laplace}(3, j)\end{aligned}$$

for all $i, j \in \mathbf{3}$. It is important to note, however, that expansions along the diagonals of A do not in general produce the correct value of $\det A$.

EXAMPLE 3.2 The Determinant of a 2 × 2 Real Matrix

Use the definition to calculate the determinant of the matrix

$$A = \begin{bmatrix} 3 & 5 \\ 2 & 4 \end{bmatrix}$$

Solution. The determinant of a 2×2 matrix $A = [a_{ij}]$ is

$$\det \begin{bmatrix} a_{11} & a_{12} \\ a_{21} & a_{22} \end{bmatrix} = a_{11}a_{22} - a_{12}a_{21}$$

Hence $\det A = 3(4) - 2(5) = 2$. ▲

EXAMPLE 3.3 The Determinant of a 2 x 2 Polynomial Matrix

Use the definition to calculate the determinant of the matrix

$$A = \begin{bmatrix} 3-t & 5+6t \\ 2 & 4-t+t^2 \end{bmatrix}$$

Solution. If we label the entries of A as

$$\begin{bmatrix} 3-t & 5+6t \\ 2 & 4-t+t^2 \end{bmatrix} = \begin{bmatrix} a_{11} & a_{12} \\ a_{21} & a_{22} \end{bmatrix}$$

then we know from the definition of the determinants of 2×2 matrices that

$$\det A = a_{11}a_{22} - a_{12}a_{21} = (3-t)\left(4-t+t^2\right) - (2)(5+6t)$$

As we can see, $\det A$ is the polynomial $2 - 19t + 4t^2 - t^3$. ▲

EXAMPLE 3.4 The Determinant of a 3 x 3 Matrix

Verify that the determinant of the 3×3 matrix

$$A = \begin{bmatrix} a_{11} & a_{12} & a_{13} \\ a_{21} & a_{22} & a_{23} \\ a_{31} & a_{32} & a_{33} \end{bmatrix}$$

is $a_{11}a_{22}a_{33} - a_{11}a_{23}a_{32} - a_{12}a_{21}a_{33} + a_{12}a_{23}a_{31} + a_{13}a_{21}a_{32} - a_{13}a_{22}a_{31}$.

Solution. By the definition of the Laplace expansion of the determinant of a 3×3 matrix A along the first row of A,

$$\det A = a_{11}C_{11} + a_{12}C_{12} + a_{13}C_{13}$$

Since the cofactors C_{ij} of A are

$$C_{11} = (-1)^{1+1} \det \begin{bmatrix} a_{22} & a_{23} \\ a_{32} & a_{33} \end{bmatrix} = +(a_{22}a_{33} - a_{23}a_{32})$$

$$C_{12} = (-1)^{1+2} \det \begin{bmatrix} a_{21} & a_{23} \\ a_{31} & a_{33} \end{bmatrix} = -(a_{21}a_{33} - a_{23}a_{31})$$

$$C_{13} = (-1)^{1+3} \det \begin{bmatrix} a_{21} & a_{22} \\ a_{31} & a_{32} \end{bmatrix} = +(a_{21}a_{32} - a_{22}a_{31})$$

the Laplace expansion of det A along the first row of A is

$$\det A = a_{11}\left(a_{22}a_{33} - a_{23}a_{32}\right) - a_{12}\left(a_{21}a_{33} - a_{23}a_{31}\right) + a_{13}\left(a_{21}a_{32} - a_{22}a_{21}\right)$$
$$= a_{11}a_{22}a_{33} - a_{11}a_{23}a_{32} - a_{12}a_{21}a_{33} + a_{12}a_{23}a_{31} + a_{13}a_{21}a_{32} - a_{13}a_{22}a_{31}$$

as expected. ▲

EXAMPLE 3.5 Using *Maple* to Calculate the Determinant of a Matrix

Use *Maple* to calculate the determinant of the matrix

$$A = \begin{bmatrix} t-21 & t^2 & 1 & t^3 \\ 19 & t & 0 & 6 \\ t+1 & 3 & t-19 & 2 \\ 0 & 0 & 8 & t-5 \end{bmatrix}$$

Solution. We first define A and then apply the **det** function in the **linalg** package.

```
[> A:=matrix([
   [t-21,t^2,1,t^3],
   19,t,0,6],
   [t+1,3,t-19,2],
   [0,0,8,t-5]]);
```

```
[> linalg[det](A);
```

$$-10t^4 - 94t^3 - 962t^2 - 1453t - 3309 + 8t^5$$

Thus, the determinant of A is a polynomial in t. ▲

EXERCISES 3.1

1. Use the definition to calculate the determinants of the following matrices.

a. $A = \begin{bmatrix} -85 & -55 \\ -37 & -35 \end{bmatrix}$ b. $A = \begin{bmatrix} 97 & 50 & 79 \\ 56 & 49 & 63 \\ 57 & -59 & 45 \end{bmatrix}$ c. $A = \begin{bmatrix} 0 & 0 \\ 0 & 0 \end{bmatrix}$

d. $A = \begin{bmatrix} 3 & 9 & 4 & 1 \\ 0 & 8 & 15 & 4 \\ 0 & 0 & 9 & 12 \\ 0 & 0 & 0 & 9 \end{bmatrix}$ e. $A = \begin{bmatrix} 6 & 0 & 0 & 0 \\ 0 & 21 & 0 & 0 \\ 0 & 4 & 3 & 0 \\ 5 & 0 & 0 & 19 \end{bmatrix}$ f. $A = \begin{bmatrix} 1 & 0 & 0 & 0 \\ 0 & 2 & 0 & 0 \\ 0 & 0 & 3 & 0 \\ 0 & 0 & 0 & 2 \end{bmatrix}$

2. Find a 3×3 matrix A with real entries for which

$$\det A \neq a_{11}C_{11} + a_{22}C_{22} + a_{33}C_{33}$$

3. Find a 3×3 matrix A with polynomial entries for which

$$\det A \neq a_{11}C_{11} + a_{22}C_{22} + a_{33}C_{33}$$

4. Find two 3×3 matrices A and B for which $\det(A + B) \neq \det A + \det B$ and $\det 5A \neq 5 \det A$.

5. (*Maple V*) Verify that $\det AB = \det BA$ for all 3×3 matrices.

6. (*Maple V*) Calculate the determinant of A, where

$$A = \begin{bmatrix} 54 & -5 & 99 & -61 & -50 \\ -12 & -18 & 31 & -26 & -62 \\ 1 & -47 & -91 & -47 & -61 \\ 41 & -58 & -90 & 53 & -1 \\ 94 & 83 & -86 & 23 & -84 \end{bmatrix}$$

7. (*Maple V, 6*) Explore the functions **linalg[det]** and **LinearAlgebra[Determinant]** and discuss their differences.

Properties of Determinants

Determinants and permutations

If we take a closer look at the formulas for determinants, we notice that the determinant of an $n \times n$ matrix A is a sum of certain products $e_1 \cdots e_n$ of entries of the matrix. Each product $e_1 \cdots e_n$ consists of n entries. Moreover, each row and column of A is represented in every one of these products exactly once. These specialized products of entries of A are called the ***elementary products*** of the entries of A.

Elementary products can be described in terms of patterns using the locations ij of the entries a_{ij}. For each row index i, the corresponding column index j can be thought of as the value of a function with input i and output j. We introduce a language that allows us to speak about these patterns. We recall that a function $f : A \to B$ is ***one–one*** if for all $x, y \in A$, the equality $f(x)$ and $f(y)$ of two values of f implies that x and y must have been equal.

DEFINITION 3.5 *For any indexing set* $\mathbf{n} = [1, 2, \ldots, n]$, *a* ***permutation*** *of* $\mathbf{n}$ *is a one–one function* $\pi : \mathbf{n} \to \mathbf{n}$.

Since the domain $\mathbf{n}$ of a permutation π is ordered, we can abbreviate the notation for π and write it simply as $[\pi(1), \pi(2), \ldots, \pi(n)]$. If $\mathbf{n} = \mathbf{5}$, for example, the list

$$[\pi(1), \pi(2), \pi(3), \pi(4), \pi(5)] = [3, 4, 1, 2, 5]$$

denotes the permutation π defined by $\pi(1) = 3$, $\pi(2) = 4$, $\pi(3) = 1$, $\pi(4) = 2$, and $\pi(5) = 5$.

DEFINITION 3.6 *An **inversion** in a permutation π is any pair $(i, j) \in \mathbf{n} \times \mathbf{n}$ for which $i < j$ and $\pi(i) > \pi(j)$.*

For example, if $\pi(1) = 3$, $\pi(2) = 2$, $\pi(3) = 1$, and $\pi(4) = 4$ is a permutation π of the list **4**, then $\pi(1) > \pi(2)$, $\pi(1) > \pi(3)$, and $\pi(2) > \pi(3)$ are the three instances where a larger integer precedes a smaller one. Hence π has three inversions. The inversions can be calculated from the order in which the values of a permutation are listed.

If $\pi = [3, 2, 1, 4]$, for example, and we compare the listed values of π from left to right, we obtain three inequalities: $3 > 2$, $3 > 1$, and $2 > 1$. Hence π has three inversions.

If $\pi = [1, 5, 4, 3, 2]$, for example, and we compare the listed values of π from left to right, we obtain six inequalities $5 > 4$, $5 > 3$, $5 > 2$, $4 > 3$, $4 > 2$, and $3 > 2$. Hence π has six inversions.

DEFINITION 3.7 *A permutation is called **even** if it has an even number of inversions and **odd** otherwise.*

We can use the idea of inversions to attach a sign $\text{sgn}(\pi)$ to a permutation. This sign is called the ***parity*** of the permutation.

DEFINITION 3.8 *The **parity** $\text{sgn}(\pi)$ of a permutation π is 1 if π has an even number of inversions and is -1 otherwise.*

For any square $n \times n$ matrix A, the parity $\text{sgn}(\pi)$ of the permutation π is used to determine the sign of the elementary product $a_{1\pi(1)} \cdots a_{n\pi(n)}$ in the expansion of the determinant of A.

The set of all permutations of **n** is denoted by S_n.The **combinat** package makes it easy to calculate S_n. Let us illustrate the method by calculating the permutations of **2** and **3**.

EXAMPLE 3.6 Using *Maple* to Calculate Permutations

Use *Maple* to find the permutations of **2** and **3**.

Solution. We apply the **combinat** function to the indexing sets $\mathbf{2} = [1, 2]$ and $\mathbf{3} = [1, 2, 3]$.

```
[> combinat[permute]([1,2]);
                    [[1, 2], [2, 1]]
```

This tells us that $S_2 = \{[1, 2], [2, 1]\}$. We can tell by inspection that the permutation $[1, 2]$ has no inversions and that the permutation $[2, 1]$ has one inversion. Hence $[1, 2]$ is even and $[2, 1]$ is odd. Table 1 records these results.

TABLE 1 PARITIES FOR n = 2.

PERMUTATION	PARITY	sgn
$\pi_1 = [1, 2]$	Even	1
$\pi_2 = [2, 1]$	Odd	-1

Next we calculate the permutations of **3**.

```
[> combinat[permute]([1,2,3]);
```

$$[[1, 2, 3], [1, 3, 2], [2, 1, 3], [2, 3, 1], [3, 1, 2], [3, 2, 1]]$$

Therefore, $S_3 = \{[1, 2, 3], [1, 3, 2], [2, 1, 3], [2, 3, 1], [3, 1, 2], [3, 2, 1]\}$. We can tell by inspection that the permutation [1, 2, 3] has no inversions, that [1, 3, 2] and [2, 1, 3] have one inversion, [2, 3, 1] and [3, 1, 2] have two inversions, and [3, 2, 1] has three inversions. ▲

We record the results of these calculations in Table 2.

TABLE 2 PARITIES FOR n = 3.

PERMUTATION	PARITY	sgn
$\pi_1 = [1, 2, 3]$	Even	1
$\pi_2 = [1, 3, 2]$	Odd	-1
$\pi_3 = [2, 1, 3]$	Odd	-1
$\pi_4 = [2, 3, 1]$	Even	1
$\pi_5 = [3, 1, 2]$	Even	1
$\pi_6 = [3, 2, 1]$	Odd	-1

Using permutations and their parities, we can describe determinants without reference to submatrices. An induction on $n \geq 2$ shows that the determinant of an $n \times n$ matrix A is

$$\det A = \sum_{\pi \in S_n} \operatorname{sgn}(\pi) a_{1\pi(1)} \cdots a_{n\pi(n)}$$

Let us illustrate this formula with an example.

EXAMPLE 3.7 The Determinant of a 3 x 3 Matrix

Use permutations to find the determinant of a general 3×3 matrix.

Solution. Suppose that

$$A = \begin{bmatrix} a_{11} & a_{12} & a_{13} \\ a_{21} & a_{22} & a_{23} \\ a_{31} & a_{32} & a_{33} \end{bmatrix}$$

is a general 3×3 matrix, and let $S_3 = \{\pi_1, \pi_2, \pi_3, \pi_4, \pi_5, \pi_6\}$ be the set

$$\{[1, 2, 3], [1, 3, 2], [2, 1, 3], [2, 3, 1], [3, 1, 2], [3, 2, 1]\}$$

of permutations of **3**. Then Table 2 shows that

a. $\text{sgn}(\pi_1) = 1$	b. $\text{sgn}(\pi_2) = -1$	c. $\text{sgn}(\pi_3) = -1$
d. $\text{sgn}(\pi_4) = 1$	e. $\text{sgn}(\pi_5) = 1$	f. $\text{sgn}(\pi_6) = -1$

Therefore,

$$\begin{aligned} \det A &= \sum_{\pi \in S_3} \text{sgn}(\pi) a_{1\pi(1)} a_{2\pi(2)} a_{3\pi(3)} \\ &= \text{sgn}(\pi_1)\, a_{11}a_{22}a_{33} + \text{sgn}(\pi_2)\, a_{11}a_{23}a_{32} \\ &\quad + \text{sgn}(\pi_3)\, a_{12}a_{21}a_{33} + \text{sgn}(\pi_4)\, a_{12}a_{23}a_{31} \\ &\quad + \text{sgn}(\pi_5)\, a_{13}a_{21}a_{32} + \text{sgn}(\pi_6)\, a_{13}a_{22}a_{31} \end{aligned}$$

This means that

$$\det A = a_{11}a_{22}a_{33} - a_{11}a_{23}a_{32} - a_{12}a_{21}a_{33} + a_{12}a_{23}a_{31} + a_{13}a_{21}a_{32} - a_{13}a_{22}a_{31}$$

Each permutations π in S_3 determines one of the terms of $\det A$. ▲

We now use permutations to discuss how the basic properties of determinants can be proved. We limit the discussion to 2×2 and 3×3 matrices.

Determinants of identity matrices

Consider the identity matrix

$$I = \begin{bmatrix} a_{11} & a_{12} & a_{13} \\ a_{21} & a_{22} & a_{23} \\ a_{31} & a_{32} & a_{33} \end{bmatrix} = \begin{bmatrix} 1 & 0 & 0 \\ 0 & 1 & 0 \\ 0 & 0 & 1 \end{bmatrix}$$

The calculations

a. $a_{11}a_{22}a_{33} = (1)(1)(1) = 1$		b. $a_{11}a_{23}a_{32} = (1)(0)(0) = 0$
c. $a_{12}a_{21}a_{33} = (0)(0)(1) = 0$		d. $a_{12}a_{23}a_{31} = (0)(0)(0) = 0$
e. $a_{13}a_{21}a_{32} = (0)(0)(0) = 0$		f. $a_{13}a_{22}a_{31} = (0)(1)(0) = 0$

show that with the exception of $a_{11}a_{22}a_{33}$, all elementary products of I are zero and $a_{11}a_{22}a_{33} = 1$. Hence $\det A = \text{sgn}([1, 2, 3])a_{11}a_{22}a_{33} = 1$. The next theorem shows that this is typical for identity matrices.

THEOREM 3.1 *The determinant* $\det I_n$ *of the* $n \times n$ *identity matrix* I_n *is* 1.

Proof. Since all terms of the form $a_{1\pi(1)} \cdots a_{n\pi(n)}$ except for the term $a_{11} \cdots a_{nn}$ are 0,

$$\det I_n = \sum_{\pi \in S_n} \operatorname{sgn}(\pi) a_{1\pi(1)} \cdots a_{n\pi(n)} = \operatorname{sgn}(\pi) a_{11} \cdots a_{nn} = a_{11} \cdots a_{nn}$$

where π is the identity permutation $[1, \ldots, n]$. Hence $\det I_n = 1$. ■

Determinants of matrices with zero rows

Consider square matrices with zero rows such as

$$A = \begin{bmatrix} a_{11} & a_{12} & a_{13} \\ a_{21} & a_{22} & a_{23} \\ a_{31} & a_{32} & a_{33} \end{bmatrix} = \begin{bmatrix} 0 & 0 & 0 \\ a_{21} & a_{22} & a_{23} \\ a_{31} & a_{32} & a_{33} \end{bmatrix}$$

The calculations

a. $a_{11}a_{22}a_{33} = 0a_{22}a_{33} = 0$ b. $a_{11}a_{23}a_{32} = 0a_{23}a_{32} = 0$
c. $a_{12}a_{21}a_{33} = 0a_{21}a_{33} = 0$ d. $a_{12}a_{23}a_{31} = 0a_{23}a_{31} = 0$
e. $a_{13}a_{21}a_{32} = 0a_{21}a_{32} = 0$ f. $a_{13}a_{22}a_{31} = 0a_{22}a_{31} = 0$

show that all elementary products of A are zero. Hence $\det A = 0$. This is true in general.

THEOREM 3.2 *If A is an $n \times n$ matrix with a zero row, then the determinant of A is* 0.

Proof. Suppose that $A = \left[a_{ij}\right]$ and that $a_{i1} = \cdots = a_{in} = 0$. Then the terms

$$a_{1\pi(1)} \cdots a_{i\pi(i)} \cdots a_{n\pi(n)}$$

are 0 for all permutations π. Hence $\det A = 0$. ■

Determinants and elementary row operations

We now explore the effect of elementary row operations on determinants.

THEOREM 3.3 *If B is a matrix resulting from an $n \times n$ matrix A by the elementary row operation $\left(R_i \rightleftharpoons R_j\right)$ and $i \neq j$, then* $\det B = -\det A$.

Proof. We sketch a proof for $n = 2$. Let $S_2 = \{\sigma, \tau\} = \{[1, 2], [2, 1]\}$ and suppose that

$$A = \begin{bmatrix} a_{11} & a_{12} \\ a_{21} & a_{22} \end{bmatrix}$$

Consider the matrix

$$B = \begin{bmatrix} a_{21} & a_{22} \\ a_{11} & a_{12} \end{bmatrix} = \begin{bmatrix} a_{2\tau(2)} & a_{2\sigma(2)} \\ a_{1\sigma(1)} & a_{1\tau(1)} \end{bmatrix}$$

obtained from A by the row interchange $(R_1 \rightleftharpoons R_2)$.

By definition,

$$\begin{aligned}
\det B &= a_{21}a_{12} - a_{11}a_{22} \\
&= -\operatorname{sgn}(\tau)\, a_{2\tau(2)}a_{1\tau(1)} - \operatorname{sgn}(\sigma)\, a_{1\sigma(1)}a_{2\sigma(2)} \\
&= -\left(\operatorname{sgn}(\sigma)\, a_{1\sigma(1)}a_{2\sigma(2)} + \operatorname{sgn}(\tau)\, a_{2\tau(2)}a_{1\tau(1)}\right) \\
&= -\left(a_{1\sigma(1)}a_{2\sigma(2)} - a_{2\tau(2)}a_{1\tau(1)}\right) \\
&= -\det A
\end{aligned}$$

The general proof is analogous. ■

COROLLARY 3.4 *If $A = \left[a_{ij}\right]$ is an $n \times n$ matrix containing two identical rows, then the determinant of A is 0.*

Proof. Suppose $\operatorname{row}_i(A) = \operatorname{row}_j(A)$, with $i \neq j$, and let B be the matrix obtained from A by the elementary row operation $\left(R_i \rightleftharpoons R_j\right)$. Then it follows from Theorem 3.3 that $\det B = -\det A$. But $B = A$. Hence $\det A = -\det A$. This can only hold if $\det A = 0$. ■

Next we consider the effect of the operation $(R_i \to sR_i)$ on the determinant of a matrix.

THEOREM 3.5 *If B is a matrix resulting from an $n \times n$ matrix A by the elementary row operation $(R_i \to sR_i)$, then $\det B = s \det A$.*

Proof. We sketch the proof for $n = 3$. The underlying idea of the proof is very simple. Every elementary product in the expansion of $\det B$ contains the factor s exactly once. If we factor out s, we are left with $\det A$. Suppose that B is obtained from A by applying the operation $(R_2 \to sR_2)$. Then

$$\begin{aligned}
\det B &= \det \begin{bmatrix} a_{11} & a_{12} & a_{13} \\ sa_{21} & sa_{22} & sa_{23} \\ a_{31} & a_{32} & a_{33} \end{bmatrix} \\
&= a_{11}sa_{22}a_{33} - a_{11}sa_{23}a_{32} - sa_{21}a_{12}a_{33} + sa_{21}a_{13}a_{32} + a_{31}a_{12}sa_{23} - a_{31}a_{13}sa_{22} \\
&= s(a_{11}a_{22}a_{33} - a_{11}a_{23}a_{32} - a_{12}a_{21}a_{33} + a_{12}a_{23}a_{31} + a_{13}a_{21}a_{32} - a_{13}a_{22}a_{31}) \\
&= s \det A
\end{aligned}$$

since every elementary product of B contains s as a factor. The general proof is analogous. ■

EXAMPLE 3.8 Determinants and Scalar Multiplication

Use *Maple* to show that if A is the matrix

$$A = \begin{bmatrix} 0 & 5 & -3 & 2 \\ -2 & 2 & 2 & -1 \\ 1 & 4 & 1 & -3 \\ 3 & 5 & 5 & -2 \end{bmatrix}$$

then $\det(9A) = 9^4 \det A$.

Solution. We recall that we get the matrix $9A$ by multiplying each entry of A by 9. This is equivalent to multiplying each row by 9. In terms of elementary row operations, $9A$ results from A by the sequence of operations

$$A\ (R_1 \to 9R_1)\ B\ (R_2 \to 9R_2)\ C\ (R_3 \to 9R_3)\ D\ (R_4 \to 9R_4)\ E$$

Let us use *Maple* to calculate $\det A$ and $\det E$ and then compare the results.

```
[> A:=matrix([[0,5,-3,2],[-2,2,2,-1],[1,4,1,-3],
    [3,5,5,-2]]):
[> linalg[det](A);
                              369
```

```
[> E:=matrix([
    [9*0,9*5,9*-3,9*2],
    [9*-2,9*2,9*2,9*-1],
    [9*1,9*4,9*1,9*-3],
    [9*3,9*5,9*5,9*-2]]):
[> linalg[det](E);
                              2421009

[> 2421009/(9^4)
                              369
```

This shows that $\det(9A) = 9^4 \det A$. ▲

EXAMPLE 3.9 Determinants and the Addition of Rows

Use *Maple* to show that if the matrix

$$A = \begin{bmatrix} 0 & 5 & -3 & 2 \\ -2 & 2 & 2 & -1 \\ 1 & 4 & 1 & -3 \\ 3 & 5 & 5 & -2 \end{bmatrix}$$

is related to the matrix B by $A\ (R_3 \to R_3 + 7R_2)\ B$, then $\det B = \det A$.

Solution. We calculate $\det A$ and $\det B$ and compare the results.

```
[> A:=matrix([[0,5,-3,2],-2,2,2,-1],[1,4,1,-3],
    [3,5,5,-2]]):
[> linalg[det](A);
```

$$369$$

```
[> B:=matrix([
    [0,5,-3,2],
    [-2,2,2,-1],
    [1+7*(-2),4+7*(2),1+7*(2),-3+7*(-1)],
    [3,5,5,-2]]):
[> linalg[det](B);
```

$$369$$

This shows that $\det A = \det B$. ▲

Finally, we consider the case where B results from A by the addition of a multiple of one row to another row.

THEOREM 3.6 *If B is a matrix resulting from an $n \times n$ matrix A by the elementary row operation $\left(R_i \to R_i + sR_j\right)$, then* $\det B = \det A$.

Proof. We sketch the proof for $n = 3$. Consider the effect of the operation $(R_1 \to R_1 + sR_2)$ on the matrix

$$A = \begin{bmatrix} a_{11} & a_{12} & a_{13} \\ a_{21} & a_{22} & a_{23} \\ a_{31} & a_{32} & a_{33} \end{bmatrix}$$

The result is the matrix

$$B = \begin{bmatrix} a_{11} + sa_{21} & a_{12} + sa_{22} & a_{13} + sa_{23} \\ a_{21} & a_{22} & a_{23} \\ a_{31} & a_{32} & a_{33} \end{bmatrix}$$

Therefore,

$$\det B = \det \begin{bmatrix} a_{11} + sa_{21} & a_{12} + sa_{22} & a_{13} + sa_{23} \\ a_{21} & a_{22} & a_{23} \\ a_{31} & a_{32} & a_{33} \end{bmatrix}$$

Using the Laplace expansion of $\det B$ along the first row, we get

$$\begin{aligned}\det B &= (a_{11} + sa_{21})\, M_{11} - (a_{12} + sa_{22})\, M_{12} + (a_{13} + sa_{23})\, M_{13} \\ &= a_{11}M_{11} - a_{12}M_{12} + a_{13}M_{13} + sa_{21}M_{11} - sa_{22}M_{12} + sa_{23}M_{13} \\ &= \det A + s \det C\end{aligned}$$

where

$$M_{11} = \det\begin{bmatrix} a_{22} & a_{23} \\ a_{32} & a_{33} \end{bmatrix} \quad M_{12} = \det\begin{bmatrix} a_{21} & a_{23} \\ a_{31} & a_{33} \end{bmatrix} \quad M_{13} = \det\begin{bmatrix} a_{21} & a_{22} \\ a_{31} & a_{32} \end{bmatrix}$$

and

$$C = \begin{bmatrix} a_{21} & a_{22} & a_{23} \\ a_{21} & a_{22} & a_{23} \\ a_{31} & a_{32} & a_{33} \end{bmatrix}$$

However, the determinant of the matrix C is zero since C has two identical rows. Hence $\det B = \det A$. The general proof is analogous. ■

COROLLARY 3.7 *If E is an elementary matrix obtained from I_n by $(R_i \rightleftharpoons R_j)$ and $i \neq j$, then* $\det E = -1$.

Proof. Since $\det I_n = 1$, Theorem 3.6 implies that $\det E = -\det I_n = -1$. ■

COROLLARY 3.8 *If E is an elementary matrix obtained from I_n by $(R_i \to sR_i)$, then* $\det E = s$.

Proof. Since $\det I_n = 1$, Theorem 3.6 implies that $\det E = s \det I_n = s$. ■

COROLLARY 3.9 *If E is an elementary matrix obtained from I_n by $\left(R_i \to R_i + sR_j\right)$, then* $\det E = \det I_n = 1$.

Proof. This result follows from Theorem 3.6 and the fact that $\det I_n = 1$. ■

COROLLARY 3.10 *If $A = \left[a_{ij}\right]$ is an $n \times n$ matrix and E is an $n \times n$ elementary matrix, then* $\det EA = \det E \det A$.

Proof. If $E = I_n$, the result is obvious. In the other cases the result follows from Theorem 3.6 and its corollaries. ■

Determinants of elementary matrices

We can use the result in the previous section to describe the effect of elementary row operations on determinants.

THEOREM 3.11 *If E is an elementary matrix and $B = EA$ for some $n \times n$ matrix A, then the following statements hold.*

1. If E results from I_n by the operation $(R_i \rightleftharpoons R_j)$ and $i \neq j$, then $\det B = -\det A$.

2. If E results from I_n by the operation $(R_i \to sR_i)$ then $\det B = s \det A$.

3. If E results from I_n by the operation $(R_i \to R_i + sR_j)$, then $\det B = \det A$.

Proof. Since every elementary row operation corresponds to a left-multiplication by an elementary matrix, the equalities follow from Theorems 3.3, 3.5, and 3.6. ■

THEOREM 3.12 *If two $n \times n$ matrices A and B are row equivalent, then* $\det A = 0$ *if and only if* $\det B = 0$.

Proof. The matrices A and B are row equivalent if and only if there exist elementary matrices operations $E_1, \ldots, E_p$ for which $B = E_1 \cdots E_p A$. Using Corollaries 3.10, 3.7, 3, and 3.9, we can conclude that $\det B = s(-1)^r \det A$, where s is the product of nonzero scalars determined by the matrices among $E_1, \ldots, E_p$ corresponding to operations of the form $(R_i \to sR_i)$, and r is the number of matrices among $E_1, \ldots, E_p$ corresponding to operations of the form $(R_i \rightleftharpoons R_j)$. Thus if $\det A = 0$, then $s(-1)^r \det A = 0 = \det B$. Conversely, if $\det B = 0$, then $s(-1)^r \det A = 0$. But since $s(-1)^r$ is not equal to 0, we have $\det A = 0$. ■

Determinants of invertible matrices

Determinants provide a useful test for the invertibility of real matrices.

THEOREM 3.13 (Zero determinant theorem) *A real $n \times n$ matrix A is invertible if and only if* $\det A \neq 0$.

Proof. Suppose that A is invertible. If $A = I_n$, then $\det A = 1$. If $A \neq I_n$, then we know from Corollary 2.19 that there exist elementary row operations $E_1, \ldots, E_p$ such that

$$E_1 \cdots E_p A = I_n$$

Therefore $\det E_1 \cdots E_p A = \det I_n = 1 = s(-1)^r \det A$. It follows that $\det A \neq 0$.

Conversely, suppose that A is not invertible. Then the reduced row echelon form B of A must have at least one zero row. Hence every term $b_{1j_1} b_{2j_2} \cdots b_{nj_n}$ in the determinant of B must have a zero factor. Therefore, $\det B = 0$. Since $\det A$ is a multiple of $\det B$, it follows that $\det A$ is also 0. ■

Since a real number a is invertible if and only $a \neq 0$, we can reword Theorem 3.13 by saying that "A matrix A is an invertible element of $\mathbb{R}^{n \times n}$ if and only if its determinant $\det A$ is an invertible element of $\mathbb{R}$."

Determinants and matrix multiplication

The next theorem provides us with a property of determinants that will be used over and over again in this text.

THEOREM 3.14 (Determinant product theorem) *If A and B are two $n \times n$ matrices, then* $\det AB = \det A \det B$.

Proof. Suppose that A is not invertible. Then AB is not invertible either since $(AB)^{-1}$ would have to factor into $B^{-1}A^{-1}$. But A^{-1} does not exist. Hence Theorem 3.13 implies that $\det A = \det AB = 0$.

Suppose, conversely, that A is invertible. Then Corollary 2.19 tells us that A is a product $E_1 \cdots E_p$ of elementary matrices. It therefore follows from Corollary 3.10 and an induction on p that

$$\begin{aligned} \det AB &= \det\left(E_1 \cdots E_p B\right) \\ &= \left(\det E_1 \det E_2 \cdots \det E_p\right) \det B \\ &= \det\left(E_1 \cdots E_p\right) \det B \\ &= \det A \det B \end{aligned}$$

This shows that the determinant of a product is the product of the determinants. ■

EXAMPLE 3.10 The Determinant of a Matrix Product

Use *Maple* to show that the determinant of the product of the matrices

$$A = \begin{bmatrix} 0 & 5 & -3 & 2 \\ -2 & 2 & 2 & -1 \\ 1 & 4 & 1 & -3 \\ 3 & 5 & 5 & -2 \end{bmatrix} \quad \text{and} \quad B = \begin{bmatrix} -9 & 8 & 4 & -5 \\ 10 & -2 & 5 & 7 \\ 7 & -7 & 4 & -8 \\ 2 & 8 & 7 & 8 \end{bmatrix}$$

is equal to the product of the determinants of A and B.

Solution. We use *Maple* to calculate $\det AB$ and $\det A \det B$ and compare the results.

```
[>with(linalg):
[>A:=matrix([[0,5,-3,2],[-2,2,2,-1],[1,4,1,-3],[3,5,5,-2]):
[>B:=matrix([[-9,8,4,-5],[10,-2,5,7],[7,-7,4,-8],
    [2,8,7,8]]):
[> [det(A)*det(B),det(A&*B)];

                    [574533, 574533]
```

This shows that $\det A \det B = \det AB$. ▲

Another important property of determinants is the fact that the determinant of the inverse of a matrix is the reciprocal of the determinant of the matrix.

EXAMPLE 3.11 The Determinant of the Inverse of a Matrix

Use *Maple* to show that the determinant of the inverse of the matrix

$$A = \begin{bmatrix} -1 & -2 & 3 \\ 0 & 6 & 6 \\ 0 & 0 & -7 \end{bmatrix}$$

is the reciprocal of the determinant of A.

Solution. We use *Maple* to calculate $\det A$ and $\det A^{-1}$ and show that $\det A^{-1} = 1/\det A$.

```
[> with(linalg):
[> A:=matrix([[-1,-2,3],[0,6,6],[0,0,-7]]):
[> [det(A),det(inverse(A))];
```

$$\left[42, \tfrac{1}{42}\right]$$

This shows that $\det A^{-1} = 1/\det A$. ▲

In the next theorem we prove that this equality always holds.

THEOREM 3.15 *For any invertible matrix* A, $\det A^{-1} = 1/\det A$.

Proof. We know from Theorem 3.1 that $\det AA^{-1} = \det I = 1$ and from Theorem 3.14 that $\det AA^{-1} = \det A \det A^{-1}$. Hence $1 = \det A \det A^{-1}$. Since $\det A \neq 0$, we can solve for $\det A^{-1}$ and obtain $1/\det A$. ■

It is important to note that in contrast to matrix products, where the determinant of a product of two matrices is the product of the determinants of the matrices, the determinant of the sum of two matrices is usually not equal to the sum of the two determinants. For example, if

$$A = \begin{bmatrix} 1 & 1 \\ 1 & 1 \end{bmatrix} \quad \text{and} \quad B = \begin{bmatrix} 0 & 2 \\ 2 & 0 \end{bmatrix}$$

then $\det A = 0$ and $\det B = -4$. Therefore, $\det A + \det B = -4$. On the other hand,

$$\det(A + B) = \det \begin{bmatrix} 1 & 3 \\ 3 & 1 \end{bmatrix} = -8$$

Therefore, $\det A + \det B \neq \det(A + B)$.

Determinants and homogeneous linear systems

Among the many useful applications of this theorem is a description of homogeneous linear systems with nontrivial solutions.

THEOREM 3.16 *For any real square matrix A, the homogeneous linear system $A\mathbf{x} = \mathbf{0}$ has a nonzero solution $\mathbf{x}$ if and only if* $\det A = 0$.

Proof. Suppose $A\mathbf{x} = \mathbf{0}$ and $\det A \neq 0$. By Theorem 3.13, the matrix A is invertible. We therefore have $A^{-1}A\mathbf{x} = A^{-1}\mathbf{0}$. This means that $\mathbf{x} = \mathbf{0}$.

Conversely, suppose that $A\mathbf{x} = \mathbf{0}$ and $\det A = 0$. We know from our previous work that the solutions of $A\mathbf{x} = \mathbf{0}$ coincide with those of the homogeneous linear system $B\mathbf{x} = \mathbf{0}$, determined by the reduced row echelon form of A. Since $\det A = 0$, the matrix A is not invertible. Theorem 2.17 therefore tells us that the matrix B is not an identity matrix. Since A is square, it follows that one of its columns is not a pivot column. Hence the system $A\mathbf{x} = \mathbf{0}$ has a free variable $\mathbf{x}_i$ to which we can assign a nonzero value a_i, so the vector $\mathbf{x} = (0, \ldots, 0, a_i, 0, \ldots, 0)$ is a nonzero solution. ■

Determinants of triangular matrices

THEOREM 3.17 *If $A = \left[a_{ij}\right]$ is an $n \times n$ triangular matrix, then the determinant of A is the product $a_{11} \cdots a_{nn}$ of the diagonal entries of A.*

Proof. It is sufficient to prove the theorem for upper-triangular matrices. The proof for lower-triangular matrices is analogous. Let

$$A = \begin{bmatrix} a_{11} & a_{12} & a_{13} \\ a_{21} & a_{22} & a_{23} \\ a_{31} & a_{32} & a_{33} \end{bmatrix} = \begin{bmatrix} a_{11} & a_{12} & a_{13} \\ 0 & a_{22} & a_{23} \\ 0 & 0 & a_{33} \end{bmatrix}$$

be an arbitrary lower-triangular 3×3 matrix. Then we know from the definition and Example 3.7 that

$$\begin{aligned} \det A &= \sum_{\pi \in S_3} \operatorname{sgn}(\pi)\, a_{1\pi(1)} a_{2\pi(2)} a_{3\pi(3)} \\ &= a_{11}a_{22}a_{33} - a_{11}a_{23}a_{32} - a_{12}a_{21}a_{33} + a_{12}a_{23}a_{31} + a_{13}a_{21}a_{32} - a_{13}a_{22}a_{31} \\ &= a_{11}a_{22}a_{33} - a_{11}a_{23}(0) - a_{12}(0)\,a_{33} + a_{12}a_{23}(0) + a_{13}(0)(0) - a_{13}a_{22}(0) \\ &= a_{11}a_{22}a_{33} \end{aligned}$$

If $\pi(i) < i$ for some $i \in \mathbf{3}$, then $a_{i\pi(i)} = 0$ since A is upper triangular. Hence

$$a_{1\pi(1)} a_{2\pi(2)} a_{3\pi(3)} = 0$$

On the other hand, if $\pi(i) \geq i$ for all $i \in \mathbf{3}$, then $\pi(3) = 3$ since 3 is the largest element of $\mathbf{3}$. Therefore, $\pi(3-1) = 3-1 = 2$ since π is one–one and since $3 - 1 = 2$ is the largest remaining element of $\mathbf{3}$. Continuing in this way, we conclude that $\pi(i) = i$ for all

$i \in \mathbf{3}$. Hence the only nonzero term of the determinant of A is $a_{11} \cdots a_{33}$. The general proof is analogous. ■

In Table 3, we summarize the process of calculating the determinant of a matrix using triangular matrices and elementary row reduction.

TABLE 3 FINDING A DETERMINANT BY ROW REDUCTION.

Step 1. Use elementary row operations to reduce a matrix A to a triangular matrix B.
Step 2. Calculate $\det B$ by forming the product of the diagonal entries of B.
Step 3. For each $(R_i \rightleftharpoons R_j)$ used in the reduction, multiply $\det B$ by -1.
Step 4. For each $(R_i \to sR_i)$ used in the reduction, multiply $\det B$ by s.
Step 5. The determinant of A is the value obtained in Steps 2–4.

We conclude this section by establishing a connection between determinants, transposes, and inverses.

Determinants and matrix transposition

An important property of determinants is the fact that they do not change under matrix transposition.

EXAMPLE 3.12 The Determinant of the Transpose of a Matrix

Use *Maple* to show that the determinant of the matrix

$$A = \begin{bmatrix} -10 & 5 & -4 \\ -8 & -1 & -10 \\ 2 & 3 & -1 \end{bmatrix}$$

is equal to the determinant of A^T.

Solution. We use *Maple* to calculate $\det A$ and $\det A^T$ and compare the results.

```
[>with(linalg):
[>A:=matrix([[-10,5,-4],[-8,-1,-10],[2,3,-1]]):
[> [det(A),det(transpose(A))];
                    [-362, -362]
```

This shows that $\det A = \det A^T$. ▲

We now prove that this equality holds in general.

THEOREM 3.18 *For any $n \times n$ matrix A,* $\det A = \det A^T$.

Proof. We prove the theorem for $n = 3$. Consider the matrix

$$A = \begin{bmatrix} a_{11} & a_{12} & a_{13} \\ a_{21} & a_{22} & a_{23} \\ a_{31} & a_{32} & a_{33} \end{bmatrix}$$

and its transpose

$$A^T = \begin{bmatrix} a_{11} & a_{21} & a_{31} \\ a_{12} & a_{22} & a_{32} \\ a_{13} & a_{23} & a_{33} \end{bmatrix} = \begin{bmatrix} b_{11} & b_{12} & b_{13} \\ b_{21} & b_{22} & b_{23} \\ b_{31} & b_{32} & b_{33} \end{bmatrix}$$

From the definition of determinants it follows that

$$\begin{aligned} \det A &= \sum_{\pi \in S_3} \operatorname{sgn}(\pi) a_{1\pi(1)} a_{2\pi(2)} a_{3\pi(3)} \\ &= \sum_{\pi^{-1} \in S_3} \operatorname{sgn}(\pi^{-1}) a_{\pi^{-1}(1)1} a_{\pi^{-1}(2)2} a_{\pi^{-1}(3)3} \\ &= \sum_{\pi \in S_3} \operatorname{sgn}(\pi) b_{1\pi(1)} b_{2\pi(2)} b_{3\pi(3)} \\ &= \det A^T \end{aligned}$$

This shows that $\det A = \det A^T$. The general proof is analogous. ■

EXERCISES 3.2

1. Use elementary row operations to simplify the calculation of the determinants of the following matrices, if helpful, and then calculate the determinants.

a. $\begin{bmatrix} 1 & 3 & 2 \\ 0 & 2 & 4 \\ 0 & 0 & 3 \end{bmatrix}$ b. $\begin{bmatrix} 1 & 3 & 2 \\ 0 & 0 & 4 \\ 0 & 0 & 3 \end{bmatrix}$ c. $\begin{bmatrix} 0 & -3 & 2 \\ 4 & 0 & 4 \\ 2 & -1 & 0 \end{bmatrix}$

d. $\begin{bmatrix} 1 & 3 & 2 & 3 \\ 0 & 2 & 4 & 6 \\ 0 & 0 & 0 & 0 \\ 1 & 1 & 1 & 1 \end{bmatrix}$ e. $\begin{bmatrix} 7 & 5 & 0 & 0 \\ 9 & 0 & 57 & -59 \\ 4 & -8 & 0 & 92 \\ 3 & -2 & 7 & 0 \end{bmatrix}$ f. $\begin{bmatrix} 5 & 0 & 0 & 0 \\ 0 & 0 & -3 & 2 \\ 0 & 4 & 0 & 4 \\ 0 & 2 & -1 & 0 \end{bmatrix}$

2. Show that for each of the following matrices, $\det aA = a^n \det A$ for all scalars a.

a. $A = \begin{bmatrix} 0 & 0 & 0 \\ 0 & 0 & 0 \\ 0 & 0 & 0 \end{bmatrix}$ b. $A = \begin{bmatrix} 4 & 2 & 3 \\ -3 & 5 & 2 \\ 7 & 0 & 4 \end{bmatrix}$ d. $A = \begin{bmatrix} \frac{1}{\sqrt{2}} & \frac{1}{\sqrt{2}} \\ -\frac{1}{\sqrt{2}} & \frac{1}{\sqrt{2}} \end{bmatrix}$

3. Prove that for all permutation matrices P, $\det P = \pm 1$.

4. (*Maple V*) Generate a random 3×3 matrix $A = \left[a_{ij}\right]$ and calculate all elementary products $e_1 e_2 e_3$ of A.

5. (*Maple V*) Generate a random 3×3 matrix $A = \left[a_{ij}\right]$ and calculate the signed elementary products $\operatorname{sgn}(\sigma)\, a_{1\sigma(1)} a_{2\sigma(2)} a_{3\sigma(3)}$ of A for the permutations $\sigma = [3, 1, 2]$ and $\tau = [2, 1, 3]$.

6. (*Maple V*) Generate a random 3×3 matrix $A = \left[a_{ij}\right]$ and calculate all minors M_{ij} of A.

7. (*Maple V*) Generate a random 3×3 matrix $A = \left[a_{ij}\right]$ and calculate all cofactors C_{ij} of A.

8. (*Maple V*) Use the Laplace expansion to calculate the determinant of A, where

$$A = \begin{bmatrix} 1 & 0 & 7 & 3 \\ 6 & 8 & 5 & 8 \\ 1 & 9 & 5 & 3 \\ 7 & 0 & 4 & 5 \end{bmatrix}$$

9. (*Maple V*) Use the determinant to test the following matrices for invertibility.

$$\text{a.} \begin{bmatrix} 1 & 1 & 2 \\ 2 & 1 & 3 \\ 3 & 1 & 4 \end{bmatrix} \quad \text{b.} \begin{bmatrix} 1 & 1 & 2 \\ 2 & 1 & 3 \\ 3 & 1 & 6 \end{bmatrix} \quad \text{c.} \begin{bmatrix} 1 & 1 & 2 & 2 \\ 2 & 1 & 3 & 3 \\ 3 & 1 & 6 & 6 \\ 0 & 7 & 4 & 1 \end{bmatrix}$$

10. (*Maple V*) Calculate the determinant of A, where

$$A = \begin{bmatrix} 54 & -5 & 99 & -61 & -50 \\ -12 & -18 & 31 & -26 & -62 \\ 1 & -47 & -91 & -47 & -61 \\ 41 & -58 & -90 & 53 & -1 \\ 94 & 83 & -86 & 23 & -84 \end{bmatrix}$$

11. (*Maple V*) Generate a random 25×25 matrix with integer entries, and verify that its determinant is equal to the determinant of its transpose.

12. (*Maple V*) Use the **array** function to generate a general 4×4 matrix A and show that $\det A = \det A^T$.

13. (*Maple 6*) Show that $\det A^{-1} = 1/\det A$ for the following matrices.

$$\text{a. } A = \begin{bmatrix} 4 & 0 & 0 \\ 0 & 5 & 0 \\ 0 & 0 & 4 \end{bmatrix} \quad \text{b. } A = \begin{bmatrix} 4 & 2 & 3 \\ 0 & 5 & 2 \\ 0 & 0 & 4 \end{bmatrix} \quad \text{c. } A = \begin{bmatrix} \frac{1}{\sqrt{2}} & \frac{1}{\sqrt{2}} & 0 \\ -\frac{1}{\sqrt{2}} & \frac{1}{\sqrt{2}} & 0 \\ 0 & 0 & 1 \end{bmatrix}$$

14. (*Maple 6*) Use the **RandomMatrix** function to generate two random 3×3 matrices A and B with real entries and show that $\det AB = \det A \det B$.

APPLICATIONS

Cramer's Rule

A well-known procedure for solving linear systems is ***Cramer's rule,*** which uses determinants. If $A\mathbf{x} = \mathbf{b}$ is a linear system and A is an $n \times n$ invertible matrix, then Cramer's rule says that for each $1 \le i \le n$,

$$x_i = \frac{\det A\,(\mathbf{c}_i/\mathbf{b})}{\det A}$$

where $A\,(\mathbf{c}_i/\mathbf{b})$ is the matrix obtained from A by replacing the ith column of A by $\mathbf{b}$. We illustrate the use of Cramer's rule in the case of a 3×3 matrix A.

EXAMPLE 3.13 Using Cramer's Rule to Solve a Linear System in Three Variables

Use Cramer's rule to solve the 3×3 linear system

$$\begin{cases} x + y - z = 4 \\ 2x - y + 2z = 18 \\ x - y + z = 5 \end{cases}$$

Solution. The matrix form $A\mathbf{x} = \mathbf{b}$ of the given system is

$$\begin{bmatrix} 1 & 1 & -1 \\ 2 & -1 & 2 \\ 1 & -1 & 1 \end{bmatrix} \begin{bmatrix} x \\ y \\ z \end{bmatrix} = \begin{bmatrix} 4 \\ 18 \\ 5 \end{bmatrix}$$

Since $\det A = 2 \neq 0$, Cramer's rule applies. It tells us that

$$x = \frac{A\,(\mathbf{c}_1/\mathbf{b})}{\det A} = \frac{\det \begin{bmatrix} 4 & 1 & -1 \\ 18 & -1 & 2 \\ 5 & -1 & 1 \end{bmatrix}}{\det \begin{bmatrix} 1 & 1 & -1 \\ 2 & -1 & 2 \\ 1 & -1 & 1 \end{bmatrix}} = \frac{9}{2}$$

and that

$$y = \frac{A\left(\mathbf{c}_2/\mathbf{b}\right)}{\det A} = \frac{\det \begin{bmatrix} 1 & 4 & -1 \\ 2 & 18 & 2 \\ 1 & 5 & 1 \end{bmatrix}}{\det \begin{bmatrix} 1 & 1 & -1 \\ 2 & -1 & 2 \\ 1 & -1 & 1 \end{bmatrix}} = 8$$

and that

$$z = \frac{A\left(\mathbf{c}_3/\mathbf{b}\right)}{\det A} = \frac{\det \begin{bmatrix} 1 & 1 & 4 \\ 2 & -1 & 18 \\ 1 & -1 & 5 \end{bmatrix}}{\det \begin{bmatrix} 1 & 1 & -1 \\ 2 & -1 & 2 \\ 1 & -1 & 1 \end{bmatrix}} = \frac{17}{2}$$

Hence $(x, y, z) = (9/2, 8, 17/2)$ is the solution of the system. ▲

Cramer's rule provides an excellent framework for elegant definitions in calculus and differential equations. However, its usefulness for solving linear systems is minimized by the fact that, in general, it requires significantly more arithmetic operations than do other methods.

EXERCISES 3.3

1. Convert the following 2×2 linear systems to the matrix form $A\mathbf{x} = \mathbf{b}$. Solve the systems by Cramer's rule given $\det A \neq 0$.

$$\text{a. } \begin{cases} 2x + y = 2 \\ x - 3y = 1 \end{cases} \quad \text{b. } \begin{cases} 2x + y = 2 \\ -2x - y = 1 \end{cases} \quad \text{c. } \begin{cases} 2x + y = 0 \\ x - 3y = 0 \end{cases}$$

2. Convert the following 3×3 linear systems to the matrix form $A\mathbf{x} = \mathbf{b}$. Solve the systems by Cramer's rule given $\det A \neq 0$.

$$\text{a. } \begin{cases} 2x + y - z = 2 \\ x - 3y + z = 1 \\ 2x + y - 2z = 9 \end{cases} \quad \text{b. } \begin{cases} 2x + y - z = 2 \\ z - y + x = 1 \\ 2x - 2y + z = 9 \end{cases} \quad \text{c. } \begin{cases} 2x + y - z = 0 \\ x - 3y + z = 0 \\ 2x + y - 4z = 0 \end{cases}$$

Adjoint Matrices and Inverses

We conclude this section by describing the link between determinants and matrix inverses. For this purpose, we let $A\left(\mathbf{r}_i/\mathbf{r}_j\right)$ be the matrix obtained from a matrix A by replacing the ith row of A by its jth row. Our description is based on two simple observations:

1. If $i = j$, then $A(\mathbf{r}_i/\mathbf{r}_j) = A$ and therefore $\det A(\mathbf{r}_i/\mathbf{r}_j) = \det A$.
2. If $i \neq j$, then $A(\mathbf{r}_i/\mathbf{r}_j)$ has two identical rows and, by Corollary 3.4, $\det A(\mathbf{r}_i/\mathbf{r}_j) = 0$.

If we use the ith row of $A(\mathbf{r}_i/\mathbf{r}_j)$ for the Laplace expansion of $\det A(\mathbf{r}_i/\mathbf{r}_j)$, we get

$$\det A(\mathbf{r}_i/\mathbf{r}_j) = a_{j1}C_{i1} + \cdots + a_{jn}C_{in} = \begin{bmatrix} a_{j1} & \cdots & a_{jn} \end{bmatrix} \begin{bmatrix} C_{i1} \\ \vdots \\ C_{in} \end{bmatrix}$$

Therefore, $\det A(\mathbf{r}_i/\mathbf{r}_j)$ is the ijth entry of the matrix

$$AC = \begin{bmatrix} a_{11} & \cdots & a_{1n} \\ & \vdots & \\ a_{n1} & \cdots & a_{nn} \end{bmatrix} \begin{bmatrix} C_{11} & \cdots & C_{n1} \\ & \vdots & \\ C_{1n} & \cdots & C_{nn} \end{bmatrix}$$

We call the matrix C the ***adjoint matrix*** of A and denote it by adj A.

We can see that the adjoint matrix is the transpose of the matrix whose entries are the cofactors of A. Moreover, it follows from the stated properties of $\det A(\mathbf{r}_i/\mathbf{r}_j)$ that $A \operatorname{adj} A = (\det A) I_n$. Therefore,

$$I_n = A\frac{\operatorname{adj} A}{\det A} = AA^{-1}$$

Solving for A^{-1}, we get the formula

$$A^{-1} = \frac{\operatorname{adj} A}{\det A}$$

for the inverse of A, expressed entirely in terms of determinants of submatrices of A.

EXAMPLE 3.14 Using Adjoint Matrices to Find the Inverse of a Matrix

Use the adjoint matrix to find the inverse of the matrix

$$A = \begin{bmatrix} -2 & 3 & 4 \\ 4 & -3 & -3 \\ 4 & -1 & 0 \end{bmatrix}$$

Solution. By definition, the entries of the adjoint matrix of A are the cofactors $C_{ij} = (-1)^{i+j} \det M_{ij}$ of A. We must therefore calculate $C_{11}, \ldots, C_{33}$.

We leave it as an exercise to verify that $C_{11} = -3$, $C_{12} = -12$, $C_{13} = 8$, $C_{21} = -4$, $C_{22} = -16$, $C_{23} = 10$, $C_{31} = 3$, $C_{32} = 10$, and $C_{33} = -6$. Therefore,

$$\operatorname{adj} A = \begin{bmatrix} -3 & -4 & 3 \\ -12 & -16 & 10 \\ 8 & 10 & -6 \end{bmatrix}$$

Moreover, it is easily seen that det $A = 2$. By the derived formula,

$$A^{-1} = \frac{1}{2}\begin{bmatrix} -3 & -4 & 3 \\ -12 & -16 & 10 \\ 8 & 10 & -6 \end{bmatrix} = \begin{bmatrix} -\frac{3}{2} & -2 & \frac{3}{2} \\ -6 & -8 & 5 \\ 4 & 5 & -3 \end{bmatrix}$$

is the inverse of A.

Let us use *Maple* to verify this fact.

```
[> with(LinearAlgebra):
[> A:=Matrix([[-2,3,4],[4,-3,-3],[4,-1,0]]):
[> B:=Matrix([[-3/2,-2,3/2],[-6,-8,5],[4,5,-3]]):
[> I3:=IdentityMatrix(3):
[> Equal(A.B,I3);
                              true
```

This confirms that $B = A^{-1}$. ▲

EXERCISES 3.4

1. Verify that

$$\begin{bmatrix} 6 & 3 \\ 4 & 4 \end{bmatrix}^{-1} = \begin{bmatrix} \frac{1}{3} & -\frac{1}{4} \\ -\frac{1}{3} & \frac{1}{2} \end{bmatrix} = \frac{1}{\det A}\begin{bmatrix} C_{11} & C_{21} \\ C_{12} & C_{22} \end{bmatrix}$$

2. Verify that

$$\begin{bmatrix} 1 & 0 & 2 \\ 2 & 0 & 3 \\ 4 & 1 & 0 \end{bmatrix}^{-1} = \begin{bmatrix} -3 & 2 & 0 \\ 12 & -8 & 1 \\ 2 & -1 & 0 \end{bmatrix} = \frac{1}{\det A}\begin{bmatrix} C_{11} & C_{21} & C_{31} \\ C_{12} & C_{22} & C_{32} \\ C_{13} & C_{23} & C_{33} \end{bmatrix}$$

3. (*Maple V*) Repeat Exercises 1 and 2 using the functions **det** and **minor** in the **linalg** package.

4. (*Maple V*) Repeat Exercises 1 and 2 using the function **adj** in the **linalg** package.

5. (*Maple V, 6*) Explore the connection between functions **minor** in the **linalg** package and the function **Minor** in the **LinearAlgebra** package.

Determinants and the LU Decomposition

In this section, we establish a useful connection between determinants and the LU decomposition. We use this result to develop strategies for calculating determinants. We recall from

Theorem 3.17 that if $A = [a_{ij}]$ is an $n \times n$ triangular matrix, then the determinant of A is the product $a_{11} \cdots a_{nn}$ of the diagonal elements of A. When combined with the LU decomposition discussed in Chapter 2, this theorem provides us with a simplified procedure for calculating determinants.

EXAMPLE 3.15 Determinants and the LU Decomposition

Use the **LUDecomposition** function to find the determinant of the matrix

$$A = \begin{bmatrix} 4 & 5 & 0 & 0 \\ 2 & 2 & 2 & 4 \\ 5 & 2 & 5 & 2 \\ 4 & 1 & 3 & 0 \end{bmatrix}$$

Solution. We use the **LinearAlgebra** package.

```
[> with(LinearAlgebra):
[> A:=Matrix([[4,5,0,0],[2,2,2,4],[5,2,5,2],[4,1,3,0]]):
[> P,L,U:=LUDecomposition(A);
```

$$P, L, U := \begin{bmatrix} 1 & 0 & 0 & 0 \\ 0 & 1 & 0 & 0 \\ 0 & 0 & 1 & 0 \\ 0 & 0 & 0 & 1 \end{bmatrix}, \begin{bmatrix} 1 & 0 & 0 & 0 \\ \frac{1}{2} & 1 & 0 & 0 \\ \frac{5}{4} & \frac{17}{2} & 1 & 0 \\ 1 & 8 & \frac{13}{12} & 1 \end{bmatrix}, \begin{bmatrix} 4 & 5 & 0 & 0 \\ 0 & \frac{-1}{2} & 2 & 4 \\ 0 & 0 & -12 & -32 \\ 0 & 0 & 0 & \frac{8}{3} \end{bmatrix}$$

Since $A = PLU$, it follows from Theorem 3.14 that $\det A = \det P \det L \det U$. Moreover, we know from Theorem 3.17 that the determinant of a triangular matrix is simply the product of the diagonal entries of the matrix. Hence $\det A = 4\,(-1/2)\,(-12)\,(8/3) = 64$. ▲

EXERCISES 3.5

1. Find an LU decomposition of the matrix

$$A = \begin{bmatrix} 3 & 2 \\ 1 & 4 \end{bmatrix}$$

and show that $\det A = \det U$.

2. Find a PLU decomposition of the matrix

$$A = \begin{bmatrix} 0 & 2 & 1 \\ 1 & 4 & 1 \\ 3 & 0 & 1 \end{bmatrix}$$

and show that $\det A = (-1) \det U$.

3. (*Maple V*) Use Theorems 3.14 and 3.17 to calculate the determinant of the matrix

$$A = \begin{bmatrix} 8 & 4 & -5 & -5 \\ 3 & -5 & 8 & 5 \\ -1 & 0 & 3 & -4 \\ -5 & -2 & -1 & -9 \end{bmatrix}$$

4. (*Maple 6*) Generate a random 25×25 matrix and use Theorems 3.14 and 3.17 to calculate the determinant of the matrix.

5. (*Maple 6*) Let f be the function on $\mathbf{n} \times \mathbf{n}$ defined by $f(i, j) = \max(i, j)$ if $i \leq j$ and $\min(i, j)$ otherwise. Use the command

```
A:=RandomMatrix(100,100,generator=f)
```

to generate a 100×100 matrix, calculate an LU decomposition of PA, verify that P is an identity matrix, and show that $\det A = \det U$. Do not display the matrices involved. Compare the length of time required by *Maple* to calculate $\det A$ and $\det U$.

Determinants and Gaussian Elimination

The next theorem shows that we can also find the determinant of a matrix by first reducing the matrix to echelon form. In practice, this method is usually an efficient way of calculating a determinant. We motivate the theorem with an example that shows how pivots can be used to calculate determinants.

EXAMPLE 3.16 Determinants and Pivots

Use pivots to calculate the determinant of the matrix

$$A = \begin{bmatrix} 0 & 5 & 0 \\ 5 & 2 & 5 \\ 2 & 2 & 3 \end{bmatrix}$$

Solution. We first use *Maple* to calculate the echelon form of A.

```
[>with(linalg):
[>A:=matrix([[0,5,0],[5,2,5],[2,2,3]]):
[>A1:=swaprow(A,1,2);
```

$$A1 := \begin{bmatrix} 5 & 2 & 5 \\ 0 & 5 & 0 \\ 2 & 2 & 3 \end{bmatrix}$$

```
[> A2:=addrow(A,1,3,-2/5);
```

$$A2 := \begin{bmatrix} 5 & 2 & 5 \\ 0 & 5 & 0 \\ 0 & \frac{6}{5} & 1 \end{bmatrix}$$

```
[> A3:=addrow(A,2,3,-6/25);
```

$$A3 := \begin{bmatrix} 5 & 2 & 5 \\ 0 & 5 & 0 \\ 0 & 0 & 1 \end{bmatrix}$$

This shows that the pivots of A are 5, 5, and 1. Since we had to use one row interchange to obtain the reduced row-echelon form $A3$ of A, the next theorem tells us that $\det A = (-1) \times 5 \times 5 \times 1 = -25$. ▲

THEOREM 3.19 *For every invertible $n \times n$ real matrix A, $\det A = (-1)^m \det B$, where B is a row echelon matrix obtained from A by elementary row operations of types $(R_i \rightleftharpoons R_j)$, with $i \neq j$, and $(R_i \to R_i + sR_j)$, and m is the number of row interchanges involved.*

Proof. We know from our work on echelon matrices that every square matrix can be reduced to row echelon form using only operations of types $(R_i \rightleftharpoons R_j)$ and $(R_i \to R_i + sR_j)$. Theorem 3.6 tells us that the operation $(R_i \to R_i + sR_j)$ leaves the determinant of a matrix unchanged, and Corollary 3.7 guarantees that if $i \neq j$, then $(R_i \rightleftharpoons R_j)$ changes the sign of the determinant. ■

COROLLARY 3.20 *For every invertible $n \times n$ real matrix A, $\det A = (-1)^m p_1 \cdots p_n$, where $p_1, \ldots, p_n$ are the pivots of A.*

Proof. Since the row echelon matrix B in Theorem 3.19 is upper triangular, its determinant is simply the product of its diagonal entries. It is clear from our work in Chapter 1 that these entries are the pivots of the matrix A. ■

EXERCISES 3.6

1. Show that

$$\det \begin{bmatrix} 1 & 0 & 0 \\ a & 1 & 0 \\ b & c & 1 \end{bmatrix} = \det \begin{bmatrix} 1 & a & b \\ 0 & 1 & c \\ 0 & 0 & 1 \end{bmatrix} = 1$$

for all a, b, c.

2. Suppose that A is an $n \times n$ diagonal matrix. Find a formula for $\det A^p$ in terms of the diagonal entries of A.

3. Find the determinant of the following matrices by inspection.

$$\text{a. } A = \begin{bmatrix} 4 & 0 & 0 & 1 & 1 \\ 0 & 5 & 0 & -4 & 3 \\ 0 & 0 & 4 & 7 & 2 \\ 0 & 0 & 0 & 0 & 0 \\ 0 & 0 & 0 & 0 & 0 \end{bmatrix} \qquad \text{b. } B = \begin{bmatrix} 4 & 0 & 0 & 1 & 1 \\ 0 & 5 & 0 & -4 & 3 \\ 0 & 0 & 4 & 7 & 2 \\ 0 & 0 & 0 & 1 & 0 \\ 0 & 0 & 0 & 0 & 2 \end{bmatrix}$$

4. Let A and B be the matrices in Exercise 3. Use inspection to find the determinant of the matrix AB.

5. Find a 3×3 matrix with nonzero diagonal entries whose trace is equal to its determinant.

Determinants in Geometry

The determinant has many beautiful geometric applications. The following examples illustrate some of them.

EXAMPLE 3.17 The Straight Line Determined by Two Points

Use determinants to find an equation for the straight line determined by two distinct points (x_1, y_1) and (x_2, y_2) in $\mathbb{R}^2$.

Solution. The required equation is of the form $ax+by+c=0$, with a, b, c not all equal to 0. The given points must satisfy this equation, so we have $ax_1+by_1+c=0$ and $ax_2+by_2+c=0$. These equations determine a homogeneous linear system

$$\begin{cases} ax + by + c = 0 \\ ax_1 + by_1 + c = 0 \\ ax_2 + by_2 + c = 0 \end{cases}$$

whose matrix form $A\mathbf{x} = \mathbf{0}$ is

$$\begin{bmatrix} x & y & 1 \\ x_1 & y_1 & 1 \\ x_2 & y_2 & 1 \end{bmatrix} \begin{bmatrix} a \\ b \\ c \end{bmatrix} = \begin{bmatrix} 0 \\ 0 \\ 0 \end{bmatrix}$$

Since a, b, c are not all 0, the system has a nontrivial solution. It follows from Theorem 3.16 that

$$\det\begin{bmatrix} x & y & 1 \\ x_1 & y_1 & 1 \\ x_2 & y_2 & 1 \end{bmatrix} = 0$$

This means that $ax' + by' + c = 0$ if and only if

$$\det\begin{bmatrix} x' & y' & 1 \\ x_1 & y_1 & 1 \\ x_2 & y_2 & 1 \end{bmatrix} = 0$$

Let us suppose that $(x_1, y_1) = (1, 2)$ and $(x_2, y_2) = (3, 5)$. Then the straight line determined by these points is given by the equation

$$\det\begin{bmatrix} x & y & 1 \\ 1 & 2 & 1 \\ 3 & 5 & 1 \end{bmatrix} = 0$$

Let us use *Maple* to extract the desired equation.

```
[> A:=matrix([[x,y,1],[1,2,1],[3,5,1]]);
```

$$A := \begin{bmatrix} x & y & 1 \\ 1 & 2 & 1 \\ 3 & 5 & 1 \end{bmatrix}$$

```
[> linalg[det](A);
```

$$-3x + 2y - 1$$

The desired equation is therefore $-3x + 2y - 1 = 0$. ▲

EXAMPLE 3.18 The Parabola Determined by Three Noncollinear Points

Use determinants to find an equation for the parabola determined by three noncollinear points (x_1, y_1), (x_2, y_2), and (x_3, y_3) in $\mathbb{R}^2$.

Solution. The method is analogous to that used in Example 3.17. Since the three points are assumed to be noncollinear, we may assume that the desired parabola is of the form $ax^2 + bx + cy + d = 0$, with a, b, c, d not all equal to 0. Then the given points must satisfy this equation. The homogeneous linear system determined by this equation and by the three given points is

$$\begin{cases} ax^2 + bx + cy + d = 0 \\ ax_1^2 + bx_1 + cy_1 + d = 0 \\ ax_2^2 + bx_2 + cy_2 + d = 0 \\ ax_3^2 + bx_3 + cy_3 + d = 0 \end{cases}$$

The matrix form $A\mathbf{x} = \mathbf{0}$ of this equation is

$$\begin{bmatrix} x^2 & x & y & 1 \\ x_1^2 & x_1 & y_1 & 1 \\ x_2^2 & x_2 & y_2 & 1 \\ x_3^2 & x_3 & y_3 & 1 \end{bmatrix} \begin{bmatrix} a \\ b \\ c \\ d \end{bmatrix} = \begin{bmatrix} 0 \\ 0 \\ 0 \\ 0 \end{bmatrix}$$

Since $\mathbf{x}$ is assumed to be nonzero, Corollary 2.22 tells us that the matrix A must be singular. By Theorem 3.13, we therefore have

$$\det \begin{bmatrix} x^2 & x & y & 1 \\ x_1^2 & x_1 & y_1 & 1 \\ x_2^2 & x_2 & y_2 & 1 \\ x_3^2 & x_3 & y_3 & 1 \end{bmatrix} = 0$$

Let us suppose that $(x_1, y_1) = (-1, 1)$, $(x_2, y_2) = (0, 0)$, and $(x_3, y_3) = (1, 1)$. Then the parabola determined by these points is given by

$$\det \begin{bmatrix} x^2 & x & y & 1 \\ 1 & -1 & 1 & 1 \\ 0 & 0 & 0 & 1 \\ 1 & 1 & 1 & 1 \end{bmatrix} = 0$$

Calculations similar to those in Example 3.17 show that the desired equation is $2x^2 - 2y = 0$, which simplifies to $y = x^2$. ▲

EXAMPLE 3.19 The Circle Determined by Three Noncollinear Points

Use determinants to find an equation for the circle defined by three noncollinear points (x_1, y_1), (x_2, y_2), and (x_3, y_3) in $\mathbb{R}^2$.

Solution. The method is similar to that employed in the two previous examples. Suppose that the desired circle is of the form $a\left(x^2 + y^2\right) + bx + cy + d = 0$, with a, b, c, d not all equal to 0. Then the given points must satisfy this equation. The homogeneous linear system determined by this equation and the three given points is

$$\begin{cases} a\left(x^2+y^2\right)+bx+cy+d=0 \\ a\left(x_1^2+y_1^2\right)+bx_1+cy_1+d=0 \\ a\left(x_2^2+y_2^2\right)+bx_2+cy_2+d=0 \\ a\left(x_3^2+y_3^2\right)+bx_3+cy_3+d=0 \end{cases}$$

The matrix form $A\mathbf{x}=\mathbf{0}$ of this equation is

$$\begin{bmatrix} x^2+y^2 & x & y & 1 \\ x_1^2+y_1^2 & x_1 & y_1 & 1 \\ x_2^2+y_2^2 & x_2 & y_2 & 1 \\ x_3^2+y_3^2 & x_3 & y_3 & 1 \end{bmatrix}\begin{bmatrix} a \\ b \\ c \\ d \end{bmatrix}=\begin{bmatrix} 0 \\ 0 \\ 0 \\ 0 \end{bmatrix}$$

Hence the determinant equation for the desired circle is

$$\det\begin{bmatrix} x^2+y^2 & x & y & 1 \\ x_1^2+y_1^2 & x_1 & y_1 & 1 \\ x_2^2+y_2^2 & x_2 & y_2 & 1 \\ x_3^2+y_3^2 & x_3 & y_3 & 1 \end{bmatrix}=0$$

Let us suppose that $(x_1, y_1)=(-1,1)$, $(x_2, y_2)=(0,0)$, and $(x_3, y_3)=(1,1)$. Then the circle defined by these points is $x^2+(y-1)^2=1$. This equation defines the circle of radius 1 centered at (0, 1), as expected. ▲

For the next examples, we recall some definitions from geometry in $\mathbb{R}^2$ and $\mathbb{R}^3$. The basic operation on which these definitions are built is the dot product $\mathbf{x}\cdot\mathbf{y}=\mathbf{x}^T\mathbf{y}$.

DEFINITION 3.9 *The **length** $\|\mathbf{x}\|$ of a vector $\mathbf{x}=(x_1, x_2)$ in $\mathbb{R}^2$ is*

$$\sqrt{\mathbf{x}^T\mathbf{x}}=\sqrt{x_1^2+x_2^2}$$

DEFINITION 3.10 *The **length** $\|\mathbf{x}\|$ of a vector $\mathbf{x}=(x_1, x_2, x_3)$ in $\mathbb{R}^3$ is*

$$\sqrt{\mathbf{x}^T\mathbf{x}}=\sqrt{x_1^2+x_2^2+x_3^2}$$

The properties of the length function $\|\mathbf{x}\|$ are the basis for our work in Chapter 7. One of these properties is the fact that for any real number a and any vector $\mathbf{x}=(x_1, x_2, x_3)\in\mathbb{R}^3$,

$$\begin{aligned} \|a\mathbf{x}\| &= \|(ax_1, ax_2, ax_3)\| \\ &= \sqrt{a^2\left(x_1^2+x_2^2+x_3^2\right)} \\ &= |a|\sqrt{x_1^2+x_2^2+x_3^2}=|a|\,\|\mathbf{x}\| \end{aligned}$$

DEFINITION 3.11 *The **distance** $d(\mathbf{x},\mathbf{y})$ between two vectors $\mathbf{x}$ and $\mathbf{y}$ in $\mathbb{R}^2$ and $\mathbb{R}^3$ is the length $\|\mathbf{x}-\mathbf{y}\|$ of the vector $\mathbf{x}-\mathbf{y}$.*

DEFINITION 3.12 *The* ***cosine*** *of the angle θ (with $0 \leq \theta < \pi$) between two nonzero vectors* $\mathbf{x}$ *and* $\mathbf{y}$ *in* $\mathbb{R}^2$ *and* $\mathbb{R}^3$ *is* $\mathbf{x}^T\mathbf{y}/(\|\mathbf{x}\|\,\|\mathbf{y}\|)$.

We can rewrite this definition in dot product notation as $\mathbf{x}\cdot\mathbf{y} = \|\mathbf{x}\|\,\|\mathbf{y}\|\cos\theta$.

Given two nonzero direction arrows $\mathbf{x}$ and $\mathbf{y}$ in $\mathbb{R}^2$ and $\mathbb{R}^3$, we can always drop a perpendicular from the tip of $\mathbf{x}$ onto the line L through $\mathbf{y}$ and form the direction arrow $\mathbf{z}$ on L whose tip lies at the intersection of L and the perpendicular. By construction, the vectors $\mathbf{z}$ and $\mathbf{x}-\mathbf{z}$ are perpendicular. Their dot product will be zero. Moreover,

$$\mathbf{x} = \mathbf{z} + (\mathbf{x} - \mathbf{z})$$

We have therefore decomposed the vector $\mathbf{x}$ into the sum of two perpendicular vectors. The vector $\mathbf{z}$ is called the ***orthogonal projection*** of $\mathbf{x}$ on $\mathbf{y}$, and we denote it by $\text{proj}\,(\mathbf{x}\to\mathbf{y})$. The vector $\mathbf{x}-\mathbf{z}$ is called the ***vector component*** of $\mathbf{x}$ orthogonal to $\mathbf{y}$, and we denote it by $\text{perp}\,(\mathbf{x}\perp\mathbf{y})$. In Chapter 8, we verify that

$$\mathbf{z} = \text{proj}\,(\mathbf{x}\to\mathbf{y}) = \frac{\mathbf{y}\cdot\mathbf{x}}{\|\mathbf{y}\|^2}\mathbf{y}$$

and $\text{perp}\,(\mathbf{x}\perp\mathbf{y}) = \mathbf{x} - \text{proj}\,(\mathbf{x}\to\mathbf{y})$ by definition. The length $\|\mathbf{z}\|$ of $\mathbf{z}$ is $|\mathbf{y}\cdot\mathbf{x}|\,/\,\|\mathbf{y}\|$.

In addition to using orthogonal projections to create orthogonal vectors, we can also use a construction known as the ***cross product***. However, while orthogonal projections can be generalized from $\mathbb{R}^2$ and $\mathbb{R}^3$ to other spaces, the cross product only makes sense in $\mathbb{R}^3$.

DEFINITION 3.13 *The* ***cross product*** $\mathbf{x}\times\mathbf{y}$ *of two vectors* $\mathbf{x} = (x_1, x_2, x_3)$ *and* $\mathbf{y} = (y_1, y_2, y_3)$ *in* $\mathbb{R}^3$ *is the vector* (z_1, z_2, z_3) *with components*

$$z_1 = \det\begin{bmatrix} x_2 & x_3 \\ y_2 & y_3 \end{bmatrix}, \quad z_2 = -\det\begin{bmatrix} x_1 & x_3 \\ y_1 & y_3 \end{bmatrix}, \quad z_3 = \det\begin{bmatrix} x_1 & x_2 \\ y_1 & y_2 \end{bmatrix}$$

The dot and cross products of vectors in $\mathbb{R}^3$ are related by the following laws and are easily verified. If $\mathbf{x}$, $\mathbf{y}$, and $\mathbf{z}$ are three vectors in $\mathbb{R}^3$, then the following relationships holds. We recall from Chapter 1 that two vectors $\mathbf{x}$ and $\mathbf{y}$ in $\mathbb{R}^2$ and $\mathbb{R}^3$ are defined to be orthogonal if $\mathbf{x}\cdot\mathbf{y} = 0$.

a. $\mathbf{x}\cdot(\mathbf{x}\times\mathbf{y}) = 0$ (Orthogonality)

b. $\mathbf{y}\cdot(\mathbf{x}\times\mathbf{y}) = 0$ (Orthogonality)

c. $(\mathbf{x}\times\mathbf{y})\times\mathbf{z} = (\mathbf{x}\cdot\mathbf{z})\,\mathbf{y} - (\mathbf{y}\cdot\mathbf{z})\,\mathbf{x}$ (Link between cross and dot products)

d. $\mathbf{x}\times(\mathbf{y}\times\mathbf{z}) = (\mathbf{x}\cdot\mathbf{z})\,\mathbf{y} - (\mathbf{x}\cdot\mathbf{y})\,\mathbf{z}$ (Link between cross and dot products)

e. $\|\mathbf{x}\times\mathbf{y}\|^2 = \|\mathbf{x}\|^2\,\|\mathbf{y}\|^2 - (\mathbf{x}\cdot\mathbf{y})^2$ (Lagrange's identity)

With each triple of vectors $\mathbf{x}$, $\mathbf{y}$, and $\mathbf{z}$ in $\mathbb{R}^3$, we can associate a useful scalar by combining the dot– and cross–product operations.

DEFINITION 3.14 *The* ***scalar triple product*** *of vectors* $\mathbf{x} = (x_1, x_2, x_3)$, $\mathbf{y} = (y_1, y_2, y_3)$, *and* $\mathbf{z} = (z_1, z_2, z_3)$ *in* $\mathbb{R}^3$ *is the scalar* $\mathbf{x}\cdot(\mathbf{y}\times\mathbf{z})$ *given by the formula*

$$\det \begin{bmatrix} x_1 & x_2 & x_3 \\ y_1 & y_2 & y_3 \\ z_1 & z_2 & z_3 \end{bmatrix}$$

It is clear from the properties of determinants that $\mathbf{x} \cdot (\mathbf{y} \times \mathbf{z}) = \mathbf{z} \cdot (\mathbf{z} \times \mathbf{y}) = \mathbf{y} \cdot (\mathbf{z} \times \mathbf{x})$.

Using Lagrange's identity and the definition of the cosine of an angle, we can give an interesting geometric interpretation to the cross product of two vectors.

$$\begin{aligned} \|\mathbf{x} \times \mathbf{y}\|^2 &= \|\mathbf{x}\|^2 \|\mathbf{y}\|^2 - (\mathbf{x} \cdot \mathbf{y})^2 \\ &= \|\mathbf{x}\|^2 \|\mathbf{y}\|^2 - \|\mathbf{x}\|^2 \|\mathbf{y}\|^2 \cos^2 \theta \\ &= \|\mathbf{x}\|^2 \|\mathbf{y}\|^2 \left(1 - \cos^2 \theta\right) \\ &= \|\mathbf{x}\|^2 \|\mathbf{y}\|^2 \sin^2 \theta \end{aligned}$$

Thus $\|\mathbf{x} \times \mathbf{y}\| = \|\mathbf{x}\| \|\mathbf{y}\| \sin \theta$ for $0 \leq \theta < \pi$. Since we know form geometry that $\|\mathbf{y}\| \sin \theta$ is the altitude of the parallelogram determined by the vectors $\mathbf{x}$ and $\mathbf{y}$, the area of the parallelogram is given by the formula

$$\text{Area} = \text{Base} \times \text{Height} = \|\mathbf{x}\| \|\mathbf{y}\| \sin \theta = \|\mathbf{x} \times \mathbf{y}\|$$

The cross product of two vectors $\mathbf{x} = (x_1, x_2, x_3)$ and $\mathbf{y} = (y_1, y_2, y_3)$ in $\mathbb{R}^3$ has a beautiful formulation using determinants and the standard unit vectors $\mathbf{i} = (1, 0, 0)$, $\mathbf{j} = (0, 1, 0)$, and $\mathbf{k} = (0, 0, 1)$ in $\mathbb{R}^3$. Since

$$\mathbf{x} = (x_1, x_2, x_3) = x_1\mathbf{i} + x_2\mathbf{j} + x_3\mathbf{k}$$

and

$$\mathbf{y} = (y_1, y_2, y_3) = y_1\mathbf{i} + y_2\mathbf{j} + y_3\mathbf{k}$$

we can use the determinant notation and write

$$\begin{aligned} \mathbf{x} \times \mathbf{y} &= \det \begin{bmatrix} \mathbf{i} & \mathbf{j} & \mathbf{k} \\ x_1 & x_2 & x_3 \\ y_1 & y_2 & y_3 \end{bmatrix} \\ &= \det \begin{bmatrix} x_2 & x_3 \\ y_2 & y_3 \end{bmatrix} \mathbf{i} - \det \begin{bmatrix} x_1 & x_3 \\ y_1 & y_3 \end{bmatrix} \mathbf{j} + \det \begin{bmatrix} x_1 & x_2 \\ y_1 & y_2 \end{bmatrix} \mathbf{k} \end{aligned}$$

Moreover, the cross product of the vectors $\mathbf{i}$, $\mathbf{j}$, and $\mathbf{k}$ are easily seen to be linked by the following equations:

a.	$\mathbf{i} \times \mathbf{i} = \mathbf{0}$	d.	$\mathbf{i} \times \mathbf{j} = \mathbf{k}$	g.	$\mathbf{j} \times \mathbf{i} = -\mathbf{k}$
b.	$\mathbf{j} \times \mathbf{j} = \mathbf{0}$	e.	$\mathbf{j} \times \mathbf{k} = \mathbf{i}$	h.	$\mathbf{k} \times \mathbf{j} = -\mathbf{i}$
c.	$\mathbf{k} \times \mathbf{k} = \mathbf{0}$	f.	$\mathbf{k} \times \mathbf{i} = \mathbf{j}$	i.	$\mathbf{i} \times \mathbf{k} = -\mathbf{j}$

These properties of the vectors **i**, **j**, and **k** are the basis for the definition of a ***quaternion*** structure on $\mathbb{R}^4$ in Chapter 4, used to calculate the rotations of vectors in $\mathbb{R}^3$.

The following theorem about orthogonal vectors is the motivation behind the definitions of length and distance.

THEOREM 3.21 (Pythagorean theorem) *If* $\mathbf{x}$ *and* $\mathbf{y}$ *are two orthogonal vectors in* $\mathbb{R}^2$ *or* $\mathbb{R}^3$, *then* $\|\mathbf{x}+\mathbf{y}\|^2 = \|\mathbf{x}\|^2 + \|\mathbf{y}\|^2$.

Distance, length, and cosines are related by the following well-known formula.

THEOREM 3.22 (Law of cosines theorem) *If* θ *is an angle determined by two vectors* $\mathbf{x}$ *and* $\mathbf{y}$ *in* $\mathbb{R}^2$ *or* $\mathbb{R}^3$, *then* $\|\mathbf{x}-\mathbf{y}\|^2 = \|\mathbf{x}\|^2 + \|\mathbf{y}\|^2 - 2\|\mathbf{x}\|\,\|\mathbf{y}\|\cos\theta$.

EXAMPLE 3.20 The Volume of a Parallelepiped

Show that the volume V of the parallelogram determined by the vectors $\mathbf{x} = (x_1, x_2, x_3)$, $\mathbf{y} = (y_1, y_2, y_3)$, and $\mathbf{z} = (z_1, z_2, z_3)$ in $\mathbb{R}^3$ is the absolute values of the determinant

$$\det \begin{bmatrix} x_1 & x_2 & x_2 \\ y_1 & y_2 & y_2 \\ z_1 & z_2 & z_3 \end{bmatrix}$$

Solution. If we use the vectors $\mathbf{x}$ and $\mathbf{y}$ to define the base of the parallelepiped, then $\|\mathbf{x}\times\mathbf{y}\|$ is the area B of the base. Moreover, it is clear from the definition of orthogonal projections that the height H of the parallelepiped is the length of the orthogonal projection $\text{proj}\,(\mathbf{z}\to\mathbf{x}\times\mathbf{y})$. Hence

$$H = \|\text{proj}\,(\mathbf{z}\to\mathbf{x}\times\mathbf{y})\| = \frac{|(\mathbf{x}\times\mathbf{y})\cdot\mathbf{z}|}{\|\mathbf{x}\times\mathbf{y}\|}$$

This tells us that

$$V = BH = \|\mathbf{x}\times\mathbf{y}\| \frac{|(\mathbf{x}\times\mathbf{y})\cdot\mathbf{z}|}{\|\mathbf{x}\times\mathbf{y}\|} = |(\mathbf{x}\times\mathbf{y})\cdot\mathbf{z}|$$

Since

$$(\mathbf{x}\times\mathbf{y})\cdot\mathbf{z} = \det \begin{bmatrix} x_1 & x_2 & x_2 \\ y_1 & y_2 & y_2 \\ z_1 & z_2 & z_3 \end{bmatrix}$$

we have found the required formula. ▲

EXERCISES 3.7

1. Use determinants to find equations for the straight lines containing the following points.

a. $(3, 4), (2, 7)$ b. $(3, 4), (2, -7)$
c. $(3, 4), (-2, 7)$ d. $(3, 4), (-2, -7)$

2. Use determinants to find equations for the parabolas containing the following points.

a. (3, 4), (2, 7), (9, 4) b. (3, 4), (2, 7), (9, −4)
c. (3, 4), (2, 7), (−9, 4) d. (3, 4), (2, 7), (−9, −4)

3. Use determinants to find equations for the circles containing the points in Exercise 2.

4. Use determinants to find the areas of the parallelograms determined by the following points.

a. (3, 4), (2, 7), (9, 4) b. (3, 4), (2, 7), (−7, 6)
c. (3, 4), (2, 7), (−2, 2) d. (3, 4), (2, 7), (−12, −15)

5. Use the fact that

$$\det\begin{bmatrix} a & c \\ b & d \end{bmatrix} = \det\begin{bmatrix} a & c & 0 \\ b & d & 0 \\ 1 & 1 & 1 \end{bmatrix}$$

to show that the area of a parallelogram in $\mathbb{R}^2$ is preserved by the homogeneous coordinate embedding of $\mathbb{R}^2$ in $\mathbb{R}^3$.

6. For the vectors $\mathbf{x} = (1, 2, 3)$, $\mathbf{y} = (-1, 3, 5)$, and $\mathbf{z} = (7, 0, 6)$ in $\mathbb{R}^3$, calculate the following:

a. The lengths of the vectors $\|\mathbf{x}\|$, $\|\mathbf{y}\|$, and $\|\mathbf{z}\|$.

b. The distances between the three vectors.

c. The cosines of the angles between the three vectors.

d. The orthogonal projections determined by any two of the three vectors.

e. The vector components determined by any two of the three vectors.

f. The cross products determined by any two of the three vectors.

g. The scalar triple product determined by the three vectors.

h. The area of the parallelogram determined by the vectors $\mathbf{u} = (1, 2, 0)$, $\mathbf{v} = (-1, 3, 0)$.

i. The volume of the parallelepiped determined by $\mathbf{x}$, $\mathbf{y}$, and $\mathbf{z}$.

Determinants in Calculus

Determinants can be used in calculus to solve a variety of problems that include the computation of partial derivatives, problems involving multiple integrals, the identification of maxima and minima, and the proof that a set of functions is linearly independent.

Jacobians

Suppose that $F(u, v, x, y) = 0$ and $G(u, v, x, y) = 0$ are two differentiable functions in some region of $\mathbb{R}^4$ and that u and v are differentiable functions of x and y. Determinants and Cramer's rule can be used to find the partial derivatives u_x, v_x, u_y, and v_y.

We proceed as follows. Since $F(u, v, x, y) = G(u, v, x, y) = 0$, the differentials dF and dG are 0. This means that

$$\begin{aligned} 0 = dF &= F_u\,du + F_v\,dv + F_x\,dx + F_y\,dy \\ 0 = dG &= G_u\,du + G_v\,dv + G_x\,dx + G_y\,dy \end{aligned}$$

Furthermore, $du = u_x\,dx + u_y\,dy$ and $dv = v_x\,dx + v_y\,dy$ since u and v are differentiable functions of x and y. Therefore,

$$\begin{aligned} 0 &= dF \\ &= F_u\,du + F_v\,dv + F_x\,dx + F_y\,dy \\ &= F_u\left(u_x\,dx + u_y\,dy\right) + F_v\left(v_x\,dx + v_y\,dy\right) + F_x\,dx + F_y\,dy \\ &= \left(F_u u_x + F_v v_x + F_x\right)dx + \left(F_u u_y + F_v v_y + F_y\right)dy \end{aligned}$$

and

$$\begin{aligned} 0 &= dG \\ &= G_u du + G_v dv + G_x dx + G_y dy \\ &= G_u\left(u_x dx + u_y dy\right) + G_v\left(v_x dx + v_y dy\right) + G_x dx + G_y dy \\ &= \left(G_u u_x + G_v v_x + G_x\right)dx + \left(G_u u_y + G_v v_y + G_y\right)dy \end{aligned}$$

Since x and y are independent variables, the coefficients of dx and dy in these equations must be zero. Hence we obtain the two linear systems

$$\begin{cases} F_u u_x + F_v v_x = -F_x \\ G_u u_x + G_v v_x = -G_x \end{cases} \quad \text{and} \quad \begin{cases} F_u u_y + F_v v_y = -F_y \\ G_u u_y + G_v v_y = -G_y \end{cases}$$

By Cramer's rule, the formulas

$$u_x = -\frac{\det\begin{bmatrix} F_x & F_v \\ G_x & G_v \end{bmatrix}}{\det\begin{bmatrix} F_u & F_v \\ G_u & G_v \end{bmatrix}} \quad \text{and} \quad v_x = -\frac{\det\begin{bmatrix} F_u & F_x \\ G_u & G_x \end{bmatrix}}{\det\begin{bmatrix} F_u & F_v \\ G_u & G_v \end{bmatrix}}$$

and

$$u_y = -\frac{\det\begin{bmatrix} F_y & F_v \\ G_y & G_v \end{bmatrix}}{\det\begin{bmatrix} F_u & F_v \\ G_u & G_v \end{bmatrix}} \quad \text{and} \quad v_y = -\frac{\det\begin{bmatrix} F_u & F_y \\ G_u & G_y \end{bmatrix}}{\det\begin{bmatrix} F_u & F_v \\ G_u & G_v \end{bmatrix}}$$

are valid at all points (x, y) at which $(F_u G_v - F_v G_u) \neq 0$. The function $(F_u G_v - F_v G_u)$ is usually denoted by

$$J\left(\frac{F, G}{u, v}\right) \quad \text{or} \quad \frac{\partial(F, G)}{\partial(u, v)}$$

and is called the ***Jacobian*** of F and G with respect to u and v. The determinants

$$J\left(\frac{F, G}{x, v}\right) = \det\begin{bmatrix} F_x & F_v \\ G_x & G_v \end{bmatrix} \quad J\left(\frac{F, G}{y, v}\right) = \det\begin{bmatrix} F_y & F_v \\ G_y & G_v \end{bmatrix}$$

$$J\left(\frac{F, G}{u, x}\right) = \det\begin{bmatrix} F_u & F_x \\ G_u & G_x \end{bmatrix} \quad J\left(\frac{F, G}{u, y}\right) = \det\begin{bmatrix} F_u & F_y \\ G_u & G_y \end{bmatrix}$$

are interpreted analogously. As such, they are the Jacobians of F and G with respect to x and v, y and v, u and x, and u and y, respectively.

Jacobians and Cramer's rule can be used to develop analogous formulas for partial derivatives involving other combinations of variables and functions.

EXAMPLE 3.21 The Jacobian of the Polar Coordinate Transformation

Find the Jacobian $J\left(\frac{F,G}{r,\theta}\right)$ of the transformation

$$\begin{cases} F(r, \theta) &= r\cos\theta \\ G(r, \theta) &= r\sin\theta \end{cases}$$

known as the ***polar coordinate transformation***.

Solution. By definition,

$$J\left(\frac{F, G}{r, \theta}\right) = \det\begin{bmatrix} \frac{\partial}{\partial r}(r\cos\theta) & \frac{\partial}{\partial\theta}(r\cos\theta) \\ \frac{\partial}{\partial r}(r\sin\theta) & \frac{\partial}{\partial\theta}(r\sin\theta) \end{bmatrix} = \det\begin{bmatrix} \cos\theta & -r\sin\theta \\ \sin\theta & r\cos\theta \end{bmatrix} = r$$

is the Jacobian of the given transformation. ▲

EXAMPLE 3.22 Using Jacobians to Calculate Partial Derivatives

Use *Maple* and Jacobians to find the partial derivative u_x, u_y, v_x, and v_y of the functions u and v defined by the system

$$\begin{cases} F(u, v, x, y) &= x^2 + y^3 x + u^2 + v^3 = 0 \\ G(u, v, x, y) &= x^2 + 3xy + u^4 - v^2 = 0 \end{cases}$$

assuming that u and v are differentiable functions of x and y.

Solution. We begin by representing the functions F and G in *Maple*.

```
[> with(linalg):
[> F:=x^2+y^3*x+u^2+v^3;
```

$$F := x^2 + y^3x + u^2 + v^3$$

```
[> G:=x^2+3*y*x+u^4-v^2;
```

$$G := x^2 + 3yx + u^4 - v^2$$

Next we use *Maple* to calculate the five required Jacobians.

```
[> J(FGxv):=det(jacobian([F,G],[x,v]));
```

$$J(FGxv) := -4vx - 2vy^3 - 6v^2x - 9v^2y$$

```
[> J(FGuv):=det(jacobian([F,G],[u,v]));
```

$$J(FGuv) := -4uv - 12v^2u^3$$

```
[> J(FGux):=det(jacobian([F,G],[u,x]));
```

$$J(FGux) := 4ux + 6uy - 8u^3x - 4u^3y^3$$

```
[> J(FGyv):=det(jacobian([F,G],[y,v]));
```

$$J(FGyv) := -6y^2xv - 9v^2x$$

```
[> J(FGuy):=det(jacobian([F,G],[u,y]));
```

$$J(FGuy) := 6ux - 12y^2xu^3$$

We now combine the Jacobians and use Cramer's rule to calculate the required partial derivatives.

```
[> u[x]:=simplify(-J(FGxv)/J(FGuv));
```

$$u_x = -\frac{1}{4}\frac{4x + 2y^3 + 6vx + 9vy}{u\left(1 + 3vu^2\right)}$$

```
[> v[x]:=simplify(-J(FGux)/J(FGuv));
```

$$v_x = -\frac{1}{2}\frac{-2x - 3y + 4u^2x + 2u^2y^3}{v\left(1 + 3vu^2\right)}$$

```
[>u[y]:=simplify(-J(FGyv)/J(FGuv));
```

$$u_y = \frac{3\,x\,(2y^2+3v)}{4\,u\,(1+3vu^2)}$$

```
[>v[y]:=simplify(-J(FGuy)/J(FGuv));
```

$$v_y = -\frac{3}{2}\frac{x\,(-1+2y^2u^2)}{v\,(1+3vu^2)}$$

The functions produced in steps 8–11 are u_x, v_x, u_y, and v_y, respectively. ▲

Jacobian determinants also play a role in the theory of multiple integration. Suppose, for example, that $f : \mathbb{R}^2 \to \mathbb{R}$ is an integrable function on some domain D, that $f(x, y) = f(\varphi(u, v), \psi(u, v))$ on D, that φ and ψ have partial derivatives on D, and that $T : \mathbb{R}^2 \to \mathbb{R}^2$ is an appropriate transformation of the xy-plane into the uv-plane given by $T(u, v) = (\varphi(u, v), \psi(u, v))$. If $E = T^{-1}(D)$ is the set of all (u, v) in the uv-plane for which $T(u, v) \in D$, then

$$\int\int_D f(x, y)\,dx\,dy = \int\int_E f(\varphi(u, v), \psi(u, v)) \left|\det J\left(\frac{\varphi, \psi}{u, v}\right)\right| du\,dv$$

The number

$$\left|\det J\left(\frac{\varphi, \psi}{u, v}\right)\right|$$

is the local ratio of the areas of the domains D and E.

Hessians

Determinants can be used to classify critical points of differentiable functions. Suppose, for example, that $f : \mathbb{R}^2 \to \mathbb{R}$ is a function with continuous second partial derivatives f_{xx}, f_{xy}, f_{yx}, and f_{yy}. Let H_f be the matrix

$$H_f = \begin{bmatrix} f_{xx} & f_{xy} \\ f_{yx} & f_{yy} \end{bmatrix}$$

H_f is called a ***Hessian matrix***. The determinant $D = \det H_f(\mathbf{v})$ of H_f is called the ***discriminant*** of f. We know from calculus that if $\mathbf{v} = (x, y) \in \mathbb{R}^2$ is a critical point of f, in other words, if $f_x(\mathbf{v}) = f_y(\mathbf{v}) = 0$, and $D > 0$, then $f(\mathbf{v})$ is a local minimum of f if $f_{xx}(\mathbf{v}) > 0$ and is a local maximum if $f_{xx}(\mathbf{v}) < 0$. If $D < 0$, then $\mathbf{v}$ is a saddle point of f. If $D = 0$, the test fails.

EXAMPLE 3.23 A Minimax Problem

Show that a box of a given volume and minimal surface area is a cube.

Solution. Let V be the volume of a cubic solid, and let x, y, and z be, respectively, the length, width, and height of the box B enclosing V. Then the surface area of B is $S(x, y, z) = 2(xy + xz + yz)$ and $V = xyz$, with $x, y, z > 0$. Hence

$$z = \frac{V}{xy} \quad \text{and} \quad S(x, y, z) = 2\left(xy + \frac{V}{y} + \frac{V}{x}\right)$$

The critical points of S are therefore the solutions of the equations

$$S_x = 2y - \frac{2V}{x^2} = 0 \quad \text{and} \quad S_y = 2x - \frac{2V}{y^2} = 0$$

Therefore, $V = xy^2 = x^2y$. Since x and y are nonzero, we must have $x = y$. Therefore, $2x - (2V)/x^2 = 0$. This implies that $V = x^3$.

Let us use the discriminant to verify that V is a minimum. We first calculate the second derivatives $S_{xy} = 2$, $S_{yx} = 2$, $S_{xx} = 4(V/x^3)$, $S_{yy} = 4(V/y^3)$ and form the Hessian matrix

$$H_S = \begin{bmatrix} S_{xx} & S_{xy} \\ S_{yx} & S_{yy} \end{bmatrix} = \begin{bmatrix} 4\left(\frac{V}{x^3}\right) & 2 \\ 2 & 4\left(\frac{V}{y^3}\right) \end{bmatrix}$$

Its determinant at $x = y$ is

$$D = \det H_S = 4\left(\frac{4V^2 - x^6}{x^6}\right)$$

The fact that $V = x^3$ implies that

$$D = 4\left(\frac{4x^6 - x^6}{x^6}\right) = 12 > 0$$

Hence S has a minimum at $x = y = z$ since $S_{xx} > 0$. ▲

Wronskians

Matrices whose entries are partial derivatives are useful for describing the solutions of differential equations. Suppose, for example, that $S = \{f_1, \ldots, f_n\}$ is a set of solutions of an nth–order linear homogeneous differential equation on an interval $[a, b]$. Then the determinant $W(f_1, \ldots, f_n)$ of the matrix of functions

$$\begin{bmatrix} f_1 & \cdots & f_n \\ \vdots & \ddots & \vdots \\ f_1^{(n-1)} & \cdots & f_n^{(n-1)} \end{bmatrix}$$

can be used to test whether the set S is ***linearly independent***, where S is linearly independent provided that $a_1 f_1 + \cdots + a_n f_n = 0$ if and only if $a_1 = \cdots = a_n = 0$. The determinant

$W(f_1, \dots, f_n)$ is called the ***Wronskian*** of S. It is proved in calculus that S is linearly independent if and only if there exists a point $x \in \mathbb{R}$ for which $W(f_1, \dots, f_n)(x) \neq 0$.

EXAMPLE 3.24 A Wronskian Determinant

Use Wronskian determinants to show that the set of functions

$$S = \left\{ f_1 = x^5, f_2 = e^x, f_3 = e^{2x} \right\}$$

is linearly independent.

Solution. By definition, the Wronskian of S is

$$W(f_1, f_2, f_3) = \det \begin{bmatrix} x^5 & e^x & e^{2x} \\ 5x^4 & e^x & 2e^{2x} \\ 20x^3 & e^x & 4e^{2x} \end{bmatrix} = 2x^5 e^x e^{2x} - 15x^4 e^x e^{2x} + 20x^3 e^x e^{2x}$$

Let $x = 1$. Then $W(f_1, f_2, f_3)(1) = 7e^3$. Since $7e^3$ is not equal to 0, the set S is linearly independent. ▲

EXERCISES 3.8

1. Suppose that the variables u and v are defined implicitly as functions of x. Use Jacobians to find their first partial derivatives, where

$$F(u, v, x) = x^2 + 2uv - 1 = 0 \quad \text{and} \quad G(u, v, x) = x^3 - u^3 + v^3 - 1 = 0$$

2. Suppose that the variables u and v are defined implicitly as functions of the remaining variables. Use Jacobians to find their first partial derivatives, where $F(u, v, x, y) = u + e^v - x - y = 0$ and $G(u, v, x, y) = e^u + v - x + y = 0$.

3. Suppose that the variables u and v are defined implicitly as functions of x, y and z. Use Jacobians to find their first partial derivatives with respect to x, y and z, where $F(x, y, z, u, v) = u + z + y - x + v = 0$ and $G(u, v, w, x, y) = uv + vz - y^2 + x^2 = 0$.

4. Use Hessians, if possible, to classify the critical points of the following functions.

   a. $f(x, y) = x^4 + y^2 - 8x - 9y$
   b. $f(x, y) = x^3 - 6x^2 - 7xy - 4y - 5$
   c. $f(x, y) = x^3 - 6x^2 + 7x - 4y - 5$

5. Find x, y, z for which $xyz = 1$ and $x^2 + y^2 + z^2$ is minimal.

6. Find the dimensions of a rectangular box with an open top whose volume equals 100 cubic feet and whose surface area is minimal.

7. Use the Wronskian to test the following sets of functions for linear independence.

   a. $S_1 = \{e^x, e^{2x}, e^{3x}\}$ b. $S_2 = \{x, x^2, \sin x\}$ c. $S_3 = \{1, (x-2), (x-2)^3\}$

8. (*Maple V*) Explore the properties of the function **jacobian** in the **linalg** package.

9. (*Maple V*) Repeat Exercise 4 using the function **hessian** in the **linalg** package.

10. (*Maple V*) Repeat Exercise 7 using the function **wronskian** in the **linalg** package.

REVIEW

KEY CONCEPTS ▶ Define and discuss each of the following.

Square Matrices

Adjoint matrix of a square matrix, cofactors of a square matrix, elementary product of square matrix entries, minors of a square matrix.

Permutations

Elementary product of matrix entries, even permutation, inversion of the elements of a permutation, odd permutation, parity of a permutation.

Determinant Functions

Laplace expansion of a determinant, permutation form of a determinant.

Linear Systems

Cramer's rule.

KEY FACTS ▶ Explain and illustrate each of the following.

1. The determinant $\det I_n$ of the $n \times n$ identity matrix I_n is 1.
2. If A is an $n \times n$ matrix with a zero row, then the determinant of A is 0.
3. If B is a matrix resulting from an $n \times n$ matrix A by the elementary row operation $(R_i \rightleftharpoons R_j)$ and $i \neq j$, then $\det B = -\det A$.
4. If $A = [a_{ij}]$ is an $n \times n$ matrix containing two identical rows, then the determinant of A is 0.
5. If B is a matrix resulting from an $n \times n$ matrix A by the elementary row operation $(R_i \to sR_i)$, then $\det B = s \det A$.
6. If B is a matrix resulting from an $n \times n$ matrix A by the elementary row operation $(R_i \to R_i + sR_j)$, then $\det B = \det A$.
7. If E is an elementary matrix obtained from I_n by $(R_i \rightleftharpoons R_j)$ and $i \neq j$, then $\det E = -1$.
8. If E is an elementary matrix obtained from I_n by $(R_i \to sR_i)$, then $\det E = s$.
9. If E is an elementary matrix obtained from I_n by $(R_i \to R_i + sR_j)$, then $\det E = \det I_n = 1$.
10. If $A = [a_{ij}]$ is an $n \times n$ matrix and E is an $n \times n$ elementary matrix, then $\det EA = \det E \det A$.

11. If E is an elementary matrix and $B = EA$ for some $n \times n$ matrix A, then the following statements hold.

    a. $\det B = -\det A$ if E results from I_n by the operation $(R_i \rightleftharpoons R_j)$, provided that $i \neq j$.

    b. $\det B = s \det A$ if E results from I_n by the operation $(R_i \to sR_i)$.

    c. $\det B = \det A$ if E results from I_n by the operation $\left(R_i \to R_i + sR_j\right)$.

12. If two $n \times n$ matrices A and B are row equivalent, then $\det A = 0$ if and only if $\det B = 0$.

13. If A and B are two $n \times n$ matrices, then $\det AB = \det A \det B$.

14. A real $n \times n$ matrix A is invertible if and only if $\det A \neq 0$.

15. A homogeneous linear system $A\mathbf{x} = \mathbf{0}$ with real coefficients has a nonzero solution $\mathbf{x}$ if and only if $\det A = 0$.

16. If $A = \left[a_{ij}\right]$ is an $n \times n$ triangular matrix, then the determinant of A is the product $a_{11} \cdots a_{nn}$ of the diagonal entries of A.

17. For any $n \times n$ matrix A and any scalar s, $\det sA = s^n \det A$.

18. For any matrix A, $\det A = \det A^T$.

19. For any invertible matrix A, $\det A^{-1} = 1/\det A$.

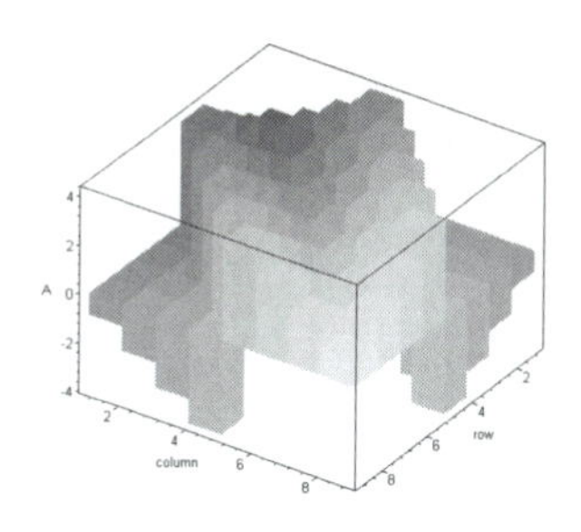

4

VECTOR SPACES

What do real numbers, matrices, continuous functions, solutions of certain differential equations, polynomials, and many other mathematical objects have in common? The answer is that they can all be *added* to produce new objects of the same kind. They can also be *multiplied by constants* to produce objects of the same kind. For example, if a and b are real numbers, then so are $a + b$ and ab. If $\mathbf{x}$ and $\mathbf{y}$ are two column vectors with real entries and the same number of rows, then $\mathbf{x} + \mathbf{y}$ is a column vector with the same number of rows and so is the multiple $a\mathbf{x}$, for any real number a. If A and B are two $m \times n$ matrices with real entries, then so are $A + B$ and aA, for any real number a.

In linear algebra, these different kinds of objects are referred to as ***vectors*** belonging to different ***vector spaces***. Thus vectors can be real numbers, continuous functions, matrices, and many other objects. The constants used to multiply vectors to produce new vectors are called ***scalars***. In this text, scalars are either real or complex numbers. More specifically, they will be real numbers unless otherwise specified.

REAL VECTOR SPACES

In this chapter, we lay the foundation for the study of ***real vector spaces***. The vectors will vary: They will be columns of real numbers, polynomials, matrices, or other objects. The scalars, however, will always be the real numbers. We examine the spaces, their bases, and their subspaces. The pivotal concept for this endeavor will be that of a ***linearly independent*** set of vectors. The sets $\mathbb{R}^n$ of real column vectors of height n will play an essential role in guiding our intuition.

DEFINITION 4.1 *A real* ***vector space*** V *consists of a set of objects called* ***vectors****, a distinguished vector* $\mathbf{0}$ *called the* ***zero vector****, and the set* $\mathbb{R}$ *of real numbers, whose elements are called* ***scalars****. The space* V *is equipped with two operations, called* ***vector addition,*** *denoted by* $\mathbf{u} + \mathbf{v}$*, and* ***scalar multiplication****, denoted by* $a\mathbf{v}$*. Table 1 lists the eight axioms satisfied by these operations.*

TABLE 1 THE REAL VECTOR SPACE AXIOMS.

Axiom 1.	Vector addition	For all $\mathbf{u}, \mathbf{v}, \mathbf{w} \in V$, $\mathbf{u} + (\mathbf{v} + \mathbf{w}) = (\mathbf{u} + \mathbf{v}) + \mathbf{w}$.
Axiom 2.	Vector addition	For all $\mathbf{u}, \mathbf{v} \in V$, $\mathbf{u} + \mathbf{v} = \mathbf{v} + \mathbf{u}$.
Axiom 3.	Vector addition	For all $\mathbf{u} \in V$, $\mathbf{u} + \mathbf{0} = \mathbf{u}$.
Axiom 4.	Vector addition	For all $\mathbf{u} \in V$ there exists a unique $-\mathbf{u} \in V$ for which $\mathbf{u} + (-\mathbf{u}) = \mathbf{0}$.
Axiom 5.	Scalar multiplication	For all $a \in \mathbb{R}$ and all $\mathbf{u}, \mathbf{v} \in V$, $a(\mathbf{u} + \mathbf{v}) = a\mathbf{u} + a\mathbf{v}$.
Axiom 6.	Scalar multiplication	For all $a, b \in \mathbb{R}$ and all $\mathbf{u} \in V$, $(a + b)\,\mathbf{u} = a\mathbf{u} + b\mathbf{u}$.
Axiom 7.	Scalar multiplication	For all $a, b \in \mathbb{R}$ and all $\mathbf{u} \in V$, $(ab)\,\mathbf{u} = a\,(b\mathbf{u})$.
Axiom 8.	Scalar multiplication	For all $a \in \mathbb{R}$ and all $\mathbf{u} \in V$, $1\mathbf{u} = \mathbf{u}$.

The first four axioms express the facts that vector addition is associative and commutative, that adding the zero vector to a vector leaves the vector unchanged, and that each vector has an additive inverse. The next three axioms describe how matrix addition and scalar multiplication interact. The last axiom indicates that multiplying a vector by the real number 1 leaves the vector unchanged.

The statement that vector addition and scalar multiplication are ***operations*** on V is equivalent to saying that if $\mathbf{u}$ and $\mathbf{v}$ are any two vectors in V and a and b are any two scalars, then $a\mathbf{u} + b\mathbf{v} \in V$. We express this fact by saying that V is ***closed under vector addition and scalar multiplication***. We follow the usual practice and write $\mathbf{u} + (-\mathbf{v})$ as $\mathbf{u} - \mathbf{v}$ and refer to this operation as the ***subtraction*** of vectors.

If we replace the set $\mathbb{R}$ of real numbers by the set $\mathbb{C}$ of complex numbers in Definition 4.1, we get a ***complex vector space.*** More general vector spaces can be defined by varying the set of scalars of the spaces.

In all vector spaces, the zero vector and the zero scalar are connected by the following basic property.

THEOREM 4.1 *For any scalar a and any vector $\mathbf{v}$, we have $\mathbf{0} = a\mathbf{0} = 0\mathbf{v}$.*

Proof. Since $\mathbf{0} = \mathbf{0} + \mathbf{0}$, it follows that $a\mathbf{0} = a\mathbf{0} + a\mathbf{0}$ for all scalars a. By subtracting $a\mathbf{0}$ from both sides of this equation, we get $\mathbf{0} = a\mathbf{0} - a\mathbf{0} = a\mathbf{0} + a\mathbf{0} - a\mathbf{0} = a\mathbf{0}$. Moreover, since $0 = 0 + 0$, we have $0\mathbf{v} = 0\mathbf{v} + 0\mathbf{v}$ for all vectors $\mathbf{v}$. By subtracting the vector $0\mathbf{v}$ from both sides of this equation, we get $\mathbf{0} = 0\mathbf{v} - 0\mathbf{v} = 0\mathbf{v} + 0\mathbf{v} - 0\mathbf{v} = 0\mathbf{v}$. ■

We can strengthen this theorem by noting that the following is true.

THEOREM 4.2 *For any scalar a and any vector $\mathbf{v}$, we have $a\mathbf{v} = \mathbf{0}$ if and only if $a = 0$ or $\mathbf{v} = \mathbf{0}$.*

Proof. In view of Theorem 4.1, it suffices to prove that if $a\mathbf{v} = \mathbf{0}$ and $a \neq 0$, then $\mathbf{v} = \mathbf{0}$. Since $a \neq 0$, there exists a nonzero inverse a^{-1} such that $\mathbf{0} = a^{-1}\mathbf{0} = a^{-1}a\mathbf{v} = 1\mathbf{v} = \mathbf{v}$. The theorem therefore follows. ■

EXERCISES 4.1

1. Show that in a vector space, $(-a)\mathbf{u} = a(-\mathbf{u}) = -(a\mathbf{u})$ for all scalars a and all vectors $\mathbf{u}$.
2. Show that in a vector space, $(a+b)(\mathbf{u}+\mathbf{v}) = a\mathbf{u} + b\mathbf{u} + a\mathbf{v} + b\mathbf{v}$ for all scalars a, b and all vectors $\mathbf{u}$, $\mathbf{v}$.
3. Show that the set of points $(x, y) \in \mathbb{R}^2$ with $(x, y) + (x', y') = (x + x', y + y')$ and $a(x, y) = (ax, ay)$ is a vector space.
4. Define vector space operations on the set of all points $(x, y, z) \in \mathbb{R}^3$ analogous to the operations in Exercise 3 and show that the result is a real vector space.
5. Explain why the set of points on the line $y = 3x + 5$ is not a vector space.
6. Use the operations as defined in Exercises 3 and 4 to show that the following sets contain the zero vector and are closed under vector addition and scalar multiplication.

   a. The set of all points on the line $y = 3x$

   b. The set of all points on the plane $x + 2y + 3z = 0$

   c. The set $V = \{0\}$ whose only vector is the number 0

   d. The set of all real upper-triangular $n \times n$ matrices

   e. The set of all real diagonal $n \times n$ matrices

   f. The set of all real symmetric $n \times n$ matrices

7. Let $V = \{a_0 + a_1 t + a_2 t^5 + t^3 : a_0, a_1, a_2 \in \mathbb{R}\}$. Explain why V is not a vector space.
8. Show that the definition $3(a, b) = (0, 3b)$ is consistent with the first seven axioms of a real vector space. Explain why this shows that axiom 8 is independent of the other axioms.
9. Explain why the set of all 2×2 matrices with rational entries is not a real vector space.
10. Show that the solutions of the equation $y'(t) + y(t) = t$ do not form a vector space.

In this text, we group vector spaces into four kinds: coordinate spaces, matrix spaces, polynomial spaces, and others. To show that a given set of objects is a real vector space, we must show that it is closed under the defined operations of vector addition and scalar multiplication, that it contains a zero vector, and that the operations satisfy the vector space axioms. In all but one case, we leave the verification of the axioms as an exercise.

Coordinate Spaces

EXAMPLE 4.1 The Space $\mathbb{R}$

If $n = 1$, we can think of $\mathbb{R}^1$ as $\mathbb{R}$ by identifying the column vectors $[x]$ with the real numbers x. The usual addition and multiplication of real numbers determine a vector space structure on $\mathbb{R}$. ▲

EXAMPLE 4.2 The Space $\mathbb{R}^n$

The space $\mathbb{R}^n$ of all $n \times 1$ column vectors $A : \mathbf{n} \times \mathbf{1} \to \mathbb{R}$, equipped with matrix addition and the scalar multiplication of matrices, is a real vector space. ▲

By defining the elements of $\mathbb{R}^n$ as column vectors, we are able to describe the two-dimensional link between the vector space operations in matrix terms. For example, by the commutativity of the multiplication of real numbers, the linear combination of vectors

$$3\begin{bmatrix} 5 \\ 6 \end{bmatrix} + 2\begin{bmatrix} 8 \\ 7 \end{bmatrix} = \begin{bmatrix} 3(5) \\ 3(6) \end{bmatrix} + \begin{bmatrix} 2(8) \\ 2(7) \end{bmatrix} = \begin{bmatrix} 3(5)+2(8) \\ 3(6)+2(7) \end{bmatrix}$$

can be expressed as the matrix product

$$\begin{bmatrix} 5 & 8 \\ 6 & 7 \end{bmatrix}\begin{bmatrix} 3 \\ 2 \end{bmatrix} = \begin{bmatrix} 5(3)+8(2) \\ 6(3)+7(2) \end{bmatrix}$$

The identity

$$\begin{bmatrix} 3(5)+2(8) \\ 3(6)+2(7) \end{bmatrix} = \begin{bmatrix} 5(3)+8(2) \\ 6(3)+7(2) \end{bmatrix}$$

holds because $3(5) = 5(3)$, $2(8) = 8(2)$, $3(6) = 6(3)$, and $2(7) = 7(2)$.

EXAMPLE 4.3 Verifying the Vector Space Axioms for $\mathbb{R}^3$

Use *Maple* to verify that the addition of vectors and the multiplication of vectors by scalars in $\mathbb{R}^3$ satisfy the axioms of a vector space.

Solution. We must verify eight axioms.

```
[> [a1,a2,a3]+([b1,b2,b3]+[c1,c2,c3]):
[> ([a1,a2,a3]+[b1,b2,b3])+[c1,c2,c3]:
[> equal(%,%%);
                         true
```

```
[> [a1,a2,a3]+[b1,b2,b3]:
[> [b1,b2,b3]+[a1,a2,a3]:
[> equal(%,%%);
```

true

```
[> [a1,a2,a3]+[0,0,0]:
[> [a1,a2,a3]:
[> equal(%,%%);
```

true

```
[> [a1,a2,a3]+[-a1,-a2,-a3]:
[> [0,0,0]:
[> equal(%,%%);
```

true

```
[> a*([a1,a2,a3] + [b1,b2,b3]):
[> a*[a1,a2,a3] + a*[b1,b2,b3]:
[> equal(expand(%),expand(%%));
```

true

```
[> (a+b)*[a1,a2,a3]:
[> a*[a1,a2,a3]+b*[a1,a2,a3]:
[> equal(expand(%),expand(%%));
```

true

```
[> (a*b)*[a1,a2,a3]:
[> a*(b*[a1,a2,a3]):
[> equal(%,%%);
```

true

```
[> 1*[a1,a2,a3]:
[> [a1,a2,a3]:
[> equal(%,%%);
                                    true
```

This shows that $\mathbb{R}^3$ satisfies the axioms of a vector space. In these calculations, it is tacitly assumed that the constants a, b, a_i, b_i, and c_i are real numbers. ▲

EXERCISES 4.2

1. Use matrix multiplication to calculate the linear combinations

$$\text{a. } 4\begin{bmatrix} 2 \\ -8 \end{bmatrix} + 5\begin{bmatrix} 1 \\ 2 \end{bmatrix} \quad \text{b. } 4\begin{bmatrix} 2 \\ -8 \end{bmatrix} + 5\begin{bmatrix} 1 \\ 2 \end{bmatrix} + 8\begin{bmatrix} 9 \\ 6 \end{bmatrix}$$

$$\text{c. } 4\begin{bmatrix} 8 \\ -2 \\ 5 \end{bmatrix} - 7\begin{bmatrix} 1 \\ 1 \\ 1 \end{bmatrix} \quad \text{d. } 4\begin{bmatrix} 1 \\ 2 \\ 3 \end{bmatrix} + 5\begin{bmatrix} 4 \\ 5 \\ 6 \end{bmatrix} - 7\begin{bmatrix} 7 \\ 8 \\ 9 \end{bmatrix}$$

2. Sketch a proof of the fact that the vector space axioms hold for the coordinate spaces $\mathbb{R}^4$.
3. Let m and n range over the indexing sets **1**, **2**, and **3**. Find three nonzero vectors each in the coordinate spaces $\mathbb{R}^{mn}$.
4. (*Maple 6*) Verify that the scalar multiplication axioms hold in $\mathbb{R}^3$.

Matrix Spaces

EXAMPLE 4.4 The Space $\mathbb{R}^{m\times n}$

The space $\mathbb{R}^{m\times n}$ consists of all $m \times n$ matrices A with real entries, equipped with matrix addition and the scalar multiplication of matrices as vector space operations. We refer to this space as the space of real $m \times n$ matrices. ▲

EXERCISES 4.3

1. Explain the difference between the vectors in the following real vector spaces.

   a. The space $\mathbb{R}$ and the space $\mathbb{R}^{1\times 1}$

   b. The spaces $\mathbb{R}^3$, $\mathbb{R}^{3\times 1}$, and $\mathbb{R}^{1\times 3}$

   c. The spaces $\mathbb{R}^{2\times 3}$ and $\mathbb{R}^{3\times 2}$

   d. The spaces $\mathbb{R}^{2\times 3}$ and $\mathbb{R}^6$

2. Show that the vector space axioms hold for $\mathbb{R}^{2\times 3}$.
3. Show that the set of real 3×2 matrices with nonnegative entries is not a real vector space.

Polynomial Spaces

EXAMPLE 4.5 The Space $\mathbb{R}[t]$

If $p(t) = a_0 + a_1 t + \cdots + a_n t^n$ and $q(t) = b_0 + b_1 t + \cdots + b_n t^n$ are two polynomials in $\mathbb{R}[t]$, then we can think of them as vectors. The definitions

$$p(t) + q(t) = (a_0 + b_0) + (a_1 + b_1)t + \cdots + (a_n + b_n)t^n$$
$$ap(t) = aa_0 + aa_1 t + \cdots + aa_n t^n$$
$$\mathbf{0} = 0$$

turn $\mathbb{R}[t]$ into a real vector space. If $p(t) = a_0 + a_1 t + \cdots + a_n t^n$ and $q(t) = b_0 + b_1 t + \cdots + b_m$ and $m \neq n$, we simply add the required number of zero terms to the polynomial of lower degree so we can apply the definition of $p(t) + q(t)$. For example, if $p[t] = 1 + t$ and $q(t) = 7 - 3t^2$, we put $p[t] = 1 + t + 0t^2$ and $q(t) = 7 + 0t - 3t^2$ and form the sum $p(t) + q(t) = 8 + t - 3t^2$. ▲

EXERCISES 4.4

1. Describe the vector space operations for all polynomials in $\mathbb{R}[t]$ of degree 3 or less.
2. Show that the vector space axioms hold for $\mathbb{R}[t]$.
3. Show that the set of all polynomials in $\mathbb{R}[t]$ of degree 4 is not closed under vector addition and not closed under scalar multiplication.
4. Show that the set of all polynomials in $\mathbb{R}[t]$ of degree 4 or less is closed under vector addition and scalar multiplication.

Other Spaces

In each of the following examples, we describe the zero vector and discuss closure under linear combinations. We leave the verification of the real vector space axioms as exercises.

EXAMPLE 4.6 The Space $\mathbb{R}_n[t]$

The set of real polynomials of degree n or less can be made into a vector space by using the vector space operations on $\mathbb{R}[t]$. Since the sum of two polynomials of degree n may be a polynomial of degree less than n, it is essential to include in the vectors of this space all polynomials of degree less than n. ▲

EXAMPLE 4.7 The Space $\mathbb{R}^{\mathbb{N}}$

The set $\mathbb{R}^{\mathbb{N}}$ of real-valued infinite sequences $(x_1, \ldots, x_n, \ldots)$ can be turned into a vector space by defining vector addition by

$$(v_1, \ldots, v_n, \ldots) + (w_1, \ldots, w_n, \ldots) = (v_1 + w_1, \ldots, v_n + w_n, \ldots)$$

and scalar multiplication and the zero vector by

$$a(v_1, \ldots, v_n, \ldots) = (av_1, \ldots, av_n, \ldots)$$
$$\mathbf{0} = (0, \ldots, 0, \ldots)$$

In this example, it would not be helpful to try to visualize the vectors as columns since such columns would have to be of infinite height. ▲

EXAMPLE 4.8 The Space $\mathbb{R}^\infty$

The set $\mathbb{R}^\infty$ of all infinite sequences $(x_1, \ldots, x_n, \ldots)$ of real numbers for which

$$|x_1|^2 + \cdots + |x_n|^2 + \cdots < \infty$$

is a real vector space. Vector addition and scalar multiplication are defined as in the previous example. ▲

EXAMPLE 4.9 The Solution Space of $A\mathbf{x} = \mathbf{0}$

The set of solutions of a homogeneous linear system with real coefficients forms a vector space. We know from Theorems 1.4 and 2.7 that this set contains the zero vector and that if **x** and **y** are solutions and a and b are two scalars, the $ax + by$ is a solution. ▲

EXAMPLE 4.10 The Solution Space of $y'(t) + ay(t) = 0$

Suppose that $y(t)$ is a differentiable function $\mathbb{R} \to \mathbb{R}$, that $y'(t)$ is the derivative of $y(t)$, and that $a \in \mathbb{R}$. Then it is shown in calculus that

$$V = \left\{ y(t) : y(t) = ce^{-at} \text{ and } c \in \mathbb{R} \right\}$$

is the set of solutions of the equation $y'(t) + ay(t) = 0$. This set is closed under addition and the multiplication by scalars. The constant function $y(t) = 0e^{-at} = 0$ is the **0** vector. ▲

EXAMPLE 4.11 The Solution Space of $y''(t) + p(t)y'(t) + q(t)y(t) = 0$

Suppose that $p(t)$ and $q(t)$ are two continuous functions on some interval (r, s), that $y(t)$ is a twice-differentiable function on (r, s), and that $y'(t)$ and $y''(t)$ are the first and second derivatives of $y(t)$, respectively. Suppose further that $y_1(t)$ and $y_2(t)$ are two solutions of the given equation on (r, s). Then it can be shown in calculus that the set

$$V = \{ay_1(t) + by_2(t) : a, b \in \mathbb{R}\}$$

is the solution space of the equation if and only if the Wronskian determinant

$$W(y_1(t), y_2(t)) = \det \begin{bmatrix} y_1(t) & y_2(t) \\ y_1'(t) & y_2'(t) \end{bmatrix}$$

is nonzero for some $t \in (r, s)$. The constant function $f(t) = 0$ is the **0** vector of the space. ▲

EXAMPLE 4.12 The Space $D^\infty(\mathbb{R}, \mathbb{R})$

Show that the set $D^\infty(\mathbb{R}, \mathbb{R})$ of infinitely differentiable functions $f : \mathbb{R} \to \mathbb{R}$ contains a zero vector and is closed under linear combinations.

Solution. Suppose that f and g belong to $D^{\infty}(\mathbb{R}, \mathbb{R})$. Then

$$\frac{d}{dt}(af(t) + bg(t)) = a\frac{d}{dt}(f(t)) + b\frac{d}{dt}(g(t))$$

This shows that $D^{\infty}(\mathbb{R}, \mathbb{R})$ is closed under vector addition and scalar multiplication. Moreover, since

$$\frac{d}{dt}(0) = 0$$

the zero function $f(t) = 0$ for all $t \in \mathbb{R}$ belongs to $D^{\infty}(\mathbb{R}, \mathbb{R})$. ▲

EXAMPLE 4.13 The Zero Space

If $\mathbf{0}$ is the zero vector of a real vector space V, then the set $\{\mathbf{0}\}$ is a real vector space. Since $a\mathbf{0} + b\mathbf{0} = \mathbf{0}$ for all real numbers a and b, it is clear that $\{\mathbf{0}\}$ contains the zero vector and is closed under vector addition and scalar multiplication. ▲

What the zero vector in a space looks like depends on the nature of the vectors in the space. As Table 2 shows, the zero vector of $\mathbb{R}^3$, for example, consists of a column vector of height 3 whose coordinates are all zero, whereas the zero vector of the space $\mathbb{R}^{3\times 2}$ of 3×2 real matrices is the 3×2 matrix all of whose entries are zero. Nevertheless, we usually speak of *the* ***zero space,*** since it is always clear from the context which zero vector is involved.

TABLE 2 TWO ZERO VECTORS.

Vector Space	Zero Vector	Vector Space	Zero Vector
$\mathbb{R}^3$	$\mathbf{0} = \begin{bmatrix} 0 \\ 0 \\ 0 \end{bmatrix}$	$\mathbb{R}^{3\times 2}$	$\mathbf{0} = \begin{bmatrix} 0 & 0 \\ 0 & 0 \\ 0 & 0 \end{bmatrix}$

EXERCISES 4.5

1. Show that the set of real polynomials of the form $a_0 + a_1 t + a_2 t^2$ is a real vector space.
2. Show that $\mathbb{R}^{\mathbb{N}}$ is a real vector space.
3. Show that the space $D^{\infty}(\mathbb{R}, \mathbb{R})$ is a real vector space.
4. Explain the difference between the vector space $\mathbb{R}^{\mathbb{N}}$ and the vector space $\mathbb{R}^{\infty}$.
5. Show that the set of all lines in the plane parallel to a given line forms a real vectors.
6. Describe the zero vectors in the spaces $\mathbb{R}^2$, $\mathbb{R}_4[t]$, $\mathbb{R}^{2\times 3}$, $\mathbb{R}^6$, $D^{\infty}(\mathbb{R}, \mathbb{R})$, and $\mathbb{R}^{\mathbb{N}}$.
7. (*Maple V*) Show that the set of all solutions of the linear system

$$\begin{cases} 3x + 7y - z = 0 \\ x + y + z = 0 \end{cases}$$

is a real vector space.

8. (*Maple V*) Show that the set of all solutions of the linear system

$$\begin{cases} 3x + 7y - z = 0 \\ x + y + z = 0 \\ 2x - y + 3z = 0 \end{cases}$$

is a real vector space.

9. (*Maple V*) Show that the solution space of the differential equation $y'(t) + 3y(t) = 0$ is a real vector space.

10. (*Maple V*) Show that the solution space of the differential equation $y^{(2)}(t) + 3y^{(1)}(t) + 6y(t) = 0$ is a real vector space.

BASES AND DIMENSION

Although infinite, a vector space may have the property that all of its vectors can be built up from a fixed set of finitely many of its vectors using vector addition and scalar multiplication. This is similar to a language being based on a finite alphabet in which the infinite set of words of the language can be written. It is this feature that will allow us to use matrices to study the properties of certain functions on real vector spaces. It will be essential for our work that all sets of vectors used in this context are ***ordered***. The ordered lists *tea* and *eat,* for example, are different *words* determined by the unordered set of letters {a, e, t}. Similarly, the lists

$$[\mathbf{x}_2, \mathbf{x}_3, \mathbf{x}_1] = [(5,7), (3,8), (2,2)] \quad \text{and} \quad [\mathbf{x}_1, \mathbf{x}_3, \mathbf{x}_2] = [(2,2), (3,8), (5,7)]$$

are different ***ordered sets of vectors,*** determined by the vectors $\mathbf{x}_1$, $\mathbf{x}_2$, and $\mathbf{x}_3$. However, since all sets of vectors in this text are assumed to be ordered, we usually speak of ***sets of vectors*** and write them in set notation as $\{\mathbf{x}_2, \mathbf{x}_3, \mathbf{x}_1\}$ and $\{\mathbf{x}_1, \mathbf{x}_3, \mathbf{x}_2\}$.

Linear Combinations

Much of the work that follows is based on bases and linear combinations. We therefore motivate these concepts with an example.

EXAMPLE 4.14 Linear Combinations of Vectors

On CRT computer monitors, color is usually represented as a vector in **RGB** space, where **R** is the red component, **G** is the green component, and **B** is the blue component. **RGB** space is a subset of the vector space $\mathbb{R}^3$ obtained from the ***basis***

$$\begin{aligned} \mathbf{R} &= (1, 0, 0) \quad \text{(pure red)} \\ \mathbf{G} &= (0, 1, 0) \quad \text{(pure green)} \\ \mathbf{B} &= (0, 0, 1) \quad \text{(pure blue)} \end{aligned}$$

by taking all ***linear combinations*** of **R**, **G**, and **B**, using nonnegative scalars. In the context of color, these basis vectors are referred to as primaries. We know that any color C in **RGB** space can be represented as a linear combination of red, green, and blue:

$$C = \alpha_1 \mathbf{R} + \alpha_2 \mathbf{G} + \alpha_3 \mathbf{B}$$

Here are some examples of colors in **RGB** space:

(0, 0, 0)	black
(1, 1, 1)	white
(1, 0, 1)	magenta
(0, 1, 1)	cyan
(1, 1, 0)	yellow

Along with red, green, and blue, these colors define the eight corners of a unit cube known as the **RGB** *color cube*. The main diagonal of the cube, which extends from white to black, is called the *grayscale*, because colors that fall on this diagonal are gray, or colorless. The process of converting an arbitrary color to grayscale is the following. Consider an arbitrary color $C = (C_x, C_y, C_z)$. The grayscale representation C_G of C is given by

$$C_G = \begin{bmatrix} .299C_x + .587C_y + .114C_z \\ .299C_x + .587C_y + .114C_z \\ .299C_x + .587C_y + .114C_z \end{bmatrix}$$

Each of its components is a *linear combination* of the colors C_x, C_y, and C_z. Observe that all **R**, **G**, and **B** components are now identical. ▲

It is interesting to note that the green weight is higher than the red and the blue weight. Much of image processing is done in grayscale because images appear in high contrast when displayed in this way. The greater importance of green in the grayscale formula seems to have a historical explanation. Apparently, the human eye sees greater contrast in green than in the other primaries because as predators in prehistoric times, humans needed to see prey that were well camouflaged in green forests. Experiments have shown that if we can take an arbitrary image and look at its red, green, and blue channels individually, our eyes will see the most contrast when looking at the green channel and the least contrast in the blue channel. This explains the choice of the experimentally established weights $\alpha_1 = .299$, $\alpha_2 = .587$, and $\alpha_3 = .114$. It also explains the low weight attached to blue in the grayscale formula.

The idea of a linear combination of vectors is fundamental to the study of vector spaces.

DEFINITION 4.2 *Let $\mathbf{v}_1, \ldots, \mathbf{v}_n$ be vectors in a vector space V and let $a_1, \ldots, a_n$ be corresponding scalars. Then the vector $a_1\mathbf{v}_1 + \cdots + a_n\mathbf{v}_n$ is a* ***linear combination*** *of the given vectors and scalars.*

EXAMPLE 4.15 A Linear Combination of Column Vectors

The vectors in $\mathbb{R}^n$ are columns of real numbers. We can therefore think of a linear combination $a_1\mathbf{v}_1 + \cdots + a_n\mathbf{v}_n$ of vectors in $\mathbb{R}^n$ as a matrix product

$$a_1\mathbf{v}_1 + \cdots + a_n\mathbf{v}_n = A\begin{bmatrix} a_1 \\ \vdots \\ a_n \end{bmatrix}$$

where $A = [\mathbf{v}_1 \ \cdots \ \mathbf{v}_n]$ is the matrix whose columns are the vectors $\mathbf{v}_i$. ▲

DEFINITION 4.3 *The* ***span*** *of a nonempty subset $\mathcal{S}$ of vectors of a vector space V is the set of all linear combinations in V that can be formed with the vectors in $\mathcal{S}$. If $\mathcal{S}$ is empty, then the span of $\mathcal{S}$ in V is the set $\{\mathbf{0}\}$, containing only the zero vector of V.*

We write span $(\mathcal{S})$ for the span of $\mathcal{S}$. If $\mathcal{S} = \{\mathbf{x}_1, \ldots, \mathbf{x}_n\}$ is a finite set of vectors, we also write span $\{\mathbf{x}_1, \ldots, \mathbf{x}_n\}$ for the span of $\mathcal{S}$. It is important to notice that the span of a set $\mathcal{S}$ of vectors depends on the set of scalars of V.

EXAMPLE 4.16 The Span of a Set of Column Vectors

Show that the set

$$\mathcal{S} = \left\{ \begin{bmatrix} 1 \\ 2 \\ 3 \end{bmatrix}, \begin{bmatrix} 0 \\ 4 \\ 4 \end{bmatrix}, \begin{bmatrix} 1 \\ 2 \\ 0 \end{bmatrix}, \begin{bmatrix} 3 \\ 2 \\ 1 \end{bmatrix} \right\}$$

spans $\mathbb{R}^3$.

Solution. We use *Maple* to show that every vector in $\mathbb{R}^3$ can be written as a linear combination of vectors in $\mathcal{S}$. Since there are infinitely many such possible linear combinations, these sums are not unique. We begin by forming a general linear combination of the form

$$a\begin{bmatrix} 1 \\ 2 \\ 3 \end{bmatrix} + b\begin{bmatrix} 0 \\ 4 \\ 4 \end{bmatrix} + c\begin{bmatrix} 1 \\ 2 \\ 0 \end{bmatrix} + d\begin{bmatrix} 3 \\ 2 \\ 1 \end{bmatrix} = \begin{bmatrix} x \\ y \\ z \end{bmatrix}$$

We then evaluate the left-hand side of this equation.

```
[> a*[1,2,3]+b*[0,4,4]+c*[1,2,0]+d*[3,2,1]:
[> expand(%);
```

$$[3d + c + a, 2d + 2c + 4b + 2a, d + 4b + 3a]$$

We therefore solve the equation

$$\begin{bmatrix} a+c+3d \\ 2a+4b+2c+2d \\ 3a+4b+d \end{bmatrix} = \begin{bmatrix} x \\ y \\ z \end{bmatrix}$$

For any given x, y, z, we have a system of three equations in the four unknowns a, b, c, and d. If we choose a, b, and c as basic variables and d as a free variable, *Maple* finds a, b, and c in terms of d.

```
[> solve({3*d+c+a=x,2*d+2*c+4*b+2*a=y,d+4*b+3*a=z},{a,b,c});
```

$$\{a = \frac{-5}{3}d + \frac{2}{3}x - \frac{1}{3}y + \frac{1}{3}z, b = d - \frac{1}{2}x + \frac{1}{4}y, c = -\frac{4}{3}d + \frac{1}{3}x + \frac{1}{3}y - \frac{1}{3}z\}$$

As we can see, we can combine the given coordinates x, y, z with infinitely many $d \in \mathbb{R}$ to build a, b, and c. Hence we can span $\mathbb{R}^3$ with infinitely many different linear combinations of the given vectors. ▲

DEFINITION 4.4 *A subset $\mathcal{S}$ of a vector space V is* ***linearly dependent*** *if there exist vectors $\mathbf{v}_1, \ldots, \mathbf{v}_n \in \mathcal{S}$ and scalars $a_1, \ldots, a_n$, not all 0, such that $a_1\mathbf{v}_1 + \cdots + a_n\mathbf{v}_n = \mathbf{0}$.*

DEFINITION 4.5 *A subset $\mathcal{S}$ of a vector space V is* ***linearly independent*** *if it is not linearly dependent.*

In Table 3, we summarize the process for verifying whether a finite set of real columns vectors is linearly independent.

TABLE 3 VERIFYING LINEAR INDEPENDENCE.

Step 1. Take the set $\mathcal{S} = \{\mathbf{x}_1, \ldots, \mathbf{x}_p\}$ of column vectors of $\mathbb{R}^n$.
Step 2. Form the linear system $\begin{bmatrix} \mathbf{x}_1 & \cdots & \mathbf{x}_p \end{bmatrix} \begin{bmatrix} a_1 \\ \vdots \\ a_p \end{bmatrix} = \begin{bmatrix} 0 \\ \vdots \\ 0 \end{bmatrix}$.
Step 3. Find the solutions of this homogeneous linear system.
Step 4. The set $\mathcal{S}$ is linearly independent if and only if the linear system has only the trivial solution.

EXAMPLE 4.17 Linearly Independent Column Vectors

Use *Maple* to show that the set

$$\mathcal{S} = \left\{ \begin{bmatrix} 1 \\ 2 \\ 3 \end{bmatrix}, \begin{bmatrix} 0 \\ 4 \\ 4 \end{bmatrix}, \begin{bmatrix} 1 \\ 2 \\ 0 \end{bmatrix} \right\}$$

is linearly independent.

Solution. First we express the zero vector as a linear combination of the vectors in $\mathcal{S}$:

$$a \begin{bmatrix} 1 \\ 2 \\ 3 \end{bmatrix} + b \begin{bmatrix} 0 \\ 4 \\ 4 \end{bmatrix} + c \begin{bmatrix} 1 \\ 2 \\ 0 \end{bmatrix} = \begin{bmatrix} 0 \\ 0 \\ 0 \end{bmatrix}$$

We then evaluate the left-hand side of this equation.

```
[> a*[1,2,3]+b*[0,4,4]+c*[1,2,0]:
[> expand(%);
```

$$[c + a, 2c + 4b + 2a, 4b + 3a]$$

We therefore solve the equation

$$\begin{bmatrix} a + c \\ 2a + 4b + 2c \\ 3a + 4b \end{bmatrix} = \begin{bmatrix} 0 \\ 0 \\ 0 \end{bmatrix}$$

```
[> solve({c+a=0, 2*c+4*b+2*a=0, 4*b+3*a=0},{a,b,c});
```

$$\{b = 0, c = 0, a = 0\}$$

This shows that the linear combination of the vectors in $\mathcal{S}$ is $\mathbf{0}$ if and only if $a = b = c = 0$. The set $\mathcal{S}$ is therefore linearly independent. ▲

Here are some important facts about linearly dependent and linearly independent sets.

1. The empty set $\emptyset$ is linearly independent. It will allow us to assign the dimension zero to the zero space.

2. Any set $\{\mathbf{v}\}$ containing a single nonzero vector is linearly independent, since $a\mathbf{v} = \mathbf{0}$ only if $a = 0$.

3. The set $\{\mathbf{0}\}$ is linearly dependent, since $a\mathbf{0} = \mathbf{0}$ for all scalars a.

4. A set $\{\mathbf{v}, \mathbf{w}\}$ containing two nonzero vectors is linearly independent if $\mathbf{v}$ is not a scalar multiple of $\mathbf{w}$, since $a\mathbf{v} + b\mathbf{w} = \mathbf{0}$ is equivalent to saying that $a\mathbf{v} = -b\mathbf{w}$. If $a \neq 0$, then we can solve this equation for $\mathbf{v}$ and write $\mathbf{v} = -(b/a)\,\mathbf{w}$.

5. All vectors in a nonempty linearly independent set $\mathcal{S}$ are nonzero. For suppose $\mathbf{v} \in \mathcal{S}$ and $\mathbf{v} = \mathbf{0}$. Then $a\mathbf{v} = \mathbf{0}$ for all scalars a, and $\mathcal{S}$ is therefore linearly dependent.

Matrix Test for Linear Independence

Corollary 2.22 provides an easy test for the linear independence of a set $\{\mathbf{x}_1, \ldots, \mathbf{x}_n\}$ of n column vectors in $\mathbb{R}^n$. For example, if $\mathcal{S} = \{\mathbf{x}_1 = (1, 2), \mathbf{x}_2 = (3, 4)\}$ is a set consisting of two column vectors in $\mathbb{R}^2$ and A is the 2×2 matrix determined by $\mathcal{S}$, then

$$\begin{bmatrix} 1 & 3 \\ 2 & 4 \end{bmatrix}\begin{bmatrix} a \\ b \end{bmatrix} = \begin{bmatrix} 0 \\ 0 \end{bmatrix} \quad \text{if and only if} \quad a\begin{bmatrix} 1 \\ 2 \end{bmatrix} + b\begin{bmatrix} 3 \\ 4 \end{bmatrix} = \begin{bmatrix} 0 \\ 0 \end{bmatrix}$$

Therefore $\mathcal{S}$ is linearly independent if and only if

$$\begin{bmatrix} 1 & 3 \\ 2 & 4 \end{bmatrix}\begin{bmatrix} a \\ b \end{bmatrix} = \begin{bmatrix} 0 \\ 0 \end{bmatrix}$$

has only the trivial solution. We know from Corollary 2.22 that this is the case if and only if the matrix A is invertible. Hence the set $\mathcal{S}$ is linearly independent if and only if the vectors in $\mathcal{S}$ determine an invertible matrix. We formulate these observations in the form of a theorem.

THEOREM 4.3 (Matrix test for linear independence) *A set $\{\mathbf{x}_1, \ldots, \mathbf{x}_n\}$ of n column vectors of height n is linearly independent if and only if the matrix $A = [\mathbf{x}_1 \cdots \mathbf{x}_n]$ is invertible.*

Proof. If we combine the vectors $\mathbf{x}_1, \ldots, \mathbf{x}_n$ into an $n \times n$ matrix $A = [\mathbf{x}_1 \cdots \mathbf{x}_n]$ and use the vector $\mathbf{x} = [a_1, \ldots, a_n]$ to form the homogeneous equation $A\mathbf{x} = \mathbf{0}$, then Corollary 2.22 tells us that A is invertible if and only if the equation $A\mathbf{x} = \mathbf{0}$ has only the trivial solution. Since

$$A\mathbf{x} = \begin{bmatrix} \mathbf{x}_1 & \cdots & \mathbf{x}_n \end{bmatrix}\begin{bmatrix} a_1 \\ \vdots \\ a_n \end{bmatrix} = a_1\mathbf{x}_1 + \cdots + a_n\mathbf{x}_n$$

this statement is equivalent to saying that $A\mathbf{x} = \mathbf{0}$ has only the trivial solution if and only if the vectors $\mathbf{x}_1, \ldots, \mathbf{x}_n$ are linearly independent. ■

COROLLARY 4.4 *A set $\{\mathbf{x}_1, \ldots, \mathbf{x}_n\}$ of column vectors of height n is linearly independent if and only if the matrix $A = [\mathbf{x}_1 \cdots \mathbf{x}_n]$ has n pivots.*

Proof. By Corollary 2.20, A is invertible if and only if it has n pivots. ■

As noted, we refer to Theorem 4.3 as the *matrix test for linear independence*. The theorem also provides us with an infinite set of linearly independent subsets of $\mathbb{R}^n$ since the columns of invertible real $n \times n$ matrices are linearly independent sets.

EXAMPLE 4.18 A Linearly Independent Subset of $\mathbb{R}^4$

Use *Maple* and Theorem 4.3 to try to generate a linearly independent subset $\{\mathbf{x}_1, \mathbf{x}_2, \mathbf{x}_3, \mathbf{x}_4\}$ of $\mathbb{R}^4$.

Solution. We use *Maple* to generate a random 4×4 matrix A and test it for invertibility. If it is invertible, we use its columns as the required vectors. If A is not invertible, we stop, since we have no guarantee that *Maple* will produce an invertible matrix in a finite number of attempts.

```
[>with(linalg):
[>r:=rand(0..5):   random:= proc() r() end:
[>A:=randmatrix(4,4,entries = random );
```

$$A := \begin{bmatrix} 0 & 3 & 3 & 5 \\ 0 & 0 & 0 & 1 \\ 3 & 1 & 0 & 2 \\ 5 & 2 & 5 & 3 \end{bmatrix}$$

Next we ask *Maple* to invert the matrix A, without requesting an explicit display of A^{-1}. If A is not invertible, *Maple* will respond to our command with a message indicating that A is not invertible.

```
[>inverse(A):
```

Since *Maple* carried out the required operation without producing an output, we know that A is invertible. Hence its columns are linearly independent. Therefore, the set

$$\{\mathbf{x}_1 = (0, 0, 3, 5), \mathbf{x}_2 = (3, 0, 1, 2), \mathbf{x}_3 = (3, 0, 0, 5), \mathbf{x}_4 = (5, 1, 2, 3)\}$$

is the required set of vectors. ▲

So far, we have defined linear dependence and independence as a property of sets. However, we can also speak of the linear dependence and independence as a property of vectors. Consider the linearly dependent subset

$$S_1 = \{\mathbf{x}_1 = (1, 2), \mathbf{x}_2 = (3, 4), \mathbf{x}_3 = (-1, 6)\}$$

of $\mathbb{R}^2$. In this case, there exist nonzero scalars $a = -11$, $b = 4$, and $c = 1$ such that

$$a\mathbf{x}_1 + b\mathbf{x}_2 + c\mathbf{x}_3 = \mathbf{0}$$

Hence the set S_1 is linearly dependent. However, each of the three vectors is linearly independent of each of the remaining two other vectors. The vector $\mathbf{x}_1$ is linearly independent of the vector $\mathbf{x}_2$ since it is not a multiple of $\mathbf{x}_2$, and it is linearly independent of $\mathbf{x}_3$ since it is not a multiple of $\mathbf{x}_3$. Similarly, the vectors $\mathbf{x}_2$ and $\mathbf{x}_3$ are linearly independent of each other since neither is a multiple of the other. This implies that each of the three vectors can be written as a linear combination of the remaining two. On the other hand, consider the linearly dependent set

$$S_2 = \{\mathbf{x}_1 = (1, 2), \mathbf{x}_2 = (2, 4), \mathbf{x}_3 = (-1, 6)\}$$

of vectors in $\mathbb{R}^2$. The set S_2 is linearly dependent since the equation

$$a\begin{bmatrix} 1 \\ 2 \end{bmatrix} + b\begin{bmatrix} 2 \\ 4 \end{bmatrix} + c\begin{bmatrix} -1 \\ 6 \end{bmatrix} = \begin{bmatrix} 0 \\ 0 \end{bmatrix}$$

has the nontrivial solution $a = -2$, $b = 1$ and $c = 0$. But since c must be zero, the vector $\mathbf{x}_3$ cannot be written as a linear combination of $\mathbf{x}_1$ and $\mathbf{x}_2$. This is due to the fact that $\mathbf{x}_2$ and $\mathbf{x}_3$ are linearly dependent. However, $\mathbf{x}_2$ can be written as $2\mathbf{x}_1$.

This shows that the linear dependence of a set does not imply that all vectors can be written as linear combinations of the remaining ones. We formulate this observation as a theorem.

THEOREM 4.5 *A nonempty set $\mathcal{S} = \{\mathbf{v}_1, \ldots, \mathbf{v}_n\}$ is linearly dependent if and only if one of the vectors $\mathbf{v}_i$ can be written as a linear combination of the remaining vectors.*

Proof. Suppose that $\mathcal{S}$ is linearly dependent. Then there exist scalars $a_1, \ldots, a_n$ that are not all 0 for which $a_1\mathbf{v}_1 + \cdots + a_n\mathbf{v}_n = \mathbf{0}$. By suitably renumbering the vectors in $\mathcal{S}$, we may assume that $a_1 \neq 0$. Therefore $a_1\mathbf{v}_1 = -a_2\mathbf{v}_2 - \cdots - a_n\mathbf{v}_n$. Hence $\mathbf{v}_1$ can be written as a linear combination of $\mathbf{v}_2, \ldots, \mathbf{v}_n$ using the coefficients $-a_2/a_1$ to $-a_n/a_1$. Conversely, suppose that $\mathbf{v}_1 = b_2\mathbf{v}_2 + \cdots + b_n\mathbf{v}_n$. Then

$$\begin{aligned}\mathbf{0} &= 1\mathbf{v}_1 - b_2\mathbf{v}_2 - \cdots - b_n\mathbf{v}_n \\ &= a_1\mathbf{v}_1 + a_2\mathbf{v}_2 + \cdots + a_n\mathbf{v}_n\end{aligned}$$

and not all a_i are 0. Hence $\mathcal{S}$ is linearly dependent. ■

COROLLARY 4.6 *A nonempty set $\mathcal{S} = \{\mathbf{v}_1, \ldots, \mathbf{v}_n\}$ is linearly independent if and only if none of the vectors $\mathbf{v}_i$ can be written as a linear combination of the remaining vectors.*

Proof. A statement of the form "φ if and only if ψ" is true when the corresponding statement "not φ if and only if not ψ" is true. Hence the corollary follows from Theorem 4.5. ■

Using Theorem 4.5 and its corollary, we say that a vector $\mathbf{x} \in V$ is ***linearly independent*** of the vectors $\mathbf{v}_1, \ldots, \mathbf{v}_n$ if $\mathbf{x}$ is not expressible as a linear combination of $\mathbf{v}_1, \ldots, \mathbf{v}_n$. Equivalently, we say that a vector $\mathbf{x} \in V$ is ***linearly independent*** of a set $\mathcal{S}$ of vectors if $\mathbf{x}$ is not in the span of $\mathcal{S}$. The idea of ***linearly dependent vectors*** is analogous.

EXAMPLE 4.19 Linearly Independent Vectors

Show that the vector $\mathbf{x} = (1, 2, 3)$ is linearly independent of the set

$$\mathcal{S} = \text{span}\{(1, 2, 0), (0, 3, 4)\}$$

Solution. Suppose that

$$\begin{bmatrix} 1 \\ 2 \\ 3 \end{bmatrix} = a \begin{bmatrix} 1 \\ 2 \\ 0 \end{bmatrix} + b \begin{bmatrix} 0 \\ 3 \\ 4 \end{bmatrix} = \begin{bmatrix} a \\ 2a + 3b \\ 4b \end{bmatrix}$$

Then $3 = 4b$, $a = 1$, and $2 = 2a + 3b$. Hence we must have both $b = 3/4$ and $b = 0$. This is obviously impossible. ▲

EXERCISES 4.6

1. Use the given vectors and scalars to construct all possible linear combinations of the given vectors.

a.	Scalars: 7, 9, −4	Vectors: (3, 5), (−8, 2)
b.	Scalar: $\pi, \pi^2, 0$	Vectors: $\begin{bmatrix} 1 & 2 \\ 3 & 4 \end{bmatrix}, \begin{bmatrix} 1 & 2 \\ 3 & 4 \end{bmatrix}, \begin{bmatrix} 1 & 2 \\ 3 & 4 \end{bmatrix}$

2. Describe the spans of the vectors in Exercise 1 in $\mathbb{R}^2$ and $\mathbb{R}^{2\times2}$.

3. Describe the span of the set of polynomials $\mathcal{S} = \{t - 7, t^3, t^5, t^6 + 1\}$.

4. Show that the pivot columns of a matrix in reduced row echelon form are linearly independent.

5. Test the following sets of vectors for linear dependence.

   a. $\mathcal{S} = \{(3, 5), (-8, 2)\}$
   b. $\mathcal{S} = \{(3, 5, 4), (-8, 2, 3), (0, 1, 1)\}$
   c. $\mathcal{S} = \{(3, 5), (-8, 2), (0, 1)\}$
   d. $\mathcal{S} = \{(3, 5, 4), (-8, 2, 3), (0, 1, 1), (6, 6, 6)\}$

6. Use linear independence to test the following matrices for invertibility.

   a. $\begin{bmatrix} 3 & -8 & 0 \\ 5 & 0 & 1 \\ 4 & 3 & 6 \end{bmatrix}$ b. $\begin{bmatrix} 3 & -8 & 1 \\ 0 & 0 & 0 \\ 4 & 3 & 1 \end{bmatrix}$ c. $\begin{bmatrix} 3 & -8 & -5 \\ 5 & 0 & 5 \\ 4 & 3 & 7 \end{bmatrix}$

7. Use pivots to test the columns of the following matrices for linear independence.

$$\text{a. } \begin{bmatrix} 3 & -8 & 0 \\ 5 & 0 & 1 \\ 4 & 3 & 6 \end{bmatrix} \quad \text{b. } \begin{bmatrix} 3 & -8 & 1 & 2 & 0 \\ 0 & 0 & 0 & 1 & 0 \\ 4 & 3 & 1 & 0 & 5 \\ 1 & 4 & 0 & 0 & 0 \\ 2 & 5 & 0 & 0 & 1 \end{bmatrix} \quad \text{c. } \begin{bmatrix} 3 & -8 & -5 \\ 5 & 0 & 5 \\ 4 & 3 & 7 \end{bmatrix}$$

8. (*Maple V*) Generate a random 3×4 matrix A and show that at least one of the columns of A can be written as a linear combination of the remaining ones.

9. (*Maple V*) Modify entries the matrix in Exercise 8, if necessary, so that precisely two columns of the new matrix cannot be written as linear combinations of the remaining ones.

Bases

We now show that all vectors in a vector space can be written as unique linear combinations of special sets of vectors.

DEFINITION 4.6 *A **basis** for a vector space V is any linearly independent ordered subset $\mathcal{B}$ of V that spans V.*

It is perhaps surprising that every vector space has a basis since, in the general case, the proof of this fact requires a powerful axiom of set theory. It is perhaps less surprising to find that if $\mathcal{B}$ and $\mathcal{C}$ are two finite bases for the same vector space, then $\mathcal{B}$ and $\mathcal{C}$ have the same number of elements. We will use this fact later to define the dimension of a vector space.

EXAMPLE 4.20 A Basis for $\mathbb{R}^3$

Use *Maple* to show that the set $\mathcal{B} = \{\mathbf{x}_1 = (1, 2, 3), \mathbf{x}_2 = (0, 5, 2), \mathbf{x}_3 = (1, 2, 0)\}$ is a basis for $\mathbb{R}^3$.

Solution. We first show that $\mathcal{B}$ spans $\mathbb{R}^3$. We show that for any vector $\mathbf{x} = (x, y, z)$ there exist scalars a, b, c such that $a\mathbf{x}_1 + b\mathbf{x}_2 + c\mathbf{x}_3 = \mathbf{x}$. The scalars a, b, c are determined by the scalars x, y, z. The following linear system expresses their relationship.

$$a \begin{bmatrix} 1 \\ 2 \\ 3 \end{bmatrix} + b \begin{bmatrix} 0 \\ 5 \\ 2 \end{bmatrix} + c \begin{bmatrix} 1 \\ 2 \\ 0 \end{bmatrix} = \begin{bmatrix} x \\ y \\ z \end{bmatrix}$$

To find a, b, and c, we solve the equation

$$\begin{bmatrix} a + c \\ 2a + 5b + 2c \\ 3a + 2b \end{bmatrix} = \begin{bmatrix} x \\ y \\ z \end{bmatrix}$$

```
[> solve({a+c=x,2*a+5*b+2*c=y,3*a+2*b=z},{a,b,c});
```

$$\{a = \tfrac{4}{15}x - \tfrac{2}{15}y + \tfrac{1}{3}z, c = \tfrac{11}{15}x + \tfrac{2}{15}y - \tfrac{1}{3}z, b = -\tfrac{2}{5}x + \tfrac{1}{5}y\}$$

Since a, b, c depend only on x, y, z, the set $\mathcal{B}$ spans $\mathbb{R}^3$. It remains to verify that $\mathcal{B}$ is linearly independent. We do so by showing that if $x = y = z = 0$, then $a = b = c = 0$.

```
[> solve({a+c=0,2*a+5*b+2*c=0,3*a+2*b=0},{a,b,c});
```

$$\{b = 0, c = 0, a = 0\}$$

Therefore, $\mathcal{B}$ is a basis for $\mathbb{R}^3$. ▲

EXAMPLE 4.21 The Basis of the Zero Space

By definition, the span of the empty set $\emptyset$ of vectors of a vector space is the space $\{\mathbf{0}\}$. As we pointed out earlier, the set $\emptyset$ is linearly independent. Hence it satisfies both parts of the definition of a basis. ▲

Finite bases have an easy description in terms of linear combinations.

THEOREM 4.7 (Unique combination theorem) *A subset $\mathcal{B} = \{\mathbf{x}_1, \ldots, \mathbf{x}_n\}$ of a vector space V is a basis for V if and only if every vector $\mathbf{x} \in V$ can be written as a unique linear combination $a_1\mathbf{x}_1 + \cdots + a_n\mathbf{x}_n$.*

Proof. Suppose that $\mathcal{B}$ is a basis and that $\mathbf{x} = a_1\mathbf{x}_1 + \cdots + a_n\mathbf{x}_n = b_1\mathbf{x}_1 + \cdots + b_n\mathbf{x}_n$. Then

$$\begin{aligned}\mathbf{0} &= \mathbf{x} - \mathbf{x} \\ &= (a_1 - b_1)\,\mathbf{x}_1 + \cdots + (a_n - b_n)\,\mathbf{x}_n\end{aligned}$$

Since the basis vectors $\mathbf{x}_i$ are linearly independent, $a_i - b_i = 0$ for all i. Hence the linear combination is unique.

Conversely, suppose that every $\mathbf{x} \in V$ can be written as a unique linear combination $a_1\mathbf{x}_1 + \cdots + a_n\mathbf{x}_n$. Then $\mathcal{B}$ obviously spans V. Now suppose that $a_1\mathbf{x}_1 + \cdots + a_n\mathbf{x}_n = \mathbf{0}$. Then

$$\begin{aligned}\mathbf{0} &= 0\mathbf{x}_1 + \cdots + 0\mathbf{x} \\ &= a_1\mathbf{x}_1 + \cdots + a_n\mathbf{x}_n\end{aligned}$$

Since the representation of $\mathbf{0}$ as a linear combination of vectors in $\mathcal{B}$ is unique, it follows that $a_1 = \cdots = a_n = 0$. Hence $\mathcal{B}$ is linearly independent. Therefore, the set $\mathcal{B}$ is a basis for V. ■

EXERCISES 4.7

1. Describe all possible bases of the space $\mathbb{R}$ of real numbers.

2. Use invertible matrices to construct two different bases for the space $\mathbb{R}^4$.

3. Show that the set

$$\mathcal{B} = \left\{ \begin{bmatrix} 1 & 2 \\ 3 & 4 \end{bmatrix}, \begin{bmatrix} 2 & 0 \\ 0 & 5 \end{bmatrix}, \begin{bmatrix} 0 & 2 \\ 5 & 0 \end{bmatrix}, \begin{bmatrix} 1 & 1 \\ 1 & 1 \end{bmatrix} \right\}$$

is a basis for the space $\mathbb{R}^{2\times 2}$ of 2×2 matrices.

4. The fact that the linear system

$$\begin{cases} a + 2b - c = 0 \\ 2a + 4b + 2c + d = 0 \\ 3a + 7b + 5c + d = 0 \\ 4a + 5b + c + 3d = 0 \end{cases}$$

has only the trivial solution ensures that the set

$$\mathcal{B} = \left\{ \begin{bmatrix} 1 & 2 \\ 3 & 4 \end{bmatrix}, \begin{bmatrix} 2 & 4 \\ 7 & 5 \end{bmatrix}, \begin{bmatrix} -1 & 2 \\ 5 & 1 \end{bmatrix}, \begin{bmatrix} 0 & 1 \\ 1 & 3 \end{bmatrix} \right\}$$

is a basis for $\mathbb{R}^{2\times 2}$. Explain why.

5. Find two different bases for the space $\mathbb{R}_5[t]$ of polynomials of degree 5 or less.

6. Find a basis for the solution space of the homogeneous linear system

$$\begin{bmatrix} 2 & 1 & 4 & 2 \\ 3 & 7 & 4 & 5 \\ 4 & 8 & 4 & 1 \end{bmatrix} \begin{bmatrix} w \\ x \\ y \\ z \end{bmatrix} = \begin{bmatrix} 0 \\ 0 \\ 0 \end{bmatrix}$$

7. (*Maple V*) Find a basis for the solution space of the differential equation $y'(t) - y(t) = 0$.

Standard Bases

Among the bases of some vector spaces are bases that are particularly simple. The vectors involved are often built from the scalars 0 and 1, but may also have other special properties. These bases are known as ***standard bases***. Except for the order in which the vectors are listed, standard bases are unique.

We can describe the standard bases of coordinate spaces and matrix spaces using the functions $\delta : \mathbf{m} \times \mathbf{n} \to \{0, 1\}$, defined by

$$\delta_{ij} = \begin{cases} 1 & \text{if} \quad i = j \\ 0 & \text{if} \quad i \neq j \end{cases}$$

These functions are known as the ***Kronecker delta*** functions. We can use a procedure to define these functions in *Maple* by putting

```
Delta:=proc(i,j) if i=j then 1 else 0 fi; end
```

As is clear from the definition, the input `Delta(i,j)` produces the output 1 if $i = j$ and 0 otherwise. On a single argument, we can define the function by putting

```
Delta:=proc(i) if i=1 then 1 else 0 fi; end
```

The command `Delta(i)` produces 1 if $i = 1$ and 0 otherwise.

EXAMPLE 4.22 The Standard Basis of $\mathbb{R}^2$

The standard basis of the coordinate space $\mathbb{R}^2$ can be defined using the Kronecker delta function $\delta : \mathbf{2} \times \mathbf{2} \to \{0, 1\}$. The set $\mathcal{E}$ consisting of the two vectors $\mathbf{e}_1 = (\delta_{11}, \delta_{12}) = (1, 0)$ and $\mathbf{e}_2 = (\delta_{21}, \delta_{22}) = (0, 1)$ is the standard basis of $\mathbb{R}^2$. ▲

EXAMPLE 4.23 The Standard Basis of $\mathbb{R}^3$ in *Maple*

The standard basis $\mathcal{E}$ of $\mathbb{R}^3$ consists of the three vectors

$$\begin{cases} \mathbf{e}_1 = (\delta_{11}, \delta_{12}, \delta_{13}) = (1, 0, 0) \\ \mathbf{e}_2 = (\delta_{21}, \delta_{22}, \delta_{23}) = (0, 1, 0) \\ \mathbf{e}_3 = (\delta_{31}, \delta_{32}, \delta_{33}) = (0, 0, 1) \end{cases}$$

discussed in Chapter 3 in connection with the cross product of vectors in $\mathbb{R}^3$. At that point, we denoted them by **i**, **j**, and **k**. We can use the **delta** functions to generate these vectors in *Maple*.

```
[> delta:=proc(i,j) if i=j then 1 else 0 fi; end:
[> e1:=matrix(3,1,[delta(1,1),delta(1,2),delta(1,3)]);
```

$$e1 := \begin{bmatrix} 1 \\ 0 \\ 0 \end{bmatrix}$$

```
[> e2:=matrix(3,1,[delta(2,1),delta(2,2),delta(2,3)]);
```

$$e2 := \begin{bmatrix} 0 \\ 1 \\ 0 \end{bmatrix}$$

```
[> e3:=matrix(3,1,[delta(3,1),delta(3,2),delta(3,3)]);
```

$$e3 := \begin{bmatrix} 0 \\ 0 \\ 1 \end{bmatrix}$$

As we can see, **e1**, **e2**, and **e3** are the required vectors. ▲

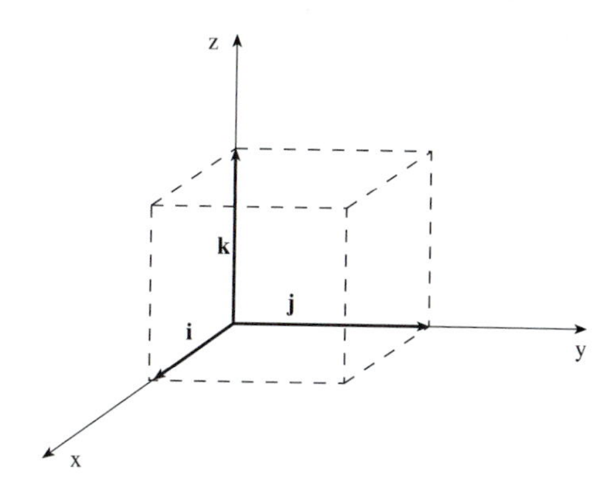

FIGURE 1 The Standard Basis of $\mathbb{R}^3$.

Figure 1 shows the vectors **i**, **j**, and **k** represented graphically. At the end of this chapter, we will provide yet another perspective on the vectors **i**, **j**, and **k** by viewing them as special ***quaternion*** vectors in $\mathbb{R}^4$.

EXAMPLE 4.24 The Standard Basis of $\mathbb{R}^{2\times 2}$

The standard basis of the matrix space $\mathbb{R}^{2\times 2}$ can also be defined using the Kronecker delta function $\delta : \mathbf{4} \times \mathbf{4} \to \{0, 1\}$. The set

$$\mathcal{E} = \left\{ \begin{bmatrix} \delta_{11} & \delta_{12} \\ \delta_{13} & \delta_{14} \end{bmatrix}, \begin{bmatrix} \delta_{21} & \delta_{22} \\ \delta_{23} & \delta_{24} \end{bmatrix}, \begin{bmatrix} \delta_{31} & \delta_{32} \\ \delta_{33} & \delta_{34} \end{bmatrix}, \begin{bmatrix} \delta_{41} & \delta_{42} \\ \delta_{43} & \delta_{44} \end{bmatrix} \right\}$$

is the standard basis of $\mathbb{R}^{2\times 2}$. ▲

EXAMPLE 4.25 The Standard Basis of $\mathbb{R}^n$

The standard basis of the coordinate space $\mathbb{R}^n$ is defined using the Kronecker delta function $\delta : \mathbf{n} \times \mathbf{n} \to \{0, 1\}$. The set $\mathcal{E} = \{\mathbf{e}_i = (\delta_{i1}, \ldots, \delta_{in}) : 1 \leq i \leq n\}$ is the standard basis for the coordinate space $\mathbb{R}^n$. ▲

EXAMPLE 4.26 The Standard Basis of $\mathbb{R}_n[t]$

The set $\mathcal{E} = \{1, t, \ldots, t^n\}$ is the standard basis of the polynomial space $\mathbb{R}_n[t]$. ▲

EXAMPLE 4.27 The Standard Basis of $\mathbb{R}[t]$

The set $\mathcal{E} = \{1, t, \ldots, t^n, \ldots\}$ is the standard basis of the polynomial space $\mathbb{R}[t]$. It is infinite since it contains the polynomials t^n for all $n \in \mathbb{N}$. ▲

EXAMPLE 4.28 The Standard Basis of a Homogeneous System

Consider the homogeneous system $A\mathbf{x} = \mathbf{0}$, determined by the linear equations

$$\begin{cases} x_1 + 3x_2 - x_5 + x_6 = 0 \\ 2x_1 + x_3 - x_4 + 4x_5 = 0 \\ x_2 + x_3 + x_4 - x_6 = 0 \end{cases}$$

We use the reduced row echelon form of the augmented matrix

$$[A \mid \mathbf{0}] = \begin{bmatrix} 1 & 3 & 0 & 0 & -1 & 1 & 0 \\ 2 & 0 & 1 & -1 & 4 & 0 & 0 \\ 0 & 1 & 1 & 1 & 0 & -1 & 0 \end{bmatrix}$$

of $A\mathbf{x} = \mathbf{0}$ to define the standard basis of the solution space of $A\mathbf{x} = \mathbf{0}$. We expedite the calculations by using *Maple* to row-reduce the matrix $[A \mid \mathbf{0}]$.

```
[>A:=matrix([[1,3,0,0,-1,1,0],[2,0,1,-1,4,0,0],
   [0,1,1,1,0,-1,0]]):
[>B:=linalg[rref](A);
```

$$B := \begin{bmatrix} 1 & 0 & 0 & \frac{-6}{7} & \frac{11}{7} & \frac{4}{7} & 0 \\ 0 & 1 & 0 & \frac{2}{7} & \frac{-6}{7} & \frac{1}{7} & 0 \\ 0 & 0 & 1 & \frac{5}{7} & \frac{6}{7} & \frac{-8}{7} & 0 \end{bmatrix}$$

We can tell from the matrix B that the variables of the system $A\mathbf{x} = \mathbf{0}$ are related by the equations

$$\begin{cases} x_1 = \frac{6}{7}x_4 - \frac{11}{7}x_5 - \frac{4}{7}x_6 \\ x_2 = -\frac{2}{7}x_4 + \frac{6}{7}x_5 - \frac{1}{7}x_6 \\ x_3 = -\frac{5}{7}x_4 - \frac{6}{7}x_5 + \frac{8}{7}x_6 \end{cases}$$

Therefore the solutions of the homogeneous linear system $A\mathbf{x} = \mathbf{0}$ are the vectors of the form

$$\begin{bmatrix} \frac{6}{7}x_4 - \frac{11}{7}x_5 - \frac{4}{7}x_6 \\ -\frac{2}{7}x_4 + \frac{6}{7}x_5 - \frac{1}{7}x_6 \\ -\frac{5}{7}x_4 - \frac{6}{7}x_5 + \frac{8}{7}x_6 \\ x_4 \\ x_5 \\ x_6 \end{bmatrix} = x_4 \begin{bmatrix} \frac{6}{7} \\ -\frac{2}{7} \\ -\frac{5}{7} \\ 1 \\ 0 \\ 0 \end{bmatrix} + x_5 \begin{bmatrix} -\frac{11}{7} \\ \frac{6}{7} \\ -\frac{6}{7} \\ 0 \\ 1 \\ 0 \end{bmatrix} + x_6 \begin{bmatrix} -\frac{4}{7} \\ -\frac{1}{7} \\ \frac{8}{7} \\ 0 \\ 0 \\ 1 \end{bmatrix}$$

written as unique linear combinations of the set

$$\mathcal{E} = \left\{ \left(\tfrac{6}{7}, -\tfrac{2}{7}, -\tfrac{5}{7}, 1, 0, 0\right), \left(-\tfrac{11}{7}, \tfrac{6}{7}, -\tfrac{6}{7}, 0, 1, 0\right), \left(-\tfrac{4}{7}, -\tfrac{1}{7}, \tfrac{8}{7}, 0, 0, 1\right) \right\}$$

By construction, the set $\mathcal{E}$ spans the solution space of $A\mathbf{x} = \mathbf{0}$. By looking at the last three components of the vectors in $\mathcal{E}$, we see that $\mathcal{E}$ is linearly independent. The set $\mathcal{E}$ is therefore a basis for the solution space of the system $A\mathbf{x} = \mathbf{0}$. This basis is called a *standard basis* since it is uniquely determined by the reduced row echelon form of $[A \mid \mathbf{0}]$. ▲

EXERCISES 4.8

1. Use the Kronecker delta function to generate the standard basis of $\mathbb{R}^4$.

2. Use the Kronecker delta function to generate the standard basis of $\mathbb{R}^{2\times 3}$.

3. Find the standard bases of the following real vector spaces.

   a. The space of real numbers

   b. The space of real upper-triangular 3×3 matrices

   c. The space of real diagonal 4×4 matrices

   d. The space of real symmetric 3×3 matrices

4. Find the standard basis of the solution space of the homogeneous linear system

$$\begin{cases} 2w + x + 4y + 2z = 0 \\ 3w + 7x + 4y + 5z = 0 \end{cases}$$

5. (*Maple V*) Discuss the relevance of the **KroneckerDelta** function, defined by the procedure

```
KroneckerDelta:=proc(i) if i=1 then 1 else 0 fi; end;
```

to the construction of standard bases using *Maple*.

Bases and Linear Systems

If $\mathbf{x}$ is any vector in the vector space $\mathbb{R}^n$ and $\mathcal{B} = \{\mathbf{v}_1, \ldots, \mathbf{v}_n\}$ is a basis for $\mathbb{R}^n$, then $\mathbf{x}$ can be written uniquely in the form $\mathbf{x} = a_1\mathbf{v}_1 + \cdots + a_n\mathbf{v}_n$ in the basis $\mathcal{B}$. The coefficients $a_1, \ldots, a_n$ can be found by solving a system of linear equations. Here is a numerical example.

Consider the basis $\mathcal{B}$ for $\mathbb{R}^4$ consisting of the vectors $\mathbf{v}_1 = (1, 0, 2, 3)$, $\mathbf{v}_2 = (1, 0, 0, 1)$, $\mathbf{v}_3 = (0, 3, -1, 0)$, $\mathbf{v}_4 = (1, 1, 1, 1)$, and let $\mathbf{x} = (1, 2, 3, 4) \in \mathbb{R}^4$. We show how the vector $\mathbf{x}$ can be written as a linear combination of the vectors $\mathbf{v}_1$, $\mathbf{v}_2$, $\mathbf{v}_3$ and $\mathbf{v}_4$. Since $\mathcal{B}$ is a basis, there exist scalars a, b, c, d for which

$$\begin{bmatrix} 1 \\ 2 \\ 3 \\ 4 \end{bmatrix} = a \begin{bmatrix} 1 \\ 0 \\ 2 \\ 3 \end{bmatrix} + b \begin{bmatrix} 1 \\ 0 \\ 0 \\ 1 \end{bmatrix} + c \begin{bmatrix} 0 \\ 3 \\ -1 \\ 0 \end{bmatrix} + d \begin{bmatrix} 1 \\ 1 \\ 1 \\ 1 \end{bmatrix}$$

We can find the scalars a, b, c, and d by solving the linear system

$$\begin{cases} a + b + d = 1 \\ 3c + d = 2 \\ 2a - c + d = 3 \\ 3a + b + d = 4 \end{cases}$$

```
[> solve({a+b+d=1,3*c+d=2,2*a-c+d=3,3*a+b+d=4},{a,b,c,d});
```

$$\{b = -1, c = \tfrac{1}{2}, a = \tfrac{3}{2}, d = \tfrac{1}{2}\}$$

Therefore $a = 3/2$, $b = -1$, $c = 1/2$, and $d = 1/2$ are the unique scalars required to represent the vector $\mathbf{x}$ in the basis $\mathcal{B}$. We cast these observations in the form of a theorem.

THEOREM 4.8 *If $\mathcal{B} = \{\mathbf{v}_1, \ldots, \mathbf{v}_n\}$ is any basis for $\mathbb{R}^n$ and $\mathbf{x}$ is any vector in $\mathbb{R}^n$, then the coefficients $a_1, \ldots, a_n$ in the unique linear combination $\mathbf{x} = a_1\mathbf{v}_1 + \cdots + a_n\mathbf{v}_n$ can be found by solving a system of linear equations.*

Proof. Let $i \in \mathbf{n}$ and suppose that the vectors $\mathbf{v}_i = (v_{i1}, \ldots, v_{in})$ constitute a basis for $\mathbb{R}^n$. Suppose further that $\mathbf{x} = (x_1, \ldots, x_n)$ is an arbitrary vector in $\mathbb{R}^n$. Then

$$\begin{bmatrix} x_1 \\ \vdots \\ x_n \end{bmatrix} = a_1 \begin{bmatrix} v_{11} \\ \vdots \\ v_{1n} \end{bmatrix} + \cdots + a_n \begin{bmatrix} v_{n1} \\ \vdots \\ v_{nn} \end{bmatrix} = \begin{bmatrix} a_1 v_{11} \\ \vdots \\ a_1 v_{1n} \end{bmatrix} + \cdots + \begin{bmatrix} a_n v_{n1} \\ \vdots \\ a_n v_{nn} \end{bmatrix}$$

and

$$\begin{bmatrix} a_1 v_{11} \\ \vdots \\ a_1 v_{1n} \end{bmatrix} + \cdots + \begin{bmatrix} a_n v_{n1} \\ \vdots \\ a_n v_{nn} \end{bmatrix} = \begin{bmatrix} a_1 v_{11} + \cdots + a_n v_{n1} \\ \vdots \\ a_1 v_{1n} + \cdots + a_n v_{nn} \end{bmatrix}$$

Therefore,

$$\begin{bmatrix} a_1 v_{11} + \cdots + a_n v_{n1} \\ \vdots \\ a_1 v_{1n} + \cdots + a_n v_{nn} \end{bmatrix} = \begin{bmatrix} x_1 \\ \vdots \\ x_n \end{bmatrix}$$

This shows that the linear system

$$\begin{cases} a_1 v_{11} + \cdots + a_n v_{n1} = x_1 \\ \vdots \\ a_1 v_{1n} + \cdots + a_n v_{nn} = x_n \end{cases}$$

determines the coefficients $a_1, \ldots, a_n$. ■

EXERCISES 4.9

1. (*Maple V*) Find the solutions of the linear systems $A\mathbf{x} = \mathbf{b}_1$, $A\mathbf{x} = \mathbf{b}_2$, and $A\mathbf{x} = \mathbf{b}_3$, where A is the 3×3 matrix determined by the basis $\mathcal{B} = \{(4, 5, -9), (4, -5, 4), (2, 9, 0)\}$ for $\mathbb{R}^3$ and $\mathbf{b}_1 = (3, -2, 1)$, $\mathbf{b}_2 = (7, 5, 3)$, and $\mathbf{b}_3 = (1, 1, 1)$. Explain why the solutions are unique.

2. (*Maple V*) Verify that the linear system

$$\begin{cases} 6w + 2x - 6y + 6z = 3 \\ 2w - 3x + 5y + 2z = 2 \\ 2w - 5x - 3y + 4z = 0 \\ 4w + 2x + 6y - 4z = 1 \end{cases}$$

has a unique solution. Use this fact to explain why the columns of the coefficient matrix are linearly independent.

3. (*Maple V*) Construct and solve the linear system that determines the coefficients a_1, a_2, a_3, and a_4 for which

$$\begin{bmatrix} 1 \\ 2 \\ 3 \\ 4 \end{bmatrix} = a_1 \begin{bmatrix} 8 \\ 3 \\ -1 \\ -5 \end{bmatrix} + a_2 \begin{bmatrix} 4 \\ -5 \\ 0 \\ -2 \end{bmatrix} + a_3 \begin{bmatrix} -5 \\ 8 \\ 3 \\ -1 \end{bmatrix} + a_4 \begin{bmatrix} -5 \\ 5 \\ -4 \\ -9 \end{bmatrix}$$

Explain why the coefficients a_1, a_2, a_3, and a_4 are unique.

Dimension

A vector space is ***finite-dimensional*** if it has a finite basis. A space that is not finite-dimensional is called ***infinite-dimensional***. All coordinate spaces $\mathbb{R}^n$ and all matrix spaces $\mathbb{R}^{m\times n}$ are finite-dimensional. So are the polynomial spaces $\mathbb{R}_n[t]$. The polynomial space $\mathbb{R}[t]$, on the other hand, is infinite-dimensional. So are the sequence spaces $\mathbb{R}^N$ and $\mathbb{R}^\infty$. However, most spaces studied in this text are finite-dimensional.

The main theorem in this section establishes that all bases of a finite-dimensional vector space V have the same number of elements. We can use this property to define the ***dimension*** of V.

THEOREM 4.9 *Let V be a vector space spanned by the linearly independent vectors $\mathbf{v}_1, \ldots, \mathbf{v}_m$. Then any linearly independent set of vectors in V is finite and contains no more than m elements.*

Proof. It is sufficient to show that any set containing more than m elements is linearly dependent. Let $\mathcal{S}$ be such a set, containing n distinct vectors $\mathbf{w}_1, \ldots, \mathbf{w}_n$, where $n > m$. We show that the set $\{\mathbf{w}_1, \ldots, \mathbf{w}_n\}$ is linearly dependent. Let $b_1\mathbf{w}_1 + \cdots + b_n\mathbf{w}_n = \mathbf{0}$. Since the vectors $\mathbf{v}_1, \ldots, \mathbf{v}_m$ span V, there exist scalars a_{ij} such that $\mathbf{w}_j = \sum_{i\in\mathbf{m}} a_{ij}\mathbf{v}_i$. Therefore,

$$\mathbf{0} = b_1\mathbf{w}_1 + \cdots + b_n\mathbf{w}_n = \sum_{j\in\mathbf{n}} b_j \left(\sum_{i\in\mathbf{m}} a_{ij}\mathbf{v}_i\right) = \sum_{i\in\mathbf{m}} \left(\sum_{j\in\mathbf{n}} b_j a_{ij}\right)\mathbf{v}_i$$

Since the vectors $\mathbf{v}_i$ are linearly independent, the coefficients

$$\sum_{j\in\mathbf{n}} b_j a_{ij}$$

must all be zero. Hence we get a homogeneous linear system

$$\sum_{j\in\mathbf{n}} b_j a_{ij} = 0 \quad (1 \le i \le m)$$

for which $n > m$. Theorem 1.5 guarantees the existence of a nontrivial solution $(b_1, \ldots, b_n)$. Therefore $\mathcal{S}$ is linearly dependent. ■

Proof. This corollary follows immediately from Theorem 4.9. ■

THEOREM 4.10 (Dimension theorem) *If $\mathcal{B}$ and $\mathcal{C}$ are two bases of a finite-dimensional vector space V, then $\mathcal{B}$ and $\mathcal{C}$ have the same number of elements.*

Proof. Suppose that $\mathcal{B} = \{\mathbf{x}_1, \ldots, \mathbf{x}_n\}$ and $\mathcal{C} = \{\mathbf{y}_1, \ldots, \mathbf{y}_m\}$ are two bases for V. Since $\mathcal{B}$ spans V, Theorem 4.9 implies that the linearly independent set $\mathcal{C}$ has at most n elements. Therefore, $m \le n$. On the other hand, $\mathcal{C}$ is also a basis and spans V. Hence the linearly independent set $\mathcal{B}$ has at most m elements, so $n \le m$. If follows that $n = m$. ■

This theorem tells us that the following definition makes sense.

DEFINITION 4.7 *If V is a vector space with a finite basis $\mathcal{B}$, then the number of elements of $\mathcal{B}$ is the* **dimension** *of V.*

EXAMPLE 4.29 The Dimension of the Coordinate Space $\mathbb{R}^n$

The space $\mathbb{R}^n$ has dimension n since its standard basis $\mathcal{E} = \{\mathbf{e}_1, \ldots, \mathbf{e}_n\}$, for example, has n elements. ▲

EXAMPLE 4.30 The Dimension of the Matrix Space $\mathbb{R}^{m\times n}$

The space $\mathbb{R}^{m\times n}$ of $m \times n$ matrices over $\mathbb{R}$ has dimension nm since its standard basis, for example, has mn elements. ▲

Contracting Spanning Sets

If a set $\mathcal{S}$ of vectors of a finite-dimensional vector space V spans V, then there are procedures for extracting a basis $\mathcal{B}$ for V from $\mathcal{S}$.

EXAMPLE 4.31 Extracting a Basis from a Finite Spanning Set

Use *Maple* and the fact that $\mathbb{R}^3$ has dimension 3 to extract a basis for $\mathbb{R}^3$ from the spanning set

$$\mathcal{S} = \left\{ \mathbf{v}_1 = \begin{bmatrix} 1 \\ 2 \\ 1 \end{bmatrix}, \mathbf{v}_2 = \begin{bmatrix} 4 \\ 4 \\ 4 \end{bmatrix}, \mathbf{v}_3 = \begin{bmatrix} 1 \\ 0 \\ 1 \end{bmatrix}, \mathbf{v}_4 = \begin{bmatrix} 2 \\ 4 \\ 2 \end{bmatrix}, \mathbf{v}_5 = \begin{bmatrix} 0 \\ 1 \\ 1 \end{bmatrix} \right\}$$

Solution. Since $\mathcal{S}$ spans $\mathbb{R}^3$, we need to extract a linearly independent subset of $\mathcal{S}$ consisting of three linearly independent vectors. We construct this step by step.

Let $\mathcal{S}_1 = \{\mathbf{v}_1\}$. We test whether $\mathbf{v}_2$ is linearly independent of $\mathbf{v}_1$. If this is so, we form the set $\mathcal{S}_2 = \{\mathbf{v}_1, \mathbf{v}_2\}$. If not, we discard $\mathbf{v}_2$ and test the linear independence of $\mathbf{v}_1$ and $\mathbf{v}_3$, and so on.

```
[>with(linalg):
[>v1:=<1,2,1>:   v2:=< 4,4,4>:   v3:=<1,0,1>:
[>v4:=< 2,4,2>:   v5:=<0,1,1>:   z:=< 0,0,0>:
[>A12:=concat(v1,v2):
[>linsolve(A12,z);

                         [0, 0]
```

This shows that $\mathbf{v}_1$ and $\mathbf{v}_2$ are linearly independent. We therefore form the set $\mathcal{S}_2 = \{\mathbf{v}_1, \mathbf{v}_2\}$.

Next we test the linear independence of $\mathbf{v}_3$ from $\mathbf{v}_1$ and $\mathbf{v}_2$. If this is so we form $\mathcal{S}_3 = \{\mathbf{v}_1, \mathbf{v}_2, \mathbf{v}_3\}$. If not, we discard $\mathbf{v}_3$ and examine whether $\mathbf{v}_4$ is linearly independent of $\mathbf{v}_1$ and $\mathbf{v}_2$, and so on.

```
[> A123:=concat(v1,v2,v3):
[> linsolve(A123,z);
```

$$[-2\_t[1], \_t[1], -2\_t[1]]$$

The fact that we have found a nonzero solution of this homogeneous linear systems tells us that the three vectors are linearly dependent. Hence we discard $\mathbf{v}_3$ and test $\mathbf{v}_4$.

```
[> A124:=concat(v1,v2,v4):
[> linsolve(A124,z);
```

$$[-2\_t[1], 0, \_t[1]]$$

Therefore, $\mathbf{v}_1$, $\mathbf{v}_2$, and $\mathbf{v}_4$ are also linearly dependent. We are therefore left with $\mathbf{v}_5$. Since $\mathcal{S}$ spans $\mathbb{R}^3$, the vector $\mathbf{v}_5$ must be linearly dependent of $\mathbf{v}_1$ and $\mathbf{v}_2$. Hence the set $\mathcal{S}' = \{\mathbf{v}_1, \mathbf{v}_2, \mathbf{v}_5\}$ is a basis for $\mathbb{R}^3$. We use *Maple* to confirm the linear independence.

```
[> A125:=concat(v1,v2,v5):
[> linsolve(A125,z);
```

$$[0, 0, 0]$$

Hence the set $\mathcal{S}' = \{\mathbf{v}_1, \mathbf{v}_2, \mathbf{v}_5\}$ is a basis for $\mathbb{R}^3$. ▲

If the dimension of a vector space equals the number of vectors in a subset of finite-dimensional space, we can simplify the verification that the set of vectors is a basis. In that case, spanning implies linear independence, and linear independence implies spanning.

THEOREM 4.11 *If V is an n-dimensional vector space and $\mathcal{S} = \{\mathbf{x}_1, \dots, \mathbf{x}_n\}$ is a subset of V that spans V, then $\mathcal{S}$ is a basis for V.*

Proof. We must show that $\mathcal{S}$ is linearly independent. Suppose that it is not. By renumbering the basis vectors, we may assume that $\mathbf{x}_n$ is linearly dependent on $\mathbf{x}_1, \dots, \mathbf{x}_{n-1}$. On the other hand, the set $\mathcal{S}' = \{\mathbf{x}_1, \dots, \mathbf{x}_{n-1}\}$ spans a vector space of dimension at most $n - 1$. If $\mathcal{B} = \{\mathbf{y}_1, \dots, \mathbf{y}_n\}$ is another basis for V, then some $\mathbf{y}_i$ cannot be a linear combination of $\mathbf{x}_1, \dots, \mathbf{x}_{n-1}$. This contradicts the fact that $\mathcal{S}$ is a basis. Hence $\mathcal{S}$ is linearly independent. ■

THEOREM 4.12 *If V is an n-dimensional vector space and $\mathcal{S} = \{\mathbf{x}_1, \dots, \mathbf{x}_n\}$ is a linearly independent subset of V, then $\mathcal{S}$ is a basis for V.*

Proof. We must show that $\mathcal{S}$ spans V. Suppose it does not. Then there exists a vector $\mathbf{y}$ in V that is linearly independent of $\mathcal{S}$. Therefore $\mathcal{S} = \{\mathbf{x}_1, \ldots, \mathbf{x}_n, \mathbf{y}\}$ is a linearly independent subset of V containing $n + 1$ elements. This contradicts the fact that V is of dimension n. ■

A spanning set of a finite-dimensional vector space may be both finite and infinite. If the set is infinite, a procedure for extracting a finite linearly independent spanning set may require some extra knowledge about the given infinite set. If the elements of the set are listed in a fixed order $S = \{\mathbf{x}_1, \mathbf{x}_2, \mathbf{x}_3, \mathbf{x}_4, \mathbf{x}_5, \ldots\}$, we can build a finite linearly independent spanning set $S' = \{\mathbf{y}_1, \ldots, \mathbf{y}_n\}$ constructively by letting $\mathbf{y}_1 = \mathbf{x}_1$, letting $\mathbf{y}_2$ be the first vector $\mathbf{x}_i$ that is linearly independent of $\mathbf{x}_1$, letting $\mathbf{y}_3$ the next vector that is linearly independent of $\mathbf{y}_1$ and $\mathbf{y}_2$, and so on. Since we require only n vectors, for a fixed n, the process will terminate after finitely many steps.

Another method is to take a basis $\mathcal{B} = \{\mathbf{b}_1, \ldots, \mathbf{b}_n\}$ and write each basis vector $\mathbf{b}_i$ as a linear combination of vectors from the spanning set. This method is nonconstructive if the vectors in the spanning set are not algorithmically specified.

EXAMPLE 4.32 Extracting a Basis from an Infinite Spanning Set

Let $\mathcal{S}$ be an infinite spanning set of a three-dimensional real vector space V. Show that S has a linearly independent subset $\mathcal{S}' = \{\mathbf{b}_1, \mathbf{b}_2, \mathbf{b}_3\}$ that spans V.

Solution. Let $\mathcal{B} = \{\mathbf{x}_1, \mathbf{x}_2, \mathbf{x}_3\}$ be a basis for V. Since $\mathcal{S}$ spans V, each basis vector can be written as a linear combination of vectors in $\mathcal{S}$. Let

$$\begin{cases} \mathbf{x}_1 = a_1\mathbf{u}_1 + \cdots + a_r\mathbf{u}_r \\ \mathbf{x}_2 = b_1\mathbf{v}_1 + \cdots + b_s\mathbf{v}_s \\ \mathbf{x}_3 = c_1\mathbf{w}_1 + \cdots + c_t\mathbf{w}_t \end{cases}$$

Then $V = \operatorname{span}\{\mathbf{u}_1, \ldots, \mathbf{u}_r, \mathbf{v}_1, \ldots, \mathbf{v}_s, \mathbf{w}_1, \ldots, \mathbf{w}_t\}$, since $\mathcal{B}$ is a basis for V. Now we proceed as in Example 4.31 and extract the linearly independent spanning set $\mathcal{S}' = \{\mathbf{b}_1, \mathbf{b}_2, \mathbf{b}_3\}$ from the finite set $\{\mathbf{u}_1, \ldots, \mathbf{u}_r, \mathbf{v}_1, \ldots, \mathbf{v}_s, \mathbf{w}_1, \ldots, \mathbf{w}_t\}$. By construction, the set $\mathcal{S}'$ is a basis for V. ▲

We now apply these ideas in the proof of the theorem that guarantees that every spanning set of a finite-dimensional vector space V contains a basis for V.

THEOREM 4.13 (Basis contraction theorem) *Suppose that V is a finite-dimensional vector space and that $\mathcal{S} \subseteq V$ spans V. Then $\mathcal{S}$ has a linearly independent subset $\mathcal{S}'$ that is a basis for V.*

Proof. Since V is finite-dimensional, it has a finite basis $\mathcal{B} = \{\mathbf{v}_1, \ldots, \mathbf{v}_n\}$. Since $\operatorname{span}(\mathcal{S}) = V$, we can write each $\mathbf{v}_i \in \mathcal{B}$ as a finite linear combination of vectors in $\mathcal{S}$. The vector $\mathbf{v}_1$ is a linear combination of vectors in a finite set $\mathcal{S}_1 = \{\mathbf{w}_{11}, \ldots, \mathbf{w}_{1p_1}\} \subseteq \mathcal{S}$, the vector $\mathbf{v}_2$ is a linear combination of vectors in a finite set $\mathcal{S}_2 = \{\mathbf{w}_2, \ldots, \mathbf{w}_{2p_1}\} \subseteq \mathcal{S}$, and so on.

Continuing in this way, we express each $\mathbf{v}_i$ as a linear combination of vectors of a finite set $\mathcal{S}_i = \{\mathbf{w}_{i1}, \ldots, \mathbf{w}_{ip_i}\} \subseteq \mathcal{S}$. Then

$$\mathcal{S}_0 = \{\mathbf{w}_{11}, \ldots, \mathbf{w}_{1p_1}, \mathbf{w}_{21}, \ldots, \mathbf{w}_{2p_1}, \mathbf{w}_{n1}, \ldots, \mathbf{w}_{np_n}\}$$

is a finite subset of V that spans V since $\mathcal{B}$ is a basis. We now extract a basis $\mathcal{C}$ from $\mathcal{S}_0$ using the method described in Example 4.32. ■

Extending Linearly Independent Sets

Bases are optimal subsets of vector spaces. They are large enough to span a space and yet small enough to be linearly independent. Theorem 4.13 explains how to contract a spanning set to a minimal spanning set. In the next theorem we describe a procedure for extending a linearly independent set to a maximally linearly independent set. We begin with an example.

EXAMPLE 4.33 Extending a Linearly Independent Set to a Basis

Use *Maple* to extend the linearly independent set $\mathcal{S} = \{\mathbf{v}_1 = (1, 0, 3), \mathbf{v}_2 = (0, 0, 1)\}$ to a basis for $\mathbb{R}^3$.

Solution. The idea behind the construction is to take a basis $\mathcal{B} = \{\mathbf{w}_1, \mathbf{w}_2, \mathbf{w}_3\}$ for $\mathbb{R}^3$ and append one of the vectors from $\mathcal{B}$ to $\mathcal{S}$. Since $\mathcal{B}$ is a basis and therefore spans $\mathbb{R}^3$, and since $\mathcal{S}$ is linearly independent, there must be a vector in $\mathcal{S}$ that is linearly independent of $\mathbf{v}_1$ and $\mathbf{v}_2$.

Let $\mathcal{B}$ be the standard basis $\{\mathbf{w}_1 = (1, 0, 0), \mathbf{w}_2 = (0, 1, 0), \mathbf{w}_3 = (0, 0, 1)\}$ for $\mathbb{R}^3$. We test for linear independence as in Example 4.31. The *Maple* dialog

```
[>matrix([[1,0,3],[0,0,1],[1,0,0]]);
[>linalg[inverse](%);
Error, (in inverse) singular matrix
```

tells us that $\mathbf{w}_1$ is linearly dependent on $\mathbf{v}_1$ and $\mathbf{v}_2$. We therefore discard it.

On the other hand, the calculation

```
[>A:=matrix([[1,0,3],[0,0,1],[0,1,0]]);
[>B:=linalg[inverse](%);
```

$$B := \begin{bmatrix} 1 & -3 & 0 \\ 0 & 0 & 1 \\ 0 & 1 & 0 \end{bmatrix}$$

produces the inverse B of the matrix A. The vector $\mathbf{w}_2$ is therefore linearly independent of $\mathbf{v}_1$ and $\mathbf{v}_2$. Hence the set $\mathcal{C} = \{\mathbf{v}_1, \mathbf{v}_2, \mathbf{w}_2\}$ is a basis for $\mathbb{R}^3$. ▲

We adapt the idea used in the proof of Theorem 4.13 to extend a linearly independent set of vectors to a basis.

THEOREM 4.14 (Basis extension theorem) *Any linearly independent set of vectors $\mathcal{S}$ of a finite-dimensional vector space V can be extended to a basis for V.*

Proof. Suppose that $\dim V = n$ and that $\mathcal{S} = \{\mathbf{w}_1, \ldots, \mathbf{w}_p\} \subseteq V$ is linearly independent, with $p \leq n$. If $p = n$, then $\mathcal{S}$ spans V and is therefore a basis for V. If $p < n$, then $\mathcal{S}$ does not span V. In that case, we take an arbitrary basis $\mathcal{B} = \{\mathbf{v}_1, \ldots, \mathbf{v}_n\}$ and form the list of vectors

$$\mathbf{w}_1, \ldots, \mathbf{w}_p, \mathbf{v}_1, \ldots, \mathbf{v}_n$$

Since $\mathcal{B}$ is a basis, the list $\mathcal{S}' = \{\mathbf{w}_1, \ldots, \mathbf{w}_p, \mathbf{v}_1, \ldots, \mathbf{v}_n\}$ spans V. By the construction in the proof of Theorem 4.13, there exists a linearly independent list

$$\mathcal{S}'' = \{\mathbf{w}_1, \ldots, \mathbf{w}_p, \mathbf{v}_1, \ldots\}$$

of $\mathcal{S}'$ that also spans V. The list $\mathcal{S}''$ is the required basis for V. ■

EXERCISES 4.10

1. Extend the set $\mathcal{S} = \{(1, 2)\}$ to a basis for $\mathbb{R}^2$.
2. Extend the set $\mathcal{S} = \{(1, 2, 3)\}$ to a basis for $\mathbb{R}^3$.
3. Extend the set $\mathcal{S} = \{(1, 2, 3, 4), (5, 6, 7, 8)\}$ to a basis for $\mathbb{R}^4$.
4. Select a linearly independent subset $\mathcal{S}_0$ from the set
$$\mathcal{S} = \{(1, 2, 3), (2, 4, 6), (-1, -2, -3)\}$$
and extend $\mathcal{S}_0$ to a basis for $\mathbb{R}^3$.
5. Select a linearly independent subset $\mathcal{S}_0$ from the set
$$\mathcal{S} = \{(1, 2, 3), (4, 5, 6), (-10, -11, -12)\}$$
and extend $\mathcal{S}_0$ to a basis for $\mathbb{R}^3$.
6. Contract the set $\mathcal{S} = \{(1, 2), (8, 16), (2, 1)\}$ to a basis for $\mathbb{R}^2$.
7. Contract the set $\mathcal{S} = \{(1, 2, 3), (1, 0, 1), (-8, 5, 7), (4, 4, 4)\}$ to a basis for $\mathbb{R}^3$.
8. Extend the set $\mathcal{S} = \{t, t^2\}$ to a basis for $\mathbb{R}_2[t]$.
9. Extend the set $\mathcal{S} = \{1, t^2\}$ to a basis for $\mathbb{R}_4[t]$.
10. Contract the set $\mathcal{S} = \{1, t, t^5, t^2, t + t^2, t^3, t^4\}$ to a basis for $\mathbb{R}_5[t]$.
11. Contract the set $\mathcal{S} = \{1, t, t^5, 3t^2, t + t^2, t^3, 8t^4\}$ to a basis for $\mathbb{R}_5[t]$.

12. Show that the set $\mathcal{B} = \{(1, 0, 1), (0, 2, 0), (0, 0, 5)\}$ is a basis for $\mathbb{R}^3$.

13. Consider the subsets $\mathcal{S}_1 = \{(1, 2, 3), (2, 2, 2)\}$ and $\mathcal{S}_2 = \{(3, 6, 9), (2, 4, 6), (1, 0, 0)\}$ of $\mathbb{R}^3$. The linearly independent set $\mathcal{S}_1$ is too small to span $\mathbb{R}^3$, and the set $\mathcal{S}_2$ is not linearly independent. Find a linearly independent subset of $\mathcal{S}_1 \cup \mathcal{S}_2$ that spans $\mathbb{R}^3$.

14. Use Theorem 4.9 to prove that if V is a vector space spanned by n vectors, then any subset of V with more than n elements is linearly dependent.

15. Extend the set

$$\mathcal{S} = \left\{ \begin{bmatrix} 1 & 0 \\ 1 & 2 \end{bmatrix} \right\}$$

to a basis for the space $\mathbb{R}^{2\times 2}$ of real 2×2 matrices.

16. Extend the set

$$\mathcal{S} = \left\{ \begin{bmatrix} 1 & 2 \\ 3 & 4 \end{bmatrix}, \begin{bmatrix} 0 & 0 \\ 1 & 0 \end{bmatrix}, \begin{bmatrix} 1 & 1 \\ 0 & 0 \end{bmatrix} \right\}$$

to a basis for $\mathbb{R}^{2\times 2}$.

17. (*Maple V*) Verify that the set of pivot columns of the matrix

$$A := \begin{bmatrix} 1 & 4 & 1 & 2 & 0 \\ 2 & 4 & 0 & 4 & 1 \\ 1 & 4 & 1 & 2 & 1 \end{bmatrix}$$

is a basis for $\mathbb{R}^3$.

18. Let $\mathcal{S}$ be the set

$$\left\{ \begin{bmatrix} 1 & 0 \\ 1 & 2 \end{bmatrix}, \begin{bmatrix} 0 & 0 \\ 0 & 0 \end{bmatrix}, \begin{bmatrix} 1 & 0 \\ 0 & 1 \end{bmatrix}, \begin{bmatrix} 0 & 0 \\ 1 & 0 \end{bmatrix}, \begin{bmatrix} 0 & 1 \\ 0 & 0 \end{bmatrix}, \begin{bmatrix} 1 & 0 \\ 0 & 0 \end{bmatrix} \right\}$$

Verify that the dimension of the span of $\mathcal{S}$ is 4.

19. (*Maple V*) Show that the span of the set

$$\mathcal{S} = \{(1, 2, 1), (4, 4, 4), (5, 6, 5), (6, 8, 6), (10, 12, 10)\}$$

is a vector space of dimension 2.

20. (*Maple 6*) Generate a random 4×2 matrix A of rank 2. Use a columns of a random invertible 4×4 matrix B to extend the set of columns of A to a 4×3 matrix A_1 of rank 3, and extend A_1 to a 4×4 matrix A_2 of rank 4. Explain why this construction always works and why it illustrates the proof of Theorem 4.14.

Bases and Invertible Matrices

The coordinate spaces $\mathbb{R}^n$ have an abundant supply of bases determined by invertible matrices.

THEOREM 4.15 (Bases and invertible matrix theorem) *Every invertible real $n \times n$ matrix determines a basis for $\mathbb{R}^n$.*

Proof. Let $A = [\mathbf{x}_1 \ \cdots \ \mathbf{x}_n]$ be an invertible real $n \times n$ matrix with columns $\mathbf{x}_i$. Then Corollary 2.20 tells us that A has n pivots. By Corollary 4.4, the set of column vectors $\mathcal{B} = \{\mathbf{x}_1, \ldots, \mathbf{x}_n\}$ is linearly independent. Since the dimension of $\mathbb{R}^n$ is n, it is clear that $\mathcal{B}$ also spans $\mathbb{R}^n$. ■

EXAMPLE 4.34 A Basis Built from an Invertible Matrix

Use *Maple* and an invertible matrix to construct a basis for $\mathbb{R}^4$.

Solution. We use the **RandomMatrix** function to generate a random invertible matrix by generating a lower-triangular matrix whose determinant is nonzero.

```
[> with(LinearAlgebra):
[> A:=RandomMatrix(4,4,generator=1..9,
     outputoptions=[shape=triangular[lower]]);
```

$$A := \begin{bmatrix} 1 & 0 & 0 & 0 \\ 2 & 7 & 0 & 0 \\ 8 & 2 & 7 & 0 \\ 4 & 5 & 8 & 5 \end{bmatrix}$$

Since the matrix A is invertible, the set of four columns

$$\mathcal{B} = \left\{ \begin{bmatrix} 1 \\ 2 \\ 8 \\ 4 \end{bmatrix}, \begin{bmatrix} 0 \\ 7 \\ 2 \\ 5 \end{bmatrix}, \begin{bmatrix} 0 \\ 0 \\ 7 \\ 8 \end{bmatrix}, \begin{bmatrix} 0 \\ 0 \\ 0 \\ 5 \end{bmatrix} \right\}$$

of A is linearly independent and therefore forms a basis for $\mathbb{R}^4$. ▲

EXERCISES 4.11

1. (*Maple V*) Use Theorem 4.3 to show that the columns of the matrix

$$A = \begin{bmatrix} -8 & -5 & 7 & 7 & -9 \\ -5 & 4 & -9 & -8 & 4 \\ 7 & -9 & 5 & 3 & 0 \\ 7 & -8 & 3 & 5 & -1 \\ -9 & 4 & 0 & -1 & -5 \end{bmatrix}$$

form a basis for $\mathbb{R}^5$.

2. (*Maple V*) Verify that the matrix

$$A = \begin{bmatrix} 19 & 15 & 21 & 28 \\ 0 & 34 & 3 & 35 \\ 0 & 0 & 8 & 9 \\ 0 & 0 & 0 & 35 \end{bmatrix}$$

is invertible by showing that the columns of A form a basis for $\mathbb{R}^4$.

3. (*Maple 6*) Restart *Maple* and use the **RandomMatrix** function in the **LinearAlgebra** package to generate an invertible 4×4 real matrix A. Show that the columns of A and the columns of A^{-1} are bases for $\mathbb{R}^4$.

Coordinate Vectors

To link linear algebra with matrix algebra, we must find a way of representing vectors numerically. This can be done by choosing a basis and writing each vector as a linear combination of basis vectors. The coefficients arising in this way are the required scalars.

DEFINITION 4.8 *The **coordinate vector** of a vector* $\mathbf{x}$ *in an* n*-dimensional real vector space* V *relative to a basis* $\mathcal{B} = \{\mathbf{v}_1, \ldots, \mathbf{v}_n\}$ *for* V *is the unique column vector*

$$[\mathbf{x}]_{\mathcal{B}} = \begin{bmatrix} a_1 \\ \vdots \\ a_n \end{bmatrix} \in \mathbb{R}^n$$

for which $\mathbf{x} = a_1\mathbf{v}_1 + \cdots + a_n\mathbf{v}_n$.

TABLE 4 FINDING COORDINATE VECTORS.

Step 1. Take a basis $\mathcal{B} = \{\mathbf{v}_1, \ldots, \mathbf{v}_n\}$ for a vector space V.

Step 2. Write the vector $\mathbf{x} \in V$ as a linear combination $a_1\mathbf{v}_1 + \cdots + a_n\mathbf{v}_n$ of the basis vectors.

Step 3. Use the scalars $a_1, \ldots, a_n$ in the linear combination to form the vector $\begin{bmatrix} a_1 \\ \vdots \\ a_n \end{bmatrix} \in \mathbb{R}^n$.

Step 4. The column vector constructed in Steps 2–3 is the coordinate vector $[\mathbf{x}]_{\mathcal{B}}$.

Table 4 describes the steps involved in the coordinate vector construction.

Since the coordinate vector $[\mathbf{x}]_{\mathcal{B}}$ of a vector $\mathbf{x}$ in a given basis is unique, we can recover $\mathbf{x}$ from $[\mathbf{x}]_{\mathcal{B}}$ in an easy way. Each column vector

$$\mathbf{y} = \begin{bmatrix} b_1 \\ \vdots \\ b_n \end{bmatrix} \in \mathbb{R}^n$$

determines a unique linear combination $\widehat{\mathbf{y}}_{\mathcal{B}} = b_1\mathbf{v}_1 + \cdots + b_n\mathbf{v}_n \in V$. Figure 2 illustrates the link between coordinate vectors and linear combinations in $\mathbb{R}^2$.

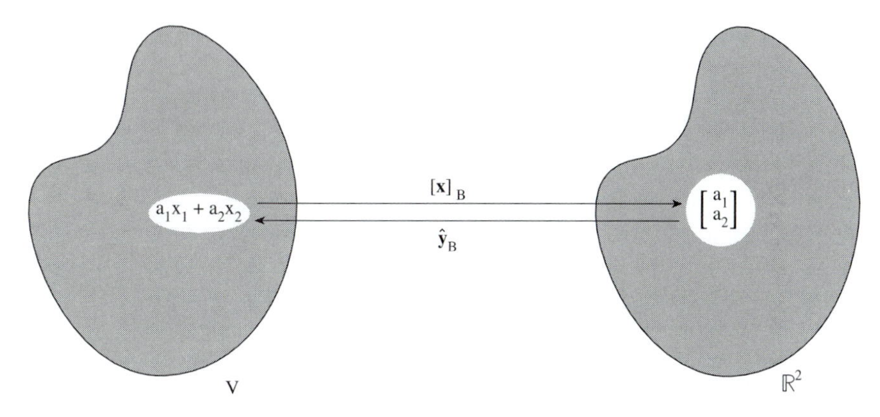

FIGURE 2 Coordinate Vector and Linear Combination.

The function $\mathbf{x} \to [\mathbf{x}]_{\mathcal{B}}$ that assigns to each vector $\mathbf{x} \in V$ the coordinate vector $[\mathbf{x}]_{\mathcal{B}} \in \mathbb{R}^n$ is the ***coordinate-vector function*** from V to $\mathbb{R}^n$ determined by the basis $\mathcal{B}$ for V. Conversely, the function $\mathbf{y} \to \widehat{\mathbf{y}}_{\mathcal{B}}$ that assigns to each column vector $\mathbf{y} \in \mathbb{R}^n$ the linear combination $\widehat{\mathbf{y}}_{\mathcal{B}} \in V$ is the ***linear-combination function*** from $\mathbb{R}^n$ to V determined by the basis $\mathcal{B}$ for V.

THEOREM 4.16 (Coordinate function theorem) *The functions $\mathbf{x} \to [\mathbf{x}]_{\mathcal{B}}$ and $\mathbf{y} \to \widehat{\mathbf{y}}_{\mathcal{B}}$ are inverses of each other.*

Proof. The calculations required to prove this theorem are easy. Let

$$\mathbf{x} = a_1\mathbf{v}_1 + \cdots + a_n\mathbf{v}_n \quad \text{and} \quad \mathbf{y} = \begin{bmatrix} a_1 \\ \vdots \\ a_n \end{bmatrix}$$

Then

$$a_1\mathbf{v}_1 + \cdots + a_n\mathbf{v}_n \to [\mathbf{x}]_{\mathcal{B}} = \begin{bmatrix} a_1 \\ \vdots \\ a_n \end{bmatrix} = \mathbf{y} \to \widehat{\mathbf{y}}_{\mathcal{B}} = a_1\mathbf{v}_1 + \cdots + a_n\mathbf{v}_n$$

and

$$\begin{bmatrix} a_1 \\ \vdots \\ a_n \end{bmatrix} \to \widehat{\mathbf{y}}_{\mathcal{B}} = a_1\mathbf{v}_1 + \cdots + a_n\mathbf{v}_n \to [a_1\mathbf{v}_1 + \cdots + a_n\mathbf{v}_n]_{\mathcal{B}} = \begin{bmatrix} a_1 \\ \vdots \\ a_n \end{bmatrix}$$

Therefore, the coordinate-vector functions are inverses of each other. ■

The coordinate-vector and linear-combination functions are useful because they are inverses of each other and because they preserve vector addition and scalar multiplication in the sense of the following theorem.

THEOREM 4.17 *The coordinate-vector function $\mathbf{x} \to [\mathbf{x}]_{\mathcal{B}}$ from V to $\mathbb{R}^n$ has the property that $[\mathbf{x}+\mathbf{y}]_{\mathcal{B}} = [\mathbf{x}]_{\mathcal{B}} + [\mathbf{y}]_{\mathcal{B}}$ and that $[a\mathbf{x}]_{\mathcal{B}} = a[\mathbf{x}]_{\mathcal{B}}$.*

Proof. Let $\mathbf{x} = a_1\mathbf{v}_1 + \cdots + a_n\mathbf{v}_n$ and $\mathbf{y} = b_1\mathbf{v}_1 + \cdots + b_n\mathbf{v}_n$ be two vectors in V and let

$$\mathbf{x}+\mathbf{y} = (a_1+b_1)\mathbf{v}_1 + \cdots + (a_n+b_n)\mathbf{v}_n$$

be their sum. Then

$$[\mathbf{x}+\mathbf{y}]_{\mathcal{B}} = \begin{bmatrix} a_1+b_1 \\ \vdots \\ a_n+b_n \end{bmatrix} = \begin{bmatrix} a_1 \\ \vdots \\ a_n \end{bmatrix} + \begin{bmatrix} b_1 \\ \vdots \\ b_n \end{bmatrix} = [\mathbf{x}]_{\mathcal{B}} + [\mathbf{y}]_{\mathcal{B}}$$

Moreover, for any scalar a,

$$a\mathbf{x} = a(a_1\mathbf{v}_1 + \cdots + a_n\mathbf{v}_n) = (aa_1)\mathbf{v}_1 + \cdots + (aa_n)\mathbf{v}_n$$

so

$$[a\mathbf{x}]_{\mathcal{B}} = \begin{bmatrix} aa_1 \\ \vdots \\ aa_n \end{bmatrix} = a\begin{bmatrix} a_1 \\ \vdots \\ a_n \end{bmatrix} = a[\mathbf{x}]_{\mathcal{B}}$$

This proves the theorem. ■

COROLLARY 4.18 *The linear-combination function $\mathbf{y} \to \widehat{\mathbf{y}}_{\mathcal{B}}$ from $\mathbb{R}^n$ to V has the properties that $\widehat{\mathbf{x}+\mathbf{y}}_{\mathcal{B}} = \widehat{\mathbf{x}}_{\mathcal{B}} + \widehat{\mathbf{y}}_{\mathcal{B}}$ and that $\widehat{a\mathbf{x}}_{\mathcal{B}} = a\widehat{\mathbf{x}}_{\mathcal{B}}$.*

Proof. By reversing the conversion steps in the proof of Theorem 4.17, we obtain the stated equalities. ■

A coordinate-vector function is an example of a ***linear transformation.*** Its key property is the fact that it "preserves" linear combinations of vectors. The obvious question to ask is what connection exists between the coordinate vectors $[\mathbf{x}]_{\mathcal{B}}$ and $[\mathbf{x}]_{\mathcal{C}}$ of a given vector $\mathbf{x}$ relative to two different bases $\mathcal{B}$ and $\mathcal{C}$ for V. The answer turns out to be quite simple and uses the idea of coordinate conversion. Table 5 lists the components required for the construction of a coordinate conversion matrix.

TABLE 5 FINDING A COORDINATE CONVERSION MATRIX.

Step 1. Take two bases $\mathcal{B} = \{\mathbf{v}_1, \ldots, \mathbf{v}_n\}$ and $\mathcal{C} = \{\mathbf{w}_1, \ldots, \mathbf{w}_n\}$ for $\mathbb{R}^n$.

Step 2. Write the vectors $\mathbf{v}_i$ as linear combinations $a_{1i}\mathbf{w}_1 + \cdots + a_{ni}\mathbf{w}_n$ in $\mathcal{C}$.

Step 3. Form the coordinate vectors $[\mathbf{v}_i]_{\mathcal{C}} = \begin{bmatrix} a_{1i} \\ \vdots \\ a_{ni} \end{bmatrix}$ in $\mathbb{R}^n$.

Step 4. Use the coordinate vectors $[\mathbf{v}_i]_{\mathcal{C}}$ as the columns of a matrix $P = [[\mathbf{v}_1]_{\mathcal{C}} \ \cdots \ [\mathbf{v}_n]_{\mathcal{C}}]$.

Step 5. The matrix P is the coordinate conversion matrix from $\mathcal{B}$ to $\mathcal{C}$.

Step 6. For each vector $\mathbf{x}$ and $\mathbf{y}$ in $\mathbb{R}^n$, it holds that $P\,[\mathbf{x}]_{\mathcal{B}} = [\mathbf{x}]_{\mathcal{C}}$ and $P^{-1}\,[\mathbf{y}]_{\mathcal{C}} = [\mathbf{y}]_{\mathcal{B}}$.

THEOREM 4.19 (Coordinate conversion theorem) *Let V be an n-dimensional real vector space, let $\mathcal{B}$ and $\mathcal{C}$ be two bases for V, and let $\mathbf{x}$ be a vector in V with coordinate vectors $[\mathbf{x}]_{\mathcal{B}}$ and $[\mathbf{x}]_{\mathcal{C}}$. Then there exists an invertible matrix P for which $P[\mathbf{x}]_{\mathcal{B}} = [\mathbf{x}]_{\mathcal{C}}$ and $P^{-1}[\mathbf{x}]_{\mathcal{C}} = [\mathbf{x}]_{\mathcal{B}}$.*

Proof. Let $\mathcal{B} = \{\mathbf{v}_1, \ldots, \mathbf{v}_n\}$ and $\mathcal{C} = \{\mathbf{w}_1, \ldots, \mathbf{w}_n\}$, and suppose that $\mathbf{x} = a_1\mathbf{v}_1 + \cdots + a_n\mathbf{v}_n$ is any vector in V, written as a linear combination in the basis $\mathcal{B}$. From the fact that the coordinate-vector function $\mathbf{x} \to [\mathbf{x}]_{\mathcal{C}}$ preserves vector addition and scalar multiplication, it follows that

$$
\begin{aligned}
[\mathbf{x}]_{\mathcal{C}} &= [a_1\mathbf{v}_1 + \cdots + a_n\mathbf{v}_n]_{\mathcal{C}} = a_1\,[\mathbf{v}_1]_{\mathcal{C}} + \cdots + a_n\,[\mathbf{v}_n]_{\mathcal{C}} \\
&= \begin{bmatrix} [\mathbf{v}_1]_{\mathcal{C}} & \cdots & [\mathbf{v}_n]_{\mathcal{C}} \end{bmatrix} \begin{bmatrix} a_1 \\ \vdots \\ a_n \end{bmatrix} = \begin{bmatrix} [\mathbf{v}_1]_{\mathcal{C}} & \cdots & [\mathbf{v}_n]_{\mathcal{C}} \end{bmatrix} [\mathbf{x}]_{\mathcal{B}} \\
&= P\,[\mathbf{x}]_{\mathcal{B}}
\end{aligned}
$$

Conversely, let $\mathbf{y} = b_1\mathbf{w}_1 + \cdots + b_n\mathbf{w}_n$ be any vector in V, written as a linear combination in the basis $\mathcal{C}$. From the fact that the coordinate-vector function $\mathbf{y} \to [\mathbf{y}]_{\mathcal{B}}$ preserves vector addition and scalar multiplication, it follows that

$$[\mathbf{y}]_{\mathcal{B}} = [b_1\mathbf{w}_1 + \cdots + b_n\mathbf{w}_n]_{\mathcal{B}} = b_1 [\mathbf{w}_1]_{\mathcal{B}} + \cdots + b_n [\mathbf{w}_n]_{\mathcal{B}}$$

$$= \begin{bmatrix} [\mathbf{w}_1]_{\mathcal{B}} & \cdots & [\mathbf{w}_n]_{\mathcal{B}} \end{bmatrix} \begin{bmatrix} b_1 \\ \vdots \\ b_n \end{bmatrix} = \begin{bmatrix} [\mathbf{w}_1]_{\mathcal{B}} & \cdots & [\mathbf{w}_n]_{\mathcal{B}} \end{bmatrix} [\mathbf{y}]_{\mathcal{C}}$$

$$= Q\,[\mathbf{y}]_{\mathcal{C}}$$

This means that $Q = P^{-1}$. ■

DEFINITION 4.9 *The $n \times n$ matrices P and P^{-1} with the property that $P[\mathbf{x}]_{\mathcal{B}} = [\mathbf{x}]_{\mathcal{C}}$ and $P^{-1}[\mathbf{x}]_{\mathcal{C}} = [\mathbf{x}]_{\mathcal{B}}$ are the* ***coordinate conversion matrices*** *from $\mathcal{B}$ to $\mathcal{C}$ and from $\mathcal{C}$ to $\mathcal{B}$, respectively.*

The next theorem explains why every coordinate space $\mathbb{R}^n$ has an unlimited supply of such matrices.

THEOREM 4.20 *Every invertible real $n \times n$ matrix is a coordinate conversion matrix for the coordinate space $\mathbb{R}^n$.*

Proof. Let $A = [\mathbf{x}_1 \ \cdots \ \mathbf{x}_n]$ be in invertible real $n \times n$ matrix with columns $\mathbf{x}_i$ and let $\mathbf{x}$ be any vector in $\mathbb{R}^n$. By Theorem 4.15, the associated set of vectors $\mathcal{B} = \{\mathbf{x}_1, \ldots, \mathbf{x}_n\}$ is a basis for $\mathbb{R}^n$. Therefore, there exist unique scalars $a_1, \ldots, a_n$ for which $\mathbf{x} = a_1\mathbf{x}_1 + \cdots + a_n\mathbf{x}_n$. This means that

$$\mathbf{x} = \begin{bmatrix} \mathbf{x}_1 & \cdots & \mathbf{x}_n \end{bmatrix} \begin{bmatrix} a_1 \\ \vdots \\ a_n \end{bmatrix} = A\,[\mathbf{x}]_{\mathcal{B}}$$

Since A is invertible, $[\mathbf{x}]_{\mathcal{B}} = A^{-1}\mathbf{x}$. Therefore, A is a coordinate conversion matrix from $\mathcal{B}$ to the standard basis for $\mathbb{R}^n$. ■

EXAMPLE 4.35 A Coordinate Conversion in $\mathbb{R}^2$

Discuss the coordinate conversion on $\mathbb{R}^2$ determined by the invertible matrix

$$A = \begin{bmatrix} 2 & 1 \\ 1 & 3 \end{bmatrix}$$

Solution. Since the determinant of A is not zero, A is an invertible matrix. By Theorem 4.20, the matrix A therefore determines a coordinate conversion from the basis $\mathcal{B} = \{\mathbf{u} = (2, 1), \mathbf{v} = (1, 3)\}$ to the standard basis $\mathcal{E} = \{\mathbf{x} = (1, 0), \mathbf{y} = (0, 1)\}$ of $\mathbb{R}^2$.

Figure 3 illustrates the link between the coordinate vectors

$$\begin{bmatrix} 3 \\ 4 \end{bmatrix}_{\mathcal{E}} \quad \text{and} \quad \begin{bmatrix} 1 \\ 1 \end{bmatrix}_{\mathcal{B}}$$

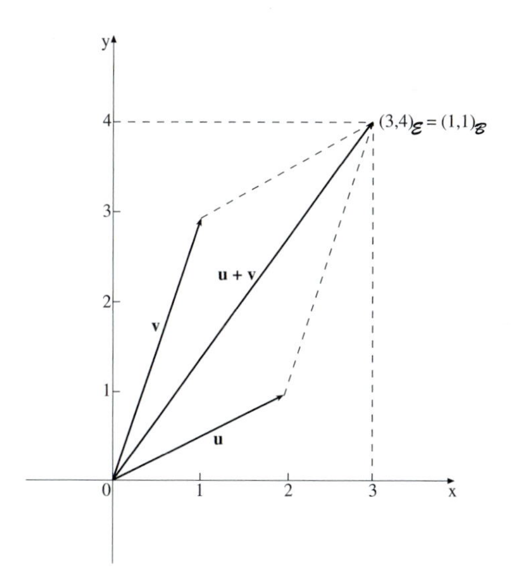

FIGURE 3 A Coordinate Conversion.

created by the matrix A. We can see geometrically that

$$\begin{bmatrix} 1 \\ 1 \end{bmatrix}_{\mathcal{B}} = \mathbf{u} + \mathbf{v} = \begin{bmatrix} 2 \\ 1 \end{bmatrix}_{\mathcal{E}} + \begin{bmatrix} 1 \\ 3 \end{bmatrix}_{\mathcal{E}} = \begin{bmatrix} 3 \\ 4 \end{bmatrix}_{\mathcal{E}}$$

The matrix equation

$$\begin{bmatrix} 2 & 1 \\ 1 & 3 \end{bmatrix} \begin{bmatrix} 1 \\ 1 \end{bmatrix} = \begin{bmatrix} 3 \\ 4 \end{bmatrix}$$

confirms algebraically that this relationship holds. ▲

EXAMPLE 4.36 A Coordinate Conversion in $\mathbb{R}^3$

Find the coordinate conversion matrices from $\mathcal{B}$ to $\mathcal{C}$ and $\mathcal{C}$ to $\mathcal{B}$, where

$$\mathcal{B} = \{\mathbf{v}_1 = (1, 2, 3)\,,\ \mathbf{v}_2 = (0, 1, 0)\,,\ \mathbf{v}_3 = (1, 0, 1)\}$$

and

$$\mathcal{C} = \{\mathbf{w}_1 = (0, 1, 1)\,,\ \mathbf{w}_2 = (1, 1, 0)\,,\ \mathbf{w}_3 = (1, 0, 1)\}$$

are two bases for $\mathbb{R}^3$.

Solution. Let P be the coordinate conversion matrix from $\mathcal{B}$ to $\mathcal{C}$. Then

$$P = \begin{bmatrix} [\mathbf{v}_1]_{\mathcal{C}} & [\mathbf{v}_2]_{\mathcal{C}} & [\mathbf{v}_3]_{\mathcal{C}} \end{bmatrix}$$

We therefore need to calculate the coordinate vectors $[\mathbf{v}_1]_C$, $[\mathbf{v}_2]_C$, and $[\mathbf{v}_3]_C$. Let

$$[\mathbf{v}_1]_C = \begin{bmatrix} 1 \\ 2 \\ 3 \end{bmatrix}_C = \begin{bmatrix} a_{11} \\ a_{12} \\ a_{13} \end{bmatrix}$$

Then

$$\begin{bmatrix} 1 \\ 2 \\ 3 \end{bmatrix} = a_{11} \begin{bmatrix} 0 \\ 1 \\ 1 \end{bmatrix} + a_{12} \begin{bmatrix} 1 \\ 1 \\ 0 \end{bmatrix} + a_{13} \begin{bmatrix} 1 \\ 0 \\ 1 \end{bmatrix} = \begin{bmatrix} a_{12} + a_{13} \\ a_{11} + a_{12} \\ a_{11} + a_{13} \end{bmatrix}$$

We use *Maple* to solve the associated linear system

$$\{1 = a_{12} + a_{13}, 2 = a_{11} + a_{12}, 3 = a_{11} + a_{13}\}$$

to find the coordinates a_{11}, a_{12}, and a_{13}.

```
[> solve({a12+a13=1,a11+a12=2,a11+a13=3},{a11,a12,a13});
```

$$\{a13 = 1, a11 = 2, a12 = 0\}$$

Therefore,

$$[\mathbf{v}_1]_C = \begin{bmatrix} 1 \\ 2 \\ 3 \end{bmatrix}_C = \begin{bmatrix} a_{11} \\ a_{12} \\ a_{13} \end{bmatrix} = \begin{bmatrix} 2 \\ 0 \\ 1 \end{bmatrix}$$

Similar calculations yield

$$[\mathbf{v}_2]_C = \begin{bmatrix} 0 \\ 1 \\ 0 \end{bmatrix}_C = \begin{bmatrix} a_{21} \\ a_{22} \\ a_{23} \end{bmatrix} = \begin{bmatrix} \frac{1}{2} \\ \frac{1}{2} \\ -\frac{1}{2} \end{bmatrix}, \quad [\mathbf{v}_3]_C = \begin{bmatrix} 1 \\ 0 \\ 1 \end{bmatrix}_C = \begin{bmatrix} a_{31} \\ a_{32} \\ a_{33} \end{bmatrix} = \begin{bmatrix} 0 \\ 0 \\ 1 \end{bmatrix}$$

where

$$\begin{bmatrix} 0 \\ 1 \\ 0 \end{bmatrix} = a_{21} \begin{bmatrix} 0 \\ 1 \\ 1 \end{bmatrix} + a_{22} \begin{bmatrix} 1 \\ 1 \\ 0 \end{bmatrix} + a_{23} \begin{bmatrix} 1 \\ 0 \\ 1 \end{bmatrix}$$

and

$$\begin{bmatrix} 1 \\ 0 \\ 1 \end{bmatrix} = a_{31} \begin{bmatrix} 0 \\ 1 \\ 1 \end{bmatrix} + a_{32} \begin{bmatrix} 1 \\ 1 \\ 0 \end{bmatrix} + a_{33} \begin{bmatrix} 1 \\ 0 \\ 1 \end{bmatrix}$$

By concatenating the column vectors $[\mathbf{v}_1]_\mathcal{C}$, $[\mathbf{v}_2]_\mathcal{C}$, and$[\mathbf{v}_3]_\mathcal{C}$, we get

$$P = \begin{bmatrix} [\mathbf{v}_1]_\mathcal{C} & [\mathbf{v}_2]_\mathcal{C} & [\mathbf{v}_3]_\mathcal{C} \end{bmatrix} = \begin{bmatrix} 2 & \frac{1}{2} & 0 \\ 0 & \frac{1}{2} & 0 \\ 1 & -\frac{1}{2} & 1 \end{bmatrix}$$

We conclude by calculating the coordinate conversion matrix $Q = P^{-1}$ from $\mathcal{C}$ to $\mathcal{B}$.

```
[> Q:=linalg[inverse](matrix([[2,1/2,0], [0,1/2,0],
   [1,-1/2,1]]));
```

$$Q := \begin{bmatrix} \frac{1}{2} & -\frac{1}{2} & 0 \\ 0 & 2 & 0 \\ -\frac{1}{2} & \frac{3}{2} & 1 \end{bmatrix}$$

This completes the calculation. ▲

EXERCISES 4.12

1. Find the coordinate vectors in $\mathbb{R}^4$ of the polynomials $3 + 7t - t^2$, $4 + t^3$, $21t^2$, $1 + t + t^2$, and $3t^3 + 2t$ in $\mathbb{R}_3[t]$ in the standard bases.

2. Let E be the standard basis for $\mathbb{R}^3$ and let $\mathcal{B} = \{(0, 0, 1), (0, 1, 1), (1, 1, 1)\}$ be a second basis.

   a. Find the coordinate conversion matrix that converts coordinate vectors in the basis $\mathcal{E}$ to coordinate vectors in the basis $\mathcal{B}$.

   b. Convert the coordinate vectors

   $$\begin{bmatrix} 1 \\ 2 \\ 3 \end{bmatrix}, \begin{bmatrix} -1 \\ 1 \\ -21 \end{bmatrix}, \begin{bmatrix} 2 \\ 2 \\ 2 \end{bmatrix}, \begin{bmatrix} 0 \\ 0 \\ 0 \end{bmatrix}, \begin{bmatrix} 3 \\ -7 \\ 1 \end{bmatrix}$$

   from vectors in the standard basis to vectors in the basis $\mathcal{B}$.

   c. Find the coordinate conversion matrix that converts coordinate vectors in the basis $\mathcal{B}$ to coordinate vectors in the basis $\mathcal{E}$.

   d. Convert the coordinate vectors

   $$\begin{bmatrix} 1 \\ 2 \\ 3 \end{bmatrix}, \begin{bmatrix} -1 \\ 1 \\ -21 \end{bmatrix}, \begin{bmatrix} 2 \\ 2 \\ 2 \end{bmatrix}, \begin{bmatrix} 0 \\ 0 \\ 0 \end{bmatrix}, \begin{bmatrix} 3 \\ -7 \\ 1 \end{bmatrix}$$

   from vectors in the basis $\mathcal{B}$ to vectors in the basis $\mathcal{E}$.

3. Find the basis $\mathcal{B}$ for $\mathbb{R}^3$ for which

$$\begin{bmatrix} 2 & 0 & 0 \\ 0 & 0 & 2 \\ 0 & 2 & 2 \end{bmatrix}$$

is the coordinate conversion matrix from the standard basis $\mathcal{E}$ to $\mathcal{B}$.

4. Explain why there is no coordinate conversion matrix from the standard basis for the polynomial space $\mathbb{R}[t]$ to any coordinate space $\mathbb{R}^n$.

5. Let $A = \left[a_{ij}\right]$ be an invertible 3×3 real matrix, let $\mathcal{B}$ be the basis for $\mathbb{R}^3$ determined by A, and let $\mathcal{E}$ be the standard basis of $\mathbb{R}^3$. Show that

$$\begin{bmatrix} a_{11} + a_{12} + a_{13} \\ a_{21} + a_{22} + a_{23} \\ a_{31} + a_{32} + a_{33} \end{bmatrix}_{\mathcal{E}} = \begin{bmatrix} 1 \\ 1 \\ 1 \end{bmatrix}$$

Explain this result geometrically.

6. (*Maple V*) Find the coordinate conversion matrix $\mathcal{P}$ from $\mathcal{B}$ to $\mathcal{C}$ and the coordinate conversion matrix $\mathcal{Q}$ from $\mathcal{C}$ to $\mathcal{B}$ for the bases $\mathcal{B} = \{\mathbf{v}_1 = (1, 2), \mathbf{v}_2 = (3, 4)\}$ and $\mathcal{C} = \{\mathbf{w}_1 = (1, 0), \mathbf{w}_2 = (0, 1)\}$ for $\mathbb{R}^2$.

7. (*Maple 6*) Use the standard bases to define a coordinate-vector function $[x]_{\mathcal{B}}$ from the space of real 2×3 matrices $\mathbb{R}^{2\times 3}$ to $\mathbb{R}^6$ and find the associated coordinate vectors of the following matrices.

a. $\begin{bmatrix} 4 & 1 & 5 \\ 4 & 3 & -4 \end{bmatrix}$ b. $\begin{bmatrix} 0 & 0 & 0 \\ 0 & 1 & 0 \end{bmatrix}$ c. $\begin{bmatrix} 1 & -1 & 1 \\ -1 & \frac{3}{5} & \pi \end{bmatrix}$

SUBSPACES

Certain subsets of a vector space are vector spaces in their own right, with the same operations as those of the ambient space. For example, the coordinate planes

$$\begin{cases} \mathcal{S}_{xy} = \{(x, y, 0) : x, y \in \mathbb{R}\} \\ \mathcal{S}_{xz} = \{(x, 0, z) : x, z \in \mathbb{R}\} \\ \mathcal{S}_{yz} = \{(0, y, z) : y, z \in \mathbb{R}\} \end{cases}$$

shown in Figure 4 are subspaces of $\mathbb{R}^3$.

The vector space operations of these spaces are those of $\mathbb{R}^3$, restricted to $\mathcal{S}_{xy}$, $\mathcal{S}_{xz}$, and $\mathcal{S}_{xy}$. The operations on $\mathcal{S}_{xy}$, for example, are

$$\begin{bmatrix} x_1 \\ y_1 \\ 0 \end{bmatrix} + \begin{bmatrix} x_2 \\ y_2 \\ 0 \end{bmatrix} = \begin{bmatrix} x_1 + x_2 \\ y_1 + y_2 \\ 0 \end{bmatrix} \quad \text{and} \quad a \begin{bmatrix} x \\ y \\ 0 \end{bmatrix} = \begin{bmatrix} ax \\ ay \\ 0 \end{bmatrix}$$

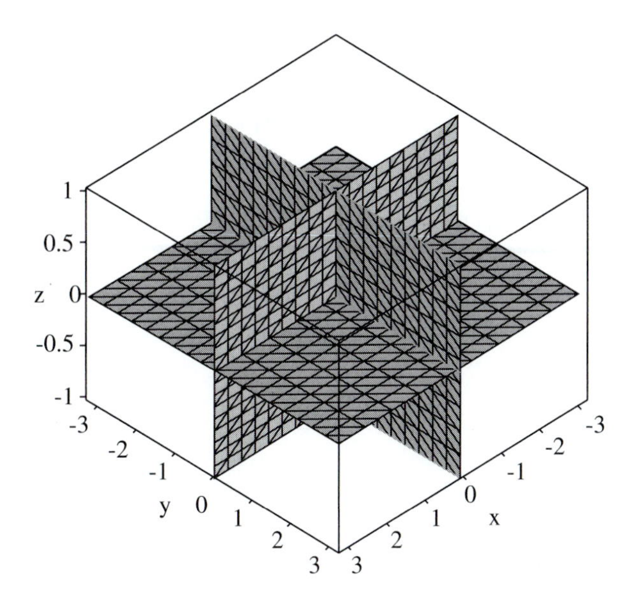

FIGURE 4 The Coordinate-Plane Subspaces of $\mathbb{R}^3$.

Since the set $\mathcal{S}_{xy}$ is a proper subset of $\mathbb{R}^3$, it does not contain all vectors of $\mathbb{R}^3$. Nevertheless, the sum of two vectors $\mathbf{x} + \mathbf{y}$ in $\mathcal{S}_{xy}$ is the same as the sum of these vectors in $\mathbb{R}^3$. Moreover, any scalar multiple $a\mathbf{x}$ of a vector $\mathbf{x}$ in $\mathcal{S}_{xy}$ is the same as the scalar multiple $a\mathbf{x}$ in $\mathbb{R}^3$. The set $\mathcal{S}_{xy}$ is an example of a *subspace* of $\mathbb{R}^3$.

DEFINITION 4.10 *A nonempty subset W of a vector space V is a* ***subspace*** *of V if for all vectors $\mathbf{x}$ and $\mathbf{y}$ in W and all scalars a and b, the linear combination $a\mathbf{x} + b\mathbf{y}$ belongs to W.*

We note that the fact that subspaces are nonempty implies that they contain the zero vector of V. If $\mathbf{x}$ is any vector in W, the linear combination $0\mathbf{x}+0\mathbf{x} = \mathbf{0}$ belongs to W. Furthermore, the vector $(-1)\,\mathbf{x}+0\mathbf{x} = -\mathbf{x}$ belongs to W for all $\mathbf{x} \in W$. Hence W contains the zero vector and is closed under inverses. Moreover, for any two vectors $\mathbf{x}$ and $\mathbf{y}$ in W, the linear combination $(1)\,\mathbf{x} + (1)\,\mathbf{y} = \mathbf{x} + \mathbf{y}$ is in W. Hence W is also closed under vector addition. Since the vectorspace operations of W are restrictions of the vector-space operations of V, the space W automatically satisfies the vector space axioms. The space W is therefore a vector space in its own right.

EXAMPLE 4.37 The Zero Subspace

For any vector space V, the space $\{\mathbf{0}\} \subseteq V$ is a subspace of V. ▲

EXAMPLE 4.38 The Ambient Space

It obvious from the definition that every vector space is a subspace of itself. ▲

Every subspace of a vector space V other than V itself is called a ***proper subspace***.

EXAMPLE 4.39 A Proper Subspace of $\mathbb{R}^2$

Use *Maple* to graph the subspace of $\mathbb{R}^2$ determined by the equation $y = 17x$.

Solution. We enter the expression $y = 17x$ and invoke the **plot** function.

```
[>plot(17*x,x=-10..10);
```

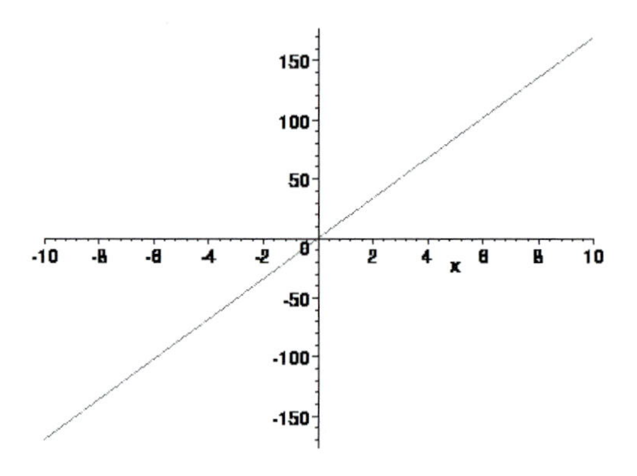

As we can see, the space W consists of a line through the origin of $\mathbb{R}^2$. ▲

Here is a general method for constructing subspaces.

THEOREM 4.21 (Subspace theorem) *The span of a subset W of a vector space V is a subspace of V.*

Proof. Suppose $W = \emptyset$. Then span W is the zero subspace of V. Now suppose $W \neq \emptyset$, that $\mathbf{x}$ and $\mathbf{y}$ belong to span W, and that a and b are two scalars. Since span W is closed under linear combinations, it follows that $0\mathbf{x} + 0\mathbf{y} = \mathbf{0} \in \text{span } W$ and $a\mathbf{x} + b\mathbf{y} \in \text{span } W$. Hence span W is a subspace of V. ■

Another method for constructing subspaces is to intersect them.

THEOREM 4.22 (Intersection theorem) *If U and W are subspaces of a vector space V, then so is $U \cap W$.*

Proof. First we observe that since U and W are subspaces, $\mathbf{0} \in U \cap W$. Suppose that in addition, $\mathbf{x}, \mathbf{y} \in U \cap W$. Then $\mathbf{x}, \mathbf{y} \in U$ and $\mathbf{x}, \mathbf{y} \in W$. Since U and W are subspaces, $\mathbf{x}+\mathbf{y} \in U$ and for any scalar a, $a\mathbf{x} \in U$. Similarly $\mathbf{x} + \mathbf{y} \in W$ and $a\mathbf{x} \in W$. Hence $\mathbf{x} + \mathbf{y} \in U \cap W$ and $a\mathbf{x} \in U \cap W$ for all scalars a. ■

If $U \cap W = \{\mathbf{0}\}$, we say that U and W are ***disjoint subspaces.*** For example, the space $U = \{(x, 0) : x \in \mathbb{R}\}$ and $W = \{(0, y) : y \in \mathbb{R}\}$ are disjoint subspaces of $\mathbb{R}^2$. Here is an example of the intersection of two subspaces of $\mathbb{R}^3$.

EXAMPLE 4.40 The Intersection of Two Subspaces

Use *Maple* to plot the two subspaces

$$U = \{(x, y, z) \in \mathbb{R}^3 : z = 3x + 17y\} \quad \text{and} \quad V = \{(x, y, z) \in \mathbb{R}^3 : z = x - y\}$$

of $\mathbb{R}^3$. Explain why U and V are not disjoint.

Solution. We use the **plot3d** function. The commands

a. `plot3d(3*x+17*y,x=-10..10,y=-10..10,axes=boxed)`

b. `plot3d(x-y,x=-10..10,y=-10..10,axes=boxed)`

c. `plot3d({3*x+17*y,x-y},x=-10..10,y=-10..10,axes=boxed)`

produce the graphs in Figures 5, 6, and 7.

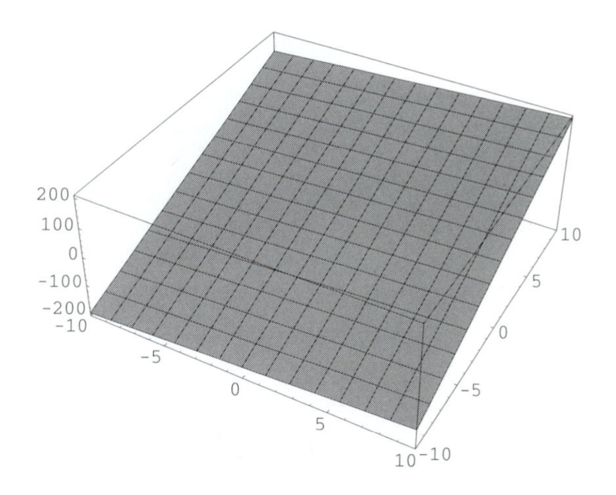

FIGURE 5 The Subspace U of $\mathbb{R}^3$.

The graph in Figure 5 illustrates the subspace of $\mathbb{R}^3$ determined by the equation $z = 3x + 17y$. The graph in Figures 6 and 7 on Page 328, on the other hand, depict, repectively, the subspace V of $\mathbb{R}^3$ determined by the equation $z = x - y$, and the subspace

$$L = U \cap V = \left\{(x, y, z) \in \mathbb{R}^3 : 3x + 17y = x - y\right\} = \left\{(x, -\tfrac{1}{9}x, z) : x, z \in \mathbb{R}\right\}$$

consisting of the line determined by the intersection of U and V.

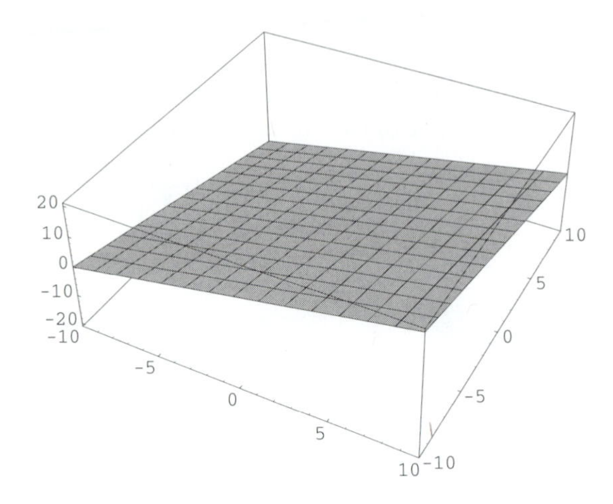

FIGURE 6 The Subspace V of $\mathbb{R}^3$.

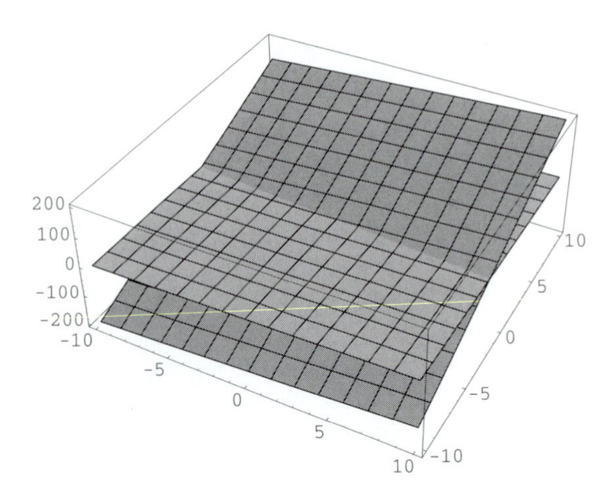

FIGURE 7 The Subspace $U \cap V$ of $\mathbb{R}^3$.

As we can see, the solutions of the linear system

$$\begin{cases} 3x + 17y - z = 0 \\ \quad x - y - z = 0 \end{cases}$$

form the line corresponding to the intersection of the spaces U and V. ▲

EXAMPLE 4.41 The Subspaces of $\mathbb{R}^3$

Since all subspaces of $\mathbb{R}^3$ must contain the zero vector, it is clear from geometry that the only possible subspaces are lines and planes passing through the origin **0**, together with the zero subspace and $\mathbb{R}^3$ itself. ▲

The situation where a subspace W of a finite-dimensional vector space V is all of V has a very simple description in terms of dimensions.

THEOREM 4.23 *If W is a subspace of a finite-dimensional vector space V, and W and V have the same dimension, then $W = V$.*

Proof. We prove this theorem by deriving a contradiction from the assumption that $W \neq V$.

Suppose that span $\{W\} \neq V$ and that $\mathcal{B} = \{\mathbf{w}_1, \ldots, \mathbf{w}_n\}$ is a basis for W. Then there exists a vector $\mathbf{v} \in V$ not in W that is linearly independent of the vectors in $\mathcal{B}$. Therefore the set $\mathcal{B}' = \{\mathbf{w}_1, \ldots, \mathbf{w}_n, \mathbf{v}\}$, consisting of $n + 1$ vectors, is a linearly independent subset of V. This contradicts the fact that W and V have the same dimension. ■

EXERCISES 4.13

1. Show that the proper subspaces of $\mathbb{R}^2$ are the zero subspace and all lines through the origin.

2. Give a geometric description of the proper subspaces of $\mathbb{R}^3$.

3. Find bases for the following subspaces of $\mathbb{R}^3$.

   a. The plane $2x - y + 3z = 0$
   b. $W = \{(a, a - b, b) : a, b \in \mathbb{R}\}$
   c. The line $x = t$, $y = -2t$, $z = 3t$
   d. The planes $2x - y + 3z = 0$ and $x + y - z = 0$

4. Prove that if $A\mathbf{x} = \mathbf{0}$ is a homogeneous linear system with real coefficients consisting of m equations in n unknowns, then the set of solution vectors is a subspace of $\mathbb{R}^n$.

5. Determine whether the set of real polynomials of even degrees is a subspace of the polynomial space $\mathbb{R}[t]$. Justify your answer.

6. Prove that if $\mathbf{u} \in U$ and $\mathbf{v} \in V$ are two nonzero vectors in two disjoint subspaces U and V of a vector space W, then $\mathbf{u}$ and $\mathbf{v}$ are linearly independent.

7. Prove that the set of symmetric real 3×3 matrices is a subspace of $\mathbb{R}^{3\times 3}$.

8. Determine whether the set of orthogonal $n \times n$ matrices is a subspace of $\mathbb{R}^{n\times n}$. Justify your answer.

9. Explain why the set of invertible $n \times n$ matrices is not a subspace of $\mathbb{R}^{n\times n}$.

10. Use Theorem 4.9 to prove that if W is a subspace of an n-dimensional vector space V, then $\dim W \leq n$.

11. Determine whether the set of real polynomials of even degrees less than or equal to 6 is a subspace of the polynomial space $\mathbb{R}_{10}[t]$. Justify your answer.

12. (*Maple V*) Generate a random 5×5 invertible matrix A and describe the different possible pairs of disjoint subspaces of $\mathbb{R}^5$ determined by the columns of A.

13. (*Maple 6*) Construct two 4×4 matrices A and B so that the columns of A are linearly independent of the columns of B, and A and B have exactly two linearly independent columns. Show that the columns of A and B span two disjoint two-dimensional subspaces of $\mathbb{R}^4$.

Fundamental Subspaces

Let A be an $m \times n$ matrix over $\mathbb{R}$. Then A determines four important vector spaces, known as the ***fundamental subspaces determined by*** A. They are the column space, the null space, the row space, and the left null space. Two of the spaces are subspaces of $\mathbb{R}^m$ and two of them are subspaces of $\mathbb{R}^n$. It left as an exercise to show that the defined sets of vectors are subspaces.

DEFINITION 4.11 *The* ***column space*** *of a real* $m \times n$ *matrix* A *is the subspace* Col A *of* $\mathbb{R}^m$ *spanned by the columns of* A.

The dimension of the column space is called the ***column rank*** of A. It counts the number of linearly independent columns of A.

DEFINITION 4.12 *The* ***null space*** *of a real* $m \times n$ *matrix* A *is the subspace* Nul A *of* $\mathbb{R}^n$ *consisting of all vectors* $\mathbf{x} \in \mathbb{R}^n$ *for which* $A\mathbf{x} = \mathbf{0}$.

The fact that Nul A is a subspace of $\mathbb{R}^n$ was essentially shown in Example 4.9. The dimension of Nul A is called the ***nullity*** of A.

Although the emphasis in this book is on column spaces, we sometimes need the space Col A^T, whose columns are the rows of A.

DEFINITION 4.13 *The* ***row space*** *of a real* $m \times n$ *matrix* A *is the subspace* Col A^T *of* $\mathbb{R}^n$ *spanned by the columns of* A^T.

We sometimes write Row A for Col A^T to highlight the fact the vectors in the Col A^T are obtained by transposing the rows of A. The dimension of the row space of A is called the ***row rank*** of A. It counts the number of linearly independent rows of A.

DEFINITION 4.14 *The* ***left null space*** *of a real* $m \times n$ *matrix* A *is the subspace* Nul A^T *of* $\mathbb{R}^m$ *consisting of all vectors* $\mathbf{y} \in \mathbb{R}^n$ *for which* $A^T\mathbf{y} = \mathbf{0}$.

These spaces can be described in terms of interesting bases. Alternative descriptions, based on the singular value decomposition of a matrix, are given in Chapter 9. The geometric relationship between the fundamental subspaces is discussed in Chapter 7.

The Column Space

Column spaces have a straightforward description in terms of pivot columns.

THEOREM 4.24 (Column space basis theorem) *The pivot columns of an $m \times n$ matrix A over $\mathbb{R}$ form a basis for* Col A.

Proof. Let B be the reduced row echelon form of an $m \times n$ real matrix A. By Theorem 1.6, $A\mathbf{x} = \mathbf{0}$ if and only if $B\mathbf{x} = \mathbf{0}$. Hence a set of columns of A is linearly dependent if and only if the corresponding set of columns of B is linearly dependent. It suffices therefore to prove the theorem for the matrix B.

Suppose $\mathbf{c}_1 B, \ldots, \mathbf{c}_n B$ are the columns of B and that $\mathbf{d}_1 B, \ldots, \mathbf{d}_k B$ are the pivot columns. We know from Exercise 4.6.4, that the columns $\mathbf{d}_1 B, \ldots, \mathbf{d}_k B$ are linearly independent. Moreover, it is clear from the definition of reduced row echelon matrices that

$$\mathbf{d}_i B = (\delta_{1i}, \ldots, \delta_{ii}, \ldots, \delta_{im}) = \mathbf{e}_i$$

is the ith vector in the standard basis E of $\mathbb{R}^m$. Since B is in reduced row echelon form, the last $(m - k)$ entries of the columns $\mathbf{c}_1 B, \ldots, \mathbf{c}_n B$ are zero. Hence any column $\mathbf{c}_j B$ of B is a linear combination

$$\mathbf{c}_j B = a_1 \mathbf{e}_1 + \cdots + a_k \mathbf{e}_k = a_1 \mathbf{d}_1 B + \cdots + a_k \mathbf{d}_k B$$

of the vectors $\mathbf{e}_1, \ldots, \mathbf{e}_k$. This means that Col $B \subseteq$ span $\{\mathbf{e}_1, \ldots, \mathbf{e}_k\}$. By Exercise 4.13.10, the dimension of Col B can therefore be no greater than k. However, since B contains the k linearly independent columns $\mathbf{e}_1, \ldots, \mathbf{e}_k$, the dimension of Col B must be at least k. Hence dim (Col B) $= k$. The pivot columns of B are therefore a basis for B. ■

EXAMPLE 4.42 A Basis for a Column Space

Use *Maple* to find a basis for the column space of the matrix

$$A = \begin{bmatrix} 3 & 1 & 0 & 2 & 4 \\ 1 & 1 & 0 & 0 & 2 \\ 5 & 2 & 0 & 3 & 7 \end{bmatrix}$$

Solution. According to Theorem 4.24, the pivot columns of A form a basis for Col A. We can find these columns by converting A^T to reduced row echelon form.

```
[>with(linalg):
[>A:=matrix([[3,1,0,2,4],[1,1,0,0,2],[5,2,0,3,7]]):
[>rref(A);
```

$$\begin{bmatrix} 1 & 0 & 0 & 1 & 1 \\ 0 & 1 & 0 & -1 & 1 \\ 0 & 0 & 0 & 0 & 0 \end{bmatrix}$$

Hence $\mathcal{B} = \{(1, 0, 0), (0, 1, 0)\}$ is a basis for Col A consisting of the pivot columns of A. We could also have used the **colspace** command in the **linalg** package to find a basis.

```
[> colspace(A);
```

$$\{[1, 0, \tfrac{3}{2}], [0, 1, \tfrac{1}{2}]\}$$

Therefore, the set $\mathcal{C} = \{(1, 0, 3/2), (0, 1, 1/2)\}$ is a second basis for the column space of A. However, it does not consist of the pivot columns of A. ▲

The Null Space

Using the concept of free variables discussed in Chapter 1, we can give an easy description of a basis for Nul A. The idea is to produce the standard basis of the solution space of the system $A\mathbf{x} = \mathbf{0}$, as described in Example 4.28.

THEOREM 4.25 (Null space basis theorem) *The set of column vectors obtained by writing the general solution of the system $A\mathbf{x} = \mathbf{0}$ as a linear combination of the free variables of A is a basis for* Nul A.

Proof. If we convert the matrix A to reduced row echelon form and write the general solution of the system as a linear combination of the free variables of the system $A\mathbf{x} = \mathbf{0}$, then it follows from the discussion in Example 4.28 that the set of column vectors arising in this way is a basis for Nul A. ■

EXAMPLE 4.43 A Basis for a Null Space

Use *Maple* to find a basis for the null space of the matrix

$$A = \begin{bmatrix} 3 & 1 & 0 & 2 & 4 \\ 1 & 1 & 0 & 0 & 2 \\ 5 & 2 & 0 & 3 & 7 \end{bmatrix}$$

Solution. We begin by finding the general solution of the system $A\mathbf{x} = \mathbf{0}$.

```
[> with(LinearAlgebra):
[> A:=Matrix([[3,1,0,2,4],[1,1,0,0,2],[5,2,0,3,7]]):
[> z:=ZeroVector(3):
```

```
[> LinearSolve(A,z);
```

$$\begin{bmatrix} \_t8_1 \\ \_t8_1 + 2\_t8_4 \\ \_t8_3 \\ \_t8_4 \\ -\_t8_1 - \_t8_4 \end{bmatrix}$$

This tells us that the general solution $\mathbf{x}$ of $A\mathbf{x} = \mathbf{0}$ is

$$\begin{bmatrix} a \\ a+2b \\ c \\ b \\ -a-b \end{bmatrix} = \begin{bmatrix} a \\ a \\ 0 \\ 0 \\ -a \end{bmatrix} + \begin{bmatrix} 0 \\ 2b \\ 0 \\ b \\ -b \end{bmatrix} + \begin{bmatrix} 0 \\ 0 \\ c \\ 0 \\ 0 \end{bmatrix} = a \begin{bmatrix} 1 \\ 1 \\ 0 \\ 0 \\ -1 \end{bmatrix} + b \begin{bmatrix} 0 \\ 2 \\ 0 \\ 1 \\ -1 \end{bmatrix} + c \begin{bmatrix} 0 \\ 0 \\ 1 \\ 0 \\ 0 \end{bmatrix}$$

By Theorem 4.25, the set of vectors

$$\mathcal{B} = \{(1, 1, 0, 0, -1), (0, 2, 0, 1, -1), (0, 0, 1, 0, 0)\}$$

is a basis for the null space of A. ▲

The Row Space

The next theorem provides an easy description of a basis for the row space of A. The proof of the theorem uses the following lemma.

LEMMA 4.26 *The nonzero rows of a matrix in reduced row echelon form are linearly independent.*

Proof. Let A be an $m \times n$ matrix in reduced row echelon form and let $\mathbf{r}_1 A, \ldots, \mathbf{r}_k A$ be the nonzero rows of A. Suppose that

$$a_1 \mathbf{r}_1 A + \cdots + a_k \mathbf{r}_k A = (0, \ldots, 0)$$

Since A is in row echelon form, all entries below the leading entry of $\mathbf{r}_1 A$ are zero. Hence $a_1 = 0$ since $a_1 \mathbf{r}_1 A$ would otherwise contribute a nonzero entry to the zero vector $(0, \ldots, 0)$. A similar argument shows that the coefficient a_2 must be zero, and so on. Hence $a_1 = \cdots = a_k = 0$. This shows that the rows $\mathbf{r}_1 A, \ldots, \mathbf{r}_k A$ must be linearly independent. ■

We use this lemma to construct a basis for the row space of A.

THEOREM 4.27 (Row space basis theorem) *The nonzero rows of the reduced row echelon form B of an $m \times n$ matrix A are a basis for* Row A.

Proof. Suppose that A is an $m \times n$ matrix and C is obtained from A by an elementary row operation. We show that Row A = Row C.

Let $\mathbf{r}_1 A, \ldots, \mathbf{r}_n A$ be the nonzero rows of A and suppose that C results from A by the row operation $(R_i \to sR_i)$. Then each vector $\mathbf{x} \in$ Row C is of the form

$$\begin{aligned}\mathbf{x} &= a_1 \mathbf{r}_1 A + \cdots + a_i (s\mathbf{r}_i A) + \cdots + a_n \mathbf{r}_n A \\ &= a_1 \mathbf{r}_1 A + \cdots + (a_i s)\, \mathbf{r}_i A + \cdots + a_n \mathbf{r}_n A\end{aligned}$$

and is therefore also in the row space of A. Conversely, since $s \neq 0$, each vector $\mathbf{y} \in$ Row A is of the form

$$\begin{aligned}\mathbf{y} &= a_1 \mathbf{r}_1 A + \cdots + a_i \mathbf{r}_i A + \cdots + a_n \mathbf{r}_n A \\ &= a_1 \mathbf{r}_1 A + \cdots + \left(\frac{a_i}{s}\right)(s\mathbf{r}_i A) + \cdots + a_n \mathbf{r}_n A\end{aligned}$$

and is also in the row space of C. The calculations for the other two types are analogous. Hence Row A = Row C. Since the reduced row echelon matrix B results from A by a finite number of applications of elementary row operations, we have

$$\text{Row } A = \text{Row } C_1 = \cdots = \text{Row } C_k = \text{Row } B$$

where C_{i+1} is the matrix obtained from C_i be a single elementary row operation. By Lemma 4.26, the rows of B are linearly independent and therefore form a basis for Row B. Since Row B = Row A, they also form a basis for Row A. ■

We note that while elementary row operations preserve row spaces, they do not preserve column spaces. For example, if

$$A = \begin{bmatrix} 1 & 2 \\ 3 & 6 \end{bmatrix} \quad \text{and} \quad B = \begin{bmatrix} 1 & 2 \\ 0 & 0 \end{bmatrix}$$

then B results from A by the operation $(R_2 \to R_2 - 3R_1)$, but Col $A \neq$ Col B since

$$\text{Col } A = \left\{ \begin{bmatrix} a \\ 3a \end{bmatrix} : a \in \mathbb{R} \right\} \quad \text{and} \quad \text{Col } B = \left\{ \begin{bmatrix} a \\ 0 \end{bmatrix} : a \in \mathbb{R} \right\}$$

Theorem 4.27 provides us with an alternative description of a basis for the column space of A.

COROLLARY 4.28 *The nonzero columns of the reduced row echelon form B of an $m \times n$ matrix A are a basis for* Col A.

Proof. In terms of columns, Theorem 4.27 says that the nonzero columns of the reduced row echelon form B of an $m \times n$ matrix A^T are a basis for Col A^T. If we replace A by A^T and use the fact that $A^{TT} = A$, we get the column version of the theorem. ■

EXAMPLE 4.44 A Basis for a Row Space

Use *Maple* to find a basis for the row space of the matrix

$$A = \begin{bmatrix} 3 & 1 & 0 & 2 & 4 \\ 1 & 1 & 0 & 0 & 2 \\ 5 & 2 & 0 & 3 & 7 \end{bmatrix}$$

Solution. We begin by finding the reduced row echelon form B of the matrix A.

```
[> with(linalg):
[> A:=matrix([[3,1,0,2,4],[1,1,0,0,2],[5,2,0,3,7]]);
[> B:=rref(A);
```

$$B := \begin{bmatrix} 1 & 0 & 0 & 1 & 1 \\ 0 & 1 & 0 & -1 & 1 \\ 0 & 0 & 0 & 0 & 0 \end{bmatrix}$$

```
[> transpose(B);
```

$$\begin{bmatrix} 1 & 0 & 0 \\ 0 & 1 & 0 \\ 0 & 0 & 0 \\ 1 & -1 & 0 \\ 1 & 1 & 0 \end{bmatrix}$$

According to Theorem 4.27, the set $\mathcal{B} = \{(1, 0, 0, 1, 1), (0, 1, 0, -1, 1)\}$ of the nonzero columns of B^T is a basis for Col A^T. We can use the **rowspace** function to confirm this fact.

```
[> rowspace(A);
```

$$\{[1, 0, 0, 1, 1], [0, 1, 0, -1, 1]\}$$

This shows that $\mathcal{B}$ is a basis for the row space of A. ▲

The Left Null Space

Next we describe the construction of a basis for a left null space. For this purpose we use the LU decomposition of rectangular matrices discussed in Chapter 2.

THEOREM 4.29 (Left null space basis theorem) *If A is an $m \times n$ matrix of rank k, and if $PA = LU$ is the permutation-free LU decomposition of A, then the last $m - k$ rows of $L^{-1}P$ are a basis for* Nul A^T.

Proof. Suppose that $PA = LU$ is the permutation-free LU decomposition of PA. We know from Theorems 2.15 and 2.16 that L is invertible. Therefore $L^{-1}PA = U$. Since U is upper triangular, all of its zero rows occur at the bottom. Moreover, since $L^{-1}P$ is invertible, the matrix A and the matrix U have the same rank. Therefore U has rank k. The last $m - k$ rows of U are therefore the zero rows of U. Hence its entries $\mathbf{r}_i\left(L^{-1}P\right)\mathbf{c}_j A$ are zero for all $i > k$. Therefore the vectors $\mathbf{r}_{k+1}\left(L^{-1}P\right), \ldots, \mathbf{r}_m\left(L^{-1}P\right)$ span Nul A^T. Since $L^{-1}P$ is invertible, its rows are linearly independent by Theorem 4.3. This tells us that the set $\left\{\mathbf{r}_{k+1}\left(L^{-1}P\right), \ldots, \mathbf{r}_m\left(L^{-1}P\right)\right\}$ is a basis for Nul A^T. ■

EXAMPLE 4.45 A Basis for a Left Null Space

Use *Maple* and Theorem 4.29 to find a basis for the left null space of the matrix

$$A = \begin{bmatrix} -1 & -2 & 39 & 8 \\ 0 & 0 & 0 & 0 \\ 0 & -7 & 2 & 0 \\ 0 & 1 & 2 & 12 \end{bmatrix}$$

Solution. We note that $m = 4$, since A is a 4×4 matrix, and that

$$\begin{bmatrix} 1 & 0 & 0 & \frac{775}{4} \\ 0 & 1 & 0 & \frac{3}{2} \\ 0 & 0 & 1 & \frac{21}{4} \\ 0 & 0 & 0 & 0 \end{bmatrix}$$

is the reduced row echelon form of A. This tells us that the rank k of A is 3. Hence $m - k = 1$. Moreover, the matrix

$$PA = \begin{bmatrix} 1 & 0 & 0 & 0 \\ 0 & 0 & 0 & 1 \\ 0 & 0 & 1 & 0 \\ 0 & 1 & 0 & 0 \end{bmatrix} \begin{bmatrix} -1 & -2 & 39 & 8 \\ 0 & 0 & 0 & 0 \\ 0 & -7 & 2 & 0 \\ 0 & 1 & 2 & 12 \end{bmatrix}$$

has the permutation-free LU decomposition

$$\begin{bmatrix} -1 & -2 & 39 & 8 \\ 0 & 1 & 2 & 12 \\ 0 & -7 & 2 & 0 \\ 0 & 0 & 0 & 0 \end{bmatrix} = \begin{bmatrix} 1 & 0 & 0 & 0 \\ 0 & 1 & 0 & 0 \\ 0 & -7 & 1 & 0 \\ 0 & 0 & 0 & 1 \end{bmatrix} \begin{bmatrix} -1 & -2 & 39 & 8 \\ 0 & 1 & 2 & 12 \\ 0 & 0 & 16 & 84 \\ 0 & 0 & 0 & 0 \end{bmatrix}$$

By Theorem 4.29, the 4th row [0, 1, 0, 0] of the matrix

$$L^{-1}P = \begin{bmatrix} 1 & 0 & 0 & 0 \\ 0 & 1 & 0 & 0 \\ 0 & -7 & 1 & 0 \\ 0 & 0 & 0 & 1 \end{bmatrix}^{-1} \begin{bmatrix} 1 & 0 & 0 & 0 \\ 0 & 0 & 0 & 1 \\ 0 & 0 & 1 & 0 \\ 0 & 1 & 0 & 0 \end{bmatrix} = \begin{bmatrix} 1 & 0 & 0 & 0 \\ 0 & 0 & 0 & 1 \\ 0 & 0 & 1 & 7 \\ 0 & 1 & 0 & 0 \end{bmatrix}$$

determines a basis for Nul A^T. We can verify directly that $A^T\mathbf{x}_0 = \mathbf{0}$.

```
[> with(linalg):
[> A:=matrix([[-1,-2,39,8],[0,0,0,0],[0,-7,2,0],
   [0,1,2,12]]):
[> transpose(A);
```

$$\begin{bmatrix} -1 & 0 & 0 & 0 \\ -2 & 0 & -7 & 1 \\ 39 & 0 & 2 & 2 \\ 8 & 0 & 0 & 12 \end{bmatrix}$$

```
[> evalm(transpose(A)&*(array([0,1,0,0]))):
[> array([0,0,0,0]):
[> equal(%%,%);
```

true

Hence $\mathbf{x}$ is in the null space of A^T. This confirms that the set $\mathcal{B} = \{(0, 1, 0, 0)\}$ is a basis for the left null space of A. ▲

In the next example, we use *Maple* to illustrate the calculation of the left null space of a rectangular matrix.

EXAMPLE 4.46 A Basis for the Left Null Space of a Rectangular Matrix

Use *Maple* to find a basis for the left null space of the matrix

$$A = \begin{bmatrix} 3 & 1 & 0 & 2 & 4 \\ 1 & 1 & 0 & 0 & 2 \\ 5 & 2 & 0 & 3 & 7 \end{bmatrix}$$

Solution. We use the **NullSpace** function to find the required basis.

```
[> with(LinearAlgebra):
[> A:=Transpose(Matrix([[3,1,0,2,4],[1,1,0,0,2],
    [5,2,0,3,7]])):
[> NullSpace(A);
```

$$\left\{\begin{bmatrix} 3 \\ 1 \\ -2 \end{bmatrix}\right\}$$

Maple is telling us that the set $\mathcal{B} = \{(3, 1, -2)\}$ is a basis for Nul A^T. ▲

Rank and Fundamental Subspaces

It is clear that if A is an $m \times n$ matrix over $\mathbb{R}$ and $m \neq n$, then Col A and Col A^T are subspaces of column spaces of different dimension. It is remarkable, therefore, that they nevertheless have the same dimension.

THEOREM 4.30 (Rank theorem) *The spaces* Col A *and* Col A^T *have the same dimension.*

Proof. Let R be the reduced row echelon form of A. Then Theorem 4.27 tells us that Row $R =$ Row A. Hence dim (Row R) $=$ dim (Row A) . Furthermore, Theorem 4.24 tells us that the pivot columns of A form a basis for Col A. Since R and A have the same pivot columns, dim (Col R) $=$ dim (Col A) . Moreover, the nonzero rows of R are precisely the rows in which the pivots of R occur. Hence the number of nonzero rows of R is the same as the number of pivot columns of R. Hence dim (Row R) $=$ dim (Col R) . It follows that dim (Row A) $=$ dim (Col A) . ■

Since the dimension of the column space of a matrix equals the dimension of its row space, we can use either columns or rows to give a name to this dimension. We opt for columns.

DEFINITION 4.15 *The* ***rank*** rank A *of an* $m \times n$ *matrix* A *is the dimension of the column space of* A.

An $n \times n$ matrix is said to be of ***full rank*** if it has rank n. Otherwise it is said to be ***rank deficient***. Although we cannot use a built-in *Maple* function to calculate the rank of a matrix, we can use the conclusion of the next theorem to define one.

THEOREM 4.31 *For any real* $m \times n$ *matrix* A, *the sum of the rank and nullity of* A *is* n.

Proof. Let A be a real $m \times n$ matrix. Then A has n columns and the homogeneous linear system $A\mathbf{x} = \mathbf{0}$ therefore consists of m equations in n variables. By Theorem 4.25, the number of free variables of A determines the nullity of A, and by Theorem 4.24, the number of basic variables of A determines the rank of A. It follows that nullity A + rank $A = n$. ■

As Theorems 4.30 and 4.31 show, the rank of a matrix uniquely determines the dimensions of the fundamental subspaces. Table 6 lists the dimensions of these subspaces for an $m \times n$ matrix A of rank r.

TABLE 6 SUBSPACE DIMENSIONS.

Spaces	Col A	Col A^T	Nul A	Nul A^T
Dimensions	r	r	$n - r$	$m - r$

We use these facts to define *Maple* functions that calculate the rank of a matrix as well as a basis for the row space of that matrix.

EXAMPLE 4.47 The Rank of a Matrix

Use *Maple* to calculate the rank of the matrix

$$A = \begin{bmatrix} 3 & 1 & 0 & 2 & 4 \\ 1 & 1 & 0 & 0 & 2 \\ 5 & 2 & 0 & 3 & 7 \end{bmatrix}$$

Solution. We use the **Rank** function in the **LinearAlgebra** package.

```
[> with(LinearAlgebra):
[> A:=Matrix([[3,1,0,4],[1,1,0,0,2],[5,2,0,3,7]]):
[> Rank(A);
                              2
```

As expected, *Maple* is telling us that the rank of A is 2. ▲

EXAMPLE 4.48 The Row Space of a Matrix

Use *Maple* to calculate the row space of the matrix

$$A = \begin{bmatrix} 3 & 1 & 0 & 2 & 4 \\ 1 & 1 & 0 & 0 & 2 \\ 5 & 2 & 0 & 3 & 7 \end{bmatrix}$$

Solution. We use the **rowspace** function in the **linalg** package to find a basis for the row space of A.

```
[>A:=matrix([[3,1,0,2,4],[1,1,0,0,2],[5,2,0,3,7]]);
[>B:=linalg[rowspace](A);
```

$$B := \{[1, 0, 0, 1, 1], [0, 1, 0, -1, 1]\}$$

Our calculation shows that the set $\mathcal{B}$ is a basis for the row space of A. We could have obtained the same result by reducing the matrix A to row echelon form.

```
[>linalg[gaussjord](A);
```

$$A := \begin{bmatrix} 1 & 0 & 0 & 1 & 1 \\ 0 & 1 & 0 & -1 & 1 \\ 0 & 0 & 0 & 0 & 0 \end{bmatrix}$$

As we can see, the rows of A coincide with the basis vectors in $\mathcal{B}$. ▲

Invertibility and Fundamental Subspaces

There are several useful connections between the invertibility of a matrix and the fundamental subspaces it determines.

THEOREM 4.32 *A real $n \times n$ matrix A is invertible if and only if* $\text{Col } A = \mathbb{R}^n$.

Proof. By Corollary 2.22, the matrix A is invertible if and only if the system $A\mathbf{x} = \mathbf{0}$ has only the trivial solution. In other words, A is invertible if and only if the nullity of A is zero. By Theorem 4.31, this happens if and only if the column space of A is an n-dimensional subspace of $\mathbb{R}^n$. Theorem 4.23 tells us that this is so if and only if $\text{Col } A = \mathbb{R}^n$. ■

Having proved that the columns of A are linearly independent and span $\mathbb{R}^n$, we have also established the following corollary.

COROLLARY 4.33 *A real $n \times n$ matrix A is invertible if and only if it has rank n.*

Proof. Since $\dim \mathbb{R}^n = n$ and $\text{Col } A = \mathbb{R}^n$, the result follows from Theorem 4.32. ■

The next corollary is a restatement of Theorem 4.3. The new proof is less elementary since it refers to bases, column spaces, and dimensions.

COROLLARY 4.34 *A real $n \times n$ matrix A is invertible if and only if the columns of A are linearly independent.*

Proof. This corollary follows from Corollary 4.33. The dimension of the column space of an $n \times n$ matrix is n if and only if the columns form a basis for A. This holds if and only if the columns of A are linearly independent. ■

EXAMPLE 4.49 A Basis for $\mathbb{R}^3$

Use *Maple* to find an invertible 3×3 matrix, and use the columns of this matrix to construct a basis for $\mathbb{R}^3$.

Solution. We use the **randmatrix** function to find an invertible matrix.

```
[>with(linalg):
[>r:=rand(-5..5):   random:= proc() r() end:
[>A:=randmatrix(3,3,entries=random);
```

$$A := \begin{bmatrix} 4 & -1 & -4 \\ -1 & -1 & -4 \\ -2 & 3 & 4 \end{bmatrix}$$

```
[>inverse(A);
```

$$\begin{bmatrix} \frac{1}{5} & -\frac{1}{5} & 0 \\ \frac{3}{10} & \frac{1}{5} & \frac{1}{2} \\ -\frac{1}{8} & -\frac{1}{4} & -\frac{1}{8} \end{bmatrix}$$

Since the matrix A has an inverse, its columns are linearly independent, and we can use them to form the basis $\mathcal{B} = \{(4, -1, -2), (-1, -1, 3), (-4, -4, 4)\}$ for $\mathbb{R}^3$. ▲

EXAMPLE 4.50 Matrices and Linear Dependence

Use *Maple* to show that the set $\mathcal{S} = \{(-3, -3, 3), (5, 8, -8), (17, 17, -17)\}$ is linearly dependent and therefore fails to form a basis for $\mathbb{R}^3$.

Solution. We begin by constructing the required matrix.

```
[>A:=matrix([[-3,-3,3],[5,8,-8],[17,17,-17]]):
```

Next we try to calculate the inverse of A.

```
[>linalg[inverse](A);
Error, (in linalg[inverse]) singular matrix
```

Maple is telling us that A is not invertible. By Corollary 4.34, the columns of $\mathcal{S}$ are therefore linearly dependent and cannot form a basis for $\mathbb{R}^3$. ▲

By reversing the role of the rows and columns of A, we get the analogous result for the row space of A.

COROLLARY 4.35 *A real $n \times n$ matrix A is invertible if and only if the dimension of* Col A^T *is n.*

Proof. By Theorem 4.32, the matrix A^T is invertible if and only if Col $A^T = \mathbb{R}^n$. Moreover, we know from Theorem 2.27 that the matrix A^T is invertible if and only if its transpose $A^{TT} = A$ is invertible. Hence the corollary follows. ■

COROLLARY 4.36 *A real $n \times n$ matrix A is invertible if and only if* Nul $A = \{\mathbf{0}\}$.

Proof. By Theorem 4.32, the matrix A is invertible if and only if Col $A = \mathbb{R}^n$. In other words, A is invertible if and only if the dimension of its column space is n. Since the dimension of the null space of A is $n - \dim \text{Col } A = n - n = 0$, it follows that Nul $A = \{\mathbf{0}\}$. ■

EXERCISES 4.14

1. Find the column space, row space, null space, and left null space of the matrix

$$A = \begin{bmatrix} 1 & 2 & 4 \\ 0 & 0 & 1 \end{bmatrix}$$

and give a geometric description of these subspaces of $\mathbb{R}^2$ and $\mathbb{R}^3$.

2. Find the column space, row space, null space, and left null space of the matrix

$$A = \begin{bmatrix} 1 & 2 & 4 & 3 \\ 0 & 0 & 1 & 3 \end{bmatrix}$$

3. Construct 3×4 matrices of rank 0, 1, 2, and 3.
4. Show that the rank of a matrix is equal to the rank of the transpose of the matrix.
5. Prove that Gaussian elimination preserves rank.
6. Prove that the rank of a matrix is equal to the number of nonzero rows of the reduced row echelon form of the matrix.
7. Prove that the rank of a matrix is equal to the number of pivots of the row echelon form of the matrix.
8. Show that for any real $m \times n$ matrix A, the spaces Col A, Nul A, Col A^T, and Nul A^T are subspaces and therefore satisfy the axioms of a real vector space.

9. Let A be an $m \times n$ matrix, B be an $m \times m$ invertible matrix, and C be an $n \times n$ invertible matrix. Show that the matrices A, BA, and AC have the same rank.

10. Let A be an $m \times n$ matrix and B be an $n \times p$ matrix. Show that the rank of AB is less than or equal to the rank of A.

11. Prove that a real $n \times n$ matrix is singular if and only if its nullity is greater than zero.

12. Construct 5×5 matrices of rank 0, 1, 2, 3, 4, and 5.

13. Prove that if $A = QR$ is a real matrix and R is invertible, then $\text{Col } A = \text{Col } Q$.

14. (*Maple V*) Generate 10 random 8×8 matrices and calculate their rank. Calculate the probability, based on your experiment, that a random 8×8 matrix is of full rank.

15. (*Maple V*) Find the column space, row space, null space, and left null space of the matrix

$$A = \begin{bmatrix} 1 & 2 & 4 \\ 0 & 0 & 1 \\ 3 & 6 & 12 \end{bmatrix}$$

and give a geometric description of these subspaces of $\mathbb{R}^3$.

16. (*Maple V*) Show that the intersection of the column space of the matrix

$$A = \begin{bmatrix} 1 & 2 & 4 \\ 0 & 0 & 1 \\ 3 & 6 & 12 \end{bmatrix}$$

and the null space of the matrix

$$B = \begin{bmatrix} 1 & 0 & 1 \\ 1 & 0 & 1 \\ 1 & 0 & 1 \\ 0 & 0 & 0 \end{bmatrix}$$

is a subspace of $\mathbb{R}^3$.

17. (*Maple V*) Show that for any matrix A, the function

```
ColumnRankk:=A->rowdim(transpose(A))-nops(nullspace(A))
```

calculates the column rank of A. Explain why this value is different from the value produced by the **coldim** function in the **linalg** package.

18. (*Maple V*) Show that for any matrix A, the function

```
RowRankk := A - >nops(rowspace(A))
```

calculates the row rank of A. Explain why this value is different from the value produced by the **rowdim** function in the **linalg** package.

19. (*Maple V*) Explain why the assignment

```
RowwSpace:=A->submatrix(rref(A),1..rank(A),1..coldim(A))
```

calculates a basis for the row space of a matrix.

20. (*Maple 6*) Explain why the function `Rankk:=A->nops(RowSpace(A))` calculates the rank of a matrix.

21. (*Maple V, 6*) Use the function **RowwSpace** from in Exercise 19 to define a function that calculates a basis for the column space of a random 3×4 matrix A. Compare your result with that obtained by the function **ColumnSpace(Transpose(A)).**

Direct Sums

Contrary to intersections, the union $U \cup V$ of two subspaces U and V of a vector space W may not be a subspace of W since it may not contain all linear combinations of the vectors in U and V. Let $U = \{(a, 0) : a \in \mathbb{R}\}$ and $V = \{(0, b) : b \in \mathbb{R}\}$ be two subspaces of $\mathbb{R}^2$. Then the vector $(1, 0)$ belongs to U, the vector $(0, 2)$ belongs to V, but the vector

$$\begin{bmatrix} 1 \\ 0 \end{bmatrix} + \begin{bmatrix} 0 \\ 2 \end{bmatrix} = \begin{bmatrix} 1 \\ 2 \end{bmatrix}$$

does not belong to $U \cup V$. Therefore, $U \cup V$ is not a subspace of W. To construct a subspace using U and V, we need to form the span of $U \cup V$. We call this set the ***sum*** of U and V. The space

$$U + V = \{\mathbf{x} + \mathbf{y} \mid \mathbf{x} \in U, \ \mathbf{y} \in V\}$$

is constructed from U and V by building all possible linear combinations of vectors $\mathbf{x} + \mathbf{y}$. It is therefore closed under vector addition and scalar multiplication. By Theorem 4.21, $U + V$ is then a subspace of W.

In many applications it is important to know the dimension of the sum of two subspaces U and V in terms of the dimensions of U and V. For example, if U and V are the subspace

$$\operatorname{span}\{\mathbf{x} = (1, 2, 3, 4), \mathbf{y} = (5, 6, 7, 8)\} \quad \text{and} \quad \operatorname{span}\{\mathbf{y} = (5, 6, 7, 8), \mathbf{z} = (1, 1, 1, 2)\}$$

of $\mathbb{R}^4$, then $U + V = \operatorname{span}\{\mathbf{x}, \mathbf{y}, \mathbf{z}\}$. Moreover, $\mathbf{x}, \mathbf{y}, \mathbf{z}$ are linearly independent since $a\mathbf{x} + b\mathbf{y} + c\mathbf{z} = (0, 0, 0, 0)$ if and only if $a = b = c = 0$. Now suppose that $\mathbf{u} \in U \cap V$. Then $\mathbf{u} = a_1\mathbf{x} + a_2\mathbf{y} = b_1\mathbf{y} + b_2\mathbf{z}$, so

$$a_1\mathbf{x} + a_2\mathbf{y} - b_1\mathbf{y} + b_2\mathbf{z} = a_1\mathbf{x} + (a_2 - b_1)\mathbf{y} + b_2\mathbf{z} = \mathbf{0}$$

Since the vectors **x**, **y**, **z** are linearly independent, $a_1 = a_2 - b_1 = b_2 = 0$. Hence $a_2 = b_1$. This tells us that **u** belongs to span $\{y\}$. This means that $\dim(U+V) = 3$, $\dim(U \cap V) = 1$ and $\dim U = \dim V = 2$. Therefore,

$$3 + 1 = \dim(U + V) + \dim(U \cap V) = \dim U + \dim V = 2 + 2$$

In this example, the sets of vectors $S_1 = \{\mathbf{x}, \mathbf{y}\}$ and $S_2 = \{\mathbf{y}, \mathbf{z}\}$ are both linearly independent but share the vector **y**, and the vector **z** is linearly independent of **x** and **y**. However, if we replace **z** by the vector $\mathbf{w} = (4, 3, 2, 1)$, for example, then $\mathbf{w} = (-9/4)\mathbf{x} + (5/4)\mathbf{y}$ and

$$U + V = \text{span}\{\mathbf{x}, \mathbf{y}\} = U \cap V$$

However, it is still true that

$$2 + 2 = \dim(U + V) + \dim(U \cap V) = \dim U + \dim V = 2 + 2$$

We now prove that this formula holds in general. The idea of the proof is rather interesting. We start with a basis for the intersection of the two spaces, and extend it to separate bases for the two subspaces. We then count the number of vectors in the extended bases.

THEOREM 4.37 *Let U and V be subspaces of a finite-dimensional vector space W. Then* $\dim(U + V) = \dim U + \dim V - \dim(U \cap V)$.

Proof. Suppose that the dimension of $U \cap V$ is $n \geq 0$ and that the $\dim U = (n + p)$ and $\dim V = (n + q)$. The idea of the proof is to take a basis $\mathcal{B}$ for the common part $U \cap V$ of U and V containing $n \geq 0$ elements and to extend $\mathcal{B}$ to a basis $\mathcal{C} = \mathcal{B} \cup \mathcal{S}_U$ for U and a basis $\mathcal{D} = \mathcal{B} \cup \mathcal{S}_V$ for V. Showing that the sets $\mathcal{B}$, $\mathcal{S}_U$, and $\mathcal{S}_V$ are linearly independent and that their union spans $U + V$ proves the theorem. Suppose, therefore, that $n > 0$.

Let $\mathcal{B} = \{\mathbf{x}_1, \ldots, \mathbf{x}_n\}$ be a basis for $U \cap V$, let $\mathcal{C} = \left\{\mathbf{x}_1, \ldots, \mathbf{x}_n, \mathbf{u}_1, \ldots, \mathbf{u}_p\right\}$ be a basis for U, and let $\mathcal{D} = \left\{\mathbf{x}_1, \ldots, \mathbf{x}_n, \mathbf{v}_1, \ldots, \mathbf{v}_q\right\}$ be a basis for V. Then $\dim U = (n + p)$ and $\dim V = (n + q)$. If we can show that $\mathcal{E} = \mathcal{B} \cup \mathcal{S}_U \cup \mathcal{S}_V$ is a basis for $U + V$, then

$$\dim(U + V) = n + p + q = (n + p) + (n + q) - n = \dim U + \dim V - \dim(U \cap V)$$

By construction, the set $\mathcal{E}$ has $n + p + q$ elements and spans $U + V$. It therefore remains to show that $\mathcal{E}$ is linearly independent. Suppose that

$$\underbrace{a_1\mathbf{x}_1 + \cdots + a_n\mathbf{x}_n}_{\in U \cap V} + \underbrace{b_1\mathbf{u}_1 + \cdots + b_p\mathbf{u}_p}_{\in U} + \underbrace{c_1\mathbf{v}_1 + \cdots + c_q\mathbf{v}_q}_{\in V} = \mathbf{0}$$

Then

$$\underbrace{c_1\mathbf{v}_1 + \cdots + c_q\mathbf{v}_q}_{\in V} = \underbrace{-(a_1\mathbf{x}_1 + \cdots + a_n\mathbf{x}_n + b_1\mathbf{u}_1 + \cdots + b_p\mathbf{u}_p)}_{\in U} = -\mathbf{w}$$

It follows that the vector $-\mathbf{w}$ belongs to both U and V. Since $\mathcal{B}$ is a basis for $U \cap V$, there exist scalars $d_1, \ldots, d_n$ for which $d_1\mathbf{x}_1 + \cdots + d_n\mathbf{x}_n = -\mathbf{w}$. This entails that

$$\mathbf{0} = c_1\mathbf{v}_1 + \cdots + c_q\mathbf{v}_q + \mathbf{w} = c_1\mathbf{v}_1 + \cdots + c_q\mathbf{v}_q + d_1\mathbf{x}_1 + \cdots + d_n\mathbf{x}_n$$

From the linear independence of $\mathcal{D}$ we get $c_1 = \cdots = c_q = d_1 = \cdots = d_n = 0$. Therefore,

$$\mathbf{0} = a_1\mathbf{x}_1 + \cdots + a_n\mathbf{x}_n + b_1\mathbf{u}_1 + \cdots + b_p\mathbf{u}_p$$

Moreover, the linear independence of $\mathcal{C}$ implies that $a_1 = \cdots = a_n = b_1 = \cdots = b_p = 0$. Therefore the set $\mathcal{E}$ is a basis for $U + V$. If $n = 0$, then the spaces U and V are disjoint and the basis $\mathcal{B}$ is empty. It follows at once that the vectors in $\mathcal{C}$ are linearly independent of the vectors in $\mathcal{D}$ and conversely. Hence $\mathcal{C} \cup \mathcal{D}$ is a basis for $U + V$. The dimension of $U + V$ is therefore the sum of the dimensions of U and V. ■

COROLLARY 4.38 *If U and V are disjoint subspaces of a finite-dimensional vector space W, then* $\dim(U + V) = \dim U + \dim V$.

Proof. Since U and V are disjoint, $\dim(U \cap V) = 0$. By Theorem 4.37,

$$\begin{aligned}\dim(U + V) &= \dim U + \dim V - \dim(U \cap V)\\ &= \dim U + \dim V - 0\\ &= \dim U + \dim V\end{aligned}$$

as required. ■

One of the standard techniques for studying a complex mathematical object is to break it down into simpler parts, study the simpler parts, and then reassemble the object in a transparent way. In the case of vector spaces, the simpler parts are frequently disjoint subspaces.

DEFINITION 4.16 *If U and V are disjoint subspaces of a vector space W, then $U + V$ is the* ***direct sum*** *of U and V, denoted by $U \oplus V$.*

Direct sums of subspaces are an important tool for analyzing the structure of vector spaces. They rely on the following fundamental property.

THEOREM 4.39 (Direct sum theorem) *If $W = U \oplus V$, then every vector $\mathbf{x} \in W$ can be written as a unique sum $\mathbf{u} + \mathbf{v}$, with $\mathbf{u} \in U$ and $\mathbf{v} \in V$.*

Proof. Suppose that $\mathbf{x} \in W$ can be written as $\mathbf{u}_1 + \mathbf{v}_1$ and $\mathbf{u}_2 + \mathbf{v}_2$, with $\mathbf{u}_1, \mathbf{u}_2 \in U$ and $\mathbf{v}_1, \mathbf{v}_2 \in V$. Then $\mathbf{u}_1 + \mathbf{v}_1 = \mathbf{u}_2 + \mathbf{v}_2$, so $\mathbf{u}_1 - \mathbf{u}_2 = \mathbf{v}_2 - \mathbf{v}_1$. But since $\mathbf{u}_1 - \mathbf{u}_2 \in U$ and $\mathbf{v}_2 - \mathbf{v}_1 \in V$, it follows that $\mathbf{u}_1 - \mathbf{u}_2 = \mathbf{v}_2 - \mathbf{v}_1 \in U \cap V$. Since $U \cap V = \{\mathbf{0}\}$, we have $\mathbf{u}_1 - \mathbf{u}_2 = \mathbf{v}_2 - \mathbf{v}_1 = \mathbf{0}$. Therefore, $\mathbf{u}_1 = \mathbf{u}_2$ and $\mathbf{v}_2 = \mathbf{v}_1$. ■

The following theorem expresses one of the basic properties of direct sums.

THEOREM 4.40 (Basis union theorem) *If U and V are subspaces of a vector space W with bases $\mathcal{B}$ and $\mathcal{C}$, then $W = U \oplus V$ if and only if $\mathcal{B} \cup \mathcal{C}$ is a basis for W.*

Proof. Suppose that $W = U \oplus V$ and that $\mathcal{B} = \{\mathbf{u}_1, \ldots, \mathbf{u}_p\}$ and $\mathcal{C} = \{\mathbf{v}_1, \ldots, \mathbf{v}_q\}$. Then for any $\mathbf{x} \in V$, $\mathbf{x} = \mathbf{u} + \mathbf{v}$ with $\mathbf{u} \in U$ and $\mathbf{v} \in V$. Since $\mathcal{B}$ is a basis for U and $\mathcal{C}$ is a basis for V, $\mathbf{u}$ is a linear combination of vectors in $\mathcal{B}$, and $\mathbf{v}$ is a linear combination of vectors in $\mathcal{C}$. Hence $\mathbf{x}$ is a linear combination of vectors in $\mathcal{B} \cup \mathcal{C}$. Therefore, $\mathcal{B} \cup \mathcal{C}$ spans W. It remains to show that $\mathcal{B} \cup \mathcal{C}$ is linearly independent. Suppose that

$$a_1\mathbf{u}_1 + \cdots + a_p\mathbf{u}_p + b_1\mathbf{v}_1 + \cdots + b_q\mathbf{v}_q = \mathbf{0}$$

and that $a_1\mathbf{u}_1 + \cdots + a_p\mathbf{u}_p \in U$ and $b_1\mathbf{v}_1 + \cdots + b_q\mathbf{v}_q \in V$. Since the sum $U \oplus W$ is direct, the zero vector $\mathbf{0} \in V$ is the unique sum of $\mathbf{0} \in U$ and $\mathbf{0} \in V$. It therefore follows from the unique decomposition property of direct sums that $a_1\mathbf{u}_1 + \cdots + a_p\mathbf{u}_p = \mathbf{0}$ and $b_1\mathbf{v}_1 + \cdots + b_q\mathbf{v}_q = \mathbf{0}$. Since $\mathcal{B}$ and $\mathcal{C}$ are bases, we get $a_1 = \cdots = a_p = b_1 = \cdots = b_q = 0$. This shows that $\mathcal{B} \cup \mathcal{C}$ is linearly independent.

Conversely, suppose that $\mathcal{B} \cup \mathcal{C}$ is a basis for W. Then, for any $\mathbf{x} \in W$ there exist scalars $a_1, \ldots, a_p, b_1, \ldots, b_q$ for which

$$\mathbf{x} = a_1\mathbf{u}_1 + \cdots + a_p\mathbf{u}_p + b_1\mathbf{v}_1 + \cdots + b_q\mathbf{v}_q = \mathbf{u} + \mathbf{v}$$

with $\mathbf{u} = a_1\mathbf{u}_1 + \cdots + a_p\mathbf{u}_p \in U$ and $\mathbf{v} = b_1\mathbf{v}_1 + \cdots + b_q\mathbf{v}_q \in V$. This show that $W = U + V$.

The theorem follows if we can show that U and V are disjoint. Suppose, therefore, that $\mathbf{u} \in U \cap V$. Then $\mathbf{u} = a_1\mathbf{u}_1 + \cdots + a_p\mathbf{u}_p = b_1\mathbf{v}_1 + \cdots + b_q\mathbf{v}_q$ for some scalars $a_1, \ldots, a_p$, $b_1, \ldots, b_q$. This means that $a_1\mathbf{u}_1 + \cdots + a_p\mathbf{u}_p - b_1\mathbf{v}_1 - \cdots - b_q\mathbf{v}_q = \mathbf{0}$. Since $\mathcal{B} \cup \mathcal{C}$ is a basis for W, it is a linearly independent set. Hence the scalars $a_1, \ldots, a_p, b_1, \ldots, b_q$ must all be zero. It follows that $U \cap V = \{\mathbf{0}\}$. Therefore, $W = U \oplus V$. ∎

EXAMPLE 4.51 A Direct Sum Basis

Use *Maple* to show that the space $W = U + V$ is a direct sum, where

$$U = \operatorname{span} \mathcal{B} = \operatorname{span} \{(3, 4, 1, 3)\}$$

and

$$V = \operatorname{span} \mathcal{C} = \operatorname{span} \{(2, 2, 1, 2), (3, 5, 1, 2), (5, 2, 3, 1)\}$$

Solution. We use the **SumBasis** function in the **LinearAlgebra** packageto show that the set $\mathcal{B} \cup \mathcal{C}$ is a basis for $U + V$. By Theorem 4.40, the sum $U + V$ is therefore direct.

```
[> with(LinearAlgebra):
[> v1:=<3|4|1|3>:     v2:=<2|2|1|2>:
[> v3:=<3|5|1|2>:     v4:=<5|2|3|1>:
[> SumBasis([{v1},{v2,v3,v4}]);

            {[5, 2, 3, 1], [3, 5, 1, 2], [2, 2, 1, 2], [3, 4, 1, 3]}
```

As we can see, the sum basis is $\mathcal{B} \cup \mathcal{C}$.

We could also have shown that the intersection basis is empty.

```
[> IntersectionBasis([{v1},{v2,v3,v4}]);

                              {}
```

Corollary 4.38 then tells us that the sum of U and W is direct. ▲

EXERCISES 4.15

1. Let U be the set of all solutions of the equation $y = 3x$ and W be the set of all solutions of the equation $y = -7x$ in $\mathbb{R}^2$. Show that $\mathbb{R}^2 = U \oplus W$.

2. Express $\mathbb{R}^3$ as a direct sum of subspaces in three different ways.

3. Show that $\mathbb{R}^3$ is the direct sum of the row space and the null space of the matrix

$$\begin{bmatrix} 1 & 0 & 1 \\ 1 & 0 & 1 \\ 1 & 0 & 1 \\ 0 & 0 & 0 \end{bmatrix}$$

4. Show that $\mathbb{R}^4$ is the direct sum of the column space and the left null space of the matrix

$$\begin{bmatrix} 1 & 0 & 1 \\ 1 & 0 & 1 \\ 1 & 0 & 1 \\ 0 & 0 & 0 \end{bmatrix}$$

5. Let $\mathcal{B} = \{\mathbf{v}_1, \ldots, \mathbf{v}_r, \mathbf{v}_{r+1}, \ldots, \mathbf{v}_n\}$ be a basis for a vector space V, and let U and V be the spaces span $\{\mathbf{v}_1, \ldots, \mathbf{v}_r\}$ and $W = \text{span}\,\{\mathbf{v}_{r+1}, \ldots, \mathbf{v}_n\}$. Show that if $V = U \oplus W$, then $\mathcal{C} = \{\mathbf{v}_{r+1}, \ldots, \mathbf{v}_n\}$ is a basis for W.

6. (*Maple V*) Choose three linearly independent vectors $\mathbf{x}$, $\mathbf{y}$, and $\mathbf{z}$ in $\mathbb{R}^4$ and construct the matrix 4×4 matrix $A = [\mathbf{x}\ \mathbf{y}\ \mathbf{y}\ \mathbf{z}]$. Show that the rank of A is 3, and use A to construct two subspaces U and V of dimension 2. Calculate the dimension of the intersection of U and V.

7. (*Maple 6*) Construct a 4×4 matrix A of column rank 2, and calculate the dimensions of the different subspaces of $\mathbb{R}^4$ determined by the columns of A.

8. (*Maple 6*) Construct a 4×4 matrix A of column rank 4, and form two different direct-sum decompositions of $\mathbb{R}^4$ determined by the columns of A.

9. (*Maple 6*) Let $U = \text{span}\{(3, 2, 2), (2, 0, 1)\}$ and $V = \text{span}\{(5, 3, 4), (0, 1, 0), (5, 3, 1)\}$ be two subspaces of $\mathbb{R}^3$. Use the **SumBasis** function in the **LinearAlgebra** package to show that the sum $U + V$ is not direct.

10. (*Maple 6*) Use the **IntersectionBasis** function in the **LinearAlgebra** package to show that the basis for the intersection of the two spaces in Exercise 6 is not empty.

COMPLEX VECTOR SPACES

At various times in our work we will need the complex analogs of ***coordinate spaces, matrix spaces***, and ***polynomial spaces***.

EXAMPLE 4.52 The Space $\mathbb{C}^n$

The space $\mathbb{C}^n$ of all $n \times 1$ column vectors $A : \mathbf{n} \times \mathbf{1} \to \mathbb{C}$, equipped with matrix addition and the scalar multiplication of matrices, is a complex vector space. ▲

EXAMPLE 4.53 The Space $\mathbb{C}^{m \times n}$

The space $\mathbb{C}^{m \times n}$ consists of all $m \times n$ matrices A with complex entries, equipped with matrix addition and the scalar multiplication of matrices as vector space operations. We refer to this space as the space of complex $m \times n$ matrices. ▲

EXAMPLE 4.54 The Space $\mathbb{C}[t]$

If $p(t) = a_0 + a_1 t + \cdots + a_n t^n$ and $q(t) = b_0 + b_1 t + \cdots + b_n t^n$ are two polynomials in $\mathbb{C}[t]$, then we can think of them as vectors. The definitions

$$
\begin{aligned}
p(t) + q(t) &= (a_0 + b_0) + (a_1 + b_1)t + \cdots + (a_n + b_n)t^n \\
ap(t) &= aa_0 + aa_1 t + \cdots + aa_n t^n \\
\mathbf{0} &= 0
\end{aligned}
$$

turn $\mathbb{C}[t]$ into a real vector space. ▲

EXERCISES 4.16

1. Describe the set of linear combinations $a\mathbf{x} + b\mathbf{y}$ in $\mathbb{C}^2$ determined by the set

$$\mathcal{S} = \{(1, 0), (0, 1)\}$$

and all $a, b \in \mathbb{C}$.

2. Describe the set of linear combinations $a\mathbf{x} + b\mathbf{y}$ in $\mathbb{C}^2$ determined by the set

$$\mathcal{S} = \{(1, 0), (0, 1)\}$$

and all $a, b \in \mathbb{R}$.

3. Explain why the set $\mathcal{S} = \{(1, 0), (0, i), (0, 1), (i, 0)\}$ is linearly independent in $\mathbb{C}^2$ when $\mathbb{C}^2$ is considered as a real vector space but linearly dependent when $\mathbb{C}^2$ is considered as complex vector space.

4. Find a basis for the space $\mathbb{C}[t]$ of complex polynomials, considered as a complex vector space.

5. (*Maple V*) Verify that the set $\mathbb{C}^2$ of complex column vectors of height 2 satisfies the axioms for a complex vector space.

6. (*Maple V*) Verify that the set $\mathbb{C}$ of complex numbers, together with the set $\mathbb{R}$ of real numbers as scalars, satisfies the axioms of a real vector space.

7. (*Maple V*) Verify that the set $\mathbb{C}^2$ of complex column vectors of height 2, together with the set $\mathbb{R}$ of real numbers as scalars, satisfies the axioms of a real vector space.

8. (*Maple V*) Use the **Complex** function to find all linear combinations $a\mathbf{x} + b\mathbf{y}$ in $\mathbb{C}^3$, for

$$\mathbf{x} = 3 + 4i, \quad \mathbf{y} = 3 - 4i, \quad a = 6, \quad b = 7$$

9. (*Maple 6*) Describe the vectors of $\mathbb{C}^2$ in three different ways as linear combinations of the vectors $\mathcal{S} = \{(1, 0), (0, i), (0, 1), (i, 0)\}$ and real scalars a, b, c, d.

10. (*Maple 6*) Find a basis $\mathcal{B}$ for the space $\mathbb{C}_3[t]$ of complex polynomials of degree 3 or less, considered as a complex vector space. Use *Maple* to prove that $\mathcal{B}$ is a basis.

Quaternions

The unit vectors i, j, and k, described in Chapter 3, can be used to construct a four-dimensional real vector space known as the ***quaternions***. The space is usually denoted by $\mathbb{H}$, in honor of Sir William Rowan Hamilton, who discovered the quaternions in the early 19th century. Today quaternions are used extensively in computer animation to calculate the rotation of images in $\mathbb{R}^3$. The essence of quaternions is not only that they satisfy the axioms of a vector space, but that they also come equipped with a noncommutative multiplication.

We conclude this chapter by discussing several ways in which the multiplication of quaternions can be described. The algebra of quaternions does for rotations in $\mathbb{R}^3$ what the algebra of complex numbers does for rotations in $\mathbb{R}^2$. The motivating idea that led Hamilton to quaternions is the fact that a complex number $a+bi$ can be written as a pair of real numbers (a, b), with

$$(a, b) + (c, d) = (a + c, b + d) \quad \text{and} \quad (a, b)(c, d) = (ac - bd, ad + bc)$$

This means that $(a, b) = (a, 0) + (0, b) = (a, 0) + (b, 0)(0, 1) = a + bi$, where $(0, 1)$ is represented by the symbol i. In particular, $i^2 = (0, 1)(0, 1) = (-1, 0) = -1$ and $i(a, b) = (0, 1)(a, b) = (-b, a)$. Since

$$\begin{bmatrix} \cos\frac{\pi}{2} & -\sin\frac{\pi}{2} \\ \sin\frac{\pi}{2} & \cos\frac{\pi}{2} \end{bmatrix} \begin{bmatrix} a \\ b \end{bmatrix} = \begin{bmatrix} -b \\ a \end{bmatrix}$$

we can see that, geometrically, the multiplication of a complex number by i corresponds to a counterclockwise rotation of the point (a, b) by $\pi/2$ radians.

Quaternions as vectors in $\mathbb{R}^4$

For any four real numbers a, b, c, and d, we can adapt the representation of complex numbers as ordered pairs to four dimensions and write the vector $(a, b, c, d) \in \mathbb{R}^4$ as a linear combination

$$(a, b, c, d) = a1 + bI + cJ + dK$$

in the standard basis for $\mathbb{R}^4$, where

$$1 = (1, 0, 0, 0), \quad I = (0, 1, 0, 0), \quad J = (0, 0, 1, 0), \quad K = (0, 0, 0, 1)$$

We can then define the addition and multiplication of these vectors by analogy with the addition and multiplication of complex numbers. However, the multiplication formula becomes more complicated. The determining properties of the multiplication are summarized in Table 7.

TABLE 7 QUATERNION MULTIPLICATION.

$\times$	1	I	J	K
1	1	I	J	K
I	I	-1	K	$-J$
J	J	$-K$	-1	I
K	K	J	$-I$	-1

The table shows that $I^2 = J^2 = K^2 = -1$, that $IJ = K$, $JK = I$, $KI = J$, and that $JI = -K$, $KJ = -I$, $IK = -J$. These laws are of course reminiscent of the laws governing the cross products of the standard basis vectors i, j, and k in $\mathbb{R}^3$, discussed in Chapter 3. It is for this reason that quaternions can be used to describe rotations in $\mathbb{R}^3$.

Quaternions as vectors in $\mathbb{C}^{2\times 2}$

Another description of quaternions is in terms of complex 2×2 matrices. For any two complex numbers $z = a + bi$ and $w = c + di$, we define a ***quaternion*** to be a complex matrix of the form

$$H = \begin{bmatrix} z & w \\ -\overline{w} & \overline{z} \end{bmatrix} = \begin{bmatrix} a+bi & c+di \\ -c+di & a-bi \end{bmatrix}$$

where $\overline{w} = c - di$ is the complex conjugate of w and $\overline{z} = a - bi$ is the complex conjugate of z.

We note that every quaternion can be written as a linear combination of the form

$$\begin{bmatrix} a+bi & c+di \\ -c+di & a-bi \end{bmatrix} = aU + bI + cJ + dK$$

where

$$U = \begin{bmatrix} 1 & 0 \\ 0 & 1 \end{bmatrix}, \quad I = \begin{bmatrix} i & 0 \\ 0 & -i \end{bmatrix}$$

$$J = \begin{bmatrix} 0 & 1 \\ -1 & 0 \end{bmatrix}, \quad K = \begin{bmatrix} 0 & i \\ i & 0 \end{bmatrix}$$

In this representation, quaternion multiplication is matrix multiplication.

It is quite remarkable that the complex multiples

$$(-i)\,I = \begin{bmatrix} 1 & 0 \\ 0 & -1 \end{bmatrix}, \quad (-i)\,J = \begin{bmatrix} 0 & -i \\ i & 0 \end{bmatrix}, \quad (-i)\,K = \begin{bmatrix} 0 & 1 \\ 1 & 0 \end{bmatrix}$$

are actually the Pauli spin matrices (see Exercise 2.8.8), used in quantum physics to study angular velocities of atoms.

Quaternions as scalar-vector pairs

One of the most useful representations of a quaternion $w + xi + yj + zk$ is as a pair

$$q = [w, \mathbf{v}]$$

where w is a real scalar and $\mathbf{v}$ is the vector $[x, y, z]$. In this notation, we can identify the points (x, y, z) in $\mathbb{R}^3$ with the quaternions $[0, \mathbf{v}]$. Let us call this representation of quaternions the ***scalar-vector form.***

The basic operations of quaternion can be defined elegantly using the scalar-vector, as shown in Table 8. The operation $\mathbf{v} \cdot \mathbf{v}'$ is the dot product of $\mathbf{v}$ and $\mathbf{v}'$ discussed briefly in Chapter 1, and $\mathbf{v} \times \mathbf{v}'$ is the vector cross product of $\mathbf{v}$ and $\mathbf{v}'$ discussed in Chapter 3.

TABLE 8 ALGEBRA OF QUATERNIONS.

Names	Operations
Addition	$q + q\prime = \left[w + w', \mathbf{v} + \mathbf{v}'\right]$
Multiplication	$qq' = \left[ww' - \mathbf{v} \cdot \mathbf{v}', \mathbf{v} \times \mathbf{v}' + w\mathbf{v}' + w'\mathbf{v}\right]$
Conjugate	$q^* = [w, -\mathbf{v}]$
Norm	$Norm(q) = \sqrt{qq^*} = \sqrt{w^2 + x^2 + y^2 + z^2}$
Inverse	$q^{-1} = q^*/Norm(q)$
Identities	$[1, (0, 0, 0)]$ and $[0, (0, 0, 0)]$

A quaternion q is a ***unit quaternion*** if $Norm(q) = 1$. In that case, $q^{-1} = q^*$. Unit quaternions can be used to represent rotations. It can be shown that the rotation of a vector $\mathbf{x} = (a, b, c)$ about the axis determined by unit quaternion $\mathbf{u} = [0, [x, y, z]]$ by an angle θ can be computed using the quaternion

$$q = [\cos(\theta/2), \sin(\theta/2)[x, y, z]]$$

The rotated vector is the vector part $\mathbf{w}$ of the quaternion $\mathbf{v}' = q\mathbf{v}q^* = [0, \mathbf{w}]$. The composition of rotations can be computed using the fact that if

$$q_1 = [\cos(\theta/2), \sin(\theta/2)[x, y, z]] \quad \text{and} \quad q_2 = [\cos(\varphi/2), \sin(\varphi/2)[x, y, z]]$$

then $q_2\left(q_1\mathbf{v}q_1^*\right)q_2^* = (q_2q_1)\,\mathbf{v}\left(q_1^*q_2^*\right) = (q_2q_1)\,\mathbf{v}\,(q_2q_1)^*$.

EXAMPLE 4.55 Quaternion Rotations in $\mathbb{R}^3$

Calculate the rotations in $\mathbb{R}^3$ determined by the quaternions I, J, and K.

Solution. Let $\mathbf{x} = [0, x, y, z]$ be an arbitrary quaternion representing a point in $\mathbb{R}^3$, and consider the following conjugate pairs:

Quaternion	Conjugate
$q_1 := [0, 1, 0, 0]$	$q_1^* = [0, -1, 0, 0]$
$q_2 := [0, 0, 1, 0]$	$q_2^* = [0, 0, -1, 0]$
$q_3 := [0, 0, 0, 1]$	$q_3^* = [0, 0, 0, -1]$

It is easy to verify that the rotations determined by I, J, and K are the following:

Rotation determined by I	$q_1\mathbf{x}q_1^* = [0, x, -y, -z]$
Rotation determined by J	$q_2\mathbf{x}q_2^* = [0, -x, y, -z]$
Rotation determined by K	$q_3\mathbf{x}q_3^* = [0, -x, -y, z]$

Moreover, the calculations

$t_1 = [\cos\frac{\pi}{2}, \sin\frac{\pi}{2}, 0, 0]$	$t_1\mathbf{x}t_1^* = [0, x, -y, -z]$
$t_2 = [\cos\frac{\pi}{2}, 0, \sin\frac{\pi}{2}, 0]$	$t_2\mathbf{x}t_2^* = [0, -x, y, -z]$
$t_3 = [\cos\frac{\pi}{2}, 0, 0, \sin\frac{\pi}{2}]$	$t_3\mathbf{x}t_3^* = [0, -x, -y, z]$

show that the multiplications by I, J, and K correspond to rotations by π around the coordinate axes. We can verify this fact in the usual way using the standard rotation matrices for rotating points in $\mathbb{R}^3$ about the coordinate axes. Let $\theta = \pi$, and apply the matrices

$$R_x = \begin{bmatrix} 1 & 0 & 0 \\ 0 & \cos\theta & -\sin\theta \\ 0 & \sin\theta & \cos\theta \end{bmatrix}, R_y = \begin{bmatrix} \cos\theta & 0 & \sin\theta \\ 0 & 1 & 0 \\ -\sin\theta & 0 & \cos\theta \end{bmatrix}, R_z = \begin{bmatrix} \cos\theta & -\sin\theta & 0 \\ \sin\theta & \cos\theta & 0 \\ 0 & 0 & 1 \end{bmatrix}$$

to the vector $\mathbf{x}$. We know from geometry that

$$R_x\mathbf{x} = (x, -y, -z), \quad R_y\mathbf{x} = (-x, y, -z), \quad R_z\mathbf{x} = (-x, -y, z)$$

are the vectors of obtained from $\mathbf{x}$ by rotating $\mathbf{x}$ by π radians about the x-, y-, and z-axis, respectively. As we can see, this agrees with the quaternion calculations. ▲

In aeronautics, the rotations corresponding to the quaternions

$qroll$	$=$	$[\cos(x/2), \sin(x/2), 0, 0]$
$qpitch$	$=$	$[\cos(y/2), 0, \sin(y/2), 0]$
$qyaw$	$=$	$[\cos(z/2), 0, 0, \sin(z/2)]$

are known as ***roll, pitch***, and ***yaw***.

EXAMPLE 4.56 Rotation about a Quaternion Axis

Let $(1/\sqrt{3}, 1/\sqrt{3}, 1/\sqrt{3})$ be the axis vector in $\mathbb{R}^3$ determined by the unit quaternion

$$\mathbf{u} = [0, 1/\sqrt{3}, 1/\sqrt{3}, 1/\sqrt{3}]$$

and let $\theta = \pi/2$. Calculate the quaternion $\mathbf{v}' = q\mathbf{v}q^*$, where $\mathbf{v}$ is the quaternion $[0, 2, -1, 4]$ and

$$q = [\cos(\theta/2), \sin(\theta/2)(1/\sqrt{3}, 1/\sqrt{3}, 1/\sqrt{3})]$$

and explain the result.

Solution. Using the *Maple* procedure for calculating with quaternions (see the answer to Exercise 4.17.3 in Appendix F), we continue as follows:

```
[> u:=quat(0,1/sqrt(3),1/sqrt(3),1/sqrt(3)):
[> theta:=Pi/2:
```

```
[> q:=quat(
     cos(theta/2),
     sin(theta/2)/sqrt(3),
     sin(theta/2)/sqrt(3),
     sin(theta/2)/sqrt(3)):
```

```
[> qconj:=quat(
     cos(theta/2),
     -sin(theta/2)/sqrt(3),
     -sin(theta/2)/sqrt(3),
     -sin(theta/2)/sqrt(3)):
```

```
[> v:=quat(0,2,-1,4):
[> w:=simplify(q&^v&^qconj);
```

$$w := \frac{5}{3}I + \frac{5}{3}J + \frac{5}{3}K - \sqrt{3}K - \frac{2}{3}\sqrt{3}J + \frac{5}{3}\sqrt{3}I$$

This calculation shows that the specified rotation maps the vector $(2, -1, 4)$ in $\mathbb{R}^3$ to the vector $(\frac{5}{3} + \frac{5}{3}\sqrt{3}, \frac{5}{3} - \frac{2}{3}\sqrt{3}, \frac{5}{3} - \sqrt{3})$. ▲

EXERCISES 4.17

1. Prove that for any unit quaternion $\mathbf{u} = [0, [x, y, z]]$, the rotation quaternion

$$q = [\cos(\theta/2), \sin(\theta/2)\,[x, y, z]]$$

is also a unit quaternion.

2. Prove that if $\mathbf{u} = [0, [x, y, z]]$ is a unit quaternion if

$$q_1 = [\cos(\theta/2), \sin(\theta/2)\,[x, y, z]] \quad \text{and} \quad q_2 = [\cos(\varphi/2), \sin(\varphi/2)\,[x, y, z]]$$

then

$$q_2\left(q_1\mathbf{v}q_1^*\right)q_2^* = (q_2q_1)\,\mathbf{v}\left(q_1^*q_2^*\right) = (q_2q_1)\,\mathbf{v}\,(q_1q_2)^*$$

3. (*Maple 6*) Implement the *Maple* procedure for the calculation with quaternions in the *Maple 6* Programming Guide.

4. (*Maple 6*) Verify that the quaternions

$$r = [0, 1/\sqrt{2}, 1/\sqrt{2}, 0], \quad s = [0, \sqrt{23}/6, 1/2, 1/3], \quad t = [0, 1/2, 1/2, 1/\sqrt{2}]$$

are unit quaternions, and calculate their conjugates, norms, and inverses.

5. (*Maple 6*) Find a real number a for which $q = [0, 1/3, 1/4, a]$ is a unit quaternion. Calculate the conjugate, norm, and inverse of q.

6. (*Maple 6*) Calculate the points in $\mathbb{R}^3$ obtained by rotating $(1, 2, 3)$ about the axis determined by the unit quaternion $\mathbf{x} = [0, 1/\sqrt{14}, 3/\sqrt{14}, 2/\sqrt{14}]$ and the angles $\theta = \pi/4$, $\pi/3$, and π.

7. (*Maple 6*) Show that if $\theta = 2\pi/3$, if q is the quaternion

$$\left[\left(1/\sqrt{3}\right)\cos\theta, \left(1/\sqrt{3}\right)\sin\theta, \left(1/\sqrt{3}\right)\sin\theta, \left(1/\sqrt{3}\right)\sin\theta\right]$$

and if $\mathbf{x}$ is the vector $[0, 1, 1, 1]$, then $(13/12)\,\mathbf{x} = q\mathbf{x}q^*$.

8. (*Maple 6*) Rewrite the multiplications in Example 4.55 using I, J, and K.

9. (*Maple 6*) Use the standard rotation matrices for rotations about the coordinate axes in $\mathbb{R}^3$ described in Example 4.55 and calculate the vectors obtained by rotating the vector $(4, 8, -1)$ by $\pi/3$ radian about the coordinate axes. Use quaternions to validate your results.

REVIEW

KEY CONCEPTS ▶ Define and discuss each of the following.

Bases

Finite basis, linear independence of a basis, span of a basis, standard basis.

Linear Transformations

coordinate-vector function, linear-combination function.

Sets of Vectors

Basis for a vector space, linearly dependent vectors, linearly independent vectors, standard basis of a vector space.

Subspaces

Column space, fundamental subspace, left null space, null space, row space, zero space.

Vector Operations

Linear combination of vectors, scalar multiplication, subtraction of vectors, vector addition.

Vector Space Properties

Dimension of a vector space, finite-dimensional space, infinite-dimensional space.

Vector Space Operations

Direct sum of vector spaces, subspace of a vector space, sum.

Vector Spaces

Complex vector space, coordinate space, matrix space, polynomial space, real vector space.

Vectors

Coordinate vector, linear combination of vectors, scalar multiple of a vector, zero vector.

KEY FACTS ▶ Explain and illustrate each of the following.

1. For any scalar a and any vector $\mathbf{v}$, it holds that $a\mathbf{v} = \mathbf{0}$ if and only if $a = 0$ or $\mathbf{v} = \mathbf{0}$.
2. A set $\{\mathbf{x}_1, \dots, \mathbf{x}_n\}$ of column vectors of height n is linearly independent if and only if the matrix $A = [\mathbf{x}_1 \ \cdots \ \mathbf{x}_n]$ is invertible.
3. A set $\{\mathbf{x}_1, \dots, \mathbf{x}_n\}$ of column vectors of height n is linearly independent if and only if the matrix $A = [\mathbf{x}_1 \ \cdots \ \mathbf{x}_n]$ has n pivots.
4. A nonempty set $\mathcal{S} = \{\mathbf{v}_1, \dots, \mathbf{v}_n\}$ is linearly dependent if and only if one of the vectors $\mathbf{v}_i$ can be written as a linear combination of the remaining vectors.

5. A nonempty set $\mathcal{S} = \{\mathbf{v}_1, \dots, \mathbf{v}_n\}$ is linearly independent if and only if none of the vectors $\mathbf{v}_i$ can be written as a linear combination of the remaining vectors.

6. A subset $\mathcal{B} = \{\mathbf{x}_1, \dots, \mathbf{x}_n\}$ of a vector space V is a basis for V if and only if every vector $\mathbf{x} \in V$ can be written as a unique linear combination $a_1\mathbf{x}_1 + \cdots + a_n\mathbf{x}_n$.

7. If $\mathcal{B} = \{\mathbf{v}_1, \dots, \mathbf{v}_n\}$ is any basis for $\mathbb{R}^n$ and $\mathbf{x}$ is any vector in $\mathbb{R}^n$, then the coefficients $a_1, \dots, a_n$ in the unique linear combination $\mathbf{x} = a_1\mathbf{v}_1 + \cdots + a_n\mathbf{v}_n$ can be found by solving a system of linear equations.

8. Let V be a vector space spanned by the vectors $\mathbf{v}_1, \dots, \mathbf{v}_m$. Then any linearly independent set of vectors in V is finite and contains no more than m elements.

9. If $\mathcal{B}$ and $\mathcal{C}$ are two bases of a finite-dimensional vector space V, then $\mathcal{B}$ and $\mathcal{C}$ have the same number of elements.

10. Suppose that $\mathcal{S}$ is a subset of a finite-dimensional vector space V that spans V. Then $\mathcal{S}$ has a linearly independent subset $\mathcal{S}'$ that is a basis for V.

11. Any linearly independent set of vectors $\mathcal{S}$ of a finite-dimensional vector space V can be extended to a basis for V.

12. Every invertible real $n \times n$ matrix determines a basis for $\mathbb{R}^n$.

13. The functions $\mathbf{x} \to [\mathbf{x}]_{\mathcal{B}}$ and $\mathbf{y} \to \widehat{\mathbf{y}}_{\mathcal{B}}$ are inverses of each other.

14. The coordinate-vector function $\mathbf{x} \to [\mathbf{x}]_{\mathcal{B}}$ from V to $\mathbb{R}^n$ has the property that $[\mathbf{x}+\mathbf{y}]_{\mathcal{B}} = [\mathbf{x}]_{\mathcal{B}} + [\mathbf{y}]_{\mathcal{B}}$ and that $[a\mathbf{x}]_{\mathcal{B}} = a[\mathbf{x}]_{\mathcal{B}}$.

15. The linear-combination function $\mathbf{y} \to \widehat{\mathbf{y}}_{\mathcal{B}}$ from $\mathbb{R}^n$ to V has the properties that $\widehat{\mathbf{x}+\mathbf{y}}_{\mathcal{B}} = \widehat{\mathbf{x}}_{\mathcal{B}} + \widehat{\mathbf{y}}_{\mathcal{B}}$ and that $\widehat{a\mathbf{x}}_{\mathcal{B}} = a\widehat{\mathbf{x}}_{\mathcal{B}}$.

16. Let V be an n-dimensional vector space, let $\mathcal{B}$ and $\mathcal{C}$ be two bases for V, and let $\mathbf{x}$ be a vector in V with coordinate vectors $[\mathbf{x}]_{\mathcal{B}}$ and $[\mathbf{x}]_{\mathcal{C}}$. Then there exists an invertible matrix P for which $P[\mathbf{x}]_{\mathcal{B}} = [\mathbf{x}]_{\mathcal{C}}$ and $P^{-1}[\mathbf{x}]_{\mathcal{C}} = [\mathbf{x}]_{\mathcal{B}}$.

17. Every invertible real $n \times n$ matrix is a coordinate conversion matrix for $\mathbb{R}^n$.

18. The span of a subset W of a vector space V is a subspace of V.

19. If U and W are subspaces of a vector space V, then so is $U \cap W$.

20. If W is a subspace of a finite-dimensional vector space V, and W and V have the same dimension, then $W = V$.

21. The pivot columns of an $m \times n$ matrix A over $\mathbb{R}$ form a basis for Col A.

22. The set of column vectors obtained by writing the general solution of the system $A\mathbf{x} = \mathbf{0}$ as a linear combination of the free variables of A is a basis for Nul A.

23. The nonzero rows of matrix in reduced row echelon form are linearly independent.

24. The nonzero rows of the reduced row echelon form B of an $m \times n$ matrix A are a basis for Col A^T.

25. The nonzero columns of the reduced row echelon form B of an $m \times n$ matrix A are a basis for Col A.

26. If A is an $m \times n$ matrix of rank k, and if $PA = LU$ is the permutation-free LU decomposition of A, then the last $m - k$ rows of $L^{-1}P$ are a basis for Nul A^T.

27. The spaces Col A and Col A^T have the same dimension.

28. For any real $m \times n$ matrix A, the sum of the rank and nullity of A is n.

29. A real $n \times n$ matrix A is invertible if and only if Col $A = \mathbb{R}^n$.

30. Let U and W be subspaces of a finite-dimensional vector space V. Then $\dim(U + W) = \dim U + \dim W - \dim(U \cap W)$.

31. A real $n \times n$ matrix A is invertible if and only if it has rank n.

32. A real $n \times n$ matrix A is invertible if and only if the columns of A are linearly independent.

33. A real $n \times n$ matrix A is invertible if and only if the dimension of Col A^T is n.

34. A real $n \times n$ matrix A is invertible if and only if Nul $A = \{\mathbf{0}\}$.

35. If U and W are disjoint subspaces of V, then

$$\dim(U + W) = \dim U + \dim W$$

36. If $V = U \oplus W$, then every vector $\mathbf{x} \in V$ can be written as a unique sum $\mathbf{u} + \mathbf{w}$, with $\mathbf{u} \in U$ and $\mathbf{w} \in W$.

37. If U and W are disjoint subspaces of a vector space V with bases $\mathcal{B}$ and $\mathcal{C}$, then $V = U \oplus W$ if and only if $\mathcal{B} \cup \mathcal{C}$ is a basis for V.

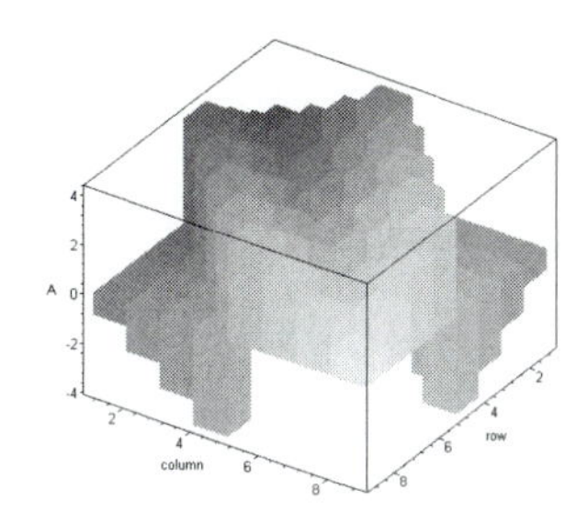

5 LINEAR TRANSFORMATIONS

In mathematics, we study both individual objects and their interactions. In linear algebra, the objects are vector spaces. The interactions between vector spaces have various names, such as linear maps, vector space homomorphisms, and linear transformations; in this text we call them ***linear transformations***. From one point of view, linear algebra deals almost exclusively with the study of the properties of linear transformations, which map the vectors in one space—called the ***domain*** of the function—to another space—called the ***codomain*** of the function (sometimes also called the ***range***). To be linear, these functions must ***preserve linear combinations***. Figure 1 illustrates the connection between the domain and the codomain of a linear transformation.

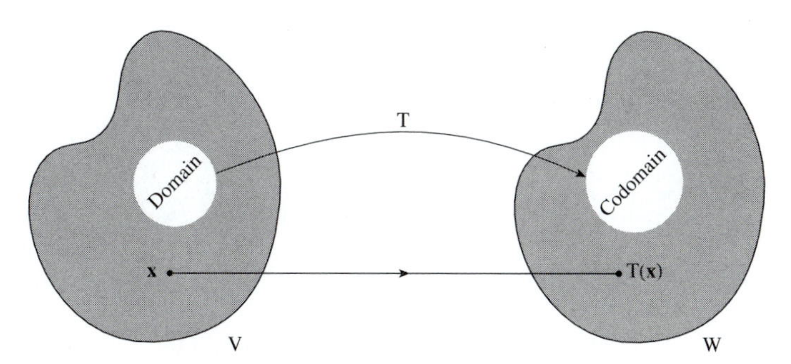

FIGURE 1 A Linear Transformation from V to W.

LINEAR TRANSFORMATIONS

We begin by defining linear transformations between real vector spaces. If we replace the word *real* by the word *complex* and $\mathbb{R}$ by $\mathbb{C}$, we get the corresponding definition for complex spaces.

DEFINITION 5.1 *If V and W are two real vector spaces, a* ***linear transformation*** *from V to W is a function $T : V \to W$ with the property that*

$$T(\mathbf{x}+\mathbf{y}) = T(\mathbf{x}) + T(\mathbf{y})$$
$$T(a\mathbf{x}) = aT(\mathbf{x})$$

for all vectors $\mathbf{x}, \mathbf{y} \in V$ and all scalars $a \in \mathbb{R}$.

We can combine the two conditions defining a linear transformation into a single equation by saying that a function $T : V \to W$ is a linear transformation if

$$T(a\mathbf{x} + b\mathbf{y}) = aT(\mathbf{x}) + bT(\mathbf{y})$$

for all $\mathbf{x}, \mathbf{y} \in V$ and all $a, b \in \mathbb{R}$. This equation expresses the idea that T ***preserves linear combinations***: The linear combination $a\mathbf{x} + b\mathbf{y} \in V$ is mapped to the linear combination $aT(\mathbf{x}) + bT(\mathbf{y}) \in W$, with the same scalars. Instead of saying that T is a linear transformation, we often simply will say that T is linear.

EXAMPLE 5.1 A Linear Transformation from $\mathbb{R}^{2\times 3}$ to $\mathbb{R}^2$

Use *Maple* to show that the function $T : \mathbb{R}^{2\times 3} \to \mathbb{R}^2$ defined by

$$T\begin{bmatrix} a_{11} & a_{12} & a_{13} \\ a_{21} & a_{22} & a_{23} \end{bmatrix} = \begin{bmatrix} a_{11} \\ a_{21} \end{bmatrix}$$

is a linear transformation.

Solution. We begin by defining the function T.

```
[> T:=(s)->array([s[1,1],s[2,1]]);
```

$$T := s- > array([s_{1,1}, s_{2,1}])$$

Next we specify two general 2×3 matrices $\mathbf{x}$ and $\mathbf{y}$.

```
[> x:=array([[p11,p12,p13],[p21,p22,p23]]);
```

$$x := \begin{bmatrix} p11 & p12 & p13 \\ p21 & p22 & p23 \end{bmatrix}$$

```
[> y:=array([[q11,q12,q13],[q21,q22,q23]]);
```

$$y := \begin{bmatrix} q11 & q12 & q13 \\ q21 & q22 & q23 \end{bmatrix}$$

We now verify that $T(a\mathbf{x}+b\mathbf{y}) = aT(\mathbf{x}) + bT(\mathbf{y})$.

```
[> evalm(T(evalm(a*x+b*y)));
```

$$[ap11 + bq11, ap21 + bq21]$$

```
[> evalm(a*T(x)+b*T(y));
```

$$[ap11 + bq11, ap21 + bq21]$$

```
[> linalg[equal](%,%%);
```

$$true$$

Since the equation $T(a\mathbf{x}+b\mathbf{y}) = aT(\mathbf{x}) + bT(\mathbf{y})$ holds for arbitrary scalars a and b and arbitrary vectors $\mathbf{x}$ and $\mathbf{y}$, the function T is linear. ▲

EXAMPLE 5.2 A Linear Transformation from $\mathbb{R}^4$ to $\mathbb{R}[t]$

Use *Maple* to show that the function $T : \mathbb{R}^4 \to \mathbb{R}[t]$ given by

$$T\begin{bmatrix} a_0 \\ a_1 \\ a_2 \\ a_3 \end{bmatrix} = a_0 + a_1 t + a_2 t^2 + a_3 t^3$$

is a linear transformation.

Solution. We define the function T taking column vectors to polynomials in t as follows:

```
[> T:=s->s[1]+s[2]*t+(s[3])*t^2+(s[4])*t^3;
```

$$T := s- > s_1 + s_2 t + s_3 t^2 + s_4 t^3$$

Next we create two arbitrary column vectors.

```
[> x:=<x1,x2,x3,x4>;
```

$$x := \begin{bmatrix} x1 \\ x2 \\ x3 \\ x4 \end{bmatrix}$$

```
[> y:=<y1,y2,y3,y4>;
```

$$y := \begin{bmatrix} y1 \\ y2 \\ y3 \\ y4 \end{bmatrix}$$

We conclude by verifying that $T(a\mathbf{x} + b\mathbf{y}) = aT(\mathbf{x}) + bT(\mathbf{y})$.

```
[> expand(T(evalm(a*x+b*y)));
```

$$a\,x1 + t\,a\,x2 + t^2\,a\,x3 + t^3\,a\,x4 + b\,y1 + b\,y2\,t + b\,y3\,t^2 + b\,y4\,t^3$$

```
[> expand(a*T(x)+b*T(y));
```

$$a\,x1 + t\,a\,x2 + t^2\,a\,x3 + t^3\,a\,x4 + b\,y1 + b\,y2\,t + b\,y3\,t^2 + b\,y4\,t^3$$

As we can tell, the two expressions are equal. The dialog

```
[> is(expand(T(evalm(a*x+b*y)))=expand(a*T(x)+b*T(y)));
```

true

confirms this fact. It shows that $T(a\mathbf{x} + b\mathbf{y}) = aT(\mathbf{x}) + bT(\mathbf{y})$. Hence T is linear. ▲

EXAMPLE 5.3 A Linear Transformation from $\mathbb{R}^4$ to $\mathbb{R}_3[t]$

Let $T : \mathbb{R}^4 \to \mathbb{R}_3[t]$ be the function defined by the same rule as the linear transformation in Example 5.2. Then the preceding calculations still apply and show that T is a linear transformation. The only difference between the linear transformations in the two examples is that the codomain of T in Example 5.2 is infinite-dimensional, whereas the codomain of T in the present example has dimension 4. ▲

EXAMPLE 5.4 A Linear Transformation from $\mathbb{R}_3[t]$ to $\mathbb{R}^4$

Use *Maple* to show that the function $T : \mathbb{R}_3[t] \to \mathbb{R}^4$ given by

$$T(a_0 + a_1 t + a_2 t^2 + a_3 t^3) = \begin{bmatrix} a_0 \\ a_1 \\ a_2 \\ a_3 \end{bmatrix}$$

is a linear transformation.

Solution. We begin by defining the function T.

```
[> T:=s->array(
    [coeff(s,x,0),coeff(s,x,1),coeff(s,x,2),coeff(s,x,3)]);
```

$$T := s- > array([coeff(s,x,0), coeff(s,x,1), coeff(s,x,2), coeff(s,x,3)])$$

We prove that T is linear by showing that it preserves linear combinations.

```
[> p:=p1+p2*x+p3*x^2+p4*x^3:
[> q:=q1+q2*x+q3*x^2+q4*x^3:
[> u:=Vector[column](T(collect(a*p+b*q,x)));
```

$$u := \begin{bmatrix} a\,p1 + b\,q1 \\ a\,p2 + b\,q2 \\ a\,p3 + b\,q3 \\ a\,p4 + b\,q4 \end{bmatrix}$$

```
[> v:=Vector[column](evalm(T(expand(a*p))+T(expand(b*q))));
```

$$v := \begin{bmatrix} a\,p1 + b\,q1 \\ a\,p2 + b\,q2 \\ a\,p3 + b\,q3 \\ a\,p4 + b\,q4 \end{bmatrix}$$

This shows that $T(a\mathbf{x} + b\mathbf{y}) = aT(\mathbf{x}) + bT(\mathbf{y})$. Hence T is linear. ▲

EXAMPLE 5.5 A Linear Transformation from $\mathbb{R}^{2\times 2}$ to $\mathbb{R}$

Use *Maple* to show that the trace function $T : \mathbb{R}^{2\times 2} \to \mathbb{R}$ defined by

$$T \begin{bmatrix} a & b \\ c & d \end{bmatrix} = a + d$$

is a linear transformation.

Solution. We first define T.

```
[> T:=s->s[1,1]+s[2,2];
```

$$T := s- > s_{1,1} + s_{2,2}$$

Next we specify two general 2×2 matrices.

```
[> with(LinearAlgebra):
[> A:=RandomMatrix(2,2,generator=g):
[> B:=RandomMatrix(2,2,generator=h):
```

Then we compare $aT(A) + bT(B)$ and $T(aA + bB)$.

```
[> T(evalm(a*A+b*B));
```

$$a\,g(1,1) + b\,h(1,1) + a\,g(2,2) + b\,h(2,2)$$

```
[> expand(a*T(A)+b*T(B));
```

$$a\,g(1,1) + b\,h(1,1) + a\,g(2,2) + b\,h(2,2)$$

The last calculation shows that $aT(\mathbf{x}) + bT(\mathbf{y}) = T(a\mathbf{x} + b\mathbf{y})$ for all vectors **x** and **y** and all scalars a and b. Hence T is linear. ▲

Matrix Transformations

Among the linear transformations are the transformations between coordinate spaces whose functional rules are determined by matrix multiplication.

DEFINITION 5.2 *A **matrix transformation** is a function $T : \mathbb{R}^n \to \mathbb{R}^m$ for which there exists a real $m \times n$ matrix A with the property that $T(\mathbf{x}) = A\mathbf{x}$.*

Since T depends on a matrix A, we denote T by T_A.

THEOREM 5.1 *Every matrix transformation is a linear transformation.*

Proof. The linearity of T_A follows from the linearity of matrix multiplication:

$$\begin{aligned} T_A(a\mathbf{x} + b\mathbf{y}) &= A(a\mathbf{x} + b\mathbf{y}) \\ &= aA\mathbf{x} + bA\mathbf{y} \\ &= aT_A(\mathbf{x}) + bT_A(\mathbf{y}) \end{aligned}$$

By Definition 5.1, T_A is a linear transformation. ■

We will see later that this theorem has a converse. All linear transformations on finite-dimensional vector spaces are, in some sense, matrix transformations. However, the matrix representation of a linear transformation $T : \mathbb{R}^m \to \mathbb{R}^n$ is virtually never unique. It varies with the bases for $\mathbb{R}^m$ and $\mathbb{R}^n$ chosen to represent T. In fact, much of our work will consist of developing strategies for finding suitable bases in which a linear transformation has a simple matrix representation. By "simple," we often mean diagonal.

EXAMPLE 5.6 A Matrix Transformation from $\mathbb{R}^2$ to $\mathbb{R}^3$

Use *Maple* to show that the function $T_A : \mathbb{R}^2 \to \mathbb{R}^3$ defined by

$$T_A \begin{bmatrix} x \\ y \end{bmatrix} = \begin{bmatrix} 1 & 2 \\ 3 & 4 \\ 5 & 6 \end{bmatrix} \begin{bmatrix} x \\ y \end{bmatrix} = \begin{bmatrix} x + 2y \\ 3x + 4y \\ 5x + 6y \end{bmatrix}$$

is linear.

Solution. We must show that

$$\begin{bmatrix} 1 & 2 \\ 3 & 4 \\ 5 & 6 \end{bmatrix} \left[a \begin{bmatrix} x_1 \\ x_2 \end{bmatrix} + b \begin{bmatrix} y_1 \\ y_2 \end{bmatrix} \right] = a \begin{bmatrix} 1 & 2 \\ 3 & 4 \\ 5 & 6 \end{bmatrix} \begin{bmatrix} x_1 \\ x_2 \end{bmatrix} + b \begin{bmatrix} 1 & 2 \\ 3 & 4 \\ 5 & 6 \end{bmatrix} \begin{bmatrix} y_1 \\ y_2 \end{bmatrix}$$

```
[>with(LinearAlgebra):
[>A:=Matrix([[1,2],[3,4],[5,6]]):
[>x:=<x1,x2>:
[>y:=<y1,y2>:
```

```
[>left:=evalm(A&*(a*x+b*y)):
[>right:=simplify(evalm(a*A&*x+b*A&*y)):
```

```
[>u:=Vector[column](3,[left[1],left[2],left[3]]):
[>v:=Vector[column](3,[right[1],right[2],right[3]]):
```

```
[>Equal(u,v);
                              true
```

This shows that $A(a\mathbf{x} + b\mathbf{y}) = aA\mathbf{x} + bA\mathbf{y}$. Hence T_A is linear. ▲

EXAMPLE 5.7 A Matrix Transformation from $\mathbb{R}^3$ to $\mathbb{R}^2$

Use *Maple* to show that the function $T_A : \mathbb{R}^3 \to \mathbb{R}^2$ defined by

$$T_A \begin{bmatrix} x \\ y \end{bmatrix} = \begin{bmatrix} 1 & 3 & 5 \\ 2 & 4 & 6 \end{bmatrix} \begin{bmatrix} x \\ y \\ z \end{bmatrix} = \begin{bmatrix} x + 3y + 5z \\ 2x + 4y + 6z \end{bmatrix}$$

is linear.

Solution. We show that the vector

$$\begin{bmatrix} 1 & 3 & 5 \\ 2 & 4 & 6 \end{bmatrix} \left(a \begin{bmatrix} x_1 \\ x_2 \\ x_3 \end{bmatrix} + b \begin{bmatrix} y_1 \\ y_2 \\ y_3 \end{bmatrix} \right)$$

is equal to the vector

$$a \begin{bmatrix} 1 & 3 & 5 \\ 2 & 4 & 6 \end{bmatrix} \begin{bmatrix} x_1 \\ x_2 \\ x_3 \end{bmatrix} + b \begin{bmatrix} 1 & 3 & 5 \\ 2 & 4 & 6 \end{bmatrix} \begin{bmatrix} y_1 \\ y_2 \\ y_3 \end{bmatrix}$$

```
[>with(LinearAlgebra):
[>A:=array([[1,3,5],[2,4,6]]):
[>x:=<x1,x2,x3>:
[>y:=<y1,y2,y3>:
```

```
[>left:=evalm(A&*(a*x+b*y)):
[>right:=simplify(evalm(a*A&*x+b*A&*y)):
```

```
[>u:=Vector[column](2,[left[1],left[2]]):
[>v:=Vector[column](2,[right[1],right[2]]):
```

```
[>Equal(u,v);
                              true
```

This shows that $A(a\mathbf{x} + b\mathbf{y}) = aA\mathbf{x} + bA\mathbf{y}$. Hence T_A is linear. ▲

Examples 5.6 and 5.7 illustrate that with every matrix transformation $T_A : \mathbb{R}^n \to \mathbb{R}^m$ we can associate a linear transformation $T_{(A^T)} : \mathbb{R}^m \to \mathbb{R}^n$ defined by

$$T_{(A^T)}(\mathbf{x}) = A^T\mathbf{x}$$

The transformation $T_{(A^T)}$ is the ***transpose*** of T_A.

One of the basic properties of a linear transformation from a vector space V to a vector space W is the fact that it maps the zero vector $\mathbf{0}_V$ of V to the zero vector $\mathbf{0}_W$ of W.

THEOREM 5.2 *If $T : V \to W$ is a linear transformation, then $T(\mathbf{0}_V) = \mathbf{0}_W$.*

Proof. Recall from Theorem 4.1 that $a\mathbf{0} = \mathbf{0}$ in all vector spaces and for any scalar a. Therefore $T(\mathbf{0}_V) = T(0\mathbf{0}_V) = 0T(\mathbf{0}_V) = \mathbf{0}_W$, where $\mathbf{0}_V$ is the zero vector of V and $\mathbf{0}_W$ is the zero vector of W. ■

This property of linear transformations prevents certain functions whose graphs are linear from being linear transformations. For example, the translation function $T : \mathbb{R}^2 \to \mathbb{R}^2$ defined by

$$T\begin{bmatrix} x \\ y \end{bmatrix} = \begin{bmatrix} x+h \\ y+k \end{bmatrix}$$

is not a linear transformation if either h or k is not zero. The failure of translations to be linear has given rise to a separate subject called ***affine geometry***, which deals with the study of transformations $T : \mathbb{R}^n \to \mathbb{R}^m$ of the form $T[\mathbf{x}] = A\mathbf{x} + \mathbf{b}$, determined by $m \times n$ matrices A and fixed translation vectors $\mathbf{b} \in \mathbb{R}^m$. As indicated in Chapter 2, these transformations are known as ***affine transformations.*** The **AffineMap** function, defined by the assignment

[> **AffineMap** := (**A**, **x**) − >
  evalm(**SubMatrix**(**A**, **1**..**2**, **1**..**2**)& ∗ **x** + **SubMatrix**(**A**, **1**..**2**, **3**))

provides a useful tool for using *Maple 6* to calculate the values of affine transformations of the plane.

EXAMPLE 5.8 An Affine Transformation $T : \mathbb{R}^2 \to \mathbb{R}^2$

Use *Maple* to calculate the values of the affine transformation $T : \mathbb{R}^2 \to \mathbb{R}^2$ defined by

$$A = \begin{bmatrix} 1 & 2 \\ 3 & 4 \end{bmatrix} \quad \text{and} \quad \mathbf{b} = \begin{bmatrix} 5 \\ 6 \end{bmatrix}$$

Solution. We begin by loading the **LinearAlgebra** package and defining the **AffineMap** functions.

```
[>with(LinearAlgebra):
[>AffineMap:=(A,x)->
   evalm(SubMatrix(A,1..2,1..2)&*x+SubMatrix(A,1..2,3));
```

Next we define the augmented matrix $[A \mid \mathbf{b}]$ and calculate $T(\mathbf{x})$ for a general vector $\mathbf{x} = (x, y)$.

```
[> A:=Matrix([[1,2,5],[3,4,6]]):
```

$$A := \begin{bmatrix} 1 & 2 & 5 \\ 3 & 4 & 6 \end{bmatrix}$$

```
[> X:=Vector([x,y]);
```

$$X := \begin{bmatrix} x \\ y \end{bmatrix}$$

```
[> AffineMap(A,X);
```

$$\begin{bmatrix} x + 2y + 5 \\ 3x + 4y + 6 \end{bmatrix}$$

As we can see, the output of the command **AffineMap(A,X)** is the vector

$$\begin{bmatrix} 1 & 2 \\ 3 & 4 \end{bmatrix} \begin{bmatrix} x \\ y \end{bmatrix} + \begin{bmatrix} 5 \\ 6 \end{bmatrix}$$

obtained by forming the vector $A\mathbf{x}$ and then adding the vector $\mathbf{b}$ to $A\mathbf{x}$. ▲

Bases and Linear Transformations

In this section, we show that every basis $\mathcal{B} = \{\mathbf{x}_1, \mathbf{x}_2, \ldots, \mathbf{x}_n\}$ for a real vector space V and every list $\{\mathbf{y}_1, \mathbf{y}_2, \ldots, \mathbf{y}_n\}$ of n not necessarily distinct vectors in a vector space W determine a unique linear transformation $T : V \to W$. We express this fact by saying *that every linear transformation is uniquely determined by its values on a basis.* We will see later that if W is an m-dimensional real vector space, then this property implies that, relative to fixed bases for $\mathbb{R}^n$ and $\mathbb{R}^m$, every linear transformation $T : V \to W$ corresponds to a unique matrix transformation $T_A : \mathbb{R}^n \to \mathbb{R}^m$. The matrix A is a real $m \times n$ matrix. The transformation T is therefore defined by mn scalars. This makes linear transformations very special functions. Although $\mathbb{R}^n$ and $\mathbb{R}^m$ consist of infinitely many vectors, and T usually has infinitely many distinct values, all of them are determined by a finite amount of information, namely, the entries of the matrix A. It is this remarkable property that allows us to use matrix algebra to study linear transformations.

THEOREM 5.3 (Linear transformation theorem) *If $\mathcal{B} = \{\mathbf{x}_1, \mathbf{x}_2, \ldots, \mathbf{x}_n\}$ is a basis for a vector space V and if $\mathbf{y}_1, \mathbf{y}_2, \ldots, \mathbf{y}_n$ is any list of vectors in a vector space W, then there exists a unique linear transformation $T : V \to W$ for which $T(\mathbf{x}_1) = \mathbf{y}_1, \ldots, T(\mathbf{x}_n) = \mathbf{y}_n$.*

Proof. Let us $T : \mathcal{B} \to W$ be the function defined by $T(\mathbf{x}_1) = \mathbf{y}_1, \ldots, T(\mathbf{x}_n) = \mathbf{y}_n$, and let $\mathbf{x}$ be any vector in V. Since $\mathcal{B}$ is a basis for V, the vector $\mathbf{x}$ is a unique linear combination

$a_1\mathbf{x}_1 + \cdots + a_n\mathbf{x}_n$ of the basis vectors $\mathbf{x}_1, \ldots, \mathbf{x}_n$. We now extend the function T from $\mathcal{B}$ to the vector space V by defining

$$\begin{aligned} T(\mathbf{x}) &= T(a_1\mathbf{x}_1 + \cdots + a_n\mathbf{x}_n) \\ &= a_1 T(\mathbf{x}_1) + \cdots + a_n T(\mathbf{x}_n) \\ &= a_1\mathbf{y}_1 + \cdots + a_n\mathbf{y}_n \end{aligned}$$

By definition, the transformation T is linear. It remains for us to show that T is unique. Suppose that $S : \mathcal{B} \to W$ is another linear transformation with the property that $S(\mathbf{x}_1) = \mathbf{y}_1, \ldots, S(\mathbf{x}_n) = \mathbf{y}_n$. Then

$$\begin{aligned} S(\mathbf{x}) &= S(a_1\mathbf{x}_1 + \cdots + a_n\mathbf{x}_n) \\ &= a_1 S(\mathbf{x}_1) + \cdots + a_n S(\mathbf{x}_n) \\ &= a_1 T(\mathbf{x}_1) + \cdots + a_n T(\mathbf{x}_n) \\ &= T(a_1\mathbf{x}_1 + \cdots + a_n\mathbf{x}_n) \\ &= T(\mathbf{x}) \end{aligned}$$

for all $\mathbf{x} \in V$. Hence $S = T$. ■

EXAMPLE 5.9 Defining a Linear Transformation $T : \mathbb{R}^3 \to \mathbb{R}^4$ on a Basis

Show that the linear transformation $T : \mathbb{R}^3 \to \mathbb{R}^4$ defined by

$$T\begin{bmatrix} a_1 \\ a_2 \\ a_3 \end{bmatrix} = \begin{bmatrix} a_1 + a_2 \\ a_2 \\ a_3 \\ a_1 - a_3 \end{bmatrix}$$

is uniquely determined by its values on the standard basis $\mathcal{E}$ of $\mathbb{R}^3$.

Solution. We recall that $\mathcal{E} = \{\mathbf{e}_1 = (1, 0, 0), \mathbf{e}_2 = (0, 1, 0), \mathbf{e}_3 = (0, 0, 1)\}$ and the values of T on $\mathcal{E}$ are

$$\begin{cases} T(\mathbf{e}_1) = (1, 0, 0, 1) \\ T(\mathbf{e}_2) = (1, 1, 0, 0) \\ T(\mathbf{e}_3) = (0, 0, 1, -1) \end{cases}$$

If $\mathbf{x} = a_1\mathbf{e}_1 + a_2\mathbf{e}_2 + a_3\mathbf{e}_3$ is any vector in $\mathbb{R}^3$, written as a linear combination in the basis $\mathcal{E}$, then the linearity of T implies that

$$T(\mathbf{x}) = T(a_1\mathbf{e}_1 + a_2\mathbf{e}_2 + a_3\mathbf{e}_3) = a_1 T(\mathbf{e}_1) + a_2 T(\mathbf{e}_2) + a_3 T(\mathbf{e}_3)$$

This shows that the three vectors $T(\mathbf{e}_1), T(\mathbf{e}_2), T(\mathbf{e}_3)$ and the three scalars a_1, a_2, a_3 contain all the information required to calculate the vector $T[\mathbf{x}]$ in the basis $\mathcal{E}$. ▲

EXAMPLE 5.10 Using *Maple* to Define a Linear Transformation

Use the **Map** command in the **LinearAlgebra** package to define the linear transformation $T : \mathbb{R}^5 \to \mathbb{R}^4$ given by $T(u, v, x, y, z) = (2x - 7z, z, u - v, y)$, and calculate the vector $T(\mathbf{x}) = T(1, 2, 3, 4, 5)$.

Solution. We define the function T by specifying its input and corresponding output.

```
[> T:=(u,v,x,y,z)->[2*x-7*z,z,u-v,y];
```

$$T := (u, v, x, y, z) \to [2 * x - 7 * z, z, u - v, y]$$

We now use the **Map** function to calculate the vector $T(\mathbf{x}) = T(1, 2, 3, 4, 5)$.

```
[> with(LinearAlgebra):
[> Map(T,(1,2,3,4,5),{u,v,x,y,z});
```

$$[-29, 5, -1, 4]$$

This shows that $T(\mathbf{x}) = (-29, 5, -1, 4) \in \mathbb{R}^4$. ▲

EXERCISES 5.1

1. Show that for all $\theta \in \mathbb{R}$, the matrix

$$A = \begin{bmatrix} \cos\theta & -\sin\theta \\ \sin\theta & \cos\theta \end{bmatrix}$$

determines a linear transformation $T_\theta : \mathbb{R}^2 \to \mathbb{R}^2$.

2. Explain why the translation function $T : \mathbb{R}^2 \to \mathbb{R}^2$ defined by

$$T\begin{bmatrix} x \\ y \end{bmatrix} = \begin{bmatrix} x + h \\ y + k \end{bmatrix}$$

is almost never linear. When is it linear?

3. Explain why the following functions $T : \mathbb{R}^1 \to \mathbb{R}^1$ are not linear transformations.

$$\text{a. } T(x) = x + 2 \quad \text{b. } T(x) = \cos x \quad \text{c. } T(x) = x^2$$
$$\text{d. } T(x) = 3 \quad \text{e. } T(x) = \exp x \quad \text{f. } T(x) = \log x$$

4. Determine which of the following functions $T : \mathbb{R}^1 \to \mathbb{R}^2$ are linear transformations.

$$\text{a. } T(x) = \begin{bmatrix} x \\ x \end{bmatrix} \quad \text{b. } T(x) = \begin{bmatrix} x \\ 2x \end{bmatrix} \quad \text{c. } T(x) = \begin{bmatrix} x \\ 0 \end{bmatrix}$$

d. $T(x) = \begin{bmatrix} 0 \\ x^2 \end{bmatrix}$ e. $T(x) = \begin{bmatrix} 0 \\ 0 \end{bmatrix}$ f. $T(x) = 3\begin{bmatrix} x \\ x \end{bmatrix}$

5. Determine which of the following functions $T : \mathbb{R}^2 \to \mathbb{R}^1$ are linear transformations.

a. $T\begin{bmatrix} x \\ y \end{bmatrix} = x$ b. $T\begin{bmatrix} x \\ y \end{bmatrix} = y$ c. $T\begin{bmatrix} x \\ y \end{bmatrix} = 0$

d. $T\begin{bmatrix} x \\ y \end{bmatrix} = xy$ e. $T\begin{bmatrix} x \\ y \end{bmatrix} = 7$ f. $T\begin{bmatrix} x \\ y \end{bmatrix} = x + y$

6. Determine which of the following functions $T : \mathbb{R}^2 \to \mathbb{R}^2$ are linear transformations.

a. $T\begin{bmatrix} x \\ y \end{bmatrix} = \begin{bmatrix} 42x + 71y \\ 18x + 44y \end{bmatrix}$ b. $T\begin{bmatrix} x \\ y \end{bmatrix} = \begin{bmatrix} x \\ y^2 \end{bmatrix}$

c. $T\begin{bmatrix} x \\ y \end{bmatrix} = \begin{bmatrix} \cos x \\ y \end{bmatrix}$ d. $T\begin{bmatrix} x \\ y \end{bmatrix} = 3\begin{bmatrix} x \\ y \end{bmatrix} - 5\begin{bmatrix} y \\ x \end{bmatrix}$

7. Determine which of the following functions are linear transformations.
   a. $T : \mathbb{R}^{n\times n} \to \mathbb{R}$ defined by $T(A) = \det A$
   b. $T : \mathbb{R}^{n\times n} \to \mathbb{R}$ defined by $T(A) = \text{trace } A$
   c. $T : \mathbb{R}^{n\times n} \to \mathbb{R}^n$ defined by $T(A) = \text{diag } A$
   d. $T : \mathbb{R}^{m\times n} \to \mathbb{R}^{n\times m}$ defined by $T(A) = A^T$
   e. $T : \mathbb{R}^{m\times n} \to \mathbb{R}^{n\times m}$ defined by $T(A) = -A^T$

8. Explain why the function $T(A) = A^{-1}$ is not a linear transformation.

9. Find the images of the standard basis vectors of the matrix transformation $T_A : \mathbb{R}^3 \to \mathbb{R}^4$ determined by the transpose of the matrix

$$\begin{bmatrix} 7 & 17 & 38 & 27 \\ 0 & 22 & 7 & 34 \\ 0 & 0 & 9 & 25 \end{bmatrix}$$

10. Explain why the distributive law $a(b + c) = ab + ac$ guarantees that the conversion formula $f(x) = 5280x$ from miles to feet is a linear transformation $f : \mathbb{R} \to \mathbb{R}$.

11. Explain why the conversion formula $F = \frac{9}{5}C + 32$ from degrees Celsius to degrees Fahrenheit is an affine transformation that is not linear.

12. Describe the images of the vertices $(-1, 0)$, $(1, 0)$, $(0, 2)$ of an isosceles triangle under the affine transformation

$$T\begin{bmatrix} x \\ y \end{bmatrix} = \begin{bmatrix} 3 & 0 \\ 0 & 5 \end{bmatrix}\begin{bmatrix} x \\ y \end{bmatrix} + \begin{bmatrix} 4 \\ 5 \end{bmatrix}$$

13. Let $T_1, T_2 : U \to V$ be two linear transformations. Show that the ***sum*** $T_1 + T_2 : U \to V$ of T_1 and T_2, defined by $(T_1 + T_2)(\mathbf{x}) = T_1(\mathbf{x}) + T_2(\mathbf{x})$, is a linear transformation.

14. Let $T : U \to V$ be a linear transformation and let s be any scalar. Show that the ***scalar multiple*** $sT : U \to V$ of T, defined by $(sT)(\mathbf{x}) = s(T(\mathbf{x}))$, is a linear transformation.

15. Verify that for any fixed vector (x_0, y_0), the function $f : \mathbb{R}^2 \to \mathbb{R}$ defined by $f(x, y) = x_0 x + y_0 y$ is linear.

16. (*Maple V*) Verify that the function $f : \mathbb{R}^2 \to \mathbb{R}$ defined by $f(x, y) = 3x + 4y$ is linear.

17. (*Maple V*) Verify that the function $f(x, y, z) = 5x + 2y - 7z$ is a linear transformation $f : \mathbb{R}^3 \to \mathbb{R}$.

18. (*Maple V*) Verify that for any fixed vector $\mathbf{x}_0 \in \mathbb{R}^n$ the function $f : \mathbb{R}^n \to \mathbb{R}$ defined by $f(\mathbf{y}) = \mathbf{x}_0^T \mathbf{y}$ is a linear transformation.

19. (*Maple V*) Write a *Maple* routine for calculating the values of f, and verify that the function $f : \mathbb{R}[t] \to \mathbb{R}$ defined by $f(p(t)) = p(0)$ is a linear transformation.

20. (*Maple V*) Use the **AffineMap** function defined by the assignment

$$(\mathbf{A}, \mathbf{x}) - >\mathbf{evalm(SubMatrix(A, 1..2, 1..2)\& * x + SubMatrix(A, 1..2, 3))}$$

to find the images $A\mathbf{x}+\mathbf{b}$ of the vertices $(0, 0)$, $(1, 0)$, $(1, 1)$, and $(0, 1)$ of the unit square for the following affine transformations.

a. $A = \begin{bmatrix} 2 & 3 \\ 1 & 6 \end{bmatrix}$, $\mathbf{b} = \begin{bmatrix} -3 \\ 4 \end{bmatrix}$ b. $A = \begin{bmatrix} 0 & 1 \\ -1 & 0 \end{bmatrix}$, $\mathbf{b} = \begin{bmatrix} 5 \\ 9 \end{bmatrix}$

21. (*Maple V*) Repeat Exercise 12 using the **AffineMap** function defined in Exercise 20.

22. (*Maple V*) Show that the affine transformation defined by a real 2×3 matrix A and the procedure in Exercise 20 determines a linear $T : \mathbb{R}^2 \to \mathbb{R}^2$ if and only if the third column of A is a zero column.

23. (*Maple V*) Let $q = \frac{1}{\sqrt{6}}i + \frac{1}{\sqrt{6}}j + \frac{2}{\sqrt{6}}k$ be a unit quaternion as defined in Chapter 4. Show that the function $T(\mathbf{x}) = q\mathbf{x}\overline{q}$ is a linear transformation from $\mathbb{R}^4$ to $\mathbb{R}^4$.

Invertible Transformations

In many applications, we require linear transformations that are *invertible*. In geometry, for example, we often require linear transformations that preserve length, distance, and angles. This will occur only if the transformations are invertible. A linear transformation $T : V \to W$ is ***invertible*** if it is paired with another linear transformation $S : W \to V$ for which $S(T(\mathbf{v})) = \mathbf{v}$ for all $\mathbf{v} \in V$ and $T(S(\mathbf{w})) = \mathbf{w}$ for all $\mathbf{w} \in W$.

In this section, we discuss the basic properties of invertible linear transformations. For reasons that will become clear later, we also refer to invertible linear transformations as ***isomorphisms.*** The prefix *iso-* comes from Greek and means "equal," and *morphism* also comes from Greek and means "form." Isomorphisms between vector spaces are functions between two mathematical objects of the same form preserving that form. We will prove that for every n-dimensional real vector space V there exists an isomorphism from V to the coordinate space $\mathbb{R}^n$. This fact is fundamental for the link between linear and matrix algebra.

The definition of invertible linear transformations is based on *identity transformations* and the *composition* of linear transformations.

DEFINITION 5.3 *The **identity transformation** $I_V : V \to V$ on a vector space V is the function defined by $I_V(\mathbf{x}) = \mathbf{x}$ for all $\mathbf{x} \in V$.*

Identity transformations are linear since $I_V(a\mathbf{x} + b\mathbf{y}) = a\mathbf{x} + b\mathbf{y} = aI_V(\mathbf{x}) + bI_V(\mathbf{y})$ for all vectors $\mathbf{x}$ and $\mathbf{y}$ and all scalars a and b. We write I in place of I_V if V is understood from the context.

DEFINITION 5.4 *The **composition** of two linear transformations $T : U \to V$ and $S : V \to W$ is the function $S \circ T : U \to W$, defined by $(S \circ T)(\mathbf{x}) = S(T(\mathbf{x}))$.*

Composite linear transformations $S \circ T$ are linear since

$$\begin{aligned}(S \circ T)(a\mathbf{x} + b\mathbf{y}) &= S(T(a\mathbf{x} + b\mathbf{y})) \\ &= S(aT(\mathbf{x}) + bT(\mathbf{y})) \\ &= a\,(S(T(\mathbf{x}))) + b\,(S(T(\mathbf{y}))) \\ &= a\,(S \circ T)(\mathbf{x}) + b\,(S \circ T)(\mathbf{y})\end{aligned}$$

for all vectors $\mathbf{x}$ and $\mathbf{y}$ and all scalars a and b. We write ST in place of $S \circ T$ when the intended meaning is clear from the context. We use identity transformations and composition to define the inverse of a linear transformation.

DEFINITION 5.5 *If $T : U \to V$ and $S : V \to U$ are linear transformations with the property that $S \circ T = I_U$ and $T \circ S = I_V$, then S is an **inverse** of T and T is an **inverse** of S.*

It is easy to see that the inverse of a linear transformation T is uniquely determined by T. We can therefore speak of ***the inverse*** of T. We write T^{-1} for the inverse of T.

DEFINITION 5.6 *If $T : U \to V$ is a linear transformation that has an inverse, T is* ***invertible****.*

It is an easy exercise to prove that the composition of two invertible linear transformations (in other words, of two isomorphisms) is an isomorphism.

The invertibility of a linear transformation can be described in terms of two basic mapping properties.

1. A linear transformation $T : V \to W$ is ***one–one*** if $T(\mathbf{x}) = T(\mathbf{y})$ implies that $\mathbf{x} = \mathbf{y}$. This is equivalent to saying that T is one–one if $\mathbf{x} \neq \mathbf{y}$ implies that $T(\mathbf{x}) \neq T(\mathbf{y})$. Figure 2 illustrates the idea of a one–one linear transformation. We note that this definition generalizes the definition of one–one functions used in Chapter 3 to define permutations of indexing sets.

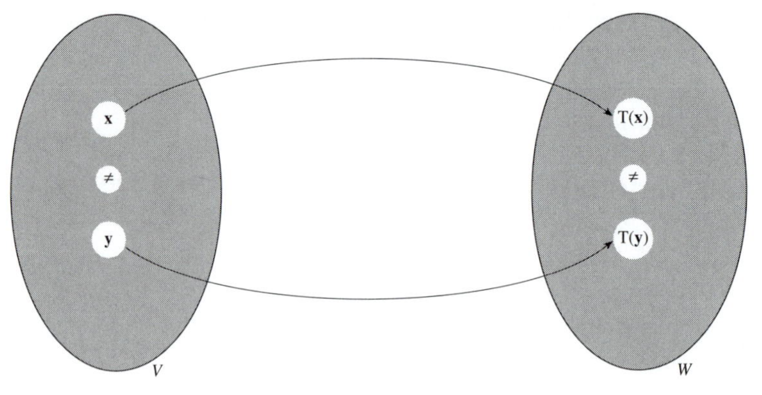

FIGURE 2 A One–One Transformation.

2. A linear transformation $T : V \to W$ is ***onto*** if for every vector $\mathbf{w} \in W$ there exists an $\mathbf{x} \in V$ such that $T(\mathbf{x}) = \mathbf{w}$. In other words, a linear transformation T is *not* onto if there exists a vector $\mathbf{w} \in W$ that is not in the image of T. Figure 3 shows graphically an onto linear transformation.

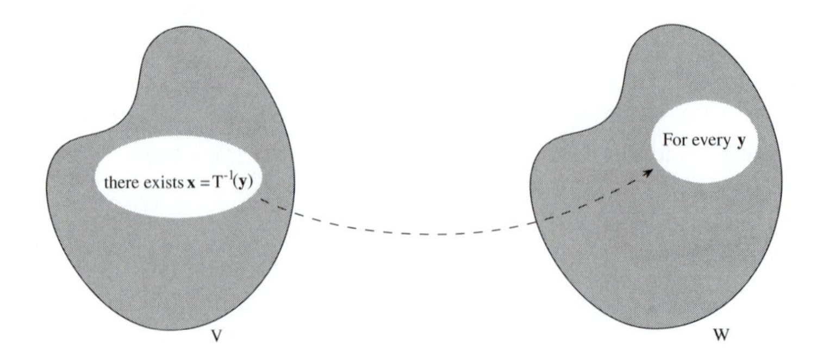

FIGURE 3 An Onto Transformation.

We now show that invertible linear transformations can be described as linear transformations that are both one–one and onto.

THEOREM 5.4 *If $T : V \to W$ is an invertible linear transformation, then T is one–one.*

Proof. We prove this theorem by showing that if $T(\mathbf{x}) = T(\mathbf{y})$, then $\mathbf{x}$ must equal $\mathbf{y}$. Suppose that T is invertible and that S is its inverse. If $T(\mathbf{x}) = T(\mathbf{y})$, then $S(T(\mathbf{x})) = S(T(\mathbf{y}))$. But since $S(T(\mathbf{x})) = (S \circ T)(\mathbf{x}) = I_V(\mathbf{x}) = \mathbf{x}$ and $S(T(\mathbf{y})) = (S \circ T)(\mathbf{y}) = I_V(\mathbf{y}) = \mathbf{y}$, it follows that $\mathbf{x} = \mathbf{y}$. ■

THEOREM 5.5 *If $T : V \to W$ is an invertible linear transformation, then T is onto.*

Proof. Suppose that T is invertible and that S is its inverse. If $\mathbf{y} \in W$, then $S(\mathbf{y}) \in V$ and $T(S(\mathbf{y})) = (T \circ S)(\mathbf{y}) = I_W(\mathbf{y}) = \mathbf{y}$. Therefore, $\mathbf{x} = S(\mathbf{y})$ is a vector in V whose value in W is $\mathbf{y}$. ■

We now prove the converse of Theorems 5.4 and 5.5. Using elementary properties of functions, we can then easily show that compositions of isomorphisms are isomorphisms.

THEOREM 5.6 *A linear transformation $T : V \to W$ is an isomorphism if and only if T is one–one and onto.*

Proof. We have just shown that every isomorphism is one–one and onto. We now prove the converse.

Suppose that the linear transformation $T : V \to W$ is one– one and onto. Since T is onto, every $\mathbf{w} \in W$ is of the form $T(\mathbf{v})$ for some $\mathbf{v} \in V$. Since T is one–one, the vector $\mathbf{v} \in V$ is unique. We therefore define a function $S : W \to V$ that associates with each $\mathbf{w} \in W$ the unique $\mathbf{v} \in V$ for which $T(\mathbf{v}) = \mathbf{w}$. Then it follows from the definition of S that $T(S(\mathbf{w})) = T(\mathbf{v}) = \mathbf{w}$ and $S(T(\mathbf{v})) = S(\mathbf{w}) = \mathbf{v}$. Hence $S = T^{-1}$.

It remains to show that S is linear. Consider two vectors $\mathbf{w}_1 = T(\mathbf{v}_1)$ and $\mathbf{w}_2 = T(\mathbf{v}_2)$ in W. By the definition of S,

$$\begin{aligned} S(\mathbf{w}_1 + \mathbf{w}_2) &= S(T(\mathbf{v}_1) + T(\mathbf{v}_2)) \\ &= S(T(\mathbf{v}_1 + \mathbf{v}_2)) \\ &= (S \circ T)(\mathbf{v}_1 + \mathbf{v}_2) \\ &= \mathbf{v}_1 + \mathbf{v}_2 \\ &= S(\mathbf{w}_1) + S(\mathbf{w}_2) \end{aligned}$$

Moreover, for any scalar a,

$$\begin{aligned} S(a\mathbf{w}_1) &= S(aT(\mathbf{v}_1)) \\ &= S(T(a\mathbf{v}_1)) \end{aligned}$$

$$\begin{aligned}&= (S \circ T)(a\mathbf{v}_1) \\ &= a\mathbf{v}_1 \\ &= aS(\mathbf{w}_1)\end{aligned}$$

Hence S is a linear transformation. ■

Next we provide examples of linear transformations that are one– one but not onto, and onto but not one–one.

EXAMPLE 5.11 A One–One Linear Transformation $T : \mathbb{R}^2 \to \mathbb{R}^3$ Not Onto

The linear transformation defined by $T(x, y) = (x, y, 0)$ is clearly one–one. But it is not onto since the vector $(x, y, 1)$, for example, is not a value of T. ▲

EXAMPLE 5.12 An Onto Linear Transformation Not One–One

The linear transformation $T : \mathbb{R}^3 \to \mathbb{R}^2$ defined by $T(x, y, z) = (x, y)$ is clearly onto. However, since $T(x, y, 1) = T(x, y, 2) = (x, y)$, the transformation T is not one–one. ▲

Example 5.12 depends on the difference between the dimensions of the domains and codomains involved. The next example shows that there are also linear transformations with identical domains and codomains that are one–one but not onto.

EXAMPLE 5.13 A One–One Linear Transformation $T : \mathbb{R}[t] \to \mathbb{R}[t]$ Not Onto

Show that the linear transformation $T : \mathbb{R}[t] \to \mathbb{R}[t]$ defined by $T(p(t)) = tp(t)$ is one–one but not onto.

Solution. The calculation

$$\begin{aligned}T(ap(t) + bq(t)) &= t(ap(t) + bq(t)) \\ &= atp(t) + btq(t) \\ &= aT(p(t)) + bT(q(t))\end{aligned}$$

shows that T is linear. In particular, $T(\mathbf{0}) = t\mathbf{0} = \mathbf{0}$. Moreover, if $T(p(t)) = T(q(t))$, then $tp(t) = tq(t)$. This implies that

$$\begin{aligned}tp(t) - tq(t) &= t(p(t) - q(t)) \\ &= p(t) - q(t) \\ &= 0\end{aligned}$$

Therefore, T is one–one. On the other hand, no nonzero constant polynomial is a value of T. Hence the transformation T is not onto. ▲

Later we will show that a one–one linear transformation $T : V \to W$ must also be onto if the vector spaces V and W have the same finite dimension. In Theorem 5.27 we will prove that if V is finite-dimensional, then every one–one linear transformation from V to V is also onto, and every onto linear transformation from V to V is also one–one.

EXAMPLE 5.14 An Isomorphism Between $\mathbb{R}_3[t]$ and $\mathbb{R}^4$

Construct an isomorphism between the matrix space $\mathbb{R}_3[t]$ and the coordinate space $\mathbb{R}^4$.

Solution. Let $T_1 : \mathbb{R}^4 \to \mathbb{R}_3[t]$ be the linear transformation defined in Example 5.2 and $T_2 : \mathbb{R}_3[t] \to \mathbb{R}^4$ be the linear transformations defined in Example 5.4. It is easy to check that $I = T_2 \circ T_1 : \mathbb{R}^4 \to \mathbb{R}^4$ is the identity transformation on $\mathbb{R}^4$ and that $I = T_1 \circ T_2 : \mathbb{R}_3[t] \to \mathbb{R}_3[t]$ is the identity transformation on $\mathbb{R}_3[t]$. Therefore, T_1 is an isomorphism between $\mathbb{R}_3[t]$ and $\mathbb{R}^4$, and so is its inverse T_2. ▲

EXAMPLE 5.15 An Isomorphism Between $\mathbb{R}^{2\times 2}$ and $\mathbb{R}^4$

Construct an isomorphism between the matrix space $\mathbb{R}^{2\times 2}$ and the coordinate space $\mathbb{R}^4$.

Solution. Let

$$A = \begin{bmatrix} a_{11} & a_{12} \\ a_{21} & a_{22} \end{bmatrix}$$

be an arbitrary matrix in $\mathbb{R}^{2\times 2}$. Then the function $T : \mathbb{R}^{2\times 2} \to \mathbb{R}^4$ defined by

$$T \begin{bmatrix} a_{11} & a_{12} \\ a_{21} & a_{22} \end{bmatrix} = \begin{bmatrix} a_{11} \\ a_{12} \\ a_{21} \\ a_{22} \end{bmatrix}$$

is an isomorphism between $\mathbb{R}^{2\times 2}$ and $\mathbb{R}^4$. ▲

EXAMPLE 5.16 An Isomorphism Between $\mathbb{R}^{2\times 3}$ and $\mathbb{R}_5[t]$

Construct an isomorphism between the matrix space $\mathbb{R}^{2\times 3}$ and the polynomial space $\mathbb{R}_5[t]$.

Solution. Let

$$A = \begin{bmatrix} a_{11} & a_{12} & a_{13} \\ a_{21} & a_{22} & a_{23} \end{bmatrix}$$

be an arbitrary matrix in $\mathbb{R}^{2\times 3}$. Then the function $T : \mathbb{R}^{2\times 3} \to \mathbb{R}_5[t]$ defined by

$$T \begin{bmatrix} a_{11} & a_{12} & a_{13} \\ a_{21} & a_{22} & a_{23} \end{bmatrix} = a_{11} + a_{12}t + a_{13}t^2 + a_{21}t^3 + a_{22}t^4 + a_{23}t^5$$

is an isomorphism between $\mathbb{R}^{2\times 3}$ and $\mathbb{R}_5[t]$. ▲

Several of the examples of linear transformations given so far are defined in terms of matrices. This suggests that matrices may be useful for defining linear transformations more generally, and this is actually the case. All real vector spaces of the same dimension d are isomorphic to the coordinate space $\mathbb{R}^d$. We can therefore study a linear transformation $T : V \to W$ from an n-dimensional vector space V to an m-dimensional vector space W by studying a corresponding linear transformation $T' : \mathbb{R}^n \to \mathbb{R}^m$. After working with T', we can go back to V and W using the given isomorphisms. Moreover, we will see later that the linear transformation T' can be represented by an $m \times n$ matrix A with the property that $T'(\mathbf{x}) = A\mathbf{x}$. This means that we will be able to find the values of T' by matrix multiplication.

THEOREM 5.7 (Isomorphism theorem) *For any n-dimensional real vector space V and any basis $\mathcal{B}$ for V, the coordinate-vector function $[-]_{\mathcal{B}} : V \to \mathbb{R}^n$ is an isomorphism.*

Proof. In Theorem 4.17 and Corollary 4.18, we proved that $[-]_{\mathcal{B}}$ is an invertible linear transformation. ■

COROLLARY 5.8 *If two real vector spaces have the same dimension, they are isomorphic.*

Proof. Suppose that V and W have dimension n. By Theorem 5.7, there exists an isomorphism $T : V \to \mathbb{R}^n$ and an isomorphism $S : W \to \mathbb{R}^n$. It is easy to check that the linear transformation $S^{-1} \circ T : V \to W$ is an isomorphism. ■

COROLLARY 5.9 *The matrix space $\mathbb{R}^{m\times n}$ is isomorphic to the coordinate space $\mathbb{R}^{mn}$.*

Proof. Since $\dim \mathbb{R}^{m\times n} = \dim \mathbb{R}^{mn} = mn$, the result follows from Corollary 5.8. ■

COROLLARY 5.10 *The polynomial space $\mathbb{R}_n[t]$ is isomorphic to the coordinate space $\mathbb{R}^{n+1}$.*

Proof. Since $\mathbb{R}_n[t]$ and $\mathbb{R}^{n+1}$ are both of dimension $n+1$, the result follows again from Corollary 5.8. ■

EXERCISES 5.2

1. Construct two linear transformations S and T for which $ST \neq TS$.
2. Find two linear transformations S and T for which $ST = TS$.
3. Show that if $S : U \to V$ and $T : V \to W$ are two isomorphisms, then $T \circ S$ is an isomorphism.
4. Show that an isomorphism is one–one and onto.
5. Show that a one–one, onto linear transformation is invertible.
6. Prove that the inverse of a linear transformation is unique.
7. Prove or refute the claim that if T is a one–one linear transformation $\mathbb{R}^n \to \mathbb{R}^m$, then $n = m$.

8. Prove or refute the claim that if T is an onto linear transformation $\mathbb{R}^n \to \mathbb{R}^m$, then $n = m$.

9. Show by means of examples that a linear transformation fails to be invertible if it is not one–one or not onto.

10. Prove or refute the claim that if T is an isomorphism $\mathbb{R}^n \to \mathbb{R}^m$, then $n = m$.

11. Prove that if T is a one–one linear transformation $\mathbb{R}^n \to \mathbb{R}^m$, then $\mathbb{R}^n$ is isomorphic to a subspace of $\mathbb{R}^m$.

12. Show that the equations

$$\begin{aligned} T_1 \circ (T_2 \circ T_3) &= (T_1 \circ T_2) \circ T_3 \\ T_1 \circ (T_2 + T_3) &= T_1 \circ T_2 + T_1 \circ T_3 \\ (T_1 + T_2) \circ T_3 &= T_1 \circ T_3 + T_2 \circ T_3 \\ s(T_1 \circ T_2) &= (sT_1) \circ T_2 = T_1 \circ (sT_2) \end{aligned}$$

hold for all compatible linear transformations T_1, T_2, T_3 and all scalars s.

13. Show that the functions defined in Examples 5.15 and 5.16 are isomorphisms.

14. Let V and W be two real vector spaces. Show that the set $\text{Hom}(V, W)$ of all linear transformations from V to W is a real vector space.

15. Show that the set $\text{Hom}(V, V)$ of all linear transformations from V to V is closed under sums, scalar multiples, and composition. Vector spaces with this structure satisfying the equations in Exercise 12 are called ***linear algebras***.

16. It follows from Exercise 14 that for any real vector space V, the set $\text{Hom}(V, \mathbb{R})$ is a real vector space. It is called the ***dual space*** of V and is denoted by V^*. The elements of V^* are called ***linear functionals.*** Show that if $V = \mathbb{R}^2$, then the projection functions $p_1(x_1, x_2) = x_1$ and $p_2(x_1, x_2) = x_2$ are linear functionals.

17. Let $\varphi : V \to V^{**}$ be the linear transformation $\varphi(\mathbf{v}) : V^* \to \mathbb{R}$ defined by $\varphi(\mathbf{v})(f) = f(\mathbf{v})$. Verify that φ is linear.

18. Let $\mathcal{B} = \{\mathbf{v}_1, \mathbf{v}_2, \mathbf{v}_3\}$ be a basis for a real vector space V, and associate with each $\mathbf{v}_i$ a linear functional $\varphi_i : V \to \mathbb{R}$ defined by

$$\varphi_i(\mathbf{v}_j) = \begin{cases} 1 \text{ if } i = j \\ 0 \text{ if } i \neq j \end{cases}$$

Show that the set $\mathcal{C} = \{\varphi_1, \varphi_2, \varphi_3\}$ is a basis for V^*. (*Hint:* Let $f : V \to \mathbb{R}$ be any linear functional determined by its three values $f(\mathbf{v}_i) = a_i \in \mathbb{R}$ on the basis $\mathcal{B}$. Show that $f = a_1\varphi_1 + a_2\varphi_2 + a_3\varphi_3$. Show that this equation holds for the basis vector $\mathbf{v}_i$.)

19. In communication theory, a *signal* is sometimes defined to be a function

$$f(t) = \frac{1}{2}a_0 + \sum_{k\in\mathbf{n}} b_k \cos 2\pi kt + \sum_{k\in\mathbf{n}} c_k \sin 2\pi kt$$

with $a_0, b_k, c_k \in \mathbb{R}$. Prove that the set of all signals $f(t)$ is isomorphic to $\mathbb{R}^{2n+1}$.

20. (*Maple V*) Suppose that $T : \mathbb{R}^3 \to \mathbb{R}^2$ is a linear transformation. Show that the associated function $T^* : \mathrm{Hom}(\mathbb{R}^2, \mathbb{R}) \to \mathrm{Hom}(\mathbb{R}^3, \mathbb{R})$ defined by $T^*(f) = f \circ T$ is a linear transformation.

21. (*Maple V*) Use *Maple* to verify that the sets $\mathrm{Hom}(\mathbb{R}, \mathbb{R})$ and $\mathrm{Hom}(\mathbb{R}^2, \mathbb{R}^2)$ are linear algebras.

Singular Transformations

A linear transformation $T : V \to W$ is ***singular*** if there exists a nonzero vector $\mathbf{x} \in V$ for which $T(\mathbf{x}) = \mathbf{0}$. A linear transformation that is not singular is called ***nonsingular***. Singular transformations cannot be one–one since $T(\mathbf{0}) = T(\mathbf{x}) = 0$ for some nonzero vector $\mathbf{x}$. Therefore, singular transformations are not invertible. On the other hand, they can certainly be onto. Let $W = \{\mathbf{0}\}$ be a zero space and let V contain nonzero vectors. Since $T(\mathbf{0}) = \mathbf{0}$, T is onto. But there also exists a nonzero vector $\mathbf{x} \in V$ for which $T(\mathbf{x}) = \mathbf{0}$. Hence T is singular. Here is an example of a singular linear transformation, called a ***projection.*** It projects the space $\mathbb{R}^3$ onto the xz-plane in $\mathbb{R}^3$.

EXAMPLE 5.17 A Projection Function

Show that the function $T : \mathbb{R}^3 \to \mathbb{R}^3$ defined by $T(x, y, z) = (x, 0, z)$ is a singular linear transformation.

Solution. First we note that

$$T\left(a\begin{bmatrix} x_1 \\ x_2 \\ x_3 \end{bmatrix} + b\begin{bmatrix} y_1 \\ y_2 \\ y_3 \end{bmatrix}\right) = T\begin{bmatrix} a\,x_1 + b\,y_1 \\ a\,x_2 + b\,y_2 \\ a\,x_3 + b\,y_3 \end{bmatrix} = \begin{bmatrix} a\,x_1 + b\,y_1 \\ 0 \\ a\,x_3 + b\,y_3 \end{bmatrix}$$

and

$$\begin{bmatrix} a\,x_1 + b\,y_1 \\ 0 \\ a\,x_3 + b\,y_3 \end{bmatrix} = \begin{bmatrix} a\,x_1 \\ 0 \\ a\,x_3 \end{bmatrix} + \begin{bmatrix} b\,y_1 \\ 0 \\ b\,y_3 \end{bmatrix} = a\begin{bmatrix} x_1 \\ 0 \\ x_3 \end{bmatrix} + b\begin{bmatrix} y_1 \\ 0 \\ y_3 \end{bmatrix}$$

Therefore

$$T\left(a\begin{bmatrix} x_1 \\ x_2 \\ x_3 \end{bmatrix} + b\begin{bmatrix} y_1 \\ y_2 \\ y_3 \end{bmatrix}\right) = aT\begin{bmatrix} x_1 \\ x_2 \\ x_3 \end{bmatrix} + bT\begin{bmatrix} y_1 \\ y_2 \\ y_3 \end{bmatrix}$$

This shows that T is linear. In addition, T is singular since $T(0, y, 0) = (0, 0, 0)$ for any $y \in \mathbb{R}$. ▲

Projections will be discussed more fully in Chapter 8.

EXERCISES 5.3

1. Show that if A is a singular $n \times n$ matrix, then the linear transformations $T : \mathbb{R}^n \to \mathbb{R}^n$ defined by $T(\mathbf{x}) = A\mathbf{x}$ is singular.

2. Show that the identity transformation $I : \mathbb{R}^n \to \mathbb{R}^n$ is a sum of singular transformations.

3. Prove or disprove the claim that the composition of two singular transformations is a singular transformation.

4. Prove or disprove the claim that the composition $T \circ S$ of a singular transformation T and a nonsingular transformation S is singular.

5. Show that if A is a square matrix and the matrix transformation T_A is nonsingular, then the matrix transformation determined by A^n is nonsingular for all n.

6. (*Maple V*) Show that the linear transformation T_A and T_B determined by the matrices

$$A = \begin{bmatrix} 1 & 0 & 0 \\ 2 & 5 & 0 \\ 3 & 0 & 0 \end{bmatrix} \quad \text{and} \quad B = \begin{bmatrix} 0 & 0 & 1 \\ 0 & 0 & 7 \\ 0 & 0 & 4 \end{bmatrix}$$

are singular. Does this imply that $T_{(A+B)}$ is singular?

7. (*Maple V*) Let A be a matrix obtained from the identity matrix I_3 by replacing one or two of its diagonal entries by 0. The matrix A is called a ***projection matrix***. Show that projection matrices are singular, and explain in what sense they determine matrix transformations that project $\mathbb{R}^3$ onto subspaces of $\mathbb{R}^3$.

MATRICES OF LINEAR TRANSFORMATIONS

We know from Theorem 5.1 that every real $m \times n$ matrix A determines a linear transformation $T_A : \mathbb{R}^n \to \mathbb{R}^m$. Now we show that the converse is also true. Every linear transformation $T : \mathbb{R}^n \to \mathbb{R}^m$ is a matrix transformation T_A for some A. The matrix A depends on a basis $\mathcal{B}$ for $\mathbb{R}^n$ and a basis $\mathcal{C}$ for $\mathbb{R}^m$. We build this dependence into our notation by writing $[T]_{\mathcal{C}}^{\mathcal{B}}$ for the matrix A. The columns of $[T]_{\mathcal{C}}^{\mathcal{B}}$ are the coordinate vectors $[T(\mathbf{x})]_{\mathcal{C}}$ of the images $T(\mathbf{x})$ in the basis $\mathcal{C}$ of the basis vectors $\mathbf{x} \in \mathcal{B}$. If $\mathcal{B} = \mathcal{C}$, we write $[T]_{\mathcal{B}}$ for $[T]_{\mathcal{C}}^{\mathcal{B}}$.

DEFINITION 5.7 *If $T : V \to W$ is a linear transformation from an n-dimensional real vector space V with basis $\mathcal{B}$ to an m-dimensional real vector space W with basis $\mathcal{C}$, then the matrix $[T]_{\mathcal{C}}^{\mathcal{B}}$ is the* ***matrix of T in the bases $\mathcal{B}$ and $\mathcal{C}$***.

The definition of a matrix of a linear transformation on complex vector spaces is analogous. If $\mathcal{B}$ and $\mathcal{C}$ are standard bases, we call $[T]_{\mathcal{C}}^{\mathcal{B}}$ the ***standard matrix of T***. We say that the matrix $[T]_{\mathcal{C}}^{\mathcal{B}}$ ***represents*** T in the bases $\mathcal{B}$ and $\mathcal{C}$. Table 1 summarizes the steps involved in the construction of a matrix of a linear transformation.

TABLE 1 FINDING THE MATRIX OF A LINEAR TRANSFORMATION .

Step 1. Take a basis $\mathcal{B} = \{\mathbf{x}_1, \ldots, \mathbf{x}_n\}$ for $\mathbb{R}^n$ and a basis $\mathcal{C} = \{\mathbf{y}_1, \ldots, \mathbf{y}_m\}$ for $\mathbb{R}^m$.
Step 2. Calculate the n image vectors $T(\mathbf{x}_i)$ in $\mathbb{R}^m$ for $T : \mathbb{R}^n \to \mathbb{R}^m$.
Step 3. Write each vector $T(\mathbf{x}_i)$ as a linear combination $a_{1i}\mathbf{y}_1 + \cdots + a_{mi}\mathbf{y}_m$ in $\mathcal{C}$.
Step 4. Form the coordinate vectors $[T(\mathbf{x}_i)]_{\mathcal{C}} = \begin{bmatrix} a_{1i} \\ \vdots \\ a_{mi} \end{bmatrix}$ in $\mathbb{R}^m$.
Step 5. Use the vectors $[T(\mathbf{x}_i)]_{\mathcal{C}}$ as the columns of a matrix $\left[[T(\mathbf{x}_1)]_{\mathcal{C}} \ \cdots \ [T(\mathbf{x}_n)]_{\mathcal{C}}\right]$.
Step 6. The matrix constructed in Steps 2-5 is the matrix $[T]_{\mathcal{C}}^{\mathcal{B}}$ of T.
Step 7. For all coordinate vectors $[\mathbf{x}]_{\mathcal{B}} \in \mathbb{R}^n$, it holds that $[T]_{\mathcal{C}}^{\mathcal{B}} [\mathbf{x}]_{\mathcal{B}} = [T(\mathbf{x})]_{\mathcal{C}}$.

EXAMPLE 5.18 The Matrix of a Linear Transformation

Let $T : \mathbb{R}^3 \to \mathbb{R}^2$ be the linear transformation defined by

$$T\begin{bmatrix} x \\ y \\ z \end{bmatrix} = \begin{bmatrix} 3x + 2y - 4z \\ x - 5y + 3z \end{bmatrix}$$

Use *Maple* to find the matrix $[T]_{\mathcal{C}}^{\mathcal{B}}$ of T in the bases

$$\mathcal{B} = \{\mathbf{x}_1 = (1, 1, 1), \mathbf{x}_2 = (1, 1, 0), \mathbf{x}_3 = (1, 0, 0)\} \quad \text{and} \quad \mathcal{C} = \{\mathbf{y}_1 = (1, 3), \mathbf{y}_2 = (2, 5)\}$$

Solution. By definition,

$$[T]_{\mathcal{C}}^{\mathcal{B}} = \left[\ [T(\mathbf{x}_1)]_{\mathcal{C}} \quad [T(\mathbf{x}_2)]_{\mathcal{C}} \quad [T(\mathbf{x}_3)]_{\mathcal{C}}\ \right]$$

We therefore begin by calculating the image vectors $T(\mathbf{x}_1)$, $T(\mathbf{x}_2)$, and $T(\mathbf{x}_3)$.

```
[> T:=x->array([3*x[1]+2*x[2]-4*x[3],x[1]-5*x[2]+3*x[3]]);
```

$$T := x-> array([3x[1] + 2x[2] - 4x[3], x[1] - 5x[2] + 3x[3]])$$

```
[> x1:=array([1,1,1]):
[> x2:=array([1,1,0]):
[> x3:=array([1,0,0]):
[> [x1,x2,x3];
```

$$[[1, 1, 1], [1, 1, 0], [1, 0, 0]]$$

```
[> T(x1):  T(x2):  T(x3):
[> [T(x1),T(x2),T(x3)];
```

$$[[1, -1], [5, -4], [3, 1]]$$

Next we need to find the coordinate vectors of $T(\mathbf{x}_1)$, $T(\mathbf{x}_2)$, and $T(\mathbf{x}_3)$ in the basis $\mathcal{C}$. We require scalars a, b, c, d, e, and f satisfying the following equations.

$$\text{a. } \begin{bmatrix} 1 \\ -1 \end{bmatrix}_{\mathcal{C}} = a\begin{bmatrix} 1 \\ 3 \end{bmatrix} + b\begin{bmatrix} 2 \\ 5 \end{bmatrix} = \begin{bmatrix} a+2b \\ 3a+5b \end{bmatrix}$$

$$\text{b. } \begin{bmatrix} 5 \\ -4 \end{bmatrix}_{\mathcal{C}} = c\begin{bmatrix} 1 \\ 3 \end{bmatrix} + d\begin{bmatrix} 2 \\ 5 \end{bmatrix} = \begin{bmatrix} c+2d \\ 3c+5d \end{bmatrix}$$

$$\text{c. } \begin{bmatrix} 3 \\ 1 \end{bmatrix}_{\mathcal{C}} = e\begin{bmatrix} 1 \\ 3 \end{bmatrix} + f\begin{bmatrix} 2 \\ 5 \end{bmatrix} = \begin{bmatrix} e+2f \\ 3e+5f \end{bmatrix}$$

To find these scalars, we solve the following linear systems.

$$\text{a. } \begin{cases} 1 = a+2b \\ -1 = 3a+5b \end{cases} \quad \text{b. } \begin{cases} 5 = c+2d \\ -4 = 3c+5d \end{cases} \quad \text{c. } \begin{cases} 3 = e+2f \\ 1 = 3e+5f \end{cases}$$

```
[> solve({1=a+2*b,-1=3*a+5*b});
```

$$\{b = 4, a = -7\}$$

```
[> solve({5=c+2*d,-4=3*c+5*d});
```

$$\{c = -33, d = 19\}$$

```
[> solve({3=e+2*f,1=3*e+5*f});
```

$$\{e = -13, f = 8\}$$

Hence

$$[T]_{\mathcal{C}}^{\mathcal{B}} = \begin{bmatrix} [T(\mathbf{x}_1)]_{\mathcal{C}} & [T(\mathbf{x}_2)]_{\mathcal{C}} & [T(\mathbf{x}_3)]_{\mathcal{C}} \end{bmatrix} = \begin{bmatrix} -7 & -33 & -13 \\ 4 & 19 & 8 \end{bmatrix}$$

is the required matrix. ▲

If $\mathcal{C}$ is the standard basis $\mathcal{E}$ for $\mathbb{R}^2$, the last few steps in Example 5.18 are redundant. In this case, the matrix $[T]_{\mathcal{E}}^{\mathcal{B}}$ is simply

$$\begin{bmatrix} [T(\mathbf{x}_1)]_{\mathcal{E}} & [T(\mathbf{x}_2)]_{\mathcal{E}} & [T(\mathbf{x}_3)]_{\mathcal{E}} \end{bmatrix} = \begin{bmatrix} 1 & 5 & 3 \\ -1 & -4 & 1 \end{bmatrix}$$

We have already pointed out that the purpose of representing linear transformations by matrices is so that we can find their values using matrix multiplication. We now show how this works.

THEOREM 5.11 (Matrix representation theorem) *If T is a linear transformation from $\mathbb{R}^n$ to $\mathbb{R}^m$ and if $\mathcal{B}$ is a basis for $\mathbb{R}^n$ and $\mathcal{C}$ is a basis for $\mathbb{R}^m$, then $[T(\mathbf{x})]_{\mathcal{C}} = [T]_{\mathcal{C}}^{\mathcal{B}} [\mathbf{x}]_{\mathcal{B}}$ for all $\mathbf{x} \in \mathbb{R}^n$.*

Proof. Suppose that $\mathcal{B} = \{\mathbf{x}_1, \ldots, \mathbf{x}_n\}$ and that $\mathbf{x} = a_1\mathbf{x}_1 + \cdots + a_n\mathbf{x}_n$ is any vector in $\mathbb{R}^n$. Then Theorem 4.17 implies that

$$\begin{aligned}[T(\mathbf{x})]_{\mathcal{C}} &= [T(a_1\mathbf{x}_1 + \cdots + a_n\mathbf{x}_n)]_{\mathcal{C}} = a_1 [T(\mathbf{x}_1)]_{\mathcal{C}} + \cdots + a_n [T(\mathbf{x}_n)]_{\mathcal{C}} \\ &= \left[[T(\mathbf{x}_1)]_{\mathcal{C}} \cdots [T(\mathbf{x}_n)]_{\mathcal{C}} \right] \begin{bmatrix} a_1 \\ \vdots \\ a_n \end{bmatrix} = [T]_{\mathcal{C}}^{\mathcal{B}} [\mathbf{x}]_{\mathcal{B}}\end{aligned}$$

This shows that the image vector $[T(\mathbf{x})]_{\mathcal{C}}$ can be computed by matrix multiplication. ■

Relative to suitable bases $\mathcal{B}$ and $\mathcal{C}$, the matrix $[T]_{\mathcal{C}}^{\mathcal{B}}$ of a linear transformation T may have a particularly simple form. It may, for example, be diagonal, triangular, or symmetric.

EXAMPLE 5.19 A Linear Transformation Represented by a Symmetric Matrix

Let $\mathcal{E}$ be the standard basis for $\mathbb{R}^3$, and let $T : \mathbb{R}^3 \to \mathbb{R}^3$ be a linear transformation defined by

$$\begin{cases} T(\mathbf{e}_1) = a_1\mathbf{e}_1 + a_2\mathbf{e}_2 + a_3\mathbf{e}_3 \\ T(\mathbf{e}_2) = a_2\mathbf{e}_1 + a_4\mathbf{e}_2 + a_5\mathbf{e}_3 \\ T(\mathbf{e}_3) = a_3\mathbf{e}_1 + a_5\mathbf{e}_2 + a_6\mathbf{e}_3 \end{cases}$$

Then the standard matrix

$$[T]_{\mathcal{E}}^{\mathcal{E}} = \begin{bmatrix} a_1 & a_2 & a_3 \\ a_2 & a_4 & a_5 \\ a_3 & a_5 & a_6 \end{bmatrix}$$

of T is symmetric. ▲

EXAMPLE 5.20 Coordinate Conversion Matrices and the Identity Transformations

Show that relative to two bases $\mathcal{B}$ and $\mathcal{C}$ for $\mathbb{R}^n$, the matrix $[I]_{\mathcal{C}}^{\mathcal{B}}$ of the identity transformation $I : \mathbb{R}^n \to \mathbb{R}^n$ is the coordinate conversion matrix from $\mathcal{B}$ to $\mathcal{C}$, and the matrix $[I]_{\mathcal{B}}^{\mathcal{C}}$ is the coordinate conversion matrix from $\mathcal{C}$ to $\mathcal{B}$.

Solution. Let $\mathcal{B} = \{\mathbf{x}_1, \ldots, \mathbf{x}_n\}$ and $\mathcal{C} = \{\mathbf{y}_1, \ldots, \mathbf{y}_n\}$ be two bases for $\mathbb{R}^n$. Then it is clear from the definition of a coordinate conversion matrix in Chapter 4 that

$$[I]_{\mathcal{C}}^{\mathcal{B}} = \left[[I(\mathbf{x}_1)]_{\mathcal{C}} \cdots [I(\mathbf{x}_n)]_{\mathcal{C}}\right] = [[\mathbf{x}_1]_{\mathcal{C}} \cdots [\mathbf{x}_n]_{\mathcal{C}}]$$

is the coordinate conversion matrix from $\mathcal{B}$ to $\mathcal{C}$, and that

$$[I]_{\mathcal{B}}^{\mathcal{C}} = \left[[I(\mathbf{y}_1)]_{\mathcal{B}} \cdots [I(\mathbf{y}_n)]_{\mathcal{B}}\right] = \left[[\mathbf{y}_1]_{\mathcal{B}} \cdots [\mathbf{y}_n]_{\mathcal{B}}\right]$$

is the coordinate conversion matrix from $\mathcal{C}$ to $\mathcal{B}$. ▲

EXAMPLE 5.21 A Coordinate Conversion

Calculate the coordinate conversion matrices $[I]_{\mathcal{C}}^{\mathcal{B}}$ and $[I]_{\mathcal{B}}^{\mathcal{C}}$ for the bases

$$\mathcal{B} = \{(0, 4, 2), (3, 0, 0), (1, 3, 4)\} \quad \text{and} \quad \mathcal{C} = \{(0, 0, 2), (3, 0, 0), (1, 1, 1)\}$$

for $\mathbb{R}^3$ and show that $[I]_{\mathcal{C}}^{\mathcal{B}} [\mathbf{x}]_{\mathcal{B}} = [\mathbf{x}]_{\mathcal{C}}$ and $[I]_{\mathcal{B}}^{\mathcal{C}} [\mathbf{x}]_{\mathcal{C}} = [\mathbf{x}]_{\mathcal{B}}$ for $\mathbf{x} = (1, 2, 3)$.

Solution. It is clear from Example 5.20 that

$$[I]_{\mathcal{C}}^{\mathcal{B}} = \left[\begin{bmatrix} 0 \\ 4 \\ 2 \end{bmatrix}_{\mathcal{C}} \begin{bmatrix} 3 \\ 0 \\ 0 \end{bmatrix}_{\mathcal{C}} \begin{bmatrix} 1 \\ 3 \\ 4 \end{bmatrix}_{\mathcal{C}} \right] = \begin{bmatrix} -1 & 0 & \frac{1}{2} \\ -\frac{4}{3} & 1 & -\frac{2}{3} \\ 4 & 0 & 3 \end{bmatrix}$$

and

$$[I]_{\mathcal{B}}^{\mathcal{C}} = \left[\begin{bmatrix} 0 \\ 0 \\ 2 \end{bmatrix}_{\mathcal{B}} \begin{bmatrix} 3 \\ 0 \\ 0 \end{bmatrix}_{\mathcal{B}} \begin{bmatrix} 1 \\ 1 \\ 1 \end{bmatrix}_{\mathcal{B}} \right] = \begin{bmatrix} -\frac{3}{5} & 0 & \frac{1}{10} \\ -\frac{4}{15} & 1 & \frac{4}{15} \\ \frac{4}{5} & 0 & \frac{1}{5} \end{bmatrix}$$

Moreover,

$$[\mathbf{x}]_{\mathcal{B}} = \begin{bmatrix} -\frac{1}{10} \\ \frac{1}{15} \\ \frac{4}{5} \end{bmatrix} \quad \text{and} \quad [\mathbf{x}]_{\mathcal{C}} = \begin{bmatrix} \frac{1}{2} \\ -\frac{1}{3} \\ 2 \end{bmatrix}$$

Therefore,

$$[I]_{\mathcal{B}}^{\mathcal{C}} [\mathbf{x}]_{\mathcal{C}} = \begin{bmatrix} -\frac{3}{5} & 0 & \frac{1}{10} \\ -\frac{4}{15} & 1 & \frac{4}{15} \\ \frac{4}{5} & 0 & \frac{1}{5} \end{bmatrix} \begin{bmatrix} \frac{1}{2} \\ -\frac{1}{3} \\ 2 \end{bmatrix} = \begin{bmatrix} -\frac{1}{10} \\ \frac{1}{15} \\ \frac{4}{5} \end{bmatrix} = [\mathbf{x}]_{\mathcal{B}}$$

and

$$[I]_{\mathcal{C}}^{\mathcal{B}} [\mathbf{x}]_{\mathcal{B}} = \begin{bmatrix} -1 & 0 & \frac{1}{2} \\ -\frac{4}{3} & 1 & -\frac{2}{3} \\ 4 & 0 & 3 \end{bmatrix} \begin{bmatrix} -\frac{1}{10} \\ \frac{1}{15} \\ \frac{4}{5} \end{bmatrix} = \begin{bmatrix} \frac{1}{2} \\ -\frac{1}{3} \\ 2 \end{bmatrix} = [\mathbf{x}]_{\mathcal{C}}$$

It is easy to show that these equations hold for any vector $\mathbf{x} \in \mathbb{R}^3$. Hence $[I]_{\mathcal{C}}^{\mathcal{B}}$ is the coordinate conversion matrix from $\mathcal{B}$ to $\mathcal{C}$ and $[I]_{\mathcal{B}}^{\mathcal{C}}$ is the coordinate conversion matrix from $\mathcal{C}$ to $\mathcal{B}$. ▲

EXAMPLE 5.22 A Matrix of an Identity Transformation

Show that in the standard basis $\mathcal{E} = \{\mathbf{e}_1, \ldots, \mathbf{e}_n\}$ for $\mathbb{R}^n$, the identity transformation $I : \mathbb{R}^n \to \mathbb{R}^n$ is represented by the identity matrix I_n.

Solution. By Theorem 5.11 it holds that $[I]_{\mathcal{E}}^{\mathcal{E}} = \left[[I(\mathbf{e}_1)]_{\mathcal{E}} \ \cdots \ [I(\mathbf{e}_n)]_{\mathcal{E}}\right]$. But $[I(\mathbf{e}_i)]_{\mathcal{E}} = \mathbf{e}_i$ for all $i \in \mathbf{n}$. Therefore, $[I]_{\mathcal{E}}^{\mathcal{E}} = [\mathbf{e}_1 \ \cdots \ \mathbf{e}_n] = I_n$. This shows that the identity matrix represents the identity transformation. ▲

EXERCISES 5.4

1. Find the standard matrix of each of the following linear transformations.

   a. $T : \mathbb{R}^2 \to \mathbb{R}^2$ defined by $T(1, 0) = (4, 2)$ and $T(0, 1) = (8, 3)$

   b. $T : \mathbb{R}^3 \to \mathbb{R}^3$ defined by $T(1, 0, 0) = (4, 2, 1)$, $T(0, 1, 0) = (1, 1, 1)$, and $T(0, 0, 1) = (8, 3, 4)$

   c. $T : \mathbb{R}^2 \to \mathbb{R}^2$ defined by $T(1, 1) = (4, 2)$ and $T(2, -1) = (8, 3)$

   d. $T : \mathbb{R}^3 \to \mathbb{R}^3$ defined by $T(1, 0, -5) = (4, 2, 1)$, $T(0, 0, 3) = (1, 1, 1)$, and $T(5, -3, 0) = (8, 3, 4)$

2. Let $\mathcal{B}$ be the basis of $\mathbb{R}^2$ determined by the invertible matrix

$$\begin{bmatrix} 1 & 3 \\ -2 & 0 \end{bmatrix}$$

and let $\mathcal{E}$ be the standard basis of $\mathbb{R}^2$. Find the matrix $[T]_{\mathcal{E}}^{\mathcal{B}}$ of the linear transformation

$$T\begin{bmatrix} x \\ y \end{bmatrix} = \begin{bmatrix} x+y \\ x \end{bmatrix}$$

3. Find the matrix $[T]_{\mathcal{B}}^{\mathcal{E}}$ of the linear transformation in Exercise 2.

4. Find the matrix $[T]_{\mathcal{B}}^{\mathcal{B}}$ of the linear transformation in Exercise 2.

5. Let $\mathcal{B}$ be the basis of $\mathbb{R}^3$ determined by the invertible matrix

$$\begin{bmatrix} -1 & -2 & 2 \\ 0 & 1 & 2 \\ 0 & 0 & 4 \end{bmatrix}$$

and let $\mathcal{E}$ be the standard matrix of $\mathbb{R}^3$. Find the matrix $[T]_{\mathcal{E}}^{\mathcal{B}}$ of the linear transformation $T(x, y, z) = (x + y, z, y)$.

6. Find the matrix $[T]_{\mathcal{B}}^{\mathcal{E}}$ of the linear transformation in Exercise 5.

7. Find the matrix $[T]_{\mathcal{B}}^{\mathcal{B}}$ of the linear transformation in Exercise 5.

8. Prove or disprove the following statements.

   a. If T_1 and T_2 are two linear transformations $\mathbb{R}^n \to \mathbb{R}^n$ and $[T_1]_{\mathcal{B}}^{\mathcal{B}} = [T_2]_{\mathcal{C}}^{\mathcal{C}}$ in two bases $\mathcal{B}$ and $\mathcal{C}$ for $\mathbb{R}^n$, then $T_1 = T_2$.

   b. If T_1 and T_2 are two linear transformations $\mathbb{R}^n \to \mathbb{R}^n$ and $[T_1]_{\mathcal{B}}^{\mathcal{B}} = [T_2]_{\mathcal{B}}^{\mathcal{B}}$ in some basis $\mathcal{B}$ for $\mathbb{R}^n$, then $T_1 = T_2$.

   c. If T is a linear transformation $\mathbb{R}^n \to \mathbb{R}^n$ and $[T]_{\mathcal{B}}^{\mathcal{B}} = [T]_{\mathcal{C}}^{\mathcal{C}}$ in two bases $\mathcal{B}$ and $\mathcal{C}$ for $\mathbb{R}^n$, then $\mathcal{B} = \mathcal{C}$.

9. Let $D_t : \mathbb{R}_3[t] \to \mathbb{R}_2[t]$ be the differentiation operation

$$D_t(a_0 + a_1t + a_2t^2 + a_3t^3) = a_1 + 2a_2t + 3a_3t^2$$

and $I_t : \mathbb{R}_2[t] \to \mathbb{R}_3[t]$ be the integration operation

$$I_t(a_0 + a_1t + a_2t^2) = a_0t + \tfrac{1}{2}a_1t^2 + \tfrac{1}{3}a_2t^3$$

Find the standard matrices $A = [D_t]$ and $B = [I_t]$ with respect to the standard bases, and show that

$$AB = \begin{bmatrix} 1 & 0 & 0 \\ 0 & 1 & 0 \\ 0 & 0 & 1 \end{bmatrix} \quad \text{and} \quad BA = \begin{bmatrix} 0 & 0 & 0 & 0 \\ 0 & 1 & 0 & 0 \\ 0 & 0 & 1 & 0 \\ 0 & 0 & 0 & 1 \end{bmatrix}$$

10. (*Maple V*) Let $\mathcal{B}$ and $\mathcal{C}$ be two bases for $\mathbb{R}^3$ determined by the invertible matrices

$$\begin{bmatrix} 5 & 1 & -3 \\ 0 & 5 & 3 \\ 0 & 0 & -2 \end{bmatrix} \quad \text{and} \quad \begin{bmatrix} 0 & 4 & 4 \\ 5 & 0 & 0 \\ 0 & 0 & 4 \end{bmatrix}$$

Find the matrices $[I]_{\mathcal{C}}^{\mathcal{B}}$ and $[I]_{\mathcal{B}}^{\mathcal{C}}$ of the identity transformation $I : \mathbb{R}^3 \to \mathbb{R}^3$.

11. (*Maple 6*) Find the coordinate conversion matrices $[I]_{\mathcal{C}}^{\mathcal{B}}$ and $[I]_{\mathcal{B}}^{\mathcal{C}}$ for the bases $\mathcal{B}$ and $\mathcal{C}$ for $\mathbb{R}^4$ determined by the columns of the following invertible matrices.

$$\begin{bmatrix} 3 & 5 & 5 & 2 \\ 5 & 2 & -2 & 0 \\ 5 & -2 & 6 & 4 \\ 2 & 0 & 4 & -3 \end{bmatrix} \quad \text{and} \quad \begin{bmatrix} 3 & 6 & 5 & 6 \\ 4 & 6 & 7 & 7 \\ 3 & 6 & 6 & 6 \\ 7 & 7 & 2 & 7 \end{bmatrix}$$

Composition of Linear Transformations and Matrix Multiplication

We now explain the connection between matrix multiplication and the composition of matrix transformations.

THEOREM 5.12 (Composition theorem) *Suppose that $T_1 : \mathbb{R}^n \to \mathbb{R}^m$ and $T_2 : \mathbb{R}^m \to \mathbb{R}^p$ are two linear transformations and that the matrices $[T_1]_{\mathcal{C}}^{\mathcal{B}}$ and $[T_2]_{\mathcal{D}}^{\mathcal{C}}$ represent the transformations T_1 and T_2 in the bases $\mathcal{B}$, $\mathcal{C}$, and $\mathcal{D}$. Then*

$$[T_2 \circ T_1]_{\mathcal{D}}^{\mathcal{B}} = [T_2]_{\mathcal{D}}^{\mathcal{C}} [T_1]_{\mathcal{C}}^{\mathcal{B}}$$

Proof. Let $\mathcal{B}$ be a basic for $\mathbb{R}^n$, let $\mathcal{C}$ be a basis for $\mathbb{R}^m$, and let $\mathcal{D}$ be a basis for $\mathbb{R}^p$. Then

$$\begin{aligned}
[T_2 \circ T_1]_{\mathcal{D}}^{\mathcal{B}} [\mathbf{x}]_{\mathcal{B}} &= [(T_2 \circ T_1)[\mathbf{x}]]_{\mathcal{D}} \\
&= [T_2(T_1[\mathbf{x}])]_{\mathcal{D}} \\
&= [T_2]_{\mathcal{D}}^{\mathcal{C}} [T_1[\mathbf{x}]]_{\mathcal{C}} \\
&= [T_2]_{\mathcal{D}}^{\mathcal{C}} \left([T_1]_{\mathcal{C}}^{\mathcal{B}} [\mathbf{x}]_{\mathcal{B}}\right) \\
&= \left([T_2]_{\mathcal{D}}^{\mathcal{C}} [T_1]_{\mathcal{C}}^{\mathcal{B}}\right) [\mathbf{x}]_{\mathcal{B}}
\end{aligned}$$

for all $\mathbf{x} \in \mathcal{B}$. It therefore follows from Theorem 5.3 that $[T_2]_{\mathcal{D}}^{\mathcal{C}} [T_1]_{\mathcal{C}}^{\mathcal{B}}$ and $[T_2 \circ T_1]_{\mathcal{D}}^{\mathcal{B}}$ are identical matrix transformations from $\mathbb{R}^n$ to $\mathbb{R}^p$ in the bases $\mathcal{B}$ and $\mathcal{D}$. Hence $[T_2 \circ T_1]_{\mathcal{D}}^{\mathcal{B}} = [T_2]_{\mathcal{D}}^{\mathcal{C}} [T_1]_{\mathcal{C}}^{\mathcal{B}}$. ■

EXAMPLE 5.23 Composition of Linear Transformations and Matrix Multiplication

Illustrate by an example that matrix multiplication corresponds to the composition of matrix transformations.

Solution. Let $T_1 : \mathbb{R}^3 \to \mathbb{R}^2$ and $T_2 : \mathbb{R}^2 \to \mathbb{R}^2$ be the transformations defined by

$$T_1 \begin{bmatrix} x \\ y \\ z \end{bmatrix} = \begin{bmatrix} 3x + 2y - 4z \\ x - 5y + 3z \end{bmatrix} \quad \text{and} \quad T_2 \begin{bmatrix} u \\ v \end{bmatrix} = \begin{bmatrix} 4u - 2v \\ 2u + v \end{bmatrix}$$

and let A be the standard matrix of T_1 and B the standard matrix of T_2. Then

$$A = \begin{bmatrix} T_1 \begin{bmatrix} 1 \\ 0 \\ 0 \end{bmatrix} & T_1 \begin{bmatrix} 0 \\ 1 \\ 0 \end{bmatrix} & T_1 \begin{bmatrix} 0 \\ 0 \\ 1 \end{bmatrix} \end{bmatrix} = \begin{bmatrix} 3 & 2 & -4 \\ 1 & -5 & 3 \end{bmatrix}$$

and

$$B = \begin{bmatrix} T_2 \begin{bmatrix} 1 \\ 0 \end{bmatrix} & T_2 \begin{bmatrix} 0 \\ 1 \end{bmatrix} \end{bmatrix} = \begin{bmatrix} 4 & -2 \\ 2 & 1 \end{bmatrix}$$

We show that $T_B \circ T_A = T_{BA}$. The composite transformation $T_B \circ T_A$, applied to an arbitrary vector $\mathbf{x} \in \mathbb{R}^3$, yields

$$(T_B \circ T_A) \begin{bmatrix} x \\ y \\ z \end{bmatrix} = T_B \left(T_A \begin{bmatrix} x \\ y \\ z \end{bmatrix} \right) = \begin{bmatrix} 4 & -2 \\ 2 & 1 \end{bmatrix} \left(\begin{bmatrix} 3 & 2 & -4 \\ 1 & -5 & 3 \end{bmatrix} \begin{bmatrix} x \\ y \\ z \end{bmatrix} \right)$$

$$= \begin{bmatrix} 4 & -2 \\ 2 & 1 \end{bmatrix} \begin{bmatrix} 3x + 2y - 4z \\ x - 5y + 3z \end{bmatrix} = \begin{bmatrix} 10x + 18y - 22z \\ 7x - y - 5z \end{bmatrix}$$

We therefore examine the matrix transformation $T_{BA} : \mathbb{R}^3 \to \mathbb{R}^2$.

Since

$$BA = \begin{bmatrix} 4 & -2 \\ 2 & 1 \end{bmatrix} \begin{bmatrix} 3 & 2 & -4 \\ 1 & -5 & 3 \end{bmatrix} = \begin{bmatrix} 10 & 18 & -22 \\ 7 & -1 & -5 \end{bmatrix}$$

we have

$$T_{BA} \begin{bmatrix} x \\ y \\ z \end{bmatrix} = \begin{bmatrix} 10 & 18 & -22 \\ 7 & -1 & -5 \end{bmatrix} \begin{bmatrix} x \\ y \\ z \end{bmatrix} = \begin{bmatrix} 10x + 18y - 22z \\ 7x - y - 5z \end{bmatrix}$$

Hence $(T_B \circ T_A)(\mathbf{x}) = T_{BA}(\mathbf{x})$ for all $\mathbf{x} \in \mathbb{R}^3$. ▲

EXERCISES 5.5

1. Decompose the invertible matrix

$$A = \begin{bmatrix} 1 & 2 \\ 7 & 8 \end{bmatrix}$$

into a product of elementary matrices $E_1, \ldots, E_n$ and explain the geometric effect of the matrix transformations $T_{E_1}, \ldots, T_{E_n}$.

2. Describe the geometric effect of the matrix transformations T_A, T_B, and T_P, where

$$AB = \begin{bmatrix} 1 & 0 & 0 \\ 0 & 0 & 1 \\ 0 & 1 & 0 \end{bmatrix} \begin{bmatrix} 0 & 0 & 1 \\ 0 & 1 & 0 \\ 1 & 0 & 0 \end{bmatrix} = \begin{bmatrix} 0 & 0 & 1 \\ 1 & 0 & 0 \\ 0 & 1 & 0 \end{bmatrix} = P$$

is a product of elementary permutation matrices.

3. (*Maple V*) Let $T_A : \mathbb{R}^2 \to \mathbb{R}^2$ be a linear transformation defined by

$$T_A \begin{bmatrix} x \\ y \end{bmatrix} = A \begin{bmatrix} x \\ y \end{bmatrix} = \begin{bmatrix} 4 & 8 \\ 2 & 3 \end{bmatrix} \begin{bmatrix} x \\ y \end{bmatrix}$$

and $T_B : \mathbb{R}^2 \to \mathbb{R}^2$ be a linear transformation defined by

$$T_B \begin{bmatrix} x \\ y \end{bmatrix} = B \begin{bmatrix} x \\ y \end{bmatrix} = \begin{bmatrix} 6 & 2 \\ 0 & 1 \end{bmatrix} \begin{bmatrix} x \\ y \end{bmatrix}$$

Show that $T_B \circ T_A = T_{BA}$ and $T_A \circ T_B = T_{AB}$.

4. (*Maple V*) Let $T_A : \mathbb{R}^3 \to \mathbb{R}^3$ be a linear transformation defined by

$$T_A \begin{bmatrix} x \\ y \\ z \end{bmatrix} = A \begin{bmatrix} x \\ y \\ z \end{bmatrix} = \begin{bmatrix} 4 & 8 & 2 \\ 2 & 3 & 6 \\ 0 & 0 & 0 \end{bmatrix} \begin{bmatrix} x \\ y \\ z \end{bmatrix}$$

and $T_B : \mathbb{R}^3 \to \mathbb{R}^3$ be a linear transformation defined by

$$T_B \begin{bmatrix} x \\ y \\ z \end{bmatrix} = B \begin{bmatrix} x \\ y \\ z \end{bmatrix} = \begin{bmatrix} 6 & 2 & 0 \\ 0 & 1 & 0 \\ 1 & 3 & 0 \end{bmatrix} \begin{bmatrix} x \\ y \\ z \end{bmatrix}$$

Show that $T_B \circ T_A = T_{BA}$ and $T_A \circ T_B = T_{AB}$.

5. (*Maple 6*) Suppose that

$$A = \begin{bmatrix} 0 & 2 & 5 \\ 1 & 5 & 2 \\ 3 & 2 & 1 \end{bmatrix} = \begin{bmatrix} 0 & 1 & 0 \\ 1 & 0 & 0 \\ 0 & 0 & 1 \end{bmatrix} \begin{bmatrix} 1 & 0 & 0 \\ 0 & 1 & 0 \\ 3 & -\frac{13}{2} & 1 \end{bmatrix} \begin{bmatrix} 1 & 5 & 2 \\ 0 & 2 & 5 \\ 0 & 0 & \frac{55}{2} \end{bmatrix} = PLU$$

Show that the linear transformation T_A is equal to the composition $T_P \circ T_L \circ T_U$.

The Inverse of a Linear Transformation and Matrix Inversion

It is easy to show that the inverse of a linear transformation is represented by the inverse of the matrix of the transformation.

THEOREM 5.13 *If $T : \mathbb{R}^n \to \mathbb{R}^n$ is an invertible linear transformation represented by a matrix A, then $\left[T^{-1}\right] = A^{-1}$.*

Proof. Let $\mathcal{B}$ and $\mathcal{C}$ be two bases for $\mathbb{R}^n$, and suppose that $A = [T]_{\mathcal{C}}^{\mathcal{B}}$ and $B = \left[T^{-1}\right]_{\mathcal{B}}^{\mathcal{C}}$. By Theorem 5.12,

$$AB = [T]_{\mathcal{C}}^{\mathcal{B}} \left[T^{-1}\right]_{\mathcal{B}}^{\mathcal{C}} = \left[T \circ T^{-1}\right]_{\mathcal{C}}^{\mathcal{C}} = [I]_{\mathcal{C}}^{\mathcal{C}} = I$$

Hence $B = A^{-1}$. ■

EXAMPLE 5.24 The Matrix of an Inverse Transformation

Show by an example that the matrix of the inverse of a linear transformation is the inverse of the matrix of the transformation.

Solution. Let $T : \mathbb{R}^2 \to \mathbb{R}^2$ be the linear transformation defined by

$$T(x, y) = (x + y, x - y)$$

It is easy to check that inverse of T is the linear transformation

$$S(x, y) = \left(\frac{1}{2}(x + y), \frac{1}{2}(x - y)\right)$$

In the $\mathcal{E}$ standard basis of $\mathbb{R}^2$, the matrices of T and S are

$$[T]_{\mathcal{E}} = \begin{bmatrix} 1 & 1 \\ 1 & -1 \end{bmatrix} \quad \text{and} \quad [S]_{\mathcal{E}} = \begin{bmatrix} \frac{1}{2} & \frac{1}{2} \\ \frac{1}{2} & -\frac{1}{2} \end{bmatrix}$$

Since

$$\begin{bmatrix} 1 & 1 \\ 1 & -1 \end{bmatrix} \begin{bmatrix} \frac{1}{2} & \frac{1}{2} \\ \frac{1}{2} & -\frac{1}{2} \end{bmatrix} = \begin{bmatrix} 1 & 0 \\ 0 & 1 \end{bmatrix}$$

it is clear that $\left[T^{-1}\right]_{\mathcal{E}} = [T]_{\mathcal{E}}^{-1}$. ▲

EXERCISES 5.6

1. Let $T_A : \mathbb{R}^2 \to \mathbb{R}^2$ be the linear transformation defined by

$$T_A \begin{bmatrix} x \\ y \end{bmatrix} = A \begin{bmatrix} x \\ y \end{bmatrix} = \begin{bmatrix} 4 & 8 \\ 2 & 3 \end{bmatrix} \begin{bmatrix} x \\ y \end{bmatrix}$$

   Find the inverse of T_A.

2. Let $T_A : \mathbb{R}^3 \to \mathbb{R}^3$ be the linear transformation defined by

$$T_A \begin{bmatrix} x \\ y \\ z \end{bmatrix} = A \begin{bmatrix} x \\ y \\ z \end{bmatrix} = \begin{bmatrix} 4 & 8 & 2 \\ 2 & 3 & 6 \\ 0 & 0 & 0 \end{bmatrix} \begin{bmatrix} x \\ y \\ z \end{bmatrix}$$

   Show that T_A does not have an inverse.

3. Prove that if $T_A : \mathbb{R}^n \to \mathbb{R}^n$ is a matrix transformation and $T_{A^{-1}} : \mathbb{R}^n \to \mathbb{R}^n$ is the matrix transformation determined by A^{-1}, then $T_{A^{-1}} \circ T_A$ and $T_A \circ T_{A^{-1}}$ are the identity transformation on $\mathbb{R}^n$.

4. Prove that if A is an orthogonal $n \times n$ matrix, then $T_A \circ T_{A^T}$ is the identity transformation on $\mathbb{R}^n$.

IMAGES AND KERNELS

Every linear transformation $T : V \to W$ determines two special subspaces: the subspace of W consisting of the values $T(\mathbf{x})$ of T, and the subspace of V consisting of all vectors $\mathbf{x} \in V$ mapped to $\mathbf{0} \in \mathbf{W}$ by T.

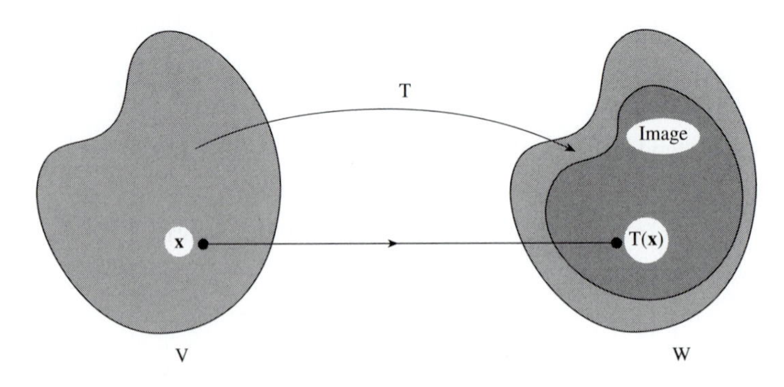

FIGURE 4 Image of a Linear Transformation.

DEFINITION 5.8 *The **image*** $\operatorname{im} T$ *of a linear transformation* $T : V \to W$ *is the set of all vectors* $\mathbf{y} \in W$ *with the property that* $\mathbf{y} = T(\mathbf{x})$ *for some* $\mathbf{x} \in V$.

Figure 4 illustrates an the idea of an image.

THEOREM 5.14 (Image theorem) *The image of a linear transformation* $T : V \to W$ *is a subspace of* W.

Proof. Since $T(0_V) = 0_W$, the image of T is nonempty. We need to show that it is also closed under linear combinations. Let **u** and **v** be two vectors in the image of T, and let a and b be any two scalars. Then $\mathbf{u} = T(\mathbf{x})$ and $\mathbf{v} = T(\mathbf{y})$ for some vectors **x** and **y** in V. By the linearity of T,

$$aT(\mathbf{x}) + bT(\mathbf{y}) = T(a\mathbf{x} + b\mathbf{y})$$

Since $T(a\mathbf{x} + b\mathbf{y}) \in \operatorname{im} T$, we can conclude that $a\mathbf{u} + b\mathbf{v} = aT(\mathbf{x}) + bT(\mathbf{y}) \in \operatorname{im} T$. The image of T is therefore closed under linear combinations. ■

If T is a matrix transformation, then the image of T has a familiar form.

THEOREM 5.15 *If* A *is a real* $m \times n$ *matrix and* $T_A : \mathbb{R}^n \to \mathbb{R}^m$ *is the associated matrix transformation, then* $\operatorname{im} T_A = \operatorname{Col} A$.

Proof. Suppose that

$$A = \begin{bmatrix} a_{11} & \cdots & a_{1n} \\ \vdots & \vdots & \vdots \\ a_{m1} & \cdots & a_{mn} \end{bmatrix} \quad \text{and} \quad \mathbf{x} = \begin{bmatrix} x_1 \\ \vdots \\ x_n \end{bmatrix} \in \mathbb{R}^n$$

In that case, the vector

$$A\mathbf{x} = \begin{bmatrix} a_{11}x_1 + \cdots + a_{1n}x_n \\ \vdots \\ a_{m1}x_1 + \cdots + a_{mn}x_n \end{bmatrix} = x_1 \begin{bmatrix} a_{11} \\ \vdots \\ a_{m1} \end{bmatrix} + \cdots + x_n \begin{bmatrix} a_{1n} \\ \vdots \\ a_{mn} \end{bmatrix}$$

is a linear combination of the columns of A. ■

EXAMPLE 5.25 An Image as a Column Space

Use *Maple* to find the image of the matrix transformation $T_A : \mathbb{R}^3 \to \mathbb{R}^2$ determined by the matrix

$$A = \begin{bmatrix} 0 & 0 & 2 \\ 0 & 1 & 0 \end{bmatrix}$$

Solution. By Theorem 5.15, the image of T_A is identical with the column space of A. We use the **colspace** function in the **linalg** package to find a basis for the image of T_A.

```
[> A:=matrix([[0,0,2],[0,1,0]]);
```

$$A := \begin{bmatrix} 0 & 0 & 2 \\ 0 & 1 & 0 \end{bmatrix}$$

```
[> linalg[colspace](A);
```

$$\{[0, 1], [1, 0]\}$$

This shows that the set $\{(1, 0), (0, 1)\}$ is a basis for the image of T_A. ▲

We can get a better understanding of the nature of the image of a linear transformation by considering how the image relates to the two defining properties of a basis.

THEOREM 5.16 *If $T : V \to W$ is a linear transformation and if $\mathcal{S} = \{\mathbf{x}_1, \ldots, \mathbf{x}_n\}$ spans V, then $\{T(\mathbf{x}_1), \ldots, T(\mathbf{x}_n)\}$ spans the image of T.*

Proof. Suppose that $\mathbf{y} \in \operatorname{im} T$. Then $\mathbf{y} = T(\mathbf{x})$ for some $\mathbf{x} \in V$. Since $\mathcal{S}$ spans V, there exist scalars $a_1, \ldots, a_n$ for which $\mathbf{x} = a_1\mathbf{x}_1 + \cdots + a_n\mathbf{x}_n$. Since T is linear,

$$T(\mathbf{x}) = T(a_1\mathbf{x}_1 + \cdots + a_n\mathbf{x}_n) = a_1T(\mathbf{x}_1) + \cdots + a_nT(\mathbf{x})$$

Therefore, $\mathbf{y}$ is a linear combination of the vectors $T(\mathbf{x}_1), \ldots, T(\mathbf{x}_n)$. ■

THEOREM 5.17 *If $T : V \to W$ is a one–one linear transformation and if the set of values $\{T(\mathbf{x}_1), \ldots, T(\mathbf{x}_n)\}$ of T spans the image of T, then $\{\mathbf{x}_1, \ldots, \mathbf{x}_n\}$ spans V.*

Proof. Let $\mathbf{x} \in V$. Then $T(\mathbf{x}) \in \operatorname{im} T$. Since $\{T(\mathbf{x}_1), \ldots, T(\mathbf{x}_n)\}$ spans the image of T, it follows that

$$T(\mathbf{x}) = a_1T(\mathbf{x}_1) + \cdots + a_nT(\mathbf{x}_n) = T(a_1\mathbf{x}_1 + \cdots + a_n\mathbf{x}_n)$$

Since T is one–one, $\mathbf{x} = a_1\mathbf{x}_1 + \cdots + a_n\mathbf{x}_n$. ■

THEOREM 5.18 *If $T : V \to W$ is a one–one linear transformation and the vectors $\mathbf{x}_1, \ldots, \mathbf{x}_n$ are linearly independent in V, then the vectors $T(\mathbf{x}_1), \ldots, T(\mathbf{x}_n)$ are linearly independent in W.*

Proof. Let $a_1T(\mathbf{x}_1) + \cdots + a_nT(\mathbf{x}_n) = \mathbf{0}$. By the linearity of T, we have

$$T(a_1\mathbf{x}_1 + \cdots + a_n\mathbf{x}_n) = \mathbf{0} = T(\mathbf{0})$$

Since T is one–one, it follows that $a_1\mathbf{x}_1 + \cdots + a_n\mathbf{x}_n = \mathbf{0}$. The linear independence of $\mathbf{x}_1, \ldots, \mathbf{x}_n$ guarantees that $a_1 = \cdots = a_n = 0$. ■

THEOREM 5.19 *If $T : V \to W$ is a linear transformation and $T(\mathbf{x}_1), \ldots, T(\mathbf{x}_n)$ are linearly independent image vectors in W, then $\mathbf{x}_1, \ldots, \mathbf{x}_n$ are linearly independent in V.*

Proof. Let $a_1\mathbf{x}_1 + \cdots + a_n\mathbf{x}_n = \mathbf{0}$. Then

$$\mathbf{0} = T(\mathbf{0}) = T(a_1\mathbf{x}_1 + \cdots + a_n\mathbf{x}_n) = a_1T(\mathbf{x}_1) + \cdots + a_nT(\mathbf{x}_n)$$

The linear independence of $T(\mathbf{x}_1), \ldots, T(\mathbf{x}_n)$ implies that $a_1 = \cdots = a_n = 0$. ■

As mentioned earlier, a linear transformation from V to W also determines an important subspace of V.

DEFINITION 5.9 *The* ***kernel*** $\ker T$ *of a linear transformation* $T : V \to W$ *is the set of all vectors* $\mathbf{x} \in V$ *with the property that* $T(\mathbf{x}) = \mathbf{0}$.

Figure 5 illustrates the idea of a kernel.

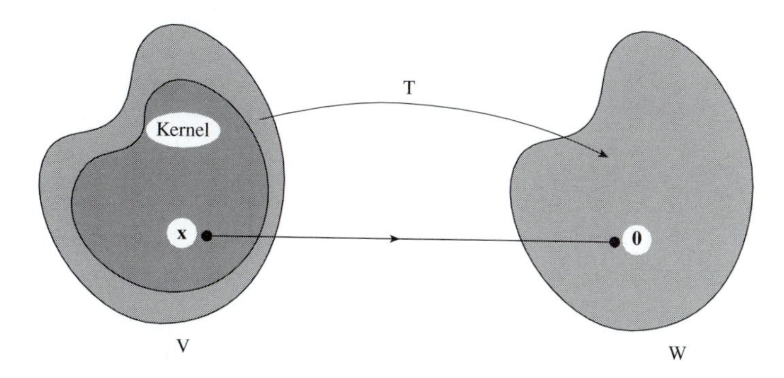

FIGURE 5 Kernel of a Linear Transformation.

THEOREM 5.20 (Kernel theorem) *The kernel of a linear transformation* $T : V \to W$ *is a subspace of* V.

Proof. Since $T(\mathbf{0}) = \mathbf{0}$, it follows that $\mathbf{0} \in \ker T$. Furthermore, the linearity of T implies that for all $\mathbf{x}_1, \mathbf{x}_2 \in \ker T$ and all scalars a and b,

$$T(a\mathbf{x}_1 + b\mathbf{x}_2) = aT(\mathbf{x}_1) + bT(\mathbf{x}_2) = a\mathbf{0} + b\mathbf{0} = \mathbf{0}$$

Hence $a\mathbf{x}_1 + b\mathbf{x}_2 \in \ker T$. Therefore the kernel of T is nonempty and is closed under linear combinations. This means that it is a subspace of V. ■

THEOREM 5.21 *The kernel of a matrix transformation* $T_A : \mathbb{R}^n \to \mathbb{R}^m$ *is the null space of* A.

Proof. We recall that the null space of the matrix A is the set of all vectors $\mathbf{x} \in \mathbb{R}^n$ for which $A\mathbf{x} = \mathbf{0}$. Moreover, the kernel of T_A consists of all vectors $\mathbf{x} \in \mathbb{R}^n$ for which $T_A(\mathbf{x}) = \mathbf{0}$. Since $T_A(\mathbf{x}) = A\mathbf{x}$, it follows that the kernel of T_A and the null space of A consist of precisely the same vectors. ■

EXAMPLE 5.26 The Kernel of a Linear Transformation

Let $T : \mathbb{R}^3 \to \mathbb{R}^3$ be the ***projection*** defined by $T(x, y, z) = (x, y, 0)$. Then $T(x, y, z) = (0, 0, 0)$ if and only if $(x, y, z) = (0, 0, z)$. Therefore, the set of vectors $\{(0, 0, z) : z \in \mathbb{R}\}$ is the kernel of T. ▲

EXAMPLE 5.27 Kernels and Homogeneous Systems

Show that the set of solutions of the homogeneous linear system

$$\begin{cases} 3x + 4y + w + z = 0 \\ x - y + 2w - z = 0 \\ x + y + w = 0 \end{cases}$$

is the kernel of a linear transformation.

Solution. The matrix form $A\mathbf{x} = \mathbf{0}$ of the given system is

$$\begin{bmatrix} 3 & 4 & 1 & 1 \\ 1 & -1 & 2 & -1 \\ 1 & 1 & 1 & 0 \end{bmatrix} \begin{bmatrix} x \\ y \\ w \\ z \end{bmatrix} = \begin{bmatrix} 0 \\ 0 \\ 0 \end{bmatrix}$$

Let $T_A : \mathbb{R}^4 \to \mathbb{R}^3$ be the matrix transformation determined by the coefficient matrix A. Then the solutions of $A\mathbf{x} = \mathbf{0}$ form the kernel of T_A. ▲

Whether or not a linear transformation is one–one can be expressed elegantly in terms of its kernel.

THEOREM 5.22 *A linear transformation T is one–one if and only if* $\ker T = \{\mathbf{0}\}$.

Proof. Suppose that T is one–one and that $\mathbf{x} \in \ker T$. Then $T(\mathbf{x}) = \mathbf{0} = T(\mathbf{0})$. Since T is one–one, we have $\mathbf{x} = \mathbf{0}$. On the other hand, if $\ker T = \{\mathbf{0}\}$ and $T(\mathbf{x}) = T(\mathbf{y})$ for some $\mathbf{x}, \mathbf{y} \in V$, then $\mathbf{0} = T(\mathbf{x}) - T(\mathbf{y}) = T(\mathbf{x} - \mathbf{y})$, so $\mathbf{x} - \mathbf{y} \in \ker T$. This can only be the case if $\mathbf{x} - \mathbf{y} = \mathbf{0}$. Therefore, $\mathbf{x} = \mathbf{y}$. ■

EXERCISES 5.7

1. Find a linear transformation $T : \mathbb{R}^2 \to \mathbb{R}^2$ whose image in the standard basis is spanned by set $\mathcal{S} = \{(-8, 2), (3, 77)\}$.

2. Suppose that $T : \mathbb{R}^3 \to \mathbb{R}^2$ is a linear transformation and that

$$T(5, 0, 1) = (-4, 7),\ T(2, 8, 0) = (1, 1),\ T(4, 5, 8) = (0, 5)$$

Show that the set $\mathcal{S} = \{(-4, 7), (1, 1), (0, 5)\}$ spans the image of T.

3. Describe the basis for the kernel of the matrix transformation T_A determined by the matrix

$$A = \begin{bmatrix} 1 & 0 \\ 0 & -1 \end{bmatrix}$$

4. Describe the kernel and image of the linear transformation $T : \mathbb{R}^{2\times 3} \to \mathbb{R}^6$ defined by

$$T\begin{bmatrix} a & b & c \\ d & e & f \end{bmatrix} = (a, b, c, 0, 0, 0)$$

5. Show that the nullity of the linear transformation $T : \mathbb{R}^{2\times 3} \to \mathbb{R}^6$ defined by

$$T\begin{bmatrix} a & b & c \\ d & e & f \end{bmatrix} = (d, e, f, a, b, c)$$

is 0 and that its rank is 6. Explain why this means that T is an isomorphism.

6. Describe the image and kernel of the differentiation operation $D : \mathbb{R}_4[t] \to \mathbb{R}_3[t]$ defined by

$$D_t(a_0 + a_1 t + a_2 t^2 + a_3 t^3 + a_4 t^4) = a_1 + 2a_2 t + 3a_3 t^2 + 4a_4 t^3$$

7. Describe the image and kernel of the integration operation $I_t : \mathbb{R}_3[t] \to \mathbb{R}[t]$ defined by

$$I_t(a_0 + a_1 t + a_2 t^2 + a_3 t^3) = a_0 t + \tfrac{1}{2}a_1 t^2 + \tfrac{1}{3}a_2 t^3 + \tfrac{1}{4}a_3 t^4$$

8. Describe the kernel of the differentiation operation $D : D^\infty(\mathbb{R}, \mathbb{R}) \to D^\infty(\mathbb{R}, \mathbb{R})$ defined by $D(f) = f'$. Show that $\ker D$ is the set of constant functions.

9. Describe the kernel of the differentiation operation $D^2 : D^\infty(\mathbb{R}, \mathbb{R}) \to D^\infty(\mathbb{R}, \mathbb{R})$ defined by $D(f) = f''$.

10. (*Maple V*) Use the procedure defined by

```
f:=proc(i,j) if i<>j then (i+j) mod 2 else 0 fi; end;
```

to generate a random 3×4 matrix A. Show that the matrix A has rank 2, and suppose that $\mathbf{x}$ is a nonzero vector in $\ker T_A$. Show that $T_A(\mathbf{x} + \mathbf{x}_0) = \mathbf{b}$ if and only if $T_A(\mathbf{x}_0) = \mathbf{b}$.

11. (*Maple V*) Find a linear transformation $T : \mathbb{R}^3 \to \mathbb{R}^3$ whose image in the standard basis is spanned by set $S = \{(-8, 2, 3), (3, 77, 5)\}$.

12. (*Maple V*) Let $T : \mathbb{R}^3 \to \mathbb{R}^2$ be the linear transformation

$$T\begin{bmatrix} x \\ y \\ z \end{bmatrix} = \begin{bmatrix} 2x + 4y - z \\ -x + y - 8z \end{bmatrix}$$

Show that the kernel of T is the solution space of the homogeneous linear system

$$\begin{cases} 2x + 4y - z = 0 \\ -x + y - 8z = 0 \end{cases}$$

and find a basis for the kernel of T.

13. (*Maple 6*) Find the image and kernel of the matrix transformations determined by the following matrices.

a. $\begin{bmatrix} 1 & 0 & -2 & 6 \\ 0 & 5 & 7 & 1 \end{bmatrix}$ b. $\begin{bmatrix} 1 & 0 \\ 0 & 5 \\ -2 & 7 \\ 6 & 1 \end{bmatrix}$ c. $\begin{bmatrix} 0 & -8 & 3 \\ 8 & 2 & 77 \\ 3 & -7 & 9 \end{bmatrix}$

d. $\begin{bmatrix} 1 & 0 & 0 \\ 0 & 3 & 0 \\ 0 & 0 & 7 \\ 0 & 0 & 0 \end{bmatrix}$ e. $\begin{bmatrix} 1 & 0 & 0 & 0 \\ 0 & 3 & 0 & 0 \\ 0 & 0 & 0 & 0 \end{bmatrix}$ f. $\begin{bmatrix} 1 & 0 & 0 & 9 \\ 0 & 0 & 0 & 0 \\ 0 & 0 & 0 & 0 \\ 0 & 0 & 0 & 0 \end{bmatrix}$

Rank and Nullity

Images and kernels of linear transformations are important for reconstructing vector spaces from their constituent parts. Since many calculations with images and kernels involve the dimensions of these spaces, these dimensions have been given special names.

DEFINITION 5.10 *The **rank** of a linear transformation T, denoted by* rank T, *is the dimension of the image of the transformation T.*

DEFINITION 5.11 *The **nullity** of T, denoted by* nullity T, *is the dimension of the kernel of the transformation T.*

One of the important connections between the rank and the nullity of a linear transformation $T : V \to W$ is the fact that if V is finite-dimensional, then $\dim V = \text{nullity}\, T + \text{rank}\, T$. This fact is proved in Theorem 5.25.

THEOREM 5.23 *The rank of a linear transformation $T_A : \mathbb{R}^n \to \mathbb{R}^m$ is the dimension of the column space of A.*

Proof. By Theorem 5.15, the image of T_A is the column space of A. The rank of T_A therefore coincides with the dimension of the column space of A. ■

THEOREM 5.24 *The nullity of a linear transformation $T_A : \mathbb{R}^n \to \mathbb{R}^m$ is the dimension of the null space of A.*

Proof. By definition, $T_A(\mathbf{x}) = A\mathbf{x}$ for all $\mathbf{x} \in \mathbb{R}^n$, and the null space of A consists of all $\mathbf{x} \in \mathbb{R}^n$ for which $A\mathbf{x} = \mathbf{0}$. The nullity of T_A therefore coincides with the dimension of the null space of A. ■

EXAMPLE 5.28 The Rank and Nullity of a Linear Transformation

Use *Maple* to find the rank and nullity of the linear transformation $T : \mathbb{R}^5 \to \mathbb{R}^4$ defined by $T(u, v, x, y, z) = (x - y + z + u, x + 2y - u, x + y + 3v - 5z, v)$.

Solution. We begin by loading the **LinearAlgebra** package and defining a function that computes the nullity of a matrix.

```
[> with(LinearAlgebra):
[> Nullity:=T->nops(NullSpace(T));
```

$$Nullity := T \to nops(NullSpace(T))$$

Next we find a matrix representation of T. In the standard bases of $\mathbb{R}^4$ and $\mathbb{R}^5$, the transformation T is represented by the matrix

$$A = \begin{bmatrix} 1 & 0 & 1 & -1 & 1 \\ -1 & 0 & 1 & 2 & 0 \\ 0 & 3 & 1 & 1 & -5 \\ 0 & 1 & 0 & 0 & 0 \end{bmatrix}$$

It remains to represent the matrix A in *Maple* and to calculate its rank and nullity.

```
[> A:=Matrix([[1,0,1,-1,1],[-1,0,1,2,0],[0,3,1,1,-5],
    [0,1,0,0,0]]):
[> Nullity(A);
```

$$1$$

```
[> Rank(A);
```

$$4$$

This shows that the nullity of T is 1 and rank of T is 4. ▲

In this example, rank T + nullity $T = 4 + 1 = 5 = \dim \mathbb{R}^5$. The corollary of the next theorem shows that this relationship between the rank and nullity of a linear transformation is no accident.

THEOREM 5.25 *If $\mathcal{B} = \{\mathbf{x}_1, \ldots, \mathbf{x}_p, \mathbf{x}_{p+1}, \ldots, \mathbf{x}_n\}$ is a basis for a vector space V and $\mathcal{C} = \{\mathbf{x}_1, \ldots, \mathbf{x}_p\}$ is a basis for the kernel of a linear transformation $T : V \to W$, then the set $\mathcal{D} = \{T(\mathbf{x}_{p+1}), \ldots, T(\mathbf{x}_n)\}$ spans the image of T.*

Proof. Suppose that

$$\mathbf{x} = a_1\mathbf{x}_1 + \cdots + a_p\mathbf{x}_p + a_{p+1}\mathbf{x}_{p+1} + \cdots + a_n\mathbf{x}_n$$

is any vector in V. Then

$$\begin{aligned} T(\mathbf{x}) &= T\left(a_1\mathbf{x}_1 + \cdots + a_p\mathbf{x}_p + a_{p+1}\mathbf{x}_{p+1} + \cdots + a_n\mathbf{x}_n\right) \\ &= a_1T(\mathbf{x}_1) + \cdots + a_pT(\mathbf{x}_p) + a_{p+1}T(\mathbf{x}_{p+1}) + \cdots + a_nT(\mathbf{x}_n) \\ &= a_{p+1}T(\mathbf{x}_{p+1}) + \cdots + a_nT(\mathbf{x}_n) \end{aligned}$$

Since every $\mathbf{y} \in \operatorname{im} T$ is of the form $T(\mathbf{x})$ for some $\mathbf{x} \in V$, the theorem follows. ■

COROLLARY 5.26 *If $T : V \to W$ is a linear transformation and V is finite-dimensional, then* $\dim V = \text{nullity}\, T + \text{rank}\, T$.

Proof. Suppose that $\mathcal{B} = \{\mathbf{x}_1, \ldots, \mathbf{x}_p, \mathbf{x}_{p+1}, \ldots, \mathbf{x}_n\}$ is a basis for V and that $\mathcal{C} = \{\mathbf{x}_1, \ldots, \mathbf{x}_p\}$ is a basis for the kernel of T. In Theorem 5.25 we showed that the set

$$\mathcal{D} = \{T(\mathbf{x}_{p+1}), \ldots, T(\mathbf{x}_n)\}$$

spans the image of T. We now show that it is also linearly independent.

Suppose that

$$\mathbf{0} = a_{p+1}T(\mathbf{x}_{p+1}) + \cdots + a_nT(\mathbf{x}_n)$$

Then the linearity of T tells us that

$$\begin{aligned} \mathbf{0} &= a_{p+1}T(\mathbf{x}_{p+1}) + \cdots + a_nT(\mathbf{x}_n) \\ &= T\left(a_{p+1}\mathbf{x}_{p+1} + \cdots + a_n\mathbf{x}_n\right) \end{aligned}$$

so the vector $a_{p+1}\mathbf{x}_{p+1} + \cdots + a_n\mathbf{x}_n$ belongs to the kernel of T. Therefore, there exist scalars $b_1, \ldots, b_p$ for which

$$a_{p+1}\mathbf{x}_{p+1} + \cdots + a_n\mathbf{x}_n = b_1\mathbf{x}_1 + \cdots + b_p\mathbf{x}_p$$

since $\mathcal{C} = \{\mathbf{x}_1, \ldots, \mathbf{x}_p\}$ is a basis for the kernel of T. This means that

$$b_1\mathbf{x}_1 + \cdots + b_p\mathbf{x}_p - a_{p+1}\mathbf{x}_{p+1} - \cdots - a_n\mathbf{x}_n = \mathbf{0}$$

However, since $\mathcal{B}$ is a basis, the vectors $x_1, \ldots, \mathbf{x}_n$ are linearly independent, so all scalars b_i and a_j must be zero. This shows that the set $\mathcal{D}$ is also linearly independent. Since $p = \dim(\ker T)$ and $n - p = \dim(\operatorname{im} T)$, and since $\dim(\ker T)$ is the nullity of T and $\dim(\operatorname{im} T)$ is the rank of T, the theorem follows. ■

We can use Corollary 5.26 to characterize isomorphisms between finite-dimensional vector spaces.

THEOREM 5.27 (Isomorphism theorem) *For any linear transformation $T : V \to V$ on a finite-dimensional vector space V, the following statements are equivalent: (1) T is nonsingular. (2) T is one–one. (3) T is onto. (4) T is an isomorphism.*

Proof. Theorem 5.22 shows that (1) is equivalent to (2). Theorem 5.6 shows that (4) implies (2) and (3) and that (2) and (3) together imply (4). It remains to show that if V is finite-dimensional, then (2) and (3) are equivalent.

We know from Corollary 5.26 that $\dim V = \text{nullity}\, T + \text{rank}\, T$. Suppose that T is one–one. Then $\ker T = \{\mathbf{0}\}$. Therefore $\text{nullity}\, T = 0$, so $\dim V = \text{rank}\, T$. This tells us that T must be onto. On the other hand, if T is onto, then $\dim V = \text{rank}\, T$. Hence $\text{nullity}\, T = 0$. This implies that $\ker T = \{\mathbf{0}\}$. By Theorem 5.22, T must be one–one. ■

EXERCISES 5.8

1. Calculate the rank and nullity of the matrix transformations determined by the following matrices. Show that in each case that $\text{rank}\, T + \text{nullity}\, T$ is equal to the dimension of the domain of T.

$$\text{a.} \begin{bmatrix} 1 & 0 & -2 & 6 \\ 0 & 5 & 7 & 1 \end{bmatrix} \quad \text{b.} \begin{bmatrix} 1 & 0 & 0 \\ 0 & 5 & 0 \\ -2 & 7 & 0 \\ 6 & 1 & 0 \end{bmatrix} \quad \text{c.} \begin{bmatrix} 0 & 0 & 0 \\ 8 & 2 & 77 \\ 3 & -7 & 9 \end{bmatrix}$$

2. Show that a matrix transformation $T_A : \mathbb{R}^n \to \mathbb{R}^n$ is an isomorphism if and only if its rank is n.

3. Show that a matrix transformation $T_A : \mathbb{R}^n \to \mathbb{R}^n$ is an isomorphism if and only if its nullity is 0.

4. (*Maple V*) Show that if

$$A = \begin{bmatrix} 5 & 1 & 0 & 0 & 1 & 1 & 2 \\ 3 & 0 & 0 & 0 & 0 & 3 & 0 \\ 8 & 1 & 0 & 0 & 1 & 4 & 2 \\ 0 & 0 & 0 & 0 & 0 & 0 & 0 \\ 3 & 0 & 0 & 0 & 1 & 0 & 2 \end{bmatrix}$$

then the nullity of the matrix transformation $T_A : \mathbb{R}^7 \to \mathbb{R}^5$ is 4. Find a basis $\mathcal{B} = \{\mathbf{x}_1, \mathbf{x}_2, \mathbf{x}_3, \mathbf{x}_4\}$ for $\ker T_A$. Extend $\mathcal{B}$ to a basis

$$\mathcal{C} = \{\mathbf{x}_1, \mathbf{x}_2, \mathbf{x}_3, \mathbf{x}_4, \mathbf{x}_5, \mathbf{x}_6, \mathbf{x}_7\}$$

for $\mathbb{R}^7$ and show that the set $\{T_A(\mathbf{x}_5), T_A(\mathbf{x}_6), T_A(\mathbf{x}_7)\}$ is a basis for $\text{im}\, T_A$.

5. (*Maple V, 6*) Calculate the rank and nullity of the matrix transformations determined by the following matrices, and show that in each case, rank T + nullity T is equal to the dimension of the domain of T.

$$\text{a.} \begin{bmatrix} 1 & 0 & 0 \\ 0 & 3 & 0 \\ 0 & 0 & 7 \\ 0 & 0 & 0 \end{bmatrix} \quad \text{b.} \begin{bmatrix} 1 & 0 & 4.5 & 0 \\ 0 & 3 & 0 & 0 \\ 0 & 0 & 0 & 0 \end{bmatrix} \quad \text{c.} \begin{bmatrix} 1 & 0 & 0 & 9 \\ 0 & 0 & 0 & 0 \\ 0 & 0 & 0 & 0 \\ 0 & 0 & 0 & 0 \end{bmatrix}$$

Compare how *Maple V* and *Maple 6* calculate the ranks of these matrices. (*Hint:* Use *Maple* Help.)

6. (*Maple 6*) Show that the function

```
Rankk:=A->ColumnDimension(A)-nops(NullSpace(A))
```

calculates the rank of any $m \times n$ matrix A.

7. (*Maple 6*) Show that the function

```
Nullityy:=A->ColumnDimension(A)-Rank(A)
```

calculates the nullity of any $m \times n$ matrix A.

8. (*Maple 6*) Generate a random 20×21 real matrix A and calculate its rank. Use the **Rank** and **Dimension** functions to explain why the matrix $B = A^T A$ is not invertible. Use the **Rankk** and **Nullityy** functions to confirm that rank B + nullity $B = 21$.

Areas and Linear Transformations

We know from Chapter 3 that the area of the parallelogram determined by two vectors

$$\mathbf{u} = \begin{bmatrix} u_1 \\ u_2 \end{bmatrix} \quad \text{and} \quad \mathbf{v} = \begin{bmatrix} v_1 \\ v_2 \end{bmatrix}$$

is the absolute value of the determinant of the matrix

$$P = \begin{bmatrix} u_1 & v_1 \\ u_2 & v_2 \end{bmatrix}$$

We now explore the change in the area of the parallelogram if the vectors **u** and **v** are modified by a matrix transformation.

EXAMPLE 5.29 The Area of the Image of a Parallelogram

Suppose that $T_A : \mathbb{R}^2 \to \mathbb{R}^2$ is a linear transformation defined by the matrix

$$A = \begin{bmatrix} a & c \\ b & d \end{bmatrix}$$

and that the columns of the matrix

$$P = \begin{bmatrix} u_1 & v_1 \\ u_2 & v_2 \end{bmatrix}$$

are two vectors determining a parallelogram in $\mathbb{R}^2$. Show that the area of the image of the parallelogram under T_A is the absolute values of $\det A \det P$.

Solution. Consider the parallelogram determined by the vectors $A\mathbf{u}$ and $A\mathbf{v}$. We know from Theorem 2.1 that the matrix $[A\mathbf{u}\ A\mathbf{v}]$, whose columns are $A\mathbf{u}$ and $A\mathbf{v}$, equals AP. Therefore the area of the parallelogram determined by $A\mathbf{u}$ and $A\mathbf{v}$ is the absolute value of

$$\det[A\mathbf{u}\ A\mathbf{v}] = \det AP = \det A \det P$$

This shows that the area of the original parallelogram is magnified by the absolute value of the determinant of the matrix A. ▲

In Example 5.29, we assumed that the parallelogram was determined by two vectors starting at the origin. If this is not the case, we can use homogeneous coordinates to find the area.

EXAMPLE 5.30 Homogeneous Coordinates and the Area of an Image

Let P be the parallelogram with vertices $(2, 3)$, $(10, -4)$, $(0, -9)$, and $(-8, -2)$. Use homogeneous coordinates to find the area of the image of P under the matrix transformation $T_A : \mathbb{R}^2 \to \mathbb{R}^2$ determined by

$$A = \begin{bmatrix} 3 & 1 \\ 0 & -2 \end{bmatrix}$$

Solution. Since any parallelogram is determined by three of its four vertices, we choose $(2, 3)$, $(10, -4)$, and $(-8, -2)$ and let

$$H = \begin{bmatrix} 2 & 10 & -8 \\ 3 & -4 & -2 \\ 1 & 1 & 1 \end{bmatrix}$$

be the matrix of the homogeneous coordinates of the chosen vertices. We translate H so that the coordinates of one of the vertices are the homogeneous coordinates of the origin of $\mathbb{R}^2$.

We have three choices and opt for the translation that moves $(2, 3)$ to $(0, 0)$.

$$A_1 = \begin{bmatrix} 1 & 0 & -2 \\ 0 & 1 & -3 \\ 0 & 0 & 1 \end{bmatrix} \begin{bmatrix} 2 & 10 & -8 \\ 3 & -4 & -2 \\ 1 & 1 & 1 \end{bmatrix} = \begin{bmatrix} 0 & 8 & -10 \\ 0 & -7 & -5 \\ 1 & 1 & 1 \end{bmatrix}$$

The last two columns of A_1 are the homogeneous coordinates of the new vectors determining the translated parallelogram.

Next we find the homogeneous coordinates of the images of the new vectors.

$$A_2 = \begin{bmatrix} 3 & 1 & 0 \\ 0 & -2 & 0 \\ 0 & 0 & 1 \end{bmatrix} \begin{bmatrix} 0 & 8 & -10 \\ 0 & -7 & -5 \\ 1 & 1 & 1 \end{bmatrix} = \begin{bmatrix} 0 & 17 & -35 \\ 0 & 14 & 10 \\ 1 & 1 & 1 \end{bmatrix}$$

It is easy to check that the area of the image of P is the absolute value of the determinant of the matrix A_2. Therefore the area of the image of P is

$$\det \begin{bmatrix} 0 & 17 & -35 \\ 0 & 14 & 10 \\ 1 & 1 & 1 \end{bmatrix} = 660$$

The areas of P and im P are connected by the determinant of A :

$$\det \begin{bmatrix} 17 & -35 \\ 14 & 10 \end{bmatrix} = 660 = \det \begin{bmatrix} 3 & 1 \\ 0 & -2 \end{bmatrix} \det \begin{bmatrix} 8 & -10 \\ -7 & -5 \end{bmatrix} = (-6) \times (-110)$$

As we can see, the magnification factor is the absolute value of the determinant of A. ▲

EXERCISES 5.9

1. Find the area of the image of the triangle determined by the vectors $(2, 3)$ and $(5, -1)$ under the matrix transformation

$$T \begin{bmatrix} x \\ y \end{bmatrix} = \begin{bmatrix} x + y \\ y \end{bmatrix}$$

2. Construct a matrix transformation $T_A : \mathbb{R}^3 \to \mathbb{R}^3$ that does not preserve linear independence. Justify your construction by exhibiting two linearly independent vectors $\mathbf{x}_1$ and $\mathbf{x}_2$ for which $T_A(\mathbf{x}_1)$ and $T_A(\mathbf{x}_2)$ are linearly dependent.

3. Find the area of the image of the parallelogram determined by the vertices

$$\mathbf{x}_1 = \begin{bmatrix} -2\sqrt{2} \\ 5\sqrt{2} \end{bmatrix}, \quad \mathbf{x}_2 = \begin{bmatrix} 0 \\ 7\sqrt{2} \end{bmatrix}, \quad \mathbf{x}_3 = \begin{bmatrix} -\sqrt{2} \\ 8\sqrt{2} \end{bmatrix}, \quad \mathbf{x}_4 = \begin{bmatrix} -3\sqrt{2} \\ 6\sqrt{2} \end{bmatrix}$$

under the linear transformation

$$T \begin{bmatrix} x \\ y \end{bmatrix} = \begin{bmatrix} x - y \\ y \end{bmatrix}$$

4. (*Maple V*) Plot the quadrilateral determined by the points $(1, 2)$, $(5, -3)$, $(7, 8)$, and $(12, 2)$ and find the area of its image under the matrix transformation

$$T\begin{bmatrix} x \\ y \end{bmatrix} = \begin{bmatrix} x - y \\ x + y \end{bmatrix}$$

5. (*Maple 6*) Find the image of the matrix transformation $T_A : \mathbb{R}^3 \to \mathbb{R}^4$ determined by the matrix

$$A = \begin{bmatrix} 1 & 0 & 2 \\ 0 & 1 & 0 \\ -6 & 0 & 1 \\ 0 & 0 & 0 \end{bmatrix}$$

SIMILARITY

We saw earlier that there are many different ways of representing a linear transformation $T : \mathbb{R}^n \to \mathbb{R}^m$ by a matrix. Different choices of bases yield different matrices. What is the connection between these matrices? If $n = m$, the answer is quite simple. Two matrices A and B represent the same transformation $T : \mathbb{R}^n \to \mathbb{R}^n$ if there exists a coordinate conversion matrix P for which $B = PAP^{-1}$. Before proving this fact, we give a name to this connection between A and B.

DEFINITION 5.12 *Two matrices A and B are* ***similar*** $B = PAP^{-1}$.

THEOREM 5.28 (Matrix similarity theorem) *Two square matrices A and B represent the same linear transformation if and only if they are similar.*

Proof. Suppose that $\mathcal{B}$ and $\mathcal{C}$ are two bases for V, that $A = [T]_{\mathcal{C}}$ and $B = [T]_{\mathcal{B}}$, and that P is the coordinate conversion matrix from $\mathcal{B}$ to $\mathcal{C}$. Then we know from Example 5.20 and Theorem 5.12 that

$$\begin{aligned} P^{-1}AP\,[\mathbf{x}]_{\mathcal{B}} &= [I]_{\mathcal{B}}^{\mathcal{C}}\,[T]_{\mathcal{C}}\,[I]_{\mathcal{C}}^{\mathcal{B}}\,[\mathbf{x}]_{\mathcal{B}} \\ &= [I]_{\mathcal{B}}^{\mathcal{C}}\,[T]_{\mathcal{C}}\,[\mathbf{x}]_{\mathcal{C}} \\ &= [I]_{\mathcal{B}}^{\mathcal{C}}\,[T(\mathbf{x})]_{\mathcal{C}} \\ &= [T(\mathbf{x})]_{\mathcal{B}} \\ &= [T]_{\mathcal{B}}\,[\mathbf{x}]_{\mathcal{B}} \\ &= B\,[\mathbf{x}]_{\mathcal{B}} \end{aligned}$$

Since this equation holds for all $\mathbf{x} \in V$, it follows that $P^{-1}AP = B$.

Conversely, suppose that $A = PBP^{-1}$ and that $A = [T_1]_{\mathcal{B}}$ and $B = [T_2]_{\mathcal{C}}$. Then

$$\begin{aligned}
[T_1(\mathbf{x})]_{\mathcal{B}} &= [T_1]_{\mathcal{B}}\,[\mathbf{x}]_{\mathcal{B}} \\
&= [I]_{\mathcal{B}}^{\mathcal{C}}\,[T_2]_{\mathcal{C}}\,[I]_{\mathcal{C}}^{\mathcal{B}}\,[\mathbf{x}]_{\mathcal{B}} \\
&= [I]_{\mathcal{B}}^{\mathcal{C}}\,[T_2]_{\mathcal{C}}\,[\mathbf{x}]_{\mathcal{C}} \\
&= [I]_{\mathcal{B}}^{\mathcal{C}}\,[T_2(\mathbf{x})]_{\mathcal{C}} \\
&= [T_2(\mathbf{x})]_{B}
\end{aligned}$$

Therefore $[T_1]_{\mathcal{B}}\,[\mathbf{x}]_{\mathcal{B}} = [T_2]_{\mathcal{B}}\,[\mathbf{x}]_{\mathcal{B}}$ for all $\mathbf{x} \in V$. Hence $T_1 = T_2$. ■

The next two theorems provide useful tests for showing that two square matrices are *not* similar.

THEOREM 5.29 (Trace similarity test) *If A and B are similar, then* trace A = trace B.

Proof. Suppose that A and B are similar. Then $A = P^{-1}BP$. We know from Theorem 2.10 that trace DE = trace ED. Therefore

$$\begin{aligned}
\text{trace } A &= \text{trace } P^{-1}BP \\
&= \text{trace } P^{-1}\,(BP) \\
&= \text{trace } (BP)\,P^{-1} \\
&= \text{trace } B(PP^{-1}) \\
&= \text{trace } B
\end{aligned}$$

This proves the theorem. ■

This theorem shows that if trace $A \neq$ trace B, then A and B are not similar. However, two nonsimilar matrices may still have the same trace.

EXAMPLE 5.31 Two Matrices with a Different Trace

Use the trace test to show that the matrices

$$A = \begin{bmatrix} 3 & 2 \\ 1 & 2 \end{bmatrix} \quad \text{and} \quad B = \begin{bmatrix} 1 & 2 \\ -1 & 2 \end{bmatrix}$$

are not similar.

Solution. The trace of A is $3 + 2 = 5$ and that of B is $1 + 2 = 3$. Therefore, A and B are not similar since they have a different trace. ▲

THEOREM 5.30 (Determinant similarity test) *If A and B are similar, then* det A = det B.

Proof. Suppose that A and B are similar. Then $A = P^{-1}BP$. We know from Theorem 3.14 that det DEF = det D det E det F.

Therefore,

$$\begin{aligned}\det A &= \det P^{-1}BP \\ &= \det P^{-1} \det B \det P \\ &= \det P^{-1} \det P \det B \\ &= \det B\end{aligned}$$

This proves the theorem. ■

This theorem shows that if $\det A \neq \det B$, then A and B are not similar. Again the test may be nonconclusive: Two nonsimilar matrices may have the same determinant.

EXAMPLE 5.32 Two Matrices with a Different Determinant

Use *Maple* and the determinant test to show that the matrices

$$A = \begin{bmatrix} 3 & 2 \\ 1 & 2 \end{bmatrix} \quad \text{and} \quad B = \begin{bmatrix} 1 & 2 \\ -1 & 1 \end{bmatrix}$$

are not similar.

Solution. We begin by defining A and B.

```
[> A:=Matrix([[3,2],[1,2]]):
[> B:=Matrix([[1,2],[-1,1]]):
```

We then use the **Determinant** function in the **LinearAlgebra** package.

```
[> LinearAlgebra[Determinant](A);
                                   4
[> LinearAlgebra[Determinant](B);
                                   3
```

Since $\det A \neq \det B$, the matrices A and B are not similar. ▲

Neither the trace test nor the determinant test allows us to conclude that two matrices are similar. The matrices

$$\begin{bmatrix} 1 & 1 & 0 \\ 3 & 3 & 2 \\ 2 & 2 & 3 \end{bmatrix} \quad \text{and} \quad \begin{bmatrix} 1 & 1 & 0 \\ 3 & 3 & 2 \\ 0 & 0 & 3 \end{bmatrix}$$

for example, have the same trace and the same determinant. However, in Example 6.13 we will show that they are in fact not similar. The trace and determinant tests establish necessary but not sufficient conditions for similarity. Sufficient conditions also exist, but their study is beyond the scope of this book. It is therefore helpful that both the **linalg** package and the **LinearAlgebra** package have a built-in test for deciding whether or not two matrices are similar.

EXAMPLE 5.33 Two Similar Matrices

Use a built-in similarity test of *Maple* to show that the following matrices are similar.

$$\begin{bmatrix} 3 & 2 & 3 \\ 5 & 4 & 2 \\ 5 & 2 & 1 \end{bmatrix} \text{ and } \begin{bmatrix} \frac{87}{17} & \frac{913}{51} & -\frac{209}{51} \\ 1 & \frac{14}{3} & -\frac{4}{3} \\ \frac{16}{17} & -\frac{91}{51} & -\frac{91}{51} \end{bmatrix}$$

Solution. We use the **IsSimilar** function in the **LinearAlgebra** package.

```
[> A:=Matrix([[3,2,3],[5,4,2],[5,2,1]]):
[> B:=Matrix([
    [87/17,913/51,-209/51],
    [1,14/3,-4/3],
    [16/17,-91/51,-91/51]]):
[> LinearAlgebra[IsSimilar](A,B);
```

true

EXERCISES 5.10

1. Show that the matrices

$$\begin{bmatrix} 3 & 0 & 0 \\ 0 & 2 & 0 \\ 0 & 0 & 2 \end{bmatrix} \text{ and } \begin{bmatrix} 2 & 0 & 0 \\ 0 & 3 & 0 \\ 0 & 0 & 2 \end{bmatrix}$$

are similar.

2. If A and B are two $n \times n$ real matrices, then Theorem 2.10 tells us that trace $AB =$ trace BA and Theorem 3.14 implies that det $AB =$ det BA. Explain why it follows from these theorems that neither the trace nor the determinant yields a sufficient condition for the similarity of AB and BA.

3. Find two similar 2×2 matrices A and B and two similar 2×2 matrices C and D with the properties that $A + C$ is not similar to $B + D$ and AC is not similar to BD.

4. Explain why two matrices with the same trace but different determinants are not similar. Use this test to construct two nonsimilar 2×2 matrices.

5. Show that the 2×2 zero matrix is not similar to any nonzero 2×2 matrix.

6. Find all matrices that are similar to the $n \times n$ identity matrix.

7. Show that if the matrices A and B are similar, then sA is similar to sB for any scalar s.

8. Explain why two real $n \times n$ matrices AB and BA are not necessarily similar.

9. (*Maple V*) Use the determinant test to show that the following three matrices are not similar.

$$\text{a. } A = \begin{bmatrix} 1 & 0 & -2 \\ 0 & 3 & 7 \\ 0 & 0 & 2 \end{bmatrix} \quad \text{b. } B = \begin{bmatrix} 6 & 0 & 0 \\ 0 & 0 & 0 \\ 0 & 0 & 0 \end{bmatrix} \quad \text{c. } C = \begin{bmatrix} 2 & -8 & 3 \\ 4 & 2 & 7 \\ 4 & -5 & 2 \end{bmatrix}$$

10. (*Maple V*) Find two real 5×5 matrices A and B for which the matrices AB and BA are not similar.

11. (*Maple V*) Use the **issimilar** function in the **linalg** package to show that the following pairs of matrices are similar.

$$\text{a. } \begin{bmatrix} 1 & 2 & 3 \\ 4 & 5 & 6 \\ 7 & 8 & 9 \end{bmatrix} \quad \text{and} \quad \begin{bmatrix} 0 & 3 & 1 \\ 9 & 15 & 2 \\ -3 & -3 & 0 \end{bmatrix}$$

$$\text{b. } \begin{bmatrix} 1 & 1 & 0 & 0 \\ 0 & 1 & 1 & 1 \\ 0 & 0 & 1 & 0 \\ 1 & 1 & 0 & 1 \end{bmatrix} \quad \text{and} \quad \begin{bmatrix} 2 & -\frac{2}{7} & \frac{4}{7} & -1 \\ 2 & \frac{1}{2} & \frac{1}{2} & -\frac{5}{2} \\ 1 & \frac{1}{2} & \frac{3}{2} & \frac{1}{2} \\ 1 & -\frac{1}{7} & \frac{2}{7} & 0 \end{bmatrix}$$

12. (*Maple V*) Use the **trace** function in the **linalg** package to show that the following three matrices are not similar.

$$\text{a. } A = \begin{bmatrix} 1 & 0 & -2 \\ 0 & 5 & 7 \\ 0 & 0 & 0 \end{bmatrix} \quad \text{b. } B = \begin{bmatrix} 0 & 0 & 0 \\ 0 & 0 & 0 \\ 0 & 0 & 0 \end{bmatrix} \quad \text{c. } C = \begin{bmatrix} 5 & -8 & 3 \\ 4 & 2 & 7 \\ 4 & -5 & 9 \end{bmatrix}$$

13. (*Maple V, 6*) Explore the properties of the **issimilar** and **IsSimilar** functions.

REVIEW

KEY CONCEPTS ▶ Define and discuss each of the following.

Components of a Linear Transformation

Codomain of a linear transformation, domain of a linear transformation, image of a linear transformation, kernel of a linear transformation, range of a linear transformation.

Matrices and Linear Transformations

Matrix of a transformation, matrix transformation, similar matrices and matrix transformations, standard matrix of a linear transformation.

Operations on Linear Transformations

Addition of linear transformations, composition of linear transformations, inversion of a transformation, scalar multiplication of a linear transformation, the transpose of a linear transformation.

Properties of a Linear Transformation

Invertible transformation, isomorphism, nullity of a linear transformation, one–one linear transformation, onto linear transformation, singular transformation.

Vectors and Linear Transformations

coordinate-vector function, linear-combination function.

KEY FACTS ▶ Explain and illustrate each of the following.

1. Every matrix transformation is a linear transformation.
2. If $T : V \to W$ is a linear transformation, then $T(\mathbf{0}_V) = \mathbf{0}_W$.
3. If $\mathcal{B} = \{\mathbf{x}_1, \mathbf{x}_2, \dots, \mathbf{x}_n\}$ is a basis for a vector space V and if $\mathbf{y}_1, \mathbf{y}_2, \dots, \mathbf{y}_n$ is any list of vectors in a vector space W, then there exists a unique linear transformation $T : V \to W$ for which $T(\mathbf{x}_1) = \mathbf{y}_1, \dots, T(\mathbf{x}_n) = \mathbf{y}_n$.
4. If $T : V \to W$ is an invertible linear transformation, then T is one–one.
5. If $T : V \to W$ is an invertible linear transformation, then T is onto.
6. A linear transformation $T : V \to W$ is an isomorphism if and only if T is one–one and onto.
7. For any basis $\mathcal{B} = \{\mathbf{x}_1, \dots, \mathbf{x}_n\}$ of a real vector space V, the coordinate-vector function $[-]_{\mathcal{B}} : V \to \mathbb{R}^n$ is an isomorphism.
8. If two real vector spaces have the same dimension, they are isomorphic.

9. The matrix space $\mathbb{R}^{m\times n}$ is isomorphic to the coordinate space $\mathbb{R}^{mn}$.

10. The polynomial space $\mathbb{R}_n[t]$ is isomorphic to the coordinate space $\mathbb{R}^{n+1}$.

11. If T is a linear transformation from $\mathbb{R}^n$ to $\mathbb{R}^m$ and if $\mathcal{B}$ is a basis for $\mathbb{R}^n$ and $\mathcal{C}$ is a basis for $\mathbb{R}^m$, then $[T(\mathbf{x})]_{\mathcal{C}} = [T]_{\mathcal{C}}^{\mathcal{B}}[\mathbf{x}]_{\mathcal{B}}$ for all $\mathbf{x} \in \mathbb{R}^n$.

12. Suppose that $T_1 : \mathbb{R}^n \to \mathbb{R}^m$ and $T_2 : \mathbb{R}^m \to \mathbb{R}^p$ are two linear transformations and that the matrices $[T_1]_{\mathcal{C}}^{\mathcal{B}}$ and$[T_2]_{\mathcal{D}}^{\mathcal{C}}$ represent the transformation T_1 and T_2 in the bases $\mathcal{B}$, $\mathcal{C}$, and $\mathcal{D}$. Then

$$[T_2 \circ T_1]_{\mathcal{D}}^{\mathcal{B}} = [T_2]_{\mathcal{D}}^{\mathcal{C}}[T_1]_{\mathcal{C}}^{\mathcal{B}}$$

13. If $T : \mathbb{R}^n \to \mathbb{R}^n$ is an invertible linear transformation represented by a matrix A, then $\left[T^{-1}\right] = A^{-1}$.

14. The image of a linear transformation $T : V \to W$ is a subspace of W.

15. If A is a real $m \times n$ matrix and $T_A : \mathbb{R}^n \to \mathbb{R}^m$ is the associated matrix transformation, then $\text{im}(T_A) = \text{Col}\, A$.

16. If $T : V \to W$ is a linear transformation and if $\mathcal{S} = \{\mathbf{x}_1, \ldots, \mathbf{x}_n\}$ spans V, then the set $\{T(\mathbf{x}_1), \ldots, T(\mathbf{x}_n)\}$ spans the image of T.

17. If $T : V \to W$ is a one–one linear transformation and if $\{T(\mathbf{x}_1), \ldots, T(\mathbf{x}_n)\}$ spans the image of T, then $\{\mathbf{x}_1, \ldots, \mathbf{x}_n\}$ spans V.

18. If $T : V \to W$ is a one–one linear transformation and the vectors $\mathbf{x}_1, \ldots, \mathbf{x}_n$ are linearly independent in V, then the vectors $T(\mathbf{x}_1), \ldots, T(\mathbf{x}_n)$ are linearly independent in W.

19. If $T : V \to W$ is a linear transformation and the vectors $T(\mathbf{x}_1), \ldots, T(\mathbf{x}_n)$ are linearly independent in W, then the vectors $\mathbf{x}_1, \ldots, \mathbf{x}_n$ are linearly independent in V.

20. The kernel of a linear transformation $T : V \to W$ is a subspace of V.

21. The kernel of a matrix transformation $T_A : \mathbb{R}^n \to \mathbb{R}^m$ is the null space of A.

22. A linear transformation T is one–one if and only if $\ker T = \{\mathbf{0}\}$.

23. The rank of a linear transformation $T_A : \mathbb{R}^n \to \mathbb{R}^m$ is the dimension of the column space of A.

24. The nullity of a linear transformation $T_A : \mathbb{R}^n \to \mathbb{R}^m$ is the dimension of the null space of A.

25. Suppose that $\mathcal{B} = \{\mathbf{x}_1, \ldots, \mathbf{x}_p\}$ is a basis for the kernel of a linear transformation $T : V \to W$ and that $\mathcal{C} = \{\mathbf{x}_1, \ldots, \mathbf{x}_p, \mathbf{x}_{p+1}, \ldots, \mathbf{x}_n\}$ is a basis for V. Then the set $\mathcal{D} = \{T(\mathbf{x}_{p+1}), \ldots, T(\mathbf{x}_n)\}$ is a basis for the image of T.

26. If $T : V \rightarrow W$ is a linear transformation and V is finite-dimensional, then $\dim V =$ nullity $T +$ rank T.

27. For any linear transformation $T : V \rightarrow V$ on a finite-dimensional vector space V, the following are equivalent: T is nonsingular, T is one–one, T is onto, T is an isomorphism.

28. Two square matrices A and B represent the same linear transformation if and only if they are similar.

29. If A and B are similar, then trace $A =$ trace B.

30. If A and B are similar, then $\det A = \det B$.

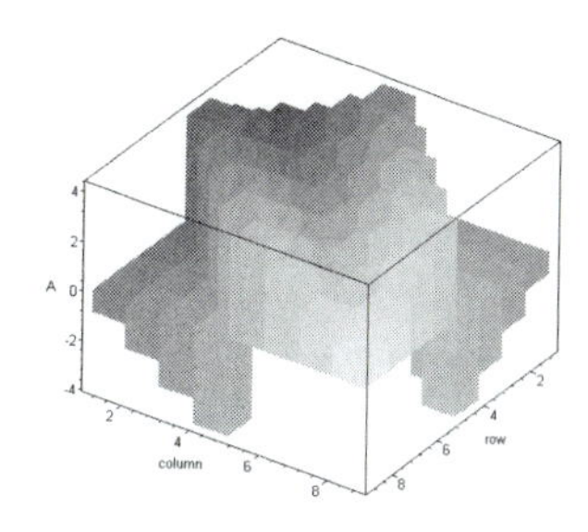

6

EIGENVALUES AND EIGENVECTORS

INTRODUCTION

In many applications involving a linear transformation $T : V \to V$ it is useful to find nonzero vectors $\mathbf{x}$ that are parallel to their images $T(\mathbf{x})$. By ***parallel*** vectors we mean vectors that are scalar multiples of each other. We are interested in finding scalars λ and nonzero vectors $\mathbf{x}$ for which $T(\mathbf{x}) = \lambda\mathbf{x}$. The scalar λ is called an ***eigenvalue*** and the associated vector $\mathbf{x}$ an ***eigenvector*** of T.

Our goal in this chapter is to describe how eigenvalues and eigenvectors can be found and to discuss conditions under which T has enough linearly independent eigenvectors to form a basis for V. Not every linear transformation has eigenvalues and eigenvectors. If V is finite-dimensional, we can reduce the problem of finding eigenvalues and eigenvectors for T to the problem of finding eigenvalues and eigenvectors for any matrix A of T.

THEOREM 6.1 *If $T : V \to V$ is a linear transformation on a finite-dimensional vector space V and $A = [T]_{\mathcal{B}}^{\mathcal{B}}$ is the matrix of T in some basis $\mathcal{B}$, then $T(\mathbf{x}) = \lambda\mathbf{x}$ if and only if $A[\mathbf{x}]_{\mathcal{B}} = \lambda[\mathbf{x}]_{\mathcal{B}}$.*

Proof. Suppose that $T(\mathbf{x}) = \lambda\mathbf{x}$. Then $[T(\mathbf{x})]_{\mathcal{B}} = [\lambda\mathbf{x}]_{\mathcal{B}}$. Since $A = [T]_{\mathcal{B}}^{\mathcal{B}}$, we know from the construction of A that $[T(\mathbf{x})]_{\mathcal{B}} = A[\mathbf{x}]_{\mathcal{B}}$. We also know from the linearity of the coordinate-vector function, that $[\lambda\mathbf{x}]_{\mathcal{B}} = \lambda[\mathbf{x}]_{\mathcal{B}}$. Hence $A[\mathbf{x}]_{\mathcal{B}} = \lambda[\mathbf{x}]_{\mathcal{B}}$. By reversing these arguments, we see that if $A[\mathbf{x}]_{\mathcal{B}} = \lambda\,[\mathbf{x}]_{\mathcal{B}}$, then $T(\mathbf{x}) = \lambda\mathbf{x}$. ■

This theorem tells us that $T(\mathbf{x}) = \lambda\mathbf{x}$ if and only if $[T]_{\mathcal{B}}^{\mathcal{B}}[\mathbf{x}]_{\mathcal{B}} = \lambda[\mathbf{x}]_{\mathcal{B}}$ for some matrix $[T]_{\mathcal{B}}^{\mathcal{B}}$. It tells us, in particular, that this relationship is independent of the specific matrix $[T]_{\mathcal{B}}^{\mathcal{B}}$ representing T. We can therefore begin by defining λ relative to matrices.

We distinguish real and complex matrices.

DEFINITION 6.1 *If A is a real $n \times n$ matrix and λ is a real number for which $A\mathbf{x} = \lambda\mathbf{x}$ for some nonzero vector $\mathbf{x} \in \mathbb{R}^n$, then λ is a **real eigenvalue** of A.*

DEFINITION 6.2 *If A is a real $n \times n$ matrix and λ is not a real eigenvalue of A, then λ is a* ***complex eigenvalue*** *of A if $A\mathbf{x} = \lambda\mathbf{x}$ for some complex number λ and some nonzero vector $\mathbf{x} \in \mathbb{C}^n$.*

If A is a complex $n \times n$ matrix for which $A\mathbf{x} = \lambda\mathbf{x}$ for some complex number λ and some nonzero vector $\mathbf{x} \in \mathbb{C}^n$, we simply say that λ is a real eigenvalue of A. We will see later that a real matrix, considered as a complex matrix, may have both real and complex eigenvalues or only complex eigenvalues. We will also see that the fundamental theorem of algebra, briefly discussed in Appendix A, implies that every $n \times n$ complex matrix has n (not necessarily distinct) complex eigenvalues. However, a real matrix may have no real eigenvalues. For example, the 2×2 rotation matrix

$$\begin{bmatrix} 0 & 1 \\ -1 & 0 \end{bmatrix}$$

has no real eigenvalues but does have the two complex eigenvalues i and $-i$. The 3×3 real matrix

$$\begin{bmatrix} 0 & 0 & 2 \\ 1 & 0 & -1 \\ 0 & 1 & 2 \end{bmatrix}$$

on the other hand, has the real eigenvalue 2 and the two complex eigenvalues i and $-i$. In this text, we are mainly concerned with real eigenvalues. Hence the term *eigenvalue* will mean *real eigenvalue* unless otherwise specified.

Theorem 6.1 guarantees that similar matrices have the same eigenvalues. Later we will confirm this fact using properties of determinants. We can therefore lift the idea of an eigenvalue from matrices to linear transformation. We can think of an eigenvalue of a matrix transformation $T : \mathbb{R}^n \to \mathbb{R}^n$ to be an eigenvalue of any one of the real matrices A representing T. Using isomorphisms, we can extend this definition from matrix transformations on $\mathbb{R}^n$ to linear transformations on n-dimensional real vector spaces.

EXAMPLE 6.1 Diagonal Eigenvalues

Show that the diagonal entries of the matrix

$$A = \begin{bmatrix} 3 & 0 \\ 0 & 0 \end{bmatrix}$$

are eigenvalues of the matrix A.

Solution. Since

$$\begin{bmatrix} 3 & 0 \\ 0 & 0 \end{bmatrix} \begin{bmatrix} 1 \\ 0 \end{bmatrix} = \begin{bmatrix} 3 \\ 0 \end{bmatrix} = 3 \begin{bmatrix} 1 \\ 0 \end{bmatrix}$$

and since

$$\begin{bmatrix} 3 & 0 \\ 0 & 0 \end{bmatrix}\begin{bmatrix} 0 \\ 1 \end{bmatrix} = \begin{bmatrix} 0 \\ 0 \end{bmatrix} = 0\begin{bmatrix} 0 \\ 1 \end{bmatrix}$$

it follows that the diagonal entries 3 and 0 of the matrix A are eigenvalues of A.▲

With each eigenvalue λ, we now associate a set of vectors for which λ is an eigenvalue.

DEFINITION 6.3 *A nonzero vector* $\mathbf{x}$ *is a **real eigenvector** of a real* $n \times n$ *matrix* A *if* $\mathbf{x} \in \mathbb{R}^n$ *and there exists a real number* λ *for which* $A\mathbf{x} = \lambda\mathbf{x}$.

DEFINITION 6.4 *A nonzero vector* $\mathbf{x}$ *is a **complex eigenvector** of a real* $n \times n$ *matrix* A *if* $\mathbf{x} \in \mathbb{C}^n$ *and there exists a complex number* λ *for which* $A\mathbf{x} = \lambda\mathbf{x}$, *provided that* $\mathbf{x}$ *is not a real eigenvector of* A.

In the case of complex matrices, we simply speak of eigenvalues and eigenvectors. We say that an eigenvector $\mathbf{x}$ ***belongs to*** or ***is associated with*** its eigenvalue λ. In the work that follows, it is essential that this relationship be specific. An eigenvector should be belong to precisely one eigenvalue. For this reason it makes sense to allow the scalar 0 be an eigenvalue but not to allow the vector $\mathbf{0}$ as an eigenvector. For any square matrix A, for example, the equation

$$A\mathbf{0} = \lambda\mathbf{0} = \mathbf{0}$$

holds for all real numbers λ. Hence there is no unique scalar λ for which $A\mathbf{0} = \lambda\mathbf{0}$.

In the language of eigenvalues and eigenvectors, Theorem 6.1 says that if $A = [T]_{\mathcal{B}}^{\mathcal{B}}$ is the matrix of a linear transformation T, then λ is an eigenvalue of T if and only if it is an eigenvalue of A, and $\mathbf{x}$ is an eigenvector of T if and only if $[\mathbf{x}]_{\mathcal{B}}$ is an eigenvector of A. Let us consider some examples.

EXAMPLE 6.2 The Eigenvalues and Eigenvectors of an Identity Transformation

Let $T : V \to V$ be the identity transformation. Then $T(\mathbf{x}) = \mathbf{x} = 1\mathbf{x}$ for all $\mathbf{x} \in V$. Therefore, 1 is an eigenvalue of T, and any nonzero vector $\mathbf{x} \in V$ is an eigenvector belonging to the eigenvalue 1. ▲

EXAMPLE 6.3 The Eigenvalues and Eigenvectors of a Zero Transformation

Let $T : V \to V$ be the zero transformation. Then $T(\mathbf{x}) = \mathbf{0} = 0\mathbf{x}$ for all $\mathbf{x} \in V$. Therefore, 0 is an eigenvalue of T, and any nonzero vector $\mathbf{x} \in V$ is an eigenvector belonging to the eigenvalue 0. ▲

The next examples are more interesting. They show that a linear transformation $T : \mathbb{R}^n \to \mathbb{R}^n$ may have any number of distinct eigenvalues. However, T cannot have more than n such values.

EXAMPLE 6.4 A Linear Transformation with Three Eigenvalues

Let $T_1 : \mathbb{R}^3 \to \mathbb{R}^3$ be the transformation defined by $T_1(x, y, z) = (x, 2y, 3z)$. Then 1, 2, and 3 are eigenvalues of T_1 since

$$\begin{cases} T_1(x,0,0) = (x,0,0) \ \ = 1(x,0,0) \\ T_1(0,y,0) = (0,2y,0) = 2(0,y,0) \\ T_1(0,0,z) = (0,0,3z) = 3(0,0,z) \end{cases}$$

for all $x, y, z \in \mathbb{R}$. ▲

As we can see, the linear transformation T_1 has three distinct eigenvalues. The transformation $T_2 : \mathbb{R}^3 \to \mathbb{R}^3$ defined by $T_2(x, y, z) = (x, 2y, 2z)$, on the other hand, has only the two distinct eigenvalues 1 and 2. The transformation $T_3 : \mathbb{R}^3 \to \mathbb{R}^3$ defined by $T_3(x, y, z) = (2x, 2y, 2z)$ has the real number 2 as its only eigenvalue. Furthermore, we will see in Example 6.7 that some linear transformations have no real eigenvalues at all.

Let us take a closer look at the eigenvectors of the linear transformations T_1, T_2, and T_3.

1. In the case of T_1, the vector $\mathbf{e}_1 = (1, 0, 0)$ is an eigenvector belonging to 1, the vector $\mathbf{e}_2 = (0, 1, 0)$ is an eigenvector belonging to 2, and the vector $\mathbf{e}_3 = (0, 0, 1)$ is an eigenvector belonging to 3. Moreover, since $\mathbf{e}_1, \mathbf{e}_2$, and $\mathbf{e}_3$ are the standard basis vector of $\mathbb{R}^3$, they are linearly independent. This is no accident. We will see later that eigenvectors belonging to distinct eigenvalues are linearly independent.

2. In the case of T_2, the vector $\mathbf{e}_1$ is an eigenvector belonging to 1, and the vectors $\mathbf{e}_2$ and $\mathbf{e}_3$ are eigenvectors belonging to 2. In this case, the two linearly independent eigenvectors $\mathbf{e}_2$ and $\mathbf{e}_3$ belong to the same eigenvalue. This tells us that, not surprisingly, linearly independent vectors can belong to the same eigenvalue.

3. In the case of T_3, the standard basis vectors $\mathbf{e}_1, \mathbf{e}_2$, and $\mathbf{e}_3$ are all eigenvectors belonging to the eigenvalue 2.

In the standard basis $\mathcal{E}$, the linear transformations T_1, T_2, and T_3 are represented by the diagonal matrices

$$\text{a.} \begin{bmatrix} 1 & 0 & 0 \\ 0 & 2 & 0 \\ 0 & 0 & 3 \end{bmatrix} \quad \text{b.} \begin{bmatrix} 1 & 0 & 0 \\ 0 & 2 & 0 \\ 0 & 0 & 2 \end{bmatrix} \quad \text{c.} \begin{bmatrix} 2 & 0 & 0 \\ 0 & 2 & 0 \\ 0 & 0 & 2 \end{bmatrix}$$

One of the fundamental questions of linear algebra is whether such diagonal representations always exist. If the number of distinct eigenvalues of a linear transformation $T : \mathbb{R}^n \to \mathbb{R}^n$ is less than n, this problem reduces to the problem of finding enough linearly independent eigenvectors belonging to the same eigenvalue. This is not always possible.

DEFINITION 6.5 *An $n \times n$ matrix is **diagonalizable** if it has n linearly independent eigenvectors.*

We will see later that this means that there exists a diagonal matrix D whose diagonal entries are eigenvalues of A and an invertible matrix P whose columns are eigenvectors of A, for which $A = PDP^{-1}$. We call PDP^{-1} an ***eigenvector-eigenvalue decomposition*** of A. In Chapter 9, we will study techniques for decomposing matrices that fail to have such decompositions.

DEFINITION 6.6 *A linear transformation $T : V \to V$ is **diagonalizable** if there exists a basis $\mathcal{B}$ for which the matrix $A = [T]_{\mathcal{B}}^{\mathcal{B}}$ is diagonal.*

EXAMPLE 6.5 A Diagonalizable Transformation

Find the eigenvalues and eigenvectors of the linear transformation $T_A : \mathbb{R}^2 \to \mathbb{R}^2$ defined by

$$T_A \begin{bmatrix} x \\ y \end{bmatrix} = \begin{bmatrix} 3 & 0 \\ 0 & 2 \end{bmatrix} \begin{bmatrix} x \\ y \end{bmatrix}$$

and discuss the geometric properties of T_A.

Solution. Since

$$\begin{bmatrix} 3 & 0 \\ 0 & 2 \end{bmatrix} \begin{bmatrix} x \\ y \end{bmatrix} = \begin{bmatrix} 3x \\ 2y \end{bmatrix} = 3 \begin{bmatrix} x \\ 0 \end{bmatrix} + 2 \begin{bmatrix} 0 \\ y \end{bmatrix}$$

it follows that $T_A(x, 0) = 3(x, 0)$ and $T_A(0, y) = 2(0, y)$ for all $x, y \in \mathbb{R}$. Hence $\lambda_1 = 3$ and $\lambda_2 = 2$ are eigenvalues of T_A and $(x, 0)$ and $(0, y)$ are their associated eigenvectors, provided x and y are not zero.

We can illustrate the geometric properties of T_A by considering its effect on the triangle determined by the points $(0, 0)$, $(1, 0)$, and $(0, 1)$. Since $T(0, 0) = (0, 0)$, $T(1, 0) = (3, 0)$, and $T(0, 1) = (0, 2)$, the transformation T stretches the triangle three units along the x-axis and two units along the y-axis. Figure 1 on Page 420 shows the corresponding triangles. ▲

Maple has an **eigenvalues** function in the **linalg** package that can be used to find eigenvalues of square matrices.

EXAMPLE 6.6 A Linear Transformation with Two Eigenvalues

Find the eigenvalues of the linear transformation $T : \mathbb{R}^2 \to \mathbb{R}^2$ defined by

$$T \begin{bmatrix} x \\ y \end{bmatrix} = \begin{bmatrix} x + y \\ x - y \end{bmatrix}$$

Solution. We must find a λ for which

$$\begin{bmatrix} x + y \\ x - y \end{bmatrix} = \lambda \begin{bmatrix} x \\ y \end{bmatrix} = \begin{bmatrix} \lambda x \\ \lambda y \end{bmatrix}$$

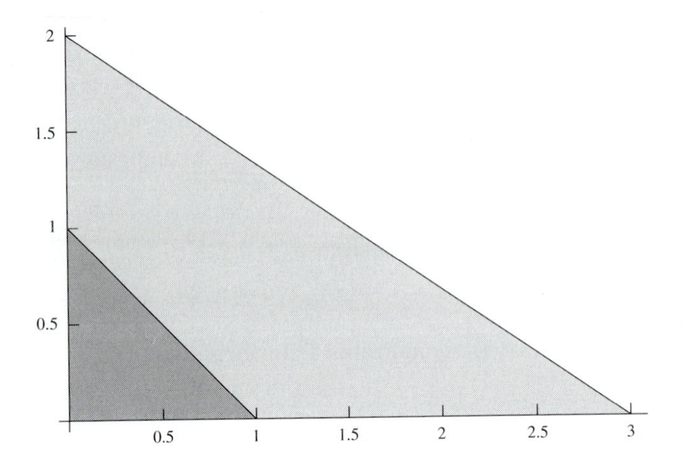

FIGURE 1 The Geometric Effect of T_A.

We begin by representing T in the standard basis $\mathcal{E} = \{\mathbf{e}_1, \mathbf{e}_2\}$ of $\mathbb{R}^2$.

$$[T]_{\mathcal{E}} = \left[[T(\mathbf{e}_1)]_{\mathcal{E}} \; [T(\mathbf{e}_2)]_{\mathcal{E}}\right] = \begin{bmatrix} 1 & 1 \\ 1 & -1 \end{bmatrix}$$

Then we use *Maple* to find the eigenvalues and associated eigenvectors of $[T]_{\mathcal{E}}$.

```
[> A:=matrix([[1,1],[1,-1]]):
[> linalg[eigenvalues](A);
```

$$\sqrt{2}, -\sqrt{2}$$

We show that $-\sqrt{2}$ is an eigenvalue by finding an associated eigenvector. The verification for $\sqrt{2}$ is analogous. If $-\sqrt{2}$ is an eigenvalue of A, then

$$\begin{bmatrix} x+y \\ x-y \end{bmatrix} = -\sqrt{2}\begin{bmatrix} x \\ y \end{bmatrix} = \begin{bmatrix} -\sqrt{2}x \\ -\sqrt{2}y \end{bmatrix}$$

This means that

$$\begin{cases} x + y = -\sqrt{2}x \\ x - y = -\sqrt{2}y \end{cases}$$

If this system has a nontrivial solution, then $-\sqrt{2}$ is an eigenvalue of A.

```
[> solve({x+y=-sqrt(2)*x, x-y=-sqrt(2)*y},{x,y});
```

$$\{x = y - \sqrt{2}y, y = y\}$$

```
[> linalg[eigenvectors](A);
```

$$[\sqrt{2}, 1, \{[1, -1 + \sqrt{2}]\}], [-\sqrt{2}, 1, \{[1, -\sqrt{2} - 1]\}]$$

The calculation

$$T\begin{bmatrix} 1-\sqrt{2} \\ 1 \end{bmatrix} = \begin{bmatrix} \left(1-\sqrt{2}\right)+1 \\ \left(1-\sqrt{2}\right)-1 \end{bmatrix} = -\sqrt{2}\begin{bmatrix} 1-\sqrt{2} \\ 1 \end{bmatrix}$$

shows that the vector $(x, y) = (1-\sqrt{2}, 1)$ is indeed an eigenvector belonging to the eigenvalue $-\sqrt{2}$. A similar calculation shows that the vector $(x, y) = (1 + \sqrt{2}, 1)$ is an eigenvector belonging to $\sqrt{2}$. Hence the matrix A has the two real eigenvalues, $-\sqrt{2}$ and $\sqrt{2}$. ▲

The next example shows that a linear transformation may have no eigenvalues at all. It also explains geometrically why this is the case.

EXAMPLE 6.7 A Linear Transformation Without Eigenvalues

Find the real eigenvalues of the linear transformation $T : \mathbb{R}^2 \to \mathbb{R}^2$ defined by

$$T\begin{bmatrix} x \\ y \end{bmatrix} = \begin{bmatrix} 0 & -1 \\ 1 & 0 \end{bmatrix}\begin{bmatrix} x \\ y \end{bmatrix}$$

if they exist.

Solution. Since

$$T\begin{bmatrix} x \\ y \end{bmatrix} = \begin{bmatrix} 0 & -1 \\ 1 & 0 \end{bmatrix}\begin{bmatrix} x \\ y \end{bmatrix} = \begin{bmatrix} -y \\ x \end{bmatrix}$$

it is clear that T rotates the points in $\mathbb{R}^2$ counterclockwise by 90°. Therefore

$$T\begin{bmatrix} x \\ y \end{bmatrix} = \lambda\begin{bmatrix} x \\ y \end{bmatrix} \quad \text{if and only if} \quad \begin{bmatrix} -y \\ x \end{bmatrix} = \lambda\begin{bmatrix} x \\ y \end{bmatrix} = \begin{bmatrix} \lambda x \\ \lambda y \end{bmatrix}$$

This means that $-y = \lambda x$ and $x = \lambda y$. These equations hold only if $(\lambda^2 + 1)y = 0$ and $(\lambda^2 + 1)x = 0$. Since the equation $\lambda^2 + 1 = 0$ has no solution in the real numbers, x and y must be 0. This shows that there are no nonzero vectors $\mathbf{x}$ for which $T(\mathbf{x}) = \lambda\mathbf{x}$. Therefore T has no eigenvalues. ▲

The existence or nonexistence of eigenvalues and eigenvectors of linear transformations depends both on the geometric properties of the transformations and on the spaces on which they act. If we were to expand the domain and codomain of the matrix transformation $T_A : \mathbb{R}^2 \to \mathbb{R}^2$ in Example 6.7 to the complex vector space $\mathbb{C}^2$, then T_A would have eigenvalues and eigenvectors. The complex roots of the equation $\lambda^2 + 1 = 0$ are $\lambda = i$ and $\lambda = -i$, and

$$T\begin{bmatrix} y \\ iy \end{bmatrix} = \begin{bmatrix} 0 & -1 \\ 1 & 0 \end{bmatrix}\begin{bmatrix} y \\ iy \end{bmatrix} = \begin{bmatrix} -iy \\ y \end{bmatrix} = -i\begin{bmatrix} y \\ iy \end{bmatrix}$$

Hence all vectors of the form (y, iy) are eigenvectors belonging to the eigenvalue $-i$. Similarly, all nonzero vectors (iy, y) are eigenvectors belonging to the eigenvalue i.

EXAMPLE 6.8 Approximate Eigenvalues of a Large Matrix

Use *Maple* to find approximate eigenvalues of a random 15×15 matrix A and extract the real eigenvalues of A, if they exist.

```
[> restart;
[> with(linalg):
[> A:=randmatrix(15,15):
[> E:=evalf(eigenvalues(A),2);

   E:=  51.,240.,100.+31.I,150.+140.I,61.+75.I,-61.+210.I,
             -81.+95.I,-77.+23.I,-240.,-77.-23.I,-81.-95.I,
                  -61.-210.I,61.-75.I,150.-140.I,100.-31.I
```

We can tell by inspection that $\lambda_1 = 51.$, $\lambda_2 = 240.$, and $\lambda_3 = -240.$ are approximate real eigenvalues of A. All other eigenvalues are complex. ▲

EXERCISES 6.1

1. Describe the geometric properties of the following linear transformations $T : \mathbb{R}^2 \to \mathbb{R}^2$.

   a. $T(x, y) = (3x, 4y)$ b. $T(x, y) = (-3x, 4y)$
   c. $T(x, y) = (3x, -4y)$ d. $T(x, y) = (-3x, -4y)$

2. Let $T : \mathbb{R}^2 \to \mathbb{R}^2$ be a linear transformation whose values on the standard basis of $\mathbb{R}^2$ are $T(\mathbf{e}_1) = 5\mathbf{e}_1$ and $T(\mathbf{e}_2) = -2\mathbf{e}_2$. Explain why 5 and -2 are eigenvalues of T, and find the eigenvectors for each eigenvalue.

3. Let $T : \mathbb{R}^3 \to \mathbb{R}^3$ be a linear transformation whose values on the standard basis of $\mathbb{R}^3$ are $T(\mathbf{e}_1) = 5\mathbf{e}_1$, $T(\mathbf{e}_2) = -2\mathbf{e}_2$, and $T(\mathbf{e}_3) = 7\mathbf{e}_3$. Explain why 5, -2, and 7 are eigenvalues of T, and find the eigenvectors for each eigenvalue.

4. Consider the matrix transformation $T_A : \mathbb{R}^3 \to \mathbb{R}^3$ determined by the matrix

$$A = \begin{bmatrix} 0 & 0 & 6 \\ 1 & 0 & -11 \\ 0 & 1 & 6 \end{bmatrix}$$

and explain the connection between the eigenvectors of T_A and the solutions of the homogeneous linear systems $(A - I)\mathbf{x} = \mathbf{0}$, $(A - 2I)\mathbf{x} = \mathbf{0}$, and $(A - 3I)\mathbf{x} = \mathbf{0}$.

5. Let $T_P : \mathbb{R}^{2\times 2} \to \mathbb{R}^{2\times 2}$ be the similarity transformation $T_P(A) : PAP^{-1}$ defined by the matrix

$$P = \begin{bmatrix} 24 & 17 \\ 9 & 6 \end{bmatrix}$$

Compare the eigenvalues of the matrices A and $T_P(A)$ for the case

$$A = \begin{bmatrix} 1 & 2 \\ 3 & 4 \end{bmatrix}$$

6. Show that if λ is an eigenvalue of a linear transformation $T : V \to V$, then λ^2 is an eigenvalue of T^2.

7. Show by an example that the existence of an eigenvalues for $T^2 : V \to V$ does not imply the existence of an eigenvalues for $T : V \to V$.

8. Explain why the set of eigenvectors belonging to an eigenvalue of a real matrix is infinite, but every set of linearly independent eigenvectors belonging to an eigenvalue is finite.

9. Show that the scalar 1 is the only eigenvalue of the identity transformation $I : \mathbb{R}^3 \to \mathbb{R}^3$ and that every nonzero vector in $\mathbb{R}^3$ is an eigenvector.

10. The eigenvalues λ of the matrix

$$A = \begin{bmatrix} 2 & 5 & 6 \\ 0 & 3 & 7 \\ 0 & 0 & 4 \end{bmatrix}$$

are 2, 3, and 4. Show that for each λ, the matrix $(A - \lambda I)$ is singular.

11. Show that the symmetric matrices AA^T and A^TA have the same nonzero eigenvalues.

12. Find the real eigenvalues of the following matrices, if they exist.

a. $\begin{bmatrix} 1 & 0 \\ 0 & 0 \end{bmatrix}$ b. $\begin{bmatrix} 0 & -1 \\ 1 & 0 \end{bmatrix}$ c. $\begin{bmatrix} 0 & 1 \\ -1 & 1 \end{bmatrix}$

d. $\begin{bmatrix} 1 & 0 & -2 \\ 0 & 5 & 7 \\ 0 & 0 & 0 \end{bmatrix}$ e. $\begin{bmatrix} 0 & -1 & 0 \\ 1 & 0 & 0 \\ 0 & 0 & 1 \end{bmatrix}$ f. $\begin{bmatrix} 2 & 0 & 0 & 0 \\ 0 & 3 & 0 & 0 \\ 0 & 0 & 2 & 0 \\ 0 & 0 & 0 & 5 \end{bmatrix}$

g. $\begin{bmatrix} 1 & 0 & 0 & 0 \\ 0 & 1 & 0 & 0 \\ 0 & 0 & 1 & 0 \\ 0 & 0 & 0 & 1 \end{bmatrix}$ h. $\begin{bmatrix} 6.0 & 3 & 6 & 5 \\ 3 & 4 & 4 & 3 \\ 6 & 4 & 2 & 2 \\ 5 & 3 & 2 & 4 \end{bmatrix}$ i. $\begin{bmatrix} 1 & 0 & 0 & 9 \\ 0 & 1 & 7 & 0 \\ 0 & 7 & 1 & 0 \\ 9 & 0 & 0 & 1 \end{bmatrix}$

13. Let $T_2 \circ T_1 : V \to V$ be a composite linear transformation, and suppose that $T_1(\mathbf{x}) = \lambda\mathbf{x}$ and $T_2(\mathbf{x}) = \mu\mathbf{x}$ for some nonzero vector $\mathbf{x} \in V$. Show that $\lambda\mu$ is an eigenvalue of $T_2 \circ T_1$.

14. (*Maple V*) Show that if $T_A : \mathbb{R}^3 \to \mathbb{R}^3$ is the matrix transformation determined by the matrix

$$A = \begin{bmatrix} 0 & 0 & 60 \\ 1 & 0 & 7 \\ 0 & 1 & -6 \end{bmatrix}$$

then -5, -4, and 3 are eigenvalues of T_A. Find two eigenvectors for each eigenvalue.

15. (*Maple V*) Generate a random 3×3 invertible matrix P and calculate its inverse P^{-1}. Show that the eigenvalues of the matrix transformation determined by PAP^{-1} are -5, -4, and 3 if A is the matrix in Exercise 14.

16. (*Maple V*) Describe the mapping properties of the following linear transformations.

a. $T(x, y, z) = (3x, 4y, 5z)$ b. $T(x, y, z) = (-3x, 4y, 5z)$
c. $T(x, y, z) = (-3x, -4y, 5z)$ d. $T(x, y, z) = (3x, 4y, -5z)$

17. (*Maple V*) Find the eigenvalues of the matrix

$$A = \begin{bmatrix} 0 & 0 & 0 & 0 & 120 \\ 1 & 0 & 0 & 0 & -274 \\ 0 & 1 & 0 & 0 & 225 \\ 0 & 0 & 1 & 0 & -85 \\ 0 & 0 & 0 & 1 & 15 \end{bmatrix}$$

Show that A has enough linearly independent eigenvectors to span $\mathbb{R}^5$.

18. (*Maple V*) Let $T_P : \mathbb{R}^{3\times3} \to \mathbb{R}^{3\times3}$ be the similarity transformation $T_P(A) : PAP^{-1}$ defined by the matrix

$$P = \begin{bmatrix} 3 & 0 & 8 \\ 0 & 1 & 3 \\ 8 & 3 & 0 \end{bmatrix}$$

Compare the eigenvalues of the matrices A and $T_P(A)$ for the case

$$A = \begin{bmatrix} 1 & 2 & 0 \\ 3 & 4 & 0 \\ 0 & 0 & 1 \end{bmatrix}$$

19. (*Maple V*) Generate a random 3×3 invertible matrix P and show that if A is the matrix in Exercise 18, then the matrix $(PAP^{-1} - \lambda I)$ is singular for every eigenvalue λ of A.

20. (*Maple V*) Verify that the matrix transformation $T_A : \mathbb{C}^3 \to \mathbb{C}^3$ defined on the complex vector space $\mathbb{C}^3$ by the matrix

$$A = \begin{bmatrix} 0 & 0 & 2 \\ 1 & 0 & 1 \\ 0 & 1 & 1 \end{bmatrix}$$

has the eigenvalues 2, $-\frac{1}{2} + \frac{1}{2}\sqrt{3}i$, and $-\frac{1}{2} - \frac{1}{2}\sqrt{3}i$. Find an eigenvector for each of the three eigenvalues.

21. (*Maple V*) Show that the matrix

$$\begin{bmatrix} 0 & 0 & 0 & 0 & 540 \\ 1 & 0 & 0 & 0 & -783 \\ 0 & 1 & 0 & 0 & 450 \\ 0 & 0 & 1 & 0 & -128 \\ 0 & 0 & 0 & 1 & 18 \end{bmatrix}$$

has only three linearly independent eigenvectors.

22. (*Maple 6*) Use `RandomMatrix(15,15)` and `evalf(Eigenvalues(-))` to calculate the approximate eigenvalues of a random 15×15 matrix A and, by inspection, discuss whether A has any approximate real eigenvalues.

CHARACTERISTIC POLYNOMIALS

In this section we show that eigenvalues of matrices are ***roots of polynomials***. Suppose that A is a real $n \times n$ matrix and that there exists a nonzero column vector $\mathbf{x} \in \mathbb{R}^n$ for which

$A\mathbf{x} = \lambda\mathbf{x}$. Using the identity matrix I, we can rewrite this equation as $A\mathbf{x} = \lambda I\mathbf{x}$ and form the associated homogeneous linear systems

$$A\mathbf{x} - \lambda I\mathbf{x} = (A - \lambda I)\mathbf{x} = (-1)(A - \lambda I)\mathbf{x} = (\lambda I - A)\mathbf{x} = \mathbf{0}$$

Finding an eigenvector λ for A is therefore corresponds to finding a nontrivial solution of the homogeneous system $(A - \lambda I)\,\mathbf{x} = \mathbf{0}$ or $(\lambda I - A)\,\mathbf{x} = \mathbf{0}$.

We know from our work on linear systems that this is possible only if the real coefficient matrix $(A - \lambda I)$ is singular. We also know that this happens precisely when the determinant of $(A - \lambda I)$ is 0. This leads us to real polynomials. For any variable t, the determinant of the matrix $(A - tI)$ is a polynomial $p_A(t) \in \mathbb{R}[t]$, since the entries of $(A - tI)$ are polynomials, and the required scalars λ will be among the roots of $p_A(t)$. Moreover, the polynomials

$$p_A(t) = \det(A - tI) \in \mathbb{R}[t] \quad \text{and} \quad q_A(t) = \det(tI - A) \in \mathbb{R}[t]$$

have the same roots since $q_A(t) = (-1)p_A(t)$. The polynomial $q_A(t)$ has the advantage that it is of the form $t^n + r(t)$, whereas $p_A(t) = (-1)^n t^n + s(t)$. *Maple* is programmed to use $q_A(t)$ rather than $p_A(t)$. For any variable t, the required scalars λ will be among the roots of $p_A(t)$ and $q_A(t)$.

We note that for the work that follows, the matrices $(A - tI)$ have the advantage over the matrices $(tI - A)$ that they preserve the signs of the entries of A. For example, if

$$A = \begin{bmatrix} 1 & 2 \\ 3 & 4 \end{bmatrix}$$

then

$$A - tI = \begin{bmatrix} 1-t & 2 \\ 3 & 4-t \end{bmatrix} \quad \text{and} \quad tI - A = \begin{bmatrix} t-1 & -2 \\ -3 & t-4 \end{bmatrix}$$

The polynomials $q_A(t) = \det(tI - A)$, on the other hand, have the advantage over the polynomials $p_A(t) = \det(A - tI)$ that the coefficients of their highest powers are always 1, whereas those of $p_A(t) = \det(A - tI)$ alternate between -1 and 1, depending on whether the dimension of A is odd or even.

DEFINITION 6.7 *For any $n \times n$ matrix A, the matrix $(A - tI)$ is called the* ***characteristic matrix*** *of A in the variable t, and the polynomial $c_A(t) = (-1)^n \det(A - tI)$ is called the* ***characteristic polynomial*** *of A in the variable t.*

The following definition provides us with the best of both worlds. In the **linalg** package of *Maple V*, the characteristic matrix of A is defined to be polynomial matrix $(tI - A)$, whereas in the **LinearAlgebra** package of *Maple 6*, the characteristic matrix of A is defined to be the polynomial matrix $(A - tI)$. In both packages, the characteristic polynomial is defined to be the polynomial $(-1)^n \det(A - tI) = \det(tI - A)$.

EXAMPLE 6.9 Calculating Characteristic Polynomials

Calculate the characteristic matrices and characteristic polynomials of the matrices

$$A := \begin{bmatrix} 1 & 2 \\ 3 & 4 \end{bmatrix} \quad \text{and} \quad B := \begin{bmatrix} 1 & 2 & 3 \\ 4 & 5 & 6 \\ 7 & 8 & 9 \end{bmatrix}$$

in three ways: (a) using the definition; (b) by using the function **charmat** and **charpoly** in the **linalg** package of *Maple V*; (c) by using the functions **CharacteristicMatrix** and **CharacteristicPolynomial** in the **LinearAlgebra** package of *Maple 6.*

Solution. By definition, the characteristic matrices of the matrices A and B in the variable t are the polynomial matrices

$$A - tI = \begin{bmatrix} 1-t & 2 \\ 3 & 4-t \end{bmatrix} \quad \text{and} \quad B - tI = \begin{bmatrix} 1-t & 2 & 3 \\ 4 & 5-t & 6 \\ 7 & 8 & 9-t \end{bmatrix}$$

The associated characteristic polynomials are

$$c_A(t) = (-1)^2 \det(A - tI) = -2 - 5t + t^2$$

and

$$c_B(t) = (-1)^3 \det(B - tI) = -18t - 15t^2 + t^3$$

We now compare these results with the results obtained by *Maple.*

```
[> with(linalg):  with(LinearAlgebra):
[> A:=matrix([[1,2],[3,4]]):
```

```
[> cmA:=charmat(A,t);
```

$$cmA := \begin{bmatrix} t-1 & -2 \\ -3 & t-4 \end{bmatrix}$$

```
[> cpA:=charpoly(A,t);
```

$$cpA := t^2 - 5t - 2$$

```
[> CMA:=CharacteristicMatrix(Matrix(A),t);
```

$$CMA := \begin{bmatrix} -t+1 & 2 \\ 3 & -t+4 \end{bmatrix}$$

```
[> CPA:=CharacteristicPolynomial(Matrix(A),t);
```

$$CPA := t^2 - 5t - 2$$

```
[> B:=matrix([[1,2,3],[4,5,6],[7,8,9]]):
[> cmB:=charmat(B,t):
```

$$cmB := \begin{bmatrix} t-1 & -2 & -3 \\ -4 & t-5 & -6 \\ -7 & -8 & t-9 \end{bmatrix}$$

```
[> CMB:=CharacteristicMatrix(Matrix(B),t):
```

$$CMB := \begin{bmatrix} -t+1 & 2 & 3 \\ 4 & 5-t & 6 \\ 7 & 8 & 9-t \end{bmatrix}$$

```
[> cpB:=charpoly(B,t):
```

$$cpB := t^3 - 15t^2 - 18t$$

```
[> CPB:=CharacteristicPolynomial(Matrix(B),t):
```

$$CPB := t^3 - 15t^2 - 18t$$

We conclude this example by calculating the characteristic matrix and characteristic polynomial of the matrix B using the **Determinant** function.

```
[> C:=B-MatrixScalarMultiply(IdentityMatrix(3),t);
```

$$C := \begin{bmatrix} 1-t & 2 & 3 \\ 4 & 5-t & 6 \\ 7 & 8 & 9-t \end{bmatrix}$$

Then we calculate the determinant of C.

```
[> (-1)^3*Determinant(C);
```

$$18t + 15t^2 - t^3$$

As expected, the characteristic matrices produced by **charmat** and **CharacteristicMatrix** differ by a factor of (-1), whereas the characteristic polynomials produced by the three methods are the same. ▲

EXAMPLE 6.10 The Characteristic Polynomial of a Random Matrix

Use the **charpoly** function in the **linalg** package to calculate the characteristic polynomial of a random real 6×6 matrix.

Solution. We begin by generating a random 6×6 matrix.

```
[> with(linalg):
[> r:=rand(-10..10):random:=proc()r()end:
[> A:=randmatrix(6,6,entries=random);
```

$$A := \begin{bmatrix} 4 & -4 & 1 & 5 & -5 & 8 \\ 7 & 7 & 3 & -2 & -8 & -10 \\ -3 & -2 & -1 & -10 & 6 & -3 \\ -6 & -3 & 7 & 6 & -7 & 7 \\ -3 & -10 & 6 & 6 & 6 & 6 \\ 4 & -5 & -8 & -9 & 9 & -6 \end{bmatrix}$$

```
[> charpoly(A,t);
```

$$-630258 + 104797t + 13345t^2 - 1698t^3 - 14t^4 - 16t^5 + t^6$$

The result is the characteristic polynomial $c_A(t)$ of A. ▲

The characteristic polynomials of certain matrices are easy to calculate.

EXAMPLE 6.11 The Characteristic Polynomial of a Triangular Matrix

Find the characteristic polynomial $c_A(t)$ of the matrix

$$A = \begin{bmatrix} 6 & 7 & 4 & 5 \\ 0 & 3 & 7 & 5 \\ 0 & 0 & 3 & 3 \\ 0 & 0 & 0 & 5 \end{bmatrix}$$

Solution. By definition, the characteristic polynomial $c_A(t)$ equals $(-1)^4 \det(A - tI)$. Since A is upper triangular, the matrix

$$A - tI = \begin{bmatrix} 6 & 7 & 4 & 5 \\ 0 & 3 & 7 & 5 \\ 0 & 0 & 3 & 3 \\ 0 & 0 & 0 & 5 \end{bmatrix} - t \begin{bmatrix} 1 & 0 & 0 & 0 \\ 0 & 1 & 0 & 0 \\ 0 & 0 & 1 & 0 \\ 0 & 0 & 0 & 1 \end{bmatrix} = \begin{bmatrix} 6-t & 7 & 4 & 5 \\ 0 & 3-t & 7 & 5 \\ 0 & 0 & 3-t & 3 \\ 0 & 0 & 0 & 5-t \end{bmatrix}$$

is also upper triangular. It therefore follows from the property of determinants that

$$\det(A - tI) = (6-t)(3-t)(3-t)(5-t)$$

By collecting identical terms, we get $c_A(t) = (-1)^4(6-t)(3-t)^2(5-t)$. ▲

The polynomial matrix version of Theorem 3.17 shows that this example is typical for all triangular matrices: The characteristic polynomial of a triangular matrix A is the product of the diagonal entries of the characteristic matrix of A.

The following theorem establishes the fundamental connection between eigenvalues and the roots of characteristic polynomials. From a theoretical point of view, the theorem provides a basic tool for identifying the eigenvalues of a matrix. In practice, however, the theorem is not particularly helpful: A remarkable reality of mathematics is that there is no general algorithm for finding the roots of polynomials of degree five or higher. In practical applications, approximation techniques are used for calculating eigenvalues.

THEOREM 6.2 (Eigenvalue theorem) *For every scalar λ and real square matrix A, the following statements are equivalent. (1) The scalar λ is an eigenvalue of A. (2) The matrix $(A - \lambda I)$ is singular. (3) The scalar λ is a root of the characteristic polynomial $c_A(t)$ of A.*

Proof. We prove the theorem by showing that the first statement implies the second, the second implies the third, the third implies the fourth, and the fourth implies the first.

1. (1) implies (2). Suppose that λ is an eigenvalue of A. Then $A\mathbf{x} = \lambda\mathbf{x}$ for some nonzero vector $\mathbf{x}$, so $A\mathbf{x} - \lambda\mathbf{x} = A\mathbf{x} - \lambda I\mathbf{x} = (A - \lambda I)\mathbf{x} = \mathbf{0}$. Hence it follows from Theorem 3.16 that $(A - \lambda I)$ is singular.

2. (2) implies (3). We know from Theorem 3.13 that if $A - \lambda I$ is singular, then $\det(A - \lambda I) = c_A(\lambda) = 0$. Hence λ is a root of the characteristic polynomial of A.

3. (3) implies (1). Suppose that λ is a root of the characteristic polynomial of A. Then $c_A(\lambda) = \det(A - \lambda I) = 0$. Therefore the homogeneous linear system $(A - \lambda I)\,\mathbf{x} = \mathbf{0}$ has a nontrivial solution $\mathbf{x}$ for which $A\mathbf{x} = \lambda\mathbf{x}$. Hence λ is an eigenvalue of A. ■

We note that this theorem does not assert that the eigenvalues of A are real eigenvalues. The eigenvalues of the rotation matrix

$$A = \begin{bmatrix} 0 & 1 \\ -1 & 0 \end{bmatrix}$$

are the complex numbers i and $-i$, with associated complex eigenvectors $(1, i)$ and $(1, -i)$.

In Table 1 we summarize the procedure for determining the eigenvalues of a real square matrix implied by Theorem 6.2. The procedure is not an algorithm since it requires us to find the roots of polynomials. Unfortunately, there is no such algorithm for arbitrary polynomials of degree five or higher.

TABLE 1 FINDING THE EIGENVALUES OF A MATRIX.

Step 1. Form the characteristic matrix $(A - tI)$ of a real $n \times n$ matrix A.
Step 2. Find the characteristic polynomial $c_A(t)$ of the matrix A.
Step 3. The real roots of $c_A(t)$ are the real eigenvalues of A.

COROLLARY 6.3 *Suppose that $T : V \to V$ is a linear transformation on a real finite-dimensional vector space V and that $c_A(t)$ is the characteristic polynomial of the matrix $A = [T]_{\mathcal{B}}^{\mathcal{B}}$. Then every eigenvalue λ of T is a real root $c_A(t)$.*

Proof. The corollary follows immediately from Theorem 6.1. ■

Characteristic Polynomials of Similar Matrices

Since similar matrices represent the same linear transformation in different bases, we would expect that similar matrices have the same characteristic polynomial. The next theorem shows that this is in fact the case.

THEOREM 6.4 (Characteristic polynomial test) *If A and B are similar matrices, then $c_A(t) = c_B(t)$.*

Proof. Suppose that A and B are two similar matrices. Then there exists an invertible matrix P such that $B = P^{-1}AP$. Hence

$$\begin{aligned} B - tI &= P^{-1}AP - tI \\ &= P^{-1}AP - P^{-1}(tI)P \\ &= P^{-1}(A - tI)P \end{aligned}$$

This means that

$$c_B(t) = \det(B - tI) = \det P^{-1}(A - tI)P = \det P^{-1} c_A(t) \det P = c_A(t)$$

The two characteristic polynomials are therefore identical. ■

EXAMPLE 6.12 The Characteristic Polynomials of A and PAP^{-1}

Use *Maple* to show that if A and P are two random 2×2 matrices and P is invertible, then A and PAP^{-1} have the same characteristic polynomial.

Solution. We begin by generating two random matrices.

```
[> with(LinearAlgebra):
[> A:=RandomMatrix(2,2);
```

$$A := \begin{bmatrix} -65 & 66 \\ 5 & -36 \end{bmatrix}$$

```
[> P:=RandomMatrix(2,2);
```

$$P := \begin{bmatrix} 55 & 26 \\ 68 & 13 \end{bmatrix}$$

For P, we need an invertible matrix. We test P for invertibility by calculating its determinant.

```
[> Determinant(P)=0;
```

$$-1053 = 0$$

Since this statement is obviously false, the determinant of P is not zero. Hence P is invertible. We can therefore compare the characteristic polynomials of A and PAP^{-1}.

```
[> pA:=CharacteristicPolynomial(A,t);
```

$$pA := 2010 + 101t + t^2$$

```
[> pPAP:=CharacteristicPolynomial(P.A.MatrixInverse(P),t);
```

$$pPAP := 2010 + 101t + t^2$$

As we can see, the matrices A and PAP^{-1} have the same characteristic polynomial. ▲

EXAMPLE 6.13 The Characteristic Polynomials of Two Nonsimilar 3 x 3 Matrices

Use characteristic polynomials to show that the following matrices are not similar.

$$A = \begin{bmatrix} 1 & 1 & 0 \\ 3 & 3 & 2 \\ 2 & 2 & 3 \end{bmatrix} \quad \text{and} \quad B = \begin{bmatrix} 1 & 1 & 0 \\ 3 & 3 & 2 \\ 0 & 0 & 3 \end{bmatrix}$$

Solution. By Theorem 6.4, the matrices A and B are not similar if the characteristic polynomial $c_B(t)$ is different from $c_A(t)$. We use the **CharacteristicPolynomial** function to find the polynomials.

```
[> A:=Matrix([[1,1,0],[3,3,2],[2,2,3]]):
[> B:=Matrix([[1,1,0],[3,3,2],[0,0,3]]):
[> cpA:=LinearAlgebra[CharacteristicPolynomial](A,t):
[> cpB:=LinearAlgebra[CharacteristicPolynomial](B,t):
[> evalb(cpA=cpB);
                              false
```

This shows that A and B are not similar. ▲

It is easy to check that in Example 6.13, $\det A = \det B$ and trace $A =$ trace B. This shows that these equalities are necessary conditions for the similarity of A and B, but they are not sufficient. *Maple 6* provides a stronger similarity test for similarity and nonsimilarity. The calculation

```
[> A:=Matrix([[1,1,0],[3,3,2],[2,2,3]]):
[> B:=Matrix([[1,1,0],[3,3,2],[0,0,3]]):
[> LinearAlgebra[IsSimilar](A,B);
                              false
```

shows that the matrices A and B are nonsimilar. We have shown that a necessary condition for similarity is not met. The calculation

```
[> C:=Matrix([[1,2],[3,4]]):
[> P:=Matrix([[1,2],[0,4]]);
[> LinearAlgebra[IsSimilar](C,P.C.MatrixInverse(P));
                              true
```

on the other hand, shows that the matrices C and PCP^{-1} are similar. We have verified that a sufficient condition for similarity is met. However, it is not apparent what that condition is since we do not know, without additional information, how the *Maple* test is implemented.

EXERCISES 6.2

1. Calculate the characteristic matrices in the variable t of the following matrices.

a. $\begin{bmatrix} 1 & 2 \\ 0 & 0 \end{bmatrix}$ b. $\begin{bmatrix} 0 & -1 \\ 1 & 0 \end{bmatrix}$ c. $\begin{bmatrix} 0 & 1 \\ -1 & 1 \end{bmatrix}$

d. $\begin{bmatrix} 1 & 0 & -2 \\ 0 & 5 & 7 \\ 0 & 0 & 0 \end{bmatrix}$ e. $\begin{bmatrix} 0 & -1 & 0 \\ 1 & 0 & 0 \\ 0 & 0 & 1 \end{bmatrix}$ f. $\begin{bmatrix} 2 & 0 & 0 & 0 \\ 0 & 3 & 0 & 0 \\ 0 & 0 & 2 & 0 \\ 0 & 0 & 0 & 5 \end{bmatrix}$

g. $\begin{bmatrix} 1 & 0 & 0 & 0 \\ 0 & 1 & 0 & 0 \\ 0 & 0 & 1 & 0 \\ 0 & 0 & 0 & 1 \end{bmatrix}$ h. $\begin{bmatrix} 6 & 3 & 6 & 5 \\ 3 & 4 & 4 & 3 \\ 6 & 4 & 2 & 2 \\ 5 & 3 & 2 & 4 \end{bmatrix}$ i. $\begin{bmatrix} 1 & 0 & 0 & 9 \\ 0 & 1 & 7 & 0 \\ 0 & 7 & 1 & 0 \\ 9 & 0 & 0 & 1 \end{bmatrix}$

2. Find two linearly independent nontrivial solutions of the homogeneous system $A\mathbf{x} = \mathbf{0}$ defined by

$$\begin{bmatrix} 5 & 2 & 7 \\ 0 & 0 & 0 \\ 0 & 0 & 0 \end{bmatrix} \begin{bmatrix} x \\ y \\ z \end{bmatrix} = \begin{bmatrix} 0 \\ 0 \\ 0 \end{bmatrix}$$

and show that they are eigenvectors of A.

3. Use the fact that the matrix

$$A = \begin{bmatrix} 5 & 2 & 7 \\ 0 & 2 & 1 \\ 0 & 0 & 3 \end{bmatrix}$$

has three distinct nonzero eigenvalues to explain why the homogeneous system $A\mathbf{x} = \mathbf{0}$ cannot have two linearly independent nontrivial solutions.

4. Use characteristic polynomials to find the eigenvalues of the following linear transformations.

a. $T(x, y) = (3x + 4y, x - y)$ b. $T(x, y, z) = (3x + y - z, x + y, x + z)$

5. Prove that if A is a matrix $\left[a_{ij}\right]$, in which $a_{i1} = 0$ if $i > 1$, and B is the submatrix of A obtained by deleting the first row and first column of A, then the characteristic polynomial $c_A(t)$ of A is $(a_{11} - t)c_B(t)$.

6. Let A and B be two $n \times n$ matrices. Show that the matrices AB and BA have the same eigenvalues.

7. The intermediate value theorem of calculus says that if f is a continuous function on a closed interval $[a, b]$ and if y is any number between $f(a)$ and $f(b)$, then there exists a number x between a and b such that $f(x) = y$. Use this theorem to prove that every real 3×3 matrix A has at least one real eigenvalue. (*Hint*: If $p(t) = a_0 + a_1 t + a_2 t^2 + t^3$ and $r_1 > 0$ is sufficiently large, the $p(r_1) > 0$, and if $-r_2 > 0$ is sufficiently large, then $p(r_2) < 0$ since $p(t)$ is of odd degree with leading coefficient 1. The intermediate value theorem therefore tells us that there exists a number r between r_1 and r_2 for which $p(r) = 0$.)

8. Use Theorem 6.4 to prove that similar matrices have the same eigenvalues.

9. (*Maple V*) Construct an example of two 3×3 matrices having the same eigenvalues but different sets of eigenvectors.

10. (*Maple V*) Find the characteristic polynomials of the matrices in Exercise 1 using the variable t.

11. (*Maple V*) Find the eigenvectors of the matrices

$$A = \begin{bmatrix} 8 & 4 & -5 & -5 \\ 3 & -5 & 8 & 5 \\ 0 & 0 & 0 & 0 \\ 0 & 0 & 0 & 0 \end{bmatrix} \quad \text{and} \quad A^T = \begin{bmatrix} 8 & 3 & 0 & 0 \\ 4 & -5 & 0 & 0 \\ -5 & 8 & 0 & 0 \\ -5 & 5 & 0 & 0 \end{bmatrix}$$

and compare your results.

12. (*Maple 6*) For each matrix A in Exercise 1, use the function **CharacteristicPolynomial** function in the **LinearAlgebra** package to show that $c_A(A)$ is the zero matrix, where $c_A(A)$ is obtained from the characteristic polynomial $c_A(t)$ by replacing the variable t by the matrix A and replacing the constant term a_0 of $c_A(t)$ by $a_0 I$.

The Cayley–Hamilton Theorem

One of the conceptually beautiful properties of characteristic polynomials is that they are satisfied by the matrices that determine them. This result is known as the Cayley–Hamilton theorem.

THEOREM 6.5 (Cayley–Hamilton theorem for diagonalizable matrices) *Every diagonalizable square matrix satisfies its characteristic polynomial.*

Proof. Let A be a diagonalizable $n \times n$ real matrix, with characteristic polynomial

$$c_A(t) = a_0 + a_1 t + \cdots + a_{n-1}t^{n-1} + t^n$$

and let $\mathcal{B} = \{\mathbf{x}_1, \ldots, \mathbf{x}_n\}$ be a basis for $\mathbb{R}^n$ consisting of eigenvectors of A.

We use $c_A(t)$ to construct the matrix

$$B = a_0 I + a_1 A + \cdots + a_{n-1}A^{n-1} + A^n$$

obtained from $c_A(t)$ by replacing the variable t by A and the constant term a_0 by $a_0 I$.

We show that $B\mathbf{x} = \mathbf{0}$ for every eigenvector $\mathbf{x} \in \mathcal{B}$. Hence B represents the zero transformation on $\mathbb{R}^n$ and must therefore be the zero matrix.

Let $\mathbf{x} \in \mathcal{B}$ and let λ be its associated eigenvalues. Then

$$\begin{aligned} B\mathbf{x} &= \left(a_0 I + a_1 A + \cdots + a_{n-1}A^{n-1} + A^n\right)\mathbf{x} \\ &= a_0\mathbf{x} + a_1\lambda\mathbf{x} + \cdots + a_{n-1}\lambda^{n-1}\mathbf{x} + \lambda^n\mathbf{x} \\ &= \left(a_0 + a_1\lambda + \cdots + a_{n-1}\lambda^{n-1} + \lambda^n\right)\mathbf{x} = 0\mathbf{x} = \mathbf{0} \end{aligned}$$

for all basis vectors. ■

THEOREM 6.6 (Cayley–Hamilton theorem) *Every square matrix satisfies its characteristic polynomial.*

Proof. Let A be a real $n \times n$ matrix, and let B be the matrix constructed in the proof of Theorem 6.5 from the characteristic polynomial $c_A(t)$ of A. If we decompose the adjugate $\text{adj}(tI - A)$ of the matrix $(tI - A)$ into the matrix sum $\left(C_{n-1}t^{n-1} + \cdots + C_1 t + C_0\right)$, where all C_i are constant matrices, we can use the equality

$$\text{adj}(tI - A)(C_{n-1}t^{n-1} + \cdots + C_1 t + C_0) = (-1)^n \det(A - t)\, I$$

to establish the theorem.

If we expand both sides of this equation, we can form n separate equations by comparing the coefficients of the powers of t. We then multiply each of the n equations by the appropriate power A^i of A required to produce the matrix $a_i A^i$ from the matrix $a_i I$. The sum of the left-hand sides of these equations is the zero matrix and the sum of their right-hand sides is the matrix $a_0 I + a_1 A + \cdots + a_{n-1}A^{n-1} + A^n$.

This proves the theorem. ■

EXAMPLE 6.14 The Zero Matrix Determined by a Characteristic Polynomial

Use *Maple* to illustrate the Cayley–Hamilton Theorem.

Solution. Let A be the matrix

$$\begin{bmatrix} 1 & 2 \\ 3 & 2 \end{bmatrix}$$

Then

$$t\begin{bmatrix} 1 & 0 \\ 0 & 1 \end{bmatrix} - \begin{bmatrix} 1 & 2 \\ 3 & 2 \end{bmatrix} = \begin{bmatrix} t-1 & -2 \\ -3 & t-2 \end{bmatrix}$$

is the characteristic matrix of A. We begin by calculating its determinant.

```
[> with(linalg):
[> det(matrix([[t-1,-2],[-3,t-2]]));
```

$$t^2 - 3t - 4$$

The Cayley–Hamilton Theorem asserts that $-4I_2 - 3A + A^2$ is the zero matrix.

```
[> A:=matrix([[1,2],[3,2]]);
```

$$A := \begin{bmatrix} 1 & 2 \\ 3 & 2 \end{bmatrix}$$

```
[> B:= evalm(-4*matrix([[1,0],[0,1]]) -3*A +A^2):
[> ZeroMatrix:= matrix([[0,0],[0,0]]):
[> equal(B,ZeroMatrix);
```

true

As we can see, the constructed matrix is the zero matrix. ▲

EXERCISES 6.3

1. The characteristic polynomial of the matrix

$$\begin{bmatrix} 0 & 2 & 1 \\ 1 & 0 & 0 \\ -1 & 0 & 3 \end{bmatrix}$$

is $c_A(t) = 6 - t - 3t^2 + t^3$. Moreover, $c_A(t)$ is the product of the polynomials $p_1(t) = (t-2)$ and $p_2(t) = (t^2 - t - 3)$. Show that neither $p_1(A)$ nor $p_2(A)$ is the zero matrix. Why does the Cayley–Hamilton theorem nevertheless guarantee that $p_1(A)p_2(A)$ is the zero matrix?

2. Show that the following statements about real $n \times n$ matrices are false.

   a. If $c_A(t)$ is the characteristic polynomial of A, then $c_A(A^2) = \mathbf{0}$.

   b. If $p(t)$ is a polynomial with real coefficients and $p(A^2) = \mathbf{0}$, then $p(A) = \mathbf{0}$.

3. Use the Cayley–Hamilton theorem to prove that the inverse of an invertible $n \times n$ matrix A is a linear combination of $I, A, A^2, \ldots, A^{n-1}$.

4. (*Maple V*) Show that the following matrices satisfy their characteristic polynomials.

$$\text{a. } A = \begin{bmatrix} 1 & 2 \\ 3 & 4 \end{bmatrix} \quad \text{b. } A = \begin{bmatrix} 97 & 50 \\ 79 & 56 \end{bmatrix} \quad \text{c. } A = \begin{bmatrix} 0 & 2 & 1 \\ 1 & 0 & 0 \\ -1 & 0 & 3 \end{bmatrix}$$

5. (*Maple 6*) Use the Cayley–Hamilton theorem to find the inverse of the matrix

$$A = \begin{bmatrix} -86 & 23 & -84 & 19 & -50 \\ 88 & -53 & 85 & 49 & 78 \\ 17 & 72 & -99 & -85 & -86 \\ 30 & 80 & 72 & 66 & -29 \\ -91 & -53 & -19 & -47 & 68 \end{bmatrix}$$

Calculating Characteristic Polynomials

Although the definition of characteristic polynomials is conceptually elegant, it is also computationally unmanageable. For general matrices, the number of terms of $\det(A - tI)$ grows too rapidly as the dimension of A increases to make a direct computation of $c_A(t)$ manually or mechanically feasible. We must therefore look for simpler matrices B for which $c_B(t)$ can easily be computed and for which $c_B(t) = c_A(t)$. Theorem 6.4 provides the clue. If B is a triangular matrix, then Example 6.11 shows that the characteristic polynomial of B is the product of the diagonal entries of the matrix $(B - tI)$. It can therefore be determined immediately from $(B - tI)$. If B is similar to A, then $c_B(t) = c_A(t)$. Unfortunately, not every square matrix is similar to a triangular matrix.

EXAMPLE 6.15 A Real Matrix Not Similar to a Triangular Matrix

Show that the real matrix

$$A = \begin{bmatrix} 0 & -1 \\ 1 & 0 \end{bmatrix}$$

is not similar to any real triangular matrix B.

Solution. Let

$$B = \begin{bmatrix} a & b \\ 0 & c \end{bmatrix}$$

be a general upper-triangular matrix. If A and B are similar, then it follows from Theorems 5.29 and 5.30 that $\det A = \det B$ and $\operatorname{trace} A = \operatorname{trace} B$. Since $\det A = 1$ and $\det B = ac$, we must have $ac = 1$. Since $\operatorname{trace} A = 0$ and $\operatorname{trace} B = a + c$, we must have $a + c = 0$. If we combine these two results, it follows that $-c^2$ must equal 1. But this is impossible for any real number c. Hence A is not similar to B for any real numbers a, b, c. ▲

The question arises whether we can find a more general substitute for triangular matrices that can be used to calculate characteristic polynomials in reasonable time. It turns out that there are such matrices. A ***Hessenberg matrix*** is a matrix $H = (h_{ij})$ for which $h_{ij} = 0$ if $i > j + 1$. Furthermore, there exist similarity transformations $M \to QMQ^{-1}$, known as ***Householder transformations***, for converting A to H. The characteristic polynomial $c_H(t)$ can be found easily either by applying the definition or by using a number of special algorithms devised for this purpose.

EXAMPLE 6.16 A Hessenberg Matrix

Explain why the matrix

$$A = \begin{bmatrix} 3 & 4 & 2 & 1 & 5 \\ 7 & 1 & 5 & 1 & 1 \\ 0 & 2 & 6 & 2 & 7 \\ 0 & 0 & 1 & 7 & 3 \\ 0 & 0 & 0 & 0 & 3 \end{bmatrix}$$

is a Hessenberg matrix.

Solution. If $i > j+1$, then $h_{ij} \in \{h_{31}, h_{41}, h_{42}, h_{51}, h_{52}, h_{53}\}$. We can see from an inspection of A that $h_{31} = h_{41} = h_{42} = h_{51} = h_{52} = h_{53} = 0$, as required. ▲

A real $n \times n$ ***Householder matrix*** Q is a matrix of the form

$$I - \frac{2}{\mathbf{u}^T\mathbf{u}}\mathbf{u}\mathbf{u}^T$$

where $\mathbf{u}^T\mathbf{u}$ is the nonzero dot product of a nonzero vector $\mathbf{u} \in \mathbb{R}^n$, whereas $\mathbf{u}\mathbf{u}^T$ is an $n \times n$ matrix. It is quite remarkable that each matrix constructed in this way is both symmetric and orthogonal: $Q = Q^T = Q^{-1}$. To see that Q is symmetric, we note that

$$\left(I - \frac{2}{\mathbf{u}^T\mathbf{u}}\mathbf{u}\mathbf{u}^T\right)^T = I^T - \frac{2}{\mathbf{u}^T\mathbf{u}}\left(\mathbf{u}\mathbf{u}^T\right)^T = I - \frac{2}{\mathbf{u}^T\mathbf{u}}\mathbf{u}^{TT}\mathbf{u}^T = I - \frac{2}{\mathbf{u}^T\mathbf{u}}\mathbf{u}\mathbf{u}^{\mathrm{T}}$$

To see that Q is orthogonal, we note that $QQ^T = QQ$ since Q is symmetric, and that $(\mathbf{u}^T\mathbf{u}) = (\mathbf{u}^T\mathbf{u})^T$, since $\mathbf{u}^T\mathbf{u}$ is a scalar. Therefore,

$$\begin{aligned} QQ &= \left(I - \frac{2}{\mathbf{u}^T\mathbf{u}}\mathbf{u}\mathbf{u}^T\right)\left(I - \frac{2}{\mathbf{u}^T\mathbf{u}}\mathbf{u}\mathbf{u}^T\right) \\ &= I - \frac{4}{\mathbf{u}^T\mathbf{u}}\mathbf{u}\mathbf{u}^T + \frac{4}{(\mathbf{u}^T\mathbf{u})(\mathbf{u}^T\mathbf{u})}\left(\mathbf{u}\mathbf{u}^T\right)\left(\mathbf{u}\mathbf{u}^T\right) \\ &= I - \frac{4}{\mathbf{u}^T\mathbf{u}}\mathbf{u}\mathbf{u}^T + \frac{4}{\mathbf{u}^T\mathbf{u}}\mathbf{u}\mathbf{u}^T = I \end{aligned}$$

If we choose the vector **u** in Q wisely, we can use a sequence of Householder transformations to reduce any real $n \times n$ matrix A to a Hessenberg matrix B. Here is how this can be done. Suppose that

$$\mathbf{x} = (x_1, \dots, x_{k-1}, x_k, x_{k+1}, \dots, x_n)$$

and

$$\mathbf{u} = (0, \dots, 0, x_k - r, x_{k+1}, \dots, x_n)$$

are two vectors in $\mathbb{R}^n$ and that r is either $\sqrt{x_k^2 + \cdots + x_n^2}$ or $-\sqrt{x_k^2 + \cdots + x_n^2}$, chosen so that $a\, x_k \leq 0$. If Q is the Householder matrix determined by the specified vector **u**, then

$$Q\mathbf{x} = (x_1, \dots, x_{k-1}, r, 0, \dots, 0)$$

is a column vector whose last components are zero. For any $n \times n$ real matrix A, we can use this recipe to find appropriate Householder transformations that convert A to a similar Hessenberg matrix H.

EXAMPLE 6.17 A Householder Conversion of a Matrix to Hessenberg Form

Use an appropriate Householder transformation to convert the matrix

$$A = \begin{bmatrix} 7 & 3 & 1 & 0 \\ 2 & 5 & 4 & 5 \\ 0 & 4 & 6 & 6 \\ 0 & 3 & 5 & 5 \end{bmatrix}$$

to a Hessenberg matrix H, and use H to find the characteristic polynomial of A.

Solution. We must find a Householder matrix Q such that $QAQ = H = (h_{ij})$ and $h_{31} = h_{41} = h_{42} = 0$. Since h_{31} and h_{41} are already 0, it suffices to find a matrix Q that converts h_{42} to 0 and leaves the remaining entries untouched. Consider the second column

$$\mathbf{x} = \begin{bmatrix} x_1 \\ x_2 \\ x_3 \\ x_4 \end{bmatrix} = \begin{bmatrix} 3 \\ 5 \\ 4 \\ 3 \end{bmatrix}$$

of A. By the definition of **u**, we have $x_4 = x_{k+1}$ and $k = 3$. Therefore,

$$a = \pm\sqrt{x_3^2 + x_4^2} = \pm\sqrt{4^2 + 3^2} = \pm 5$$

Since $x_3 - a$ must be less than or equal to 0, we take a to be -5. This means that

$$\mathbf{u} = \begin{bmatrix} 0 \\ 0 \\ x_3 - a \\ x_4 \end{bmatrix} = \begin{bmatrix} 0 \\ 0 \\ 4-(-5) \\ 3 \end{bmatrix} = \begin{bmatrix} 0 \\ 0 \\ 9 \\ 3 \end{bmatrix}$$

is the vector required to construct the matrix $Q = I - b\left(\mathbf{u}\mathbf{u}^T\right)$, with

$$b = 2/(\mathbf{u}^T\mathbf{u}) = \frac{2}{90} = \frac{1}{45}$$

Therefore,

$$Q = I - b\left(\mathbf{u}\mathbf{u}^T\right) = \begin{bmatrix} 1 & 0 & 0 & 0 \\ 0 & 1 & 0 & 0 \\ 0 & 0 & 1 & 0 \\ 0 & 0 & 0 & 1 \end{bmatrix} - \frac{1}{45}\begin{bmatrix} 0 \\ 0 \\ 9 \\ 3 \end{bmatrix}\begin{bmatrix} 0 \\ 0 \\ 9 \\ 3 \end{bmatrix}^T = \begin{bmatrix} 1 & 0 & 0 & 0 \\ 0 & 1 & 0 & 0 \\ 0 & 0 & -\frac{4}{5} & -\frac{3}{5} \\ 0 & 0 & -\frac{3}{5} & \frac{4}{5} \end{bmatrix}$$

The product

$$QAQ = \begin{bmatrix} 1 & 0 & 0 & 0 \\ 0 & 1 & 0 & 0 \\ 0 & 0 & -\frac{4}{5} & -\frac{3}{5} \\ 0 & 0 & -\frac{3}{5} & \frac{4}{5} \end{bmatrix}\begin{bmatrix} 7 & 3 & 1 & 0 \\ 2 & 5 & 4 & 5 \\ 0 & 4 & 6 & 6 \\ 0 & 3 & 5 & 5 \end{bmatrix}\begin{bmatrix} 1 & 0 & 0 & 0 \\ 0 & 1 & 0 & 0 \\ 0 & 0 & -\frac{4}{5} & -\frac{3}{5} \\ 0 & 0 & -\frac{3}{5} & \frac{4}{5} \end{bmatrix}$$

$$= \begin{bmatrix} 7 & 3 & -\frac{4}{5} & -\frac{3}{5} \\ 2 & 5 & -\frac{31}{5} & \frac{8}{5} \\ 0 & -5 & \frac{273}{25} & -\frac{39}{25} \\ 0 & 0 & -\frac{14}{25} & \frac{2}{25} \end{bmatrix}$$

is the required Hessenberg matrix H. If we use the Laplace expansion formula for det $(H - I)$ along the first column, we get

$$c_H(t) = 18 - 112t + 130t^2 - 23t^3 + t^4$$

By Theorem 6.4, $c_A(t) = c_H(t)$ is the characteristic polynomial of the matrix A. ▲

EXERCISES 6.4

1. Construct a 3×3, 4×4, and 5×5 Hessenberg matrix.

2. (*Maple 6*) Use the **HessenbergForm** function in the **LinearAlgebra** package to confirm that the matrices constructed in Exercise 1 are Hessenberg matrices.

3. (*Maple 6*) Use the **LinearAlgebra[HouseholderMatrix]** function to create a 3×3, 4×4, and 5×5 Householder matrix and show that each matrix is both symmetric and orthogonal.

4. (*Maple 6*) Use the **HouseholderMatrix** function to generate a Householder matrix Q from the vector $\mathbf{v} = (1, 2, 3)$ and let A by a random 3×3 real matrix. Verify that the matrices A and QAQ have the same characteristic polynomial.

5. (*Maple 6*) Use two Householder transformations to convert the matrix

$$A = \begin{bmatrix} 1 & 0 & 1 & 0 \\ 0 & 2 & 0 & 1 \\ 4 & 0 & 0 & 0 \\ 0 & -1 & 4 & 0 \end{bmatrix}$$

to Hessenberg form. Compare you result with that obtained by the **HessenbergForm** function in the **LinearAlgebra** package.

EIGENSPACES

Each eigenvalue λ of a linear transformation $T : V \to V$ determines a subspace $E_\lambda = \{\mathbf{x} : T(\mathbf{x}) = \lambda\mathbf{x}\}$ of V called the ***eigenspace*** of T associated with λ. If T is a matrix transformation $T_A : \mathbb{R}^n \to \mathbb{R}^n$, then we call E_λ an ***eigenspace*** of A.

THEOREM 6.7 (Eigenspace theorem) *For every eigenvalue λ of a linear transformation $T : V \to V$, the set $E_\lambda = \{\mathbf{x} : T(\mathbf{x}) = \lambda\mathbf{x}\}$ is a nonzero subspace of V.*

Proof. Since $T(\mathbf{0}) = 0 = \lambda\mathbf{0}$, the zero vector belongs to E_λ. We show that E_λ is also closed under linear combinations. Suppose that $\mathbf{x}$ and $\mathbf{y}$ belong to E_λ. Then

$$T(a\mathbf{x} + b\mathbf{y}) = T(a\mathbf{x}) + T(b\mathbf{y}) = aT(\mathbf{x}) + bT(\mathbf{y}) = a(\lambda\mathbf{x}) + b(\lambda\mathbf{y}) = \lambda(a\mathbf{x} + b\mathbf{y})$$

Hence $a\mathbf{x} + b\mathbf{y} \in E_\lambda$, so E_λ is a subspace of V. ■

EXAMPLE 6.18 The Eigenspaces of a Matrix

Describe the eigenspaces of the matrix

$$A = \begin{bmatrix} 5 & 0 & 0 & 0 \\ 0 & 5 & 0 & 0 \\ 0 & 0 & 5 & 0 \\ 0 & 0 & 0 & 7 \end{bmatrix}$$

Solution. For any vector $\mathbf{x} \in \mathbb{R}^4$,

$$A\mathbf{x} = \begin{bmatrix} 5 & 0 & 0 & 0 \\ 0 & 5 & 0 & 0 \\ 0 & 0 & 5 & 0 \\ 0 & 0 & 0 & 7 \end{bmatrix} \begin{bmatrix} w \\ x \\ y \\ z \end{bmatrix} = \begin{bmatrix} 5w \\ 5x \\ 5y \\ 7z \end{bmatrix}$$

If $z = 0$, then

$$A\mathbf{x} = \begin{bmatrix} 5 & 0 & 0 & 0 \\ 0 & 5 & 0 & 0 \\ 0 & 0 & 5 & 0 \\ 0 & 0 & 0 & 7 \end{bmatrix} \begin{bmatrix} w \\ x \\ y \\ 0 \end{bmatrix} = \begin{bmatrix} 5w \\ 5x \\ 5y \\ 0 \end{bmatrix} = 5 \begin{bmatrix} w \\ x \\ y \\ 0 \end{bmatrix} = 5\mathbf{x}$$

This means that 5 is an eigenvalue of A and the eigenspace $E_5 = \{\mathbf{x} : A\mathbf{x} = 5\mathbf{x}\}$ consists of all vectors $\mathbf{x}$ in which $z = 0$. On the other hand, if $w = x = y = 0$, then

$$A\mathbf{x} = \begin{bmatrix} 5 & 0 & 0 & 0 \\ 0 & 5 & 0 & 0 \\ 0 & 0 & 5 & 0 \\ 0 & 0 & 0 & 7 \end{bmatrix} \begin{bmatrix} 0 \\ 0 \\ 0 \\ z \end{bmatrix} = \begin{bmatrix} 0 \\ 0 \\ 0 \\ 7z \end{bmatrix} = 7 \begin{bmatrix} 0 \\ 0 \\ 0 \\ z \end{bmatrix} = 7\mathbf{x}$$

Therefore, 7 is also an eigenvalue of A. The eigenspace $E_7 = \{\mathbf{x} : A\mathbf{x} = 7\mathbf{x}\}$ consists of all vectors $\mathbf{x}$ in which $w = x = y = 0$.

Since the characteristic polynomial of A is easily seen to be $c_A(t) = (t - 5)^3(t - 7)$, we can conclude that A has no other eigenvalues and therefore no other eigenspaces. ▲

The number of repetitions of the factors $(t - 5)$ and $(t - 7)$ in the characteristic polynomial $c_A(t) = (t - 5)^3(t - 7)$ in Example 6.18 is called the ***algebraic multiplicity*** of the eigenvalues 5 and 7. Thus the eigenvalue 5 has algebraic multiplicity 3, and the eigenvalue 7 has multiplicity 1. For any $n \times n$ matrix A, the algebraic multiplicity of an eigenvalue λ of A is the largest integer m for which the characteristic polynomial $c_A(t)$ factors into the form $c_A(t) = (t - \lambda)^m p(t)$. In this particular case, these multiplicities coincide with the dimensions of the eigenspaces. These dimensions are called the ***geometric multiplicities.*** In the example, the two multiplicities coincide because the matrix is diagonal. In general, however, the algebraic multiplicity of an eigenvalue is often greater than its geometric multiplicity.

THEOREM 6.8 *The algebraic multiplicity of an eigenvalue is always greater than or equal to its geometric multiplicity.*

Proof. Suppose that λ is an eigenvalue of a linear transformation T on an n-dimensional vector space V and that the geometric multiplicity of λ is m. Then, by definition, the subspace E_λ has a basis $\mathcal{B} = \{\mathbf{x}_1, \ldots, \mathbf{x}_m\}$ consisting of eigenvectors of T. By Theorem 4.14, we can extend $\mathcal{B}$ to a basis $\mathcal{C} = \{\mathbf{x}_1, \ldots, \mathbf{x}_m, \mathbf{x}_{m+1}, \ldots, \mathbf{x}_n\}$ of V. If we represent T in the basis $\mathcal{C}$, we get a matrix of the form

$$A = \begin{bmatrix} \lambda & 0 & \cdots & 0 & a_{1(m+1)} & \cdots & a_{1n} \\ 0 & \lambda & \cdots & 0 & a_{2(m+1)} & \cdots & a_{2n} \\ \vdots & \vdots & \ddots & \vdots & \vdots & & \vdots \\ 0 & 0 & \cdots & \lambda & a_{m(m+1)} & \cdots & a_{mn} \\ \vdots & \vdots & & \vdots & \vdots & & \vdots \\ 0 & 0 & \cdots & 0 & a_{n(m+1)} & \cdots & a_{nn} \end{bmatrix}$$

for which

$$\det(A - tI) = (\lambda - t)^m \det \begin{bmatrix} a_{(m+1)(m+1)} - t & \cdots & a_{(m+1)n} \\ \vdots & \ddots & \vdots \\ a_{n(m+1)} & \cdots & a_{nn} - t \end{bmatrix}$$

Hence the algebraic multiplicity of λ is greater than or equal to m. ■

EXAMPLE 6.19 A Maximally Linearly Independent Set of Eigenvectors

Find all eigenvalues and a maximum set $\mathcal{S}$ of linearly independent eigenvectors for the matrix

$$A = \begin{bmatrix} 5 & 6 \\ 3 & -2 \end{bmatrix}$$

Solution. We use *Maple* to find the eigenvalues of A.

```
[> linalg[eigenvalues](matrix([[5,6],[3,-2]]));
```

$$-4, 7$$

Thus the eigenvalues of A are $\lambda_1 = 7$ and $\lambda_2 = -4$. To find the eigenvectors corresponding to λ_1, we form the matrix

$$(A - \lambda_1 I) = \begin{bmatrix} 5 & 6 \\ 3 & -2 \end{bmatrix} - \begin{bmatrix} 7 & 0 \\ 0 & 7 \end{bmatrix} = \begin{bmatrix} -2 & 6 \\ 3 & -9 \end{bmatrix}$$

The eigenvectors belonging to λ_1 must therefore satisfy the equation

$$\begin{bmatrix} -2 & 6 \\ 3 & -9 \end{bmatrix} \begin{bmatrix} x \\ y \end{bmatrix} = \begin{bmatrix} -2x + 6y \\ 3x - 9y \end{bmatrix} = \begin{bmatrix} 0 \\ 0 \end{bmatrix}$$

We solve the resulting homogeneous linear system.

```
[> solve({-2*x+6*y=0,3*x-9*y=0},x);
```

$$\{x = 3y\}$$

This means that $x - 3y = 0$. Therefore $\mathbf{x}_1 = (3, 1)$ is an eigenvector of λ_1. All other eigenvectors belonging to λ_1 are multiples of $\mathbf{x}_1$. To find the eigenvectors corresponding to λ_2, we form the matrix equation

$$(A - \lambda_2 I) = \begin{bmatrix} 5 & 6 \\ 3 & -2 \end{bmatrix} - \begin{bmatrix} -4 & 0 \\ 0 & -4 \end{bmatrix} = \begin{bmatrix} 9 & 6 \\ 3 & 2 \end{bmatrix}$$

The eigenvectors belonging to λ_2 must therefore satisfy the matrix equation

$$\begin{bmatrix} 9 & 6 \\ 3 & 2 \end{bmatrix}\begin{bmatrix} x \\ y \end{bmatrix} = \begin{bmatrix} 9x + 6y \\ 3x + 2y \end{bmatrix} = \begin{bmatrix} 0 \\ 0 \end{bmatrix}$$

We solve the resulting homogeneous linear system.

```
[> solve({9*x+6*y=0,3*x+2*y=0},x);
```

$$\{x = \frac{-2}{3}y\}$$

If we let $y = -3$, then $x = 2$. Therefore, the vector $\mathbf{x}_2 = (2, -3)$, for example, is an eigenvector of λ_2. All other eigenvectors belonging to λ_2 are multiples of $\mathbf{x}_2$.

We know from Chapter 3 that any maximally linearly independent subset of $\mathbb{R}^2$ can have at most two elements. If we can show that the vectors $\mathbf{x}_1$ and $\mathbf{x}_2$ are linearly independent, then $\mathcal{S} = \{\mathbf{x}_1, \mathbf{x}_2\}$ is a maximally linearly independent set of eigenvectors of A. Suppose, therefore, that

$$a\mathbf{x}_1 + b\mathbf{x}_2 = a\begin{bmatrix} 3 \\ 1 \end{bmatrix} + b\begin{bmatrix} 2 \\ -3 \end{bmatrix} = \begin{bmatrix} 3a + 2b \\ a - 3b \end{bmatrix} = \begin{bmatrix} 0 \\ 0 \end{bmatrix}$$

Then $a = 3b$, so $9b + 2b = 11b = 0$. Hence $b = 0$. This implies that a is also 0. Therefore, $\mathbf{x}_1$ and $\mathbf{x}_2$ are linearly independent. ▲

EXAMPLE 6.20 The Algebraic Multiplicities of Eigenvalues

Use *Maple* to find the algebraic multiplicities of the eigenvalues of the matrix

$$A = \begin{bmatrix} 0 & 0 & 0 & -40 \\ 1 & 0 & 0 & 68 \\ 0 & 1 & 0 & -42 \\ 0 & 0 & 1 & 11 \end{bmatrix}$$

Solution. First we calculate the characteristic polynomial of A.

```
[> A:=matrix([[0,0,0,-40],[1,0,0,68],[0,1,0,-42],
   [0,0,1,11]]):
[> p:=linalg[charpoly](A,t);
```

$$p := t^4 - 11t^3 + 42t^2 - 68t + 40$$

Next we find the roots of the characteristic polynomial.

```
[> roots(p);
```

$$[[2, 3], [5, 1]]$$

Thus A has the eigenvalues 2 and 5. The algebraic multiplicity of the eigenvalue 2 is 3, and that of the eigenvalue 5 is 1. ▲

It is elegant to express the eigenvalues of matrices as the roots of characteristic polynomials. However, a remarkable discovery made by Galois in the 19th century shows that it is impossible to find an algorithm for calculating roots of arbitrary polynomials of degree greater than 4. For polynomials of degree 5 or higher, there are no variants of the familiar quadratic formula

$$\frac{-b \pm \sqrt{b^2 - 4ac}}{2a}$$

used to find the roots of the quadratic equation $at^2 + bt + c$. This means that in applications involving large matrices, we must use numerical techniques and usually work with approximations of eigenvalues. For example, if we generate a random matrix

```
[> A:=LinearAlgebra[RandomMatrix](5,5,generator=-10..10);
```

$$A := \begin{bmatrix} -7 & -5 & -5 & 9 & -3 \\ -1 & 2 & -4 & 3 & 6 \\ 8 & 6 & -2 & -6 & -7 \\ 10 & 10 & -5 & 8 & -9 \\ 2 & -7 & 4 & -10 & 5 \end{bmatrix}$$

and ask *Maple* to calculate the eigenvalues of A, either directly, using the **Eigenvalues** function, or indirectly, using the **CharacteristicPolynomial** and **solve** functions, we get symbolic outputs from which approximate roots can be calculated.

```
[> LinearAlgebra[Eigenvalues](A);
```

$$\begin{bmatrix} RootOf(-64660+448\_Z-249\_Z^2-160\_Z^3-6\_Z^4+\_Z^5, index=1) \\ RootOf(-64660+448\_Z-249\_Z^2-160\_Z^3-6\_Z^4+\_Z^5, index=2) \\ RootOf(-64660+448\_Z-249\_Z^2-160\_Z^3-6\_Z^4+\_Z^5, index=3) \\ RootOf(-64660+448\_Z-249\_Z^2-160\_Z^3-6\_Z^4+\_Z^5, index=4) \\ RootOf(-64660+448\_Z-249\_Z^2-160\_Z^3-6\_Z^4+\_Z^5, index=5) \end{bmatrix}$$

```
[> p:=LinearAlgebra[CharacteristicPolynomial](A,t);
```

$$p=-64660+448\,t-249\,t^2-160\,t^3-6\,t^4+t^5$$

```
[> solve(p=0,t);
```

$$RootOf(-64660+448\_Z-249\_Z^2-160\_Z^3-6\_Z^4+\_Z^5, index=1),$$
$$RootOf(-64660+448\_Z-249\_Z^2-160\_Z^3-6\_Z^4+\_Z^5, index=2),$$
$$RootOf(-64660+448\_Z-249\_Z^2-160\_Z^3-6\_Z^4+\_Z^5, index=3),$$
$$RootOf(-64660+448\_Z-249\_Z^2-160\_Z^3-6\_Z^4+\_Z^5, index=4),$$
$$RootOf(-64660+448\_Z-249\_Z^2-160\_Z^3-6\_Z^4+\_Z^5, index=5)$$

In both cases, *Maple* is telling us that the eigenvalues of the matrix A are among the roots of the characteristic polynomial. We can find approximations of these values by applying the **evalf** function to the **solve** command.

```
[> evalf(solve(p=0,t),3);
```

$$17.0, 3.07+5.81I, -8.56+3.87I, -8.56-3.87I, 3.07-5.81I$$

We can also designate the entries of A as *real* numbers by placing a decimal point behind one of them.

```
[> A:=Matrix([
   [-7,-5,-5,9,-3],
   [-1,2,-4,3,6],
   [8.,6,-2,-6,-7],
   [10,10,-5,8,-9],
   [2,-7,4,-10,5]]):
```

```
[> evalf(LinearAlgebra[Eigenvalues](A),4);
```

$$\begin{bmatrix} -8.560 + 3.866I \\ -8.560 - 3.866I \\ 17.00 + 0.I \\ 3.059 + 5.800I \\ 3.059 - 5.800I \end{bmatrix}$$

As we can see, the matrix A has one real eigenvalue near 17. The remaining roots of the characteristic polynomial are the complex eigenvalues of A.

The **Eigenvectors** function in the **LinearAlgebra** package and the **eigenvectors** function in the **linalg** package provide quick ways to calculate the eigenspaces of a particular matrix. Both functions produce the eigenvalues of a matrix and for each eigenvalue, they calculate a maximally linearly independent set of eigenvectors.

EXAMPLE 6.21 The Eigenspaces of a Diagonal Matrix

Use *Maple* to find the eigenspaces of the matrix

$$A = \begin{bmatrix} 5 & 0 & 0 & 0 & 0 \\ 0 & 5 & 0 & 0 & 0 \\ 0 & 0 & 7 & 0 & 0 \\ 0 & 0 & 0 & 7 & 0 \\ 0 & 0 & 0 & 0 & 7 \end{bmatrix}$$

Solution. Since the matrix A is diagonal, we know that its eigenvalues are its diagonal entries 5 and 7. We use the **eigenvalues** function to produce two bases for the associated eigenspaces.

```
[> A:=matrix([
    [5,0,0,0,0],
    [0,5,0,0,0],
    [0,0,7,0,0],
    [0,0,0,7,0],
    [0,0,0,0,7]]):
```

```
[> linalg[eigenvectors](A);
```

[5, 2, {[1, 0, 0, 0, 0], [0, 1, 0, 0, 0]}], [7, 3, {[0, 0, 1, 0, 0], [0, 0, 0, 0, 1], [0, 0, 0, 1, 0]}]

TABLE 2 EIGENVALUES AND EIGENVECTORS.

Eigenvalues	Associated Eigenvectors
5	(1, 0, 0, 0, 0) , (0, 1, 0, 0, 0)
7	(0, 0, 1, 0, 0) , (0, 0, 0, 0, 1) , (0, 0, 0, 1, 0)

For each of the five occurrences of eigenvalues of the matrix A, *Maple* has produced a matching eigenvector. Table 2 summarizes the results.

It shows that the eigenspace E_5 of the eigenvalue 5 has a basis consisting of two vectors, and the eigenspace E_7 of the eigenvalue 7 has a basis consisting of three vectors. ▲

EXAMPLE 6.22 The Eigenspaces of a Nondiagonalizable Matrix

Use *Maple* to find the eigenspaces of the matrix

$$A = \begin{bmatrix} 2 & 1 & 4 & 5 & 0 \\ 0 & 2 & 3 & 0 & 9 \\ 0 & 0 & 3 & 6 & 0 \\ 0 & 0 & 0 & 3 & 0 \\ 0 & 0 & 0 & 0 & 3 \end{bmatrix}$$

Solution. Since the matrix A is upper triangular, we know that its eigenvalues are its diagonal entries 2 and 3. We use the **eigenvectors** function to produce bases for the associated eigenspaces.

```
[> A:=matrix([
    [2,1,4,5,0],
    [0,2,3,0,9],
    [0,0,3,6,0],
    [0,0,0,3,0],
    [0,0,0,0,3]]):
```

```
[> linalg[eigenvectors](A);
```

$$[2, 2, \{[1, 0, 0, 0, 0]\}], [3, 3, \{[9, 9, 0, 0, 1], [7, 3, 1, 0, 0]\}]$$

For each of two eigenvalues of the matrix A, *Maple* has produced a maximal number of linearly independent eigenvectors. The match-up of eigenvalues and eigenvectors of the matrix is shown in Table 3.

TABLE 3 EIGENVALUES AND EIGENVECTORS.

Eigenvalues	Associated Eigenvectors
2	(1, 0, 0, 0, 0)
3	(9, 9, 0, 0, 1) , (7, 3, 1, 0, 0)

As we can see, the eigenspace E_2 of the eigenvalue 2 has a basis consisting of one vector, and the eigenspace E_3 of the eigenvalue 3 has a basis consisting of two vectors. ▲

We note that the **Eigenvectors** function in the **LinearAlgebra** package produces a different result.

```
[> B:=Matrix(A):
```

```
[> LinearAlgebra[Eigenvectors](B);
```

$$\begin{bmatrix} 2 \\ 2 \\ 3 \\ 3 \\ 3 \end{bmatrix}, \begin{bmatrix} 1 & 0 & 9 & 7 & 0 \\ 0 & 0 & 9 & 3 & 0 \\ 0 & 0 & 0 & 1 & 0 \\ 0 & 0 & 0 & 0 & 0 \\ 0 & 0 & 1 & 0 & 0 \end{bmatrix}$$

The column vector on the left lists the eigenvalues of A, repeated according to their algebraic multiplicities. The columns of the matrix on the right consist of associated linearly independent eigenvectors, together with the zero vector. Some caution is therefore required when interpreting results of calculations based on the **Eigenvectors** function in *Maple 6*.

The **eigenvectors** function can also be evaluated at matrices without real eigenvalues. The calculation

```
[> linalg[eigenvectors](matrix([[0,-1],[1,0]]));
```

$$[I, 1, \{[1, -I]\}], [-I, 1, \{[1, I]\}]$$

for example, shows that the rotation matrix

$$A = \begin{bmatrix} 0 & -1 \\ 1 & 0 \end{bmatrix}$$

has no real eigenvalues. Considered as a complex matrix, however, it has the eigenvalues $\pm i$, with associated eigenvectors $(-i, 1)$ and $(i, 1)$.

EXERCISES 6.5

1. Let λ be an eigenvalue of an $n \times n$ real matrix A, and suppose that the matrix $(A - \lambda I)$ has rank r. Find the dimension of the eigenspace E_λ.

2. Suppose that λ is an eigenvalue of a matrix A and that the vector $\mathbf{x}$ belongs to the eigenspace E_λ. Show that for all $n \geq 2$, the vectors $A^n\mathbf{x}$ belong to E_{λ^n}.

3. Show that a nondiagonal 3×3 real matrix cannot have an eigenspace E_λ of dimension 3.

4. Show that for each eigenvalue λ of a real matrix A, the eigenspace E_λ is $\text{Nul}(A - \lambda I)$.

5. Show that if the geometric multiplicity of an eigenvalue λ of an $n \times n$ real matrix A is n, then $A = \lambda I_n$.

6. Show that if E_λ and E_μ are eigenspaces of a linear transformation $T : V \to V$ and $\lambda \neq \mu$, then E_λ and E_μ are disjoint.

7. (*Maple 5*) Construct a nondiagonal 3×3 real matrix whose eigenspaces E_λ and E_μ have dimensions 1 and 2.

8. (*Maple 6*) Find bases for the eigenspaces of the following real matrices, if they exist. Restrict your calculations to real eigenvalues.

a. $\begin{bmatrix} 1 & 2 \\ 0 & 0 \end{bmatrix}$ b. $\begin{bmatrix} 1 & -1 \\ 1 & 1 \end{bmatrix}$ c. $\begin{bmatrix} 1 & 2 \\ 3 & 2 \end{bmatrix}$

d. $\begin{bmatrix} 0 & 0 & 0 & -225 \\ 1 & 0 & 0 & -60 \\ 0 & 1 & 0 & 26 \\ 0 & 0 & 1 & 4 \end{bmatrix}$ e. $\begin{bmatrix} 6 & 3 & 6 & 5 \\ 3 & 4 & 4.0 & 3 \\ 6 & 4 & 2 & 2 \\ 5 & 3 & 2 & 4 \end{bmatrix}$ f. $\begin{bmatrix} 1 & 0 & 0 & 9 \\ 0 & 1 & 7 & 0 \\ 0 & 7 & 1 & 0 \\ 9 & 0 & 0 & 1 \end{bmatrix}$

Invertibility and Eigenvalues

The invertibility of a matrix is intimately connected to the nature of its eigenvalues.

THEOREM 6.9 *A real square matrix A is invertible if and only if the number 0 is not an eigenvalue of A.*

Proof. Suppose that A is invertible and that λ is an eigenvalue of A. Then there exists a nonzero vector $\mathbf{x}$ such that $A\mathbf{x} = \lambda\mathbf{x}$. Therefore,

$$\mathbf{x} = (A^{-1}A)(\mathbf{x}) = A^{-1}(A\mathbf{x}) = A^{-1}(\lambda\mathbf{x}) = \lambda A^{-1}(\mathbf{x})$$

This can only be so if $\lambda \neq 0$. Conversely, suppose that there exists a nonzero vector $\mathbf{x}$ for which $A\mathbf{x} = 0\mathbf{x} = \mathbf{0}$. Then the invertibility of A would lead to the contradiction that

$$\mathbf{x} = (A^{-1}A)\mathbf{x} = A^{-1}(0\mathbf{x}) = 0A^{-1}(\mathbf{x}) = \mathbf{0}$$

This proves the equivalence. ■

EXAMPLE 6.23 An Invertible Matrix and Its Eigenvalues

Show that 0 cannot be an eigenvalue of the invertible matrix

$$A = \begin{bmatrix} 1 & 0 & 0 \\ 3 & 2 & 2 \\ 0 & 0 & 3 \end{bmatrix}$$

Solution. Suppose that $A\mathbf{x} = 0\mathbf{x} = \mathbf{0}$. We use the **NullSpace** function in the **LinearAlgebra** package to show that $\mathbf{x}$ must be $\mathbf{0}$.

```
[> with(LinearAlgebra):
[> A:=Matrix([[1,0,0],[3,2,2],[0,0,3]]):
[> NullSpace(A);
                              {}
```

Since the null space of A has the empty basis, $A\mathbf{x} = \mathbf{0}$ if and only if $\mathbf{x} = \mathbf{0}$. Since eigenvectors are nonzero, the vector $\mathbf{x}$ cannot be an eigenvector of A. Therefore, the scalar 0 cannot be an eigenvalue of A. ▲

We can use this theorem to relate the eigenvalues of T^{-1} to those of T.

THEOREM 6.10 (Inverse eigenvalue theorem) *If λ is an eigenvalue of an invertible linear transformation T on a real vector space V, then λ^{-1} is an eigenvalue of T^{-1}.*

Proof. Suppose that λ is an eigenvalue of T. Then there exists a nonzero vector $\mathbf{x} \in V$ for which $T(\mathbf{x}) = \lambda\mathbf{x}$. This means that $\mathbf{x} = T^{-1}(T(\mathbf{x})) = T^{-1}(\lambda\mathbf{x}) = \lambda T^{-1}(\mathbf{x})$. Hence $\lambda \neq 0$, and we can rewrite the equation $\mathbf{x} = \lambda T^{-1}(\mathbf{x})$ as $\lambda^{-1}\mathbf{x} = T^{-1}(\mathbf{x})$. This shows that λ^{-1} is an eigenvalue of T^{-1}. ■

EXAMPLE 6.24 The Eigenvalues of the Inverse of a Linear Transformation

Use *Maple* to show that the eigenvalues of the inverse of the matrix transformation

$$T_A \begin{bmatrix} x \\ y \\ z \end{bmatrix} = \begin{bmatrix} 1 & 0 & 0 \\ 3 & 2 & 2 \\ 0 & 0 & 3 \end{bmatrix} \begin{bmatrix} x \\ y \\ z \end{bmatrix} = \begin{bmatrix} x \\ 3x + 2y + 2z \\ 3z \end{bmatrix}$$

are the reciprocals of the eigenvalues of T.

Solution. By Corollary 6.3, the eigenvalues of T_A are the roots of the characteristic polynomial of T_A. By definition, the characteristic polynomial $c_T(t)$ of T is $c_A(t)$, and we saw in Example 6.23 that the roots of $c_A(t)$ are 1, 2, and 3. By Theorem 5.13, the transformation T^{-1} is represented by A^{-1}. Hence its eigenvalues are those of A^{-1}. Let us find the eigenvalues of this matrix.

```
[>with(linalg):
[>A:=matrix([[1,0,0],[3,2,2],[0,0,3]]):
[>eigenvalues(inverse(A));
```

$$1, \frac{1}{2}, \frac{1}{3}$$

This shows that the eigenvalues of T^{-1} are the reciprocals of the eigenvalues of T_A. ▲

EXERCISES 6.6

1. (*Maple V*) Show that the characteristic polynomial of the matrix

$$A = \begin{bmatrix} 0 & 0 & 0 & 8 \\ 1 & 0 & 0 & -4 \\ 0 & 1 & 0 & -6 \\ 0 & 0 & 1 & 5 \end{bmatrix}$$

is $p(t) = (t+1)(t-2)^3$. Use this fact to explain why the matrix A is invertible.

2. (*Maple V*) Show that the following matrices are invertible by showing that 0 is not an eigenvalue.

a. $\begin{bmatrix} 1 & 2 \\ 3 & 0 \end{bmatrix}$ b. $\begin{bmatrix} 4 & 0 & 0 \\ 0 & 4 & 0 \\ 0 & 0 & 3 \end{bmatrix}$ c. $\begin{bmatrix} 1 & 2 & 1 & 2 \\ 3 & 0 & 2 & 2 \\ 1 & 0 & 1 & 2 \\ 2 & 2 & 2 & 2 \end{bmatrix}$

d. $\begin{bmatrix} 0 & -1 \\ 1 & 0 \end{bmatrix}$ e. $\begin{bmatrix} 1 & 2 & 1 \\ 3 & 0 & 2 \\ 1 & 0 & 1 \end{bmatrix}$ f. $\begin{bmatrix} 1 & 2 & 1 & 2 & 0 \\ 3 & 0 & 2 & 2 & 0 \\ 1 & 0 & 1 & 2 & 0 \\ 2 & 2 & 2 & 2 & 0 \\ 0 & 0 & 0 & 0 & 3 \end{bmatrix}$

3. (*Maple 6*) Find the eigenvalues of the inverses of the following matrices without calculating the inverse matrices.

a. $\begin{bmatrix} 3 & 0 \\ 4 & -4 \end{bmatrix}$ b. $\begin{bmatrix} -4 & -3 \\ -3 & -1 \end{bmatrix}$ c. $\begin{bmatrix} 2 & 0 & 0 & 0 \\ 1 & 2 & 0 & 0 \\ 0 & \frac{3}{2} & 3 & 0 \\ 0 & 0 & 1 & 4 \end{bmatrix}$

d. $\begin{bmatrix} 7 & 7 \\ 0 & -7 \end{bmatrix}$ e. $\begin{bmatrix} 8 & 4 & -8 \\ 0 & 7 & 2 \\ 0 & 0 & 8 \end{bmatrix}$ f. $\begin{bmatrix} 0 & 0 & 0 & 0 & 2250 \\ 1 & 0 & 0 & 0 & -2850 \\ 0 & \frac{3}{2} & 0 & 0 & 2130 \\ 0 & 0 & 1 & 0 & -522 \\ 0 & 0 & 0 & \frac{1}{3} & 21 \end{bmatrix}$

Linearly Independent Eigenvectors

In many applications we require linearly independent eigenvectors. It is therefore fortunate that eigenvalues belonging to distinct eigenvalues are always linearly independent. This fact is proved by mathematical induction. The following example explains the idea behind the proof.

EXAMPLE 6.25 Linearly Independent Eigenvectors

Suppose $\mathbf{x}_1, \mathbf{x}_2, \mathbf{x}_3$, and $\mathbf{x}_4$ are eigenvectors belonging to the distinct nonzero eigenvalues $\lambda_1, \lambda_2, \lambda_3$, and λ_4 of a linear transformation $T : V \to V$, and suppose that $\mathbf{x}_1, \mathbf{x}_2$, and $\mathbf{x}_3$ are linearly independent. Then $\mathbf{x}_1, \mathbf{x}_2, \mathbf{x}_3$, and $\mathbf{x}_4$ are also linearly independent. The following calculation proves this fact. Let $a\mathbf{x}_1 + b\mathbf{x}_2 + c\mathbf{x}_3 + d\mathbf{x}_4 = \mathbf{0}$. Then

$$\begin{aligned} \mathbf{0} &= T(a\mathbf{x}_1 + b\mathbf{x}_2 + c\mathbf{x}_3 + d\mathbf{x}_4) \\ &= aT(\mathbf{x}_1) + bT(\mathbf{x}_2) + cT(\mathbf{x}_3) + dT(\mathbf{x}_4) \\ &= a\lambda_1\mathbf{x}_1 + b\lambda_2\mathbf{x}_2 + c\lambda_3\mathbf{x}_3 + d\lambda_4\mathbf{x}_4 \end{aligned}$$

Moreover, $a\lambda_4\mathbf{x}_1 + b\lambda_4\mathbf{x}_2 + c\lambda_4\mathbf{x}_3 + d\lambda_4\mathbf{x}_4 = \mathbf{0}$. Therefore,

$$a(\lambda_1 - \lambda_4)\mathbf{x}_1 + b(\lambda_2 - \lambda_4)\mathbf{x}_2 + c(\lambda_3 - \lambda_4)\mathbf{x}_3 + d(\lambda_4 - \lambda_4)\mathbf{x}_4 = \mathbf{0}$$

Since $d(\lambda_4 - \lambda_4)\mathbf{x}_4 = \mathbf{0}$, the preceding equation reduces to

$$a(\lambda_1 - \lambda_4)\mathbf{x}_1 + b(\lambda_2 - \lambda_4)\mathbf{x}_2 + c(\lambda_3 - \lambda_4)\mathbf{x}_3 + \mathbf{0} = \mathbf{0}$$

Since the vectors $\mathbf{x}_1, \mathbf{x}_2$, and $\mathbf{x}_3$ are linearly independent, the scalars $a(\lambda_1 - \lambda_4)$, $b(\lambda_2 - \lambda_4)$, and $c(\lambda_3 - \lambda_4)$ must be zero. Furthermore, the eigenvalues $\lambda_1, \lambda_2, \lambda_3$, and λ_4 are distinct, so the scalars $(\lambda_1 - \lambda_4)$, $(\lambda_2 - \lambda_4)$, and $(\lambda_3 - \lambda_4)$ are all nonzero. Therefore a, b, and c must be zero. This implies that $d\mathbf{x}_4 = \mathbf{0}$. However, since $\mathbf{x}_4$ is an eigenvector and is therefore nonzero, the scalar d must be 0. ▲

The proof of the next theorem is modeled after this calculation, somewhat disguised in a general induction.

THEOREM 6.11 (Eigenvector theorem) *Eigenvectors belonging to distinct eigenvalues of a linear transformation $T : V \to V$ are linearly independent.*

Proof. Let $\lambda_1, \ldots, \lambda_n$ be a sequence of distinct eigenvalues of T, and let $\mathbf{x}_1, \ldots, \mathbf{x}_n$ be a sequence of corresponding eigenvectors. We must show that if $a_1\mathbf{x}_1 + \cdots + a_n\mathbf{x}_n = 0$, then $a_1 = \cdots = a_n = 0$. The proof is by induction on n.

Let $n = 1$ and $a_1\mathbf{x}_1 = 0$. Since $\mathbf{x}_1$ is an eigenvector and is therefore nonzero, the scalar a_1 must be zero. Now suppose that $n > 1$, assume that the theorem holds for $(n - 1)$, and consider the equation $a_1\mathbf{x}_1 + \cdots + a_n\mathbf{x}_n = 0$. We first observe that

$$\begin{aligned}\mathbf{0} &= T(\mathbf{0}) \\ &= T(a_1\mathbf{x}_1 + \cdots + a_n\mathbf{x}_n) \\ &= a_1T(\mathbf{x}_1) + \cdots + a_nT(\mathbf{x}_n) \\ &= a_1\lambda_1\mathbf{x}_1 + \cdots + a_n\lambda_n\mathbf{x}_n\end{aligned}$$

Therefore $a_1\lambda_1\mathbf{x}_1 + \cdots + a_n\lambda_n\mathbf{x}_n = \mathbf{0}$. Next we multiply the zero vector $a_1\mathbf{x}_1 + \cdots + a_n\mathbf{x}_n$ by the last eigenvalue λ_n and note that

$$a_1\lambda_n\mathbf{x}_1 + \cdots + a_{n-1}\lambda_n\mathbf{x}_{n-1} + a_n\lambda_n\mathbf{x}_n = \mathbf{0}$$

We now subtract the last two equations and get

$$a_1(\lambda_1 - \lambda_n)\mathbf{x}_1 + \cdots + a_{n-1}(\lambda_{n-1} - \lambda_n)\mathbf{x}_{n-1} + a_n(\lambda_n - \lambda_n)\mathbf{x}_n = \mathbf{0}$$

Therefore, $a_1(\lambda_1 - \lambda_n)\mathbf{x}_1 + \cdots + a_{n-1}(\lambda_{n-1} - \lambda_n)\mathbf{x}_{n-1} = \mathbf{0}$. By the induction hypothesis, the vectors $\mathbf{x}_1, \ldots, \mathbf{x}_{n-1}$ are linearly independent. Therefore, the scalars $a_i(\lambda_i - \lambda_n)$ must all be zero for $1 \le i \le n - 1$. However, since the given eigenvalues are assumed to be distinct, $\lambda_i - \lambda_n \ne 0$ for $1 \le i \le n - 1$. Hence the scalars $a_1, \ldots, a_{n-1}$ must be zero. This in turn implies that

$$\begin{aligned}\mathbf{0} &= a_1\mathbf{x}_1 + \cdots + a_{n-1}\mathbf{x}_{n-1} + a_n\mathbf{x}_n \\ &= \mathbf{0} + a_n\mathbf{x}_n \\ &= a_n\mathbf{x}_n\end{aligned}$$

Since $\mathbf{x}_n$ is an eigenvector and is therefore nonzero, we must have $a_n = 0$. ■

EXERCISES 6.7

1. Show that if $\mathbf{x}$ is an eigenvector associated with an eigenvalue λ and a is a nonzero scalar, then the vector $a\mathbf{x}$ is an eigenvector associated with λ.

2. Show that if $\mathbf{x}_1$ and $\mathbf{x}_2$ are eigenvectors associated with the same eigenvalue λ, and a and b are two scalars for which $a\mathbf{x}_1 + b\mathbf{x}_2 \neq \mathbf{0}$, then the linear combination $a\mathbf{x}_1 + b\mathbf{x}_2$ is an eigenvector associated with λ.

3. Prove or disprove the statement that if $\mathbf{x}_1$ and $\mathbf{x}_2$ are two eigenvectors of a real matrix A, associated with two different eigenvalues λ_1 and λ_2, then the product $\mathbf{x}_1^T\mathbf{x}_2$ is zero.

4. Prove or disprove the statement that if two eigenvectors $\mathbf{x}$ and $\mathbf{y}$ are linearly independent, then they are associated with distinct eigenvalues.

5. Show that if a real $n \times n$ matrix A has n distinct eigenvalues, then the eigenvectors associated with these eigenvalues form a basis for $\mathbb{R}^n$.

6. (*Maple V*) Find an eigenvector for each eigenvalue of the following matrices.

a. $\begin{bmatrix} 1 & 0 \\ 0 & 0 \end{bmatrix}$ b. $\begin{bmatrix} 0 & 1 \\ 1 & 0 \end{bmatrix}$ c. $\begin{bmatrix} 1 & 1 \\ 0 & 1 \end{bmatrix}$

d. $\begin{bmatrix} 1 & 0 & -2 & 6 \\ 0 & 5 & 7 & 1 \\ 0 & 0 & 0 & 1 \\ 0 & 0 & 0 & 1 \end{bmatrix}$ e. $\begin{bmatrix} 0 & 0 & 0 & 0 \\ 0 & 0 & 0 & 0 \\ 0 & 0 & 0 & 0 \\ 0 & 0 & 0 & 0 \end{bmatrix}$ f. $\begin{bmatrix} 1 & 0 & 0 & 9 \\ 0 & 1 & 7 & 0 \\ 0 & 7 & 1 & 0 \\ 9 & 0 & 0 & 1 \end{bmatrix}$

7. (*Maple 6*) Show that the eigenvectors of the matrices in Exercise 6 associated with distinct eigenvalues are linearly independent.

DIAGONALIZING SQUARE MATRICES

At the beginning of this chapter, we discussed some of the conceptual and computational benefits of diagonal matrices. Some of these benefits transfer to nondiagonal matrices, as long as the matrices involved are similar to diagonal ones.

DEFINITION 6.8 *A square matrix A is **diagonalizable** if there exists a diagonal matrix D and an invertible matrix P such that* $A = PDP^{-1}$.

In this section, we will show that an $n \times n$ real matrix A is ***diagonalizable*** if and only if A has n linearly independent eigenvectors. If the matrix A is diagonalizable, we say that D and P ***diagonalize*** A. To stress the nature of the matrices P and D involved in a diagonalization, we call a product PDP^{-1} an ***eigenvector-eigenvalue decomposition*** of A.

If D and P diagonalize the matrix A, then the diagonal entries of D are the eigenvalues of A, and the columns of the matrix P are eigenvectors belonging to the matching eigenvalues. If $A = PDP^{-1}$, then $AP = PD$.

Therefore,

$$AP = A\,[\mathbf{x}_1 \ \cdots \ \mathbf{x}_n] = [A\mathbf{x}_1 \ \cdots \ A\mathbf{x}_n]$$

and

$$PD = P\,\text{diag}(\lambda_1, \ldots, \lambda_n) = [\lambda\mathbf{x}_1 \ \cdots \ \lambda_n\mathbf{x}_n]$$

Since $AP = PD$, it follows that $A\mathbf{x}_i = \lambda\mathbf{x}_1$. Table 4 lists the elements involved in using similarity to link diagonal and nondiagonal matrices. We will show later that such links exists only for matrices with enough linearly independent eigenvectors.

TABLE 4 DIAGONALIZING A SQUARE MATRIX.

Step 1. Take a diagonalizable $n \times n$ matrix A.
Step 2. Find the n (not necessarily distinct) eigenvalues $\lambda_1, \ldots, \lambda_n$ of A.
Step 3. For each eigenvalue λ_i, calculate an eigenvector $\mathbf{x}_i$ belonging to λ_i with the property that the set $\{\mathbf{x}_1, \ldots, \mathbf{x}_n\}$ is linearly independent.
Step 4. Form the matrices $D = \text{diag}\,(\lambda_1, \ldots, \lambda_n)$ and $P = [\mathbf{x}_1 \ \cdots \ \mathbf{x}_n]$, and calculate P^{-1}.
Step 5. The matrix D diagonalizes A in the sense that $A = PDP^{-1}$.

The next theorem explains why a matrix is diagonalizable only if it has enough linearly independent eigenvectors.

THEOREM 6.12 (Diagonalizability theorem) *An $n \times n$ matrix A is similar to a diagonal matrix D if and only if A has n linearly independent eigenvectors.*

Proof. Suppose that A is diagonalizable and that $A = PDP^{-1}$. Then by right-multiplication we have $AP = PD$. Hence, as observed earlier,

$$\begin{bmatrix} A\mathbf{x}_1 & A\mathbf{x}_2 & \cdots & A\mathbf{x}_n \end{bmatrix} = \begin{bmatrix} \lambda_1\mathbf{x}_1 & \lambda_2\mathbf{x}_2 & \cdots & \lambda_n\mathbf{x}_n \end{bmatrix}$$

By equating columns, we get

$$A\mathbf{x}_1 = \lambda_1\mathbf{x}_1,\ A\mathbf{x}_2 = \lambda_2\mathbf{x}_2, \ldots, A\mathbf{x}_n = \lambda_n\mathbf{x}_n$$

Since P is invertible, the column vectors $\mathbf{x}_1, \mathbf{x}_2, \ldots, \mathbf{x}_n$ must be linearly independent. They must also be nonzero. Therefore, $\lambda_1, \lambda_2, \ldots, \lambda_n$ are eigenvalues and $\mathbf{x}_1, \mathbf{x}_2, \ldots, \mathbf{x}_n$ are the corresponding eigenvectors.

Suppose, conversely, that $\mathbf{x}_1, \mathbf{x}_2, \ldots, \mathbf{x}_n$ are linearly independent eigenvectors. We can use them as the columns of an invertible matrix P, and let $D = \text{diag}(\lambda_1, \ldots, \lambda_n)$, where λ_1, $\lambda_2, \ldots, \lambda_n$ are the eigenvalues associated with $\mathbf{x}_1, \mathbf{x}_2, \ldots, \mathbf{x}_n$. It follows that $AP = PD$. Since the eigenvectors are linearly independent, the matrix P is invertible. We therefore get $A = PDP^{-1}$. ■

EXAMPLE 6.26 A Diagonalizable 3 x 3 Matrix

Use *Maple* to verify that the matrices

$$D = \begin{bmatrix} 3 & 0 & 0 \\ 0 & 3 & 0 \\ 0 & 0 & 5 \end{bmatrix} \quad \text{and} \quad P = \begin{bmatrix} 1 & 1 & 1 \\ -1 & 0 & 2 \\ 0 & 1 & 1 \end{bmatrix}$$

diagonalize the matrix

$$A = \begin{bmatrix} 4 & 1 & -1 \\ 2 & 5 & -2 \\ 1 & 1 & 2 \end{bmatrix}$$

Solution. We begin by representing A, P, and D.

```
[> A:=matrix([[4,1,-1],[2,5,-2],[1,1,2]]):
[> P:=matrix([[1,1,1],[-1,0,2],[0,1,1]]):
[> Dm:=matrix([[3,0,0],[0,3,0],[0,0,5]]):
```

Next we show that $A = PDP^{-1}$.

```
[> with(linalg):
[> equal(A,P&*Dm&*inverse(P));
                              true
```

This shows that the matrices D and P diagonalize the matrix A. ▲

EXAMPLE 6.27 A Diagonalizable Linear Transformation

Let $T : \mathbb{R}^2 \to \mathbb{R}^2$ be the linear transformation defined by

$$T(x, y) = \begin{bmatrix} 6x - y \\ 3x + 2y \end{bmatrix}$$

Use *Maple* to find an eigenvector-eigenvalue decomposition for the matrix $[T]_{\mathcal{E}}$.

Solution. Let $A = [T]_{\mathcal{E}}$. It is easy to verify that

$$A = \begin{bmatrix} 6 & -1 \\ 3 & 2 \end{bmatrix}$$

We use the **eigenvectors** function to find an eigenvector-eigenvalue decomposition of A. This provides us with both the eigenvalues of A and the corresponding eigenvectors at the same time.

```
[> with(linalg):
[> A:=matrix([[6,-1],[3,2]]):
[> eigenvectors(A);
```

$$[3, 1, \{[1, 3]\}], [5, 1, \{[1, 1]\}]$$

This tells us that $\{3, 5\}$ is the set of eigenvalues of A, that $(1, 3)$ is an eigenvector belonging to the eigenvalue 3, and that $(1, 1)$ is an eigenvector belonging to the eigenvalue 5. It follows that the matrices

$$D = \begin{bmatrix} 3 & 0 \\ 0 & 5 \end{bmatrix} \quad \text{and} \quad P = \begin{bmatrix} 1 & 1 \\ 3 & 1 \end{bmatrix}$$

yield an eigenvector-eigenvalue decomposition of A. We now verify that $A = PDP^{-1}$.

```
[> Dm:=matrix([[3,0],[0,5]]):
[> P:=matrix([[1,1],[3,1]]):
[> equal(A,P&*Dm&*inverse(P));
```

$$\mathit{true}$$

Therefore, $A = PDP^{-1}$. ▲

EXAMPLE 6.28 A Diagonalizable Matrix with Repeated Eigenvalues

Use *Maple* to diagonalize the matrix

$$A = \begin{bmatrix} 4 & 1 & -1 \\ 2 & 5 & -2 \\ 1 & 1 & 2 \end{bmatrix}$$

Solution. We use the **eigenvectors** function in the **linalg** package to calculate the required eigenvalues and eigenvectors of A.

```
[> with(linalg):
[> A:=matrix([[4,1,-1],[2,5,-2],[1,1,2]]):
[> evA:=eigenvectors(A);
```

$$evA := [3, 2, \{[1, 0, 1], [0, 1, 1]\}], [5, 1, \{[1, 2, 1]\}]$$

We use this information to build the matrices P and D.

```
[> P:=matrix([[1,0,1],[0,1,2],[1,1,1]]):
[> Dm:=diag(3,3,5):
[> B:=evalm(P&*Dm&*inverse(P));
```

$$B := \begin{bmatrix} 4 & 1 & -1 \\ 2 & 5 & -2 \\ 1 & 1 & 2 \end{bmatrix}$$

As we can see, the matrix B is the required diagonalization PDP^{-1}. ▲

The next example shows that not all square matrices have enough linearly independent eigenvectors to be diagonalizable.

EXAMPLE 6.29 A Nondiagonalizable Matrix

Use *Maple* to show that the matrix

$$A = \begin{bmatrix} -3 & 1 & -1 \\ -7 & 5 & -1 \\ -6 & 6 & -2 \end{bmatrix}$$

is not diagonalizable.

Solution. Let us first find the eigenvalues of A.

```
[> with(linalg):
[> A:=matrix([[-3,1,-1],[-7,5,-1],[-6,6,-2]]):
[> eigenvalues(A);
```

$$4, -2, -2$$

As we can see, the matrix A has the two eigenvalues $\lambda_1 = -2$ and $\lambda_2 = 4$. From

$$\left(\begin{bmatrix} -3 & 1 & -1 \\ -7 & 5 & -1 \\ -6 & 6 & -2 \end{bmatrix} + 2 \begin{bmatrix} 1 & 0 & 0 \\ 0 & 1 & 0 \\ 0 & 0 & 1 \end{bmatrix} \right) \begin{bmatrix} x \\ y \\ z \end{bmatrix} = \begin{bmatrix} 1 & -1 & 1 \\ 7 & -7 & 1 \\ 6 & -6 & 0 \end{bmatrix} \begin{bmatrix} x \\ y \\ z \end{bmatrix}$$

we get the homogeneous linear system

$$\begin{cases} x - y + z = 0 \\ 7x - 7y + z = 0 \\ 6x - 6y = 0 \end{cases}$$

We use *Maple* to solve this system. Let us solve the system for y and z.

```
[> solve({x-y+z=0,7*x-7*y+z=0,6*x-6*y=0},{y,z});
```

$$\{z = 0, y = x\}$$

This shows that x is the only free variable of the system. Therefore we cannot find two linearly independent eigenvectors belonging to the eigenvalue 2. Hence the matrix A is not diagonalizable. ▲

EXERCISES 6.8

All eigenvalues in the following exercises must be real numbers, and all eigenvectors must have real entries.

1. Show that the following real matrices have two linearly independent eigenvectors.

$$\text{a.} \begin{bmatrix} 1 & 1 \\ 1 & 1 \end{bmatrix} \quad \text{b.} \begin{bmatrix} 1 & 1 \\ 0 & 3 \end{bmatrix} \quad \text{c.} \begin{bmatrix} 1 & 5 \\ 5 & 3 \end{bmatrix} \quad \text{d.} \begin{bmatrix} -5 & 5 \\ 5 & 3 \end{bmatrix}$$

2. Show that the following real matrices are not diagonalizable by showing that they fail to have two linearly independent eigenvectors.

$$\text{a.} \begin{bmatrix} 2 & 1 \\ 0 & 2 \end{bmatrix} \quad \text{b.} \begin{bmatrix} 1 & 1 \\ 0 & 1 \end{bmatrix} \quad \text{c.} \begin{bmatrix} 1 & 0 \\ -1 & 1 \end{bmatrix} \quad \text{d.} \begin{bmatrix} 0 & -1 \\ 1 & 0 \end{bmatrix}$$

3. Show that not every invertible matrix is diagonalizable and that not every diagonalizable matrix is invertible.

4. Find all real numbers a for which the following matrices are diagonalizable.

$$\text{a.} \begin{bmatrix} 1 & 1 \\ a & 1 \end{bmatrix} \quad \text{b.} \begin{bmatrix} 1 & a \\ 0 & 3 \end{bmatrix} \quad \text{c.} \begin{bmatrix} 2 & 0 & 0 \\ 0 & 2 & 0 \\ a & 0 & 2 \end{bmatrix} \quad \text{d.} \begin{bmatrix} 2 & 0 & a \\ 0 & 2 & 0 \\ 1 & 0 & 2 \end{bmatrix}$$

5. Use Theorem 6.12 to show that if a real $n \times n$ matrix A has an eigenspace E_λ of dimension n, then A is diagonal.

6. Show that if A is diagonalized by an $n \times n$ matrix D, then the determinant of A is the product $\lambda_1 \lambda_2 \cdots \lambda_n$, where λ_i are the diagonal entries of D.

7. (*Maple V, 6*) Show that the following real matrices are diagonalizable by showing that they have three linearly independent eigenvectors.

$$\text{a.} \begin{bmatrix} 2 & 0 & -1 \\ 0 & 1 & 0 \\ -1 & \frac{1}{2} & 1 \end{bmatrix} \quad \text{b.} \begin{bmatrix} 2 & 0 & 0 \\ 5 & 0 & 5 \\ 0 & \frac{1}{2} & 1 \end{bmatrix} \quad \text{c.} \begin{bmatrix} 2 & -\frac{1}{2} & 0 \\ 0 & 1 & 0 \\ -1 & \frac{1}{2} & 1 \end{bmatrix}$$

8. (*Maple V, 6*) Show that the following real matrices are not diagonalizable by showing that they do not have three linearly independent eigenvectors.

$$\text{a. } \begin{bmatrix} 2 & 3 & 4 \\ 0 & 1 & 5 \\ 0 & 0 & 1 \end{bmatrix} \quad \text{b. } \begin{bmatrix} 2 & -\frac{1}{2} & 0 \\ 0 & 1 & 0 \\ -1 & 0 & 1 \end{bmatrix} \quad \text{c. } \begin{bmatrix} 2 & 3 & 4 \\ 0 & 2 & 5 \\ 0 & 0 & 2 \end{bmatrix}$$

9. (*Maple 6*) Show that each of the following real matrices has four linearly independent eigenvectors.

$$\text{a. } \begin{bmatrix} 0 & 0 & 0 & 0 \\ 0 & 0 & 0 & 0 \\ 0 & 0 & 0 & 0 \\ 0 & 0 & 0 & 0 \end{bmatrix} \quad \text{b. } \begin{bmatrix} 3 & 0 & 0 & 0 \\ 0 & 3 & 0 & 0 \\ 0 & 0 & 3 & 0 \\ 0 & 0 & 0 & 3 \end{bmatrix} \quad \text{c. } \begin{bmatrix} 2 & 0 & 0 & 0 \\ -2 & 3 & 0 & 0 \\ 0 & 0 & 3 & 0 \\ -1 & 0 & 0 & 3 \end{bmatrix}$$

10. (*Maple 6*) Show that if A has a diagonal decomposition PDP^{-1} and $n \geq 2$, then $A^n = PD^nP^{-1}$ and use this fact to calculate A^{12}, where

$$A = \begin{bmatrix} 1 & 0 & 3 & 0 & 0 \\ 0 & 1 & 0 & 0 & 0 \\ 0 & 0 & 0 & 0 & 0 \\ 0 & 0 & 2 & 3 & 0 \\ 1 & 0 & 0 & 0 & 3 \end{bmatrix}$$

11. (*Maple 6*) Every polynomial of the form $p(t) = t^n + a_{n-1}t^{n-1} + \cdots + a_1 t + a_0 \in \mathbb{R}[t]$ has an associated $n \times n$ real matrix

$$A = \begin{bmatrix} 0 & 0 & \cdots & 0 & -a_0 \\ 1 & 0 & \cdots & 0 & -a_1 \\ 0 & 1 & \cdots & 0 & -a_2 \\ \vdots & \vdots & \vdots & \vdots & \vdots \\ 0 & 0 & \cdots & 1 & -a_{n-1} \end{bmatrix}$$

called the ***companion matrix*** of $p(t)$, with the property that $p(t)$ is the characteristic polynomials $c_A(t)$ of A. If $p(t)$ has n distinct real roots, then A is diagonalizable. Construct the companion matrix of $p(t) = (t-2)(t-3)(t-4)$, and find a diagonal matrix D and an invertible matrix P that diagonalize A.

12. (*Maple 6*) Repeat Exercise 11 for polynomials of degree 4 with the following real roots.

a. $\lambda_1 = 1; \lambda_2 = 2; \lambda_3 = 3; \lambda_4 = 4$ b. $\lambda_1 = 3; \lambda_2 = 5; \lambda_3 = -3; \lambda_4 = 7$

c. $\lambda_1 = 0; \lambda_2 = 2; \lambda_3 = 4; \lambda_4 = 5$ d. $\lambda_1 = 5; \lambda_2 = 6; \lambda_3 = -5; \lambda_4 = 12$

13. (*Maple 6*) Repeat Exercise 12 using the **LinearAlgebra[CompanionMatrix]** function.

14. (*Maple V, 6*) Compare the functions **companion** and **CompanionMatrix** in the **linalg** and **LinearAlgebra** packages and discuss their properties.

APPLICATIONS

Calculus

Let V be the real vector space of functions $f : \mathbb{R} \to \mathbb{R}$ spanned by the set $S = \{\sin x, \cos x\}$, and let D_x be the differentiation operator on V. Since $D_x(af + bg) = aD_x f(x) + bD_x g(x)$ preserves linear combinations, it is a linear transformation on V. We show that D_x has no real eigenvalues.

EXAMPLE 6.30 The Eigenvalues of D_x

Use *Maple* to show that D_x has no real eigenvalues.

Solution. First we find the values of D_x for the basis vectors.

```
[> diff(sin(x),x);
```

$$\cos(x)$$

```
[> diff(cos(x),x);
```

$$-\sin(x)$$

This means that

$$D_x \sin x = \cos x = 0 \sin x + 1 \cos x$$

and

$$D_x \cos x = -\sin x = -1 \sin x + 0 \cos x$$

It follows that

$$\begin{aligned} D_x (a \sin x + b \cos x) &= aD_x \sin x + bD_x \cos x \\ &= a \cos x - b \sin x \\ &= -b \sin x + a \cos x \end{aligned}$$

On the other hand, if $\mathbf{x} = a \sin x + b \cos x$ is an eigenvector of D_x, then $\mathbf{x} \neq 0$, and

$$\begin{aligned} D_x (a \sin x + b \cos x) &= \lambda (a \sin x + b \cos x) \\ &= (\lambda a) \sin x + (\lambda b) \cos x \end{aligned}$$

This implies that $(\lambda a)\sin x + (\lambda b)\cos x = -b\sin x + a\cos x$. Since $\sin x$ and $\cos x$ are linearly independent functions, it follows that $\lambda a = -b$ and $\lambda b = a$. Since either a or b must be nonzero, these equations can hold only if $\lambda^2 = -1$. Hence D_x has no real eigenvalues. ▲

Differential Equations

Suppose that $\mathbf{y}' = A\mathbf{y}$ denotes the system

$$\begin{cases} y_1' &= a_{11}y_1 + \cdots + a_{1n}y_n \\ \vdots & \vdots \qquad\qquad \vdots \\ y_n' &= a_{n1}y_1 + \cdots + a_{nn}y_n \end{cases}$$

of linear differential equations, where the y_i are functions of t, and let $\mathbf{y} = P\mathbf{z}$ for some invertible matrix P. Then $P\mathbf{z}' = \mathbf{y}'$.

Therefore, $\mathbf{z}' = P^{-1}\mathbf{y}' = P^{-1}A\mathbf{y} = P^{-1}AP\mathbf{z}$. If A is diagonalizable, we can choose P to be a matrix whose columns are eigenvectors of A. This means that $P^{-1}AP$ is a diagonal matrix D whose diagonal entries are the eigenvalues of A. Therefore,

$$\mathbf{z}' = \begin{bmatrix} z_1(t)' \\ \vdots \\ z_n(t)' \end{bmatrix} = \operatorname{diag}(\lambda_1, \ldots, \lambda_n) \begin{bmatrix} z_1(t) \\ \vdots \\ z_n(t) \end{bmatrix} = \begin{bmatrix} \lambda_1 z_1(t) \\ \vdots \\ \lambda_n z_n(t) \end{bmatrix}$$

Moreover, we know from calculus that the general solution of the differential equation $z_i'(t) = \lambda_i z_i(t)$ is $z_i(t) = a_1 e^{\lambda_i t}$, for an arbitrary scalar a_i.

Therefore,

$$\mathbf{y} = \begin{bmatrix} y_1(t) \\ \vdots \\ y_n(t) \end{bmatrix} = P\mathbf{z} = P \begin{bmatrix} a_1 e^{\lambda_i t} \\ \vdots \\ a_n e^{\lambda_n t} \end{bmatrix}$$

We apply this technique to solve two types of differential equations. In Example 6.31, we show how a higher-order homogeneous differential equation can be converted to a system of first-order differential equations and solved by diagonalization. In Example 6.32 we show how a problem in population dynamics can be solved by diagonalization.

EXAMPLE 6.31 A Third-Order Differential Equation

Solve the homogeneous third-order differential equation

$$y''' - 2y'' - y' + 2y = 0$$

by reducing the equation to a system of first-order equations.

Solution. We introduce two new functions u and v connected by $y' = u$ and $u' = v$. Using these functions, we rewrite the given equation as

$$y''' = 2y'' + y' - 2y = 2v + u - 2y$$

We now write this equation as the linear system $\{y' = u, u' = v, v' = -2y + u + 2v\}$ and convert this system to matrix form:

$$\begin{bmatrix} y' \\ u' \\ v' \end{bmatrix} = \begin{bmatrix} 0 & 1 & 0 \\ 0 & 0 & 1 \\ -2 & 1 & 2 \end{bmatrix} \begin{bmatrix} y \\ u \\ v \end{bmatrix}$$

Next we use *Maple* to diagonalize the coefficient matrix.

```
[> with(linalg):
[> A:=matrix([[0,1,0],[0,0,1],[-2,1,2]]):
[> eigenvectors(A);
```

$$[1, 1, \{[1, 1, 1]\}], [2, 1, \{[1, 2, 4]\}], [-1, 1, \{[1, -1, 1]\}]$$

Therefore, the eigenvalues of A are -1, 1, and 2 and the vectors $\mathbf{x}_1 = (-1, 1, -1)$, $\mathbf{x}_2 = (1, 1, 1)$, and $\mathbf{x}_3 = (1, 2, 4)$ are corresponding eigenvectors. We use this information to construct an eigenvector-eigenvalue decomposition PDP^{-1} of A. Let

$$P = \begin{bmatrix} -1 & 1 & 1 \\ 1 & 1 & 2 \\ -1 & 1 & 4 \end{bmatrix} \quad \text{and} \quad D = \begin{bmatrix} -1 & 0 & 0 \\ 0 & 1 & 0 \\ 0 & 0 & 2 \end{bmatrix}$$

It is clear from our introductory remarks that $\mathbf{z}(t)' = P^{-1}AP\mathbf{z}(t) = D\mathbf{z}(t)$, where

$$\mathbf{y} = \begin{bmatrix} y \\ u \\ v \end{bmatrix} = P\mathbf{z}$$

In other words,

$$\begin{bmatrix} z_1(t)' \\ z_2(t)' \\ z_3(t)' \end{bmatrix} = \begin{bmatrix} -1 & 0 & 0 \\ 0 & 1 & 0 \\ 0 & 0 & 2 \end{bmatrix} \begin{bmatrix} z_1(t) \\ z_2(t) \\ z_3(t) \end{bmatrix} = \begin{bmatrix} -z_1(t) \\ z_2(t) \\ 2z_3(t) \end{bmatrix}$$

Moreover, we know from calculus that

$$\begin{bmatrix} -z_1(t) \\ z_2(t) \\ 2z_3(t) \end{bmatrix} = \begin{bmatrix} ae^{-t} \\ be^{t} \\ ce^{2t} \end{bmatrix}$$

Therefore,

$$\mathbf{y}(t) = \begin{bmatrix} y(t) \\ u(t) \\ v(t) \end{bmatrix} = \begin{bmatrix} -1 & 1 & 1 \\ 1 & 1 & 2 \\ -1 & 1 & 4 \end{bmatrix} \begin{bmatrix} ae^{-t} \\ be^{t} \\ ce^{2t} \end{bmatrix} = \begin{bmatrix} -ae^{-t} + be^{t} + ce^{2t} \\ ae^{-t} + be^{t} + 2ce^{2t} \\ -ae^{-t} + be^{t} + 4ce^{2t} \end{bmatrix}$$

We conclude by using *Maple* to verify that the calculated functions satisfy the given differential equation.

```
[> y:=-a*exp(-t)+b*exp(t)+c*exp(2*t);
```

$$y := -a\,e^{(-t)} + b\,e^{t} + c\,e^{(2t)}$$

```
[> evalb(diff(y,t$3) = 2*diff(y,t$2)+diff(y,t)-2*y);
```

true

As we can see, the equation holds. ▲

In population biology, the rates of change of certain populations are sometimes given by linear combinations of other populations. If the growth rate of each species, measured over time, is a function of the size of the different species involved, the resulting model is a system of first-order differential equations. We now show how the growth rates of the individual species can be calculated when the coefficient matrix of the system is diagonalizable.

EXAMPLE 6.32 Biological Rates of Change

Use *Maple* to calculate the populations $y_1(t)$ and $y_2(t)$ of two interacting species determined by the system

$$\begin{cases} y_1' = 5y_1 + 2y_2 \\ y_2' = 3y_1 + 4y_2 \end{cases}$$

if $y_1(0) = 500$ and $y_2(0) = 700$.

Solution. First we write the given system of differential equations in matrix form.

$$\begin{bmatrix} y_1' \\ y_2' \end{bmatrix} = \begin{bmatrix} 5 & 2 \\ 3 & 4 \end{bmatrix} \begin{bmatrix} y_1 \\ y_2 \end{bmatrix}$$

Next we diagonalize the coefficient matrix by finding the matrices P and D.

```
[> A:=matrix([[5,2],[3,4]]):
[> linalg[eigenvectors](A);
```

$$[2, 1, \{[1, \tfrac{-3}{2}]\}], [7, 1, \{[1, 1]\}]$$

Hence the eigenvalues of A are 2 and 7, and two corresponding eigenvectors are $\{-2, 3\}$ and $\{1, 1\}$. We therefore form the matrices

$$P = \begin{bmatrix} -2 & 1 \\ 3 & 1 \end{bmatrix} \quad \text{and} \quad D = \begin{bmatrix} 2 & 0 \\ 0 & 7 \end{bmatrix}$$

We know from the introduction to this section that

$$\begin{bmatrix} z_1(t)' \\ z_2(t)' \end{bmatrix} = \begin{bmatrix} 2 & 0 \\ 0 & 7 \end{bmatrix} \begin{bmatrix} z_1(t) \\ z_2(t) \end{bmatrix} = \begin{bmatrix} 2z_1(t) \\ 7z_2(t) \end{bmatrix}$$

so $z_1 = ae^{2t}$, $z_2 = be^{7t}$, and

$$\begin{bmatrix} y_1(t) \\ y_2(t) \end{bmatrix} = \begin{bmatrix} -2 & 1 \\ 3 & 1 \end{bmatrix} \begin{bmatrix} ae^{2t} \\ be^{7t} \end{bmatrix} = \begin{bmatrix} -2ae^{2t} + be^{7t} \\ 3ae^{2t} + be^{7t} \end{bmatrix}$$

We now use the size of the populations at time $t = 0$ to calculate the scalars a and b.

$$\begin{cases} y_1(0) = -2a + b = 500 \\ y_2(0) = 3a + b = 700 \end{cases}$$

```
[> solve({-2*a+b=500,3*a+b=700},{a,b});
```

$$\{a = 40, b = 580\}$$

Hence $y_1(t) = -80e^{2t} + 580e^{7t}$ and $y_2(t) = 120e^{2t} + 580e^{7t}$ are the populations of the two species. ▲

EXERCISES 6.9

1. Suppose that $D : D^\infty(\mathbb{R}, \mathbb{R}) \to D^\infty(\mathbb{R}, \mathbb{R})$ is the differentiation operation. Then λ is an eigenvalue of D if and only if $D(f) = f' = \lambda f$. The eigenvectors of D belonging to λ are therefore the solutions of the differential equation $y' - \lambda y = 0$. Use this information to show that every real number λ is an eigenvalue of D, and find the eigenvectors belonging to λ.

2. Find the eigenvalues and eigenvectors of the differentiation operation $D : \mathbb{R}_2[t] \to \mathbb{R}_2[t]$. Explain why D is not diagonalizable.

3. Diagonalize the linear transformation $T : \mathbb{R}_3[t] \to \mathbb{R}_3[t]$ defined by $T(p(t)) = p(0)$.

4. Suppose that $D : D^\infty(\mathbb{R}, \mathbb{R}) \to D^\infty(\mathbb{R}, \mathbb{R})$ is the differentiation operation. Find the eigenvalues and eigenvectors of each of the following linear transformations.

   a. $D^2 : D^\infty(\mathbb{R}, \mathbb{R}) \to D^\infty(\mathbb{R}, \mathbb{R})$ b. $D^2 + D : D^\infty(\mathbb{R}, \mathbb{R}) \to D^\infty(\mathbb{R}, \mathbb{R})$

5. Suppose that $D : D^{\infty}(\mathbb{R}, \mathbb{R}) \to D^{\infty}(\mathbb{R}, \mathbb{R})$ is the differentiation operation. Show that a vector $f \in D^{\infty}(\mathbb{R}, \mathbb{R})$ may be an eigenvector of $D^2 : D^{\infty}(\mathbb{R}, \mathbb{R}) \to D^{\infty}(\mathbb{R}, \mathbb{R})$ without being an eigenvector of D.

6. Use the isomorphism $\varphi(a \cos x + b \sin x) = (a, b)$ and the fact that the matrix

$$\begin{bmatrix} 0 & -1 \\ 1 & 0 \end{bmatrix}$$

has no real eigenvalues to show that the differentiation operator in Example 6.30 has no real eigenvalues.

7. Find the population for an ecosystem made up of three species whose populations are related by

$$\begin{bmatrix} y_1(t)' \\ y_3(t)' \\ y_3(t)' \end{bmatrix} = \begin{bmatrix} 9 & 0 & 2 \\ 6 & 5 & 3 \\ -4 & 0 & 3 \end{bmatrix} \begin{bmatrix} y_1(t) \\ y_3(t) \\ y_3(t) \end{bmatrix}$$

if $y_1(0) = y_2(0) = y_3(0) = 400$.

Discrete Dynamical Systems

Sequences of vectors generated by a vector and a matrix are frequently used to model physical processes. In this section, we discuss some of the mathematical ideas involved. Although the range of applications is more general, we limit ourselves to the diagonalizable case.

DEFINITION 6.9 *Let $\mathbf{x}_0$ be a fixed vector in $\mathbb{R}^n$ and let A be a diagonalizable $n \times n$ real matrix. Then the sequence of vectors*

$$\mathbf{x}_0, A\mathbf{x}_0, A^2\mathbf{x}_0, \ldots, A^n\mathbf{x}_0, \ldots$$

*is a **discrete dynamical system**.*

We can rewrite the system in terms a family of equations $\mathbf{x}_{k+1} = A\mathbf{x}_k$, with $k \in \mathbb{N}$ ranging over discrete intervals of time such as days, months, years, and so on. Hence the system is called discrete. It is called dynamical because at each stage $k + 1$, the system models an evolving process determined only by its condition at stage k and by the matrix A. If $\mathbf{x}_0$ is an eigenvector A belonging to an eigenvalue λ, then the system $\mathbf{x}_0, A\mathbf{x}_0, A^2\mathbf{x}_0, \ldots, A^n\mathbf{x}_0, \ldots$ becomes $\mathbf{x}_0, \lambda\mathbf{x}_0, \lambda^2\mathbf{x}_0, \ldots, \lambda^n\mathbf{x}_0, \ldots$. The long-term behavior of the system therefore can be analyzed in terms of the limit of $\lambda^n\mathbf{x}_0$, as $n \to \infty$, if such a limit exists.

EXAMPLE 6.33 A Discrete Dynamical System

Describe the long-term behavior of the discrete dynamical system determined by

$$A = \begin{bmatrix} \frac{9}{10} & \frac{5}{100} \\ \frac{1}{10} & \frac{95}{100} \end{bmatrix} \quad \text{and} \quad \mathbf{x}_0 = \begin{bmatrix} 30 \\ 20 \end{bmatrix}$$

Solution. We begin by showing that A has two linearly independent eigenvectors. We can then decompose the vector $\mathbf{x}_0$ into a linear combination of eigenvectors and calculate the required limits. The calculation

```
[> A:=matrix([[9/10,5/100],[1/10,95/100]]):
[> linalg[eigenvectors](A);
```

$$\left[\frac{17}{20}, 1, \{[1, -1]\}\right], [1, 1, \{[1, 2]\}]$$

shows that $17/20$ and 1 are the eigenvalues of A and that the vectors $\mathbf{x}_1 = (-1, 1)$ and $\mathbf{x}_2 = (1/2, 1)$ are eigenvectors of A belonging to the distinct eigenvalues $17/20$ and 1. They are therefore linearly independent and form a basis for $\mathbb{R}^2$. Hence there exist unique scalars a and b for which

$$\begin{bmatrix} 30 \\ 20 \end{bmatrix} = a \begin{bmatrix} -1 \\ 1 \end{bmatrix} + b \begin{bmatrix} \frac{1}{2} \\ 1 \end{bmatrix}$$

Now we find the scalars a and b.

```
[> solve({-a+b/2=30,a+b=20},{a,b});
```

$$\{a = \frac{-40}{3}, b = \frac{100}{3}\}$$

Therefore, $a = -40/3$ and $b = 100/3$. This means that long-term behavior of the system $\mathbf{x}_{k+1} = A\mathbf{x}_k$ is described by the following limit:

$$\begin{aligned} L &= \lim_{n\to\infty} \left(\begin{bmatrix} \frac{9}{10} & \frac{5}{100} \\ \frac{1}{10} & \frac{95}{100} \end{bmatrix}^n \begin{bmatrix} 30 \\ 20 \end{bmatrix} \right) \\ &= \lim_{n\to\infty} \left(\begin{bmatrix} \frac{9}{10} & \frac{5}{100} \\ \frac{1}{10} & \frac{95}{100} \end{bmatrix}^n \left(-\frac{40}{3} \begin{bmatrix} -1 \\ 1 \end{bmatrix} + \frac{100}{3} \begin{bmatrix} \frac{1}{2} \\ 1 \end{bmatrix} \right) \right) \\ &= -\frac{40}{3} \lim_{n\to\infty} \left(\frac{17}{20}\right)^n \begin{bmatrix} -1 \\ 1 \end{bmatrix} + \frac{100}{3} \lim_{n\to\infty} (1)^n \begin{bmatrix} \frac{1}{2} \\ 1 \end{bmatrix} \\ &= \frac{100}{3} \begin{bmatrix} \frac{1}{2} \\ 1 \end{bmatrix} = \begin{bmatrix} \frac{50}{3} \\ \frac{100}{3} \end{bmatrix} \end{aligned}$$

Hence the limiting vector L of the system is $(50/3, 100/3)$. ▲

EXERCISES 6.10

1. Explain why the compound interest formula $P_{n+1} = 1.08P_n$ for investing a fixed amount P at 8% per year for $n + 1$ years determines a dynamical system.

2. A population of 1000 gulls on a secluded island is attacked by a communicable disease. The disease is not hereditary. A gull becomes immune to the disease if it has survived the first month of infection. Each month, one-quarter of the healthy gulls contract the disease and one-fifth of them die after a month. Nevertheless, the gull population increases by 10% each month, after births and deaths have been taken into account. Find the dynamical system $\mathbf{x}_{n+1} = A\mathbf{x}_n$ that describes the growth of the gull population, and calculate the population after 12 months.

3. Find the dynamical system $\mathbf{x}_{n+1} = A\mathbf{x}_n$ that yields the value of four investments certificates after $n + 1$ years if the first certificate is invested at 6%, the second at 4.5%, the third at 8%, and the fourth certificate at 5%.

4. (*Maple V*) Create a diagonalizable nondiagonal 3×3 matrix A and use it to construct the dynamical system $\mathbf{x}_{k+1} = A\mathbf{x}_k$. Calculate $\mathbf{x}_{10}$ for an arbitrary initial vector $\mathbf{x}_1$.

5. (*Maple V*) Show that for a random integer a between 100 and 1000, **fibonacci(a+2)** is the sum of **fibonacci(a+1)** and **fibonacci(a)**.

6. (*Maple V*) Explain why the function **seq(fibonacci(i), i=0..n)** generates the first $n+1$ terms of the sequence of Fibonacci numbers.

7. (*Maple V*) Show that the coordinates of the vector x_{101} of dynamical system $\mathbf{x}_{i+1} = A\mathbf{x}_i$ determined by

$$A = \begin{bmatrix} 1 & 1 \\ 1 & 0 \end{bmatrix} \quad \text{and} \quad \mathbf{x} = \begin{bmatrix} 1 \\ 1 \end{bmatrix}$$

are **fibonacci(102)** and **fibonacci(101)**.

8. (*Maple 6*) Build discrete dynamical systems $\mathbf{x}_{k+1} = A\mathbf{x}_k$ with the 4×4 diagonal matrices determined by the following eigenvalues

a. $\lambda_1 = 1; \lambda_2 = 2; \lambda_3 = 3; \lambda_4 = 4$ b. $\lambda_1 = 3; \lambda_2 = 3; \lambda_3 = -3; \lambda_4 = 7$

c. $\lambda_1 = 0; \lambda_2 = 0; \lambda_3 = 4; \lambda_4 = 5$ d. $\lambda_1 = 5; \lambda_2 = 5; \lambda_3 = 5; \lambda_4 = 5$

and calculate the vector $\mathbf{x}_{100}$ for the initial vector $\mathbf{x} = (1, 2, 3, 4)$.

Markov Chains

Frequently a discrete dynamical system is a sequence of probability vectors determined by a stochastic matrix. A ***probability vector*** is a vector $\mathbf{u} = (u_1, \ldots, u_n) \in \mathbb{R}^n$ with the property that $0 \leq u_1, \ldots, u_n$ and $u_1 + \cdots + u_n = 1$, and a ***stochastic matrix*** is an $n \times n$ matrix

$A = [\mathbf{u}_1 \cdots \mathbf{u}_n]$ whose columns are probability vectors. The matrix A is called the ***transition matrix*** of the chain. It is proved in probability theory that if $\mathbf{u}$ is a probability vector and A is a transition matrix, then $A^n\mathbf{x}$ is a probability vector for all n. A stochastic matrix A is ***regular*** if there exists an $n \in \mathbb{N}$ for which all entries of A^n are nonzero.

EXAMPLE 6.34 Probability Vectors

Show that the vectors $\mathbf{x}_1 = (0, 1, 0)$, $\mathbf{x}_2 = (\frac{1}{4}, \frac{1}{4}, \frac{1}{4}, \frac{1}{4})$, $\mathbf{x}_3 = (\frac{1}{3}, 0, \frac{2}{3})$, and $\mathbf{x}_4 = (\frac{1}{4}, \frac{3}{8}, \frac{1}{8}, \frac{1}{4})$ are probability vectors.

Solution. Since the entries of all vectors are nonnegative, and since the sums of their entries are 1, each vector is a probability vector. ▲

EXAMPLE 6.35 Stochastic Matrices

Show that the matrices

$$A = \begin{bmatrix} 1 & 0 & 0 \\ 0 & 1 & 0 \\ 0 & 0 & 1 \end{bmatrix}, B = \begin{bmatrix} \frac{1}{3} & 0 & 0 \\ \frac{1}{3} & \frac{1}{2} & 0 \\ \frac{1}{3} & \frac{1}{2} & 1 \end{bmatrix}, C = \begin{bmatrix} \frac{1}{4} & \frac{1}{3} & \frac{1}{3} \\ \frac{1}{2} & \frac{1}{3} & \frac{1}{3} \\ \frac{1}{4} & \frac{1}{3} & \frac{1}{3} \end{bmatrix}, D = \begin{bmatrix} \frac{1}{2} & \frac{1}{2} & \frac{1}{3} \\ \frac{1}{2} & 0 & \frac{1}{3} \\ 0 & \frac{1}{2} & \frac{1}{3} \end{bmatrix}$$

are stochastic.

Solution. Since each of the four matrices is square and since the columns of all four matrices are probability vectors, the matrices are stochastic. ▲

EXAMPLE 6.36 Regular Stochastic Matrices

Determine which of the stochastic matrices in Example 6.35 are regular.

Solution. Since the powers of a diagonal matrix are diagonal and therefore contain non-diagonal zero entries, the identity matrix A_1 is not regular. Moreover, the product of two lower-triangular matrices is lower triangular. Hence the powers of A_2 contain zero entries above the diagonal. The matrix A_2 is therefore not regular. The matrix A_3 is regular since it contains no zero entries. A direct calculation shows that

$$A_4^2 = \begin{bmatrix} \frac{1}{2} & \frac{1}{2} & \frac{1}{3} \\ \frac{1}{2} & 0 & \frac{1}{3} \\ 0 & \frac{1}{2} & \frac{1}{3} \end{bmatrix} \begin{bmatrix} \frac{1}{2} & \frac{1}{2} & \frac{1}{3} \\ \frac{1}{2} & 0 & \frac{1}{3} \\ 0 & \frac{1}{2} & \frac{1}{3} \end{bmatrix} = \begin{bmatrix} \frac{1}{2} & \frac{5}{12} & \frac{4}{9} \\ \frac{1}{4} & \frac{5}{12} & \frac{5}{18} \\ \frac{1}{4} & \frac{1}{6} & \frac{5}{18} \end{bmatrix}$$

Hence A_4 is regular. ▲

A ***steady-state vector*** of a stochastic matrix A is a probability vector $\mathbf{s}$ for which $A\mathbf{s} = \mathbf{s}$. It can be shown that every stochastic matrix has a steady-state vector.

EXAMPLE 6.37 Steady-State Vectors

Find steady-state vectors for the stochastic matrices in Example 6.35.

Solution. For each of the matrices in Example 6.35 we find a stochastic eigenvector associated with the eigenvalue 1.

1. The vector $\mathbf{s} = (x, y, z) = (0, 0, 1)$ is a steady-state vector of the identity matrix A since

$$A\mathbf{s} = \begin{bmatrix} 1 & 0 & 0 \\ 0 & 1 & 0 \\ 0 & 0 & 1 \end{bmatrix} \begin{bmatrix} x \\ y \\ z \end{bmatrix} = \begin{bmatrix} x \\ y \\ z \end{bmatrix}$$

for all (x, y, z) and since $\mathbf{s}$ is stochastic if $0 \leq x, y, z$ and $x + y + z = 1$.

2. The vector $\mathbf{s} = (x, y, z) = (0, 0, 1)$ is also a steady-state vector of the matrix B since

$$B\mathbf{s} = \begin{bmatrix} \frac{1}{3} & 0 & 0 \\ \frac{1}{3} & \frac{1}{2} & 0 \\ \frac{1}{3} & \frac{1}{2} & 1 \end{bmatrix} \begin{bmatrix} x \\ y \\ z \end{bmatrix} = \begin{bmatrix} \frac{1}{3}x \\ \frac{1}{3}x + \frac{1}{2}y \\ \frac{1}{3}x + \frac{1}{2}y + z \end{bmatrix} = \begin{bmatrix} x \\ y \\ z \end{bmatrix}$$

implies that $x = y = 0$ and $z = 1$.

3. The vector $\mathbf{s} = (x, y, z) = \left(\frac{4}{13}, \frac{5}{13}, \frac{4}{13}\right)$ is a steady-state vector of the matrix C since the equation

$$C\mathbf{s} = \begin{bmatrix} \frac{1}{4} & \frac{1}{3} & \frac{1}{3} \\ \frac{1}{2} & \frac{1}{3} & \frac{1}{3} \\ \frac{1}{4} & \frac{1}{3} & \frac{1}{3} \end{bmatrix} \begin{bmatrix} x \\ y \\ z \end{bmatrix} = \begin{bmatrix} \frac{1}{4}x + \frac{1}{3}y + \frac{1}{3}z \\ \frac{1}{2}x + \frac{1}{3}y + \frac{1}{3}z \\ \frac{1}{4}x + \frac{1}{3}y + \frac{1}{3}z \end{bmatrix} = \begin{bmatrix} x \\ y \\ z \end{bmatrix}$$

holds if $z = x$ and $y = \frac{5}{4}x$ and since $\mathbf{s}$ is stochastic if $0 \leq x, y, z$ and

$$x + y + z = x + 5/4x + x = \tfrac{13}{4}x = 1$$

4. The vector $\mathbf{s} = (x, y, z) = \left(\frac{6}{13}, \frac{4}{13}, \frac{3}{13}\right)$ is a steady-state vector of the matrix D since the equation

$$D\mathbf{s} = \begin{bmatrix} \frac{1}{2} & \frac{1}{2} & \frac{1}{3} \\ \frac{1}{2} & 0 & \frac{1}{3} \\ 0 & \frac{1}{2} & \frac{1}{3} \end{bmatrix} \begin{bmatrix} x \\ y \\ z \end{bmatrix} = \begin{bmatrix} \frac{1}{2}x + \frac{1}{2}y + \frac{1}{3}z \\ \frac{1}{2}x + \frac{1}{3}z \\ \frac{1}{2}y + \frac{1}{3}z \end{bmatrix} = \begin{bmatrix} x \\ y \\ z \end{bmatrix}$$

holds if $x = 2z$ and $y = \frac{4}{3}z$ and since $\mathbf{s}$ is stochastic if $0 \leq x, y, z$ and

$$x + y + z = 2z + \frac{4}{3}z + z = \tfrac{13}{3}z = 1$$

We have found at least one steady-state vector for each of the matrices A, B, C, and D. ▲

DEFINITION 6.10 *A **Markov chain** is a sequence of probability vectors*

$$\mathbf{x}_1, A\mathbf{x}_1, A^2\mathbf{x}_1 \ldots, A^n\mathbf{x}_1, \ldots$$

determined by a stochastic matrix A and an initial probability vector $\mathbf{x}_1$.

A steady-state vector of the matrix A is a steady-state vector of the chain. A Markov chain is ***regular*** if there exists an n for which all entries of the matrix power A^n are nonzero. The next theorem explains why regularity is important.

THEOREM 6.13 *If A is a regular stochastic matrix, then the powers of A approach a stochastic matrix $P = [\mathbf{x} \cdots \mathbf{x}]$ with identical columns and positive entries.* ■

This theorem has the following important corollary.

COROLLARY 6.14 *If $\mathbf{x}_0$ is any initial probability vector and $\mathbf{x}_0, A\mathbf{x}_0, \ldots, A^n\mathbf{x}_0, \ldots$ is a regular Markov chain for which the powers of A approach the stochastic matrix $P = [\mathbf{x} \cdots \mathbf{x}]$, then $A\mathbf{x} = \mathbf{x}$ and $\lim_{n\to\infty} (A^n\mathbf{x}_0) = \mathbf{x}$.* ■

Many practical problems can be solved using regular Markov chains. One of the basic examples of a Markov chain deals with population movements.

EXAMPLE 6.38 Markov Chains

Assume that the population of an island is divided into two distinct groups, city dwellers and country dwellers. Every year, 10% of the city dwellers move to the country and 5% of the country dwellers move to the city. If the total population in the year 2000 is 50,000, with 30,000 living in the city and 20,000 living in the country, how many people will live in the city and how many will live in the country between the years 2000 and 2003?

Solution. To solve this problem, we note that the percentage changes in the two populations can be expressed by the linear system

$$\begin{cases} C = .90x + .05y \\ S = .10x + .95y \end{cases}$$

where C is the percentage of the population living in the city and S is the percentage of population living in the country at the end of one year. The variables x and y range over the city and country population percentages at the beginning of a year. Since initially 60% of the population live in the city and 40% live in the country, the initial probability vector of the chain is $\mathbf{x}_1 = (.6, .4)$ and we have the Markov chain $\mathbf{x}_1, A\mathbf{x}_1, A^2\mathbf{x}_1, A^3\mathbf{x}_1, \ldots$, determined by the matrix

$$A = \begin{bmatrix} .90 & .05 \\ .10 & .95 \end{bmatrix}$$

The next three vectors of the chain,

$$A\mathbf{x}_1 = \begin{bmatrix} .56 \\ .44 \end{bmatrix}, A^2\mathbf{x}_1 = \begin{bmatrix} .526 \\ .474 \end{bmatrix}, A^3\mathbf{x}_1 = \begin{bmatrix} .497\,1 \\ .502\,9 \end{bmatrix}$$

TABLE 5 POPULATION DISTRIBUTION.

	2000	2001	2002	2003
City	30	28	26.3	24.855
Country	20	22	23.7	25.145

determine the population distributions of the island at the beginning of 2001, 2002, and 2003, respectively. Table 5 lists the actual distributions, measured in thousands.

Since the matrix A is regular, we can also calculate the long-term population distribution of the island. We first calculate the steady-state vector **s** of A.

```
[> A:=matrix([[.90,.05],[.10,.95]]):
[> linalg[eigenvectors](A);
```

$$[1.000000000, 1, \{[-.4714045208, -.9428090416]\}],$$
$$[.8500000000, 1, \{[-.7071067812, .7071067812]\}]$$

Therefore,

$$\mathbf{q} = \begin{bmatrix} -.4714045208 \\ -.9428090416 \end{bmatrix}$$

is an eigenvector belonging to the eigenvalue 1. But it is obviously not a probability vector. The desired steady-state vector **s** will be a scalar multiple of $a\mathbf{q}$. An easy calculation shows that if

$$a\begin{bmatrix} -.4714045208 \\ -.9428090416 \end{bmatrix} = \begin{bmatrix} p \\ 1-p \end{bmatrix}$$

then $a = -0.707\,11$ and $p = 0.333\,33$. Therefore

$$a\mathbf{q} = (-.707\,11)\begin{bmatrix} -.4714045208 \\ -.9428090416 \end{bmatrix} = \begin{bmatrix} .333\,34 \\ .666\,67 \end{bmatrix} = \mathbf{s}$$

is the required probability vector **s**. It follows that in the long run, one-third of the island's population will live in the city and two-thirds will live in the country.

Let us calculate the population percentages of the island for the year 2103 and estimate how close to its final state it will have come at that time.

```
[> evalm((A^103)&*[.6,.4]);
```

$$[.3333333474, .6666666518]$$

Therefore, $\mathbf{x}_{104} = (.333333, .666667)$. As we can see, the population of the island will have essentially reached its final distribution at the beginning of the year 2103. ▲

EXERCISES 6.11

1. Prove that for all probabilities p and q, the vector (q, p) is a steady-state vector of the matrix

$$A = \begin{bmatrix} 1-p & q \\ p & 1-q \end{bmatrix}$$

2. Determine which of the following matrices are stochastic.

$$\text{a.} \begin{bmatrix} .3 & .8 \\ .7 & .2 \end{bmatrix} \qquad \text{b.} \begin{bmatrix} 0 & 1 \\ 1 & 0 \end{bmatrix}$$

$$\text{c.} \begin{bmatrix} .1 & .5 & 0 \\ .1 & .2 & 1 \\ .8 & .3 & 0 \end{bmatrix} \qquad \text{d.} \begin{bmatrix} .3 & .5 & .2 \\ .3 & .2 & .2 \\ .4 & .3 & .5 \end{bmatrix}$$

3. A student has the following study habits. If he studies one evening, he is 60% certain of not studying the next evening. Moreover, the probability that he studies two nights in a row is also 60%. How often does he study in the long run?

4. Suppose that a businessman manages three branches of a large corporation. The matrix

$$A = \begin{bmatrix} 0 & \frac{1}{2} & \frac{1}{2} \\ 1 & 0 & \frac{1}{2} \\ 0 & \frac{1}{2} & 0 \end{bmatrix}$$

describes the probabilities of his visiting patterns. Describe these patterns in words, and calculate how often he visits each branch in the long run.

5. Suppose that a large circle is drawn on the floor and that five points are marked off on the circle in counterclockwise order. Imagine that a girl performs a random walk on the circle by moving from point to point. Let p be the probability that she moves clockwise from one point to the next and $q = 1 - p$ be the probability that she moves counterclockwise. Calculate the transition matrix of this Markov chain.

6. (*Maple V*) Use the **rand** function to find two probabilities p and q and form the matrix

$$A = \begin{bmatrix} 1-p & q \\ p & 1-q \end{bmatrix}$$

Show that the scalar 1 is an eigenvalue of A and find a steady-state vector of A.

7. (*Maple V*) Find the steady-state vectors of the following stochastic matrices.

a. $\begin{bmatrix} .5 & .3 \\ .5 & .7 \end{bmatrix}$ b. $\begin{bmatrix} .4 & .6 \\ .6 & .4 \end{bmatrix}$ c. $\begin{bmatrix} .1 & .5 & .2 \\ .1 & .2 & .5 \\ .8 & .3 & .3 \end{bmatrix}$

8. (*Maple V*) Show that the scalar 1 is an eigenvalue of the stochastic matrix

$$A = \begin{bmatrix} \frac{1}{4} & \frac{1}{3} & \frac{1}{3} \\ \frac{1}{2} & \frac{1}{3} & \frac{1}{3} \\ \frac{1}{4} & \frac{1}{3} & \frac{1}{3} \end{bmatrix}$$

and find a steady-state vector for A.

9. (*Maple V*) Show that the scalar 1 is an eigenvalue of the stochastic matrix

$$A = \begin{bmatrix} \frac{1}{2} & \frac{1}{2} & \frac{1}{3} \\ \frac{1}{2} & 0 & \frac{1}{3} \\ 0 & \frac{1}{2} & \frac{1}{3} \end{bmatrix}$$

and find a steady-state vector for A.

10. (*Maple V*) A businesswoman is the manager of branches A, B, and C of a large corporation. She never visits the same branch on consecutive days. If she visits branch A one day, she visits branch B the next day. If she visits either branch B or C that day, then the next day she is twice as likely to visit branch A as to visit branch B or C. How often does she visit each branch in the long run?

11. (*Maple V*) Show that the following stochastic matrices are regular.

a. $\begin{bmatrix} \frac{1}{10} & \frac{3}{10} & \frac{7}{10} & 0 \\ \frac{1}{10} & \frac{2}{10} & 0 & \frac{8}{10} \\ 0 & \frac{5}{10} & 0 & \frac{2}{10} \\ \frac{8}{10} & 0 & \frac{3}{10} & 0 \end{bmatrix}$ b. $\begin{bmatrix} 0 & 1 & \frac{4}{10} & \frac{2}{10} \\ 1 & 0 & \frac{2}{10} & \frac{5}{10} \\ 1 & 0 & \frac{2}{10} & \frac{1}{10} \\ 0 & 0 & \frac{2}{10} & \frac{3}{10} \end{bmatrix}$

c. $\begin{bmatrix} \frac{1}{10} & \frac{5}{10} & 0 \\ \frac{1}{10} & 0 & 1 \\ \frac{8}{10} & \frac{5}{10} & 0 \end{bmatrix}$ d. $\begin{bmatrix} \frac{3}{10} & \frac{5}{10} & \frac{2}{10} \\ \frac{3}{10} & \frac{2}{10} & \frac{2}{10} \\ \frac{4}{10} & \frac{3}{10} & \frac{6}{10} \end{bmatrix}$

12. (*Maple V*) A car rental company in Montreal has three locations: Dorval Airport, Old Montreal, and the Dominion Square. Of the cars rented at Dorval Airport, 60% are returned to Dorval Airport, 10% are returned to Old Montreal, and 30% are returned to Dominion Square. Of the cars rented in Old Montreal, 30% are returned to Old Montreal,

50% are returned to Dorval Airport, and 20% are returned to Dominion Square. Of the cars rented in Dominion Square, 30% are returned to Dominion Square, 10% are returned to Old Montreal, and 60% are returned to Dorval Airport. Construct the transition matrix for this Markov chain, and determine the distribution of the cars among the three locations in the long run.

REVIEW

KEY CONCEPTS ▶ Define and discuss each of the following.

Eigenspaces

Eigenvector basis.

Eigenvalues

Algebraic multiplicity, geometric multiplicity.

Eigenvectors

Eigenvector belonging to an eigenvalue.

Linear Transformations

Diagonalizable transformation, Householder transformation.

Matrices

Characteristic matrix, diagonalizable matrix, Hessenberg matrix.

Polynomials

Characteristic polynomial, root of a characteristic polynomial.

KEY FACTS ▶ Explain and illustrate each of the following.

1. If $T : V \to V$ is a linear transformation on a finite-dimensional vector space V and $A = [T]_{\mathcal{B}}^{\mathcal{B}}$ is the matrix of T in some basis $\mathcal{B}$, then $T(\mathbf{x}) = \lambda\mathbf{x}$ if and only if $A[\mathbf{x}]_{\mathcal{B}} = \lambda[\mathbf{x}]_{\mathcal{B}}$.
2. For any scalar λ and any real square matrix A, the following are equivalent: (a) The scalar λ is an eigenvalue of A. (b) The matrix $(A - \lambda I)$ is singular. (c) The scalar λ is a root of the characteristic polynomial $c_A(t)$ of A.
3. Suppose that $T : V \to V$ is a linear transformation on a real finite-dimensional vector space V and that $c_A(t)$ is the characteristic polynomial of the matrix $A = [T]_{\mathcal{B}}^{\mathcal{B}}$. Then every eigenvalue λ of T is a real root $c_A(t)$.
4. If A and B are similar matrices, then $c_A(t) = c_B(t)$.
5. Every matrix satisfies its characteristic polynomial.
6. For every eigenvalue λ of a linear transformation T on a real vector space V, the set $E_\lambda = \{\mathbf{x} : T(\mathbf{x}) = \lambda\mathbf{x}\}$ is a subspace of V.
7. The algebraic multiplicity of an eigenvalue is always greater than or equal to its geometric multiplicity.
8. A real square matrix A is invertible if and only if the number 0 is not an eigenvalue of A.

9. If λ is an eigenvalue of an invertible linear transformation T on a real vector space V, then λ^{-1} is an eigenvalue of T^{-1}.

10. Eigenvectors belonging to distinct eigenvalues of a linear transformation $T : V \to V$ are linearly independent.

11. An $n \times n$ matrix A is similar to a diagonal matrix D if and only if A has n linearly independent eigenvectors.

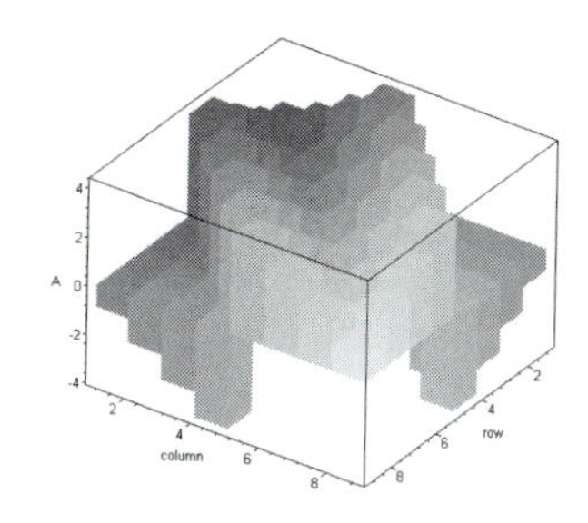

7

NORMS AND INNER PRODUCTS

We now lay the foundation for doing geometry in vector spaces. The underlying spaces will be the coordinate spaces. Our understanding of Euclidean geometry in $\mathbb{R}^2$ and $\mathbb{R}^3$ motivates most ideas developed in this chapter.

The three basic geometric concepts needed in our work are length, distance, and angle. Let us recall how these concepts are defined in ordinary geometry.

1. **Length.** Let $\mathbf{x} \in \mathbb{R}^2$ be the vector determined by the line segment from the origin $(0, 0)$ to the point (x_1, x_2). The length of $\mathbf{x}$ is measured by the Pythagorean formula

$$\|\mathbf{x}\| = \sqrt{x_1^2 + x_2^2}$$

Since the length of a vector is always a nonnegative real number, the assignment $\mathbf{x} \mapsto \|\mathbf{x}\|$ is a function from $\mathbb{R}^2$ to the set $[0, \infty) = \{x \in \mathbb{R} : 0 \leq x\}$ of nonnegative real numbers.

2. **Distance.** If $\mathbf{y} \in \mathbb{R}^2$ is a second vector determined by the line segment from $(0, 0)$ to the point (y_1, y_2), then the distance between $\mathbf{x}$ and $\mathbf{y}$ is measured by the formula

$$d(\mathbf{x}, \mathbf{y}) = \|\mathbf{x} - \mathbf{y}\| = \sqrt{(x_1 - y_1)^2 + (x_2 - y_2)^2}$$

The measure $d(\mathbf{x}, \mathbf{y}) = \|\mathbf{x} - \mathbf{y}\|$ is a function from $\mathbb{R}^2 \times \mathbb{R}^2$ to $[0, \infty)$.

3. **Angles.** Angles are defined indirectly through their cosines. The cosine of the angle θ between $\mathbf{x}$ and $\mathbf{y}$ is given by the formula

$$\cos\theta = \frac{\mathbf{x}^T\mathbf{y}}{\|\mathbf{x}\| \, \|\mathbf{y}\|}$$

The assignment $(\mathbf{x}, \mathbf{y}) \mapsto \cos\theta$ is a function from $\mathbb{R}^2 \times \mathbb{R}^2$ to the real numbers between -1 and 1.

Analogous formulas are used to measure length, distance, and angles in $\mathbb{R}^3$.

1. The length of a vector $\mathbf{x}$ from the origin $(0, 0, 0)$ to the point (x_1, x_2, x_3) is measured by the formula

$$\|\mathbf{x}\| = \sqrt{x_1^2 + x_2^2 + x_3^2}$$

2. The distance between $\mathbf{x}$ and a second vector $\mathbf{y}$ from (x_1, x_2, x_3) to the point (y_1, y_2, y_3) is measured by the formula

$$\|\mathbf{x} - \mathbf{y}\| = \sqrt{(x_1 - y_1)^2 + (x_2 - y_2)^2 + (x_3 - y_3)^2}$$

3. The cosine of the angle θ between $\mathbf{x}$ and $\mathbf{y}$ is measured by the formula

$$\cos\theta = \frac{\mathbf{x}^T\mathbf{y}}{\|\mathbf{x}\|\,\|\mathbf{y}\|}$$

The scalar $\mathbf{x}^T\mathbf{y}$ in the cosine formulas is of course the familiar ***dot product*** $\mathbf{x} \cdot \mathbf{y}$ of $\mathbf{x}$ and $\mathbf{y}$. One of the objectives of this chapter is to generalize the dot product and thereby expand the applicability of geometric ideas.

A remarkable fact about the dot product is that length, distance, and angles can all be defined in terms of it. Consider two points $\mathbf{x} = (x_1, x_2)$ and $\mathbf{y} = (y_1, y_2)$ in $\mathbb{R}^2$. The products $\mathbf{x} \cdot \mathbf{x}$ and $\mathbf{y} \cdot \mathbf{y}$ are equal to $x_1x_1 + x_2x_2 = x_1^2 + x_2^2$ and $y_1y_1 + y_2y_2 = y_1^2 + y_2^2$. This means that the length of $\mathbf{x}$ is $\sqrt{\mathbf{x} \cdot \mathbf{x}}$ and that of $\mathbf{y}$ is $\sqrt{\mathbf{y} \cdot \mathbf{y}}$. The cosine

$$\frac{\mathbf{x} \cdot \mathbf{y}}{\|\mathbf{x}\|\,\|\mathbf{y}\|} = \frac{\mathbf{x} \cdot \mathbf{y}}{\sqrt{\mathbf{x} \cdot \mathbf{x}}\sqrt{\mathbf{y} \cdot \mathbf{y}}}$$

of the angle between $\mathbf{x}$ and $\mathbf{y}$ is therefore completely determined by the dot product. Moreover, the distance $d(\mathbf{x}, \mathbf{y})$ between $\mathbf{x}$ and $\mathbf{y}$ is simply the square root of the dot product

$$(\mathbf{x} - \mathbf{y}) \cdot (\mathbf{x} - \mathbf{y})$$

EUCLIDEAN NORMS AND DOT PRODUCTS

We begin our discussion of geometry in vector spaces with the Euclidean case. The definitions of the basic geometric concepts familiar from $\mathbb{R}^2$ and $\mathbb{R}^3$ discussed in the introduction to this chapter can extended to $\mathbb{R}^n$ for any $n > 3$. The result is called a ***Euclidean space***. Moreover, the concepts of length and the dot product operation can be generalized to add other geometric structures to vector spaces. In this broader context, the length of a vector is called a ***norm***, and the functions that generalize (and include) the dot products are called ***inner products***. A norm built from an inner product is then called an ***inner product norm***. As we will see, not all norms are inner product norms. A real vector space equipped with an inner product is called a ***real inner product space***. We sometimes omit the word *real* since it is usually clear from the context. The ideas presented in this chapter apply both to the coordinate spaces and to matrix spaces. We therefore distinguish further between ***vector norms*** and ***matrix norms***. A matrix norm is a vector norm with an additional property that relates the norm of a matrix product AB to the norms of the matrices A and B.

The Euclidean Vector Norm

DEFINITION 7.1 *For any vector $\mathbf{x} = (x_1, \ldots, x_n)$ in $\mathbb{R}^n$, the* ***Euclidean norm*** *$\|\mathbf{x}\|_2$ is the square root $(\mathbf{x}^T\mathbf{x})^{1/2}$ of the dot product $\mathbf{x}^T\mathbf{x} = x_1^2 + \cdots + x_n^2$ of $\mathbf{x}$ with itself.*

The Euclidean norm is also called the ***two-norm*** since it is a special case of the p-norms, defined by $\|\mathbf{x}\|_p = (\mathbf{x}^T\mathbf{x})^{1/p}$ for suitable values of p. If it is clear from the context that we are talking about the two-norm of a vector, we often simply refer to it as the *norm* of the vector. The basic properties of a vector norm, as a function from $\mathbb{R}^n$ to $\mathbb{R}$, are summarized in the following definition.

DEFINITION 7.2 *A* ***vector norm*** *is a function $\mathbf{x} \mapsto \|\mathbf{x}\|$ on a vector space V that assigns a nonnegative real number $\|\mathbf{x}\|$ to every vector $\mathbf{x} \in V$ and has the following properties.*

1. $\|\mathbf{x}\| > 0$ *if* $\mathbf{x} \neq \mathbf{0}$
2. $\|\mathbf{x}\| = 0$ *if* $\mathbf{x} = \mathbf{0}$
3. $\|a\mathbf{x}\| = |a| \, \|\mathbf{x}\|$
4. $\|\mathbf{x} + \mathbf{y}\| \leq \|\mathbf{x}\| + \|\mathbf{y}\|$

The inequality $\|\mathbf{x} + \mathbf{y}\| \leq \|\mathbf{x}\| + \|\mathbf{y}\|$ is known as the ***triangle inequality.***

THEOREM 7.1 *The two-norm on $\mathbb{R}^n$ is a vector norm.*

Proof. Let $\mathbf{x} = (x_1, \ldots, x_n)$ be a vector in $\mathbb{R}^n$ and let $\mathbf{x}^T\mathbf{x} = x_1^2 + \cdots + x_n^2$ be the dot product of $\mathbf{x}$ with itself. The first two axioms of a vector norm are satisfied since we are taking the nonnegative square root of the sum $(x_1^2 + \cdots + x_n^2)$ of squares, and since $x_1^2 + \cdots + x_n^2 = 0$ if and only if all coordinates x_i are zero. Moreover, since $a\mathbf{x} = (ax_1, \ldots, ax_n)$, we have

$$\begin{aligned} \|a\mathbf{x}\|_2 &= \sqrt{(ax_1)^2 + \cdots + (ax_n)^2} \\ &= \sqrt{a^2x_1^2 + \cdots + a^2x_n^2} \\ &= |a| \sqrt{x_1^2 + \cdots + x_n^2} \\ &= |a| \, \|\mathbf{x}\|_2 \end{aligned}$$

Hence the third axiom is satisfied. Theorem 7.9 guarantees that the triangle inequality also holds. ■

The coordinate space $\mathbb{R}^n$ equipped with the Euclidean norm is called the ***Euclidean n-space.*** A ***normed vector space*** is a vector space equipped with a vector norm.

The Euclidean distance

We can use the two-norm on $\mathbb{R}^n$ to define the distance between two vectors.

DEFINITION 7.3 *For any two vectors* $\mathbf{x} = (x_1, \ldots, x_n)$ *and* $\mathbf{y} = (y_1, \ldots, y_n)$ *in* $\mathbb{R}^n$, *the real number*

$$d(\mathbf{x}, \mathbf{y}) = \|\mathbf{x} - \mathbf{y}\|_2 = \sqrt{(x_1 - y_1)^2 + \cdots + (x_n - y_n)^2}$$

is the ***Euclidean distance*** *between* $\mathbf{x}$ *and* $\mathbf{y}$.

The Euclidean distance is a special case of a distance function, also known as a ***metric***.

DEFINITION 7.4 *A* ***distance function*** *on a vector space* V *is a function that assigns a nonnegative real number* $d(\mathbf{x}, \mathbf{y})$ *to every pair* $(\mathbf{x}, \mathbf{y})$ *of vectors in* V *and has the following properties.*

1. $d(\mathbf{x}, \mathbf{y}) \geq 0$
2. $d(\mathbf{x}, \mathbf{y}) = d(\mathbf{y}, \mathbf{x})$
3. $d(\mathbf{x}, \mathbf{y}) \leq d(\mathbf{x}, \mathbf{z}) + d(\mathbf{z}, \mathbf{y})$
4. $d(\mathbf{x}, \mathbf{y}) = 0$ *if and only if* $\mathbf{x} = \mathbf{y}$

THEOREM 7.2 *The Euclidean distance is a distance function on* $\mathbb{R}^n$.

Proof. Since the Euclidean distance $\|\mathbf{x} - \mathbf{y}\|_2$ is defined by a norm, it is nonnegative and therefore satisfies the first axiom. Moreover,

$$\begin{aligned}\|\mathbf{x} - \mathbf{y}\|_2 &= \sqrt{(x_1 - y_1)^2 + \cdots + (x_n - y_n)^2} \\ &= \sqrt{(y_1 - x_1)^2 + \cdots + (y_n - x_n)^2} \\ &= \|\mathbf{y} - \mathbf{x}\|_2\end{aligned}$$

Hence the Euclidean distance also satisfies the second axiom. It is also clear that

$$\sqrt{(x_1 - y_1)^2 + \cdots + (x_n - y_n)^2} = 0$$

if and only if $x_i = y_i$ for all i. Hence $\|\mathbf{x} - \mathbf{y}\|_2$ also satisfies the fourth axiom. It therefore remains to verify the triangle inequality. Since $\|-\|_2$ is a norm, it satisfies the triangle inequality. Hence

$$\begin{aligned}d(\mathbf{x}, \mathbf{y}) &= \|\mathbf{x} - \mathbf{y}\|_2 \\ &= \|\mathbf{x} - \mathbf{z} + \mathbf{z} - \mathbf{y}\|_2 \\ &\leq \|\mathbf{x} - \mathbf{z}\|_2 + \|\mathbf{z} - \mathbf{y}\|_2 \\ &= d(\mathbf{x}, \mathbf{z}) + d(\mathbf{z}, \mathbf{y})\end{aligned}$$

This shows that $d(\mathbf{x}, \mathbf{y})$ is a distance function. ■

EXAMPLE 7.1 The Euclidean Distance in $\mathbb{R}^2$

Use *Maple* to find the Euclidean distance between the vectors $\mathbf{x} = (1, 2)$ and $\mathbf{y} = (3, 4)$.

Solution. By definition, $d(\mathbf{x}, \mathbf{y}) = \|\mathbf{x} - \mathbf{y}\| = \sqrt{(1-3)^2 + (2-4)^2}$.

```
[> x:=vector(2,[1,2]):  y:=vector(2,[3,4]):
[> d:=(x,y)->sqrt((x[1]-y[1])^2+(x[2]-y[2])^2):
[> d(x,y);
```

$$2\sqrt{2}$$

Thus the Euclidean distance between $\mathbf{x}$ and $\mathbf{y}$ is $2\sqrt{2}$. ▲

EXERCISES 7.1

1. Calculate the Euclidean norm of the following vectors in $\mathbb{R}^2$.

   a. $\mathbf{x} = (3, 4)$ b. $\mathbf{x} = (0, 0)$ c. $\mathbf{x} = (-3, 4)$
   d. $\mathbf{x} = (1, 0)$ e. $\mathbf{x} = (-9, 11)$ f. $\mathbf{x} = (-9, -11)$

2. Show that the function

$$d(\mathbf{x}, \mathbf{y}) = \begin{cases} 1 \text{ if } \mathbf{x} \neq \mathbf{y} \\ 0 \text{ if } \mathbf{x} = \mathbf{y} \end{cases}$$

   satisfies the axioms of a distance function.

3. Calculate the Euclidean distance between following pairs of vectors in $\mathbb{R}^2$.

   a. $\mathbf{x} = (4, 5)$ and $\mathbf{y} = (1, 2)$ b. $\mathbf{x} = (4, 5)$ and $\mathbf{y} = (-1, 2)$
   c. $\mathbf{x} = (4, 5)$ and $\mathbf{y} = (0, -2)$ d. $\mathbf{x} = (3, 0)$ and $\mathbf{y} = (4, 5)$
   e. $\mathbf{x} = (1, 2)$ and $\mathbf{y} = (0, 0)$ f. $\mathbf{x} = (1, 0)$ and $\mathbf{y} = (0, 1)$

4. Show that the attempt to use the distance function $d(\mathbf{x}, \mathbf{y})$ in Exercise 3 to define a vector norm by putting $\|\mathbf{x}\| = d(\mathbf{x}, \mathbf{0})$ fails.

5. (*Maple V*) Calculate the two-norm of the following vectors in $\mathbb{R}^3$.

   a. $\mathbf{x} = (3, 4, 5)$ b. $\mathbf{x} = (0, 0, 0)$ c. $\mathbf{x} = (-3, 4, 5)$
   d. $\mathbf{x} = (1, 0, 1)$ e. $\mathbf{x} = (0, 0, 1)$ f. $\mathbf{x} = (-9, -11, -5)$

6. (*Maple 6*) Calculate the Euclidean distance between the following pairs of vectors in $\mathbb{R}^3$.

   a. $\mathbf{x} = (1, 2, 3)$ and $\mathbf{y} = (4, 5, 7)$ b. $\mathbf{x} = (4, 5, 7)$ and $\mathbf{y} = (-1, 2, 0)$
   c. $\mathbf{x} = (4, 5, 7)$ and $\mathbf{y} = (0, -2, -3)$ d. $\mathbf{x} = (1, 0, 0)$ and $\mathbf{y} = (4, 5, 7)$
   e. $\mathbf{x} = (1, 2, 3)$ and $\mathbf{y} = (0, 0, 0)$ f. $\mathbf{x} = (1, 0, 0)$ and $\mathbf{y} = (0, 0, 1)$

The Frobenius Matrix Norm

We now extend the idea of a norm on the coordinate space $\mathbb{R}^n$ to a norm on the matrix space $\mathbb{R}^{m \times n}$. The resulting function will be a special case of a ***matrix norm***. It is known as the ***Frobenius norm***.

DEFINITION 7.5 *The **Frobenius norm** $\|A\|_F$ of any $m \times n$ real matrix A is the square root of the trace of the matrix $A^T A$.*

EXAMPLE 7.2 The Frobenius Norm of a General Matrix

Use the definition to calculate the Frobenius norm of a general 2×3 matrix.

Solution. Let

$$A = \begin{bmatrix} a_{11} & a_{12} & a_{13} \\ a_{21} & a_{22} & a_{23} \end{bmatrix}$$

be a general 2×3 matrix. Then

$$\begin{aligned} A^T A &= \begin{bmatrix} a_{11} & a_{12} & a_{13} \\ a_{21} & a_{22} & a_{23} \end{bmatrix}^T \begin{bmatrix} a_{11} & a_{12} & a_{13} \\ a_{21} & a_{22} & a_{23} \end{bmatrix} \\ &= \begin{bmatrix} a_{11}^2 + a_{21}^2 & a_{11}a_{12} + a_{21}a_{22} & a_{11}a_{13} + a_{21}a_{23} \\ a_{11}a_{12} + a_{21}a_{22} & a_{12}^2 + a_{22}^2 & a_{12}a_{13} + a_{22}a_{23} \\ a_{11}a_{13} + a_{21}a_{23} & a_{12}a_{13} + a_{22}a_{23} & a_{13}^2 + a_{23}^2 \end{bmatrix} \end{aligned}$$

The trace of this matrix is $(a_{11}^2 + a_{21}^2) + (a_{12}^2 + a_{22}^2) + (a_{13}^2 + a_{23}^2)$. Hence

$$\|A\|_F = \sqrt{(a_{11}^2 + a_{21}^2) + (a_{12}^2 + a_{22}^2) + (a_{13}^2 + a_{23}^2)}$$

is the Frobenius norm of A. ▲

As we can see from Example 7.2, the Frobenius norm of A is the square root of the squares of the Euclidean norms of the columns of A. But by regrouping terms we realize that $\|A\|_F$ is also the square root of the sum of the squares of the entries of A. Moreover, we proved in Corollary 5.9 that the matrix space $\mathbb{R}^{m \times n}$ is isomorphic to the coordinate space $\mathbb{R}^{mn}$. This means that the formula for the Euclidean vector norm on $\mathbb{R}^{mn}$ produces the Frobenius norm on $\mathbb{R}^{m \times n}$.

EXAMPLE 7.3 The Frobenius Norm of a Matrix

Use *Maple* to find the Frobenius norm $\|A\|_F$ of the matrix

$$A = \begin{bmatrix} 9 & -3 \\ -2 & 1 \end{bmatrix}$$

Solution. We first calculate the trace of $A^T A$.

```
[> with(LinearAlgebra):
[> A:=Matrix([[9,-3],[-2,1]]):
[> Trace(Transpose(A).A);
                              95
```

The Frobenius norm is therefore $\sqrt{95}$. We can confirms this conclusion using the **Norm** function in the **LinearAlgebra** package with the **Frobenius** option.

```
[> Norm(A,Frobenius);
                              √95
```

Therefore, $\|A\|_F = \sqrt{95}$ is the Frobenius norm of A. ▲

In addition to being identical with the Euclidean vector norm on $\mathbb{R}^{mn}$, the Frobenius norm also satisfies the inequality

$$\|AB\|_F \leq \|A\|_F \, \|B\|_F$$

what is known as the ***consistency condition.*** In that sense, the Frobenius norm on $\mathbb{R}^{m \times n}$ is a matrix norm. Since the composition of matrix transformations, and therefore matrix multiplication, plays a major role in linear algebra, it is important that any measure of the size of a matrix should be stable for matrix products. The size $\|AB\|$ of the product of two matrices should be no larger than the product $\|A\| \, \|B\|$ of the size $\|A\|$ of A and the size $\|B\|$ of B. This consideration leads us to the definition of a matrix norm.

DEFINITION 7.6 *A **matrix norm** on a space V of square matrices is a function $A \to \|A\|$ that assigns to each matrix $A \in V$ a real number $\|A\|$ and satisfies the following properties:*

1. $\|A\| \geq 0$
2. $\|A\| = 0$ *only if* $A = \mathbf{0}$
3. $\|cA\| = |c| \, \|A\|$
4. $\|A + B\| \leq \|A\| + \|B\|$
5. $\|AB\| \leq \|A\| \, \|B\|$

THEOREM 7.3 *The Frobenius norm on $\mathbb{R}^{m \times n}$ is a matrix norm.*

Proof. We first show that for an vector $\mathbf{x} \in \mathbb{R}^n$, the inequality $\|A\mathbf{x}\|_2 \leq \|A\|_F \, \|\mathbf{x}\|_2$ holds.

Let $\mathbf{r}_i = \mathbf{r}_i A$ be the ith row of A. Then

$$A\mathbf{x} = \begin{bmatrix} \mathbf{r}_1^T \mathbf{x} \\ \vdots \\ \mathbf{r}_m^T \mathbf{x} \end{bmatrix}$$

Therefore,

$$\|A\mathbf{x}\|_2^2 = (\mathbf{r}_1^T\mathbf{x})^2 + \cdots + (\mathbf{r}_m^T\mathbf{x})^2$$

and by the Cauchy–Schwarz inequality, proved in Theorem 7.7,

$$\|A\mathbf{x}\|_2^2 \le \|\mathbf{x}\|_2^2 \left(\|\mathbf{r}_1\|_2^2 + \cdots + \|\mathbf{r}_m\|_2^2\right) = \|\mathbf{x}\|_2^2 \|A\|_F^2$$

Now let B be a real $n \times p$ matrix and consider the product AB. Let $\mathbf{c}_j = \mathbf{c}_j B$ be the jth column of B. Then we know from Theorem 2.1 that

$$\begin{aligned}
\|AB\|_F^2 &= \left\|\begin{bmatrix} A\mathbf{c}_1 & \cdots & A\mathbf{c}_p \end{bmatrix}\right\|_F^2 \\
&= \|A\mathbf{c}_1\|_2^2 + \cdots + \left\|A\mathbf{c}_p\right\|_2^2 \\
&\le \|A\|_F^2 \left(\|\mathbf{c}_1\|_2^2 + \cdots + \left\|\mathbf{c}_p\right\|_2^2\right) \\
&= \|A\|_F^2 \|B\|_F^2
\end{aligned}$$

The Frobenius norm is therefore a matrix norm. ■

Using the Frobenius norm, we can extend the idea of the distance between two vectors in $\mathbb{R}^n$ to the distance between two matrices in $\mathbb{R}^{m\times n}$.

DEFINITION 7.7 *For any two matrices A and B in $\mathbb{R}^{m\times n}$, the real number $\|A - B\|_F$ is the **Frobenius distance** between A and B.*

EXAMPLE 7.4 The Frobenius Distance between Two Matrices

Use *Maple* to calculate the Frobenius distance between the matrices

$$A = \begin{bmatrix} 1 & 0 & 2 & 3 \\ 1 & 3 & 0 & 3 \end{bmatrix} \quad \text{and} \quad B = \begin{bmatrix} 0 & 3 & 0 & -4 \\ 2 & 4 & 4 & -3 \end{bmatrix}$$

Solution. We define the matrices A and B and use the **Frobenius** option of the **Norm** function in the **LinearAlgebra** package to calculate the norm of the matrix $(A - B)$.

```
[> with(LinearAlgebra):
[> A:=Matrix([[1,0,2,3],[1,3,0,3]]):
[> B:=Matrix([[0,3,0,-4],[2,4,4,-3]]):
[> fnA:=Norm(A-B,Frobenius);
```

$$fnA := 3\sqrt{13}$$

Since

$$A - B = \begin{bmatrix} 1 & -3 & 2 & 7 \\ -1 & -1 & -4 & 6 \end{bmatrix}$$

we can see that $3\sqrt{13}$ is the square root of the sum of the squares of the entries of the matrix $(A - B)$, as expected. ▲

EXERCISES 7.2

1. Calculate the Frobenius norm of the following matrices.

a. $\begin{bmatrix} 3 & 4 \\ -1 & 2 \end{bmatrix}$ b. $\begin{bmatrix} 3 & 0 \\ 0 & 2 \end{bmatrix}$ c. $\begin{bmatrix} 1 & 0 \\ 0 & 1 \end{bmatrix}$

d. $\begin{bmatrix} 1 & -1 & 0 \\ 0 & 1 & 0 \\ 0 & 0 & 1 \end{bmatrix}$ e. $\begin{bmatrix} -3 & 2 & 0 \\ 2 & -3 & 0 \\ 0 & 0 & 4 \end{bmatrix}$ f. $\begin{bmatrix} -3 & 2 & 4 \\ 2 & -3 & 1 \\ 1 & 0 & 4 \end{bmatrix}$

2. Show that the Frobenius norm of the matrix

$$A = \begin{bmatrix} a & 2 \\ 3 & 4 \end{bmatrix}$$

is less than the Frobenius norm of the matrix

$$B = \begin{bmatrix} b & 2 \\ 3 & 4 \end{bmatrix}$$

if and only if $|a| < |b|$.

3. Use the Frobenius norm to calculate the distance between the matrices

$$\begin{bmatrix} 1.4 & 2 \\ 3 & 4 \end{bmatrix} \text{ and } \begin{bmatrix} 1.13 & 2 \\ 3 & 4 \end{bmatrix}$$

4. (*Maple V*) Repeat Exercises 1–3 using the **norm** function in the **linalg** package.

5. (*Maple 6*) Repeat Exercises 1–3 using the **Norm** function in the **LinearAlgebra** package.

Cosines and Angles

In Euclidean geometry, the ***law of cosines*** says that the angle between two vectors $\mathbf{x}$ and $\mathbf{y}$ in the plane is related to the length of $\mathbf{x}$ and $\mathbf{y}$ by the formula

$$\|\mathbf{x} - \mathbf{y}\|_2^2 = \|\mathbf{x}\|_2^2 + \|\mathbf{y}\|_2^2 - 2\,\|\mathbf{x}\|_2\,\|\mathbf{y}\|_2 \cos\theta$$

By the definition of the two-norm and Theorem 2.9, this equation converts to

$$\mathbf{x}^T\mathbf{x} - 2\mathbf{x}^T\mathbf{y} + \mathbf{y}^T\mathbf{y} - \mathbf{x}^T\mathbf{x} - \mathbf{y}^T\mathbf{y} = -2\,\|\mathbf{x}\|_2\,\|\mathbf{y}\|_2 \cos\theta$$

If we simplify the equation and solve for $\cos\theta$, we get $\cos\theta = \mathbf{x}^T\mathbf{y}/(\|\mathbf{x}\|_2\,\|\mathbf{y}\|_2)$, provided that $\mathbf{x}$ and $\mathbf{y}$ are nonzero vectors.

EXAMPLE 7.5 The Cosine of an Angle in $\mathbb{R}^2$

Use *Maple* to find the cosine of the angle θ between the vectors $\mathbf{x} = (0, 5)$ and $\mathbf{y} = (1, 1)$.

Solution. We use the definition $\cos\theta = \mathbf{x}^T\mathbf{y}/\left(\|\mathbf{x}\|_2\ \|\mathbf{y}\|_2\right)$.

```
[>with(linalg):
[>x:=vector([0,5]):   y:=vector([1,1]):
[>a:=evalm(transpose(x)&*y):
[>b:=sqrt(evalm(transpose(x)&*x)):
[>c:=sqrt(evalm(transpose(y)&*y)):
[>cosine:=a/(b*c);
```

$$cosine := \frac{1}{2}\sqrt{2}$$

Therefore, $\cos\theta = \sqrt{2}/2$. ▲

We can adapt this formula to $\mathbb{R}^n$ for any $n \geq 2$.

DEFINITION 7.8 *The* ***cosine of the angle*** *between two nonzero vectors* **x** *and* **y** *in* $\mathbb{R}^n$ *is*

$$\frac{\mathbf{x}^T\mathbf{y}}{\|\mathbf{x}\|_2\ \|\mathbf{y}\|_2}$$

EXAMPLE 7.6 The Cosine of an Angle in $\mathbb{R}^5$

Use *Maple* to find the cosine of the angle θ between the vectors $\mathbf{x} = (1, 2, 3, 4, 5)$ and $\mathbf{y} = (3, 3, 3, 3, 3)$.

Solution. We use **dotprod** function in the **linalg** package.

```
[>with(linalg):
[>x:=vector([1,2,3,4,5]):
[>y:=vector([3,3,3,3,3]):
[>cosine:=dotprod(x,y)/sqrt(dotprod(x,x)*dotprod(y,y));
```

$$cosine := \frac{3}{11}\sqrt{11}$$

Therefore, $\cos\theta = (3/11)\sqrt{11}$. ▲

Using the fact that on the interval $0 \leq x < \pi$ the cosine function has a unique inverse, we can define the angle between two nonzero vectors.

DEFINITION 7.9 *The **angle** between two nonzero vectors* $\mathbf{x}$ *and* $\mathbf{y}$ *in* $\mathbb{R}^n$ *is the unique real number* θ *for which*

$$\cos\theta = \frac{\mathbf{x}^T\mathbf{y}}{\|\mathbf{x}\|_2\,\|\mathbf{y}\|_2} \qquad (0 \le \theta < \pi)$$

EXAMPLE 7.7 An Angle in $\mathbb{R}^3$

Use *Maple* to find the angle θ between the vectors $\mathbf{x} = (1, 2, 3)$ and $\mathbf{y} = (-2, 8, 1)$.

Solution. We first calculate the cosine of θ.

```
[>with(linalg):
[>x:=vector([1,2,3]):
[>y:=vector([-2,8,1]):
```

```
[>cosTheta:=dotprod(x,y)/sqrt(dotprod(x,x)*dotprod(y,y));
```

$$cosTheta := \frac{17}{966}\sqrt{966}$$

```
[>theta:=evalf(arccos(cosTheta));
```

$$\theta := .9920605370$$

Thus the angle between the two given vectors is approximately 1 radian. ▲

We could have calculated the angle θ in Example 7.7 directly using the **angle** function in the **linalg** package. The calculation

```
[>with(linalg):
[>x:=vector([1,2,3]):
[>y:=vector([-2,8,1]):
[>evalf(angle(x,y));
```

$$.9920605370$$

produces the same result.

EXERCISES 7.3

1. Find the cosine of the angles between the following pairs of vectors in $\mathbb{R}^2$, if they exist.

a. $\mathbf{x} = (1, 2)$ and $\mathbf{y} = (4, 5)$ b. $\mathbf{x} = (4, 5)$ and $\mathbf{y} = (-1, 2)$

c. $\mathbf{x} = (1, 0)$ and $\mathbf{y} = (4, 5)$ d. $\mathbf{x} = (4, 5)$ and $\mathbf{y} = (0, -2)$
e. $\mathbf{x} = (1, 2)$ and $\mathbf{y} = (0, 0)$ f. $\mathbf{x} = (1, 0)$ and $\mathbf{y} = (0, 1)$

2. Find the angles between the pairs of vectors in Exercise 1, if possible.
3. (*Maple V*) Find the cosine of the angles between the following pairs of vectors in $\mathbb{R}^3$, if they exist.

   a. $\mathbf{x} = (1, 2, 3)$ and $\mathbf{y} = (4, 5, 7)$ b. $\mathbf{x} = (4, 5, 7)$ and $\mathbf{y} = (-1, 2, 0)$
   c. $\mathbf{x} = (1, 0, 0)$ and $\mathbf{y} = (4, 5, 7)$ d. $\mathbf{x} = (4, 5, 7)$ and $\mathbf{y} = (0, -2, -3)$
   e. $\mathbf{x} = (1, 2, 3)$ and $\mathbf{y} = (0, 0, 0)$ f. $\mathbf{x} = (1, 0, 1)$ and $\mathbf{y} = (0, 1, 0)$

4. (*Maple V*) Find the angles between the pairs of vectors in Exercise 3, if they exist.
5. (*Maple V*) Repeat Exercises 2 and 4, using the function **linalg[angle]**.

NON-EUCLIDEAN NORMS

In this section we discuss some non-Euclidean vector and matrix norms.

Vector Norms

We recall from Definition 7.2 that a vector norm is a real-valued function that assigns to every vector $\mathbf{x}$ in a real vector space V a nonnegative real number $\|\mathbf{x}\|$ so that the following conditions are satisfied: The norm $\|\mathbf{x}\|$ is zero if and only if $\mathbf{x}$ is the zero vector, the norm of a scalar multiple $a\mathbf{x}$ is equal to the absolute value of a times the norm of $\mathbf{x}$, and the norm $\|\mathbf{x} + \mathbf{y}\|$ of the sum of two vectors $\mathbf{x}$ and $\mathbf{y}$ is no larger than the sum of the norms of $\mathbf{x}$ and $\mathbf{y}$. We saw earlier that the Euclidean norm is a vector norm. We now give some examples of other norms.

EXAMPLE 7.8 The One-Norm on $\mathbb{R}^n$

The function defined by $\|\mathbf{x}\|_1 = |x_1| + \cdots + |x_n|$ on the vector $\mathbf{x} = (x_1, \ldots, x_n) \in \mathbb{R}^n$ is a vector norm on $\mathbb{R}^n$. The absolute values $|x_i|$ are zero if $x_i = 0$ and positive otherwise. Hence $\|\mathbf{x}\|_1$ satisfies the first and second axioms. The properties of the absolute value function also imply that

$$\begin{aligned}
\|a\mathbf{x}\|_1 &= |ax_1| + \cdots + |ax_n| \\
&= |a| \left[|x_1| + \cdots + |x_n|\right] \\
&= |a| \, \|\mathbf{x}\|_1
\end{aligned}$$

Therefore, the function $\|\mathbf{x}\|_1$ satisfies the third axiom. It remains to verify the triangle inequality. Consider any two vectors $\mathbf{x} = (x_1, \ldots, x_n)$ and $\mathbf{y} = (y_1, \ldots, y_n)$. By the definition of the function $\|\cdot\|_1$ and by the triangle inequality for absolute values,

$$\begin{aligned}\|\mathbf{x}+\mathbf{y}\|_1 &= |x_1+y_1| + \cdots + |x_n+y_n| \\ &\le |x_1| + \cdots + |x_n| + |y_1| + \cdots + |y_n| \\ &= \|\mathbf{x}\|_1 + \|\mathbf{y}\|_1\end{aligned}$$

Therefore the function $\mathbf{x} \mapsto \|\mathbf{x}\|_1$ is a vector norm on $\mathbb{R}^n$. ▲

EXAMPLE 7.9 The Infinity-Norm on $\mathbb{R}^n$

Let $\mathbf{x} = (x_1, \ldots, x_n)$ be a vector in $\mathbb{R}^n$. Then the function $\|\mathbf{x}\|_\infty = \max(|x_1|, \ldots, |x_n|)$ is a vector norm on $\mathbb{R}^n$. The properties of the absolute value function guarantee that the function $\|\mathbf{x}\|_\infty$ satisfies the first two axioms. Moreover, $a\mathbf{x} = (ax_1, \ldots, ax_n)$. Therefore

$$\begin{aligned}\|a\mathbf{x}\|_\infty &= \max(|ax_1|, \ldots, |ax_n|) \\ &= \max(|a|\ |x_1|, \ldots, |a|\ |x_n|) \\ &= |a| \max(|x_1|, \ldots, |x_n|) \\ &= |a|\ \|\mathbf{x}\|_\infty\end{aligned}$$

Hence the third axiom holds. In addition, it follows from the triangle inequality for absolute values and the properties of the max function that for all vectors $\mathbf{x} = (x_1, \ldots, x_n)$ and $\mathbf{y} = (y_1, \ldots, y_n)$ in $\mathbb{R}^n$,

$$\begin{aligned}\max(|x_1+y_1|, \ldots, |x_n+y_n|) &\le \max(|x_1|+|y_1|, \ldots, |x_n|+|y_n|) \\ &\le \max(|x_1|, \ldots, |x_n|) + \max(|y_1|, \ldots, |y_n|)\end{aligned}$$

Hence the triangle inequality $\|\mathbf{x}+\mathbf{y}\|_\infty \le \|\mathbf{x}\|_\infty + \|\mathbf{y}\|_\infty$ holds. ▲

We can get an idea of the geometric properties of these norms by examining the shapes of the unit spheres determined by them. A ***unit vector*** is any vector $\mathbf{x}$ for which $\|\mathbf{x}\| = 1$, and the ***unit sphere*** determined by a norm is the set of all unit vectors of that space. In the two-dimensional case, we usually refer to a unit sphere as a ***unit circle***.

EXAMPLE 7.10 Three Unit Circles in $\mathbb{R}^2$

Draw the unit circles $\{\mathbf{x} : \|\mathbf{x}\|_1 = 1\}$, $\{\mathbf{x} : \|\mathbf{x}\|_2 = 1\}$, and $\{\mathbf{x} : \|\mathbf{x}\|_\infty = 1\}$ in $\mathbb{R}^2$.

Solution. We use *Maple* to plot the sets for which

a. $|x| + |y| = 1$ b. $x^2 + y^2 = 1$ c. $\max(|x|, |y|) = 1$

The calculation

```
[>with(plots):
[>implicitplot({
   x^2+y^2=1,abs(x)+abs(y)=1,max(abs(x),abs(y))=1},
   x=-1..1,y=-1..1):
```

produces the following result:

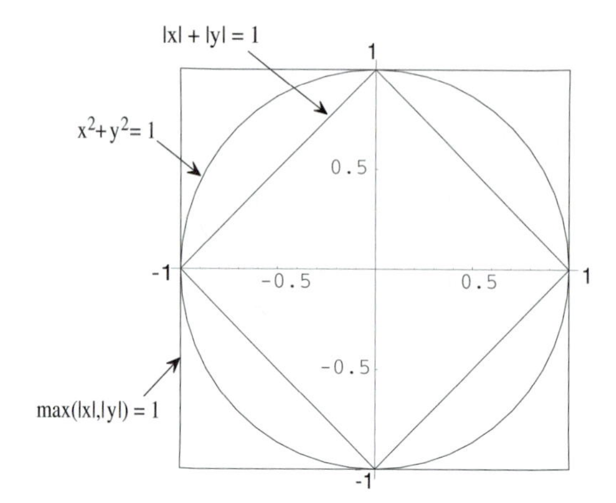

The *unit circles* are three familiar geometric objects: a square centered at the origin, the usual unit circle determined by the Euclidean norm, and a diamond, inside the unit square.

EXERCISES 7.4

1. Calculate the one-norm and the infinity-norm of the following vectors in $\mathbb{R}^2$.

   a. $\mathbf{x} = (3, 4)$ b. $\mathbf{x} = (0, 0)$ c. $\mathbf{x} = (-3, 4)$
   d. $\mathbf{x} = (1, 0)$ e. $\mathbf{x} = (-9, 11)$ f. $\mathbf{x} = (-9, -11)$

2. Calculate the one-norm and the infinity-norm of the following vectors in $\mathbb{R}^3$.

   a. $\mathbf{x} = (3, 4, 5)$ b. $\mathbf{x} = (0, 0, 0)$ c. $\mathbf{x} = (-3, 4, 5)$
   d. $\mathbf{x} = (1, 0, 1)$ e. $\mathbf{x} = (0, 0, 1)$ f. $\mathbf{x} = (-9, -11, -5)$

3. (*Maple V*) Repeat Exercise 1 using the **linalg[norm]** function.

4. (*Maple 6*) Repeat Exercise 2 using the **LinearAlgebra[Norm]** function.

5. (*Maple V, 6*) Discuss the difference between the function **norm** in the **linalg** package and the function **Norm** in the **LinearAlgebra** package.

Distance

We recall from Definition 7.4 that a distance function on a vector space V is a function that assigns a nonnegative real number $d(\mathbf{x}, \mathbf{y})$ to every pair of vectors $\mathbf{x}$ and $\mathbf{y}$ in V and satisfies the following conditions: The distance between $\mathbf{x}$ and $\mathbf{y}$ is zero if and only if $\mathbf{x} = \mathbf{y}$, the distance between $\mathbf{x}$ and $\mathbf{y}$ is the same as the distance between $\mathbf{y}$ and $\mathbf{x}$, and the distance between $\mathbf{x}$ and $\mathbf{y}$ is no greater than the sum of the distances from $\mathbf{x}$ and $\mathbf{y}$ to any other vector $\mathbf{z}$. Earlier we saw that in the Euclidean case, the distance $d(\mathbf{x}, \mathbf{y})$ between two vectors **x** and **y** can be defined as the norm of their difference $\mathbf{x} - \mathbf{y}$. This is true in general.

DEFINITION 7.10 *For any two vectors* **x** *and* **y** *in a normed vector space* V, *the* ***distance*** *between* **x** *and* **y** *is the real number* $d(\mathbf{x}, \mathbf{y}) = \|\mathbf{x} - \mathbf{y}\|$.

EXAMPLE 7.11 The One-Norm Distance between Two Vectors in $\mathbb{R}^2$

Use *Maple* to find the one-norm distance between the vectors $\mathbf{x} = (1, 2)$ and $\mathbf{y} = (3, 5)$ in $\mathbb{R}^2$.

Solution. By definition, $d(\mathbf{x}, \mathbf{y}) = \|\mathbf{x} - \mathbf{y}\|_1 = |1 - 3| + |2 - 5|$.

```
[>x:=vector(2,[1,2]):  y:=vector(2,[3,5]):
[>d:=(x,y)->abs(x[1]-y[1])+abs(x[2]-y[2]):
[>d(x,y);

                        5
```

Thus the distance between **x** and **y** in the one-norm is 5. ▲

EXAMPLE 7.12 The Infinity-Norm Distance between Two Vectors in $\mathbb{R}^2$

Use *Maple* to find the distance between the vectors $\mathbf{x} = (1, 2)$ and $\mathbf{y} = (3, 5)$ in the infinity-norm.

Solution. By definition, $d(\mathbf{x}, \mathbf{y}) = \|\mathbf{x} - \mathbf{y}\|_\infty = \max[|1 - 3|, |2 - 5|]$.

```
[>x:=vector(2,[1,2]):  y:=vector(2,[3,4]):
[>d:=(x,y)->max(abs(x[1]-y[1]),abs(x[2]-y[2])):
[>d(x,y);

                        3
```

Thus the distance between **x** and **y** in the infinity-norm is 3. ▲

EXERCISES 7.5

1. Calculate the one-norm distance between following pairs of vectors in $\mathbb{R}^2$.

a. $\mathbf{x} = (4, 5)$ and $\mathbf{y} = (1, 2)$ b. $\mathbf{x} = (4, 5)$ and $\mathbf{y} = (-1, 2)$

c. $\mathbf{x} = (4, 5)$ and $\mathbf{y} = (0, -2)$ d. $\mathbf{x} = (3, 0)$ and $\mathbf{y} = (4, 5)$

e. $\mathbf{x} = (1, 2)$ and $\mathbf{y} = (0, 0)$ f. $\mathbf{x} = (1, 0)$ and $\mathbf{y} = (0, 1)$

2. Calculate the infinity-norm distance between the pairs of vectors in Exercise 1.

3. Calculate the one-norm distance between the following pairs of vectors in $\mathbb{R}^3$.

   a. $\mathbf{x} = (1, 2, 3)$ and $\mathbf{y} = (4, 5, 7)$
   b. $\mathbf{x} = (4, 5, 7)$ and $\mathbf{y} = (-1, 2, 0)$
   c. $\mathbf{x} = (4, 5, 7)$ and $\mathbf{y} = (0, -2, -3)$
   d. $\mathbf{x} = (1, 0, 0)$ and $\mathbf{y} = (4, 5, 7)$
   e. $\mathbf{x} = (1, 2, 3)$ and $\mathbf{y} = (0, 0, 0)$
   f. $\mathbf{x} = (1, 0, 0)$ and $\mathbf{y} = (0, 0, 1)$

4. In his special relativity theory, Einstein defined a dependence between space, $\mathbb{R}^3$, and time, $\mathbb{R}$, by combining the two quantities in the vector space $\mathbb{R}^4$. The theory is based on the star function $\star$, defined by $\mathbf{x} \star \mathbf{y} = -xx' - yy' - zz' + tt'$ for any two points $\mathbf{x} = (x, y, z, t)$ and $\mathbf{y} = (x', y', z', t')$ in $\mathbb{R}^4$. Let

$$\|\mathbf{x}\| = \sqrt{|\mathbf{x} \star \mathbf{x}|} \quad \text{and} \quad d(\mathbf{x}, \mathbf{y}) = \|\mathbf{x} - \mathbf{y}\|$$

   Discuss to what extent $d(\mathbf{x}, \mathbf{y})$ is a distance function and in what sense $\|\mathbf{x}\|$ is a vector norm.

5. (*Maple V*) Use the infinity-norm to calculate the distance between the pairs of vectors in Exercise 3.

6. (*Maple V*) Repeat Exercises 1 and 2 using the **norm** function in the **linalg** package.

7. (*Maple 6*) Repeat Exercises 3 and 4 using the **Norm** function in the **LinearAlgebra** package.

Matrix Norms

We recall from Definition 7.6 that a matrix norm is vector norm that assigns to every matrix A in a matrix space V a nonnegative number $\|A\|$ and satisfies the conditions of a vector norm, with the additional condition that $\|AB\| \leq \|A\| \, \|B\|$ for all matrices A and B in V. As we saw, the Frobenius norm is a matrix norm on $\mathbb{R}^{m \times n}$. Using the one-norm and the infinity-norm for vectors, we now describe two other matrix norms. They will provide us with estimates of the ratios $\|A\mathbf{x}\| \, / \, \|\mathbf{x}\|$ that are independent of $\mathbf{x}$.

Some general properties of vector norm will be needed. To prove them, we require a fact established in advanced texts: for each matrix A and each nonzero vector $\mathbf{x}$, the set of real numbers $\left\{ \frac{\|A\mathbf{x}\|}{\|\mathbf{x}\|} : \|\mathbf{x}\| \neq 0 \right\}$ has a maximum.

THEOREM 7.4 *Let A be a real $n \times n$ matrix, and let $\mathbf{x} \in \mathbb{R}^n$. Then*

$$\max \left\{ \frac{\|A\mathbf{x}\|}{\|\mathbf{x}\|} : \|\mathbf{x}\| \neq 0 \right\} = \max \left\{ \|A\mathbf{x}\| : \|\mathbf{x}\| = 1 \right\}$$

Proof.

$$
\begin{aligned}
\max\left\{\frac{\|A\mathbf{x}\|}{\|\mathbf{x}\|} : \|\mathbf{x}\| \neq 0\right\} &= \max\left\{\frac{1}{\|\mathbf{x}\|}\|A\mathbf{x}\| : \|\mathbf{x}\| \neq 0\right\} \\
&= \max\left\{\left\|\frac{1}{\|\mathbf{x}\|}(A\mathbf{x})\right\| : \|\mathbf{x}\| \neq 0\right\} \\
&= \max\left\{\left\|A\frac{\mathbf{x}}{\|\mathbf{x}\|}\right\| : \|\mathbf{x}\| \neq 0\right\} \\
&= \max\{\|A\mathbf{x}\| : \|\mathbf{x}\| = 1\}
\end{aligned}
$$

This proves the theorem. ■

Theorem 7.4 provides us with a tool for constructing matrix norms.

THEOREM 7.5 (Matrix norm theorem) *The function* $\|A\| = \max\{\|A\mathbf{x}\| : \|\mathbf{x}\| = 1\}$ *is a matrix norm.*

Proof. Since $\|A\mathbf{x}\|$ is a vector norm, $\|A\mathbf{x}\| \geq 0$. Moreover, since $\|\mathbf{x}\| = 1$, the quantity $\|A\mathbf{x}\| = 0$ only if $A\mathbf{x} = \mathbf{0}$. Hence the first two axioms are satisfied. The third axiom is satisfied since

$$
\begin{aligned}
\|cA\| &= \max\{\|(cA)(\mathbf{x})\| : \|\mathbf{x}\| = 1\} \\
&= \max\{\|c(A\mathbf{x})\| : \|\mathbf{x}\| = 1\} \\
&= \max\{|c|\,\|A\mathbf{x}\| : \|\mathbf{x}\| = 1\} \\
&= |c| \max\{\|A\mathbf{x}\| : \|\mathbf{x}\| = 1\} \\
&= |c|\,\|A\|
\end{aligned}
$$

Furthermore, $\|A + B\| = \max\{\|(A + B)\mathbf{x}\| : \|\mathbf{x}\| = 1\}$. Let $\mathbf{y}$ be a vector of norm 1 that causes the right-hand side to attain a maximum. Then

$$
\begin{aligned}
\|A + B\| &= \|(A + B)\mathbf{y}\| \\
&= \|Ay + By\| \\
&\leq \|A\mathbf{y}\| + \|B\mathbf{y}\| \\
&\leq \|A\|\,\|\mathbf{y}\| + \|B\|\,\|\mathbf{y}\| \\
&= \|A\| + \|B\|
\end{aligned}
$$

Hence the fourth axiom is satisfied.

Finally consider the quantity $\|AB\| = \max\{\|(AB)\mathbf{x}\| : \|\mathbf{x}\| = 1\}$ and suppose that $\|(AB)\mathbf{x}\|$ is a maximum at some vector $\mathbf{y}$ of norm 1. Then

$$
\begin{aligned}
\|AB\| &= \|(AB)\mathbf{y}\| \\
&= \|A[B\mathbf{y}]\|
\end{aligned}
$$

$$\leq \|A\| \, \|B\mathbf{y}\|$$
$$\leq \|A\| \, \|B\| \, \|\mathbf{y}\| = \|A\| \, \|B\|$$

Therefore, the function $\max\{\|A\mathbf{x}\| : \|\mathbf{x}\| = 1\}$ is a matrix norm. ■

A geometric interpretation of this result is that the value $\max\{\|A\mathbf{x}\| \,/\, \|\mathbf{x}\| : \|\mathbf{x}\| \neq \mathbf{0}\}$ is the maximum stretching factor obtained when multiplying the vectors on the unit sphere by the matrix A.

EXAMPLE 7.13 Stretching a Unit Circle in $\mathbb{R}^2$

Use *Maple* to graph the set $W = \left\{A\mathbf{x} \in \mathbb{R}^2 : \|\mathbf{x}\|_2 = 1\right\}$ for

$$A = \begin{bmatrix} 2 & 0 \\ 0 & 3 \end{bmatrix}$$

Solution. Let $\mathbf{x} = (x, y)$ be a point in $\mathbb{R}^2$ whose Euclidean norm is 1. Then $x^2 + y^2 = 1$. Furthermore, the image $A\mathbf{x}$ of $\mathbf{x}$ is

$$\begin{bmatrix} 2 & 0 \\ 0 & 3 \end{bmatrix}\begin{bmatrix} x \\ y \end{bmatrix} = \begin{bmatrix} 2x \\ 3y \end{bmatrix}$$

This means that

$$W = \left\{\begin{bmatrix} 2x \\ 3y \end{bmatrix} : x^2 + y^2 = 1\right\} = \left\{\begin{bmatrix} x \\ y \end{bmatrix} : 9x^2 + 4y^2 = 36\right\}$$

We use *Maple* to plot W.

```
[> with(plots):
[> implicitplot(9*x^2+4*y^2=36,x=-2..2,y=-3..3);
```

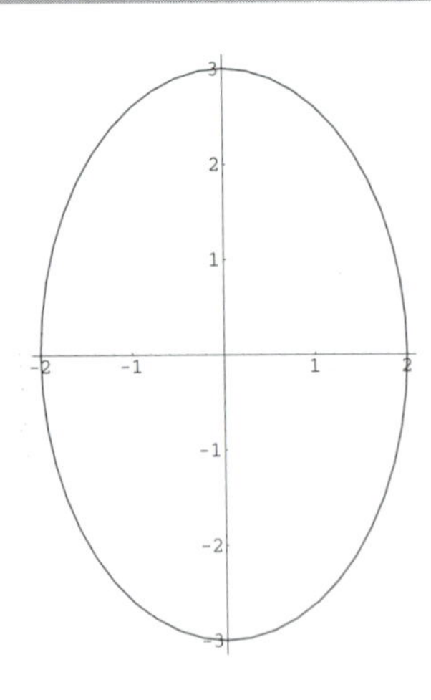

As we can see, the set W forms an ellipse. The norm of the longest vector in W is three units. Thus

$$\|A\| = \left\| \begin{bmatrix} 2 & 0 \\ 0 & 3 \end{bmatrix} \right\| = 3$$

In this case, the norm of A corresponds to the maximum stretch of the unit circle in the direction of the y-axis. ▲

We now define the three non-Euclidean matrix norms that are directly linked to the three vector norms discussed. They are called the one-norm $\|A\|_1$, the two-norm $\|A\|_2$, and the infinity-norm $\|A\|_\infty$. As we will see, the matrix two-norm is not the immediate analog of the vector two-norm. We omit the verification that these functions are indeed matrix norms.

1. The ***one-norm*** $\|A\|_1$ of A is the maximum of the absolute value sums of the columns of A. If A is a one-column vector, then $\|A\|_1$ corresponds to the one-norm of the vector A.
2. The ***two-norm*** $\|A\|_2$ of A is the largest singular value of A. By definition, a ***singular value*** of a matrix A is the positive square root $\sqrt{\lambda}$ of an eigenvalue λ of A^TA. Chapter 9, which deals in depth with such values, shows that the eigenvalues of symmetric matrices of the form A^TA are always nonnegative.
3. The ***infinity-norm*** $\|A\|_\infty$ of A is the maximum of the absolute-value sums of the rows of the matrix A. If A is a one-column matrix, then $\|A\|_\infty$ corresponds to the infinity vector norm of A.

Table 1 illustrates the difference between corresponding vector and matrix norms for vectors in $\mathbb{R}^{2\times 1}$ and matrices in $\mathbb{R}^{2\times 2}$. To be able to compare the vector and matrix forms of these norms, we treat column vectors of height 2 as 2×1 matrices. Suppose that

$$A = \begin{bmatrix} a_{11} & a_{12} \\ a_{21} & a_{22} \end{bmatrix}$$

is a given 2×2 matrix and that $\mathbf{x}$ is the first column of A. We tabulate the one-, two-, and infinity-norms for $\mathbf{x}$ and the corresponding matrix norms for A.

TABLE 1 VECTOR AND MATRIX NORMS.

Vector Norms on $\mathbb{R}^{2\times 1}$	**Matrix Norms on $\mathbb{R}^{2\times 2}$**
$\|\mathbf{x}\|_1 = \|a_{11}\| + \|a_{21}\|$	$\|A\|_1 = \max\{\|a_{11}\| + \|a_{21}\|, \|a_{12}\| + \|a_{22}\|\}$
$\|\mathbf{x}\|_2 = \left(a_{11}^2 + a_{21}^2\right)^{1/2}$	$\|A\|_2 = \max\left\{\sqrt{\lambda_1}, \sqrt{\lambda_2}\right\}$ (where λ_i is an eigenvalue of A^TA)
$\|\mathbf{x}\|_3 = \max\{\|a_{11}\|, \|a_{21}\|\}$	$\|A\|_\infty = \max\{\|a_{11}\| + \|a_{12}\|, \|a_{21}\| + \|a_{22}\|\}$

The general rules for calculating these matrix norms are analogous to those described for the 2×2 case.

EXAMPLE 7.14 The Two-Norm of a Matrix

Use *Maple* to calculate the two-norm of the matrix

$$A = \begin{bmatrix} 1 & 3 \\ 0 & 2 \end{bmatrix}$$

Solution. We use the **Norm** function in the **LinearAlgebra** package.

```
[> with(LinearAlgebra):
[> A:=Matrix([[1,3],[0,2]]):
[> TwoNorm:=allvalues(Norm(A,2));
```

$$TwoNorm := \sqrt{7+3\sqrt{5}}$$

The value $\sqrt{7+3\sqrt{5}}$ is the square root of the largest eigenvalue of the matrix $A^T A$, in other words, the largest singular value of the matrix A. ▲

EXAMPLE 7.15 The One-Norm and the Infinity-Norm of a Matrix

Calculate $\|A\|_1$ and $\|A\|_\infty$ for the matrix

$$A = \begin{bmatrix} 3 & -1 & 0 \\ 2 & 0 & 1 \\ -3 & 5 & 1 \end{bmatrix}$$

Solution. It is clear from our discussion that

$$\begin{aligned} \|A\|_1 &= \max\{|3| + |2| + |-3|, |-1| + |0| + |5|, |0| + |1| + |1|\} \\ &= \max\{8, 6, 2\} = 8 \end{aligned}$$

and

$$\begin{aligned} \|A\|_\infty &= \max\{|3| + |-1| + |0|, |2| + |0| + |1|, |-3| + |5| + |1|\} \\ &= \max\{4, 3, 9\} = 9 \end{aligned}$$

We can see that for nonsymmetric matrices, the one-norm and the infinity-norm are usually different. ▲

EXERCISES 7.6

All two-norms in the following exercises are assumed to be matrix norms.

1. Let $S_1 = \{\mathbf{x}_1 = (2, 3), \mathbf{x}_2 = (5, -6), \mathbf{x}_3 = (1, 1), \mathbf{x}_4 = (0, 0), \mathbf{x}_5 = (9, -2)\}$ be a set of vectors in $\mathbb{R}^2$ and let

$$A = \begin{bmatrix} 3 & 4 \\ -1 & 2 \end{bmatrix}$$

Calculate $\max \{\|A\mathbf{x}\|_1 : \mathbf{x} \in S_1\}$ in $\mathbb{R}^2$.

2. Repeat Exercise 1 using the Euclidean norm.

3. Repeat Exercise 1 using the infinity-norm.

4. Find the one-, two-, and infinity-norm of the following matrices.

a. $\begin{bmatrix} 3 & 4 \\ -1 & 2 \end{bmatrix}$ b. $\begin{bmatrix} 3 & 0 \\ 0 & 2 \end{bmatrix}$ c. $\begin{bmatrix} 1 & 0 \\ 0 & 1 \end{bmatrix}$

d. $\begin{bmatrix} 1 & -1 & 0 \\ 0 & 1 & 0 \\ 0 & 0 & 1 \end{bmatrix}$ e. $\begin{bmatrix} -3 & 2 & 0 \\ 2 & -3 & 0 \\ 0 & 0 & 4 \end{bmatrix}$ f. $\begin{bmatrix} -3 & 2 & 4 \\ 2 & -3 & 1 \\ 1 & 0 & 4 \end{bmatrix}$

5. Use the definition of the one-, two-, and infinity matrix norms to calculate the distances between the matrices (a)–(c) in Exercise 4.

6. (*Maple V*) Use the definition of the one-, two-, and infinity matrix norms to calculate the distance between the matrices (d)–(f) in Exercise 4.

7. (*Maple 6*) Repeat Exercise 4 using the **Norm** function in the **LinearAlgebra** package.

8. (*Maple V, 6*) Explore the properties of the **norm** function in the **linalg** package and the **Norm** function in the **LinearAlgebra** package.

REAL INNER PRODUCTS

We now begin the study of inner products and show how they can be used to link the concepts of length, distance, and angles. In the case of matrices, the trace will play a key role in the specification of an important inner product.

DEFINITION 7.11 *An* ***inner product*** *on a real vector space V is a real-valued function $(\mathbf{x}, \mathbf{y}) \mapsto \langle \mathbf{x}, \mathbf{y} \rangle$ satisfying the following four axioms.*

1. $\langle a\mathbf{x} + b\mathbf{y}, \mathbf{z} \rangle = a \langle \mathbf{x}, \mathbf{z} \rangle + b \langle \mathbf{y}, \mathbf{z} \rangle$
2. $\langle \mathbf{x}, \mathbf{y} \rangle = \langle \mathbf{y}, \mathbf{x} \rangle$
3. $\langle \mathbf{x}, \mathbf{x} \rangle \geq 0$
4. $\langle \mathbf{x}, \mathbf{x} \rangle = 0$ *if and only if* $\mathbf{x} = \mathbf{0}$

Inner products are ***bilinear*** because they are linear in each of their two variables. The linearity $\langle a\mathbf{x}+b\mathbf{y}, \mathbf{z}\rangle = a\langle \mathbf{x}, \mathbf{z}\rangle + b\langle \mathbf{y}, \mathbf{z}\rangle$ in the first variable is part of the definition. The linearity in the second variable follows from their ***symmetry***. The fact that $\langle \mathbf{x}, \mathbf{y}\rangle = \langle \mathbf{y}, \mathbf{x}\rangle$ entails that

$$\begin{aligned}\langle \mathbf{u}, a\mathbf{v}+b\mathbf{w}\rangle &= \langle a\mathbf{v}+b\mathbf{w}, \mathbf{u}\rangle \\ &= a\langle \mathbf{v}, \mathbf{u}\rangle + b\langle \mathbf{w}, \mathbf{u}\rangle \\ &= a\langle \mathbf{u}, \mathbf{v}\rangle + b\langle \mathbf{u}, \mathbf{w}\rangle\end{aligned}$$

Hence $\langle \mathbf{x}, \mathbf{y}\rangle$ is linear in $\mathbf{y}$.

DEFINITION 7.12 *A **real inner product space** is a real vector space V equipped with a real inner product* $(\mathbf{x}, \mathbf{y}) \mapsto \langle \mathbf{x}, \mathbf{y}\rangle$.

EXAMPLE 7.16 The Dot Product on $\mathbb{R}^n$

Show that the dot product function

$$\mathbf{x}^T\mathbf{y} = \begin{bmatrix} x_1 & \cdots & x_n \end{bmatrix}\begin{bmatrix} y_1 \\ \vdots \\ y_n \end{bmatrix} = x_1y_1 + \cdots + x_ny_n$$

is an inner product on $\mathbb{R}^n$.

Solution. Let $\mathbf{x}$ and $\mathbf{y}$ be two vectors in $\mathbb{R}^n$. Then

$$\begin{aligned}\langle a\mathbf{x}+b\mathbf{y}, \mathbf{z}\rangle &= (a\mathbf{x}+b\mathbf{y})^T\,\mathbf{z} \\ &= \left((a\mathbf{x})^T + (b\mathbf{y})^T\right)\mathbf{z} \\ &= \left(a\mathbf{x}^T + b\mathbf{y}^T\right)\mathbf{z} \\ &= a\mathbf{x}^T\mathbf{z} + b\mathbf{y}^T\mathbf{z} \\ &= a\,\langle \mathbf{x}, \mathbf{z}\rangle + b\,\langle \mathbf{y}, \mathbf{z}\rangle\end{aligned}$$

Therefore, $\mathbf{x}^T\mathbf{y}$ is linear in the first variable. The calculation

$$\begin{aligned}\mathbf{x}^T\mathbf{y} &= x_1y_1 + \cdots + x_ny_n \\ &= y_1x_1 + \cdots + y_nx_n \\ &= \mathbf{y}^T\mathbf{x} = (\mathbf{x}^T\mathbf{y})^T\end{aligned}$$

shows that $\mathbf{x}^T\mathbf{y}$ is also symmetric. Moreover, $\langle \mathbf{x}, \mathbf{x}\rangle = \mathbf{x}^T\mathbf{x} = x_1^2 + \cdots + x_n^2 \geq 0$ for all $\mathbf{x}$ and is 0 if and only if $x_i = 0$ for all i. Hence the function $\mathbf{x}^T\mathbf{y}$ is an inner product on $\mathbb{R}^n$. ▲

The inner product in Example 7.16 has various names. In addition to being called the ***dot product***, it is also known as the ***standard inner product*** and the ***Euclidean inner product***. In *Maple,* it is called the dot product.

EXAMPLE 7.17 A Nonstandard Inner Product on $\mathbb{R}^2$

Show that the function $\langle \mathbf{x}, \mathbf{y} \rangle = 2x_1y_1 + 3x_2y_2$ is an inner product on $\mathbb{R}^2$, where $\mathbf{x} = (x_1, x_2)$ and $\mathbf{y} = (y_1, y_2)$.

Solution. Let $\mathbf{z} = (z_1, z_2)$ be an arbitrary third vector in $\mathbb{R}^2$, and consider the arbitrary functional value $\langle a\mathbf{x}+b\mathbf{y}, \mathbf{z} \rangle$, with

$$a\mathbf{x} + b\mathbf{y} = \begin{bmatrix} ax_1 \\ ax_2 \end{bmatrix} + \begin{bmatrix} by_1 \\ by_2 \end{bmatrix} = \begin{bmatrix} ax_1 + by_1 \\ ax_2 + by_2 \end{bmatrix}$$

By the laws of arithmetic,

$$\begin{aligned} \langle a\mathbf{x} + b\mathbf{y}, \mathbf{z} \rangle &= 2\,(ax_1 + by_1)\,z_1 + 3\,(ax_2 + by_2)\,z_2 \\ &= 2ax_1z_1 + 3ax_2z_2 + 2by_1z_1 + 3by_2z_2 \\ &= a\,\langle \mathbf{x}, \mathbf{z} \rangle + b\,\langle \mathbf{y}, \mathbf{z} \rangle \end{aligned}$$

Hence $\langle \mathbf{x}, \mathbf{y} \rangle$ is linear in $\mathbf{x}$. The symmetry of $\langle \mathbf{x}, \mathbf{y} \rangle$ follows from the fact that

$$\langle \mathbf{x}, \mathbf{y} \rangle = 2x_1y_1 + 3x_2y_2 = 2y_1x_1 + 3y_2x_2 = \langle \mathbf{y}, \mathbf{x} \rangle$$

Moreover, $\langle \mathbf{x}, \mathbf{x} \rangle = 2x_1x_1 + 3x_2x_2 \geq 0$ and is zero if and only if $x_1 = x_2 = 0$. This shows that the function $\langle \mathbf{x}, \mathbf{y} \rangle$ is an inner product. ▲

In the next example, we present an important inner product for matrix spaces. We saw earlier how it was used to define the Frobenius norm on $\mathbb{R}^{m\times n}$.

EXAMPLE 7.18 The Trace as an Inner Product

Let $A = \left[a_{ij}\right]$ and $B = \left[b_{ij}\right]$ be two real $m \times n$ matrices. Then the function

$$\langle A, B \rangle = \text{trace}\, B^T A$$

is an inner product on $\mathbb{R}^{m\times n}$. It is easy to check that

$$\langle A, B \rangle = \sum_{i\in\mathbf{m}} \sum_{j\in\mathbf{n}} a_{ij} b_{ij}$$

is the sum of the products of the corresponding entries of A and B. Therefore $\langle A, A \rangle \geq 0$ and it is 0 if and only if $A = \mathbf{0}$. Moreover, the distributive and commutative laws $a(b + c) = ab + ac$ and $ab = ba$ for the multiplication and addition of real numbers imply that $\langle A, B \rangle$ is both linear in A and symmetric. Hence $\langle A, B \rangle$ is an inner product. ▲

EXAMPLE 7.19 The Trace Inner Product on $\mathbb{R}^{2\times 3}$

Use *Maple* to calculate the inner product trace $\left(B^T A\right)$ of the matrices

$$A = \begin{bmatrix} a_{11} & a_{12} & a_{13} \\ a_{21} & a_{22} & a_{23} \end{bmatrix} \quad \text{and} \quad B = \begin{bmatrix} b_{11} & b_{12} & b_{13} \\ b_{21} & b_{22} & b_{23} \end{bmatrix}$$

Solution. We use the **linalg** package.

```
[> with(linalg):
[> A:=matrix(2,3,a);
```

$$A := \begin{bmatrix} a(1,1) & a(1,2) & a(1,3) \\ a(2,1) & a(2,2) & a(2,3) \end{bmatrix}$$

```
[> B:=matrix(2,3,b);
```

$$B := \begin{bmatrix} b(1,1) & b(1,2) & b(1,3) \\ b(2,1) & b(2,2) & b(2,3) \end{bmatrix}$$

```
[> Bt:=transpose(B);
```

$$BT := \begin{bmatrix} b(1,1) & b(2,1) \\ b(1,2) & b(2,2) \\ b(1,3) & b(2,3) \end{bmatrix}$$

```
[> t:=trace(evalm(Bt&*A));
```

$$t := b(1,1)a(1,1) + b(2,1)a(2,1) + b(1,2)a(1,2)$$
$$+ b(2,2)a(2,2) + b(1,3)a(1,3) + b(2,3)a(2,3)$$

This shows that $\langle A, B\rangle$ is $\sum_{i\in\mathbf{2}} \sum_{j\in\mathbf{3}} a_{ij}b_{ij}$. ▲

In our discussion of matrix norms we implicitly used the fact that every nonzero vector $\mathbf{x}$ in an inner product space can be converted to a unit vector by dividing it by its norm $\|\mathbf{x}\|$. This process is called ***normalization.*** Unit vectors are therefore also called ***normal vectors.***

THEOREM 7.6 (Normalization theorem) *For every nonzero vector* $\mathbf{x}$ *in an inner product space* V*, the vector* $\mathbf{y} = \mathbf{x}/\|\mathbf{x}\|$ *is a unit vector.*

Proof. By definition,

$$\|\mathbf{y}\|^2 = \langle \mathbf{y}, \mathbf{y}\rangle = \left\langle \frac{\mathbf{x}}{\|\mathbf{x}\|}, \frac{\mathbf{x}}{\|\mathbf{x}\|} \right\rangle = \frac{1}{\|\mathbf{x}\|}\left\langle \mathbf{x}, \frac{\mathbf{x}}{\|\mathbf{x}\|} \right\rangle = \frac{1}{\|\mathbf{x}\|^2}\langle \mathbf{x}, \mathbf{x}\rangle = \frac{\langle \mathbf{x}, \mathbf{x}\rangle}{\langle \mathbf{x}, \mathbf{x}\rangle} = 1$$

Hence $\mathbf{y}$ is a unit vector. ■

The next theorem relates the size of the values of inner products to the size of the vectors involved. It is known as the ***Cauchy–Schwarz inequality***. Its proof uses all four of the defining properties of inner products. We let $\|\mathbf{x}\|$ be the scalar $\sqrt{\langle \mathbf{x}, \mathbf{x}\rangle}$ determined by $\langle \mathbf{x}, \mathbf{y}\rangle$.

THEOREM 7.7 (Cauchy–Schwarz inequality) *If* $(\mathbf{x}, \mathbf{y}) \mapsto \langle \mathbf{x}, \mathbf{y}\rangle$ *is an inner product on a real vector space* V*, then* $|\langle \mathbf{x}, \mathbf{y}\rangle| \leq \|\mathbf{x}\|\,\|\mathbf{y}\|$ *for all* $\mathbf{x}, \mathbf{y} \in V$.

Proof. Suppose first that either $\mathbf{x}$ or $\mathbf{y}$ is 0. Then $\langle \mathbf{x}, \mathbf{y}\rangle = 0$ and the inequality holds.

Suppose next that neither $\mathbf{x}$ nor $\mathbf{y}$ is 0. It follows that $\|\mathbf{x}\| > 0$ and $\|\mathbf{y}\| > 0$. Since $\langle \mathbf{z}, \mathbf{z}\rangle \geq 0$ for all $\mathbf{z} \in V$, the bilinearity and symmetry of inner products therefore implies that

$$
\begin{aligned}
0 &\leq \left\langle \frac{\mathbf{x}}{\|\mathbf{x}\|} + \frac{\mathbf{y}}{\|\mathbf{y}\|}, \frac{\mathbf{x}}{\|\mathbf{x}\|} + \frac{\mathbf{y}}{\|\mathbf{y}\|} \right\rangle = \left\langle \frac{\mathbf{x}}{\|\mathbf{x}\|}, \frac{\mathbf{x}}{\|\mathbf{x}\|} \right\rangle + 2\left\langle \frac{\mathbf{x}}{\|\mathbf{x}\|}, \frac{\mathbf{y}}{\|\mathbf{y}\|} \right\rangle + \left\langle \frac{\mathbf{y}}{\|\mathbf{y}\|}, \frac{\mathbf{y}}{\|\mathbf{y}\|} \right\rangle \\
&= \frac{1}{\|\mathbf{x}\|^2}\langle \mathbf{x}, \mathbf{x}\rangle + \frac{2}{\|\mathbf{x}\|\,\|\mathbf{y}\|}\langle \mathbf{x}, \mathbf{y}\rangle + \frac{1}{\|\mathbf{y}\|^2}\langle \mathbf{y}, \mathbf{y}\rangle \\
&= \frac{1}{\|\mathbf{x}\|^2}\|\mathbf{x}\|^2 + \frac{2}{\|\mathbf{x}\|\,\|\mathbf{y}\|}\langle \mathbf{x}, \mathbf{y}\rangle + \frac{1}{\|\mathbf{y}\|^2}\|\mathbf{y}\|^2 \\
&= 1 + \frac{2}{\|\mathbf{x}\|\,\|\mathbf{y}\|}\langle \mathbf{x}, \mathbf{y}\rangle + 1
\end{aligned}
$$

Hence $-\|\mathbf{x}\|\,\|\mathbf{y}\| \leq \langle \mathbf{x}, \mathbf{y}\rangle$. Similarly,

$$
\begin{aligned}
0 &\leq \left\langle \frac{\mathbf{x}}{\|\mathbf{x}\|} - \frac{\mathbf{y}}{\|\mathbf{y}\|}, \frac{\mathbf{x}}{\|\mathbf{x}\|} - \frac{\mathbf{y}}{\|\mathbf{y}\|} \right\rangle = \left\langle \frac{\mathbf{x}}{\|\mathbf{x}\|}, \frac{\mathbf{x}}{\|\mathbf{x}\|} \right\rangle - 2\left\langle \frac{\mathbf{x}}{\|\mathbf{x}\|}, \frac{\mathbf{y}}{\|\mathbf{y}\|} \right\rangle + \left\langle \frac{\mathbf{y}}{\|\mathbf{y}\|}, \frac{\mathbf{y}}{\|\mathbf{y}\|} \right\rangle \\
&= \frac{1}{\|\mathbf{x}\|^2}\langle \mathbf{x}, \mathbf{x}\rangle - \frac{2}{\|\mathbf{x}\|\,\|\mathbf{y}\|}\langle \mathbf{x}, \mathbf{y}\rangle + \frac{1}{\|\mathbf{y}\|^2}\langle \mathbf{y}, \mathbf{y}\rangle \\
&= \frac{1}{\|\mathbf{x}\|^2}\|\mathbf{x}\|^2 - \frac{2}{\|\mathbf{x}\|\,\|\mathbf{y}\|}\langle \mathbf{x}, \mathbf{y}\rangle + \frac{1}{\|\mathbf{y}\|^2}\|\mathbf{y}\|^2 \\
&= 1 - \frac{2}{\|\mathbf{x}\|\,\|\mathbf{y}\|}\langle \mathbf{x}, \mathbf{y}\rangle + 1
\end{aligned}
$$

Hence $\langle \mathbf{x}, \mathbf{y}\rangle \leq \|\mathbf{x}\|\,\|\mathbf{y}\|$. By combining the two inequalities, we obtain the theorem. ■

EXERCISES 7.7

1. Use the one-norm, the Euclidean norm, and infinity-norm to normalize the following vectors in $\mathbb{R}^2$, if possible.

   a. $\mathbf{x} = (3, 4)$ b. $\mathbf{x} = (0, 0)$ c. $\mathbf{x} = (-3, 4)$

   d. $\mathbf{x} = (1, 0)$ e. $\mathbf{x} = (-9, 11)$ f. $\mathbf{x} = (-9, -11)$

2. Calculate the standard inner products $\mathbf{x}^T\mathbf{y}$ for the following pairs of vectors.

   a. $\mathbf{x} = (1, 2)$ and $\mathbf{y} = (4, 5)$ b. $\mathbf{x} = (-1, 2)$ and $\mathbf{y} = (4, 5)$

   c. $\mathbf{x} = (0, -2)$ and $\mathbf{y} = (4, 5)$ d. $\mathbf{x} = (1, 2)$ and $\mathbf{y} = (0, 0)$

3. Calculate the nonstandard inner products $\langle \mathbf{x}, \mathbf{y} \rangle = 5x_1y_1 + 7x_2y_2$ of the pairs of vectors $\mathbf{x} = (x_1, x_2)$ and $\mathbf{y} = (y_1, y_2)$ in Exercise 2.

4. Use the norm determined by the inner product in Exercise 3 to normalize the vectors in Exercise 2.

5. (*Maple V*) Use the dot product to verify the Cauchy–Schwarz inequality for the pairs of vectors in Exercise 2.

6. (*Maple V*) Calculate $\mathbf{x}^T\mathbf{y}$ for the following pairs of vectors.

   a. $\mathbf{x} = (1, 2, 3)$ and $\mathbf{y} = (4, 5, 7)$ b. $\mathbf{x} = (-1, 2, 0)$ and $\mathbf{y} = (4, 5, 7)$
   c. $\mathbf{x} = (0, -2, -3)$ and $\mathbf{y} = (4, 5, 7)$ d. $\mathbf{x} = (1, 0, 0)$ and $\mathbf{y} = (4, 5, 7)$

7. (*Maple 6*) Use the one-norm, Euclidean norm, and infinity-norm to normalize the following vectors in $\mathbb{R}^3$, if possible.

   a. $\mathbf{x} = (3, 4, 5)$ b. $\mathbf{x} = (0, 0, 0)$ c. $\mathbf{x} = (-3, 4, 5)$
   d. $\mathbf{x} = (1, 0, 1)$ e. $\mathbf{x} = (0, 0, 1)$ f. $\mathbf{x} = (-9, -11, -5)$

8. (*Maple 6*) Calculate the inner products $\langle \mathbf{x}, \mathbf{y} \rangle = 5x_1y_1 + 7x_2y_2 + 4x_3y_3$ of the pairs of vectors $\mathbf{x} = (x_1, x_2, x_3)$ and $\mathbf{y} = (y_1, y_2, y_3)$ in Exercise 6.

9. (*Maple 6*) Calculate the inner product trace B^TA for the following pairs of matrices.

   a. $A = \begin{bmatrix} 5 & 0 & -3 \\ 0 & 0 & -3 \end{bmatrix}$ and $B = \begin{bmatrix} -1 & 4 & 5 \\ 5 & -2 & -2 \end{bmatrix}$

   b. $A = \begin{bmatrix} 5 & 0 & 1 \\ 6 & 2 & 6 \\ 1 & 4 & -4 \\ 4 & 2 & 6 \end{bmatrix}$ and $B = \begin{bmatrix} 3 & 1 & 4 \\ -6 & 3 & -2 \\ 1 & 3 & 0 \\ 1 & -8 & 3 \end{bmatrix}$

10. (*Maple 6*) Use the inner product defined in Exercise 8 to verify the Cauchy–Schwarz inequality for the pairs of vectors in Exercise 6.

11. (*Maple 6*) Define two general vectors **x:=<r*x1,x2>** and **y:=<s*y1,y2>** and find a condition on r and s for which the function **DotProduct(x,y)** is an inner product on $\mathbb{R}^2$.

Positive Definite Matrices

We saw in the last section that both the function $f(x_1, x_2, y_1, y_2) = x_1y_1 + x_2y_2$ and the function $g(x_1, y_1, x_2, y_2) = 2x_1y_1 + 3x_2y_2$ define an inner product on $\mathbb{R}^2$. In matrix notation, we can write these functions as

$$f(\mathbf{x},\mathbf{y}) = \begin{bmatrix} x_1 \\ x_2 \end{bmatrix}^T \begin{bmatrix} 1 & 0 \\ 0 & 1 \end{bmatrix} \begin{bmatrix} y_1 \\ y_2 \end{bmatrix} \quad \text{and} \quad g(\mathbf{x},\mathbf{y}) = \begin{bmatrix} x_1 \\ x_2 \end{bmatrix}^T \begin{bmatrix} 2 & 0 \\ 0 & 3 \end{bmatrix} \begin{bmatrix} y_1 \\ y_2 \end{bmatrix}$$

In other words, both inner products can be written in the form $\mathbf{x}^T A\mathbf{y}$, for some 2×2 matrix A. This leads us to search for conditions under which an inner product can be defined by a matrix. Since $\langle \mathbf{x}, \mathbf{x} \rangle$ must be positive for nonzero vectors $\mathbf{x}$, we certainly require that $\mathbf{x}^T A\mathbf{x} > 0$ for all nonzero $\mathbf{x} \in \mathbb{R}^n$. In the next theorem, we prove that the symmetry of $\langle \mathbf{x}, \mathbf{y} \rangle$ also requires that the matrix A is symmetric. We combine these two properties and define a real $n \times n$ symmetric matrix A as ***positive definite*** if $\mathbf{x}^T A\mathbf{x} > 0$ for all nonzero $\mathbf{x} \in \mathbb{R}^n$.

THEOREM 7.8 (Inner product theorem) *If A is a positive definite symmetric matrix, then the function $\langle \mathbf{x}, \mathbf{y} \rangle = \mathbf{x}^T A\mathbf{y}$ is an inner product on $\mathbb{R}^n$.*

Proof. The calculation

$$\begin{aligned} \langle a\mathbf{u} + b\mathbf{v}, \mathbf{y} \rangle &= (a\mathbf{u} + b\mathbf{v})^T A\mathbf{y} \\ &= (a\mathbf{u}^T + b\mathbf{v}^T) A\mathbf{y} \\ &= (a\mathbf{u})^T A\mathbf{z} + (b\mathbf{v})^T A\mathbf{y} \\ &= a\,\langle \mathbf{u}, \mathbf{y} \rangle + b\,\langle \mathbf{v}, \mathbf{y} \rangle \end{aligned}$$

shows that the function $\mathbf{x}^T A\mathbf{y}$ is linear in the first variable. Moreover, the symmetry of A and the fact that every real number z can be considered as a 1×1 symmetric matrix $[z]$ imply that $\mathbf{x}^T A\mathbf{y} = \left(\mathbf{x}^T A\mathbf{y}\right)^T$. It follows that $\langle \mathbf{x}, \mathbf{y} \rangle$ is symmetric since

$$\langle \mathbf{x}, \mathbf{y} \rangle = \mathbf{x}^T A\mathbf{y} = (\mathbf{x}^T A\mathbf{y})^T = \mathbf{y}^T A^T \mathbf{x} = \mathbf{y}^T A\mathbf{x} = \langle \mathbf{y}, \mathbf{x} \rangle$$

The fact that A is positive definite means that $\langle \mathbf{x}, \mathbf{x} \rangle = \mathbf{x}^T A\mathbf{x} > 0$ for all nonzero $\mathbf{x} \in \mathbb{R}^n$. Furthermore, $\langle \mathbf{0}, \mathbf{0} \rangle = \mathbf{0}^T A\mathbf{0} = 0$. Therefore $\langle \mathbf{x}, \mathbf{y} \rangle$ is an inner product on $\mathbb{R}^n$. ■

In Chapter 8, we will show that all real symmetric matrices are diagonalizable. It is an easy consequence of this fact that a symmetric matrix is positive definite if and only if its eigenvalues are positive. We call this theorem the ***eigenvalue test*** for positive definiteness.

EXAMPLE 7.20 A Positive Definite Matrix

Use the eigenvalue test to show that the symmetric matrix

$$A = \begin{bmatrix} 9 & 0 & -2 \\ 0 & 7 & 0 \\ -2 & 0 & 1 \end{bmatrix}$$

is positive definite.

Solution. The following calculation shows that the eigenvalues of A are positive.

```
[> A:=matrix([[9,0,-2],[0,7,0],[-2,0,1]]):
[> linalg[eigenvalues](A);
```

$$7, 5+2\sqrt{5}, 5-2\sqrt{5}$$

The first and second eigenvalues are clearly positive, and since $20 < 25$, it follows that $2\sqrt{5} < 5$. Therefore, $0 < 5 - 2\sqrt{5}$. This shows that A is positive definite. ▲

The **LinearAlgebra** package has a built-in test function for deciding whether a matrix is positive definite matrix or not. The matrix

$$\begin{bmatrix} 9 & 0 & -2 \\ 0 & 7 & 0 \\ -2 & 0 & 1 \end{bmatrix}$$

for example, can be shown to be positive definite using the following commands:

```
[> A:=Matrix([[9,0,-2],[0,7,0],[-2,0,1]]):
[> LinearAlgebra[IsDefinite](A);
```

true

EXERCISES 7.8

1. Show that no 2×2 matrix A^TA is positive definite if A is a 1×2 matrix.

2. Show that no 3×3 matrix A^TA is positive definite if A is a 1×3 matrix.

3. Find conditions on a, b, c, d such that A^TA is positive definite, where

$$A = \begin{bmatrix} a & b \\ c & d \end{bmatrix}$$

4. Show in detail that the products $\mathbf{x}^TA\mathbf{y}$ determined by the positive definite matrices in Exercise 1 are inner products.

5. (*Maple V*) Show that if A is the matrix

$$\begin{bmatrix} 50 & 79 \\ 56 & 49 \end{bmatrix}$$

then the symmetric matrix A^TA is positive definite and $\mathbf{x}A^TA\mathbf{y}$ is an inner product on $\mathbb{R}^2$. Find the norm of the vectors $\mathbf{x}_1 = (1, -1)$, $\mathbf{x}_2 = (45, 46)$, and $\mathbf{x}_3 = (10, 0)$ in this inner product. Find the distance between $\mathbf{x}_1$ and $\mathbf{x}_2$, between $\mathbf{x}_1$ and $\mathbf{x}_3$, and between $\mathbf{x}_2$ and $\mathbf{x}_3$.

6. (*Maple 6*) Use the eigenvalue test to determine which of the following symmetric matrices are positive definite.

$$\text{a.} \begin{bmatrix} 4 & 1 \\ 1 & -1 \end{bmatrix} \quad \text{b.} \begin{bmatrix} 1 & 1 & 0 \\ 1 & 2 & 0 \\ 0 & 0 & 4 \end{bmatrix} \quad \text{c.} \begin{bmatrix} 1 & -2 \\ -2 & 7 \end{bmatrix}$$

7. (*Maple 6*) Generate a random 3×1 matrix A, a random 2×2 matrix B, and a random 2×3 matrix C. Form the symmetric matrices AA^T, BB^T, and CC^T, and determine which of these matrices, if any, is positive definite.

8. (*Maple 6*) Show that if A is the matrix

$$\begin{bmatrix} 1 & 0 & -1 \\ 0 & 2 & 0 \\ -1 & 0 & 3 \end{bmatrix}$$

then $\mathbf{x}A^T A\mathbf{y}$ is an inner product on $\mathbb{R}^3$. Find the norm of the vectors $\mathbf{x}_1 = (1, -1, 3)$, $\mathbf{x}_2 = (45, 46, 47)$, and $\mathbf{x}_3 = (10, 0, -9)$ in this inner product. Find the distance between $\mathbf{x}_1$ and $\mathbf{x}_2$, between $\mathbf{x}_1$ and $\mathbf{x}_3$, and between $\mathbf{x}_2$ and $\mathbf{x}_3$.

9. (*Maple 6*) Generate three random 3×3 matrices A, B, and C. Determine whether the functions $\mathbf{x}A^T A\mathbf{y}$, $\mathbf{x}B^T B\mathbf{y}$, and $\mathbf{x}C^T C\mathbf{y}$ are inner products on $\mathbb{R}^3$.

Inner Product Norms

We are now in a position to unify the study of length, distance, and angles by using vector norms determined by inner products. The following theorem explains the construction of a norm from an inner product.

THEOREM 7.9 (Inner product norm theorem) *If V is a real vector space with an inner product $\langle \mathbf{x}, \mathbf{y} \rangle$, then the function $\|\mathbf{x}\| = \sqrt{\langle \mathbf{x}, \mathbf{x} \rangle}$ is a norm on V.*

Proof. Since $\langle \mathbf{x}, \mathbf{x} \rangle = 0$ if $\mathbf{x}$ is $\mathbf{0}$ and is greater than 0 otherwise, $\|\mathbf{x}\| = \sqrt{\langle \mathbf{x}, \mathbf{x} \rangle} > 0$ for all $\mathbf{x} \neq \mathbf{0}$. In addition, $\sqrt{\langle \mathbf{0}, \mathbf{0} \rangle} = 0$. Hence the first axiom of norms is satisfied. Furthermore,

$$\|a\mathbf{x}\| = \sqrt{\langle a\mathbf{x}, a\mathbf{x} \rangle} = \sqrt{a^2 \langle \mathbf{x}, \mathbf{x} \rangle} = |a| \sqrt{\langle \mathbf{x}, \mathbf{x} \rangle} = |a| \, \|\mathbf{x}\|$$

Therefore the second axiom is also satisfied. It remains to verify the third axiom. For this we use Theorem 7.7, together with the bilinearity and symmetry of the inner product:

$$\begin{aligned} \|\mathbf{x} + \mathbf{y}\|^2 &= \langle \mathbf{x} + \mathbf{y}, \mathbf{x} + \mathbf{y} \rangle \\ &= \langle \mathbf{x}, \mathbf{x} \rangle + 2 \langle \mathbf{x}, \mathbf{y} \rangle + \langle \mathbf{y}, \mathbf{y} \rangle \end{aligned}$$

$$\leq \|\mathbf{x}\|^2 + 2\|\mathbf{x}\|\,\|\mathbf{y}\| + \|\mathbf{y}\|^2$$
$$= (\|\mathbf{x}\| + \|\mathbf{y}\|)^2$$

If we take the square root of each side of this inequality, we get

$$\|\mathbf{x} + \mathbf{y}\| \leq \|\mathbf{x}\| + \|\mathbf{y}\|$$

Hence the function $\mathbf{x} \mapsto \sqrt{\langle \mathbf{x}, \mathbf{x} \rangle}$ is a norm on V. ■

DEFINITION 7.13 *If $\langle \mathbf{x}, \mathbf{y} \rangle$ is an inner product on a real vector space V, then the function $\|\mathbf{x}\| = \sqrt{\langle \mathbf{x}, \mathbf{x} \rangle}$ is an* ***inner product norm*** *on V.*

Although every inner product determines a norm in this way, there are norms that are not determined by an inner product. The infinity norm on $\mathbb{R}^2$, for example, is not an inner product norm. We will show this by showing that it fails to satisfy the ***parallelogram law***, a law satisfied by all inner product norms. The law takes its name from ordinary geometry, where it says that the sum of the squares on the diagonals of a parallelogram is equal to twice the sum of the squares on its sides.

THEOREM 7.10 (Parallelogram law) *If $\|\cdot\|$ is an inner product norm on a real vector space V, then $\|\mathbf{x} + \mathbf{y}\|^2 + \|\mathbf{x} - \mathbf{y}\|^2 = 2\left(\|\mathbf{x}\|^2 + \|\mathbf{y}\|^2\right)$ for all $\mathbf{x}, \mathbf{y} \in V$.*

Proof. Since $\|\cdot\|$ is an inner product norm, $\|\mathbf{z}\| = \sqrt{\langle \mathbf{z}, \mathbf{z} \rangle}$ and $\|\mathbf{z}\|^2 = \langle \mathbf{z}, \mathbf{z} \rangle$ for all $\mathbf{z} \in V$.

Therefore,

$$\|\mathbf{x} + \mathbf{y}\|^2 = \langle \mathbf{x} + \mathbf{y}, \mathbf{x} + \mathbf{y} \rangle = \langle \mathbf{x}, \mathbf{x} \rangle + 2\langle \mathbf{x}, \mathbf{y} \rangle + \langle \mathbf{y}, \mathbf{y} \rangle$$

and

$$\|\mathbf{x} - \mathbf{y}\|^2 = \langle \mathbf{x} - \mathbf{y}, \mathbf{x} - \mathbf{y} \rangle = \langle \mathbf{x}, \mathbf{x} \rangle - 2\langle \mathbf{x}, \mathbf{y} \rangle + \langle \mathbf{y}, \mathbf{y} \rangle$$

If we add the two equations, we get

$$\|\mathbf{x} + \mathbf{y}\|^2 + \|\mathbf{x} - \mathbf{y}\|^2 = 2\left(\langle \mathbf{x}, \mathbf{x} \rangle + \langle \mathbf{y}, \mathbf{y} \rangle\right) = 2\left(\|\mathbf{x}\|^2 + \|\mathbf{y}\|^2\right)$$

for all $\mathbf{x}, \mathbf{y} \in V$. ■

Figure 1 illustrates the geometric connection between the vectors linked by the parallelogram law.

EXAMPLE 7.21 The Infinity Norm on $\mathbb{R}^2$

Show that the infinity norm $\|(x_1, x_2)\|_\infty = \max(|x_1|, |x_2|)$ is not an inner product norm.

Solution. It suffices to find two vectors in $\mathbb{R}^2$ for which the norm fails to satisfy the parallelogram law. Consider the vectors $\mathbf{x} = (1, 0)$ and $\mathbf{y} = (0, 1)$. Since $\mathbf{x} + \mathbf{y} = (1, 1)$ and $\mathbf{x} - \mathbf{y} = (1, -1)$, we get

$$\|\mathbf{x} + \mathbf{y}\|^2 + \|\mathbf{x} - \mathbf{y}\|^2 = \max(|1|, |1|)^2 + \max(|1|, |-1|)^2 = 1 + 1 = 2$$

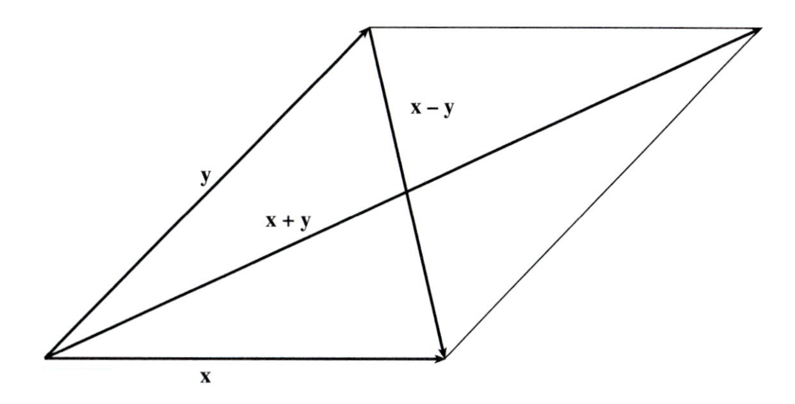

FIGURE 1 Diagonals of a Parallelogram.

and

$$2(\|\mathbf{x}\|^2 + \|\mathbf{y}\|^2) = 2(\max(|1|, |0|)^2 + \max(|0|, |1|)^2) = 2(1+1) = 4$$

Hence the parallelogram law fails. ▲

EXERCISES 7.9

1. Use the standard inner product $\mathbf{x}^T\mathbf{y}$ on $\mathbb{R}^2$ to find the norm $\|\mathbf{x}\|$ of the following vectors.

   a. $\mathbf{x} = (4, 5)$ b. $\mathbf{x} = (-1, 2)$ c. $\mathbf{x} = (0, 0)$
   d. $\mathbf{x} = (-4, 5)$ e. $\mathbf{x} = (1, 0)$ f. $\mathbf{x} = (-4, -5)$

2. Repeat Exercise 1 using the inner product $\mathbf{x}^T A\mathbf{y}$ determined by the matrix

$$A = \begin{bmatrix} 3 & 0 \\ 0 & 5 \end{bmatrix}$$

3. Prove that in any inner product space, the ***polar identity***

$$\langle \mathbf{x}, \mathbf{y} \rangle = \tfrac{1}{4}(\|\mathbf{x}+\mathbf{y}\|^2 - \|\mathbf{x}-\mathbf{y}\|^2)$$

   expresses the inner product in terms of the norm.

4. (*Maple V*) Use the standard inner product $\mathbf{x}^T\mathbf{y}$ on $\mathbb{R}^3$ to find the norm $\|\mathbf{x}\|$ of the following vectors.

   a. $\mathbf{x} = (4, 5, 7)$ b. $\mathbf{x} = (4, -5, -7)$ c. $\mathbf{x} = (1, 0, 0)$
   d. $\mathbf{x} = (-1, 2, 0)$ e. $\mathbf{x} = (0, 1, 1)$ f. $\mathbf{x} = (-1, -2, -3)$

5. (*Maple V*) Show that the symmetric matrix

$$A = \begin{bmatrix} 3 & 0 & 1 \\ 0 & 5 & 0 \\ 1 & 0 & 6 \end{bmatrix}$$

is positive definite, and use it to calculate the norms of the vectors in Exercise 4, determined by the inner product $\mathbf{x}^T A\mathbf{y}$.

6. (*Maple V*) Repeat Exercise 5 using the inner product $\mathbf{x}^T A\mathbf{y}$ determined by the matrix

$$A = \begin{bmatrix} 3 & 0 & 0 \\ 0 & 5 & 0 \\ 0 & 0 & 6 \end{bmatrix}$$

7. (*Maple V*) Verify the parallelogram law for the following pairs of vectors using the standard inner product on $\mathbb{R}^3$

a. $\mathbf{x} = (1, 2, -2)$ and $\mathbf{y} = (4, 5, 7)$ b. $\mathbf{x} = (-1, 2, 0)$ and $\mathbf{y} = (4, 5, 7)$
c. $\mathbf{x} = (1, 0, 0)$ and $\mathbf{y} = (0, 5, 7)$ d. $\mathbf{x} = (0, -2, -5)$ and $\mathbf{y} = (4, 5, -2)$
e. $\mathbf{x} = (1, 2, 3)$ and $\mathbf{y} = (0, 0, 0)$ f. $\mathbf{x} = (4, 5, 7)$ and $\mathbf{y} = (-4, -5, -7)$

8. (*Maple 6*) Find the pairs of vectors $\mathbf{x}$ and $\mathbf{y}$ in Exercise 6 for which the equation $\|\mathbf{x}+\mathbf{y}\|^2 = \|\mathbf{x}\|^2 + \|\mathbf{y}\|^2$ holds.

ANGLES

The idea of angles extends to non-Euclidean geometry and is based on the law of cosines. As we observed earlier, this law says that the angle between two vectors $\mathbf{x}$ and $\mathbf{y}$ in the plane is related to the length of $\mathbf{x}$ and $\mathbf{y}$ by the formula

$$\|\mathbf{x}-\mathbf{y}\|^2 = \|\mathbf{x}\|^2 + \|\mathbf{y}\|^2 - 2\,\|\mathbf{x}\|\,\|\mathbf{y}\|\cos\theta$$

Figure 2 on Page 513 illustrates this relationship. If we express $\|\mathbf{x}-\mathbf{y}\|^2$, $\|\mathbf{x}\|^2$, and $\|\mathbf{y}\|^2$ in terms of inner products, we get

$$\langle \mathbf{x}-\mathbf{y}, \mathbf{x}-\mathbf{y}\rangle = \langle \mathbf{x}, \mathbf{x}\rangle + \langle \mathbf{y}, \mathbf{y}\rangle - 2\,\|\mathbf{x}\|\,\|\mathbf{y}\|\cos\theta$$

Therefore,

$$\langle \mathbf{x}-\mathbf{y}, \mathbf{x}-\mathbf{y}\rangle - \langle \mathbf{x}, \mathbf{x}\rangle - \langle \mathbf{y}, \mathbf{y}\rangle = -2\,\|\mathbf{x}\|\,\|\mathbf{y}\|\cos\theta$$

Since inner products are linear in both variables, this equation converts to

$$\langle \mathbf{x}, \mathbf{x}\rangle - 2\,\langle \mathbf{x}, \mathbf{y}\rangle + \langle \mathbf{y}, \mathbf{y}\rangle - \langle \mathbf{x}, \mathbf{x}\rangle - \langle \mathbf{y}, \mathbf{y}\rangle = -2\,\|\mathbf{x}\|\,\|\mathbf{y}\|\cos\theta$$

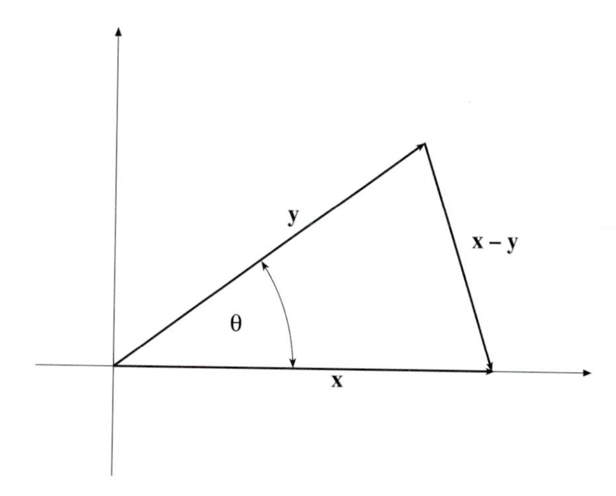

FIGURE 2 An Angle between Two Vectors.

If we simplify and solve for $\cos\theta$, we get $\cos\theta = \langle \mathbf{x}, \mathbf{y}\rangle \,/\, (\|\mathbf{x}\|\,\|\mathbf{y}\|)$. The right-hand side of this equation involves only the inner product $\langle \mathbf{x}, \mathbf{y}\rangle$ and the norms $\|\mathbf{x}\|$ and $\|\mathbf{y}\|$. Moreover, the norms $\|\mathbf{x}\| = \sqrt{\langle \mathbf{x}, \mathbf{x}\rangle}$ and $\|\mathbf{y}\| = \sqrt{\langle \mathbf{y}, \mathbf{y}\rangle}$ are both defined by the inner product $\langle \mathbf{x}, \mathbf{y}\rangle$. This tells us how angles should be defined in terms of inner products. As in the Euclidean case, the idea is to define the cosine of an angle first and then define the angle θ as the inverse of $\cos\theta$.

DEFINITION 7.14 *The **cosine of the angle** θ $(0 \leq \theta < \pi)$ between two nonzero vectors* $\mathbf{x}$ *and* $\mathbf{y}$ *in a real inner product space* V *is*

$$\cos\theta = \frac{\langle \mathbf{x}, \mathbf{y}\rangle}{\|\mathbf{x}\|\,\|\mathbf{y}\|}$$

It follows at once from the Cauchy–Schwarz inequality that $|\langle \mathbf{x}, \mathbf{y}\rangle| \leq \|\mathbf{x}\|\,\|\mathbf{y}\|$. Therefore, $|\cos\theta| \leq 1$. Moreover, we know from our introductory remarks that in the Euclidean case, there exists a unique angle θ between 0 and π for which $\cos\theta = \langle \mathbf{x}, \mathbf{y}\rangle / \|\mathbf{x}\|\,\|\mathbf{y}\|$.

EXAMPLE 7.22 The Cosine of an Angle

Use the inner product

$$\langle \mathbf{x}, \mathbf{y}\rangle = \begin{bmatrix} x & y \end{bmatrix} \begin{bmatrix} 3 & 0 \\ 0 & 5 \end{bmatrix} \begin{bmatrix} x \\ y \end{bmatrix}$$

to calculate the cosine of the angle between the vectors $\mathbf{x} = (0, 5)$ and $\mathbf{y} = (1, 1)$.

Solution. By definition, the cosine of the angle between $\mathbf{x}$ and $\mathbf{y}$ is the ratio of $\langle \mathbf{x}, \mathbf{y}\rangle$ and $\|\mathbf{x}\|\,\|\mathbf{y}\|$. We use *Maple* to calculate this ratio.

```
[> with(LinearAlgebra):
[> x:=<0,5>:  y:=<1,1>:
[> A:=Matrix([[3,0],[0,5]]):
[> ipXY:=Transpose(x).A.y:
[> normX:=sqrt(Transpose(x).A.x):
[> normY:=sqrt(Transpose(y).A.y):
```

```
[> cosTheta:=ipXY/(normX*normY);
```

$$cosTheta := \frac{1}{4}\sqrt{5}\sqrt{2}$$

Therefore, in the nonstandard inner product $\langle \mathbf{x}, \mathbf{y} \rangle$, the cosine of the angle between the vectors $\mathbf{x}$ and $\mathbf{y}$ is $(1/4)\sqrt{5}\sqrt{2}$. ▲

DEFINITION 7.15 *The **angle** between two nonzero vectors* $\mathbf{x}$ *and* $\mathbf{y}$ *in a real inner product space* V *is the unique real number* θ *in the interval* $(0 \leq x < \pi)$ *for which* $\cos\theta = \langle \mathbf{x}, \mathbf{y} \rangle / (\|\mathbf{x}\| \, \|\mathbf{y}\|)$.

EXAMPLE 7.23 Two Angles in $\mathbb{R}^3$

Compare the angles between the vectors $\mathbf{x} = (1, 2, 3)$ and $\mathbf{y} = (5, -1, 2)$ determined by the standard inner product $\langle \mathbf{x}, \mathbf{y} \rangle = \mathbf{x}^T \mathbf{y}$ and the nonstandard inner product

$$\langle \mathbf{x}, \mathbf{y} \rangle = \mathbf{x}^T \begin{bmatrix} 2 & 0 & 0 \\ 0 & 3 & 0 \\ 0 & 0 & 7 \end{bmatrix} \mathbf{y}$$

Solution. We use the **LinearAlgebra** package in combination with the **linalg[angle]** function.

```
[> with(LinearAlgebra):
[> x:=<1,2,3>:
[> y:=<5,-1,2>:
[> evalf(linalg[angle](x,y));
```

$$1.116138382$$

```
[> A:=matrix([[2,0,0],[0,3,0],[0,0,7]]):
[> a:=evalm(Transpose(x).A.y):
[> b:=sqrt(evalm(Transpose(x).A.x)):
[> c:=sqrt(evalm(Transpose(y).A.y)):
```

```
[> cosTheta:=a/(b*c);
```

$$cosTheta := \frac{46}{693}\sqrt{77}$$

```
[> theta:=evalf(arccos(cosTheta));
```

$$\theta := .9490381935$$

This shows that in the standard inner product, the angle between the two vectors is approximately 1 radian, whereas in the nonstandard inner product, it is .95 radians. ▲

EXERCISES 7.10

Use the standard inner product in Exercises 1–6.

1. Find the cosine of the angles between the following pairs of vectors in $\mathbb{R}^2$, if they exist.

a. $\mathbf{x} = (1, 2)$ and $\mathbf{y} = (4, 5)$ b. $\mathbf{x} = (4, 5)$ and $\mathbf{y} = (-1, 2)$
c. $\mathbf{x} = (1, 0)$ and $\mathbf{y} = (4, 5)$ d. $\mathbf{x} = (4, 5)$ and $\mathbf{y} = (0, -2)$
e. $\mathbf{x} = (1, 2)$ and $\mathbf{y} = (0, 0)$ f. $\mathbf{x} = (1, 0)$ and $\mathbf{y} = (0, 1)$

2. (*Maple V*) Find the cosine of the angles between the following pairs of vectors in $\mathbb{R}^3$, if they exist.

a. $\mathbf{x} = (1, 2, 3)$ and $\mathbf{y} = (4, 5, 7)$ b. $\mathbf{x} = (4, 5, 7)$ and $\mathbf{y} = (-1, 2, 0)$
c. $\mathbf{x} = (1, 0, 0)$ and $\mathbf{y} = (4, 5, 7)$ d. $\mathbf{x} = (4, 5, 7)$ and $\mathbf{y} = (0, -2, -3)$
e. $\mathbf{x} = (1, 2, 3)$ and $\mathbf{y} = (0, 0, 0)$ f. $\mathbf{x} = (1, 0, 1)$ and $\mathbf{y} = (0, 1, 0)$

3. (*Maple V*) Find the angles between the pairs of vectors in Exercise 1, if possible.

4. (*Maple V*) Find the angles between the pairs of vectors in Exercise 2, if they exist.

5. (*Maple V*) Repeat Exercise 3, using the function **linalg[angle]**.

6. (*Maple V*) Repeat Exercise 4, using the function **LinearAlgebra[VectorAngle]**.

7. (*Maple 6*) Use the trace inner product to find the angle between the matrices

$$\mathbf{x} = \begin{bmatrix} 4 & 1 \\ -4 & -1 \end{bmatrix} \quad \text{and} \quad \mathbf{y} = \begin{bmatrix} 1 & 2 \\ -2 & 7 \end{bmatrix}$$

Angles in Statistics

In this section, we discuss an interesting conceptual connection between inner products and statistics. We show that ***covariances*** and ***standard deviations*** can be computed using dot products. We then use this information to interpret the ***Pearson correlation coefficient*** of two data sets as the cosine of the angle between these sets. In this interpretation, statistically independent data correspond to orthogonal vectors.

In statistics, the covariance is used to measure the extent to which two ranges of data $\mathbf{x}$ and $\mathbf{y}$ vary together: whether large entries of $\mathbf{x}$ correspond to large entries of $\mathbf{y}$ (positive covariance), or whether small entries of $\mathbf{x}$ correspond to large entries of $\mathbf{y}$ (negative covariance), or whether the entries of $\mathbf{x}$ and $\mathbf{y}$ are unrelated (covariance near zero). The standard deviations of $\mathbf{x}$ and $\mathbf{y}$, on the other hand, measure to what extent the data in $\mathbf{x}$ and $\mathbf{y}$ deviate from the arithmetic mean of their components. We begin with an example.

EXAMPLE 7.24 Variances and Standard Deviations

Suppose that $\mathbf{x} = (x_1, x_2, x_3)$ and $\mathbf{y} = (y_1, y_2, y_3)$ are two numerical data vectors and that $\widehat{x} = \frac{1}{3}(x_1 + x_2 + x_3)$ and $\widehat{y} = \frac{1}{3}(y_1 + y_2 + y_3)$ are their arithmetic means. Then the values $(x_i - \widehat{x})$ are the deviations of the data x_i from the mean $\widehat{x}$. Similarly, the quantities $(y_i - \widehat{y})$ are the deviations of the data y_i from the mean $\widehat{y}$. The arithmetic means

$$\frac{1}{3}\left[(x_1 - \widehat{x})^2 + (x_2 - \widehat{x})^2 + (x_3 - \widehat{x})^2\right] \quad \text{and} \quad \frac{1}{3}\left[(y_1 - \widehat{y})^2 + (y_2 - \widehat{y})^2 + (y_3 - \widehat{y})^2\right]$$

of these quantities are called the ***variances*** of $\mathbf{x}$ and $\mathbf{y}$, and the square roots

$$\sigma_{\mathbf{x}} = \sqrt{\frac{1}{3}\left[(x_1 - \widehat{x})^2 + (x_2 - \widehat{x})^2 + (x_3 - \widehat{x})^2\right]}$$

and

$$\sigma_{\mathbf{y}} = \sqrt{\frac{1}{3}\left[(y_1 - \widehat{y})^2 + (y_2 - \widehat{y})^2 + (y_3 - \widehat{y})^2\right]}$$

of the variances are called the ***standard deviations*** of $\mathbf{x}$ and $\mathbf{y}$. ▲

In some applications, a distinction is required between the variance and standard deviation over an entire set of n data and over a proper subset. In the latter case, the variance and standard deviation are averaged over $(n - 1)$ instead of n.

EXAMPLE 7.25 Covariances

The ***covariance*** of two data sets is calculated by taking the average of the sum of the products of the point-by-point deviations of the data from the mean. The covariance

$$\operatorname{cov}(\mathbf{x}, \mathbf{y}) = \frac{1}{3}\left[(x_1 - \widehat{x})(y_1 - \widehat{y}) + (x_2 - \widehat{x})(y_2 - \widehat{y}) + (x_3 - \widehat{x})(y_3 - \widehat{y})\right]$$

of $\mathbf{x}$ and $\mathbf{y}$ is also denoted by $\sigma_{\mathbf{xy}}$. ▲

The Pearson correlation coefficient r combines covariances and standard deviations into a single formula that measures the degree to which two sets of data are correlated.

DEFINITION 7.16 *Let* $\mathbf{x}$ *and* $\mathbf{y}$ *be two finite sets of data. Then the* ***Pearson correlation coefficient*** *of* $\mathbf{x}$ *and* $\mathbf{y}$ *is*

$$r = \frac{\operatorname{cov}(\mathbf{x}, \mathbf{y})}{\sigma_{\mathbf{x}} \sigma_{\mathbf{y}}}$$

The coefficient r ranges from -1 to 1. If we write out the formula for r in full, we get

$$r = \frac{\frac{1}{n} \sum_{i \in \mathbf{n}} (x_i - \widehat{x})(y_i - \widehat{y})}{\sqrt{\frac{\sum_{i \in \mathbf{n}} (x_i - \widehat{x})^2}{n}} \sqrt{\frac{\sum_{i \in \mathbf{n}} (y_i - \widehat{y})^2}{n}}}$$

Since the factor $1/n$ in the denominator and numerator of r cancels out, the coefficient r becomes

$$\frac{\sum_{i \in \mathbf{n}} (x_i - \widehat{x})(y_i - \widehat{y})}{\sqrt{\sum_{i \in \mathbf{n}} (x_i - \widehat{x})^2} \sqrt{\sum_{i \in \mathbf{n}} (y_i - \widehat{y})^2}}$$

This allows us to rewrite r is the cosine of an angle. Here are the calculations.

Let $\mathbf{u} = (x_1 - \widehat{x}, \ldots, x_n - \widehat{x})$ and $\mathbf{v} = (y_1 - \widehat{y}, \ldots, y_n - \widehat{y})$ be the vectors listing the deviations from the means of $\mathbf{x} = (x_1, \ldots, x_n)$ and $\mathbf{y} = (y_1, \ldots, y_n)$. Then the dot product of $\mathbf{u}$ and $\mathbf{v}$ and their two-norms are the following.

1. The dot product of $\mathbf{u}$ and $\mathbf{v}$ is

$$\mathbf{u} \cdot \mathbf{v} = \sum_{i \in \mathbf{n}} (x_i - \widehat{x})(y_i - \widehat{y})$$

2. The two-norms of $\mathbf{u}$ and $\mathbf{v}$ are

$$\|\mathbf{u}\| = \sqrt{\mathbf{u} \cdot \mathbf{u}} = \sqrt{\sum_{i \in \mathbf{n}} (x_i - \widehat{x})(x_i - \widehat{x})} = \sqrt{\sum_{i \in \mathbf{n}} (x_i - \widehat{x})^2}$$

and

$$\|\mathbf{v}\| = \sqrt{\mathbf{v} \cdot \mathbf{v}} = \sqrt{\sum_{i \in \mathbf{n}} (y_i - \widehat{y})(y_i - \widehat{y})} = \sqrt{\sum_{i \in \mathbf{n}} (y_i - \widehat{y})^2}$$

Therefore,

$$r = \frac{\mathbf{u} \cdot \mathbf{v}}{\|\mathbf{u}\| \, \|\mathbf{v}\|}$$

is the cosine of the angle θ between the vectors $\mathbf{u}$ and $\mathbf{v}$ in $\mathbb{R}^n$.

EXAMPLE 7.26 The Pearson Correlation Coefficient of a Set of Points

Use *Maple* to find the Pearson correlation coefficient for the points $(1, 2)$, $(2, 2)$, $(3, 5)$, $(4, 6)$ in $\mathbb{R}^2$.

Solution. Using the given points we let $x_1 = 1$, $x_2 = 2$, $x_3 = 3$, $x_4 = 4$, $y_1 = 2$, $y_2 = 2$, $y_3 = 5$, and $y_4 = 6$. Then $\widehat{x} = (1 + 2 + 3 + 4)/4 = 5/2$ and $\widehat{y} = (2 + 2 + 5 + 6)/4 = 15/4$.

We use $\widehat{x}$ and $\widehat{y}$ to define

$$\mathbf{u} = (1 - \tfrac{5}{2}, 2 - \tfrac{5}{2}, 3 - \tfrac{5}{2}, 4 - \tfrac{5}{2}) \quad \text{and} \quad \mathbf{v} = (2 - \tfrac{15}{4}, 2 - \tfrac{15}{4}, 5 - \tfrac{15}{4}, 6 - \tfrac{15}{4})$$

Next we calculate the cosine of the angle between **u** and **v**. First we input **u** and **x**.

```
[> with(linalg):
[> u:=vector(4,[1-5/2,2-5/2,3-5/2,4-5/2]):
[> v:=vector(4,[2-15/4,2-15/4,5-15/4,6-15/4]):
```

We then calculate the cosine of the angle between **u** and **x**.

```
[> r:=evalf(
   dotprod(u,v)/(sqrt(dotprod(u,u))*sqrt(dotprod(v,v))));

                    r := .9393364370
```

This shows that the coefficient r of the given set of points is approximately .93936. ▲

Let us now consider the problem of calculating the Pearson correlation coefficient for two sets of data arranged in the form of two matrices. Suppose $A = [a_{ij}]$ and $B = [b_{ij}]$ are two real $m \times n$ matrices. We show that r can also be expressed as the cosine of an angle, this time an angle determined by the trace inner product and the Frobenius norm. By definition, the Pearson correlation coefficient $r(A, B)$ between A and B is

$$\frac{\sum_{i \in \mathbf{m}} \sum_{j \in \mathbf{n}} (a_{ij} - \widehat{a})(b_{ij} - \widehat{b})}{\sqrt{\sum_{i \in \mathbf{m}} \sum_{j \in \mathbf{n}} (a_{ij} - \widehat{a})^2} \sqrt{\sum_{i \in \mathbf{m}} \sum_{j \in \mathbf{n}} (b_{ij} - \widehat{b})^2}}$$

where $\widehat{a}$ is the arithmetic mean taken over all entries of the matrix A and $\widehat{b}$ is the arithmetic mean taken over all entries of the matrix B.

Let $\widehat{A} = (\widehat{a})$ be the constant $m \times n$ matrix with entries $\widehat{a}$, and let $\widehat{B} = (\widehat{b})$ be the constant $m \times n$ matrix with entries $\widehat{b}$. If we let $C = A - \widehat{A}$ and $D = B - \widehat{B}$, then the trace inner product of C and D is

$$\begin{aligned} \langle C, D \rangle &= \operatorname{trace} C^T D \\ &= \operatorname{trace} \left(A - \widehat{A}\right)^T \left(B - \widehat{B}\right) \\ &= \sum_{i \in \mathbf{m}} \sum_{j \in \mathbf{n}} (a_{ij} - \widehat{a})(b_{ij} - \widehat{b}) \end{aligned}$$

and the squares of the Frobenius norms of C and D are

$$\|C\|_F^2 = \left\|A - \widehat{A}\right\|_F^2 = \sum_{i \in \mathbf{m}} \sum_{j \in \mathbf{n}} (a_{ij} - \widehat{a})^2$$

and

$$\|D\|_F^2 = \|B - \widehat{B}\|_F^2 = \sum_{i \in \mathbf{m}} \sum_{j \in \mathbf{n}} (b_{ij} - \widehat{b})^2$$

Therefore,

$$r(A, B) = \frac{\langle C, D \rangle}{\|C\|_F \|D\|_F}$$

is the cosine of the angle determined by the matrices C and D in $\mathbb{R}^{m \times n}$.

Maple has built-in functions for calculating Pearson correlation coefficients, covariances, and standard deviations. The measures are organized in matrix form. If the given data set consist of n vectors in $\mathbb{R}^m$, the ***correlation matrix*** $R = [r_{ij}]$ is a symmetric matrix in which r_{ij} is the Pearson correlation coefficient of the ith and jth columns of the data set. The ***covariance matrix*** $C = [c_{ij}]$ is a symmetric matrix in which c_{ij} is the covariance of the ith and jth columns of the data set.

EXAMPLE 7.27 A Correlation Matrix for Three Data Sets

Use *Maple* to calculate the Pearson correlation coefficients for the vectors

$$\mathbf{x} = (1, 2, 1, 1.5, 7), \quad \mathbf{y} = (3, -2, 1, 0, 7), \quad \text{and} \quad \mathbf{z} = (2, 2, -1, 1, -9)$$

Solution. We first construct a 5×3 matrix whose columns are the vectors **x**, **y**, and **z**. Then we apply the **describe** and **linearcorrelation** functions in the **stats** package. The result is a 3×3 matrix $A = [a_{ij}]$ in which the entry a_{ij} is the Pearson correlation coefficient of the ith and jth columns of A.

```
[> with(stats):
[> x:=[1,2.0,1,1.5,7]:
[> y:=[3.0,-2,1,0,7]:
[> z:=[2.0,2,-1,1,-9]:
[> cor:=(x,y)- > describe[linearcorrelation](x,y):
```

```
[> A:=matrix([
    [cor(x,x),cor(x,y),cor(x,z)],
    [cor(y,x),cor(y,y),cor(y,z)],
    [cor(z,x),cor(z,y),cor(z,z)]]);
```

$$A := \begin{bmatrix} 1.000000000 & .7596897616 & -.9305008558 \\ .7596897616 & 1.000000000 & -.8354177867 \\ -.9305008558 & -.8354177867 & 1.000000000 \end{bmatrix}$$

The value $a_{12} = .75969$ indicates that the vectors **x** and **y** are positively correlated; the value $a_{13} = -.930501$ indicates that **x** and **z** are negatively correlated; and the value $a_{23} = -.835418$ indicates that **y** and **z** are also negatively correlated. ▲

EXAMPLE 7.28 A Covariance Matrix for Three Data Sets

Use *Maple* to find the covariances $\text{cov}(\mathbf{x}, \mathbf{y})$, $\text{cov}(\mathbf{x}, \mathbf{z})$, and $\text{cov}(\mathbf{y}, \mathbf{z})$ of the data sets represented by the columns of the matrix

$$A = \begin{bmatrix} 1 & 3.0 & 2.0 \\ 2 & -2 & 2 \\ 1 & 1 & -1 \\ 1.5 & 0 & 1 \\ 7 & 7 & -9 \end{bmatrix}$$

where **x**, **y**, and **z** are the first, second, and third columns of A, respectively.

Solution. We use the **describe** and **covariance** functions in the **stats** package.

```
[> with(stats):
[> x:=[1,2,1,1.5,7]:
[> y:=[3.0,-2,1,0,7]:
[> z:=[2.0,2,-1,1,-9]:
[> cov:=(x,y)-> describe[covariance](x,y):
```

```
[> covA:=matrix([
    [cov(x,x),cov(x,y),cov(x,z)],
    [cov(y,x),cov(y,y),cov(y,z)],
    [cov(z,x),cov(z,y),cov(z,z)]]);
```

$$covA := \begin{bmatrix} 5.200000000 & 5.300000000 & -8.800000000 \\ 5.300000000 & 9.360000000 & -10.60000000 \\ -8.800000000 & -10.60000000 & 17.20000000 \end{bmatrix}$$

The value $covA_{12} = 5.3$ is $\text{cov}(\mathbf{x}, \mathbf{y})$; the value $covA_{13} = -8.8$ is $\text{cov}(\mathbf{x}, \mathbf{z})$; and the value $covA_{23} = -10.6$ is $\text{cov}(\mathbf{y}, \mathbf{z})$. ▲

EXAMPLE 7.29 The Standard Deviations of Three Data Sets

Use *Maple* to find the standard deviations of the data sets represented by the columns of the matrix A in Example 7.28.

Solution. We use the **describe** and **standarddeviation** functions in the **stats** package.

```
[> with(stats):
[> x:=[1,2,1,1.5,7]:
[> y:=[3.0,-2,1,0,7]:
[> z:=[2.0,2,-1,1,-9]:
```

```
[> stdev:=x->describe[standarddeviation](x):
[> [stdev(x),stdev(y),stdev(z)];
```

$$[2.280350850, 3.059411708, 4.147288271]$$

The value 2.28035 is the standard deviation of the first column; the value 3.05941 is the standard deviation of the second column; and the value 4.14729 is the standard deviation of the third column. ▲

EXERCISES 7.11

1. Use the definition to find the Pearson correlation coefficient for each of the following data sets.

   a. $(x_1, y_1) = (1, -1)$, $(x_2, y_2) = (2, 1)$, $(x_3, y_3) = (3, 7)$, $(x_4, y_4) = (4, 5)$, $(x_5, y_5) = (5, 6)$

   b. $(x_1, y_1) = (1.5, 2.7)$, $(x_2, y_2) = (-3, 4)$, $(x_3, y_3) = (3, 4)$, $(x_4, y_4) = (5, 0)$, $(x_5, y_5) = (6.75, 3)$

   c. $(x_1, y_1) = (1, 0)$, $(x_2, y_2) = (2, 3)$, $(x_3, y_3) = (3, 0)$, $(x_4, y_4) = (4, 3)$, $(x_5, y_5) = (5, 0)$

2. (*Maple V*) Use the **stats[describe,covariance]** function to find the covariances of the data sets in Exercise 1.

3. (*Maple V*) Use the **stats[describe,linearcorrelation]** function to find the Pearson correlation coefficients of the data pairs in Exercise 1.

4. (*Maple V*) Consider the following data.

	x	y	z
Observation 1	1.0	1.3	1.3
Observation 2	1.1	1.2	1.2
Observation 3	2.0	1.9	1.9
Observation 4	1.0	8.0	8.0
Observation 4	3.0	4.0	4.0
Observation 6	5.0	−6.0	6.0

a. Use the standard inner product on $\mathbb{R}^6$ to calculate the covariances $\text{cov}(\mathbf{x}, \mathbf{y})$, $\text{cov}(\mathbf{x}, \mathbf{z})$ and the standard deviations $\sigma_{\mathbf{x}}$, $\sigma_{\mathbf{y}}$, and $\sigma_{\mathbf{z}}$ for the data sets $\mathbf{x}$, $\mathbf{y}$, and $\mathbf{z}$.

b. Find a nonzero vector $\mathbf{u} \in \mathbb{R}^6$ for which $.5 \leq \text{cov}(\mathbf{x}, \mathbf{u}) \leq .75$.

c. Find a nonzero vector $\mathbf{u} \in \mathbb{R}^6$ for which $-.75 \leq \text{cov}(\mathbf{x}, \mathbf{u}) \leq -.5$.

d. Find a nonzero vector $\mathbf{u} \in \mathbb{R}^6$ for which $\sigma_{\mathbf{u}} > 4$.

e. Find a nonzero vector $\mathbf{u} \in \mathbb{R}^6$ for which $\sigma_{\mathbf{u}} < .4$.

5. (*Maple 6*) Use the definition to find the Pearson correlation coefficients for the following pairs of matrices.

a. $\begin{bmatrix} 8 & 4 & -5 \\ -5 & 3 & -5 \end{bmatrix}, \begin{bmatrix} 8 & 5 & -1 \\ 0 & 3 & -4 \end{bmatrix}$ b. $\begin{bmatrix} 1 & 2 & 3 \\ 4 & 5 & 6 \end{bmatrix}, \begin{bmatrix} 2 & 4 & 6 \\ 8 & 10 & 12 \end{bmatrix}$

c. $\begin{bmatrix} 1 & 2 & 3 \\ 4 & 5 & 6 \end{bmatrix}, \begin{bmatrix} -1 & -2 & -3 \\ -4 & -5 & -6 \end{bmatrix}$ d. $\begin{bmatrix} 1 & 2 & 3 \\ 4 & 5 & 6 \end{bmatrix}, \begin{bmatrix} 6 & 5 & 4 \\ 3 & 2 & 1 \end{bmatrix}$

6. (*Maple 6*) Use the **stats[describe,standarddeviation]** function to find the standard deviations of the data sets in Exercise 5.

QUADRATIC FORMS

Inner products are special cases of functions $f : V \times V \to \mathbb{R}$ satisfying the equations

$$f(a_1\mathbf{x}_1 + a_2\mathbf{x}_2, \mathbf{y}) = a_1 f(\mathbf{x}_1, \mathbf{y}) + a_2 f(\mathbf{x}_2, \mathbf{y})$$
$$f(\mathbf{x}, b_1\mathbf{y}_1 + b_2\mathbf{y}_2) = b_1 f(\mathbf{x}, \mathbf{y}_1) + b_2 f(\mathbf{x}, \mathbf{y}_2)$$

for all vectors $\mathbf{x}, \mathbf{y} \in V$. Such functions are known as ***bilinear forms***. A bilinear form is ***symmetric*** if $f(\mathbf{x}, \mathbf{y}) = f(\mathbf{y}, \mathbf{x})$ for all $\mathbf{x} \in V$. Thus inner products are symmetric bilinear forms. We now show that every symmetric bilinear form is determined by a symmetric matrix relative to a fixed basis $\mathcal{B}$ for V. We simplify the notation in this section by often suppressing the distinction between a vector $\mathbf{v} \in \mathbf{V}$ and its coordinate representation $[\mathbf{v}]_{\mathcal{B}} \in \mathbb{R}^n$. We apply this result to study the quadratic equations $q(\mathbf{x}) = f(\mathbf{x}, \mathbf{x})$ of standard geometric objects determined by symmetric bilinear forms.

THEOREM 7.11 (Bilinear form theorem) *For every bilinear form $f : V \times V \to \mathbb{R}$, there exists a matrix A for which $f(\mathbf{x}, \mathbf{y}) = \mathbf{x}^T A\mathbf{y}$.*

Proof. Suppose that $\mathcal{B} = \{\mathbf{v}_1, \mathbf{v}_2\}$ is a basis for V. Then bilinearity of f implies that for any two vectors $\mathbf{x} = a_1\mathbf{v}_1 + a_2\mathbf{v}_2$ and $\mathbf{y} = b_1\mathbf{v}_1 + b_2\mathbf{v}_2$, the scalar $f(\mathbf{x}, \mathbf{y})$ can be written as matrix products:

$$f(\mathbf{x}, \mathbf{y}) = \begin{bmatrix} a_1 & a_2 \end{bmatrix} \begin{bmatrix} f(\mathbf{v}_1, \mathbf{v}_1) & f(\mathbf{v}_1, \mathbf{v}_2) \\ f(\mathbf{v}_2, \mathbf{v}_1) & f(\mathbf{v}_2, \mathbf{v}_2) \end{bmatrix} \begin{bmatrix} b_1 \\ b_2 \end{bmatrix}$$
$$= \begin{bmatrix} a_1 \\ a_2 \end{bmatrix}^T \begin{bmatrix} f(\mathbf{v}_1, \mathbf{v}_1) & f(\mathbf{v}_1, \mathbf{v}_2) \\ f(\mathbf{v}_2, \mathbf{v}_1) & f(\mathbf{v}_2, \mathbf{v}_2) \end{bmatrix} \begin{bmatrix} b_1 \\ b_2 \end{bmatrix}$$

This shows that $f(\mathbf{x}, \mathbf{y}) = \mathbf{x}^T A\mathbf{y}$, where $A = \begin{bmatrix} a_{ij} \end{bmatrix}$ and $a_{ij} = f(\mathbf{v}_i, \mathbf{v}_j)$. The general case is analogous. ■

EXAMPLE 7.30 A Bilinear Form on $\mathbb{R}^3$

Represent the function $f : \mathbb{R}^3 \times \mathbb{R}^3 \to \mathbb{R}$ defined by

$$f(x_1, x_2, x_3, y_1, y_2, y_3) = 6x_1y_1 - 2x_1y_2 + 4x_2y_1 - 8x_2y_3 + 4x_3y_2 - x_3y_3$$

in the form $\mathbf{x}^T A\mathbf{y}$.

Solution. Let

$$A = \begin{bmatrix} 6 & -2 & 0 \\ 4 & 0 & -8 \\ 0 & 4 & -1 \end{bmatrix}$$

Then

$$f(\mathbf{x}, \mathbf{y}) = \begin{bmatrix} x_1 & x_2 & x_3 \end{bmatrix} \begin{bmatrix} 6 & -2 & 0 \\ 4 & 0 & -8 \\ 0 & 4 & -1 \end{bmatrix} \begin{bmatrix} y_1 \\ y_2 \\ y_3 \end{bmatrix}$$
$$= 6x_1y_1 - 2x_1y_2 + 4x_2y_1 - 8x_2y_3 + 4x_3y_2 - x_3y_3$$

Therefore, $f(\mathbf{x}, \mathbf{y}) = \mathbf{x}^T A\mathbf{y}$. ▲

It follows easily from Theorem 7.11 that symmetric bilinear forms are represented by symmetric matrices.

THEOREM 7.12 *Every symmetric bilinear form can be represented by a symmetric matrix.*

Proof. Suppose that $f : V \times V \to \mathbb{R}$ is a symmetric bilinear form and let $\mathcal{B} = \{\mathbf{v}_1, \ldots, \mathbf{v}_n\}$ be a basis for V. By Theorem 7.11, the matrix $A = [a_{ij}]$, with $a_{ij} = f(\mathbf{v}_i, \mathbf{v}_j)$, has the property that $f(\mathbf{x}, \mathbf{y}) = \mathbf{x}^T A\mathbf{y}$ for all $\mathbf{x}, \mathbf{y} \in V$. Since f is symmetric, $f(\mathbf{v}_i, \mathbf{v}_j) = f(\mathbf{v}_j, \mathbf{v}_i)$. Hence the matrix A is symmetric. ■

We now use symmetric bilinear forms to define ***quadratic forms***.

DEFINITION 7.17 *A* ***quadratic form*** $q : V \to \mathbb{R}$ *on a real inner product space* V *is a function* $q(\mathbf{x}) = f(\mathbf{x}, \mathbf{x})$ *for some symmetric bilinear form* $f : V \times V \to \mathbb{R}$.

Since every symmetric bilinear form can be represented by a symmetric matrix, it is clear that every quadratic form can also be represented by a symmetric matrix. The study of quadratic forms is thus intimately connected with the study of symmetric matrices.

EXAMPLE 7.31 The Matrix of a Quadratic Form on $\mathbb{R}^2$

Use *Maple* to find the quadratic form $q(\mathbf{x}) : \mathbb{R}^2 \to \mathbb{R}$ determined by

$$A = \begin{bmatrix} 1 & 2 \\ 2 & 3 \end{bmatrix} \quad \text{and} \quad \mathbf{x} = \begin{bmatrix} x \\ y \end{bmatrix}$$

Solution. Since *Maple* detects from the context whether a vector is a row or column vector, we can write $\mathbf{x}A\mathbf{x}$ in place of $\mathbf{x}^T A\mathbf{x}$.

```
[> with(linalg):
[> v:=vector(2,[x,y]):
[> A:=matrix([[1,2],[2,3]]):
```

```
[> simplify(evalm(v&*A&*v));
```

$$x^2 + 4xy + 3y^2$$

This shows that $q(\mathbf{x}) = \mathbf{x}^T A\mathbf{x} = x^2 + 4xy + 3y^2 = q(x, y)$. ▲

EXAMPLE 7.32 A Quadratic Form on $\mathbb{R}^2$ as a Matrix Product

Use *Maple* to convert the quadratic form $q(x, y) = 7x^2 - 9y^2$ into a matrix product.

Solution. We use *Maple* to find A so that

$$q(x, y) = q(\mathbf{x}) = \mathbf{x}^T A\mathbf{x} = \begin{bmatrix} x & y \end{bmatrix} \begin{bmatrix} a & b \\ b & c \end{bmatrix} \begin{bmatrix} x \\ y \end{bmatrix}$$

```
[> with(linalg):
[> v:=vector(2,[x,y]):
[> A:=matrix([[a,b],[b,c]]):
```

```
[> simplify(evalm(v&*A&*v));
```

$$x^2 a + 2xyb + y^2 c$$

This tells us that we must have $ax^2+2bxy+cy^2 = 7x^2-9y^2$. By comparing coefficients we conclude that $a = 7$, $b = 0$, and $c = -9$. Hence

$$q(\mathbf{x}) = \begin{bmatrix} x & y \end{bmatrix} \begin{bmatrix} 7 & 0 \\ 0 & -9 \end{bmatrix} \begin{bmatrix} x \\ y \end{bmatrix}$$

is the required matrix form. ▲

EXAMPLE 7.33 A Quadratic Form on $\mathbb{R}^3$ as a Matrix Product

Use *Maple to* convert the quadratic form

$$q(\mathbf{x}) = q(x, y, z) = 3x^2 - 2xy + 3xz + 6y^2 + 5yz - 8z^2$$

to a matrix product.

Solution. We know from the definition of quadratic forms that the required matrix is symmetric. Suppose, therefore, that

$$A = \begin{bmatrix} d & a & b \\ a & e & c \\ b & c & f \end{bmatrix}$$

is the required matrix. We use *Maple* to find $q(\mathbf{x})$.

```
[> with(linalg):
[> v:=vector(3,[x,y,z]):
[> A:=matrix([[d,a,b],[a,e,c],[b,c,f]]):
```

```
[> simplify(evalm(v&*A&*v))[1,1];
```

$$x^2d + 2xya + 2xzb + y^2e + 2yzc + z^2f$$

This tells us that the entries of the matrix must satisfy the equation

$$dx^2 + 2axy + ey^2 + 2bxz + 2cyz + fz^2 = 3x^2 - 2xy + 3xz + 6y^2 + 5yz - 8z^2$$

By comparing coefficients, we can see that $d = 3$, $a = -1$, $e = 6$, $2b = 3$, $2c = 5$, and $f = -8$. This means that

$$\begin{bmatrix} x & y & z \end{bmatrix} \begin{bmatrix} 3 & -1 & \frac{3}{2} \\ -1 & 6 & \frac{5}{2} \\ \frac{3}{2} & \frac{5}{2} & -8 \end{bmatrix} \begin{bmatrix} x \\ y \\ z \end{bmatrix}$$

is the matrix form of $q(\mathbf{x})$. ▲

EXERCISES 7.12

1. Rewrite the following bilinear forms as matrix products $\mathbf{x}^T A\mathbf{y}$.

   a. $f(u, v, x, y) = 6xu + 5xv + 5yu + 8vy$

   b. $f(u, v, x, y) = 6xu + 5xv + 8yu + 8vy$

   c. $f(u, v, x, y) = -6xu + 5xv + 8yu - 8vy$

2. Rewrite the following symmetric bilinear forms $f(u, v, x, y)$ in equational form.

   a. $\begin{bmatrix} u & v \end{bmatrix}\begin{bmatrix} 6 & 0 \\ 0 & 8 \end{bmatrix}\begin{bmatrix} x \\ y \end{bmatrix}$ b. $\begin{bmatrix} u & v \end{bmatrix}\begin{bmatrix} 6 & -1 \\ -1 & 8 \end{bmatrix}\begin{bmatrix} x \\ y \end{bmatrix}$

   c. $\begin{bmatrix} u & v \end{bmatrix}\begin{bmatrix} -6 & 2 \\ 2 & 8 \end{bmatrix}\begin{bmatrix} x \\ y \end{bmatrix}$ d. $\begin{bmatrix} u & v \end{bmatrix}\begin{bmatrix} -6 & 2 \\ 2 & -8 \end{bmatrix}\begin{bmatrix} x \\ y \end{bmatrix}$

3. Decide which of the following functions $\mathbb{R}^2 \to \mathbb{R}$ are quadratic forms. Explain your answer.

   a. $q(x, y) = x + y$ b. $q(x, y) = xy$
   c. $q(x, y) = x^2 + x + y^2$ d. $q(x, y) = 3x^2 + 2xy - y^2$
   e. $q(x, y) = 5y^2$ f. $q(x, y) = 5y^2 + 7$
   g. $q(x, y) = 0$ h. $q(x, y) = x^2 + xy$

4. Find the bilinear forms associated with the following quadratic forms.

   a. $q(x, y) = x^2 + y^2$ b. $q(x, y) = 3x^2 + 2y^2$
   c. $q(x, y) = x^2 - 2xy + y^2$ d. $q(x, y) = x^2$

5. Show that if q is a quadratic form determined by a symmetric bilinear form f, then

$$f(\mathbf{x}, \mathbf{y}) = \frac{1}{2}(q(\mathbf{x} + \mathbf{y}) - q(\mathbf{x}) - q(\mathbf{y}))$$

6. (*Maple V*) Show that the function $f : \mathbb{R}^2 \times \mathbb{R}^2 \to \mathbb{R}$ defined by

$$f(a, b, c, d) = \det\begin{bmatrix} a & c \\ b & d \end{bmatrix}$$

   is a bilinear form.

7. (*Maple 6*) Show that if $f : \mathbb{R}^2 \times \mathbb{R}^2 \to \mathbb{R}$ is a bilinear form for which $f(\mathbf{x}, \mathbf{y}) = -f(\mathbf{y}, \mathbf{x})$ for all $\mathbf{x}, \mathbf{y} \in \mathbb{R}^2$, then

$$f(a, b, c, d) = s \det \begin{bmatrix} a & c \\ b & d \end{bmatrix}$$

for some scalar s. (*Hint*: Use the standard basis of $\mathbb{R}^2$.)

8. (*Maple 6*) Find the quadratic form $q(\mathbf{x}) : \mathbb{R}^3 \to \mathbb{R}$ determined by

$$A = \begin{bmatrix} 1 & 2 & 3 \\ 2 & 4 & 5 \\ 3 & 5 & 6 \end{bmatrix} \quad \text{and} \quad \mathbf{x} = \begin{bmatrix} x \\ y \\ z \end{bmatrix}$$

The Principal Axis Theorem

In Chapter 8, we will prove that for every real symmetric matrix A there exists an orthogonal matrix Q and a diagonal matrix D for which $A = QDQ^T$. We now show how this result can be used to simplify the equations of standard geometric objects.

THEOREM 7.13 (Principal axis theorem) *For every real quadratic form $q(\mathbf{x}) = \mathbf{x}^T A\mathbf{x}$ there exists an orthogonal matrix Q and a diagonal matrix D for which $\mathbf{x} = Q\mathbf{y}$ and $q(\mathbf{y}) = \mathbf{y}^T D\mathbf{y}$.*

Proof. We know that the quadratic form $q(\mathbf{x})$ can be represented in the form $\mathbf{x}^T A\mathbf{x}$ for some symmetric matrix A. By Corollary 8.36, the matrix A is orthogonally diagonalizable. Therefore, there exists a diagonal matrix D and an orthogonal matrix Q such that $A = QDQ^T$. Let $\mathbf{y} = Q^T\mathbf{x}$. Then $\mathbf{x} = Q\mathbf{y}$ and

$$q(\mathbf{x}) = \mathbf{x}^T A\mathbf{x} = (Q\mathbf{y})^T A(Q\mathbf{y}) = \mathbf{y}^T Q^T A Q\mathbf{y} = \mathbf{y}^T D\mathbf{y}$$

Thus $q(\mathbf{y}) = \mathbf{y}^T D\mathbf{y}$ is the required quadratic form. ■

Theorem 7.13 tells us that we can always change the coordinate vector $\mathbf{x}$ to a vector

$$\mathbf{y} = \begin{bmatrix} y_1 \\ \vdots \\ y_n \end{bmatrix} = Q^T\mathbf{x}$$

such that $D = \text{diag}(\lambda_1, \ldots, \lambda_n)$ and $q(\mathbf{y}) = \mathbf{y}^T D\mathbf{y} = \lambda_1 y_1^2 + \cdots + \lambda_n y_n^2$.

EXAMPLE 7.34 A Diagonal Representation of a Quadratic Form in $\mathbb{R}^3$

Use *Maple* to represent the quadratic form $q(x_1, x_2) = 3x_1^2 - 2x_1x_2 + 3x_2^2$ by a diagonal matrix.

Solution. We begin by writing $q(\mathbf{x})$ in matrix form:

$$q(\mathbf{x}) = \mathbf{x}^T A\mathbf{x} = \begin{bmatrix} x_1 & x_2 \end{bmatrix} \begin{bmatrix} 3 & -1 \\ -1 & 3 \end{bmatrix} \begin{bmatrix} x_1 \\ x_2 \end{bmatrix}$$

Since A is a real symmetric matrix, it is orthogonally diagonalizable. This means that $A = QDQ^T$ for some diagonal matrix D and some orthogonal matrix Q. We know that the diagonal entries of D must be the eigenvalues of A and the columns of Q must be the normalized eigenvectors associated with the eigenvalues of D. Moreover, $\mathbf{x} = Q\mathbf{y}$ and $q(\mathbf{y}) = \mathbf{y}^T Q^T A Q\mathbf{y}$. To represent $q(\mathbf{x})$ in the diagonal form $q(\mathbf{y})$, we must therefore find the matrices D and Q.

```
[> A:=matrix([[3,-1],[-1,3]]):
[> linalg[eigenvectors](A);
```

$$[4, 1, \{[-1, 1]\}], [2, 1, \{[1, 1]\}]$$

Thus the scalars 2 and 4 are the eigenvalues of A, and $(1, 1)$ and $(-1, 1)$ are corresponding eigenvectors. Moreover, the dot product

$$\begin{bmatrix} 1 & 1 \end{bmatrix} \begin{bmatrix} -1 \\ 1 \end{bmatrix} = 0$$

of the returned eigenvectors is zero. Hence the vectors are orthogonal. To form the orthogonal matrix Q, we normalize these vectors in the two-norm. Since

$$\|(1, 1)\| = \sqrt{1^2 + 1^2} = \sqrt{2} = \sqrt{1^2 + [-1]^2} = \|(-1, 1)\|$$

we let $y_1 = \frac{1}{\sqrt{2}}(1, 1)$ and $y_2 = \frac{1}{\sqrt{2}}(-1, 1)$. We use these normalized vectors as the columns of the required matrix Q. Let

$$Q = \begin{bmatrix} \frac{1}{\sqrt{2}} & -\frac{1}{\sqrt{2}} \\ \frac{1}{\sqrt{2}} & \frac{1}{\sqrt{2}} \end{bmatrix} \quad \text{and} \quad Q^T = \begin{bmatrix} \frac{1}{\sqrt{2}} & \frac{1}{\sqrt{2}} \\ -\frac{1}{\sqrt{2}} & \frac{1}{\sqrt{2}} \end{bmatrix}$$

We must verify that $A = QDQ^T$.

```
[> with(linalg):
[> Q:=matrix([[1/sqrt(2),-1/sqrt(2)],[1/sqrt(2),1/sqrt(2)]]):
[> Dm:=diag(2,4):
[> Qt:=transpose(Q):
[> equal(A,evalm(Q&*Dm&*Qt));
```

true

This shows that $q(y_1, y_2) = 2y_1^2 + 4y_2^2$ is the required diagonal representation. ▲

Since the matrix Q in the transformation $A \to Q^T A Q = D$ in the previous example is orthogonal, the transformation preserves the basic geometric properties of length and angles;

it merely changes the position of graph of the form. This property explains in part the following definition: Two symmetric matrices A and B are said to be ***congruent*** if there exists an orthogonal matrix Q for which $A = Q^T BQ$. The matrices A and D in Example 7.34 are congruent.

EXAMPLE 7.35 Congruence and Quadratic Forms

We know from Example 7.34 that the matrices

$$\begin{bmatrix} 3 & -1 \\ -1 & 3 \end{bmatrix} \quad \text{and} \quad \begin{bmatrix} 2 & 0 \\ 0 & 4 \end{bmatrix}$$

are congruent. Use *Maple* to graph the quadratic equations

$$3x^2 - 2xy + y^2 = 10 \quad \text{and} \quad 2x^2 + 4y^2 = 10$$

determined by these matrices.

Solution. We load the **plots** package and use the **implicitplot** function.

```
[> with(plots):
[> implicitplot({
    3*x^2-2*x*y+3*y^2=10,
    2*x^2+4*y^2=10},
    x=-4..4, y=-4..4);
```

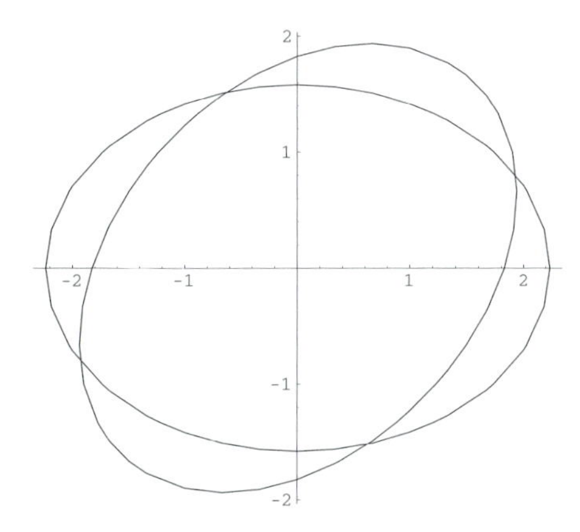

The shape and position of the two graphs suggest that we can obtain the graph of the equation $2x^2 + 4y^2 = 10$ by rotating the graph of the equation $3x^2 - 2xy + y^2 = 10$ through $\pi/4$ radians about the origin. ▲

Conic Sections

One of the benefits of the diagonal representation of quadratic forms is that the geometric interpretation of the associated quadratic equations is immediate. We know from geometry, for instance, that the equation

$$2y_1^2 + 4y_2^2 = 1$$

represents an ellipse in $\mathbb{R}^2$ since a quadratic equation of the form $ax^2 + by^2 = 1$ represents an ellipse for all $a, b > 0$ and $a \neq b$. The equation represents a circle if $a, b > 0$ and $a = b$. An interpretation of an equation of the form

$$ax^2 + bxy + cy^2 = 1$$

with an xy term is less obvious. However, we know that we can use an orthogonal change of coordinates to represent the given geometric object by an equivalent equation without an xy term.

EXAMPLE 7.36 An Ellipse in $\mathbb{R}^2$

Use *Maple* to graph the quadratic equation $3x^2 + 5y^2 = 1$.

Solution. We use the **implicitplot** function in the **plots** package.

```
[>with(plots):
[>implicitplot(
   3*x^2+5*y^2=1,
   x=-sqrt(3)..sqrt(3),
   y=-sqrt(2)..sqrt(2));
```

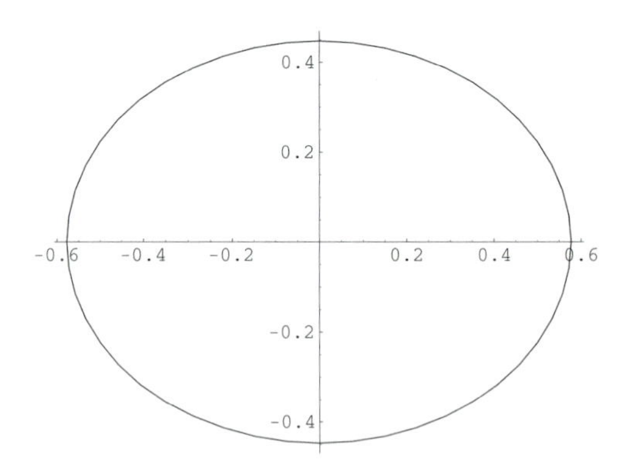

As we can see, the graph of the equation $3x^2 + 5y^2 = 1$ has the shape of an ellipse. ▲

EXAMPLE 7.37 A Hyperbola in $\mathbb{R}^2$ Centered around the *y*-Axis

Use *Maple* to graph the quadratic equation $3x^2 - 5y^2 = 1$.

Solution. We use the **implicitplot** function in the **plots** package.

```
[> with(plots):
[> implicitplot(
    3*x^2-5*y^2=1,
    x=-sqrt(3)..sqrt(3),
    y=-sqrt(2)..sqrt(2));
```

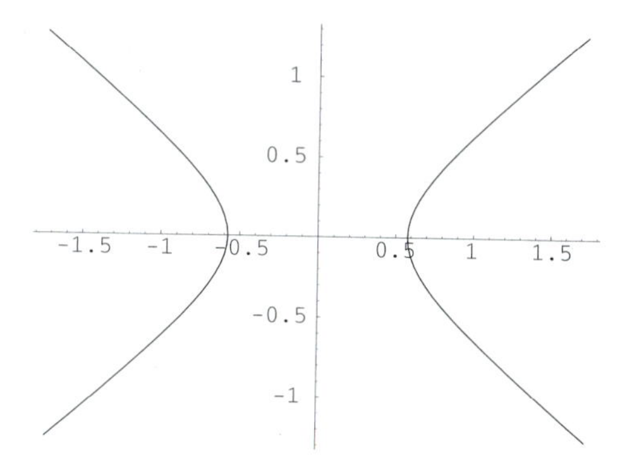

As we can see, the graph of the equation $3x^2 - 5y^2 = 1$ has the shape of a hyperbola centered around the y-axis. ▲

EXAMPLE 7.38 A Hyperbola in $\mathbb{R}^2$ Centered around the x-Axis

Use *Maple* to graph the quadratic equation $5y^2 - 3x^2 = 1$.

Solution. We use the **implicitplot** function in the **plots** package.

```
[> with(plots):
[> implicitplot(
    5*y^2-3*x^2=1,
    x=-5..5,
    y=-sqrt(5)..sqrt(5));
```

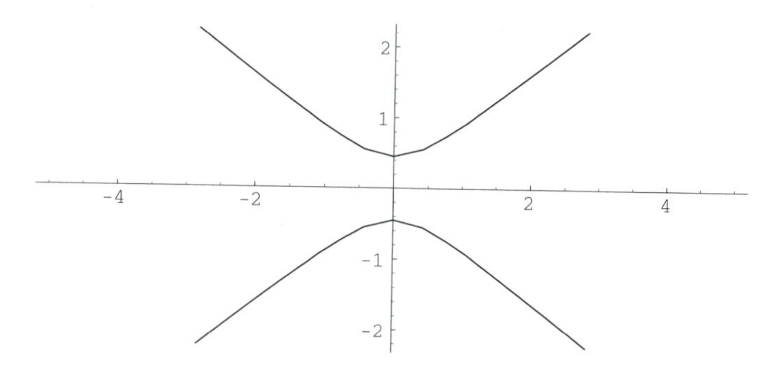

As we can see, the graph of the equation $5y^2 - 3x^2 = 1$ has the shape of a hyperbola centered around the x-axis. ▲

Quadric Surfaces

The following theorem, known as Sylvester's theorem, shows that all real quadratic forms can be written in a canonical way that reveals the properties of the geometric objects represented by the form. The theorem uses the fact that every real symmetric matrix is congruent to exactly one diagonal matrix with diagonal entries 1, -1, and 0, in which all positive diagonal entries precede all negative diagonal entries and all negative diagonal entries precede all zero entries.

The set of these diagonal matrices is a set of canonical forms for real symmetric matrices under the relation of ***congruence***. We recall from our earlier definition that two symmetric matrices A and B are ***congruent*** if there exists an orthogonal matrix Q for which $A = Q^T BQ$. Sylvester's theorem says that every real quadratic form is congruent to a diagonal matrix of the form just described. Up to congruence, the theorem therefore provides a complete finite classification of real quadratic forms.

THEOREM 7.14 (Sylvester's theorem) *If q is a real quadratic form on $\mathbb{R}^n$, then there exists integers s and r, with $s \leq r \leq n$, depending uniquely on q, such that $q(x_1, \ldots, x_n) = x_1^2 + \cdots + x_s^2 - x_{s+1}^2 - \cdots - x_r^2$ in some orthonormal basis for $\mathbb{R}^n$.*

Proof. We will show in Corollary 8.36 that there exists an orthonormal basis for $\mathbb{R}^n$ in which

$$q(y_1, \ldots, y_n) = \begin{bmatrix} y_1 & \cdots & y_n \end{bmatrix} D \begin{bmatrix} y_1 \\ \vdots \\ y_n \end{bmatrix}$$

for some diagonal matrix $D = \text{diag}(\lambda_1, \ldots, \lambda_n) \in \mathbb{R}^{n \times n}$. Since real numbers can be ordered, we can arrange the diagonal entries of D so that all positive $\lambda_1 \geq \cdots \geq \lambda_s$ precede all negative $\lambda_{s+1} \leq \cdots \leq \lambda_r$ and so that these precede all $\lambda_{r+1} = \cdots = \lambda_n = 0$. Moreover, since every positive real number λ has a square root μ, we can put $\lambda_i y_i^2 = \mu_i^2 y_i^2 = x_i^2$ for $\lambda_i > 0$ and $\lambda_i y_i^2 = -\mu_i^2 y_i^2 = -x_i^2$ for $\lambda_i < 0$. This yields

$$q(x_1, \ldots, x_n) = x_1^2 + \cdots + x_s^2 - x_{s+1}^2 - \cdots - x_r^2$$

We know that r is unique since it is the number of nonzero diagonal terms and is therefore equal to the rank of D, and diagonalization preserves rank. It therefore remains to show that s is unique. Suppose that

$$q(z_1, \ldots, z_n) = z_1^2 + \cdots + z_t^2 - z_{t+1}^2 - \cdots - z_r^2$$

is another representation of q with t positive terms. Let $s \neq t$ and assume that $t < s$. Now consider the two subspaces

$$U = \{(x_1, \ldots, x_n) \in \mathbb{R}^n : x_1 = \cdots = x_t = 0\}$$
$$W = \{(z_1, \ldots, z_n) \in \mathbb{R}^n : z_{s+1} = \cdots = z_n = 0\}$$

of $\mathbb{R}^n$. The space U is of dimension $n - t$ and W is of dimension s. Moreover, the dimension of $U + W$ is at most $\dim \mathbb{R}^n = n$. By Theorem 4.37, we therefore have

$$\begin{aligned}\dim (U \cap W) &= \dim U + \dim W - \dim (U + W) \\ &\geq (n - t) + s - n = s - t\end{aligned}$$

Since $t < s$, we can find a nonzero vector $\mathbf{x} \in U \cap W$. The definition of U and the fact that

$$q(z_1, \ldots, z_n) = z_1^2 + \cdots + z_t^2 - z_{t+1}^2 - \cdots - z_r^2$$

imply that $q(\mathbf{x}) \leq 0$. On the other hand, it follows from the definition of W and the fact that

$$q(x_1, \ldots, x_n) = x_1^2 + \cdots + x_s^2 - x_{s+1}^2 - \cdots - x_r^2$$

that $q(\mathbf{x}) > 0$. This is a contradiction. Hence we must have $s = t$. ■

Sylvester's theorem tells us that by choosing a suitable basis, we can make an ellipse look like a circle, but we can never choose a basis, for example, that makes an ellipse look like a hyperbola.

EXAMPLE 7.39 Circles and Ellipses

Use *Maple* to superimpose the graphs of two sections of the quadratic forms $q_1(x, y) = 2x^2 + 5y^2$ and $q_2(x, y) = x^2 + y^2$.

Solution. We use the **implicitplot** function in the **plots** package.

```
[>with(plots):
[>implicitplot({2*x^2+5*y^2=10,x^2+y^2=5},x=-4..4,y=-4..4);
```

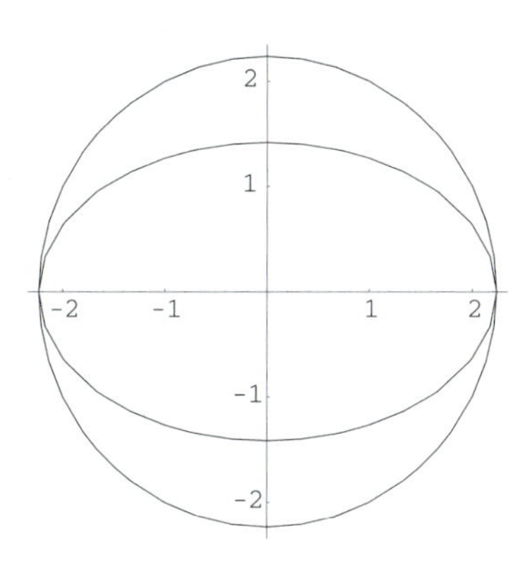

The graph consists of an ellipse inside of a circle. ▲

The number of plus and minus signs in the canonical representation of a quadratic form provided by Sylvester's theorem is often called the ***signature*** of the form. Theorem 7.14 proves that the signature of a quadratic form is unique and can be used to classify these forms. We use *Maple* to illustrate Sylvester's theorem by graphing the five possible canonical quadric surfaces determined by the signs of the eigenvalues λ_1 and λ_2. Let $q(x_1, x_n) = \lambda_1 x_1^2 + \lambda_2 x_2^2$, and consider the following five cases:

1. $\lambda_1 > 0$ and $\lambda_2 > 0$ 2. $\lambda_1 > 0$ and $\lambda_2 < 0$ 3. $\lambda_1 < 0$ and $\lambda_2 < 0$
4. $\lambda_1 > 0$ and $\lambda_2 = 0$ 5. $\lambda_1 < 0$ and $\lambda_2 = 0$

We can think of these forms as representations of three-dimensional geometric objects and use the **plot3d** function to graph them. To simplify the labeling, we will write x in place of x_1 and y in place of x_2.

EXAMPLE 7.40 A Concave-Up Circular Paraboloid in $\mathbb{R}^3$

```
[>plot3d(x^2+y^2,x=-10..10,y=-10..10);
```

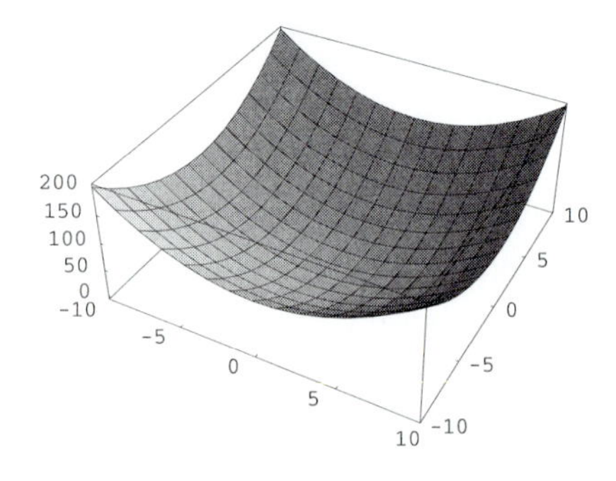

The graph is concave up and has the shape of a circular paraboloid. ▲

EXAMPLE 7.41 A Hyperbolic Paraboloid in $\mathbb{R}^3$

```
[>plot3d(x^2-y^2,x=-10..10,y=-10..10);
```

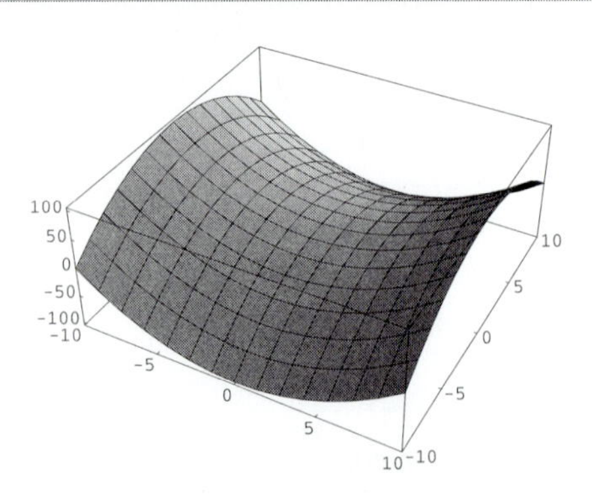

The graph has the shape of a hyperbolic paraboloid. ▲

EXAMPLE 7.42 A Concave-Down Circular Paraboloid in $\mathbb{R}^3$

```
[> plot3d(-x^2-y^2,x=-10..10,y=-10..10);
```

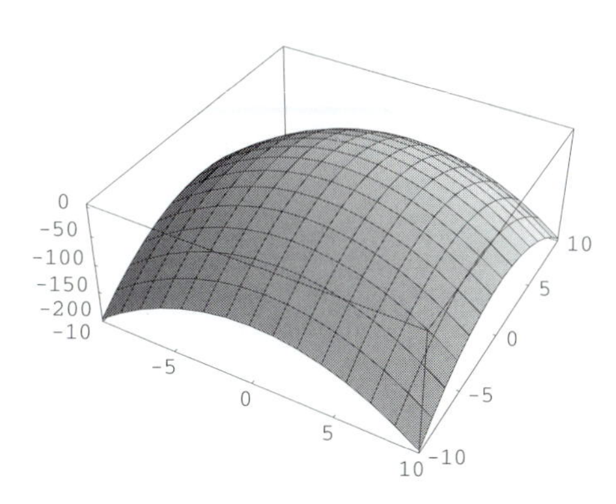

The graph is concave down and has the shape of a circular paraboloid. ▲

EXAMPLE 7.43 A Concave-Up Parabolic Cylinder in $\mathbb{R}^3$

```
[> plot3d(x^2,x=-10..10,y=-10..10);
```

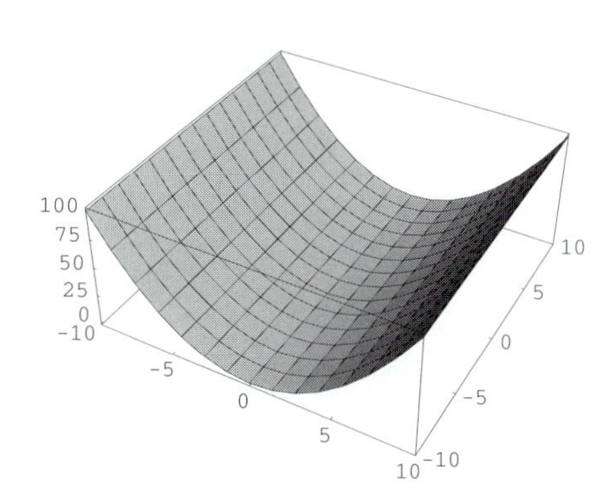

The graph is concave up and has the shape of a parabolic cylinder. ▲

EXAMPLE 7.44 A Concave-Down Parabolic Cylinder in $\mathbb{R}^3$

```
[> plot3d(-x^2,x=-10..10,y=-10..10);
```

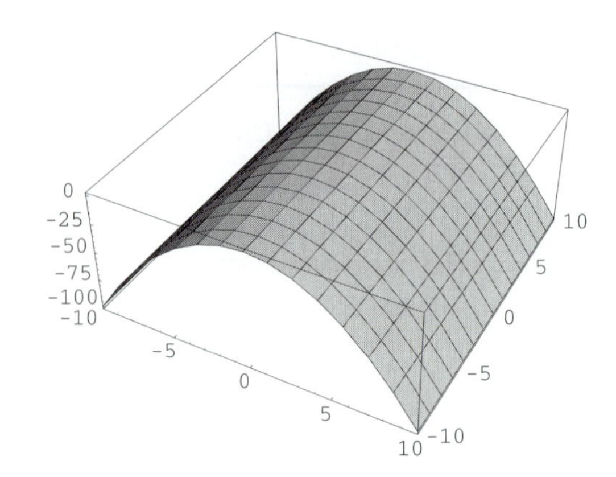

The graph is concave up and has the shape of a parabolic cylinder. ▲

EXERCISES 7.13

1. (*Maple V*) Use the **plots[implicitplot]** function to graph the following equations.

a. $4x^2 + 7y^2 = 1$	b. $4x^2 - 7y^2 = 1$
c. $-x^2 + 10xy + 2y^2 = 1$	d. $x^2 + 10xy + y^2 = 1$

2. (*Maple V*) Use quadratic forms of your choice to illustrate Sylvester's theorem for quadratic forms in two, three, and four variables.

3. (*Maple 6*) Decompose the following matrices into a product of the form QDQ^T, consisting of an orthogonal matrix Q and a diagonal matrix D.

$$\text{a. } A = \begin{bmatrix} -5 & -2 \\ -2 & -1 \end{bmatrix} \quad \text{b. } A = \begin{bmatrix} 1 & 0 & 0 \\ 0 & 2 & 4 \\ 0 & 4 & 2 \end{bmatrix} \quad \text{c. } A = \begin{bmatrix} 9 & 5 \\ 5 & 3 \end{bmatrix}$$

COMPLEX INNER PRODUCTS

Certain applications of linear algebra require complex-valued inner products. A ***complex inner product*** on a complex vector space V is a complex-valued function $(\mathbf{x}, \mathbf{y}) \mapsto \langle \mathbf{x}, \mathbf{y} \rangle$ satisfying the following four axioms.

1. $\langle a\mathbf{x} + b\mathbf{y}, \mathbf{z} \rangle = a \langle \mathbf{x}, \mathbf{z} \rangle + b \langle \mathbf{y}, \mathbf{z} \rangle$
2. $\langle \mathbf{x}, \mathbf{y} \rangle = \overline{\langle \mathbf{y}, \mathbf{x} \rangle}$
3. $\langle \mathbf{x}, \mathbf{x} \rangle \geq 0$
4. $\langle \mathbf{x}, \mathbf{x} \rangle = 0$ if and only if $\mathbf{x} = \mathbf{0}$

The scalar $\overline{\langle \mathbf{y}, \mathbf{x} \rangle}$ is the complex conjugate of $\langle \mathbf{x}, \mathbf{y} \rangle$. Complex inner products are no longer symmetric since $\langle \mathbf{x}, \mathbf{y} \rangle$ is not always equal to its complex conjugate. A ***complex inner product space*** is a complex vector space V equipped with a complex inner product.

EXAMPLE 7.45 The Standard Inner Product on $\mathbb{C}^n$

Let $\mathbf{x} = (u_1, \ldots, u_n)$ and $\mathbf{y} = (v_1, \ldots, v_n)$ be two vectors in $\mathbb{C}^n$. Then the function defined by

$$\mathbf{x}^T\overline{\mathbf{y}} = \begin{bmatrix} u_1 & \cdots & u_n \end{bmatrix} \begin{bmatrix} \overline{v}_1 \\ \vdots \\ \overline{v}_n \end{bmatrix} = u_1\overline{v}_1 + \cdots + u_n\overline{v}_n$$

is a complex inner product. ▲

The inner product in Example 7.45 is called the ***standard inner product*** on $\mathbb{C}^n$. As in the real case, we can use it to define a norm.

EXAMPLE 7.46 The Two-Norm on $\mathbb{C}^2$

Suppose that $\mathbf{z} = (z_1, z_2) = (x_1 + y_1 i, x_2 + y_2 i)$ is a vector in the coordinate space $\mathbb{C}^2$, determined by the real numbers x_1, y_1, x_2, and y_2, and that $|z_k| = \sqrt{x_k^2 + y_k^2}$ is the modulus of z_k. Then the function

$$\|\mathbf{z}\|_2 = \sqrt{|z_1|^2 + |z_2|^2} = \sqrt{x_1^2 + y_1^2 + x_2^2 + y_2^2}$$

is a vector norm on $\mathbb{C}^2$. The space $\mathbb{C}^2$, equipped with this norm, known as the ***complex Euclidean 2-space.*** It is also know as a ***unitary space.*** ▲

We will need complex inner product spaces in Chapter 8 to be able to prove that real symmetric matrices have real eigenvalues.

EXERCISES 7.14

1. Calculate the standard inner product of the following pairs of vectors in $\mathbb{C}^2$.

a. $\begin{bmatrix} 3-5i \\ -i \end{bmatrix}$ and $\begin{bmatrix} 4+i \\ 7+9i \end{bmatrix}$ b. $\begin{bmatrix} -i \\ -4+4i \end{bmatrix}$ and $\begin{bmatrix} 7+9i \\ 5-2i \end{bmatrix}$

2. Calculate the two-norm of the vectors in Exercise 1.

3. Use the two-norm to calculate the distance between the pairs of vectors in Exercise 1.

4. (*Maple V*) Find the standard inner product of the following pairs of complex vectors.

a. $\begin{bmatrix} 3 \\ 1 \\ -4 \end{bmatrix}$ and $\begin{bmatrix} 4 \\ 7 \\ 5 \end{bmatrix}$ b. $\begin{bmatrix} 3-5i \\ -i \\ -4+4i \end{bmatrix}$ and $\begin{bmatrix} 4+i \\ 7+9i \\ 5-2i \end{bmatrix}$

5. (*Maple V*) Calculate the two-norm distance between the pairs of vectors Exercise 4.

6. (*Maple 6*) Use the **LinearAlgebra[Norm]** function to calculate the two-norm distance between the pairs of vectors in Exercise 4.

REVIEW

KEY CONCEPTS ▶ Define and discuss each of the following.

Euclidean Spaces

Euclidean distance, Euclidean inner product, Euclidean n-space.

Geometry

Angle between two vectors, cosine of the angle between two vectors, distance between two vectors, length of a vector, parallelogram law, triangle inequality.

Inner Products

Complex inner product, dot product, Euclidean inner product, standard inner product.

Matrices

Positive definite matrix, symmetric matrix.

Matrix Norms

Frobenius norm of a matrix, infinity-norm of a matrix, one-norm of a matrix, two-norm of a matrix.

Vector Norms

Inner product norm of a vector, one-norm of a vector, two-norm of a vector, infinity-norm of a vector.

Vector Spaces

Normed vector space, real inner product space, unitary space.

Vectors

Unit vector.

KEY FACTS ▶ Explain and illustrate each of the following.

1. Let A be a real $n \times n$ matrix, and let $\mathbf{x} \in \mathbb{R}^n$. Then

$$\max\left\{\frac{\|A\mathbf{x}\|}{\|\mathbf{x}\|} : \|\mathbf{x}\| \neq \mathbf{0}\right\} = \max\{\|A\mathbf{x}\| : \|\mathbf{x}\| = 1\}$$

2. The function $\|A\| = \max\left\{\frac{\|A\mathbf{x}\|}{\|\mathbf{x}\|} : \|\mathbf{x}\| \neq \mathbf{0}\right\} = \max\{\|A\mathbf{x}\| : \|\mathbf{x}\| = 1\}$ is a matrix norm.
3. For every nonzero vector $\mathbf{x}$ in an inner product space V, the vector $\mathbf{y} = \mathbf{x}/\|\mathbf{x}\|$ is a unit vector.
4. If $(\mathbf{x}, \mathbf{y}) \mapsto \langle \mathbf{x}, \mathbf{y} \rangle$ is an inner product on a real vector space V, then $|\langle \mathbf{x}, \mathbf{y} \rangle| \leq \|\mathbf{x}\|\,\|\mathbf{y}\|$ for all $\mathbf{x}, \mathbf{y} \in V$.

5. If A is a positive definite symmetric matrix, then the function $\langle \mathbf{x}, \mathbf{y} \rangle = \mathbf{x}^T A \mathbf{y}$ is an inner product on $\mathbb{R}^n$.

6. If V is a real vector space with an inner product $\langle \mathbf{x}, \mathbf{y} \rangle$, then the function $\|\mathbf{x}\| = \sqrt{\langle \mathbf{x}, \mathbf{x} \rangle}$ is a norm on V.

7. If $\|\mathbf{z}\|$ is an inner product norm on a real vector space V, then $\|\mathbf{x} + \mathbf{y}\|^2 + \|\mathbf{x} - \mathbf{y}\|^2 = 2\left(\|\mathbf{x}\|^2 + \|\mathbf{y}\|^2\right)$ for all $\mathbf{x}, \mathbf{y} \in V$.

8. For every bilinear form $f : V \times V \to \mathbb{R}$ there exists a matrix A for which $f(\mathbf{x}, \mathbf{y}) = \mathbf{x}^T A \mathbf{y}$.

9. Every symmetric bilinear form can be represented by a symmetric matrix.

10. For every real quadratic form $q(\mathbf{x}) = \mathbf{x}^T A \mathbf{x}$ there exist an orthogonal matrix Q and a diagonal matrix D for which $q(\mathbf{y}) = \mathbf{y}^T D \mathbf{y}$ and $\mathbf{x} = Q\mathbf{y}$.

11. If q is a real quadratic form on $\mathbb{R}^n$, then there exists integers s and r, with $s \leq r \leq n$, depending uniquely on q, such that $q(x_1, \ldots, x_n) = x_1^2 + \cdots + x_s^2 - x_{s+1}^2 - \cdots - x_r^2$ in some orthonormal basis for $\mathbb{R}^n$.

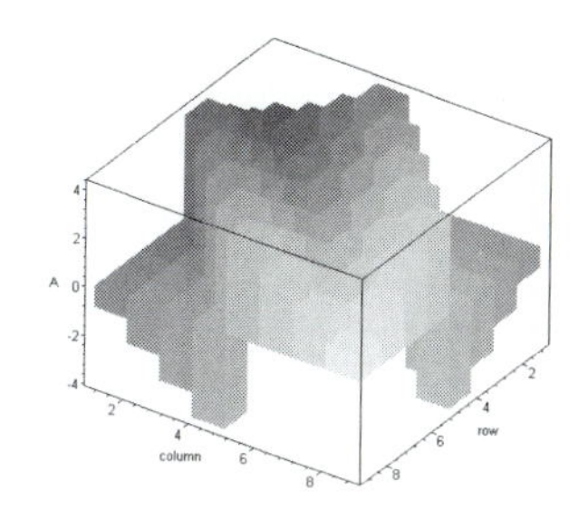

8

ORTHOGONALITY

Many applications of linear algebra are based on the geometric concept of orthogonality. Best possible approximate solutions of overdetermined linear systems, real inner products as dot products, fitting of curves to observational data, the representation of adjoint transformations using transposed matrices, and many other applications involve orthogonality in one form or another.

ORTHOGONAL VECTORS

Inner products provide us with the appropriate tool for defining and analyzing orthogonality. The definition is based on the fact that if $\mathbf{x} = (x_1, x_2)$ and $\mathbf{y} = (y_1, y_2)$ are two nonzero directional arrows in the Euclidean plane, then $\mathbf{x}$ and $\mathbf{y}$ are orthogonal if the angle between them is $\pi/2$ radians, so

$$\cos\frac{\pi}{2} = 0 = \frac{\mathbf{x}^T\mathbf{y}}{\|\mathbf{x}\|\,\|\mathbf{y}\|}$$

This holds precisely when $\mathbf{x}^T\mathbf{y} = 0$.

DEFINITION 8.1 *Two vectors* $\mathbf{x}$ *and* $\mathbf{y}$ *in a real inner product space* V *are* ***orthogonal*** *with respect to an inner product* $\langle\mathbf{x}, \mathbf{y}\rangle$ *on* V *if* $\langle\mathbf{x}, \mathbf{y}\rangle = 0$.

The most famous theorem involving orthogonality is probably the Pythagorean theorem that tells us that the square on the longest side of a right triangle equals the sum of the squares on the other two sides of the triangle. The inner product version of this theorem relates the norms of two orthogonal vectors to the norm of their sum.

THEOREM 8.1 (Pythagorean theorem) *If* $\mathbf{x}$ *and* $\mathbf{y}$ *are two orthogonal vectors in an inner product space* V, *then* $\|\mathbf{x}+\mathbf{y}\|^2 = \|\mathbf{x}\|^2 + \|\mathbf{y}\|^2$.

Proof. Since $\mathbf{x}$ and $\mathbf{y}$ are orthogonal vectors, $\langle\mathbf{x}, \mathbf{y}\rangle = \langle\mathbf{y}, \mathbf{x}\rangle = 0$. Therefore,

$$\|\mathbf{x}+\mathbf{y}\|^2 = \langle\mathbf{x}+\mathbf{y}, \mathbf{x}+\mathbf{y}\rangle$$

$$= \langle \mathbf{x}, \mathbf{x} \rangle + 2 \langle \mathbf{x}, \mathbf{y} \rangle + \langle \mathbf{y}, \mathbf{y} \rangle$$
$$= \langle \mathbf{x}, \mathbf{x} \rangle + \langle \mathbf{y}, \mathbf{y} \rangle$$
$$= \|\mathbf{x}\|^2 + \|\mathbf{y}\|^2$$

This proves the theorem. ■

EXAMPLE 8.1 Two Orthogonal Vectors in $\mathbb{R}^2$

Show that the vectors $\mathbf{x} = (1, 0)$ and $\mathbf{y} = (0, 1)$ are orthogonal in the standard inner product $\langle \mathbf{x}, \mathbf{y} \rangle = \mathbf{x}^T \mathbf{y} = x_1 y_1 + x_2 y_2$.

Solution. Since $\langle \mathbf{x}, \mathbf{y} \rangle = \mathbf{x}^T \mathbf{y} = 1 \times 0 + 0 \times 1 = 0$, the vectors are orthogonal. ▲

EXAMPLE 8.2 Two Nonorthogonal Vectors in $\mathbb{R}^2$

Show that the vectors $\mathbf{x} = (2, 1)$ and $\mathbf{y} = (-1, 2)$ are orthogonal in the standard inner product $\mathbf{x}^T \mathbf{y}$, but are not orthogonal in the inner product $\langle \mathbf{x}, \mathbf{y} \rangle = 2x_1 y_1 + 3x_2 y_2$.

Solution. The vectors $\mathbf{x}$ and $\mathbf{y}$ are orthogonal in the standard inner product since

$$\mathbf{x}^T \mathbf{y} = \begin{bmatrix} 2 & 1 \end{bmatrix} \begin{bmatrix} -1 \\ 2 \end{bmatrix} = 0$$

However, $\langle \mathbf{x}, \mathbf{y} \rangle = 2 \times 2 \times (-1) + 3 \times 1 \times 2 = 2$. Since $\langle \mathbf{x}, \mathbf{y} \rangle \neq 0$, the vectors $\mathbf{x}$ and $\mathbf{y}$ are not orthogonal in the inner product $\langle \mathbf{x}, \mathbf{y} \rangle$. ▲

The next theorem provides an important link between orthogonality and linear independence. We call it the ***orthogonality test*** for linear independence.

THEOREM 8.2 (Orthogonality test for linear independence) *Nonzero orthogonal vectors in an inner product space are linearly independent.*

Proof. Let $\mathcal{S} = \{\mathbf{x}_1, \ldots, \mathbf{x}_n\}$ be a set of nonzero orthogonal vectors in an inner product space V. Then $\langle \mathbf{x}_i, \mathbf{x}_j \rangle = 0$ for all $1 \leq i, j \leq n$ for which $i \neq j$. Now suppose that $a_1 x_1 + \cdots + a_n x_n = \mathbf{0}$. It follows from the linearity of $\langle \mathbf{x}, \mathbf{y} \rangle$ in $\mathbf{y}$ and the orthogonality of the vectors $\mathbf{x}_i$ that $\langle \mathbf{x}_i, a_1 \mathbf{x}_1 + \cdots + a_n \mathbf{x}_n \rangle = a_i \langle \mathbf{x}_i, \mathbf{x}_i \rangle = 0$ for all $i \in \mathbf{n}$. Since $\mathbf{x}_i \neq 0$, we have $\langle \mathbf{x}_i, \mathbf{x}_i \rangle \neq 0$. Hence $a_i = 0$ for all $i \in \mathbf{n}$. Therefore $\mathcal{S}$ is linearly independent. ■

COROLLARY 8.3 *If a basis $\mathcal{B} = \{\mathbf{x}_1, \ldots, \mathbf{x}_n\}$ for an inner product space V consists of orthogonal unit vectors, then $\mathbf{x} = \langle \mathbf{x}_1, \mathbf{x} \rangle \mathbf{x}_1 + \cdots + \langle \mathbf{x}_n, \mathbf{x} \rangle \mathbf{x}_n$ for any vector $\mathbf{x} \in V$.*

Proof. Since $\mathcal{B}$ is a basis, the vector $\mathbf{x}$ can be written as a unique linear combination

$$\mathbf{x} = a_1 \mathbf{x}_1 + \cdots + a_n \mathbf{x}_n$$

of the vectors in $\mathcal{B}$. Therefore,

$$\begin{aligned}\langle \mathbf{x}_i, \mathbf{x}\rangle &= \langle \mathbf{x}_i, a_1\mathbf{x}_1 + \cdots + a_n\mathbf{x}_n\rangle \\ &= a_i \langle \mathbf{x}_i, \mathbf{x}_i\rangle = a_i\end{aligned}$$

This proves the corollary. ■

Another useful consequence of Theorem 8.2 is the fact that only the zero vector can be orthogonal to all vectors of an orthogonal basis.

THEOREM 8.4 (Zero vector theorem) *If $\mathcal{B} = \{\mathbf{x}_1, \ldots, \mathbf{x}_p\}$ is an orthogonal set of non-zero vectors in an inner product space V and if $\mathbf{x}$ is a vector in the span of $\mathcal{B}$ and is orthogonal to all vectors in $\mathcal{B}$, then $\mathbf{x} = \mathbf{0}$.*

Proof. Suppose that $\mathbf{x} = a_1\mathbf{x}_1 + \cdots + a_p\mathbf{x}_p$. Since $\mathbf{x}$ is orthogonal to each $\mathbf{x}_i$, we have

$$0 = \langle \mathbf{x}, \mathbf{x}_i\rangle = \langle a_1\mathbf{x}_1 + \cdots + a_p\mathbf{x}_p, \mathbf{x}_i\rangle = a_i \langle \mathbf{x}_i, \mathbf{x}_i\rangle$$

Hence $a_i = 0$ for all i. Hence $\mathbf{x} = \mathbf{0}$. ■

EXAMPLE 8.3 Linear Independence of Orthogonal Vectors

Use *Maple* and the orthogonality test to show that the set of quaternions

$$\{q_1 = 3I + 2J - 6K, q_2 = 2I + 6J - 3K, q_3 = -6I + 3J - 2K\}$$

is linearly independent in $\mathbb{R}^4$.

Solution. By Theorem 8.2, it suffices to show that $\mathbf{x}$, $\mathbf{y}$, and $\mathbf{z}$ are orthogonal in the standard inner product. To do so, we use the **dotprod** function in the **linalg** package.

```
[>with(linalg):
[>x:=vector(4,[0,3,2,-6]):
[>y:=vector(4,[0,2,6,3]):
[>z:=vector(4,[0,-6,3,-2]):
[>is(dotprod(x,y)=0 and dotprod(x,z)=0 and dotprod(y,z)=0);
                              true
```

Since the three dot products are zero, the vectors $\mathbf{x}$, $\mathbf{y}$, and $\mathbf{z}$ are orthogonal. They are therefore also linearly independent. ▲

EXAMPLE 8.4 A Linear Combination of Orthogonal Vectors

Express the vector $\mathbf{y} = (3, 7, 1)$ as a linear combination $\mathbf{x} = a\mathbf{x}_1 + b\mathbf{x}_2 + c\mathbf{x}_3$ of the pairwise orthogonal unit vectors in the basis

$$\mathcal{B} = \left\{\mathbf{x}_1 = \left(-\tfrac{6}{19}, \tfrac{15}{19}, \tfrac{10}{19}\right), \mathbf{x}_2 = \left(\tfrac{15}{19}, \tfrac{10}{19}, -\tfrac{6}{19}\right), \mathbf{x}_3 = \left(\tfrac{10}{19}, -\tfrac{6}{19}, \tfrac{15}{19}\right)\right\}$$

and show that in the standard inner product on $\mathbb{R}^3$,

$$a = \langle \mathbf{x}_1, \mathbf{x} \rangle, \quad b = \langle \mathbf{x}_2, \mathbf{x} \rangle, \quad \text{and} \quad c = \langle \mathbf{x}_3, \mathbf{x} \rangle$$

Solution. Suppose that

$$\begin{bmatrix} 3 \\ 7 \\ 1 \end{bmatrix} = a \begin{bmatrix} -\frac{6}{19} \\ \frac{15}{19} \\ \frac{10}{19} \end{bmatrix} + b \begin{bmatrix} \frac{15}{19} \\ \frac{10}{19} \\ -\frac{6}{19} \end{bmatrix} + c \begin{bmatrix} \frac{10}{19} \\ -\frac{6}{19} \\ \frac{15}{19} \end{bmatrix}$$

Then

$$\begin{bmatrix} 3 \\ 7 \\ 1 \end{bmatrix} = \begin{bmatrix} -\frac{6}{19}a + \frac{15}{19}b + \frac{10}{19}c \\ \frac{15}{19}a + \frac{10}{19}b - \frac{6}{19}c \\ \frac{10}{19}a - \frac{6}{19}b + \frac{15}{19}c \end{bmatrix}$$

We calculate the coefficients a, b, and c by solving the associated linear system.

```
[> solve({3=-(6/19)*a+(15/19)*b+(10/19)*c,
    7=(15/19)*a+(10/19)*b-(6/19)*c,
    1=(10/19)*a-(6/19)*b+(15/19)*c},{a,b,c});
```

$$\{c = \frac{3}{19}, b = \frac{109}{19}, a = \frac{97}{19}\}$$

It remains to show that $\langle \mathbf{x}_1, \mathbf{x} \rangle = \frac{97}{19}$, $\langle \mathbf{x}_2, \mathbf{x} \rangle = \frac{109}{19}$, and $\langle \mathbf{x}_3, \mathbf{x} \rangle = \frac{3}{19}$.

```
[> with(linalg):
[> x1:=vector(3,[-6/19,15/19,10/19]):
[> x2:=vector(3,[15/19,10/19,-6/19]):
[> x3:=vector(3,[10/19,-6/19,15/19]):
[> x:=vector(3,7,1):
[> [dotprod(x1,x),dotprod(x2,x),dotprod(x3,x)];
```

$$\left[\frac{97}{19}, \frac{109}{19}, \frac{3}{19}\right]$$

Therefore, $a = \langle \mathbf{x}_1, \mathbf{x} \rangle$, $b = \langle \mathbf{x}_2, \mathbf{x} \rangle$, and $c = \langle \mathbf{x}_3, \mathbf{x} \rangle$. ▲

EXERCISES 8.1

1. Use Theorem 8.2 to prove that the following pairs of vectors are linearly independent.

a. $\mathbf{x} = (0, 24, 107, 0)$ and $\mathbf{y} = (1, 0, 0, 1)$
b. $\mathbf{x} = (1, 24, 0, 1)$ and $\mathbf{y} = (-1, 0, 7, 1)$
c. $\mathbf{x} = (0, 24, 0, 0)$ and $\mathbf{y} = (1, 0, 7, 1)$
d. $\mathbf{x} = (1, -1, 1, 1)$ and $\mathbf{y} = (1, 1, -1, 1)$

2. Use the standard inner product to verify the Pythagorean for the triangles in the plane determined by the following sets of points.

a. $\mathbf{x} = (1, 0), \mathbf{y} = (1, 2), \mathbf{z} = (5, 2)$
b. $\mathbf{x} = (1, 0), \mathbf{y} = (1, 2), \mathbf{z} = (-5, 2)$
c. $\mathbf{x} = (-8, 6), \mathbf{y} = (3, 6), \mathbf{z} = (3, -12)$
d. $\mathbf{x} = (8, 6), \mathbf{y} = (3, 6), \mathbf{z} = (3, -12)$

3. (*Maple V*) Use the standard inner product to verify the Pythagorean theorem for the triangles in the plane determined by the following sets of points.

a. $\mathbf{x} = (1, 0, 3), \mathbf{y} = (1, 2, 3), \mathbf{z} = (5, 2, 3)$
b. $\mathbf{x} = (1, 0, -3), \mathbf{y} = (1, 2, -3), \mathbf{z} = (-5, 2, -3)$
c. $\mathbf{x} = (-8, 7, 6), \mathbf{y} = (3, 7, 6), \mathbf{z} = (3, 7, -12)$
d. $\mathbf{x} = (4, 8, 6), \mathbf{y} = (4, 3, 6), \mathbf{z} = (4, 3, -12)$

4. (*Maple V*) Repeat Exercise 2 for the inner product $\mathbf{x}^T A\mathbf{y}$ determined by the diagonal matrix

$$A = \begin{bmatrix} 3 & 0 \\ 0 & 5 \end{bmatrix}$$

5. (*Maple V*) Repeat Exercise 3 for the inner product $\mathbf{x}^T A\mathbf{y}$ determined by the positive definite matrix

$$A = \begin{bmatrix} 9 & 0 & -2 \\ 0 & 7 & 0 \\ -2 & 0 & 1 \end{bmatrix}$$

Orthogonal Projections

The basic operation for constructing orthogonal vectors is that of an ***orthogonal projection***. Suppose that $\mathbf{x}$ and $\mathbf{y}$ are two vectors in an inner product space V. We would like to split the vector $\mathbf{x}$ into two orthogonal vectors $\text{proj}(\mathbf{x} \to \mathbf{y})$ and $\text{perp}(\mathbf{x} \perp \mathbf{y})$ so that the following hold.

1. $\text{proj}(\mathbf{x} \to \mathbf{y}) + \text{perp}(\mathbf{x} \perp \mathbf{y}) = \mathbf{x}$
2. $\text{proj}(\mathbf{x} \to \mathbf{y}) = a\mathbf{y}$
3. $\langle \text{perp}(\mathbf{x} \perp \mathbf{y}), \mathbf{y} \rangle = 0$

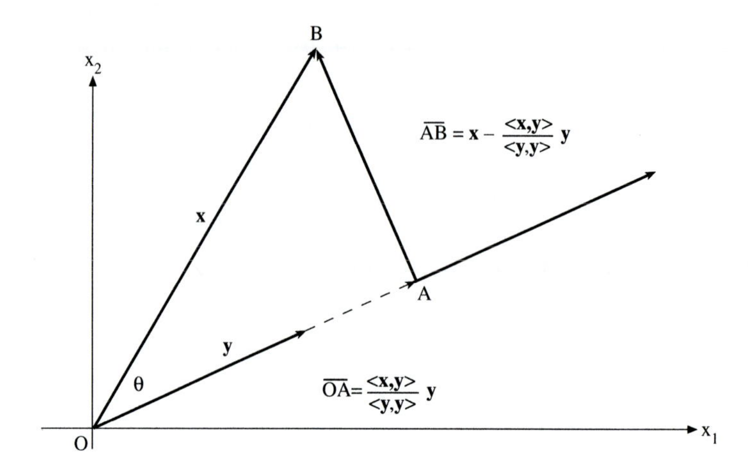

FIGURE 1 Projection of a Vector onto a Subspace.

This turns out to be quite straightforward. If the vectors $\text{proj}(\mathbf{x} \to \mathbf{y})$ and $\text{perp}(\mathbf{x} \perp \mathbf{y})$ are orthogonal, then

$$\begin{aligned}\langle \mathbf{x}, \mathbf{y} \rangle &= \langle \text{proj}(\mathbf{x} \to \mathbf{y}) + \text{perp}(\mathbf{x} \perp \mathbf{y}), \mathbf{y} \rangle \\ &= \langle \text{proj}(\mathbf{x} \to \mathbf{y}), \mathbf{y} \rangle + \langle \text{perp}(\mathbf{x} \perp \mathbf{y}), \mathbf{y} \rangle \\ &= \langle a\mathbf{y}, \mathbf{y} \rangle + 0 \\ &= a\langle \mathbf{y}, \mathbf{y} \rangle\end{aligned}$$

Therefore, a must equal $\langle \mathbf{x}, \mathbf{y} \rangle / \langle \mathbf{y}, \mathbf{y} \rangle$. Hence

$$\text{proj}(\mathbf{x} \to \mathbf{y}) = \frac{\langle \mathbf{x}, \mathbf{y} \rangle}{\langle \mathbf{y}, \mathbf{y} \rangle}\mathbf{y} \quad \text{and} \quad \text{perp}(\mathbf{x} \perp \mathbf{y}) = \mathbf{x} - \frac{\langle \mathbf{x}, \mathbf{y} \rangle}{\langle \mathbf{y}, \mathbf{y} \rangle}\mathbf{y}$$

Figure 1 illustrates this construction relative to the standard inner product. The vector $\overline{OA} = \text{proj}(\mathbf{x} \to \mathbf{y})$ is the ***orthogonal projection*** of the vector $\mathbf{x}$ onto the vector $\mathbf{y}$, and the vector $\overline{AB} = \text{perp}(\mathbf{x} \perp \mathbf{y})$ is the ***vector component*** of $\mathbf{x}$ orthogonal to $\mathbf{y}$.

We can generalize this construction and define the orthogonal projection of a vector $\mathbf{x} \in V$ onto a subspace W of V. The construction is particularly simple if the basis $\mathcal{B} = \{\mathbf{x}_1, \ldots, \mathbf{x}_n\}$ for V consists of mutually orthogonal unit vectors. In that case, we know from Corollary 8.3 that every vector $\mathbf{x} \in V$ can be written as a unique linear combination of the form

$$\mathbf{x} = \langle \mathbf{x}_1, \mathbf{x} \rangle \mathbf{x}_1 + \cdots + \langle \mathbf{x}_n, \mathbf{x} \rangle \mathbf{x}_n$$

If $\mathcal{C} = \{\mathbf{y}_1, \ldots, \mathbf{y}_p\}$ is a subset of the basis $\mathcal{B}$ and $W = \text{span}(\mathcal{C})$, then we can form the projection

$$\text{proj}(\mathbf{x} \to W) = \langle \mathbf{y}_1, \mathbf{x} \rangle \, \mathbf{y}_1 + \cdots + \langle \mathbf{y}_p, \mathbf{x} \rangle \, \mathbf{y}_p$$

of $\mathbf{x}$ onto the subspace W, obtained from $\langle \mathbf{x}_1, \mathbf{x} \rangle \mathbf{x}_1 + \cdots + \langle \mathbf{x}_n, \mathbf{x} \rangle \mathbf{x}_n$ by deleting all terms $\langle \mathbf{x}_i, \mathbf{x} \rangle \mathbf{x}_i$ for which $\mathbf{x}_i \notin \mathcal{C}$. By construction, the vector $\text{proj}(\mathbf{x} \to W)$ belongs to W, and the vector $\text{perp}(\mathbf{x} \perp W) = \mathbf{x} - \text{proj}(\mathbf{x} \to W)$ is the vector component of $\mathbf{x}$ orthogonal to every vector in W.

Therefore, the vector $\text{proj}(\mathbf{x} \to W)$ belongs to W, the inner product $\langle \text{perp}(\mathbf{x} \perp W), \mathbf{y}_i \rangle$ is zero for each i, and

$$\text{proj}(\mathbf{x} \to W) + \text{perp}(\mathbf{x} \perp \mathbf{W}) = \mathbf{x}$$

Hence the vector $\text{proj}(\mathbf{x} \to W)$ is the ***orthogonal projection*** of $\mathbf{x}$ onto the subspace W.

We note that in Theorem 8.20, we will be able to give an even simpler formula for calculating the orthogonal projections $\text{proj}(\mathbf{x} \to W)$ by showing that $\text{proj}(\mathbf{x} \to W) = PP^T\mathbf{x}$, where P is the matrix whose columns are the basis vectors $\mathbf{y}_1, \ldots, \mathbf{y}_p$.

EXAMPLE 8.5 An Orthogonal Projection onto a Subspace of $\mathbb{R}^2$

Let $\mathcal{E} = \{\mathbf{e}_1, \mathbf{e}_2\}$ be the standard basis of $\mathbb{R}^2$, let W_1 be the subspace of $\mathbb{R}^2$ spanned by $\{\mathbf{e}_1\}$, and let $\mathbf{x} = x_1\mathbf{e}_1 + x_2\mathbf{e}_2$ be any vector in $\mathbb{R}^2$. Find the vectors

a. $\text{proj}(\mathbf{x} \to W_1)$ b. $\text{perp}(\mathbf{x} \to W_1)$

c. $\text{proj}(\mathbf{x} \to W_2)$ d. $\text{perp}(\mathbf{x} \to W_2)$

and show that $\text{proj}(\mathbf{x} \to W_i)$ is orthogonal to $\text{perp}(\mathbf{x} \to W_i)$ in the standard inner product.

Solution. By definition,

$$\begin{aligned} \text{proj}(\mathbf{x} \to W_1) &= x_1\mathbf{e}_1, \quad \text{perp}(\mathbf{x} \perp W_1) = x_1\mathbf{e}_1 + x_2\mathbf{e}_2 - x_1\mathbf{e}_1 = x_2\mathbf{e}_2 \\ \text{proj}(\mathbf{x} \to W_2) &= x_2\mathbf{e}_2, \quad \text{perp}(\mathbf{x} \perp W_2) = x_1\mathbf{e}_1 + x_2\mathbf{e}_2 - x_2\mathbf{e}_2 = x_1\mathbf{e}_1 \end{aligned}$$

Furthermore,

$$\begin{bmatrix} x_1 \\ 0 \end{bmatrix}^T \begin{bmatrix} 0 \\ x_2 \end{bmatrix} = 0 \quad \text{and} \quad \begin{bmatrix} x_1 \\ 0 \end{bmatrix}^T \begin{bmatrix} 0 \\ x_2 \end{bmatrix} = 0$$

Hence $\text{proj}(\mathbf{x} \to W_i)$ is orthogonal to $\text{perp}(\mathbf{x} \perp W_i)$. ▲

EXAMPLE 8.6 An Orthogonal Projection onto a Subspace of $\mathbb{R}^3$

Let W be the subspace of $\mathbb{R}^3$ spanned by the subset $\{\mathbf{x}_2, \mathbf{x}_3\}$ of the basis

$$\mathcal{B} = \left\{\mathbf{x}_1 = \left(-\tfrac{1}{3}, \tfrac{2}{3}, \tfrac{2}{3}\right), \mathbf{x}_2 = \left(\tfrac{2}{3}, -\tfrac{1}{3}, \tfrac{2}{3}\right), \mathbf{x}_3 = \left(\tfrac{2}{3}, \tfrac{2}{3}, -\tfrac{1}{3}\right)\right\}$$

and let $\mathbf{x} = (x_1, x_2, x_3)$ be a vector in $\mathbb{R}^3$. Find $\text{proj}(\mathbf{x} \to W)$ and $\text{perp}(\mathbf{x} \perp W)$ and show that $\text{proj}(\mathbf{x} \to W)$ is orthogonal to $\text{perp}(\mathbf{x} \perp W)$ in the standard inner product on $\mathbb{R}^3$.

Solution. We begin by showing that the basis $\mathcal{B}$ consists of orthogonal unit vectors.

```
[> with(linalg):
[> x1:=vector(3,[-1/3,2/3,2/3]):
[> x2:=vector(3,[2/3,-1/3,2/3]):
[> x3:=vector(3,[2/3,2/3,-1/3]):
[> [dotprod(x1,x2),dotprod(x1,x3),dotprod(x2,x3)];
```

$$[0, 0, 0]$$

This shows that the vectors $\mathbf{x}_1$, $\mathbf{x}_2$, and $\mathbf{x}_3$ are orthogonal. Next we show that their length is 1.

```
[> [norm(x1,2),norm(x2,2),norm(x3,2)];
```

$$[1, 1, 1]$$

Hence $\mathcal{B}$ is an orthonormal basis for $\mathbb{R}^3$.

Now consider any vector $\mathbf{x} = x_1\mathbf{x}_1 + x_2\mathbf{x}_2 + x_3\mathbf{x}_3$ in $\mathbb{R}^3$. By Corollary 8.3,

$$\begin{aligned}
\text{proj}(\mathbf{x} \to W) &= \left(\begin{bmatrix} x_1 \\ x_2 \\ x_3 \end{bmatrix}^T \begin{bmatrix} \frac{2}{3} \\ -\frac{1}{3} \\ \frac{2}{3} \end{bmatrix}\right)\begin{bmatrix} \frac{2}{3} \\ -\frac{1}{3} \\ \frac{2}{3} \end{bmatrix} + \left(\begin{bmatrix} x_1 \\ x_2 \\ x_3 \end{bmatrix}^T \begin{bmatrix} \frac{2}{3} \\ \frac{2}{3} \\ -\frac{1}{3} \end{bmatrix}\right)\begin{bmatrix} \frac{2}{3} \\ \frac{2}{3} \\ -\frac{1}{3} \end{bmatrix} \\
&= \begin{bmatrix} \frac{4}{9}x_1 - \frac{2}{9}x_2 + \frac{4}{9}x_3 \\ -\frac{2}{9}x_1 + \frac{1}{9}x_2 - \frac{2}{9}x_3 \\ \frac{4}{9}x_1 - \frac{2}{9}x_2 + \frac{4}{9}x_3 \end{bmatrix} + \begin{bmatrix} \frac{4}{9}x_1 + \frac{4}{9}x_2 - \frac{2}{9}x_3 \\ \frac{4}{9}x_1 + \frac{4}{9}x_2 - \frac{2}{9}x_3 \\ -\frac{2}{9}x_1 - \frac{2}{9}x_2 + \frac{1}{9}x_3 \end{bmatrix} \\
&= \begin{bmatrix} \frac{8}{9}x_1 + \frac{2}{9}x_2 + \frac{2}{9}x_3 \\ \frac{2}{9}x_1 + \frac{5}{9}x_2 - \frac{4}{9}x_3 \\ \frac{2}{9}x_1 - \frac{4}{9}x_2 + \frac{5}{9}x_3 \end{bmatrix}
\end{aligned}$$

and

$$\text{perp}(\mathbf{x} \perp W) = \begin{bmatrix} x_1 \\ x_2 \\ x_3 \end{bmatrix} - \begin{bmatrix} \frac{8}{9}x_1 + \frac{2}{9}x_2 + \frac{2}{9}x_3 \\ \frac{2}{9}x_1 + \frac{5}{9}x_2 - \frac{4}{9}x_3 \\ \frac{2}{9}x_1 - \frac{4}{9}x_2 + \frac{5}{9}x_3 \end{bmatrix} = \begin{bmatrix} \frac{1}{9}x_1 - \frac{2}{9}x_2 - \frac{2}{9}x_3 \\ \frac{4}{9}x_2 - \frac{2}{9}x_1 + \frac{4}{9}x_3 \\ \frac{4}{9}x_2 - \frac{2}{9}x_1 + \frac{4}{9}x_3 \end{bmatrix}$$

Since

$$\begin{bmatrix} \frac{8}{9}x_1 + \frac{2}{9}x_2 + \frac{2}{9}x_3 \\ \frac{2}{9}x_1 + \frac{5}{9}x_2 - \frac{4}{9}x_3 \\ \frac{2}{9}x_1 - \frac{4}{9}x_2 + \frac{5}{9}x_3 \end{bmatrix}^T \begin{bmatrix} \frac{1}{9}x_1 - \frac{2}{9}x_2 - \frac{2}{9}x_3 \\ \frac{4}{9}x_2 - \frac{2}{9}x_1 + \frac{4}{9}x_3 \\ \frac{4}{9}x_2 - \frac{2}{9}x_1 + \frac{4}{9}x_3 \end{bmatrix} = 0$$

the vectors proj$(\mathbf{x} \to W)$ and perp $(\mathbf{x} \perp W)$ are orthogonal. ▲

Next we consider an example of the projection of a vector onto a single vector.

EXAMPLE 8.7 An Orthogonal Projection onto a Vector

Use the **dotprod** function to find the orthogonal projection of the vector $\mathbf{x} = (1, 2, 3)$ onto the vector $\mathbf{y} = (5, 0, 3)$ in the standard inner product.

Solution. We first load the **linalg** package, input the vectors **x** and **y**, and calculate the coefficient $a = \langle \mathbf{x}, \mathbf{y} \rangle / \langle \mathbf{y}, \mathbf{y} \rangle$.

```
[>with(linalg):
[>x:=vector(3,[1,2,3]):
[>y:=vector(3,[5,0,3]):
[>a:=(dotprod(x,y)/dotprod(y,y));
```

$$a := \frac{7}{17}$$

Next we calculate $\text{proj}(\mathbf{x} \rightarrow \mathbf{y})$ and $\text{perp}(\mathbf{x} \perp \mathbf{y})$.

```
[>proj[x,y]:=evalm(a*y);
```

$$proj_{x,y} := \left[\frac{35}{17}, 0, \frac{21}{17}\right]$$

```
[>perp[x,y]:=evalm(x-proj[x,y]);
```

$$perp_{x,y} := \left[\frac{-18}{17}, 2, \frac{30}{17}\right]$$

We conclude by verifying that $\text{proj}(\mathbf{x} \rightarrow \mathbf{y}) + \text{perp}(\mathbf{x} \perp \mathbf{y}) = \mathbf{x}$ and that the dot product of $\text{proj}(\mathbf{x} \rightarrow \mathbf{y})$ and $\text{perp}(\mathbf{x} \perp \mathbf{y})$ is zero.

```
[>equal(proj[x,y]+perp[x,y],x);
```

true

```
[>dotprod(proj[x,y],perp[x,y]);
```

$$0$$

This shows that $\text{proj}(\mathbf{x} \rightarrow \mathbf{y}) + \text{perp}(\mathbf{x} \perp \mathbf{y}) = \mathbf{x}$ is an orthogonal decomposition of the vector **x**. ▲

EXERCISES 8.2

1. Use the standard basis for $\mathbb{R}^2$ and the standard inner product for the following problems.

   a. Find the orthogonal projection of the vector $(1, 2)$ onto the subspace $W = \text{span}\{\mathbf{e}_1\}$.

   b. Find the orthogonal projection of the vector $(1, 2)$ onto the subspace $W = \text{span}\{\mathbf{e}_2\}$.

   c. Explain what happens when you try to find the orthogonal projection of a vector (x, y) onto $\mathbb{R}^2$.

2. Repeat Exercise 1 for the basis

$$\mathcal{B} = \left\{ \mathbf{x}_1 = \frac{1}{\sqrt{2}} \begin{bmatrix} 1 \\ 1 \end{bmatrix}, \mathbf{x}_2 = \frac{1}{\sqrt{2}} \begin{bmatrix} 1 \\ -1 \end{bmatrix} \right\}$$

   for $\mathbb{R}^2$. Use the standard inner product. The vector $(1, 2)$ is to be taken as a coordinate vector in the basis $\mathcal{B}$.

3. Use the standard basis for $\mathbb{R}^3$ and the standard inner product for the following problems.

   a. Find the orthogonal projection of the vector $(1, 2, 3)$ onto the subspace W spanned by the unit vector $\mathbf{e}_1$.

   b. Find the orthogonal projection of the vector $(1, 2, 3)$ onto the subspace W spanned by the unit vectors $\mathbf{e}_1$ and $\mathbf{e}_2$.

   c. Find the orthogonal projection of the vector $(1, 2, 3)$ onto the subspace W spanned by the unit vectors $\mathbf{e}_1$ and $\mathbf{e}_3$.

4. (*Maple V*) Find the vectors $\text{proj}(\mathbf{x} \to \mathbf{y})$ and $\text{perp}(\mathbf{x} \perp \mathbf{y})$ for the following pairs of vectors in the standard inner product.

$$\text{a. } \mathbf{x} = \begin{bmatrix} 1 \\ 1 \end{bmatrix} \text{ and } \mathbf{y} = \begin{bmatrix} 4 \\ 5 \end{bmatrix} \quad \text{b. } \mathbf{x} = \begin{bmatrix} -2 \\ 7 \end{bmatrix} \text{ and } \mathbf{y} = \begin{bmatrix} 4 \\ 5 \end{bmatrix}$$

5. (*Maple 6*) Find the vectors $\text{proj}(\mathbf{x} \to \mathbf{y})$ and $\text{perp}(\mathbf{x} \perp \mathbf{y})$ for the following pairs of vectors in the standard inner product.

$$\text{a. } \mathbf{x} = \begin{bmatrix} 1 \\ 3 \\ 1 \end{bmatrix} \text{ and } \mathbf{y} = \begin{bmatrix} 4 \\ 5 \\ 0 \end{bmatrix} \quad \text{b. } \mathbf{x} = \begin{bmatrix} 1 \\ 3 \\ 1 \end{bmatrix} \text{ and } \mathbf{y} = \begin{bmatrix} -1 \\ -9 \\ 12 \end{bmatrix}$$

ORTHOGONAL BASES

In this section, we show that every basis of a finite-dimensional inner product space can be converted to a basis whose vectors are mutually orthogonal.

Gram–Schmidt Orthogonalization

In many applications, it is essential to work with bases whose vectors are orthogonal. We now show that this can be done. Every basis in an inner product space can be converted to one in which all pairs of vectors are orthogonal. The algorithm involved is known as the ***Gram–Schmidt process.*** The process iterates the orthogonal projection construction.

DEFINITION 8.2 *A basis $\mathcal{B} = \{\mathbf{v}_1, \ldots, \mathbf{v}_n\}$ for a finite-dimensional inner product space V is **orthogonal** if $\langle \mathbf{v}_i, \mathbf{v}_j \rangle = 0$ for each $1 \le i, j \le n$.*

The process of converting a basis to an orthogonal one is known as ***orthogonalization***. We now use induction to describe this process.

THEOREM 8.5 (Gram–Schmidt orthogonalization process) *If $\mathcal{B} = \{\mathbf{x}_1, \ldots, \mathbf{x}_n\}$ is a basis for a real inner product space V with inner product $(\mathbf{x}, \mathbf{y}) \mapsto \langle \mathbf{x}, \mathbf{y} \rangle$, then the set $\mathcal{C} = \{\mathbf{y}_1, \ldots, \mathbf{y}_n\}$ is an orthogonal basis for V, where y_i are the following vectors.*

1. $\mathbf{y}_1 = \mathbf{x}_1$
2. $\mathbf{y}_2 = \mathbf{x}_2 - \dfrac{\langle \mathbf{x}_2, \mathbf{y}_1 \rangle}{\langle \mathbf{y}_1, \mathbf{y}_1 \rangle}\mathbf{y}_1$
3. $\mathbf{y}_3 = \mathbf{x}_3 - \dfrac{\langle \mathbf{x}_3, \mathbf{y}_1 \rangle}{\langle \mathbf{y}_1, \mathbf{y}_1 \rangle}\mathbf{y}_1 - \dfrac{\langle \mathbf{x}_3, \mathbf{y}_2 \rangle}{\langle \mathbf{y}_2, \mathbf{y}_2 \rangle}\mathbf{y}_2$

   $\vdots$

n. $\mathbf{y}_n = \mathbf{x}_n - \dfrac{\langle \mathbf{x}_n, \mathbf{y}_1 \rangle}{\langle \mathbf{y}_1, \mathbf{y}_1 \rangle}\mathbf{y}_1 - \dfrac{\langle \mathbf{x}_n, \mathbf{y}_2 \rangle}{\langle \mathbf{y}_2, \mathbf{y}_2 \rangle}\mathbf{y}_2 - \cdots - \dfrac{\langle \mathbf{x}_n, \mathbf{y}_{n-1} \rangle}{\langle \mathbf{y}_{n-1}, \mathbf{y}_{n-1} \rangle}\mathbf{y}_{n-1}$

Proof. We prove the theorem by induction on the number n of vectors in $\mathcal{B}$. If $n = 1$, the result is obvious. Suppose, therefore, that we have found the orthogonal vectors $\{\mathbf{y}_1, \ldots, \mathbf{y}_k\}$ and that $\mathbf{x}_{k+1}$ is linearly independent of $\{\mathbf{y}_1, \ldots, \mathbf{y}_k\}$. We form the subspace

$$\text{span}\{\mathbf{y}_1, \ldots, \mathbf{y}_k, \mathbf{x}_{k+1}\} \subseteq V$$

and look for a nonzero vector

$$\mathbf{y}_{k+1} \in \text{span}\{\mathbf{y}_1, \ldots, \mathbf{y}_k, \mathbf{x}_{k+1}\}$$

that is orthogonal to the vectors $\{\mathbf{y}_1, \ldots, \mathbf{y}_k\}$. Since $\{\mathbf{y}_1, \ldots, \mathbf{y}_k, \mathbf{y}_{k+1}\}$ is a basis for

$$\text{span}\{\mathbf{y}_1, \ldots, \mathbf{y}_k, \mathbf{x}_{k+1}\}$$

we can solve our problem by finding scalars for which

$$\mathbf{y}_{k+1} = a_1\mathbf{y}_1 + \cdots + a_k\mathbf{y}_k + \mathbf{x}_{k+1}$$

and $\langle \mathbf{y}_i, \mathbf{y}_{k+1} \rangle = 0$ for $1 \le i \le k$.

Since the vector $\mathbf{y}_{k+1}$ is orthogonal to the vectors $\{\mathbf{y}_1, \ldots, \mathbf{y}_k\}$ precisely when

$$\begin{aligned}\langle \mathbf{y}_i, \mathbf{y}_{k+1}\rangle &= \langle \mathbf{y}_i, a_1\mathbf{y}_1 + \cdots + a_k\mathbf{y}_k + \mathbf{x}_{k+1}\rangle \\ &= a_i \langle \mathbf{y}_i, \mathbf{y}_i\rangle + \langle \mathbf{y}_i, \mathbf{x}_{k+1}\rangle \\ &= 0\end{aligned}$$

it is clear that $a_i = -\langle \mathbf{y}_i, \mathbf{x}_{k+1}\rangle / \langle \mathbf{y}_i, \mathbf{y}_i\rangle$ for all $1 \leq i \leq k$. The vector $\mathbf{y}_{k+1}$ is nonzero since otherwise $\mathbf{0} = a_1\mathbf{y}_1 + \cdots + a_k\mathbf{y}_k + \mathbf{x}_{k+1}$. This means that $\mathbf{x}_{k+1}$ is a linear combination of $\mathbf{y}_1, \ldots, \mathbf{y}_k$. This contradicts the assumption that the vectors $\mathbf{y}_1, \ldots, \mathbf{y}_k, \mathbf{x}_{k+1}$ are linearly independent. ■

We note that in the Gram–Schmidt process, each vector $\mathbf{y}_{k+1}$ is constructed from the vector $\mathbf{x}_{k+1}$ and the vectors $\mathbf{y}_1, \ldots, \mathbf{y}_k$ by simply subtracting from $\mathbf{x}_{k+1}$ the orthogonal projections proj $(\mathbf{x}_{k+1} \to \mathbf{y}_i)$ of $\mathbf{x}_{k+1}$ onto the preceding vectors $\mathbf{y}_1, \ldots, \mathbf{y}_k$. We summarize the algorithm described in Theorem 8.5 in Table 1.

TABLE 1 CONVERTING A BASIS TO AN ORTHOGONAL BASIS.

Step 1. Take a finite basis $\mathcal{B} = \{\mathbf{x}_1, \ldots, \mathbf{x}_n\}$ for an inner product space and let $\mathcal{C}_1$ be the set $\{\mathbf{y}_1 = \mathbf{x}_1\}$.

Step 2. Extend $\mathcal{C}_1 = \{\mathbf{y}_1\}$ to the set $\mathcal{C}_2 = \{\mathbf{y}_1, \mathbf{y}_2\}$ by forming the scalar $a = \langle \mathbf{x}_2, \mathbf{y}_1\rangle / \langle \mathbf{y}_1, \mathbf{y}_1\rangle$ and defining the vector $\mathbf{y}_2$ to be $\mathbf{x}_2 - a\mathbf{y}_1$.

Step 3. Extend $\mathcal{C}_2 = \{\mathbf{y}_1, \mathbf{y}_2\}$ to $\mathcal{C}_3 = \{\mathbf{y}_1, \mathbf{y}_2, \mathbf{y}_3\}$ by forming the scalars $b_1 = \langle \mathbf{x}_3, \mathbf{y}_1\rangle / \langle \mathbf{y}_1, \mathbf{y}_1\rangle$ and $b_2 = \langle \mathbf{x}_3, \mathbf{y}_2\rangle / \langle \mathbf{y}_2, \mathbf{y}_2\rangle$ and defining the vector $\mathbf{y}_3$ to be $\mathbf{x}_3 - b_1\mathbf{y}_1 - b_2\mathbf{y}_2$.

Step 4. Continue in this way until the orthogonal basis $\mathcal{C} = \mathcal{C}_n$ is complete.

In the next two examples, we apply this process to convert a basis for $\mathbb{R}^2$ and a basis for $\mathbb{R}^3$ to orthogonal bases. We use *Maple* for the details of the calculations. In the third example, we use the **GramSchmidt** function of the **linalg** package for the conversion.

EXAMPLE 8.8 An Orthogonal Basis for $\mathbb{R}^2$

Use the induction steps of the Gram–Schmidt process to convert the basis

$$\mathcal{B} = \{\mathbf{x}_1 = (1, -1), \mathbf{x}_2 = (3, 2)\}$$

for $\mathbb{R}^2$ to an orthogonal basis $\mathcal{C} = \{\mathbf{y}_1, \mathbf{y}_2\}$ in the standard inner product on $\mathbb{R}^2$.

Solution. The standard inner product on $\mathbb{R}^2$ is defined by $\langle \mathbf{u}, \mathbf{x}\rangle = \mathbf{u}^T\mathbf{x}$. By the Gram–Schmidt process, the vectors

$$\mathbf{y}_1 = \mathbf{x}_1 = \begin{bmatrix} 1 \\ -1 \end{bmatrix}$$

and

$$\mathbf{y}_2 = \mathbf{x}_2 - \frac{\mathbf{x}_2^T \mathbf{y}_1}{\mathbf{y}_1^T \mathbf{y}_1} \mathbf{y}_1 = \begin{bmatrix} 3 \\ 2 \end{bmatrix} - \frac{\mathbf{x}_2^T \mathbf{y}_1}{\mathbf{y}_1^T \mathbf{y}_1} \begin{bmatrix} 1 \\ -1 \end{bmatrix}$$

are orthogonal and form the required basis $\mathcal{C}$.

We use *Maple* to calculate the coefficient $\mathbf{x}_2^T \mathbf{y}_1 / \mathbf{y}_1^T \mathbf{y}_1$.

```
[> with(linalg):
[> a:=dotprod(vector(2,[3,2]),vector(2,[1,-1])):
[> b:=dotprod(vector(2,[1,-1]),vector(2,[1,-1])):
[> a/b;
```

$$\frac{1}{2}$$

Therefore,

$$\mathbf{y}_2 = \begin{bmatrix} 3 \\ 2 \end{bmatrix} - \frac{1}{2} \begin{bmatrix} 1 \\ -1 \end{bmatrix} = \begin{bmatrix} \frac{5}{2} \\ \frac{5}{2} \end{bmatrix}$$

By construction, the set $\mathcal{C} = \{\mathbf{y}_1, \mathbf{y}_2\}$ is an orthogonal basis for $\mathbb{R}^2$. ▲

EXAMPLE 8.9 An Orthogonal Basis for $\mathbb{R}^3$

Use the Gram–Schmidt process to convert the basis

$$\mathcal{B} = \{\mathbf{x}_1 = (1, -1, 0), \mathbf{x}_2 = (3, 0, 1), \mathbf{x}_3 = (3, 2, 1)\}$$

for $\mathbb{R}^3$ to an orthogonal basis $\mathcal{C} = \{\mathbf{y}_1, \mathbf{y}_2, \mathbf{y}_3\}$ in the standard inner product on $\mathbb{R}^3$.

Solution. By the Gram–Schmidt process, the required vectors are

1. $\mathbf{y}_1 = \mathbf{x}_1$
2. $\mathbf{y}_2 = \mathbf{x}_2 - \dfrac{\mathbf{x}_2^T \mathbf{y}_1}{\mathbf{y}_1^T \mathbf{y}_1} \mathbf{y}_1$
3. $\mathbf{y}_3 = \mathbf{x}_3 - \dfrac{\mathbf{x}_3^T \mathbf{y}_1}{\mathbf{y}_1^T \mathbf{y}_1} \mathbf{y}_1 - \dfrac{\mathbf{x}_3^T \mathbf{y}_2}{\mathbf{y}_2^T \mathbf{y}_2} \mathbf{y}_2$

We begin by calculating $\mathbf{y}_2$.

```
[> with(linalg):
[> a:=dotprod(vector(3,[3,0,1]),vector(3,[1,-1,0])):
[> b:=dotprod(vector(3,[1,-1,0]),vector(3,[1,-1,0])):
[> a/b;
```

$$\frac{3}{2}$$

Therefore,

$$\mathbf{y}_2 = \begin{bmatrix} 3 \\ 0 \\ 1 \end{bmatrix} - \frac{3}{2}\begin{bmatrix} 1 \\ -1 \\ 0 \end{bmatrix} = \begin{bmatrix} \frac{3}{2} \\ \frac{3}{2} \\ 1 \end{bmatrix}$$

Next we calculate $\mathbf{y}_3$.

```
[> c:=dotprod(vector(3,[3,2,1]),vector(3,[1,-1,0])):
[> c/b;
```

$$\frac{1}{2}$$

```
[> d:=dotprod(vector(3,[3,2,1]),vector(3,[3/2,3/2,1])):
[> e:=dotprod(vector(3,[3/2,3/2,1]]),vector(3,[3/2,3/2,1])):
[> d/e;
```

$$\frac{17}{11}$$

This means that

$$\mathbf{y}_3 = \mathbf{x}_3 - \frac{\mathbf{x}_3^T\mathbf{y}_1}{\mathbf{y}_1^T\mathbf{y}_1}\mathbf{y}_1 - \frac{\mathbf{x}_3^T\mathbf{y}_2}{\mathbf{y}_2^T\mathbf{y}_2}\mathbf{y}_2$$

$$= \begin{bmatrix} 3 \\ 2 \\ 1 \end{bmatrix} - \frac{1}{2}\begin{bmatrix} 1 \\ -1 \\ 0 \end{bmatrix} - \frac{17}{11}\begin{bmatrix} \frac{3}{2} \\ \frac{3}{2} \\ 1 \end{bmatrix} = \begin{bmatrix} \frac{2}{11} \\ \frac{2}{11} \\ -\frac{6}{11} \end{bmatrix}$$

Thus the set $\mathcal{C} = \{\mathbf{y}_1, \mathbf{y}_2, \mathbf{y}_3\}$ is an orthogonal basis for $\mathbb{R}^3$. ▲

The Gram–Schmidt algorithm is built into both the **linalg** and the **LinearAlgebra** packages. We illustrate the use of the **GramSchmidt** function in the **linalg** package.

EXAMPLE 8.10 Orthogonalizing a Basis for $\mathbb{R}^3$

Use the **GramSchmidt** function in *Maple* to convert the basis

$$\mathcal{B} = \{\mathbf{x}_1 = (1, 1, 0), \mathbf{x}_2 = (0, 2, 1), \mathbf{x}_3 = (3, 0, 1)\}$$

to a basis $\mathcal{C} = \{\mathbf{y}_1, \mathbf{y}_2, \mathbf{y}_3\}$ whose vectors are mutually orthogonal unit vectors in the standard inner product.

Solution. We use the **GramSchmidt** function in the **linalg** package.

```
[> with(linalg):
[> x:=vector(3,[1,1,0]):
[> y:=vector(3,[0,2,1]):
[> z:=vector(3,[3,0,1]):
```

```
[> C:=GramSchmidt([x,y,z],normalized);
```

$$C := \left[\left[\tfrac{1}{2}\sqrt{2}, \tfrac{1}{2}\sqrt{2}, 0\right], \left[-\tfrac{1}{3}\sqrt{3}, \tfrac{1}{3}\sqrt{3}, \tfrac{1}{3}\sqrt{3}\right], \left[\tfrac{1}{30}\sqrt{25}\sqrt{6}, -\tfrac{1}{30}\sqrt{25}\sqrt{6}, \tfrac{1}{15}\sqrt{25}\sqrt{6}\right]\right]$$

The calculation

```
[> [dotprod(C[1],C[1]),dotprod(C[1],C[2]),
   dotprod(C[1],C[3]),dotprod(C[2],C[2]),
   dotprod(C[2],C[3]),dotprod(C[3],C[3])];
```

$$[1, 0, 0, 1, 0, 1]$$

shows that the vectors in the set

$$\mathcal{C} = \left\{ \begin{bmatrix} \frac{1}{\sqrt{2}} \\ \frac{1}{\sqrt{2}} \\ 0 \end{bmatrix}, \begin{bmatrix} -\frac{1}{\sqrt{3}} \\ \frac{1}{\sqrt{3}} \\ \frac{1}{\sqrt{3}} \end{bmatrix}, \begin{bmatrix} \frac{1}{6}\sqrt{6} \\ -\frac{1}{6}\sqrt{6} \\ \frac{5}{3}\sqrt{6} \end{bmatrix} \right\}$$

are unit vectors and are mutually orthogonal and have length 1. ▲

EXERCISES 8.3

1. Use the Gram–Schmidt process to convert the basis for $\mathbb{R}^2$ determined by the columns of the matrix

$$A = \begin{bmatrix} 1 & 1 \\ 1 & -1 \end{bmatrix}$$

to an orthogonal basis.

2. Use the Gram–Schmidt process with a nonstandard inner product of your choice to convert the columns of the matrix in Exercise 1 to an orthogonal basis.

3. Use the Gram–Schmidt process to convert the basis for $\mathbb{R}^3$ determined by the columns of the matrix

$$A = \begin{bmatrix} 1 & 1 & 2 \\ 1 & -1 & 0 \\ 0 & 2 & 0 \end{bmatrix}$$

to an orthogonal basis.

4. (*Maple V*) Use the vector $\mathbf{x} = (1, 1, 1)$ and the Gram–Schmidt process to extend the set $\mathcal{S} = \{(1, 1, 0), (1, -1, 2)\}$ to an orthogonal basis for $\mathbb{R}^3$.

5. (*Maple V*) Generate a random 5×5 real invertible matrix A and use the **linalg[Gram-Schmidt]** function to convert the columns of A to an orthogonal basis for $\mathbb{R}^5$.

6. (*Maple 6*) Use the **LinearAlgebra[GramSchmidt]** functions to verify your calculations in Exercise 2.

7. (*Maple V, 6*) Discuss the difference between the functions **linalg[GramSchmidt]** and **LinearAlgebra[GramSchmidt]**.

ORTHONORMAL BASES

In addition to being orthogonal, the bases in many applications must also be unit vectors.

DEFINITION 8.3 *An orthogonal basis $\mathcal{B} = \{\mathbf{v}_1, \ldots, \mathbf{v}_n\}$ for an inner product space V is* ***orthonormal*** *if $\|\mathbf{v}_i\| = 1$ for all $i \in \mathbf{n}$.*

By Theorem 7.6, every nonzero vector can be normalized. We can therefore convert every orthogonal basis to an orthonormal basis by normalizing its vectors. The **GramSchmidt** function can be used to produce orthogonal bases that are either normalized or not. The default is a normalized basis. The option `normalized=false`, on the other hand, produces a basis whose vectors may not be normal. If no option is specified, *Maple* sets the `normalized` option to `true`.

EXAMPLE 8.11 An Orthonormal Basis for $\mathbb{R}^3$

Verify that the vectors in the orthogonal basis $\mathcal{C} = \{\mathbf{x}_1, \mathbf{x}_2, \mathbf{x}_3\}$ produced by the **Gram-Schmidt** function from the basis $\mathcal{B} = \{(1, -1, 1), (2, 2, 2), (-3, 1, 3)\}$ are unit vectors.

Solution. We first load the **LinearAlgebra** package and define the basis vectors.

```
[>with(LinearAlgebra):
[>x:=Vector([1,-1,1]):
[>y:=Vector([2,2,2]):
[>z:=Vector([-3,1,3]):
```

Next we carry out the one-step conversion.

```
[> g:=GramSchmidt({x,y,z},normalized);
```

$$g := \left\{ \begin{bmatrix} \frac{1}{3}\sqrt{3} \\ \frac{1}{3}\sqrt{3} \\ \frac{1}{3}\sqrt{3} \end{bmatrix}, \begin{bmatrix} \frac{1}{6}\sqrt{6} \\ -\frac{1}{3}\sqrt{6} \\ \frac{1}{6}\sqrt{6} \end{bmatrix}, \begin{bmatrix} -\frac{1}{2}\sqrt{2} \\ 0 \\ \frac{1}{2}\sqrt{2} \end{bmatrix} \right\}$$

We use the **Normalize** function to test whether the vectors in g are normal.

```
[> e1:=Equal(Normalize(g[1],Euclidean),g[1]):
[> e2:=Equal(Normalize(g[2],Euclidean),g[2]):
[> e3:=Equal(Normalize(g[3],Euclidean),g[3]):
[> [e1,e2,e3];
```

$$[true, true, true]$$

This shows that in each case, the input vector equals the output vector. Hence the vectors are normal. ▲

In Example 8.11, the command

GramSchmidt[{{1, -1, 1}, {2, 2, 2}, { - 3, 1, 3}}, **normalized** = **false**]

would have produced the orthogonal basis

$$\{\mathbf{y}_1, \mathbf{y}_2, \mathbf{y}_3\} = \left\{\{1, -1, 1\}, \left\{\frac{4}{3}, \frac{8}{3}, \frac{4}{3}\right\}, \{-3, 0, 3\}\right\}$$

However, this basis would not normal since the norm of the vector

$$\|\mathbf{y}_1\| = \sqrt{1^2 + (-1)^2 + 1^2} = \sqrt{3} \neq 1$$

for example, is not equal to 1.

EXERCISES 8.4

1. Use the Gram–Schmidt process to convert the basis $\mathcal{B} = \{(1, 2), (1, -1)\}$ to a basis $\mathcal{C} = \{(a, b), (c, d)\}$ that is orthogonal in the standard inner product on $\mathbb{R}^2$. Convert $\mathcal{C}$ to an orthonormal basis.

2. Use the Gram–Schmidt process to convert the basis $\mathcal{B} = \{(1, 2), (1, -1)\}$ to a basis $\mathcal{C} = \{(a, b), (c, d)\}$ that is orthogonal in the inner product $\mathbf{x}A^TA\mathbf{y}$ on $\mathbb{R}^2$ determined by

the matrix

$$A = \begin{bmatrix} 50 & 79 \\ 56 & 49 \end{bmatrix}$$

Convert $\mathcal{C}$ to an orthonormal basis.

3. Let $\langle \mathbf{x}, \mathbf{y} \rangle$ be an arbitrary inner product on $\mathbb{R}^2$ and let $\mathbf{x} = (3, 4)$ and $\mathbf{y} = (2, 5)$ be two vectors in $\mathbb{R}^2$. Construct an orthonormal basis $\mathcal{B}$ for $\mathbb{R}^2$, and find the coordinates

$$[\mathbf{x}]_{\mathcal{B}} = \begin{bmatrix} a \\ b \end{bmatrix} \quad \text{and} \quad [\mathbf{y}]_{\mathcal{B}} = \begin{bmatrix} c \\ d \end{bmatrix}$$

in the basis $\mathcal{B}$. Show that $\langle \mathbf{x}, \mathbf{y} \rangle = ac + bd$.

4. (*Maple V*) Use the **GramSchmidt** function to convert the basis $\mathcal{B} = \{(3, 4), (4, 3)\}$ for $\mathbb{R}^2$ to an orthogonal basis whose vectors may not be normal.

5. (*Maple V*) Let $\langle \mathbf{x}, \mathbf{y} \rangle$ be an arbitrary inner product on $\mathbb{R}^3$ and let $\mathbf{x} = (3, 4, -2)$ and $\mathbf{y} = (2, 5, 7)$ be two vectors in $\mathbb{R}^3$. Construct an orthonormal basis $\mathcal{B}$ for $\mathbb{R}^3$, and calculate the coordinate vectors

$$[\mathbf{x}]_{\mathcal{B}} = \begin{bmatrix} a \\ b \\ c \end{bmatrix} \quad \text{and} \quad [\mathbf{y}]_{\mathcal{B}} = \begin{bmatrix} d \\ e \\ f \end{bmatrix}$$

in the basis $\mathcal{B}$. Show that $\langle \mathbf{x}, \mathbf{y} \rangle = ad + be + cf$.

6. (*Maple V*) Use the **normalize** function in the **linalg** package to normalize the basis produced in Exercise 1.

7. (*Maple 6*) Use the Gram–Schmidt process to convert the basis

$$\mathcal{B} = \{(1, 2, 1), (1, -1, 0), (0, 1, 0)\}$$

to a basis $\mathcal{C} = \{(a, b, c), (d, e, f), (g, h, i)\}$ that is orthogonal in the standard inner product on $\mathbb{R}^2$. Convert $\mathcal{C}$ to an orthonormal basis.

8. (*Maple 6*) Use the Gram–Schmidt process to convert the basis

$$\mathcal{B} = \{(1, 2, 1), (1, -1, 0), (0, 1, 0)\}$$

to a basis $\mathcal{C} = \{(a, b, c), (d, e, f), (g, h, i)\}$ that is orthogonal in the inner product $\mathbf{x}A^T A\mathbf{y}$ on $\mathbb{R}^3$ determined by the matrix

$$A = \begin{bmatrix} 1 & 0 & 0 \\ 0 & 2 & 0 \\ 0 & 0 & 3 \end{bmatrix}$$

Convert $\mathcal{C}$ to an orthonormal basis.

9. (*Maple 6*) Use the **GramSchmidt** function in the **LinearAlgebra** package to convert the basis $\mathcal{B} = \{(8, -5, 8), (4, 3, 5), (-5, -5, -1)\}$ for $\mathbb{R}^3$ to an orthogonal basis whose vectors may not be normal.

10. (*Maple 6*) Use the **Normalize** function in the **LinearAlgebra** package to normalize the basis produced in Exercise 8.

11. (*Maple 6*) Use the **GramSchmidt** function in the **LinearAlgebra** package to convert the basis in Exercise 8 to an orthonormal basis in one step.

Real Inner Products and Orthonormal Bases

It is quite remarkable that relative to orthonormal bases, all inner products on real inner product spaces can be computed as dot products. The difficult arithmetic involved in calculating inner products is now hidden in the arithmetic required to convert a given basis to an orthonormal one.

Suppose $\mathcal{B} = \{\mathbf{x}_1, \mathbf{x}_2\}$ is a basis for an inner product space V, and let

$$\mathbf{x} = a_1\mathbf{x}_1 + a_2\mathbf{x}_2 \quad \text{and} \quad \mathbf{y} = b_1\mathbf{x}_1 + b_2\mathbf{x}_2$$

be two vectors in V. Then it follows from the linearity of $\langle \mathbf{x}, \mathbf{y} \rangle$ in both $\mathbf{x}$ and $\mathbf{y}$ that

$$\begin{aligned} \langle \mathbf{x}, \mathbf{y} \rangle &= \langle a_1\mathbf{x}_1 + a_2\mathbf{x}_2, b_1\mathbf{x}_1 + b_2\mathbf{x}_2 \rangle \\ &= a_1b_1 \langle \mathbf{x}_1, \mathbf{x}_1 \rangle + a_1b_2 \langle \mathbf{x}_1, \mathbf{x}_2 \rangle + a_2b_1 \langle \mathbf{x}_2, \mathbf{x}_1 \rangle + a_2b_2 \langle \mathbf{x}_2, \mathbf{x}_2 \rangle \end{aligned}$$

If $\mathcal{B}$ is orthonormal, then $\langle \mathbf{x}_1, \mathbf{x}_2 \rangle = \langle \mathbf{x}_2, \mathbf{x}_1 \rangle = 0$ and $\langle \mathbf{x}_1, \mathbf{x}_1 \rangle = \langle \mathbf{x}_2, \mathbf{x}_2 \rangle = 1$. Hence

$$\langle \mathbf{x}, \mathbf{y} \rangle = a_1b_1 \times 1 + a_1b_2 \times 0 + a_2b_1 \times 0 + a_2b_2 \times 1 = a_1b_1 + a_2b_2$$

takes the form of the dot product. The next theorem expresses this property for inner product spaces of arbitrary finite dimension.

THEOREM 8.6 (Orthonormal basis theorem) *If the function* $(\mathbf{x}, \mathbf{y}) \mapsto \langle \mathbf{x}, \mathbf{y} \rangle$ *is an inner product on a finite-dimensional real vector space V and if* $\mathcal{E} = \{\mathbf{e}_1, \ldots, \mathbf{e}_n\}$ *is an orthonormal basis for V, then* $\langle \mathbf{x}, \mathbf{y} \rangle = a_1b_1 + \cdots + a_nb_n$ *for all vectors* $\mathbf{x} = a_1\mathbf{e}_1 + \cdots + a_n\mathbf{e}_n$ *and* $\mathbf{y} = b_1\mathbf{e}_1 + \cdots + b_n\mathbf{e}_n$ *in V.*

Proof. Consider the inner product

$$\langle \mathbf{x}, \mathbf{y} \rangle = \langle a_1\mathbf{e}_1 + \cdots + a_n\mathbf{e}_n, b_1\mathbf{e}_1 + \cdots + b_n\mathbf{e}_n \rangle$$

Since $\mathcal{E}$ is orthonormal and $\langle \mathbf{x}, \mathbf{y} \rangle$ is bilinear, we get

$$\langle \mathbf{x}, \mathbf{y} \rangle = \sum_{i,j \in \mathbf{n}} a_i b_j \left\langle \mathbf{e}_i, \mathbf{e}_j \right\rangle = a_1b_1 + \cdots + a_nb_n$$

since $\left\langle \mathbf{e}_i, \mathbf{e}_j \right\rangle = 0$ if $i \neq j$ and $\left\langle \mathbf{e}_i, \mathbf{e}_j \right\rangle = 1$ otherwise. ■

COROLLARY 8.7 *If $\mathcal{B} = \{\mathbf{e}_1, \ldots, \mathbf{e}_n\}$ is an orthonormal basis for an inner product space V, then $\mathbf{x} = \langle \mathbf{x}, \mathbf{e}_1 \rangle \mathbf{e}_1 + \cdots + \langle \mathbf{x}, \mathbf{e}_n \rangle \mathbf{e}_n$ for any $\mathbf{x} \in V$.*

Proof. Let $\mathbf{x} = a_1\mathbf{e}_1 + \cdots + a_n\mathbf{e}_n$ and $\mathbf{y} = \mathbf{e}_i$. Then Theorem 8.6 shows that

$$\langle \mathbf{x}, \mathbf{e}_i \rangle = \langle a_1\mathbf{e}_1 + \cdots + a_n\mathbf{e}_n, \mathbf{e}_i \rangle = a_i$$

for all $i \in \mathbf{n}$. This proves the corollary. ■

EXERCISES 8.5

1. Find an orthonormal basis $\mathcal{B}$ for $\mathbb{R}^2$ in which the real inner product $\mathbf{x}^T A\mathbf{y}$ determined by the positive definite matrix

$$A = \begin{bmatrix} 8 & 0 \\ 0 & 3 \end{bmatrix}$$

can be computed as a dot product. Illustrate your result with a numerical example.

2. (*Maple 6*) Find an orthonormal basis $\mathcal{B}$ for $\mathbb{R}^2$ in which the real inner product $\mathbf{x}^T A\mathbf{y}$ determined by the positive definite matrix

$$A = \begin{bmatrix} 8 & 4 \\ 4 & 3 \end{bmatrix}$$

can be computed as a dot product. Illustrate your result with a numerical example.

3. (*Maple 6*) Find an orthonormal basis $\mathcal{B}$ for $\mathbb{R}^3$ in which the real inner product $\mathbf{x}^T A\mathbf{y}$ determined by the positive definite matrix

$$A = \begin{bmatrix} 8 & 0 & 0 \\ 0 & 3 & 0 \\ 0 & 0 & 5 \end{bmatrix}$$

can be computed as a dot product. Illustrate your result with a numerical example.

4. (*Maple 6*) Find an orthonormal basis $\mathcal{B}$ for $\mathbb{R}^3$ in which the real inner product $\mathbf{x}^T A\mathbf{y}$ determined by the positive definite matrix

$$A = \begin{bmatrix} 8 & 0 & 0 \\ 0 & 3 & -1 \\ 0 & -1 & 5 \end{bmatrix}$$

can be computed as a dot product. Illustrate your result with a numerical example.

THE QR DECOMPOSITION

In this section, we show how the Gram–Schmidt orthogonalization process can be used to decompose a matrix A into a specific matrix product QR. The decomposition is called a ***QR decomposition***. It has several powerful applications in numerical linear algebra. Among them are a procedure for solving linear systems $A\mathbf{x} = \mathbf{b}$ whose coefficient matrices A are invertible, a method for solving the ***normal equations*** $A^T A\mathbf{x} = A^T\mathbf{b}$ of overdetermined linear systems $A\mathbf{x} = \mathbf{b}$, and an algorithm for calculating approximations of eigenvalues and eigenvectors in situations where exact solutions are unavailable.

The idea behind the construction of the matrices Q and R is simple. Suppose, for example, that $\{\mathbf{x}_1, \mathbf{x}_2, \mathbf{x}_3\}$ are three linearly independent columns in $\mathbb{R}^3$. Then the Gram–Schmidt process yields three orthogonal vectors $\{\mathbf{y}_1, \mathbf{y}_2, \mathbf{y}_3\}$, where

$$\begin{aligned}
\mathbf{y}_1 &= \mathbf{x}_1 \\
\mathbf{y}_2 &= \mathbf{x}_2 - \frac{\langle \mathbf{x}_2, \mathbf{y}_1 \rangle}{\langle \mathbf{y}_1, \mathbf{y}_1 \rangle}\mathbf{y}_1 \\
\mathbf{y}_3 &= \mathbf{x}_3 - \frac{\langle \mathbf{x}_3, \mathbf{y}_1 \rangle}{\langle \mathbf{y}_1, \mathbf{y}_1 \rangle}\mathbf{y}_1 - \frac{\langle \mathbf{x}_3, \mathbf{y}_2 \rangle}{\langle \mathbf{y}_2, \mathbf{y}_2 \rangle}\mathbf{y}_2
\end{aligned}$$

If we solve each of these equations for $\mathbf{x}$, we get

$$\begin{aligned}
\mathbf{x}_1 &= \mathbf{y}_1 \\
\mathbf{x}_2 &= \frac{\langle \mathbf{x}_2, \mathbf{y}_1 \rangle}{\langle \mathbf{y}_1, \mathbf{y}_1 \rangle}\mathbf{y}_1 + \mathbf{y}_2 \\
\mathbf{x}_3 &= \frac{\langle \mathbf{x}_3, \mathbf{y}_1 \rangle}{\langle \mathbf{y}_1, \mathbf{y}_1 \rangle}\mathbf{y}_1 + \frac{\langle \mathbf{x}_3, \mathbf{y}_2 \rangle}{\langle \mathbf{y}_2, \mathbf{y}_2 \rangle}\mathbf{y}_2 + \mathbf{y}_3
\end{aligned}$$

If we normalize these vectors, the result are three orthonormal vectors

$$\begin{aligned}
\mathbf{u}_1 &= a_{11}\mathbf{y}_1 \\
\mathbf{u}_2 &= a_{12}\mathbf{y}_1 + a_{22}\mathbf{y}_2 \\
\mathbf{u}_3 &= a_{13}\mathbf{y}_1 + a_{23}\mathbf{y}_2 + a_{33}\mathbf{y}_3
\end{aligned}$$

for which a_{11}, a_{22}, and a_{33} are nonzero. We now write each column vectors $\mathbf{x}_i$ as linear combinations of the vectors $\mathbf{u}_1$, $\mathbf{u}_2$, and $\mathbf{u}_3$. This means that

$$\mathbf{x}_i = r_{1i}\mathbf{u}_1 + r_{2i}\mathbf{u}_2 + r_{3i}\mathbf{u}_3 = \begin{bmatrix} \mathbf{u}_1 & \mathbf{u}_2 & \mathbf{u}_3 \end{bmatrix} \begin{bmatrix} r_{1i} \\ r_{2i} \\ r_{3i} \end{bmatrix}$$

and since $A = \begin{bmatrix} \mathbf{x}_1 & \mathbf{x}_2 & \mathbf{x}_3 \end{bmatrix}$,

$$A = \begin{bmatrix} Q\begin{bmatrix} r_{11} \\ 0 \\ 0 \end{bmatrix} & Q\begin{bmatrix} r_{21} \\ r_{22} \\ 0 \end{bmatrix} & Q\begin{bmatrix} r_{31} \\ r_{32} \\ r_{33} \end{bmatrix} \end{bmatrix} = Q\begin{bmatrix} r_{11} & r_{21} & r_{31} \\ 0 & r_{22} & r_{32} \\ 0 & 0 & r_{33} \end{bmatrix} = QR$$

By construction, the columns of Q form an orthonormal set in $\mathbb{R}^3$ and R is an upper triangular matrix whose diagonal entries are nonzero. Hence R is invertible. We summarize this construction as a theorem.

THEOREM 8.8 (QR decomposition theorem) *If A is an $m \times n$ matrix whose n columns are linearly independent vectors in $\mathbb{R}^m$, then there exist an $m \times n$ matrix Q whose columns form an orthonormal set in $\mathbb{R}^m$ and an $n \times n$ upper-triangular invertible matrix R such that $A = QR$.*

Proof. Suppose that $\mathcal{B} = \{\mathbf{x}_1, \ldots, \mathbf{x}_n\}$ is the set of columns of A. Then $\mathcal{B}$ is a basis for a subspace W of $\mathbb{R}^m$. Let $\mathcal{C} = \{\mathbf{u}_1, \ldots, \mathbf{u}_n\}$ be the orthonormal basis for W obtained from $\mathcal{B}$ by the Gram–Schmidt process, and form the matrix $Q = [\mathbf{u}_1 \ \cdots \ \mathbf{u}_n]$. Then the columns of Q are an orthonormal set in $\mathbb{R}^m$. Since $\mathcal{C}$ is a basis for W, every vector $\mathbf{x}_i \in \mathcal{B}$ is a unique linear combination

$$r_{1i}\mathbf{u}_1 + \cdots + r_{ni}\mathbf{u}_n = \begin{bmatrix} \mathbf{u}_1 & \cdots & \mathbf{u}_n \end{bmatrix} \begin{bmatrix} r_{1i} \\ r_{2i} \\ r_{3i} \end{bmatrix} = \begin{bmatrix} \mathbf{u}_1 & \cdots & \mathbf{u}_n \end{bmatrix} \mathbf{r}_i$$

in which the coefficients $r_{11}, \ldots, r_{nm}$ are all nonzero. Therefore,

$$A = \begin{bmatrix} \mathbf{x}_1 & \mathbf{x}_2 & \cdots & \mathbf{x}_n \end{bmatrix} = \begin{bmatrix} \mathbf{u}_1 & \cdots & \mathbf{u}_n \end{bmatrix} \begin{bmatrix} \mathbf{r}_1 & \mathbf{r}_2 & \cdots & \mathbf{r}_n \end{bmatrix} = QR$$

By construction, the columns of Q are an orthonormal set in $\mathbb{R}^n$, and R is an $n \times n$ upper-triangular matrix. Since the diagonal entries r_{ii} are all nonzero, R is invertible. ■

In Table 2, we summarize the process of finding a QR decomposition for an $m \times n$ matrix A, described in the proof of Theorem 8.8. The algorithm requires that the columns of A are linearly independent. This condition is equivalent to requiring that the rank of A be n.

TABLE 2 FINDING A QR DECOMPOSITION OF A MATRIX.

Step 1. Take an $m \times n$ matrix $A = [\mathbf{x}_1 \ \cdots \ \mathbf{x}_n]$ of rank n and convert the set $\{\mathbf{x}_1, \ldots, \mathbf{x}_n\}$ of columns of A to a set $\{\mathbf{u}_1, \ldots, \mathbf{u}_n\}$ of orthonormal vectors using the Gram–Schmidt process.

Step 2. Write each vector $\mathbf{x}_i$ as a linear combination $r_{1i}\mathbf{u}_1 + \cdots + r_{ni}\mathbf{u}_n$ of the vectors $\mathbf{u}_j$.

Step 3. Build the vectors $\mathbf{r}_i = \begin{bmatrix} r_{1i} \\ \vdots \\ r_{ni} \end{bmatrix}$ from the linear combinations $r_{1i}\mathbf{u}_1 + \cdots + r_{ni}\mathbf{u}_n$.

Step 4. Form the matrices $Q = [\mathbf{u}_1 \ \cdots \ \mathbf{u}_n]$ and $R = [\mathbf{r}_1 \ \cdots \ \mathbf{r}_n]$.

Step 5. The product QR of the matrices Q and R is the desired decomposition of A.

The QR decomposition process is easy to state, but tedious to apply in practice. Fortunately, *Maple* has a built-in algorithm for calculating QR decompositions. For any $m \times n$ matrix A of rank n, the command `Q,R=QRDecomposition(A)` produces the desired decomposition of A.

EXAMPLE 8.12 The QR Decomposition of a Square Matrix

Use *Maple* to find the QR decomposition of the matrix

$$A = \begin{bmatrix} 1 & 3 & 0 \\ 0 & 5 & 7 \\ 2 & -8 & 4 \end{bmatrix}$$

and verify that the matrix Q in the product QR is orthogonal.

Solution. We use the **QRDecomposition** function in the **LinearAlgebra** package to decompose the matrix A into an orthogonal matrix Q and an upper-triangular matrix R.

```
[> with(LinearAlgebra):
[> A:=Matrix([[[1,3,0],[0,5,7],[2,-8,4]]):
[> Q,R:=QRDecomposition(A);
```

$$QR := \begin{bmatrix} \frac{1}{5}\sqrt{5} & \frac{28}{1605}\sqrt{1605} & -\frac{10}{321}\sqrt{321} \\ 0 & \frac{5}{321}\sqrt{1605} & \frac{14}{321}\sqrt{321} \\ \frac{2}{5}\sqrt{5} & -\frac{14}{1605}\sqrt{1605} & \frac{5}{321}\sqrt{321} \end{bmatrix}, \begin{bmatrix} \sqrt{5} & -\frac{13}{5}\sqrt{5} & \frac{8}{5}\sqrt{5} \\ 0 & \frac{1}{5}\sqrt{1605} & \frac{119}{1605}\sqrt{1605} \\ 0 & 0 & \frac{118}{321}\sqrt{321} \end{bmatrix}$$

We show that $QR[1]$ is orthogonal. This guarantees that its columns are orthonormal.

```
[> Transpose(Q).Q;
```

$$\begin{bmatrix} 1 & 0 & 0 \\ 0 & 1 & 0 \\ 0 & 0 & 1 \end{bmatrix}$$

Next we show that Q and R yield the expected decomposition.

```
[> Q.R;
```

$$\begin{bmatrix} 1 & 3 & 0 \\ 0 & 5 & 7 \\ 2 & -8 & 4 \end{bmatrix}$$

It is clear by inspection that R is an upper-triangular matrix. It therefore remains to show that R is invertible. We do so by verifying that the determinant of R is not zero.

```
[> evalb(Determinant(R)=0);
```

$$false$$

Hence the matrices Q and R yield a QR decomposition of A. ▲

EXAMPLE 8.13 The QR Decomposition of a Rectangular Matrix

Use *Maple* to find a QR decomposition of the matrix

$$A = \begin{bmatrix} 1 & 3 \\ 0 & 5 \\ 2 & -8 \end{bmatrix}$$

Solution. We use the **QRdecomp** function in the **linalg** package to decompose A into a product QR consisting of a matrix Q whose set of columns is orthonormal and an invertible upper-triangular matrix R.

```
[> with(LinearAlgebra):
[> A:=Matrix([[1,3],[0,5],[2,-8]]):
[> Q,R:=QRDecomposition(A);
```

$$Q, R := \begin{bmatrix} \frac{1}{5}\sqrt{5} & \frac{28}{1605}\sqrt{1605} \\ 0 & \frac{5}{321}\sqrt{1605} \\ \frac{2}{5}\sqrt{5} & -\frac{14}{1605}\sqrt{1605} \end{bmatrix}, \begin{bmatrix} \sqrt{5} & -\frac{13}{5}\sqrt{5} \\ 0 & \frac{1}{5}\sqrt{1065} \end{bmatrix}$$

The following calculation shows that $QR = A$.

```
[> Q.R;
```

$$\begin{bmatrix} 1 & 3 \\ 0 & 5 \\ 2 & -8 \end{bmatrix}$$

We verify that the columns of Q form an orthonormal set in the standard inner product on $\mathbb{R}^3$ by showing that $Q^T Q$ is an identity matrix.

```
[> Transpose(Q).Q;
```

$$\begin{bmatrix} 1 & 0 \\ 0 & 1 \end{bmatrix}$$

This shows that the dot products of the columns of Q are 0 if the columns are distinct and 1 if they are identical. Hence the columns are mutually orthogonal unit vectors. Furthermore, the matrix R is clearly upper triangular and is invertible since its diagonal entries are nonzero. ▲

The QR Decomposition and Orthonormal Bases

Theorem 8.8 provides us with a powerful technique for constructing orthonormal bases for column spaces.

COROLLARY 8.9 *If $A = QR$ is a QR decomposition of an $m \times n$ real matrix, then the columns of Q form an orthonormal basis for* Col A.

Proof. If QR is a QR decomposition of an $m \times n$ matrix A, then Theorem 8.8 tells us that R is invertible. By Exercise 4.14.12, the matrices A and Q therefore have the same column space. Since the columns of Q form an orthonormal set in $\mathbb{R}^m$, the corollary follows. ■

EXAMPLE 8.14 Finding an Orthonormal Basis by QR Decomposition

Use the QR decomposition to construct an orthonormal basis for the column space of the matrix

$$A = \begin{bmatrix} 3 & 2 & 3 \\ 5 & 4 & 2 \\ 5 & 2 & 1 \\ 1 & 1 & 3 \end{bmatrix}$$

Solution. We first use *Maple* to verify that the columns of A are linearly independent by showing that the rank of A is 3.

```
[> with(LinearAlgebra):
[> A:=Matrix([[3,2,3],[5,4,2],[5,2,1],[1,1,3]]):
[> Rank(A);
```

$$3$$

This shows that the columns of A are linearly independent. Any QR decomposition of A therefore yields a matrix Q whose columns form an orthonormal basis for Col A.

```
[> Q,R:=QRDecomposition(A):
[> Q;
```

$$\begin{bmatrix} \frac{1}{10}\sqrt{15} & \frac{3}{1310}\sqrt{1965} & \frac{39}{28558}\sqrt{142790} \\ \frac{1}{6}\sqrt{15} & \frac{11}{786}\sqrt{1965} & \frac{-81}{71395}\sqrt{142790} \\ \frac{1}{6}\sqrt{15} & \frac{-13}{786}\sqrt{1965} & \frac{-11}{142790}\sqrt{142790} \\ \frac{1}{30}\sqrt{15} & \frac{23}{3930}\sqrt{1965} & \frac{28}{14279}\sqrt{142790} \end{bmatrix}$$

We show that the columns of Q are mutually orthogonal and normal by showing that the matrix Q^TQ is the 3×3 identity matrix.

```
[> Equal(Transpose(Q).Q,IdentityMatrix(3));
```

true

Hence the columns of Q form an orthonormal basis for Col A. ▲

We note that the calculation

```
[> GramSchmidt([
    Column(A,1),
    Column(A,2),
    Column(A,3)],
    normalized=true);
```

$$\left[\begin{bmatrix} \frac{1}{10}\sqrt{15} \\ \frac{1}{6}\sqrt{15} \\ \frac{1}{6}\sqrt{15} \\ \frac{1}{30}\sqrt{15} \end{bmatrix}, \begin{bmatrix} \frac{3}{1310}\sqrt{1965} \\ \frac{11}{786}\sqrt{1965} \\ \frac{-13}{786}\sqrt{1965} \\ \frac{23}{3930}\sqrt{1965} \end{bmatrix}, \begin{bmatrix} \frac{39}{28558}\sqrt{142790} \\ \frac{-81}{71395}\sqrt{142790} \\ \frac{-11}{142790}\sqrt{142790} \\ \frac{28}{14279}\sqrt{142790} \end{bmatrix} \right]$$

produces the same result.

Linear Systems with Invertible Coefficient Matrices

The QR decomposition provides an efficient method for solving linear equations determined by invertible matrices. Suppose that A is invertible and we would like to solve the system $A\mathbf{x} = \mathbf{b}$. Since A is invertible, it has a QR decomposition. Therefore, $A\mathbf{x} = QR\mathbf{x} = \mathbf{b}$. If we multiply both sides by Q^T, then the orthogonality of Q implies that $Q^T Q R\mathbf{x} = R\mathbf{x} = Q^T\mathbf{b}$. It therefore remains to solve the equation $R\mathbf{x} = Q^T\mathbf{b}$. Since R is upper triangular, this equation can be solved by back substitution.

EXAMPLE 8.15 Solving a Linear System by QR Decomposition

Use the QR decomposition and back substitution to solve the linear system

$$A\mathbf{x} = \begin{bmatrix} 1 & 2 & 1 & 0 \\ 1 & 1 & 3 & 1 \\ 2 & 3 & 1 & 5 \\ 0 & 0 & 4 & 1 \end{bmatrix} \begin{bmatrix} w \\ x \\ y \\ z \end{bmatrix} = \begin{bmatrix} 2 \\ 1 \\ 0 \\ 5 \end{bmatrix} = \mathbf{b}$$

Solution. We begin by finding a QR decomposition of the matrix A.

```
[> with(LinearAlgebra):
[> A:=Matrix([[1,2,1,0],[1,1,3,1],[2,3,1,5],[0,0,4,1]]):
[> v:=Matrix([[w],[x],[y],[z]]):
[> b:=Matrix([[2],[1],[0],[5]]):
```

```
[> Q,R:=QRDecomposition(A):
[> r:=evalm(R.v):
[> s:=evalm(Transpose(Q).b):
```

```
[> z:=solve(r[4,1]=s[4,1],z):
[> y:=solve(r[3,1]=s[3,1],y):
[> x:=solve(r[2,1]=s[2,1],x):
[> w:=solve(r[1,1]=s[1,1],w):
[> [z,y,x,w];
```

$$\left[\frac{3}{19}, \frac{23}{19}, \frac{68}{19}, \frac{-121}{19}\right]$$

Therefore $w = -121/19$, $x = 68/19$, $y = 23/19$, and $z = 3/19$ is the required solution. ▲

Normal Equations of Overdetermined Linear Systems

The QR decomposition can be used to find the least-squares solution to certain linear systems.

THEOREM 8.10 *If A is an $m \times n$ matrix of rank n, then the solution of the linear system $A^T A\mathbf{x} = A^T\mathbf{b}$ is $\mathbf{x} = R^{-1}Q^T\mathbf{b}$, where R and Q are the matrices obtained from A by a QR decomposition.*

Proof. Consider the linear system $A\mathbf{x} = \mathbf{b}$. Since A is an $m \times n$ matrix of rank n, its columns must be linearly independent. By Theorem 8.8, A can be decomposed into a product QR.

Therefore,

$$A^T A\mathbf{x} = (QR)^T (QR)\mathbf{x} = (QR)^T \mathbf{b}$$

This means that

$$R^T (Q^T Q) R\mathbf{x} = R^T Q^T \mathbf{b}$$

Furthermore, the fact that the columns of Q are orthonormal entails $Q^T Q = I$. It follows that

$$R^T R\mathbf{x} = R^T Q^T \mathbf{b}$$

We also know from Theorem 8.8 that R is invertible. Hence R^T is invertible.

It follows that

$$R\mathbf{x} = (R^T)^{-1} R^T R\mathbf{x} = (R^T)^{-1} R^T Q^T \mathbf{b} = Q^T \mathbf{b}$$

By multiplying both sides by R^{-1}, we get $\mathbf{x} = R^{-1} Q^T \mathbf{b}$. ■

EXAMPLE 8.16 Solving a Normal Equation by QR Decomposition

Use QR decomposition to solve the normal equation of the linear system

$$A\mathbf{x} = \begin{bmatrix} 1 & 2 & 3 \\ 2 & 2 & 2 \\ 1 & 1 & 3 \\ 1 & 2 & 4 \end{bmatrix} \begin{bmatrix} x \\ y \\ z \end{bmatrix} = \begin{bmatrix} 2 \\ -1 \\ 4 \\ 3 \end{bmatrix} = \mathbf{b}$$

Solution. If the matrix A has rank 3, then the required solution is $\widehat{\mathbf{x}} = R^{-1} Q^T \mathbf{b}$, where Q and R are the matrices obtained from A by QR decomposition. We use *Maple* to verify that A has rank 3 and to calculate $\widehat{\mathbf{x}}$.

```
[> with(LinearAlgebra):
[> A:=Matrix([[1,2,3],[2,2,2],[1,1,3],[1,2,4]]):
[> v:=Matrix([[x],[y],[z]]):
[> b:=Matrix([[2],[-1],[4],[3]]):
[> Rank(A);
                              3
```

Since A is a 3×3 matrix of rank 3, Theorem 8.10 applies.

```
[> Q,R:=QRDecomposition(A):
[> MatrixInverse(R).Transpose(Q).b;
```

$$\begin{bmatrix} -\frac{9}{37} \\ -\frac{84}{37} \\ \frac{77}{37} \end{bmatrix}$$

It is easy to verify that

$$\begin{bmatrix} 1 & 2 & 3 \\ 2 & 2 & 2 \\ 1 & 1 & 3 \\ 1 & 2 & 4 \end{bmatrix}^T \begin{bmatrix} 1 & 2 & 3 \\ 2 & 2 & 2 \\ 1 & 1 & 3 \\ 1 & 2 & 4 \end{bmatrix} \begin{bmatrix} -\frac{9}{37} \\ -\frac{84}{37} \\ \frac{77}{37} \end{bmatrix} = \begin{bmatrix} 7 \\ 12 \\ 28 \end{bmatrix} = \begin{bmatrix} 1 & 2 & 3 \\ 2 & 2 & 2 \\ 1 & 1 & 3 \\ 1 & 2 & 4 \end{bmatrix}^T \begin{bmatrix} 2 \\ -1 \\ 4 \\ 3 \end{bmatrix}$$

Hence the vector $\mathbf{x} = (-9/37, -84/37, 77/37)$ solves the normal equation. ▲

The QR Algorithm for Approximating Eigenvalues

The QR algorithm provides an efficient method for approximating the eigenvalues of certain invertible matrices. The algorithm is based on the construction of a sequence of similar matrices $A_1, A_2, \dots, A_{k+1}$ for which the diagonal entries of A_{k+1}, for a large enough k, are good approximations of the eigenvalues of A. The sequence of matrices is defined recursively.

$$\begin{aligned} A = A_1 &= Q_1 R_1 \\ A_2 &= R_1 Q_1 = Q_2 R_2 \\ A_3 &= R_2 Q_2 = Q_3 R_3 \\ &\vdots \\ A_{k+1} &= R_k Q_k \end{aligned}$$

For a discussion of convergence properties of the QR algorithm for different types of matrices we refer to standard numerical analysis textbooks. There it is shown, for example, that if the eigenvalues of A are of distinct absolute value, the matrices A_k converge to an upper triangular matrix U in which eigenvalues of A are diagonal entries of U.

EXAMPLE 8.17 Eigenvalues and the *QR* Algorithm

Use *Maple* and the QR algorithm to find the sequence

$$A_1 = A \qquad A_2 = R_1 Q_1 \qquad \cdots \qquad A_9 = R_8 Q_8$$

of matrices obtained from the matrix

$$A = \begin{bmatrix} 0 & 1 \\ 1 & 1 \end{bmatrix}$$

by successive QR decompositions. At each step, $Q_i R_i$ is a QR decomposition of A_i. Show that the the eigenvalues of A and are close to the diagonal entries of A_9.

Solution. We begin by calculating the eigenvalues of A.

```
[> with(LinearAlgebra):
[> A:=Matrix([[0,1],[1,1]]):
[> evA:=Eigenvalues(A);
```

$$e := \begin{bmatrix} \frac{1}{2} + \frac{1}{2}\sqrt{5} \\ \frac{1}{2} - \frac{1}{2}\sqrt{5} \end{bmatrix}$$

As we can see, the matrix A has two distinct eigenvalues. By our introductory remarks, the QR algorithm therefore supplies us with approximations of these values on the diagonal of the matrices A_k. We use the QR algorithm to calculate the sequence of matrices $A_1, \ldots, A_9$.

```
[> QR:=A->QRDecomposition(A):
[> A1:=A: QR(A1):
[> A2:=QR(A1)[2].QR(A1)[1]:
[> A3:=QR(A2)[2].QR(A2)[1]:
[> A4:=QR(A3)[2].QR(A3)[1]:
[> A5:=QR(A4)[2].QR(A4)[1]:
[> A6:=QR(A5)[2].QR(A5)[1]:
[> A7:=QR(A6)[2].QR(A6)[1]:
[> A8:=QR(A7)[2].QR(A7)[1]:
[> A9:=QR(A8)[2].QR(A8)[1]:
```

We extract the sequences of corresponding diagonal entries from the matrices $A_1, \ldots, A_9$.

```
[> evalf(
   [A1[1,1],A2[1,1],A3[1,1],
   A4[1,1],A5[1,1],A6[1,1],
   A7[1,1],A8[1,1],A9[1,1]],5);
```

[0., 1., 1.5000, 1.6000, 1.6154, 1.6176, 1.6180, 1.6180, 1.6180]

Proof. We prove the theorem for $V = \mathbb{R}^3$ and $p = 2$. The general case is analogous. We let

$$\mathcal{B} = \{\mathbf{x}_1 = (a_1, a_2, a_3), \mathbf{x}_2 = (b_1, b_2, b_3)\}$$

be an orthonormal basis for a two-dimensional subspace for $\mathbb{R}^3$, and we let $\mathbf{x} = (x, y, z)$ be an arbitrary vector in $\mathbb{R}^3$. Then

$$P = \begin{bmatrix} a_1 & b_1 \\ a_2 & b_2 \\ a_3 & b_3 \end{bmatrix} \quad \text{and} \quad P^T = \begin{bmatrix} a_1 & a_2 & a_3 \\ b_1 & b_2 & b_3 \end{bmatrix}$$

Therefore,

$$P^T\mathbf{x} = \begin{bmatrix} a_1 & a_2 & a_3 \\ b_1 & b_2 & b_3 \end{bmatrix} \begin{bmatrix} x \\ y \\ z \end{bmatrix} = \begin{bmatrix} a_1x + a_2y + a_3z \\ b_1x + b_2y + b_3z \end{bmatrix} = \begin{bmatrix} \mathbf{x}_1^T\mathbf{x} \\ \mathbf{x}_2^T\mathbf{x} \end{bmatrix}$$

and

$$\begin{aligned} PP^T\mathbf{x} &= \begin{bmatrix} a_1 & b_1 \\ a_2 & b_2 \\ a_3 & b_3 \end{bmatrix} \begin{bmatrix} \mathbf{x}_1^T\mathbf{x} \\ \mathbf{x}_2^T\mathbf{x} \end{bmatrix} = \begin{bmatrix} a_1\mathbf{x}_1^T\mathbf{x} + b_1\mathbf{x}_2^T\mathbf{x} \\ a_2\mathbf{x}_1^T\mathbf{x} + b_2\mathbf{x}_2^T\mathbf{x} \\ a_3\mathbf{x}_1^T\mathbf{x} + b_3\mathbf{x}_2^T\mathbf{x} \end{bmatrix} \\ &= \begin{bmatrix} a_1\mathbf{x}_1^T\mathbf{x} \\ a_2\mathbf{x}_1^T\mathbf{x} \\ a_3\mathbf{x}_1^T\mathbf{x} \end{bmatrix} + \begin{bmatrix} b_1\mathbf{x}_2^T\mathbf{x} \\ b_2\mathbf{x}_2^T\mathbf{x} \\ b_3\mathbf{x}_2^T\mathbf{x} \end{bmatrix} = \mathbf{x}_1^T\mathbf{x} \begin{bmatrix} a_1 \\ a_2 \\ a_3 \end{bmatrix} + \mathbf{x}_2^T\mathbf{x} \begin{bmatrix} b_1 \\ b_2 \\ b_3 \end{bmatrix} \\ &= \left[\mathbf{x}_1^T\mathbf{x}\right]\mathbf{x}_1 + \left[\mathbf{x}_2^T\mathbf{x}\right]\mathbf{x}_2 = \langle \mathbf{x}_1, \mathbf{x} \rangle \mathbf{x}_1 + \langle \mathbf{x}_2, \mathbf{x} \rangle \mathbf{x}_2 \end{aligned}$$

By Corollary 8.19, $PP^T\mathbf{x} = \text{proj}(\mathbf{x} \to W)$. ■

We summarize a process for calculating an orthogonal projection onto a subspace of $\mathbb{R}^n$ established in Theorem 8.20 in Table 3.

TABLE 3 ORTHOGONAL PROJECTIONS AND ORTHONORMAL BASES.

Step 1. Take a basis $B = \{\mathbf{x}_1, \ldots, \mathbf{x}_p\}$ for a subspace W of $\mathbb{R}^n$ that is orthonormal in the standard inner product on R^n.
Step 2. Form the matrix $P = [\mathbf{x}_1 \ \cdots \ \mathbf{x}_p]$ with the vectors in B, and let $\mathbf{x}$ be any vector in $\mathbb{R}^n$.
Step 3. The vector $PP^T\mathbf{x}$ is the orthogonal projection $\text{proj}(\mathbf{x} \to W)$ of $\mathbf{x}$ onto W.

In the next section, we show that this procedure for finding orthogonal projections can be adapted to calculate orthogonal projections onto column spaces of real matrices with linearly independent columns.

The QR Decomposition and Orthogonal Projections

In combination with the QR decomposition, Theorem 8.20 provides us with a useful formula for calculating orthogonal projections of vectors into column spaces.

EXAMPLE 8.24 Finding Orthogonal Projections by QR Decomposition

Combine Theorems 8.8 and 8.20 to find the orthogonal projection of the vector $\mathbf{x} = (1, 2, 3)$ onto the column space of the matrix

$$A = \begin{bmatrix} 3 & 2 \\ 2 & 1 \\ 0 & 3 \end{bmatrix}$$

Solution. We first note that $\mathbf{x} \notin \text{Col } A$ since the linear system

$$\begin{cases} 3x + 2y = 1 \\ 2x + y = 2 \\ 3y = 3 \end{cases}$$

has no solution in $\mathbb{R}^2$. Furthermore, the columns of A are linearly independent since the homogeneous linear system

$$\begin{cases} 3a + 2b = 0 \\ 2a + b = 0 \\ 3b = 0 \end{cases}$$

has only the trivial solution. Theorem 8.8 therefore guarantees that the matrix A has a QR decomposition $QR = A$. Furthermore, we know from Exercise 4.14.13 that $\text{Col } A = \text{Col } Q$. Hence $\text{proj}(\mathbf{x} \to \text{Col } A) = \text{proj}(\mathbf{x} \to \text{Col } Q) = QQ^T\mathbf{x}$ by Theorem 8.20. We can therefore use the **QRDecomposition** function in the **LinearAlgebra** package to find $\text{proj}(\mathbf{x} \to \text{Col } A)$.

```
[> with(LinearAlgebra):
[> A:=Matrix([[3,2],[2,1],[0,3]]):
[> x:=<1,2,3>:
[> Q,R:=QRDecomposition(A):
```

$$\text{proj}\,(\mathbf{x} \to \text{Col } A) = QQ^T\mathbf{x}$$

```
[> projxQ:=Q.Transpose(Q).x;
```

$$\text{proj}\,x\,Q := \begin{bmatrix} \frac{104}{59} \\ \frac{101}{118} \\ \frac{339}{118} \end{bmatrix}$$

The vector $(104/59, 101/118, 339/118)$ is the required projection of the vector $\mathbf{x}$ onto the column space of A. ▲

We summarize the process of using the QR decomposition to calculate the orthogonal projection of a vector onto the column space of a matrix in Table 4.

TABLE 4 ORTHOGONAL PROJECTIONS AND QR DECOMPOSITION.

Step 1. Take an $m \times n$ real matrix A with n linearly independent columns.
Step 2. Calculate a QR decomposition of A.
Step 3. Use the $m \times n$ matrix Q and any vector $\mathbf{x} \in \mathbb{R}^m$ to form $QQ^T\mathbf{x}$.
Step 4. The vector $QQ^T\mathbf{x}$ is the orthogonal projection of $\mathbf{x}$ onto the column space of A.

Orthogonality of Fundamental Subspaces

One of the remarkable properties of the fundamental subspaces determined by a real matrix is the fact that they are pairwise complementary subspaces.

THEOREM 8.21 (Fundamental subspace theorem) *If A is an $m \times n$ real matrix, then* $\text{Nul}\,A \oplus \text{Col}\,A^T = \mathbb{R}^n$ *and* $\text{Nul}\,A^T \oplus \text{Col}\,A = \mathbb{R}^m$.

Proof. In Theorem 8.14, we showed that $\text{Nul}\,A$ is orthogonal to $\text{Col}\,A^T$ and that $\text{Nul}\,A^T$ is orthogonal to $\text{Col}\,A$. Moreover, we know from Theorems 4.30 and 4.31 that if A has rank r, then the dimension of $\text{Col}\,A^T$ is r and the dimension of $\text{Nul}\,A$ is $n - r$. Since these spaces are disjoint, we can form their direct sum $\text{Nul}\,A \oplus \text{Col}\,A^T$ and obtain a subspace of $\mathbb{R}^n$ of dimension $(n - r) + r = n$. This shows that $\text{Nul}\,A \oplus \text{Col}\,A^T$ must equal $\mathbb{R}^n$. An analogous argument shows that $\text{Nul}\,A^T \oplus \text{Col}\,A = \mathbb{R}^m$ since the disjoint subspaces $\text{Col}\,A$ and $\text{Nul}\,A^T$ of $\mathbb{R}^m$ have dimensions r and $m - r$. ■

Figure 2 on Page 586 illustrates Theorem 8.21. This result implies several versions of Fredholm's alternative theorem. Here is one formulated in terms of fundamental subspaces.

THEOREM 8.22 (Fredholm's alternative theorem) *For any $m \times n$ real matrix A precisely one of the following statement holds: (1) The column space of A is $\mathbb{R}^m$. (2) The null space of A^T contains a nonzero vector.*

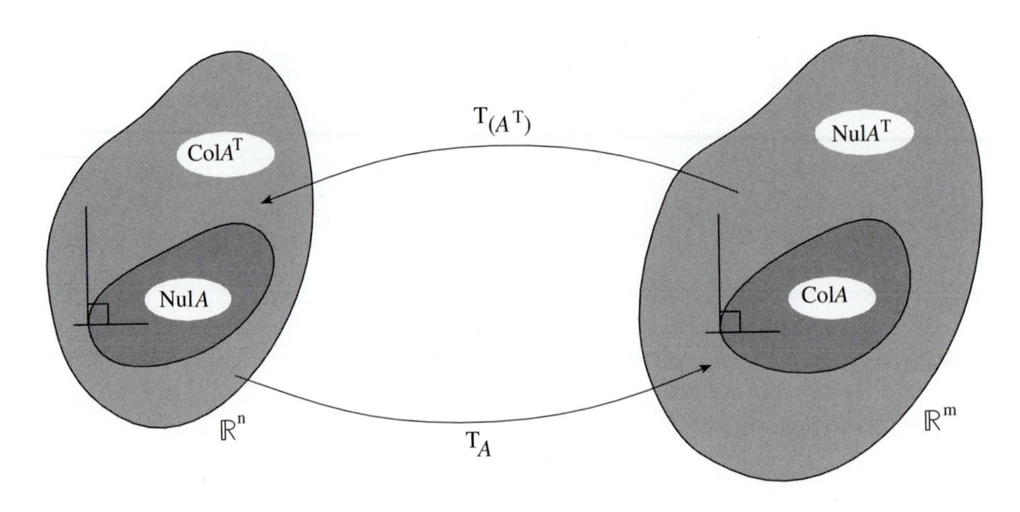

FIGURE 2 Orthogonal Complements.

Proof. By Theorem 8.21, Nul $A^T \oplus$ Col $A = \mathbb{R}^m$. Therefore

$$m = \dim \mathbb{R}^m = \dim(\text{Nul } A^T) + \dim(\text{Col } A)$$

since the direct sum of Nul A^T and Col A is $\mathbb{R}^m$. Suppose that Col $A = \mathbb{R}^m$. Then $m = \dim(\text{Nul } A^T) + m$. Therefore $\dim(\text{Nul } A^T) = 0$. This implies that Nul $A^T = \{\mathbf{0}\}$. Now suppose that there exists a nonzero vector $\mathbf{x} \in$ Nul A^T. Then $\dim(\text{Nul } A^T) \geq 1$. Hence $\dim(\text{Col } A) < m$. This means that must be an $\mathbf{x} \in \mathbb{R}^m$ that is not in Col A. ■

Rephrased in terms of solutions of linear systems, Fredholm's alternative theorem says that either an $m \times n$ linear system $A\mathbf{x} = \mathbf{b}$ with real coefficients is solvable for all $\mathbf{b} \in \mathbb{R}^m$ or the homogeneous linear system $A\mathbf{x} = \mathbf{0}$ has a nontrivial solution $\mathbf{x} \in \mathbb{R}^n$. Theorem 2.25 was the special case of this theorem for square matrices.

EXERCISES 8.9

1. Find the orthogonal complement of the set $S = \{(1, 3)\}$ in $\mathbb{R}^2$.
2. Find the orthogonal complement $S^{\perp}$ of the set $S = \{(1, 3, 2), (4, 12, 8)\}$ in $\mathbb{R}^3$. What is the dimension of $S^{\perp}$? Explain your answer.
3. Find the orthogonal complement of the set $S = \{(1, 3, 2), (2, 1, 0)\}$ in $\mathbb{R}^3$.
4. (*Maple V*) Find a basis for the orthogonal complement of the column space of the matrix

$$A = \begin{bmatrix} 5 & 0 & 5 & 0 \\ 1 & 2 & 3 & 0 \\ 0 & 2 & 2 & 0 \end{bmatrix}$$

Compare your basis with that obtained by using the function **linalg[nullspace]**.

5. (*Maple 6*) Find a basis for the orthogonal complement of the null space of the matrix

$$A = \begin{bmatrix} 3 & 1 & 0 & 2 & 4 \\ 1 & 1 & 0 & 0 & 2 \\ 5 & 2 & 0 & 3 & 7 \end{bmatrix}$$

   Compare your basis with that obtained by using the method illustrated in Example 4.44.

6. (*Maple 6*) Generate a 3×4 real matrix A and use the standard inner product to show that Fredholm's alternative theorem holds for the matrix A.

7. (*Maple 6*) Generate a 4×3 real matrix A of rank 3, and use the standard inner product to show that Fredholm's alternative theorem holds for the matrix A.

8. (*Maple V, 6*) Compare the functions **linalg[nullspace]** and **LinearAlgebra[NullSpace]** and explore their properties.

ORTHOGONAL TRANSFORMATIONS

In many applications of linear algebra such as in computer animation, we require linear transformations that preserve the shape and size of geometric objects. We now study such transformations. They are known as ***orthogonal transformations***.

We will show that a linear transformation T on a finite-dimensional real inner product space is orthogonal if and only if its matrix $[T]_{\mathcal{B}}^{\mathcal{B}}$ is orthogonal in some orthonormal basis $\mathcal{B}$. In particular, we will show that they are linear transformations that *preserve* the inner product on V. By a linear transformation $T : V \to V$ *preserving* an inner product $\langle \mathbf{x}, \mathbf{y} \rangle$ on V we mean a linear transformation T with the property that $\langle \mathbf{x}, \mathbf{y} \rangle = \langle T(\mathbf{x}), T(\mathbf{y}) \rangle$ for all $\mathbf{x}, \mathbf{y} \in V$.

We begin by showing that every T is naturally paired with a linear transformation T^* for which $\langle T(\mathbf{x}), \mathbf{y} \rangle = \langle \mathbf{x}, T^*(\mathbf{y}) \rangle$. The transformation T^* is called the ***adjoint*** of T. We will show that if T is represented by a matrix A in some orthonormal basis, then T^* is represented by A^T in the same basis. We then discuss the case of linear transformations that are equal to their adjoints, the ***self-adjoint transformations***, and use their properties to show that a linear transformation T preserves an inner product $\langle \mathbf{x}, \mathbf{y} \rangle$ if and only if $TT^* = T^*T = I$.

Adjoint Transformations

In this section we show that adjoints are represented by transposes. In order to be able to prove this fact, we need to establish a fundamental property of linear functionals. We recall from earlier work that a linear transformation f from a real vector space V to the space $\mathbb{R}$ of real numbers is called a ***linear functional***.

LEMMA 8.23 *If $f : V \to \mathbb{R}$ is a linear functional on a finite-dimensional inner product space V, then there exists a unique vector $\mathbf{y} \in V$ such that $f(\mathbf{x}) = \langle \mathbf{x}, \mathbf{y} \rangle$ for all $\mathbf{x} \in V$.*

Proof. Using the Gram–Schmidt process, we construct an orthonormal basis $\mathcal{S} = \{\mathbf{v}_1, \ldots, \mathbf{v}_n\}$ for V. We then use the functional f to define the vector $\mathbf{y}$ as a linear combination of the given basis vectors:

$$\mathbf{y} = f(\mathbf{v}_1)\mathbf{v}_1 + \cdots + f(\mathbf{v}_n)\mathbf{v}_n$$

Next we construct the linear functional $\vec{\mathbf{y}}$ that agrees with f on the orthonormal basis $\mathcal{S}$:

$$\vec{\mathbf{y}}(\mathbf{v}_i) = \langle \mathbf{v}_i, \mathbf{y} \rangle = \langle \mathbf{v}_i, f(\mathbf{v}_1)\mathbf{v}_1 + \cdots + f(\mathbf{v}_n)\mathbf{v}_n \rangle = f(\mathbf{v}_i) \langle \mathbf{v}_i, \mathbf{v}_i \rangle = f(\mathbf{v}_i).$$

Since f agrees with $\vec{\mathbf{y}}$ on the basis $\mathcal{S}$, Theorem 5.3 tells us that $f = \vec{\mathbf{y}}$.

It remains to establish the uniqueness of $\mathbf{y}$. Suppose that $\mathbf{z} \in V$ also has the property that $f(\mathbf{x}) = \langle \mathbf{x}, \mathbf{z} \rangle$ for all $\mathbf{x} \in V$. Then $\langle \mathbf{x}, \mathbf{y} \rangle = \langle \mathbf{x}, \mathbf{z} \rangle$, so

$$\langle \mathbf{x}, \mathbf{y} \rangle - \langle \mathbf{x}, \mathbf{z} \rangle = \langle \mathbf{x}, \mathbf{y} - \mathbf{z} \rangle = 0$$

Let $\mathbf{x} = \mathbf{y} - \mathbf{z}$. Then $\langle \mathbf{y} - \mathbf{z}, \mathbf{y} - \mathbf{z} \rangle = 0$. Since the inner product of a vector with itself is 0 if and only if the vector is $\mathbf{0}$, it follows that $\mathbf{y} - \mathbf{z} = \mathbf{0}$, so $\mathbf{y} = \mathbf{z}$. ■

We use this lemma to prove the existence of adjoints.

THEOREM 8.24 (Adjoint existence theorem) *If $T : V \to V$ is a linear transformation on a finite-dimensional real inner product space V, then there exists a unique linear transformation $T^* : V \to V$ for which $\langle T(\mathbf{x}), \mathbf{y} \rangle = \langle \mathbf{x}, T^*(\mathbf{y}) \rangle$.*

Proof. Let $\mathbf{y}$ be a fixed vector in V. Then the function $f(\mathbf{x}) = \langle T(\mathbf{x}), \mathbf{y} \rangle$ is a linear functional $f : V \to \mathbb{R}$. Lemma 8.23 guarantees that there exists a unique vector $\mathbf{z} \in V$ such that $f(\mathbf{x}) = \langle \mathbf{x}, \mathbf{z} \rangle$. We can therefore put $T^*(\mathbf{y}) = \mathbf{z}$ and obtain for all $\mathbf{x}, \mathbf{y} \in V$, $\langle T(\mathbf{x}), \mathbf{y} \rangle = \langle \mathbf{x}, T^*(\mathbf{y}) \rangle$. The uniqueness of $\mathbf{z}$ ensures that T^* is a function on V.

It remains to verify that T^* is linear. Since

$$\begin{aligned}
\langle \mathbf{x}, T^*(a\mathbf{u} + b\mathbf{v}) \rangle &= \langle T(\mathbf{x}), a\mathbf{u} + b\mathbf{v} \rangle \\
&= \langle T(\mathbf{x}), a\mathbf{u} \rangle + \langle T(\mathbf{x}), b\mathbf{v} \rangle \\
&= a \langle T(\mathbf{x}), \mathbf{u} \rangle + b \langle T(\mathbf{x}), \mathbf{v} \rangle \\
&= a \langle \mathbf{x}, T^*(\mathbf{u}) \rangle + b \langle \mathbf{x}, T^*(\mathbf{v}) \rangle \\
&= \langle \mathbf{x}, aT^*(\mathbf{u}) \rangle + \langle \mathbf{x}, bT^*(\mathbf{v}) \rangle \\
&= \langle \mathbf{x}, aT^*(\mathbf{u}) + bT^*(\mathbf{v}) \rangle
\end{aligned}$$

for all $\mathbf{x}, \mathbf{y} \in V$ and all $a, b \in \mathbb{R}$, the linearity of T^* follows. ■

The next theorem tells us that the two matrices A and B representing T and T^* in an orthonormal basis have a very simple connection.

THEOREM 8.25 (Adjoint representation theorem) *If A represents a linear transformation $T : V \to V$ on a real inner product space V in an orthonormal basis $\mathcal{B} = \{\mathbf{x}_1, \ldots, \mathbf{x}_n\}$, then the transpose A^T of A represents the adjoint $T^* : V \to V$ of T in the basis $\mathcal{B}$.*

Proof. Suppose that the matrix A represents T and that B represents T^* in the orthonormal basis $\mathcal{B}$. It follows from Corollary 8.3 that for each $1 \leq j \leq n$,

$$T(\mathbf{x}_j) = \langle T(\mathbf{x}_j), \mathbf{x}_1 \rangle \mathbf{x}_1 + \cdots + \langle T(\mathbf{x}_j), \mathbf{x}_n \rangle \mathbf{x}_n$$

By Theorem 5.11, the jth column of A is therefore the coordinate vector

$$T(\mathbf{x}_j) = \begin{bmatrix} \langle T(\mathbf{x}_j), \mathbf{x}_1 \rangle \\ \vdots \\ \langle T(\mathbf{x}_j), \mathbf{x}_n \rangle \end{bmatrix}$$

Similarly,

$$\begin{aligned} T^*(\mathbf{x}_j) &= \langle T^*(\mathbf{x}_j), \mathbf{x}_1 \rangle \mathbf{x}_1 + \cdots + \langle T^*(\mathbf{x}_j), \mathbf{x}_n \rangle \mathbf{x}_n \\ &= \langle \mathbf{x}_j, T(\mathbf{x}_1) \rangle \mathbf{x}_1 + \cdots + \langle \mathbf{x}_j, T(\mathbf{x}_n) \rangle \mathbf{x}_n \\ &= \langle T(\mathbf{x}_1), \mathbf{x}_j \rangle \mathbf{x}_1 + \cdots + \langle T(\mathbf{x}_n), \mathbf{x}_j \rangle \mathbf{x}_n \end{aligned}$$

Hence the jth column of B is the coordinate vector

$$T^*(\mathbf{x}_j) = \begin{bmatrix} \langle T(\mathbf{x}_1), \mathbf{x}_j \rangle \\ \vdots \\ \langle T(\mathbf{x}_n), \mathbf{x}_j \rangle \end{bmatrix}$$

This shows that the columns of A are the rows of B. Hence $B = A^T$. ■

EXAMPLE 8.25 The Adjoint of a Linear Transformation

Use *Maple* to illustrate Theorem 8.25.

Solution. We create the matrices

$$A = \begin{bmatrix} 1 & 2 & 3 \\ 4 & 5 & 6 \\ 7 & 8 & 9 \end{bmatrix} \quad \text{and} \quad A^T = \begin{bmatrix} 1 & 4 & 7 \\ 2 & 5 & 8 \\ 3 & 6 & 9 \end{bmatrix}$$

and compare the inner products

$$\left\langle A \begin{bmatrix} x \\ y \\ z \end{bmatrix}, \begin{bmatrix} u \\ v \\ w \end{bmatrix} \right\rangle \quad \text{and} \quad \left\langle \begin{bmatrix} x \\ y \\ z \end{bmatrix}, A^T \begin{bmatrix} u \\ v \\ w \end{bmatrix} \right\rangle$$

```
[> with(linalg):
[> A:=matrix([[1,2,3],[4,5,6],[7,8,9]]):
[> X:=matrix([[x],[y],[z]]):
[> Y:=matrix([[u],[v],[w]]):
[> L:=evalm(transpose(A&*X)&*Y):
[> R:=evalm(transpose(X)&*transpose(A)&*Y):
[> equal(L,R);
                              true
```

By combining these results, we see that

$$\left\langle A\begin{bmatrix} x \\ y \\ z \end{bmatrix}, \begin{bmatrix} u \\ v \\ w \end{bmatrix}\right\rangle = \left\langle \begin{bmatrix} x \\ y \\ z \end{bmatrix}, A^T\begin{bmatrix} u \\ v \\ w \end{bmatrix}\right\rangle$$

holds for all x, y, z, u, v, w. Hence the linear transformations represented by the matrices A and A^T are adjoints. ▲

EXERCISES 8.10

1. Use Theorem 8.25 to find the adjoint of the linear transformation $T : \mathbb{R}^2 \to \mathbb{R}^2$ defined by $T(x, y) = (2x - 3y, x + y)$.

2. Use Theorem 8.25 to find the adjoint of the linear transformation $T : \mathbb{R}^3 \to \mathbb{R}^3$ defined by $T(x, y, z) = (2x - 3y + z, x + y, z)$.

3. (*Maple V*) Suppose that T is the linear transformation represented by the matrix

$$A = \begin{bmatrix} 1 & 0 & 2 \\ 4 & 3 & 0 \\ 0 & 0 & 3 \end{bmatrix}$$

and T^* is the linear transformation represented by A^T in the standard basis of $\mathbb{R}^3$. Use the standard inner product to verify that $\langle T(\mathbf{x}), \mathbf{y}\rangle = \langle \mathbf{x}, T^*(\mathbf{y})\rangle$ for all $\mathbf{x}, \mathbf{y} \in \mathbb{R}^3$.

4. (*Maple 6*) Suppose that T is the linear transformation represented by the matrix

$$A = \begin{bmatrix} -3 & 3 & 5 \\ 3 & 5 & -2 \\ 5 & -2 & -3 \end{bmatrix}$$

and T^* is the linear transformation represented by A^T in the standard basis of $\mathbb{R}^3$. Use the standard inner product to verify that $\langle T(\mathbf{x}), \mathbf{y}\rangle = \langle \mathbf{x}, T^*(\mathbf{y})\rangle$ for all $\mathbf{x}, \mathbf{y} \in \mathbb{R}^3$. What is the connection between T and T^*?

Self-Adjoint Linear Transformations

If a matrix $A = [T]_{\mathcal{B}}^{\mathcal{B}}$ represents a linear transformation T on a real inner product space V in an orthonormal basis $\mathcal{B} = \{\mathbf{x}_1, \ldots, \mathbf{x}_n\}$, then we know from Theorem 8.25 that the matrix A^T represents the adjoint T^* of T in the basis $\mathcal{B}$. Therefore, $T = T^*$ if and only if $A = A^T$. For this reason we say that T is ***self-adjoint*** if it can be represented by a real symmetric matrix in some orthonormal basis. In this section, we show that a linear transformation T is orthogonal if and only if the associated transformation $TT^* - I$ is self-adjoint. We begin with two examples.

EXAMPLE 8.26 Rotations and Self-Adjointness

Show that the rotation T through the angle $\theta = \pi$ is self-adjoint.

Solution. We know that in the standard basis $\mathcal{E}$ of $\mathbb{R}^2$, the transformation T is represented by the matrix

$$A = \begin{bmatrix} \cos\pi & -\sin\pi \\ \sin\pi & \cos\pi \end{bmatrix} = \begin{bmatrix} -1 & 0 \\ 0 & -1 \end{bmatrix}$$

Moreover, by Theorem 8.25, the matrix A^T represents the adjoint T^* of T. But $A = A^T$. Hence $T = T^*$. ▲

Another important family of self-adjoint transformations are the reflections.

EXAMPLE 8.27 Reflections and Self-Adjointness

Show that the reflection T about the lines $y = x$ is a self-adjoint linear transformation.

Solution. We know that in the standard basis $\mathcal{E}$ of $\mathbb{R}^2$, the transformation T is represented by the matrix

$$A = \begin{bmatrix} 0 & 1 \\ 1 & 0 \end{bmatrix}$$

and Theorem 8.25 tells us that the adjoint T^* of T is represented by A^T. But $A = A^T$. Therefore, $T = T^*$. This means that T is self-adjoint. ▲

We now use self-adjointness to characterize orthogonal transformations.

LEMMA 8.26 *For any linear transformation T on a real inner product space V, the transformation $T^*T - I$ is self-adjoint.*

Proof. For all $\mathbf{x} \in V$ it holds that

$$\begin{aligned} \big\langle (T^*T - I)(\mathbf{x}), \mathbf{x} \big\rangle &= \big\langle (T^*T)(\mathbf{x}), \mathbf{x} \big\rangle - \langle I(\mathbf{x}), \mathbf{x} \rangle \\ &= \langle T(\mathbf{x}), T(\mathbf{x}) \rangle - \langle \mathbf{x}, I(\mathbf{x}) \rangle \end{aligned}$$

$$= \left\langle \mathbf{x}, T^*T(\mathbf{x})\right\rangle - \langle \mathbf{x}, I(\mathbf{x})\rangle$$
$$= \left\langle \mathbf{x}, (T^*T - I)(\mathbf{x})\right\rangle$$

Hence $T^*T - I$ is self-adjoint. ■

LEMMA 8.27 *If T is a self-adjoint linear transformation on a finite-dimensional real inner product space V and $\langle T(\mathbf{x}), \mathbf{x}\rangle = 0$ for all $\mathbf{x} \in V$, then $T = 0$.*

Proof. Since T is self-adjoint and $\langle T(\mathbf{x}), \mathbf{x}\rangle = \langle T(\mathbf{y}), \mathbf{y}\rangle = 0$, we have

$$\begin{aligned}
0 &= \langle T(\mathbf{x}+\mathbf{y}), \mathbf{x}+\mathbf{y}\rangle = \langle T(\mathbf{x}) + T(\mathbf{y}), \mathbf{x}+\mathbf{y}\rangle \\
&= \langle T(\mathbf{x}), \mathbf{x}+\mathbf{y}\rangle + \langle T(\mathbf{y}), \mathbf{x}+\mathbf{y}\rangle \\
&= \langle T(\mathbf{x}), \mathbf{x}\rangle + \langle T(\mathbf{x}), \mathbf{y}\rangle + \langle T(\mathbf{y}), \mathbf{x}\rangle + \langle T(\mathbf{y}), \mathbf{y}\rangle \\
&= \langle T(\mathbf{x}), \mathbf{y}\rangle + \langle T(\mathbf{y}), \mathbf{x}\rangle
\end{aligned}$$

for all $\mathbf{x}, \mathbf{y} \in V$. It follows that if $\mathbf{y} = T(\mathbf{x})$, then

$$\begin{aligned}
0 &= \langle T(\mathbf{x}), T(\mathbf{x})\rangle + \langle T\left(T(\mathbf{x})\right), \mathbf{x}\rangle \\
&= \langle T(\mathbf{x}), T(\mathbf{x})\rangle + \langle T(\mathbf{x}), T(\mathbf{x})\rangle \\
&= 2\langle T(\mathbf{x}), T(\mathbf{x})\rangle
\end{aligned}$$

Therefore $\langle T(\mathbf{x}), T(\mathbf{x})\rangle = 0$ for all $\mathbf{x} \in V$. This means that $T(\mathbf{x}) = 0$ for all $\mathbf{x} \in V$. It follows that the transformation T itself is 0. ■

We now apply Lemmas 8.26 and 8.27 to characterize orthogonal transformations.

THEOREM 8.28 (Orthogonal transformation theorem) *If T is a linear transformation on a finite-dimensional real inner product space V, then the following conditions are equivalent: (1) $\langle T(\mathbf{x}), T(\mathbf{y})\rangle = \langle \mathbf{x}, \mathbf{y}\rangle$ for all $\mathbf{x}, \mathbf{y} \in V$. (2) $\|T(\mathbf{x})\| = \|\mathbf{x}\|$ for all $\mathbf{x} \in V$. (3) $TT^* = T^*T = I$.*

Proof. We show that (1) implies (2), (2) implies (3), and (3) implies (1).

Suppose that $\langle T(\mathbf{x}), T(\mathbf{y})\rangle = \langle \mathbf{x}, \mathbf{y}\rangle$ for all $\mathbf{x}, \mathbf{y} \in V$. Then

$$\|T(\mathbf{x})\| = \sqrt{\langle T(\mathbf{x}), T(\mathbf{x})\rangle} = \sqrt{\langle \mathbf{x}, \mathbf{x}\rangle} = \|\mathbf{x}\|$$

Therefore (2) holds. Suppose now that $\|T(\mathbf{x})\| = \|\mathbf{x}\|$ for all $\mathbf{x} \in V$. Then

$$\left\langle T^*T(\mathbf{x}), \mathbf{x}\right\rangle = \langle T(\mathbf{x}), T(\mathbf{x})\rangle = \langle \mathbf{x}, \mathbf{x}\rangle = \langle I(\mathbf{x}), \mathbf{x}\rangle$$

Therefore, $\langle (T^*T - I)(\mathbf{x}), \mathbf{x}\rangle = 0$ for all $\mathbf{x} \in V$. By Lemma 8.26, $T^*T - I$ is self-adjoint. In that case, Lemma 8.27 tells us that $T^*T - I = 0$. In other words, $T^*T = I$. Since $T^*T = I$ if and only if $TT^* = I$, it follows that (3) holds. Suppose, finally, that $T^*T = I$. Then

$$\begin{aligned}\langle T(\mathbf{x}), T(\mathbf{y})\rangle - \langle \mathbf{x}, \mathbf{y}\rangle &= \left\langle \mathbf{x}, T^*T(\mathbf{y})\right\rangle - \langle \mathbf{x}, I(\mathbf{y})\rangle \\ &= \left\langle \mathbf{x}, (T^*T - I)(\mathbf{y})\right\rangle \\ &= \langle \mathbf{x}, 0(\mathbf{y})\rangle \\ &= 0\end{aligned}$$

for all $\mathbf{x}, \mathbf{y} \in V$. This means that $\langle T(\mathbf{x}), T(\mathbf{y})\rangle = \langle \mathbf{x}, \mathbf{y}\rangle$ for all $\mathbf{x}, \mathbf{y} \in V$. Hence (1) holds. ■

COROLLARY 8.29 (Orthogonal transformations and orthogonal matrices) *A matrix A represents an orthogonal transformation on a finite-dimensional real inner product space in an orthonormal basis if and only if A is orthogonal.*

Proof. Suppose A represents an orthogonal transformation T with respect to an orthonormal basis $\mathcal{B}$. By Theorem 8.25, T has an adjoint T^* represented by the transpose A^T of A. By Theorem 8.28, $TT^* = T^*T = I$, so A^T must be A^{-1}. ■

Rotations are another important family of orthogonal transformations.

EXAMPLE 8.28 Rotations about the Coordinate Axes of $\mathbb{R}^3$

Use *Maple* to show that the rotation matrices

$$R_x = \begin{bmatrix} 1 & 0 & 0 \\ 0 & \cos\theta & -\sin\theta \\ 0 & \sin\theta & \cos\theta \end{bmatrix}, R_y = \begin{bmatrix} \cos\theta & 0 & \sin\theta \\ 0 & 1 & 0 \\ -\sin\theta & 0 & \cos\theta \end{bmatrix}, R_z = \begin{bmatrix} \cos\theta & -\sin\theta & 0 \\ \sin\theta & \cos\theta & 0 \\ 0 & 0 & 1 \end{bmatrix}$$

discussed in Example 4.55 represent orthogonal transformations on $\mathbb{R}^3$.

Solution. By Corollary 8.29 it suffices to show that the matrices are orthogonal.

```
[> with(LinearAlgebra):
[> Rx:=Matrix([
    [1,0,0],
    [0,cos(theta),-sin(theta)],
    [0,sin(theta),cos(theta)]]):
[> equal(simplify(evalm(Rx.Transpose(Rx))),
    IdentityMatrix(3));
                          true
```

```
[> Ry:=Matrix([
    [cos(theta),0,sin(theta)],
    [0,1,0],
    [-sin(theta),0,cos(theta)]]):
[> map(simplify,evalm(Ry.Transpose(Ry)));
```

$$\begin{bmatrix} 1 & 0 & 0 \\ 0 & 1 & 0 \\ 0 & 0 & 1 \end{bmatrix}$$

```
[> Rz:=Matrix([
    [cos(theta),-sin(theta),0],
    [sin(theta),cos(theta),0],
    [0,0,1]]):
[> M:=map(simplify,Rz.Transpose(Rz)):
[> equal(IdentityMatrix(3),M);
```

true

Since the transpose of each matrix equals the inverse of the matrix, the three rotation matrices are orthogonal. They therefore represent orthogonal transformations from $\mathbb{R}^3$ to $\mathbb{R}^3$.

EXAMPLE 8.29 Rotation About the Vector (1, 1, 1) in $\mathbb{R}^3$

Use *Maple* to calculate the matrix A of the orthogonal transformation that rotates vectors in $\mathbb{R}^3$ counterclockwise through the angle $\theta = \pi/6$ about the axis determined by the vector $(1, 1, 1)$. Use quaternions to confirm that

$$A \begin{bmatrix} 1 \\ 2 \\ 3 \end{bmatrix} = \begin{bmatrix} 2 \\ 1 \\ 3 \end{bmatrix}$$

Solution. In the standard coordinate basis $\mathcal{E} = \{\mathbf{e}_1 = (1, 0, 0), \mathbf{e}_2 = (0, 1, 0), \mathbf{e}_3 = (0, 0, 1)\}$ for $\mathbb{R}^3$, the matrix

$$R_z = \begin{bmatrix} \cos\frac{\pi}{3} & -\sin\frac{\pi}{3} & 0 \\ \sin\frac{\pi}{3} & \cos\frac{\pi}{3} & 0 \\ 0 & 0 & 1 \end{bmatrix}$$

represents a counterclockwise rotation about the $\mathbf{e}_3$-axis by the angle $\pi/3$. To be able to use this matrix, we consider the unit vector $\mathbf{z} = \left(1/\sqrt{3}\right)(1, 1, 1)$ as one of an orthonormal set of axes for $\mathbb{R}^3$ and interpret the required rotation as a rotation about that axis. We therefore begin with the set

$$\mathcal{S} = \left\{\mathbf{z} = \frac{1}{\sqrt{3}}(1, 1, 1)\right\}$$

and extend $\mathcal{S}$ to an orthonormal basis $\mathcal{B} = \{\mathbf{x}, \mathbf{y}, \mathbf{z}\}$ for $\mathbb{R}^3$. We then calculate the coordinate conversion matrix $P = [I]_{\mathcal{E}}^{\mathcal{B}}$ for which the matrix PR_zP^{-1} represents a rotation through the angle $\theta = \pi/6$ about the axis determined by the vector $\mathbf{z}$.

Let $\mathbf{x} = (a, b, c)$ and $\mathbf{y} = (d, e, f)$. Since $\mathbf{x}$ must be orthogonal to the vector $(1, 1, 1)$, their dot product must be zero. The calculation

```
[> solve(a+b+c=0,{a,b,c});
```

$$\{a = -b - c, b = b, c = c\}$$

shows that the vector $\mathbf{u} = (-2, 1, 1)$ is orthogonal to the vector $(1, 1, 1)$. Similarly, the calculation

```
[> solve({d+e+f=0,-2*d+e+f=0},{d,e,f});
```

$$\{d = 0, f = -e, e = e\}$$

shows that the vector $\mathbf{v} = (0, -1, 1)$ is orthogonal to both $\mathbf{u}$ and $(1, 1, 1)$.

We now normalize the vectors $\mathbf{u}$ and $\mathbf{v}$.

```
[> with(linalg):
[> u:=vector(3,[-2,1,1]):   v:=vector(3,[0,-1,1]):
[> x:=normalize(u):   y:=normalize(v):
```

We now use these vectors to construct the required rotation matrix in the basis $\mathcal{B} = \{\mathbf{x}, \mathbf{y}, \mathbf{z}\}$. We know from Theorem 4.19, that the matrix $P = [I]_{\mathcal{E}}^{\mathcal{B}}$ is simply the matrix whose columns are the vectors $\mathbf{x}$, $\mathbf{y}$, $\mathbf{z}$. We use *Maple* to construct P and R_z.

```
[> P:=matrix(augment(x,y,z));
```

$$P := \begin{bmatrix} -\frac{1}{3}\sqrt{6} & 0 & \frac{1}{3}\sqrt{3} \\ \frac{1}{6}\sqrt{6} & -\frac{1}{2}\sqrt{2} & \frac{1}{3}\sqrt{3} \\ \frac{1}{6}\sqrt{6} & \frac{1}{2}\sqrt{2} & \frac{1}{3}\sqrt{3} \end{bmatrix}$$

```
[> Rz:=matrix([[cos(Pi/3),-sin(Pi/3),0],
   [sin(Pi/3),cos(Pi/3),0],[0,0,1]]);
```

$$Rz := \begin{bmatrix} \frac{1}{2} & -\frac{1}{2}\sqrt{3} & 0 \\ \frac{1}{2}\sqrt{3} & \frac{1}{2} & 0 \\ 0 & 0 & 1 \end{bmatrix}$$

The required matrix A is PR_zP^{-1}.

```
[> A:=simplify(evalm([P&*Rz&*inverse(P)));
```

$$A := \begin{bmatrix} \frac{2}{3} & \frac{-1}{3} & \frac{2}{3} \\ \frac{2}{3} & \frac{2}{3} & \frac{-1}{3} \\ \frac{-1}{3} & \frac{2}{3} & \frac{2}{3} \end{bmatrix}$$

It is easily verified that

$$\mathbf{w} = A \begin{bmatrix} 1 \\ 2 \\ 3 \end{bmatrix} = \begin{bmatrix} \frac{2}{3} & \frac{-1}{3} & \frac{2}{3} \\ \frac{2}{3} & \frac{2}{3} & \frac{-1}{3} \\ \frac{-1}{3} & \frac{2}{3} & \frac{2}{3} \end{bmatrix} \begin{bmatrix} 1 \\ 2 \\ 3 \end{bmatrix} = \begin{bmatrix} 2 \\ 1 \\ 3 \end{bmatrix}$$

We use the method of Example 4.55 to show that $\mathbf{w} = qrq^*$, where q, r, and q^* are the following quaternions:

$$\begin{aligned} q &= [\cos\tfrac{\pi}{6}, \tfrac{1}{\sqrt{3}}\sin\tfrac{\pi}{6}, \tfrac{1}{\sqrt{3}}\sin\tfrac{\pi}{6}, \tfrac{1}{\sqrt{3}}\sin\tfrac{\pi}{6}] = \tfrac{\sqrt{3}}{2} + \tfrac{\sqrt{3}}{6}I + \tfrac{\sqrt{3}}{6}J + \tfrac{\sqrt{3}}{6}K \\ r &= [0, 1, 2, 3] = I + 2J + 3K \\ q^* &= [\cos\tfrac{\pi}{6}, -\tfrac{1}{\sqrt{3}}\sin\tfrac{\pi}{6}, -\tfrac{1}{\sqrt{3}}\sin\tfrac{\pi}{6}, -\tfrac{1}{\sqrt{3}}\sin\tfrac{\pi}{6}] = \tfrac{\sqrt{3}}{2} - \tfrac{\sqrt{3}}{6}I - \tfrac{\sqrt{3}}{6}J - \tfrac{\sqrt{3}}{6}K \end{aligned}$$

We first load the procedure defined in the answer to Exercise 4.17.3 in Appendix F. Next we define the quaternions q, r, and q^*.

```
[> q:=quat(cos(Pi/6),
   sin(Pi/6)/sqrt(3),sin(Pi/6)/sqrt(3),sin(Pi/6)/sqrt(3)):
[> r:=quat(0,1,2,3):
[> qconj:=quat(cos(Pi/6),-sin(Pi/6)/sqrt(3),
   -sin(Pi/6)/sqrt(3),-sin(Pi/6)/sqrt(3)):
```

Then we evaluate qrq^*.

```
[> simplify(q&^r&^qconj);
```

$$2I + J + 3K$$

As we can see, $qrq^* = \mathbf{w}$. ▲

EXERCISES 8.11

1. Prove that the eigenvalues of an orthogonal matrix are ± 1. Show by means of a counterexample that the converse is not true: Not every matrix whose eigenvalues are ± 1 is orthogonal.

2. Prove or disprove the following statements.

   a. All rotations $T : \mathbb{R}^2 \to \mathbb{R}^2$ are self-adjoint in the standard inner product on $\mathbb{R}^2$.

   b. Some rotations $T : \mathbb{R}^2 \to \mathbb{R}^2$ are self-adjoint in the standard inner product on $\mathbb{R}^2$.

   c. All reflections $T : \mathbb{R}^2 \to \mathbb{R}^2$ are self-adjoint in the standard inner product on $\mathbb{R}^2$.

   d. Some reflections $T : \mathbb{R}^2 \to \mathbb{R}^2$ are self-adjoint in the standard inner product on $\mathbb{R}^2$.

   e. The zero transformation $T : \mathbb{R}^2 \to \mathbb{R}^2$ is self-adjoint in all inner products on $\mathbb{R}^2$.

   f. If $T_1, T_2 : \mathbb{R}^2 \to \mathbb{R}^2$ are self-adjoint linear transformations, then $T_1 + T_2 : \mathbb{R}^2 \to \mathbb{R}^2$ is a self-adjoint linear transformation.

   g. If $T : \mathbb{R}^2 \to \mathbb{R}^2$ is a self-adjoint linear transformation and $a \in \mathbb{R}$, then $(aT) : \mathbb{R}^2 \to \mathbb{R}^2$ is a self-adjoint linear transformation.

   h. The self-adjoint linear transformations (in the standard inner product) form a subspace of $\text{Hom}(\mathbb{R}^2, \mathbb{R}^2)$.

   i. All orthogonal transformations $T : \mathbb{R}^2 \to \mathbb{R}^2$ are either rotations or reflections.

   j. All orthogonal transformations $T : \mathbb{R}^3 \to \mathbb{R}^3$ are either rotations or reflections.

   k. If $T_1, T_2 : \mathbb{R}^2 \to \mathbb{R}^2$ are two rotations, then their composition $T_1 \circ T_2 : \mathbb{R}^2 \to \mathbb{R}^2$ is a rotation.

   l. If $T_1, T_2 : \mathbb{R}^2 \to \mathbb{R}^2$ are two reflections, then their composition $T_1 \circ T_2 : \mathbb{R}^2 \to \mathbb{R}^2$ is a reflection.

   m. All reflections $T : \mathbb{R}^2 \to \mathbb{R}^2$ have a real eigenvalue.

   n. Some reflections $T : \mathbb{R}^2 \to \mathbb{R}^2$ have a real eigenvalue.

   o. All rotations $T : \mathbb{R}^2 \to \mathbb{R}^2$ have a real eigenvalue.

   p. Some rotations $T : \mathbb{R}^2 \to \mathbb{R}^2$ have a real eigenvalue.

   q. Some orthogonal transformations $T : \mathbb{R}^2 \to \mathbb{R}^2$ are both a rotation and a reflection.

r. If $\mathbf{x}$ and $\mathbf{y}$ are two unit vectors in a two-dimensional real inner product space, then there exists a rotation $T : \mathbb{R}^2 \to \mathbb{R}^2$ for which $T(\mathbf{x}) = \mathbf{y}$.

3. Show that for any linear transformation $T : \mathbb{R}^n \to \mathbb{R}^n$, the transformation $T + T^* : \mathbb{R}^n \to \mathbb{R}^n$ is self-adjoint.

4. (*Maple V*) Find an angle θ for which the matrix

$$A = \begin{bmatrix} -\frac{1}{2} & -\frac{1}{2}\sqrt{3} \\ \frac{1}{2}\sqrt{3} & -\frac{1}{2} \end{bmatrix}$$

represents a rotation $T : \mathbb{R}^2 \to \mathbb{R}^2$ in the standard inner product.

5. (*Maple 6*) Use the methods in Example 8.29 to show that the vector resulting from $(1, -1, -1)$ by a counterclockwise rotation by the angle $\theta = \pi/2$ about the axis determined by the vector $\mathbf{x} = (1, 2, -1)$ is $(-\sqrt{6}/2, 0, -\sqrt{6}/2)$.

6. (*Maple 6*) Show that the function $T : \mathbb{R}^4 \to \mathbb{H}$ defined by $T(a, b, c, d) = a + bI + cJ + dK$ is an isomorphism and use T^{-1} to define a multiplication of vectors for $\mathbb{R}^4$. Show that the vector $(1, 0, 0, 0)$ is the identity element for this operation.

7. (*Maple 6*) Show that the linear transformation $T : \mathbb{R}^4 \to \mathbb{R}^4$ defined in Exercise 5.1.23 is an orthogonal transformation by showing that T preserves the quaternion norm on $\mathbb{R}^4$.

The Spectral Theorem

A real square matrix A is ***orthogonally diagonalizable*** if there exists a diagonal matrix D and an orthogonal matrix Q for which $A = QDQ^T$. In this section, we prove that every real symmetric matrix is orthogonally diagonalizable. This fact is the basis for the singular value decomposition of arbitrary rectangular real matrices studied in Chapter 9. We begin by establishing some preliminary results needed to prove the existence of D and Q.

THEOREM 8.30 (Orthogonal eigenvector theorem) *If T is a self-adjoint linear transformation and $\mathbf{x}$ and $\mathbf{y}$ are eigenvectors belonging to distinct eigenvalues λ_1 and λ_2 of T, then $\mathbf{x}$ and $\mathbf{y}$ are orthogonal.*

Proof. Suppose that $T(\mathbf{x}) = \lambda_1\mathbf{x}$ and $T(\mathbf{y}) = \lambda_2\mathbf{y}$ and that $\lambda_1 \neq \lambda_2$. Then

$$\lambda_1 \langle \mathbf{x}, \mathbf{y} \rangle = \langle \lambda_1\mathbf{x}, \mathbf{y} \rangle = \langle T(\mathbf{x}), \mathbf{y} \rangle = \langle \mathbf{x}, T(\mathbf{y}) \rangle = \langle \mathbf{x}, \lambda_2\mathbf{y} \rangle = \lambda_2 \langle \mathbf{x}, \mathbf{y} \rangle$$

Therefore $\lambda_1 \langle \mathbf{x}, \mathbf{y} \rangle - \lambda_2 \langle \mathbf{x}, \mathbf{y} \rangle = (\lambda_1 - \lambda_2) \langle \mathbf{x}, \mathbf{y} \rangle = 0$. Since $\lambda_1 - \lambda_2 \neq 0$, we must have $\langle \mathbf{x}, \mathbf{y} \rangle = 0$. Hence $\mathbf{x}$ and $\mathbf{y}$ are orthogonal. ■

COROLLARY 8.31 *The eigenvectors of a real symmetric matrix A belonging to distinct eigenvalues are orthogonal.*

Proof. Let $T_A : \mathbb{R}^n \to \mathbb{R}^n$ be the matrix transformation represented by A in the standard orthonormal basis of $\mathbb{R}^n$. Then Theorem 8.25 tells us that the adjoint T_A^* is represented by A^T. Since A is symmetric, $T_A = T_A^*$. This means that T_A is self-adjoint. The corollary therefore follows from Theorem 8.30. ■

The next result requires a brief excursion into complex inner product spaces and uses the fundamental theorem of algebra. (This theorem is discussed in Appendix A.)

THEOREM 8.32 (Symmetric matrix theorem) *The eigenvalues of a real symmetric matrix A are real.*

Proof. Let λ be a root of the characteristic polynomial $c_A(t)$ of A. We show that λ is real. By the fundamental theorem of algebra we know that there exists at least one complex root λ for which $c_A(t) = (t - \lambda)q(t)$. Since λ is an eigenvalue of A, there exists a nonzero vector $\mathbf{x} \in \mathbb{C}^n$ for which $A\mathbf{x} = \lambda\mathbf{x}$.

Let $\langle \mathbf{x}, \mathbf{y} \rangle = \mathbf{x}^T\overline{\mathbf{y}}$. Then we know from Chapter 7 that $\langle \mathbf{x}, \mathbf{y} \rangle$ is a complex inner product on $\mathbb{C}^n$. Since A is a symmetric matrix, $A = A^T$, and since A is real, $A = A^H$, where A^H is the conjugate transpose of A. Therefore

$$\langle A\mathbf{x}, \mathbf{x} \rangle = (A\mathbf{x})^T\overline{\mathbf{x}} = \mathbf{x}^T A^T \overline{\mathbf{x}} = \mathbf{x}^T A^H \overline{\mathbf{x}} = \mathbf{x}^T \overline{A\mathbf{x}} = \langle \mathbf{x}, A\mathbf{x} \rangle$$

It follows that if $\mathbf{x}$ is an eigenvector associated with an eigenvalue λ, then

$$\lambda \langle \mathbf{x}, \mathbf{x} \rangle = \langle \lambda\mathbf{x}, \mathbf{x} \rangle = \langle A\mathbf{x}, \mathbf{x} \rangle = \langle \mathbf{x}, A\mathbf{x} \rangle = \langle \mathbf{x}, \lambda\mathbf{x} \rangle = \bar{\lambda} \langle \mathbf{x}, \mathbf{x} \rangle$$

Since $\mathbf{x}$ is nonzero, we have $\langle \mathbf{x}, \mathbf{x} \rangle \neq 0$. Consequently, $\lambda = \bar{\lambda}$. The eigenvalue λ is therefore real. ■

COROLLARY 8.33 *The eigenvalues of a self-adjoint linear transformation T on a finite-dimensional real inner product space V are real.*

Proof. It is clear from our introduction to self-adjoint linear transformations that there exists an orthonormal basis for V in which the matrix of T is symmetric. The result therefore follows from Theorem 8.32. ■

Next we show that the eigenvectors of a self-adjoint linear transformation on a finite-dimensional real inner product space span the space.

THEOREM 8.34 (Orthonormal basis theorem) *Let T be a self-adjoint linear transformation on a finite-dimensional real inner product space V. Then there exists an orthonormal basis for V consisting of eigenvectors of T.*

Proof. We prove the theorem by an induction on the dimension of V.

If $\dim V = 1$, Theorem 8.32 guarantees the existence of a nonzero eigenvector $\mathbf{x}$ of T. Since every nonzero vector can be normalized, we may assume that $\mathbf{x}$ is a normal eigenvector. Therefore, $\mathcal{B} = \{\mathbf{x}\}$ is the required orthonormal basis.

Now suppose that $\dim V = n > 1$ and that the theorem holds for spaces of dimension $n - 1$. We know from Theorem 8.32 that T has at least one real eigenvalue λ. Let $\mathbf{x}$ be a normal eigenvector belonging to λ, and let $W = \text{span}(\mathbf{x})$ be the one-dimensional subspace of V spanned by $\mathbf{x}$. Then it follows from the properties of orthogonal complements that $V = W \oplus W^{\perp}$. Moreover, since T is self-adjoint,

$$\langle T(\mathbf{y}), \mathbf{x}\rangle = \langle \mathbf{y}, T(\mathbf{x})\rangle = \langle \mathbf{y}, \lambda\mathbf{x}\rangle = \lambda\langle \mathbf{y}, \mathbf{x}\rangle = \lambda 0 = 0$$

for all $\mathbf{y} \in W^{\perp}$. Therefore, $T(\mathbf{y}) \in W^{\perp}$, so T is a self-adjoint linear transformation on $W^{\perp}$. By the induction hypothesis, there therefore exists an orthonormal basis

$$\mathcal{B} = \{\mathbf{x}_2, \ldots, \mathbf{x}_n\}$$

for $W^{\perp}$ consisting of eigenvectors of T. Let $\mathcal{C}$ be the set of eigenvectors

$$\{\mathbf{x}, \mathbf{x}_2, \ldots, \mathbf{x}_n\}$$

of T. Since $W \oplus W^{\perp}$ is a direct sum, the vector $\mathbf{x}$ is linearly independent of the vectors $\mathbf{x}_2, \ldots, \mathbf{x}_n$, and since $\mathcal{B}$ is a basis for $W^{\perp}$, the set $\mathcal{C}$ is therefore a basis for $W \oplus W^{\perp} = V$. Since the vectors in $\mathcal{B}$ are normal and orthogonal eigenvectors of T, the set $\mathcal{C}$ is an orthonormal basis for the whole space consisting entirely of eigenvectors of T. ■

COROLLARY 8.35 *Every self-adjoint transformation T on a finite-dimensional real inner product space can be represented by a diagonal matrix whose diagonal entries are the eigenvalues of T.*

Proof. By Theorem 8.34, the inner product space V has an orthonormal basis $\mathcal{B} = \{\mathbf{x}_1, \ldots, \mathbf{x}_n\}$ consisting of eigenvectors of T. Let $[T(\mathbf{x}_i)]_{\mathcal{C}}$ be the coordinate vector of $T(\mathbf{x}_i)$ in the basis $\mathcal{B}$. Then $[T(\mathbf{x}_i)]_{\mathcal{C}}$ is the column vector whose ith entry is λ_i and whose other coordinates are 0. Hence the matrix $[[T(\mathbf{x}_1)]_{\mathcal{C}} \ \cdots \ [T(\mathbf{x}_n)]_{\mathcal{C}}]$ is diagonal and represents T in $\mathcal{B}$. ■

Corollary 8.35 yields the *spectral theorem* for real symmetric matrices.

COROLLARY 8.36 (Spectral theorem) *For every real symmetric matrix A, there exists an orthogonal matrix Q such that $D = Q^T A Q$ is a diagonal matrix whose entries are the eigenvalues of A.*

Proof. Consider $T : \mathbb{R}^n \to \mathbb{R}^n$ as a linear transformation defined by $T(\mathbf{x}) = A\mathbf{x}$ in some orthonormal basis $\mathcal{B}$. We know from Theorem 8.34 that there exists an orthonormal basis $\mathcal{C}$ for $\mathbb{R}^n$ consisting of eigenvectors of A. Therefore, the matrix $[T]_{\mathcal{C}} = D$ is diagonal. Let Q be the coordinate conversion matrix from $\mathcal{C}$ to $\mathcal{B}$. Then $D = Q^{-1}AQ$. By Theorem 8.13, the matrix Q is orthogonal. Hence $D = Q^T A Q$. ■

We note that if $A = QDQ^T$, then

$$A = \begin{bmatrix} \mathbf{u}_1 & \cdots & \mathbf{u}_n \end{bmatrix} \begin{bmatrix} \lambda_1 & & 0 \\ & \ddots & \\ 0 & & \lambda_n \end{bmatrix} \begin{bmatrix} \mathbf{u}_1^T \\ \vdots \\ \mathbf{u}_n^T \end{bmatrix}$$

$$= \begin{bmatrix} \lambda_1 \mathbf{u}_1 & \cdots & \lambda_n \mathbf{u}_n \end{bmatrix} \begin{bmatrix} \mathbf{u}_1^T \\ \vdots \\ \mathbf{u}_n^T \end{bmatrix}$$

$$= \lambda_1 \mathbf{u}_1 \mathbf{u}_1^T + \cdots + \lambda_n \mathbf{u}_n \mathbf{u}_n^T$$

Since the set of eigenvalues of a matrix A is called the ***spectrum*** of A, the equation

$$A = \lambda_1 \mathbf{u}_1 \mathbf{u}_1^T + \cdots + \lambda_n \mathbf{u}_n \mathbf{u}_n^T$$

is called a ***spectral decomposition*** of A.

It follows from our earlier analysis of the matrices $\mathbf{u}_i \mathbf{u}_i^T$ that for all $\mathbf{x} \in \mathbb{R}^n$, the vectors $\left(\mathbf{u}_i \mathbf{u}_i^T\right) \mathbf{x}$ are the orthogonal projections $\text{proj}(\mathbf{x} \to \text{span}\{\mathbf{u}_i\})$ of $\mathbf{x}$ onto the subspace of $\mathbb{R}^n$ spanned by $\mathbf{u}_i$.

EXAMPLE 8.30 Spectral Decomposition of a Symmetric Matrix

Use *Maple* to find a spectral decomposition of the matrix

$$A = \begin{bmatrix} 1 & 0 & 0 & 0 \\ 0 & 1 & 1 & 0 \\ 0 & 1 & 1 & 0 \\ 0 & 0 & 0 & 1 \end{bmatrix}$$

Solution. We use the **eigenvectors** function in the **linalg** package to calculate the eigenvalues and an associated full set of linearly independent eigenvectors of A.

```
[> A:=matrix([[1,0,0,0],[0,1,1,0],[0,1,1,0],[0,0,0,1]]):
[> linalg[eigenvectors](A);
```

$$[2, 1, \{[0, 1, 1, 0]\}], [1, 2, \{[0, 0, 0, 1], [1, 0, 0, 0]\}], [0, 1, \{[0, -1, 1, 0]\}]$$

This tells us that the nonzero eigenvalues of A are 2 and 1, with geometric multiplicities of 1 and 2, respectively. Thus

$$A = \mathbf{u}_1 \mathbf{u}_1^T + \mathbf{u}_2 \mathbf{u}_2^T + 2\mathbf{u}_3 \mathbf{u}_3^T + 0\mathbf{u}_3 \mathbf{u}_3^T$$

$$= (1)\begin{bmatrix} 0 \\ 0 \\ 0 \\ 1 \end{bmatrix}\begin{bmatrix} 0 \\ 0 \\ 0 \\ 1 \end{bmatrix}^T + (1)\begin{bmatrix} 1 \\ 0 \\ 0 \\ 0 \end{bmatrix}\begin{bmatrix} 1 \\ 0 \\ 0 \\ 0 \end{bmatrix}^T + 2\frac{1}{\sqrt{2}}\begin{bmatrix} 0 \\ 1 \\ 1 \\ 0 \end{bmatrix}\frac{1}{\sqrt{2}}\begin{bmatrix} 0 \\ 1 \\ 1 \\ 0 \end{bmatrix}^T$$

$$= \begin{bmatrix} 0 & 0 & 0 & 0 \\ 0 & 0 & 0 & 0 \\ 0 & 0 & 0 & 0 \\ 0 & 0 & 0 & 1 \end{bmatrix} + \begin{bmatrix} 1 & 0 & 0 & 0 \\ 0 & 0 & 0 & 0 \\ 0 & 0 & 0 & 0 \\ 0 & 0 & 0 & 0 \end{bmatrix} + \begin{bmatrix} 0 & 0 & 0 & 0 \\ 0 & 1 & 1 & 0 \\ 0 & 1 & 1 & 0 \\ 0 & 0 & 0 & 0 \end{bmatrix}$$

is a spectral decomposition of A. ▲

The spectral theorem provides us with a characterization of positive definite matrices.

THEOREM 8.37 (Positive definiteness test) *A real symmetric matrix is positive definite if and only if its eigenvalues are positive.*

Proof. Suppose that A is a positive definite real $n \times n$ matrix and that λ is an eigenvalue of A. For all eigenvectors $\mathbf{x} \in \mathbb{R}^n$ associated with λ,

$$0 < \mathbf{x}^T A\mathbf{x} = \mathbf{x}^T \lambda \mathbf{x} = \lambda \mathbf{x}^T \mathbf{x} = \lambda \, \|\mathbf{x}\|^2$$

where $\|\mathbf{x}\| = \|\mathbf{x}\|_2$ is the Euclidean norm on $\mathbb{R}^n$. Since $0 < \lambda \, \|\mathbf{x}\|^2$ and $0 < \|\mathbf{x}\|$, the eigenvalue λ must be positive.

Suppose now that all eigenvalues of A are positive. By Corollary 8.36, $A = QDQ^T$, where D is a diagonal matrix whose diagonal entries are the eigenvalues of A and Q is an orthogonal matrix whose columns are associated eigenvectors. Let $\mathbf{x} = (x_1, \ldots, x_n)$ and $\mathbf{y} = (y_1, \ldots, y_n)$ be two nonzero vectors $\mathbb{R}^n$, related by $\mathbf{x} = Q\mathbf{y}$. Then

$$\mathbf{x}^T A\mathbf{x} = (Q\mathbf{y})^T A(Q\mathbf{y}) = \mathbf{y}^T Q^T A Q\mathbf{y} = \mathbf{y}^T D\mathbf{y} = \lambda_1 y_1^2 + \cdots + \lambda_n y_n^2$$

Since all $\lambda_1, \ldots, \lambda_n$ are positive and since $\mathbf{y}$ is nonzero, the real number $\lambda_1 y_1^2 + \cdots + \lambda_n y_n^2$ must be positive. Therefore, $\mathbf{x}^T A\mathbf{x}$ is positive. ■

EXAMPLE 8.31 Orthogonal Diagonalization of a Real Symmetric Matrix

Use the **linalg** package to decompose the matrix

$$A = \begin{bmatrix} 3 & -4 \\ -4 & -3 \end{bmatrix}$$

into a product QDQ^T, where Q is an orthogonal and D is a diagonal matrix.

Solution. We begin by loading the **LinearAlgebra** package and defining the matrix A.

```
[> with(LinearAlgebra):
[> A:=Matrix([[3,-4],[-4,-3]]):
```

Since A is a real symmetric matrix, Corollary 8.36 guarantees that there exist eigenvalues and associated eigenvectors for which $A = QDQ^T$. We use *Maple* to find them.

```
[> v:=Eigenvectors(A);
```

$$\begin{bmatrix} 5 \\ -5 \end{bmatrix}, \begin{bmatrix} -2 & 1 \\ 1 & 2 \end{bmatrix}$$

The components of the vector on the left are the eigenvalues of A, and the columns of the matrix on the right are the associated eigenvectors. We therefore let

```
[> Dm:=DiagonalMatrix(v[1][1],v[1][2]);
```

$$\begin{bmatrix} 5 & 0 \\ 0 & -5 \end{bmatrix}$$

be the required diagonal matrix D. It remains to define the orthogonal matrix Q whose columns are normalized eigenvectors in the 2-norm.

```
[> n1:=Normalize(Column(v[2],1),2);
[> n2:=Normalize(Column(v[2],2),2);
[> Q:=Matrix(linalg[augment](n1,n2));
```

$$Q := \begin{bmatrix} \frac{1}{5}\sqrt{5} & -\frac{2}{5}\sqrt{5} \\ \frac{2}{5}\sqrt{5} & \frac{1}{5}\sqrt{5} \end{bmatrix}$$

We now verify that Q is orthogonal and that $QDQ^T = A$.

```
[> IsOrthogonal(Q);
```

true

```
[> Q.Dm.Transpose(Q);
```

$$\begin{bmatrix} 3 & -4 \\ -4 & -3 \end{bmatrix}$$

As intended, we have produced an orthogonal diagonalization of the matrix A. ▲

EXERCISES 8.12

1. Show that there exists a basis $\mathcal{B} = \{\mathbf{x}_1, \mathbf{x}_2\}$ for $\mathbb{R}^2$ with the property that $\mathbf{x}_1$ and $\mathbf{x}_2$ are eigenvectors of each of the matrices

$$A = \begin{bmatrix} 1 & 2 \\ 2 & 1 \end{bmatrix} \quad B = \begin{bmatrix} 1 & 1 \\ 1 & 1 \end{bmatrix} \quad C = \begin{bmatrix} 0 & 3 \\ 3 & 0 \end{bmatrix}$$

2. Calculate the rank of the symmetric matrices A^TA, B^TB, and C^TC, where

$$A = \begin{bmatrix} 2 & 3 \\ 5 & 2 \\ 4 & 2 \end{bmatrix} \quad B = \begin{bmatrix} 2 & 5 & 4 \\ 3 & 2 & 2 \end{bmatrix} \quad C = \begin{bmatrix} 2 & 5 & 4 \end{bmatrix}$$

and use this information to conclude that the columns of A^TA are linearly independent but those of B^TB, and C^TC are not. Show that it is still possible, however, to construct two bases for $\mathbb{R}^3$ consisting of eigenvectors of B^TB and C^TC.

3. Find a spectral decomposition for the real symmetric matrix

$$A = \begin{bmatrix} 0 & 2 \\ 2 & 4 \end{bmatrix}$$

4. Apply the techniques used in the proofs of Theorems 8.36 and 8.37 to write the quadratic form

$$\begin{bmatrix} x & y \end{bmatrix} \begin{bmatrix} 1 & 3 \\ 3 & 2 \end{bmatrix} \begin{bmatrix} x \\ y \end{bmatrix}$$

as a sum of squares.

5. Prove that the dimension of an eigenspace of a real symmetric matrix equals the algebraic multiplicity of the corresponding eigenvalue of the matrix.

6. (*Maple V*) Use the *Maple* commands

`charpoly(A,t)` and `evalf(solve(p=0,t))`

to find the eigenvalues of the matrix

$$A = \begin{bmatrix} 2 & 1 & 4 & 6 & 6. \\ 1 & 8 & 7 & 9 & 4 \\ 4 & 7 & 9 & 2 & 8 \\ 6 & 9 & 2 & 10 & 1 \\ 6 & 4 & 8 & 1 & 5 \end{bmatrix}$$

by calculating the roots of the characteristic polynomial of A. Find an eigenvector for each eigenvalue and show that these vectors are *approximately* orthogonal. Explain why your result is not exact.

7. (*Maple 6*) Use the method applied in Example 8.31 to diagonalize the following symmetric matrices.

$$\text{a.} \begin{bmatrix} 4 & 3 \\ 3 & 2 \end{bmatrix} \quad \text{b.} \begin{bmatrix} -2 & 0 & 0 \\ 0 & 4 & 3 \\ 0 & 3 & 2 \end{bmatrix} \quad \text{c.} \begin{bmatrix} -1 & 2 & 2 \\ 2 & -1 & 2 \\ 2 & 2 & -1 \end{bmatrix}$$

8. (*Maple 6*) Apply the techniques used in the proof of Theorems 8.36 and 8.37 to write the quadratic form

$$\begin{bmatrix} x & y & z \end{bmatrix} \begin{bmatrix} 1 & 0 & 0 \\ 0 & 2 & 4 \\ 0 & 4 & 2 \end{bmatrix} \begin{bmatrix} x \\ y \\ z \end{bmatrix}$$

as a sum of squares.

9. (*Maple 6*) Use the matrix

$$A = \begin{bmatrix} 0 & 0 & 0 & 4 \\ 1 & 0 & 0 & 11 \\ 0 & 1 & 0 & 9 \\ 0 & 0 & 1 & 1 \end{bmatrix}$$

to show that the spectral decomposition formula may not hold for a nonsymmetric real matrix.

THE METHOD OF LEAST SQUARES

In some applications of linear algebra we are faced with the problem of having to solve overdetermined linear systems. Since exact solutions are usually impossible, we look for best approximations. In this section, we explain what is meant by *best* and develop the method of least squares for finding such approximations.

Suppose A is an $m \times n$ real matrix. We know from Chapter 5 that A corresponds to a matrix transformation $T_A : \mathbb{R}^n \to \mathbb{R}^m$ defined by the rule $T_A(\mathbf{x}) = A\mathbf{x}$. Moreover, Theorem 5.15 tells us that $A\mathbf{x} \in \operatorname{Col} A$. Hence the statement that $A\mathbf{x} = \mathbf{b}$ asserts that the vector $\mathbf{b}$ belongs to $\operatorname{Col} A$. If $\mathbf{b} \notin \operatorname{Col} A$ for any $\mathbf{x} \in \mathbb{R}^n$, then the equation $A\mathbf{x} = \mathbf{b}$ has no solution. However, we can still ask whether there is a vector $\widehat{\mathbf{x}} \in \mathbb{R}^n$ for which the vector $A\widehat{\mathbf{x}} - \mathbf{b}$ is as short as possible. Figure 3 on Page 606 illustrates this situation.

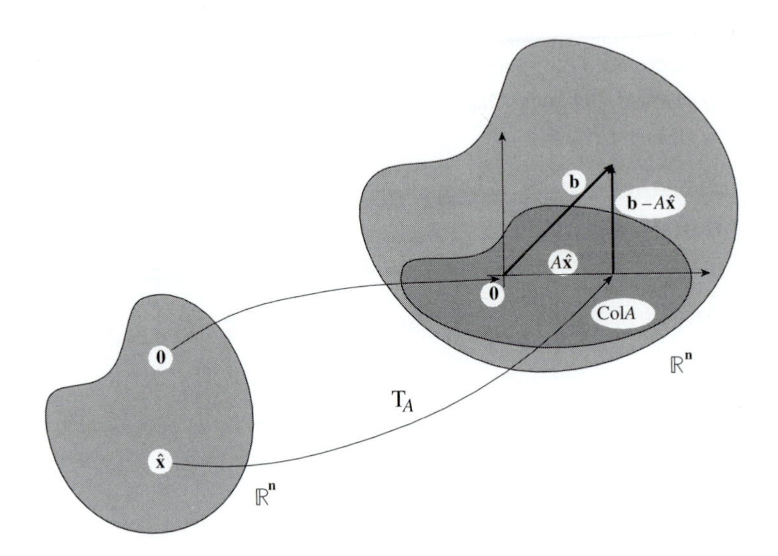

FIGURE 3 A Least-Squares Approximation.

A best possible approximation of a solution of the equation $A\mathbf{x} = \mathbf{b}$ consists of a vector $\widehat{\mathbf{x}} \in \mathbb{R}^n$ for which the distance between $A\widehat{\mathbf{x}}$ and $\mathbf{b}$ is a minimum. We use the Euclidean norm to measure this distance. The vector $\widehat{\mathbf{x}}$ is called the ***least-squares solution*** of the equation $A\mathbf{x} = \mathbf{b}$. Thus we are trying to find a vector $\widehat{\mathbf{x}} \in \mathbb{R}^n$ for which $\|A\widehat{\mathbf{x}} - \mathbf{b}\|_2$ is a minimum.

Suppose that $A\widehat{\mathbf{x}} - \mathbf{b} = (y_1, \ldots, y_m)$. Then $\|A\widehat{\mathbf{x}} - \mathbf{b}\| = \sqrt{y_1^2 + \cdots + y_m^2}$. Therefore $\|A\widehat{\mathbf{x}} - \mathbf{b}\|$ is a minimum if and only if $y_1^2 + \cdots + y_m^2$ is a minimum. This explains the term ***method of least squares***.

THEOREM 8.38 (Best approximation theorem) *Let W be a subspace of $\mathbb{R}^m$ and let $\mathbf{b}$ be a vector in $\mathbb{R}^m$. Then the vector* $\text{proj}(\mathbf{b} \to W)$ *is the best approximation of $\mathbf{b}$ by vectors in W.*

Proof. We show that $\|\mathbf{b} - \text{proj}(\mathbf{b} \to W)\| < \|\mathbf{b} - \mathbf{w}\|$ for all $\mathbf{w} \in W$, provided that $\mathbf{w} \neq \text{proj}(\mathbf{b} \to W)$. By the Pythagorean theorem and since $\mathbf{b} - \text{proj}(\mathbf{b} \to W)$ is orthogonal to W, we have

$$\|\mathbf{b} - \text{proj}(\mathbf{b} \to W)\|^2 + \|\text{proj}(\mathbf{b} \to W) - \mathbf{w}\|^2 = \|\mathbf{b} - \mathbf{w}\|^2$$

Hence $\|\mathbf{b} - \text{proj}(\mathbf{b} \to W)\|^2 < \|\mathbf{b} - \mathbf{w}\|^2$ if $\mathbf{w} \neq \text{proj}(\mathbf{b} \to W)$. ■

We can reword Theorem 8.38 in terms of matrices. If A is any $m \times n$ real matrix and $W = \text{Col}\,A$, then for any vector $\mathbf{b} \in \mathbb{R}^m$, a vector $\mathbf{x} \in \mathbb{R}^n$ minimizes the distance $\|A\mathbf{x} - \mathbf{b}\|$ if and only if $A\mathbf{x} = \text{proj}(\mathbf{b} \to \text{Col}\,A)$.

THEOREM 8.39 *For any $m \times n$ real matrix A, a vector $\widehat{\mathbf{x}} \in \mathbb{R}^n$ is a least-squares solution of the equation $A\mathbf{x} = \mathbf{b}$ if and only if $\widehat{\mathbf{x}}$ is a solution of the normal equation $A^TA\mathbf{x} = A^T\mathbf{b}$.*

Proof. Suppose that $\widehat{\mathbf{x}}$ is a least-squares solution of the equation $A\mathbf{x} = \mathbf{b}$ and let $W = \text{Col } A$. Then $A\widehat{\mathbf{x}} = \text{proj}(\mathbf{b} \rightarrow W)$. We know from Section 8 that the vector $\mathbf{b} - A\widehat{\mathbf{x}}$ is orthogonal to any vector $A\mathbf{x} \in W$. Hence $(A\mathbf{x})^T(\mathbf{b} - A\widehat{\mathbf{x}}) = 0$. Thus,

$$0 = (A\mathbf{x})^T(\mathbf{b} - A\widehat{\mathbf{x}}) = \mathbf{x}^T A^T(\mathbf{b} - A\widehat{\mathbf{x}})$$

for all $\mathbf{x} \in \mathbb{R}^n$. Therefore, $\mathbf{x}A^T(\mathbf{b} - A\widehat{\mathbf{x}}) = 0$ for all $\mathbf{x} \in \mathbb{R}^n$. It follows that $A^T(\mathbf{b} - A\widehat{\mathbf{x}}) = \mathbf{0}$. In other words, $A^T\mathbf{b} - A^TA\widehat{\mathbf{x}} = \mathbf{0}$. Hence $\widehat{\mathbf{x}}$ is a solution of the normal equation $A^TA\widehat{\mathbf{x}} = A^T\mathbf{b}$. Conversely, suppose that $A^TA\widehat{\mathbf{x}} = A^T\mathbf{b}$ for some $\widehat{\mathbf{x}} \in \mathbb{R}^n$. Then $A^T(\mathbf{b} - A\widehat{\mathbf{x}}) = \mathbf{0}$. Therefore, $\mathbf{b} - A\widehat{\mathbf{x}}$ is orthogonal to each row of A^T. Since the rows of A^T are the columns of A, the vector $\mathbf{b} - A\widehat{\mathbf{x}}$ is orthogonal to all vectors in W. Hence it is clear from Section 8 that $\mathbf{b} = A\widehat{\mathbf{x}} + (\mathbf{b} - A\widehat{\mathbf{x}})$ is a decomposition of $\mathbf{b}$, with $A\widehat{\mathbf{x}} \in W$ and $\mathbf{b} - A\widehat{\mathbf{x}} \in W^{\perp}$. By Corollary 8.18 the decomposition is unique. Hence $A\widehat{\mathbf{x}} = \text{proj}(\mathbf{b} \rightarrow W)$. ■

We summarize the process for calculating a least-squares solution of a linear system described in Theorem 8.39 in Table 5.

TABLE 5 FINDING LEAST-SQUARES SOLUTIONS.

Step 1. Form the normal equation $A^TA\dot{\mathbf{x}} = A^T\mathbf{b}$ of the linear system $A\mathbf{x} = \mathbf{b}$.
Step 2. Use one of the algorithms for solving linear systems to find a solution $\widehat{\mathbf{x}}$ of the normal equation $A^TA\mathbf{x} = A^T\mathbf{b}$.
Step 3. The vector $\widehat{\mathbf{x}}$ is a least-squares solution of $A\mathbf{x} = \mathbf{b}$.

EXAMPLE 8.32 A Least-Squares Solution

Use the 5×4 matrix A obtained from the 5×5 Hilbert matrix H by deleting the last column of H, together with the column vector $\mathbf{b} = (1, 2, 3, 4, 5)$, to show that the least-squares solution of the equation $A\mathbf{x} = \mathbf{b}$ coincides with the solution of the associated normal equation $A^TA\mathbf{x} = A^T\mathbf{b}$.

Solution. We use the **HilbertMatrix** function in the **LinearAlgebra** package.

```
[> with(LinearAlgebra):
```

Next we generate the matrix A, input the vector $\mathbf{b}$, and evaluate the functions **LeastSquares** and **LinearSolve**.

```
[> A:=SubMatrix(HilbertMatrix(5,5),1..5,1..4);
```

$$A := \begin{bmatrix} 1 & \frac{1}{2} & \frac{1}{3} & \frac{1}{4} \\ \frac{1}{2} & \frac{1}{3} & \frac{1}{4} & \frac{1}{5} \\ \frac{1}{3} & \frac{1}{4} & \frac{1}{5} & \frac{1}{6} \\ \frac{1}{4} & \frac{1}{5} & \frac{1}{6} & \frac{1}{7} \\ \frac{1}{5} & \frac{1}{6} & \frac{1}{7} & \frac{1}{8} \end{bmatrix}$$

```
[> b:=<1,2,3,4,5> :
```

```
[> LeastSquares(A,b);
```

$$\begin{bmatrix} \frac{-3028000}{33001} \\ \frac{40249620}{33001} \\ \frac{-108896760}{33001} \\ \frac{76940360}{33001} \end{bmatrix}$$

```
[> LinearSolve(Transpose(A).A,Transpose(A).b);
```

$$\begin{bmatrix} \frac{-3028000}{33001} \\ \frac{40249620}{33001} \\ \frac{-108896760}{33001} \\ \frac{76940360}{33001} \end{bmatrix}$$

As we can see, the results are identical. ▲

Theorem 8.39 implies that if the matrix $A^T A$ is invertible, the vector $\widehat{\mathbf{x}} = (A^T A)^{-1} A^T \mathbf{b}$ is the least-squares solution of $A\mathbf{x} = \mathbf{b}$. If A is the 5×5 Hilbert matrix in Example 8.32 and $\mathbf{b}$ is the column vector $(1, 2, 3, 4, 5)$, then the calculation

```
[> evalm(MatrixInverse(Transpose(A).A).Transpose(A).b);
```

$$[125, -2880, 14490, -24640, 13230]$$

produces the same result as the calculations in Example 8.32.

EXAMPLE 8.33 Fitting a Straight Line to a Set of Data

Use the method of least squares to fit a straight line to the points $(x_1, y_1) = (2, 3)$, $(x_2, y_2) = (3, 5)$, $(x_3, y_3) = (4, 3)$, and $(x_4, y_4) = (5, 6)$.

Solution. Suppose the required line is defined by the equation $y = b + mx$. Then the given points lie on this line if and only if

$$\begin{cases} y_1 = b + mx_1 \\ y_2 = b + mx_2 \\ y_3 = b + mx_3 \\ y_4 = b + mx_4 \end{cases}$$

In other words, the points satisfy the matrix vector equation $\mathbf{y} = A\mathbf{x}$, given by

$$\begin{bmatrix} y_1 \\ y_2 \\ y_3 \\ y_4 \end{bmatrix} = \begin{bmatrix} 1 & x_1 \\ 1 & x_2 \\ 1 & x_3 \\ 1 & x_4 \end{bmatrix} \begin{bmatrix} b \\ m \end{bmatrix}$$

If we think of the matrix

$$A = \begin{bmatrix} 1 & x_1 \\ 1 & x_2 \\ 1 & x_3 \\ 1 & x_4 \end{bmatrix}$$

as a matrix transformation from $A : \mathbb{R}^2 \to \mathbb{R}^4$, then the equation $\mathbf{y} = A\mathbf{x}$ asserts that $\mathbf{y}$ lies in the image of A. If $\mathbf{y} \neq A\mathbf{x}$, we would like to find an approximate solution by locating a vector $\mathbf{y} \in \mathbb{R}^4$ whose distance from the subspace $\{A\mathbf{x} : \mathbf{x} \in \mathbb{R}^2\}$ of $\mathbb{R}^2$ is a minimum. In that case, $\mathbf{y} - A\mathbf{x}$ must be orthogonal to the image of A. By the definition of the Euclidean inner product, this is so if and only if

$$(A\mathbf{z})^T(\mathbf{y} - A\mathbf{x}) = (\mathbf{z}^T A^T)(\mathbf{y} - A\mathbf{x}) = \mathbf{z}^T(A^T\mathbf{y} - A^T A\mathbf{x}) = 0$$

Since this equation holds for all $\mathbf{z}$, it follows that $A^T\mathbf{y} - A^T A\mathbf{x} = \mathbf{0}$. Therefore, $A^T\mathbf{y} = A^T A\mathbf{x}$. If $A^T A$ is invertible, we can solve this equation for $\mathbf{x}$ and obtain the required minimizing vector

$$\mathbf{x} = (A^T A)^{-1} A^T \mathbf{y}$$

Since the given points are distinct, the matrix $A^T A$ is invertible. Therefore,

$$\mathbf{x} = \left[\begin{bmatrix} 1 & 1 & 1 & 1 \\ x_1 & x_2 & x_3 & x_4 \end{bmatrix} \begin{bmatrix} 1 & x_1 \\ 1 & x_2 \\ 1 & x_3 \\ 1 & x_4 \end{bmatrix} \right]^{-1} \begin{bmatrix} 1 & 1 & 1 & 1 \\ x_1 & x_2 & x_3 & x_4 \end{bmatrix} \begin{bmatrix} y_1 \\ y_2 \\ y_3 \\ y_4 \end{bmatrix}$$

If we replace the coordinates x_i and y_j by their given values, we obtain

$$\mathbf{x} = \begin{bmatrix} \frac{9}{5} \\ \frac{7}{10} \end{bmatrix}$$

Hence $(b, m) = (9/5, 7/10)$. The required straight line is therefore $y = \frac{9}{5} + \frac{7}{10}x$. We can use the **fit[leastsquare]** function to confirm this result.

```
[> with(stats):
[> Xvalues:=[2,3,4,5]:
[> Yvalues:=[3,5,3,6]:
[> fit[leastsquare[[x,y],y=a*x+b]]([Xvalues,Yvalues]);
```

$$y = \frac{7}{10}x + \frac{9}{5}$$

As we can see, *Maple* has produced the same line. ▲

EXAMPLE 8.34 Fitting a Parabola to a Set of Data

Use the method of least squares to fit a parabola to the points $(x_1, y_1) = (2, 3)$, $(x_2, y_2) = (3, 5)$, $(x_3, y_3) = (4, 3)$, and $(x_4, y_4) = (5, 6)$.

Solution. Suppose that the required parabola is given by $y = a + bx + cx^2$. Then the given points lie on this parabola if and only if

$$\begin{cases} y_1 = a + bx_1 + cx_1^2 \\ y_2 = a + bx_2 + cx_2^2 \\ y_3 = a + bx_3 + cx_3^2 \\ y_4 = a + bx_4 + cx_4^2 \end{cases}$$

This is so if and only if

$$\begin{bmatrix} y_1 \\ y_2 \\ y_3 \\ y_4 \end{bmatrix} = \begin{bmatrix} 1 & x_1 & x_1^2 \\ 1 & x_2 & x_2^2 \\ 1 & x_3 & x_3^2 \\ 1 & x_4 & x_4^2 \end{bmatrix} \begin{bmatrix} a \\ b \\ c \end{bmatrix}$$

If we think of this equation as $\mathbf{y} = A\mathbf{x}$ and reason as in Example 8.33, we see that $\|\mathbf{y} - A\mathbf{x}\|$ is a minimum in $\mathbb{R}^4$ if and only if $\mathbf{x} = (A^T A)^{-1} A^T \mathbf{y}$. This means that

$$\mathbf{x} = \left(\begin{bmatrix} 1 & 1 & 1 & 1 \\ x_1 & x_2 & x_3 & x_4 \\ x_1^2 & x_2^2 & x_3^2 & x_4^2 \end{bmatrix} \begin{bmatrix} 1 & x_1 & x_1^2 \\ 1 & x_2 & x_2^2 \\ 1 & x_3 & x_3^2 \\ 1 & x_4 & x_4^2 \end{bmatrix} \right)^{-1} \begin{bmatrix} 1 & 1 & 1 & 1 \\ x_1 & x_2 & x_3 & x_4 \\ x_1^2 & x_2^2 & x_3^2 & x_4^2 \end{bmatrix} \begin{bmatrix} y_1 \\ y_2 \\ y_3 \\ y_4 \end{bmatrix}$$

If we replace the coordinates x_i and y_j by their given values, we obtain

$$\mathbf{x} = \begin{bmatrix} \frac{91}{20} \\ -\frac{21}{20} \\ \frac{1}{4} \end{bmatrix} = \begin{bmatrix} a \\ b \\ c \end{bmatrix}$$

The required parabola is therefore $y = \frac{91}{20} - \frac{21}{20}x + \frac{1}{4}x^2$.
We use the **fit[leastsquare]** function to confirm this result.

```
[> with(stats):
[> Xvalues:=[2,3,4,5]:
[> Yvalues:=[3,5,3,6]:
[> fit[leastsquare[[x,y],y=a*x^2+b*x+c]]([Xvalues,
   Yvalues]);
```

$$y = \frac{1}{4}x^2 - \frac{21}{20}x + \frac{91}{20}$$

As we can see, *Maple* has produced the expected equation. ▲

For certain types of numerical matrices, the round-off errors in the solution of normal equations may produce results that are very different from those obtained by the least-squares method. The numerical matrices involved are known as ***ill-conditioned matrices***. They will be studied in greater depth in Chapter 9. Hilbert matrices are ill-conditioned.

EXAMPLE 8.35 An Unstable Normal Equation

Let A be 10×10 Hilbert matrix and let $\mathbf{b} = (1, 2, 3, 4, 5, 6, 7, 8, 9, 10.)$. Show that the least-squares solution of the equation $A\mathbf{x} = \mathbf{b}$ differs from the solution.

Solution. We begin by loading the **LinearAlgebra** package.

```
[> with(LinearAlgebra):
```

Next we define the matrix A and the column vector $\mathbf{b}$.

```
[> A:=HilbertMatrix(10):
[> b:=<1,2,3,4,5,6,7,8,9,10.>:
```

Next we compare two corresponding coordinates of the least-squares solution and the solution of the associated normal equation.

```
[> LeastSquares(A,b)[4];
                         23301796.5716981515
[> LinearSolve(Transpose(A).A,Transpose(A).b)[4];
                         180158.611032684887
```

As we can see, the results are very different. The **LinearSolve** method is based on Gaussian elimination, and small numerical errors in row reductions produce relatively large errors in solution. ▲

In practice, the QR decomposition method provides an efficient way to find least-squares solutions of linear systems $A\mathbf{x} = \mathbf{b}$ if the columns of A are linearly independent. Theorem 8.8 tells us that in this case, the matrix A decomposes into a product QR consisting of an $m \times n$ matrix Q whose columns form an orthonormal set in $\mathbb{R}^m$ and an $n \times n$ upper-triangular invertible matrix R. We can therefore find the least-squares solution using back substitution.

EXAMPLE 8.36 Finding a Least-Squares Solution by QR Decomposition

Use the QR decomposition to find the least-squares solution of the linear system

$$\begin{bmatrix} 1 & 2 \\ 3 & 4 \\ 5 & 6 \end{bmatrix} \begin{bmatrix} x \\ y \end{bmatrix} = \begin{bmatrix} 6. \\ -1 \\ 3 \end{bmatrix}$$

Solution. We compare the **LeastSquares** and **QRDecomposition** functions.

```
[> A:=Matrix([[1,2],[3,4],[5,6]]):
[> b:=<6.,-1,3> :
```

We first show that the given system has no exact solution.

```
[> LinearSolve(A,b);
Error, (in LinearAlgebra:-LA_Main:-LinearSolve) inconsistent
system
```

Next we find the least-squares solution and compare it with the solution found by back substitution.

```
[> LeastSquares(A,b);
```

$$\begin{bmatrix} -5.66666666666667052 \\ 4.91666666666667051 \end{bmatrix}$$

```
[> Q,R:=QRDecomposition(A):
[> BackwardSubstitute(R,Transpose(Q).b);
```

$$\begin{bmatrix} -5.666666666 \\ 4.916666668 \end{bmatrix}$$

As we can see, the two solutions are relatively close since solving a normal equation is equivalent to finding a least-squares solution. ▲

EXERCISES 8.13

1. Convert the overdetermined linear system

$$\begin{cases} x = 1 \\ x = 2 \end{cases}$$

to the matrix form $A\mathbf{x} = \mathbf{b}$ and use the normal equation $A^TA\mathbf{x} = A^T\mathbf{b}$ to find the least-squares solution $\widehat{\mathbf{x}}$ of the given system. Explain your result.

2. Repeat Exercise 1 using the equations $x = a$ and $x = b$.

3. (*Maple V*) Express the arithmetic mean of the numbers $1, 2, 5, 6, 7$ as the least-squares solution of an overdetermined system.

4. (*Maple V*) Let y_1, y_2, and y_3 be three scalars. Find an overdetermined system whose least-squares solution is $(y_1a + y_2b + y_3c)/3$. Explain your result.

5. (*Maple V*) Convert the overdetermined linear system

$$\begin{cases} x + y = 1 \\ x - y = 2 \\ 2x + y = 7 \end{cases}$$

to the matrix form $A\mathbf{x} = \mathbf{b}$ and use the normal equation $A^TA\mathbf{x} = A^T\mathbf{b}$ to find the least-squares solution $\widehat{\mathbf{x}}$ of the given system.

6. (*Maple V*) Show that if A is an $m \times n$ real matrix whose columns form an orthonormal set in $\mathbb{R}^m$, then $\mathbf{x} = A^T\mathbf{b}$ is a least-squares solution of the linear system $A\mathbf{x} = \mathbf{b}$.

7. (*Maple V*) Let $A\mathbf{x} = \mathbf{b}$ be the linear system

$$\begin{bmatrix} 1 & -2 \\ 2 & 1 \end{bmatrix} \begin{bmatrix} x \\ y \end{bmatrix} = \begin{bmatrix} -7 \\ 5 \end{bmatrix}$$

Find the least-squares solution of the system and show that it coincides with the usual solution of the system.

8. (*Maple V*) Find the solutions of the normal equation of the linear system

$$\begin{bmatrix} 1 & 3 & 0 \\ 0 & -1 & 8 \end{bmatrix} \begin{bmatrix} x \\ y \\ z \end{bmatrix} = \begin{bmatrix} 1 \\ 2 \end{bmatrix}$$

9. (*Maple V*) Let $A\mathbf{x} = \mathbf{b}$ be the linear system

$$\begin{bmatrix} 1 & 3 & 0 \\ 0 & -1 & 8 \\ 1 & 2 & 4 \end{bmatrix} \begin{bmatrix} x \\ y \\ z \end{bmatrix} = \begin{bmatrix} 1 \\ 2 \\ 3 \end{bmatrix}$$

Find the least-squares solution $\mathbf{x} = (A^T A)^{-1} A^T \mathbf{b}$ of the system and show that it coincides with the usual solution.

10. (*Maple V*) Use the method of least squares to fit a curve $y = a + bx + cx^2 + dx^3 + ex^4$ to each of the following sets of points.

    a. $(x_1, y_1) = (1, 6)$, $(x_2, y_2) = (2, 9)$, $(x_3, y_3) = (3, 10)$, $(x_4, y_4) = (4, 21)$, $(x_5, y_5) = (5, 30)$

    b. $(x_1, y_1) = (-2, 7)$, $(x_2, y_2) = (-1, -2)$, $(x_3, y_3) = (0, -5)$, $(x_4, y_4) = (1, -1)$, $(x_5, y_5) = (2, 7)$

    c. $(x_1, y_1) = (-2, 1)$, $(x_2, y_2) = (-1, 2)$, $(x_3, y_3) = (0, 1)$, $(x_4, y_4) = (1, 2)$, $(x_5, y_5) = (2, 1)$, $(x_6, y_6) = (3, 5)$

11. (*Maple V*) Use the function **stats[fit, leastsquare]** to fit a curve $y = a + bx + cx^2 + dx^3 + ex^4$ to the set of points $(x_1, y_1) = (1, 6)$, $(x_2, y_2) = (2, 9)$, $(x_3, y_3) = (3, 10)$, $(x_4, y_4) = (4, 21)$, $(x_5, y_5) = (5, 30)$.

12. (*Maple 6*) Use the 15×15 Hilbert matrix A and a random column vector $\mathbf{b}$ with nonzero real entries to show that the least-squares solution and the solution of the normal equation of $A\mathbf{x} = \mathbf{b}$ may be very different. Then show that the QR decomposition, coupled with back substitution, yields a solution close to the least-squares solution.

REVIEW

KEY CONCEPTS ▶ Define and discuss each of the following.

Matrices

Orthogonal matrix, orthogonally diagonalizable matrix.

Orthogonal

Basis, complement, projection, subspaces, vectors.

Orthogonal Methods

Gram–Schmidt process, method of least squares, QR decomposition, perp function.

Orthogonal Transformations

Adjoint transformation, self-adjoint transformation.

Orthonormal

Basis, set.

KEY FACTS ▶ Explain and illustrate each of the following.

1. If $\mathbf{x}$ and $\mathbf{y}$ are two orthogonal vectors in an inner product space V, then $\|\mathbf{x}+\mathbf{y}\|^2 = \|\mathbf{x}\|^2 + \|\mathbf{y}\|^2$.
2. Orthogonal vectors in an inner product space are linearly independent.
3. If a basis $\mathcal{B} = \{\mathbf{x}_1, \ldots, \mathbf{x}_n\}$ for an inner product space V consists of orthogonal unit vectors, then $\mathbf{x} = \langle \mathbf{x}_1, \mathbf{x}\rangle \mathbf{x}_1 + \cdots + \langle \mathbf{x}_n, \mathbf{x}\rangle \mathbf{x}_n$ for any vector $\mathbf{x} \in V$.
4. If $\mathcal{B} = \{\mathbf{x}_1, \ldots, \mathbf{x}_p\}$ is a set of nonzero vectors in an inner product space V and if $\mathbf{x}$ is a vector in the span of $\mathcal{B}$ and is orthogonal to all vectors in $\mathcal{B}$, then $\mathbf{x} = \mathbf{0}$.
5. If $\mathcal{B} = \{\mathbf{x}_1, \ldots, \mathbf{x}_n\}$ is a basis for a real inner product space V with inner product $(\mathbf{x}, \mathbf{y}) \mapsto \langle \mathbf{x}, \mathbf{y}\rangle$ and if

$$
\begin{aligned}
&1. \quad \mathbf{y}_1 = \mathbf{x}_1 \\
&2. \quad \mathbf{y}_2 = \mathbf{x}_2 - \frac{\langle \mathbf{x}_2, \mathbf{y}_1\rangle}{\langle \mathbf{y}_1, \mathbf{y}_1\rangle}\mathbf{y}_1 \\
&3. \quad \mathbf{y}_3 = \mathbf{x}_3 - \frac{\langle \mathbf{x}_3, \mathbf{y}_1\rangle}{\langle \mathbf{y}_1, \mathbf{y}_1\rangle}\mathbf{y}_1 - \frac{\langle \mathbf{x}_3, \mathbf{y}_2\rangle}{\langle \mathbf{y}_2, \mathbf{y}_2\rangle}\mathbf{y}_2 \\
&\quad\ \vdots \\
&n. \quad \mathbf{y}_n = \mathbf{x}_n - \frac{\langle \mathbf{x}_n, \mathbf{y}_1\rangle}{\langle \mathbf{y}_1, \mathbf{y}_1\rangle}\mathbf{y}_1 - \frac{\langle \mathbf{x}_n, \mathbf{y}_2\rangle}{\langle \mathbf{y}_2, \mathbf{y}_2\rangle}\mathbf{y}_2 - \cdots - \frac{\langle \mathbf{x}_n, \mathbf{y}_{n-1}\rangle}{\langle \mathbf{y}_{n-1}, \mathbf{y}_{n-1}\rangle}\mathbf{y}_{n-1}
\end{aligned}
$$

then the set $\mathcal{C} = \{\mathbf{y}_1, \ldots, \mathbf{y}_n\}$ is an orthogonal basis for V.

6. If the function $(\mathbf{x}, \mathbf{y}) \mapsto \langle \mathbf{x}, \mathbf{y} \rangle$ is an inner product on a finite-dimensional real vector space V and if $\mathcal{E} = \{\mathbf{e}_1, \ldots, \mathbf{e}_n\}$ is an orthonormal basis for V, then $\langle \mathbf{x}, \mathbf{y} \rangle = a_1 b_1 + \cdots + a_n b_n$ for all vectors $\mathbf{x} = a_1 \mathbf{e}_1 + \cdots + a_n \mathbf{e}_n$ and $\mathbf{y} = b_1 \mathbf{e}_1 + \cdots + b_n \mathbf{e}_n$ in V.

7. If $\mathcal{B} = \{\mathbf{e}_1, \ldots, \mathbf{e}_n\}$ is an orthonormal basis for an inner product space V, then $\mathbf{x} = \langle \mathbf{x}, \mathbf{e}_1 \rangle \mathbf{e}_1 + \cdots + \langle \mathbf{x}, \mathbf{e}_n \rangle \mathbf{e}_n$ for any $\mathbf{x} \in V$.

8. If A is an $m \times n$ matrix whose n columns are linearly independent vectors in $\mathbb{R}^m$, then there exist an $m \times n$ matrix Q whose columns form an orthonormal set in $\mathbb{R}^m$ and an $n \times n$ upper-triangular invertible matrix R such that $A = QR$.

9. If A is an $m \times n$ matrix of rank n, then the solution of the linear system $A^T A\mathbf{x} = A^T \mathbf{b}$ is $\widehat{\mathbf{x}} = R^{-1} Q^T \mathbf{b}$, where R and Q are the matrices obtained from A by QR decomposition.

10. A real square matrix is orthogonal if and only if its rows and columns are orthonormal sets in the standard inner product.

11. Every invertible real matrix A is the product QR of an orthogonal matrix Q and an upper-triangular invertible matrix R.

12. Every coordinate conversion matrix between orthonormal bases of a real inner product space is orthogonal.

13. If A is an $m \times n$ real matrix, then the null space Nul A is orthogonal to the row space Row A in the standard inner product on $\mathbb{R}^n$.

14. By exchanging the matrices A and A^T in Theorem 8.14, we immediately obtain the result that the left null space Nul A^T is orthogonal to the column space Col A in the standard inner product on $\mathbb{R}^m$.

15. The orthogonal complement $S^\perp$ of a subset S of an inner product space V is a subspace of V.

16. The perp function from the subsets to subspaces of an inner product space V has the following properties: $\emptyset^\perp = V$; $V^\perp = \{\mathbf{0}\}$; $S \subseteq S^{\perp\perp}$; and if $S_1 \subseteq S_2$, then $S_2^\perp \subseteq S_1^\perp$.

17. If V is a finite-dimensional inner product space and W is a subspace of V, then $V = W \oplus W^\perp$.

18. If W is a subspace of a finite-dimensional real inner product space V and $\mathcal{B} = \{\mathbf{x}_1, \ldots, \mathbf{x}_p\}$ is an orthogonal basis for W, then every vector $\mathbf{x} \in V$ can be written in the form $\widehat{\mathbf{x}} + \mathbf{y}$, where $\mathbf{y} = \mathbf{x} - \widehat{\mathbf{x}} \in W^\perp$ and

$$\widehat{\mathbf{x}} = \frac{\langle \mathbf{x}, \mathbf{x}_1 \rangle}{\langle \mathbf{x}_1, \mathbf{x}_1 \rangle} \mathbf{x}_1 + \cdots + \frac{\langle \mathbf{x}, \mathbf{x}_p \rangle}{\langle \mathbf{x}_p, \mathbf{x}_p \rangle} \mathbf{x}_p \in W$$

19. If W is a subspace of a finite-dimensional real inner product space V and $\mathcal{B} = \{\mathbf{x}_1, \ldots, \mathbf{x}_p\}$ is an orthonormal basis for W, then every vector $\mathbf{x} \in V$ can be written in the form $\widehat{\mathbf{x}} + \mathbf{y}$, where $\widehat{\mathbf{x}} = \langle \mathbf{x}, \mathbf{x}_1 \rangle x_1 + \cdots + \langle \mathbf{x}, \mathbf{x}_p \rangle x_p$ and $\mathbf{y} = \mathbf{x} - \widehat{\mathbf{x}}$.

20. If $\mathcal{B} = \{\mathbf{x}_1, \ldots, \mathbf{x}_p\}$ is an orthonormal basis for a subspace W of $\mathbb{R}^n$ in the standard inner product and if $P = (\mathbf{x}_1 \ \cdots \ \mathbf{x}_p)$ is the matrix whose columns are the vectors in $\mathcal{B}$, then $\text{proj}(\mathbf{x} \to W) = PP^T\mathbf{x}$ for any $\mathbf{x} \in \mathbb{R}^n$.

21. If A is an $m \times n$ real matrix, then $\text{Nul}\, A \oplus \text{Col}\, A^T = \mathbb{R}^n$ and $\text{Nul}\, A^T \oplus \text{Col}\, A = \mathbb{R}^m$.

22. If $f : V \to \mathbb{R}$ is a linear functional on a finite-dimensional inner product space V, then there exists a unique vector $\mathbf{y} \in V$ such that $f(\mathbf{x}) = \langle \mathbf{x}, \mathbf{y} \rangle$ for all $\mathbf{x} \in V$.

23. If $T : V \to V$ is a linear transformation on a finite-dimensional real inner product space V, then there exists a unique linear transformation $T^* : V \to V$ for which $\langle T(\mathbf{x}), \mathbf{y} \rangle = \langle \mathbf{x}, T^*(\mathbf{y}) \rangle$.

24. If A represents a linear transformation $T : V \to V$ on a real inner product space V in an orthonormal basis $\mathcal{B} = \{\mathbf{x}_1, \ldots, \mathbf{x}_n\}$, then the matrix B representing the adjoint $T^* : V \to V$ of T in the basis $\mathcal{B}$ is A^T.

25. For any linear transformation T on a real inner product space V, the transformation $T^*T - I$ is self-adjoint.

26. If T is a self-adjoint linear transformation on a finite-dimensional real inner product space V and $\langle T(\mathbf{x}), \mathbf{x} \rangle = 0$ for all $\mathbf{x} \in V$, then $T = 0$.

27. If T is a linear transformation on a finite-dimensional real inner product space V, then the following conditions are equivalent. (1) $T(\mathbf{x}), T(\mathbf{y}) = \langle \mathbf{x}, \mathbf{y} \rangle$ for all $\mathbf{x}, \mathbf{y} \in V$. (2) $\|T(\mathbf{x})\| = \|\mathbf{x}\|$ for all $\mathbf{x} \in V$. (3) $TT^* = T^*T = I$.

28. A matrix A represents an orthogonal transformation on a finite-dimensional real inner product space in an orthonormal basis if and only if A is orthogonal.

29. If T is a self-adjoint linear transformation and $\mathbf{x}$ and $\mathbf{y}$ are eigenvectors belonging to distinct eigenvalues λ_1 and λ_2 of T, then $\mathbf{x}$ and $\mathbf{y}$ are orthogonal.

30. The eigenvectors of a real symmetric matrix A belonging to distinct eigenvalues of A are orthogonal.

31. The eigenvalues of a real symmetric matrix A are real.

32. The eigenvalues of a self-adjoint linear transformation T on a finite-dimensional real inner product space V are real.

33. Let T be a self-adjoint linear transformation on a finite-dimensional real inner product space V. Then there exists an orthonormal basis for V consisting of eigenvectors of T.

34. Every self-adjoint transformation T on a finite-dimensional real inner product space can be represented by a diagonal matrix whose diagonal entries are the eigenvalues of T.

35. For every real symmetric matrix A, there exists an orthogonal matrix Q such that $D = Q^T A Q$ is a diagonal matrix whose entries are the eigenvalues of A.

36. A real symmetric matrix is positive definite if and only if its eigenvalues are positive.

37. Let W be a subspace of $\mathbb{R}^m$ and let $\mathbf{b}$ be a vector in $\mathbb{R}^m$. Then the vector $\text{proj}(\mathbf{b} \to W)$ is the best approximation of $\mathbf{b}$ by vectors in W.

38. Suppose that A is an $m \times n$ real matrix of rank n and $\mathbf{x} \in \mathbb{R}^m$. Then the vector $\widehat{\mathbf{x}} = (A^T A)^{-1} A^T \mathbf{b}$ is a least-squares solution of the equation $A\mathbf{x} = \mathbf{b}$.

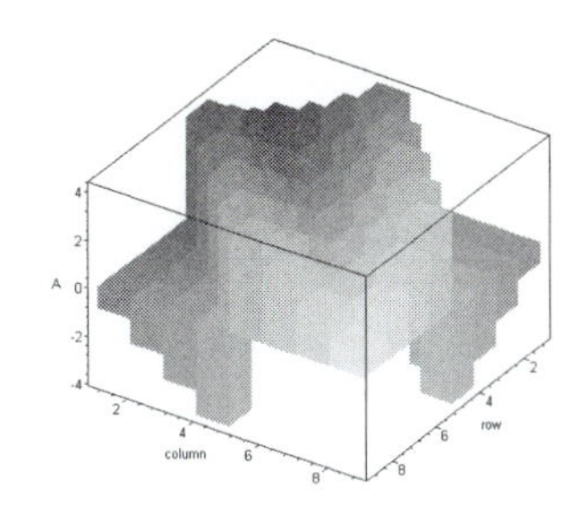

9

SINGULAR VALUES AND SINGULAR VECTORS

In this chapter, we show that every rectangular real matrix A can be decomposed into a product UDV^T of two orthogonal matrices U and V and a generalized diagonal matrix D. The product UDV^T is called the ***singular value decomposition*** of A.

The construction of UDV^T is based on the fact that for all real matrices A the matrix A^TA is symmetric and that by Corollary 8.36 there therefore exists an orthogonal matrix Q and a diagonal matrix D for which $A^TA = QDQ^T$. We know from our earlier work that the diagonal entries of D are the eigenvalues of A^TA. We now show that they are nonnegative in all cases and that their square roots, called the ***singular values*** of A, can be used to construct UDV^T. Due to the nature of the singular value decomposition algorithm, all numerical results in this chapter are approximations.

SINGULAR VALUES

We begin by showing that the eigenvalues of all symmetric matrices of the form A^TA are nonnegative.

LEMMA 9.1 *For all real $m \times n$ matrices A, the eigenvalues of A^TA are nonnegative.*

Proof. The $n \times n$ matrix A^TA is symmetric since

$$(A^TA)^T = A^TA^{TT} = A^TA$$

By Corollary 8.36, it can therefore be diagonalized relative to an orthonormal basis $\mathcal{S} = \{\mathbf{v}_1, \ldots, \mathbf{v}_n\}$ for $\mathbb{R}^n$. Let $\lambda_1, \ldots, \lambda_n$ be the associated eigenvalues. Then

$$\|A\mathbf{v}_i\|_2^2 = (A\mathbf{v}_i)^T(A\mathbf{v}_i) = \mathbf{v}_i^TA^TA\mathbf{v}_i = \mathbf{v}_i^T\lambda_i\mathbf{v}_i = \lambda_i\mathbf{v}_i^T\mathbf{v}_i = \lambda_i$$

Therefore, $\lambda_i \geq 0$ for all $i \in \mathbf{n}$. ■

Since the eigenvalues of the matrix A^TA are nonnegative real numbers, we can use their square roots to study the decomposition properties of the matrix A.

DEFINITION 9.1 *The **singular values** of a real matrix A are the square roots of the eigenvalues of the matrix A^TA.*

EXAMPLE 9.1 The Singular Values of a Square Matrix

Use *Maple* to find the singular values of the matrix

$$A = \begin{bmatrix} 1 & 3 \\ 0 & 2 \end{bmatrix}$$

Solution. The singular values of A are the square roots of the eigenvalues of A^TA. We therefore calculate these values.

```
[>with(linalg):
[>A:=matrix([[1,3],[0,2]]):
[>eigenvalues(evalm(transpose(A)&*A));
```

$$7+3\sqrt{5}, 7-3\sqrt{5}$$

This tells us that the singular values of A are $\sqrt{7-3\sqrt{5}}$ and $\sqrt{7+3\sqrt{5}}$. ▲

EXAMPLE 9.2 The Singular Values of a Rectangular Matrix

Use *Maple* to find the singular values of the matrix

$$A = \begin{bmatrix} 0 & 1 & 0 & 0 & 0 \\ -1 & 0 & 0 & 0 & 0 \\ 0 & 0 & 0 & 0 & 1 \end{bmatrix}$$

and of its transpose A^T.

Solution. We first find the eigenvalues of the matrices A^TA and $(A^T)^TA^T = AA^T$.

```
[>with(linalg):
[>A:=matrix([[0,1,0,0,0],[-1,0,0,0,0],[0,0,0,0,1]]):
[>eigenvalues(evalm(transpose(A)&*A));
```

$$1, 1, 1, 0, 0$$

```
[>eigenvalues(evalm(A&*transpose(A)));
```

$$1, 1, 1$$

Thus A has the singular values 0 and 1, whereas A^T has only the singular value 1. However, they both have the same positive singular values. ▲

We can use Lemma 9.1 and Corollary 8.36 to prove that the two-norm of a real matrix is its largest singular value.

THEOREM 9.2 (Two-norm theorem) *The two-norm of any real square matrix A is the largest singular value of A.*

Proof. Let A be a real $n \times n$ matrix. By Lemma 9.1, we can assume that the eigenvalues of A^TA are $\lambda_1 \geq \cdots \geq \lambda_n \geq 0$. Let $\mathbf{v}_1, \ldots, \mathbf{v}_n$ be eigenvectors associated with $\lambda_1, \ldots, \lambda_n$. Then

$$\|A\mathbf{v}_i\|_2^2 = (A\mathbf{v}_i)^T(A\mathbf{v}_i) = \mathbf{v}_i^T A^T A\mathbf{v}_i = \mathbf{v}_i^T \lambda_i \mathbf{v}_i = \lambda_i \mathbf{v}^T \mathbf{v}_i = \lambda_i \|\mathbf{v}_i\|_2^2$$

Therefore, $\sigma_i = \sqrt{\lambda_i} = \|A\mathbf{v}_i\|_2 / \|\mathbf{v}_i\|_2$. This implies that

$$\|A\|_2 = \max\left\{\frac{\|A\mathbf{v}\|_2}{\|\mathbf{v}\|_2} : \|\mathbf{v}\|_2 \neq 0\right\} \geq \frac{\|A\mathbf{v}_1\|_2}{\|\mathbf{v}_1\|_2} = \sigma_1$$

Let us show that $\|A\|_2 \leq \sigma_1$. By Corollary 8.36, A^TA can be diagonalized relative to an orthonormal basis $\mathcal{S} = \{\mathbf{v}_1, \ldots, \mathbf{v}_n\}$ for $\mathbb{R}^n$. Therefore, any vector $\mathbf{v} \in \mathbb{R}^n$ is a unique linear combination $a_1\mathbf{v}_1 + \cdots + a_n\mathbf{v}_n$ of orthogonal unit vectors. Moreover, the linearity of the inner product $\langle \mathbf{x}, \mathbf{y} \rangle$ implies that

$$\begin{aligned}\|\mathbf{v}\|_2^2 &= \langle \mathbf{v}, \mathbf{v} \rangle = \langle a_1\mathbf{v}_1 + \cdots + a_n\mathbf{v}_n, a_1\mathbf{v}_1 + \cdots + a_n\mathbf{v}_n \rangle \\ &= \sum_{i \in \mathbf{n}} \sum_{j \in \mathbf{n}} a_i a_j \langle \mathbf{v}_i, \mathbf{v}_j \rangle = \sum_{i \in \mathbf{n}} a_i^2\end{aligned}$$

and that

$$\begin{aligned}\|A\mathbf{v}\|_2^2 &= \langle A\mathbf{v}, A\mathbf{v} \rangle = (A\mathbf{v})^T A\mathbf{v} = \mathbf{v}A^TA\mathbf{v} = \left\langle \mathbf{v}, A^TA\mathbf{v} \right\rangle \\ &= \sum_{i \in \mathbf{n}} a_i^2 \lambda_i \leq \sum_{i \in \mathbf{n}} a_i^2 \lambda_1 = \lambda_1 \sum_{i \in \mathbf{n}} a_i^2 = \lambda_1 \|\mathbf{v}\|_2^2\end{aligned}$$

Therefore, $\|A\mathbf{v}\|_2 \leq \sqrt{\lambda_1}\, \|\mathbf{v}\|_2$. It follows that $\|A\|_2 \leq \sigma_1$. ■

EXERCISES 9.1

1. Find conditions on the entries of the matrix

$$A = \begin{bmatrix} a & b \\ c & d \end{bmatrix}$$

for which $AA^T = A^TA$.

2. Show that the singular values of the following matrices are the same as the eigenvalues of the matrices.

a. $\begin{bmatrix} 4 & 2 & 1 \\ 2 & 8 & 0 \\ 1 & 0 & 8 \end{bmatrix}$ b. $\begin{bmatrix} 4 & 0 & 1 \\ 0 & 8 & 0 \\ 1 & 0 & 8 \end{bmatrix}$ c. $\begin{bmatrix} 4 & 0 & 0 \\ 0 & 8 & 0 \\ 0 & 0 & 9 \end{bmatrix}$

3. Prove that all singular values of an orthogonal matrix are 1.

4. (*Maple V*) Use the **linalg[singularvals]** function to calculate the singular values of the matrix

$$A = \begin{bmatrix} 1 & 1 & 4. & 0. \\ 1 & 0 & 0 & 0. \\ 4 & 0 & 0 & -1 \end{bmatrix}$$

and of its transpose A^T. Compare the results.

5. (*Maple V*) Use the **eigenvalues** function in the **linalg** package to show that each of the following matrices has the same positive singular values as its transpose.

a. $\begin{bmatrix} 0 & 1 \\ -1 & 0 \end{bmatrix}$ b. $\begin{bmatrix} 0 & 1 & 0 \\ -1 & 0 & 0 \end{bmatrix}$ c. $\begin{bmatrix} 0 \\ -1 \\ 0 \end{bmatrix}$

6. (*Maple 6*) Use the function **LinearAlgebra[SingularValues]** to calculate the singular values of the following matrices and explain their accuracy.

a. $\begin{bmatrix} 2 & 4 \\ 0 & 1.0 \\ 0 & 1 \\ 1 & 1 \end{bmatrix}$ b. $\begin{bmatrix} 2 & 0 & 1 & 4 \\ 5.0 & 0 & 1 & 5 \end{bmatrix}$ c. $\begin{bmatrix} 2.0 & 3 & 0 & 0 & 0 \\ 5 & 3 & 5 & 2 & 0 \\ -1 & 1 & 1 & 0 & 2 \end{bmatrix}$

7. (*Maple 6*) Use Theorem 9.2 to calculate the two-norm of the following matrices.

a. $\begin{bmatrix} -1 & -2 & 0 \\ 1 & -1 & 1 \\ -2 & 1 & -1 \end{bmatrix}$ b. $\begin{bmatrix} -2 & 2 & -2 & 1 \\ 0 & -1 & 0 & -2 \\ 0 & 0 & -1 & 2 \\ 0 & 0 & 0 & 0 \end{bmatrix}$ c. $\begin{bmatrix} 0 & 2 & 9 \\ -2 & 0 & 0 \\ -9 & 0 & 0 \end{bmatrix}$

SINGULAR VALUE DECOMPOSITION

The following are some of the properties that make singular value decompositions useful.

1. All real matrices have singular value decompositions.

2. A real square matrix is invertible if and only if all of its singular values are nonzero.

3. If A is an $n \times n$ real invertible matrix, the ratio $\text{cond}\, A = \sigma_1/\sigma_n$ of the largest and smallest singular values tells us how close A is computationally to being singular relative to the arithmetic precision of a computing device. A large ratio indicates that a computer may not be able to detect that the matrix is invertible.

4. For any $m \times n$ real rectangular matrix A, the number of nonzero singular values of A is equal to the rank of A.

5. If $A = UDV^T$ is a singular value decomposition of an invertible matrix A, then $A^{-1} = VD^{-1}U^T$.

6. The Frobenius norm $\|A\|_F$ of a real matrix is the square root of the sum of the squares of the singular values of A.

7. For positive definite symmetric matrices, the orthogonal decomposition QDQ^T and the singular value decomposition UDV^T coincide.

THEOREM 9.3 (Singular value decomposition theorem) *For every $m \times n$ real matrix A, there exists an orthogonal $m \times m$ matrix U, an orthogonal $n \times n$ matrix V, and an $m \times n$ generalized diagonal matrix D such that $A = UDV^T$.*

Proof. Let A be a real $m \times n$ matrix, and form the real symmetric matrix A^TA. By Corollary 8.36, the matrix A^TA is an orthogonally diagonalizable matrix. It therefore has n real eigenvalues $\lambda_1, \ldots, \lambda_n$. We know from Lemma 9.1 that these eigenvalues are nonnegative. We can therefore assume that $\lambda_1 \geq \cdots \geq \lambda_n \geq 0$. We use this information to construct an $n \times n$ orthogonal matrix V.

By Theorems 8 and 7.6, we can choose the eigenvectors $\mathbf{v}_i$ for the eigenvalues λ_i in such a way that the set $\{\mathbf{v}_1, \ldots, \mathbf{v}_n\}$ is an orthonormal basis for $\mathbb{R}^n$. We then combine the column vectors $\mathbf{v}_i$ into the matrix

$$V = [\mathbf{v}_1 \ \cdots \ \mathbf{v}_n]$$

We know from Theorem 8.11 that V is orthogonal. The columns of V are called the ***right singular vectors*** of A. Now we show if A has r nonzero singular values and if $\{\mathbf{v}_1, \ldots, \mathbf{v}_n\}$ is an orthonormal basis for $\mathbb{R}^n$ consisting of eigenvectors of A^TA, then $\{A\mathbf{v}_1, \ldots, A\mathbf{v}_r\}$ is an orthogonal basis for $\text{Col}\, A$.

Since the vectors $\mathbf{v}_i$ and $\lambda_j \mathbf{v}_j$ are mutually orthogonal for all $i \neq j$, we know that

$$0 = \langle \mathbf{v}_i, \lambda_j \mathbf{v}_j \rangle = \left\langle \mathbf{v}_i, A^TA\mathbf{v}_j \right\rangle = \langle A\mathbf{v}_i, A\mathbf{v}_j \rangle$$

Hence the set $\{A\mathbf{v}_1, \ldots, A\mathbf{v}_n\}$ is orthogonal. By Lemma 9.1, $\|A\mathbf{v}_i\|_2 = \sqrt{\lambda_i} = \sigma_i$. By assumption, A has r nonzero singular values. Therefore $A\mathbf{v}_i \neq \mathbf{0}$ if and only if $i \in \mathbf{r}$. Moreover, the set $\{A\mathbf{v}_1, \ldots, A\mathbf{v}_r\}$ is linearly independent since the nonzero vectors $A\mathbf{v}_i$ are mutually orthogonal. Hence $\{A\mathbf{v}_1, \ldots, A\mathbf{v}_r\} \subseteq \text{Col}\, A$.

Now let $\mathbf{x} \in \text{Col } A$. Then $\mathbf{x} = A\mathbf{y}$ for some $\mathbf{y} = a_1\mathbf{v}_1 + \cdots + a_n\mathbf{v}_n$ and

$$\begin{aligned} A\mathbf{y} &= a_1 A\mathbf{v}_1 + \cdots + a_r A\mathbf{v}_r + a_{r+1} A\mathbf{v}_{+1} + \cdots + a_n A\mathbf{v}_n \\ &= a_1 A\mathbf{v}_1 + \cdots + a_r A\mathbf{v}_r + \mathbf{0} \\ &= a_1 A\mathbf{v}_1 + \cdots + a_r A\mathbf{v}_r \end{aligned}$$

Hence $\mathbf{y} \in \text{Col } A$. Therefore $\{A\mathbf{v}_1, \ldots, A\mathbf{v}_r\}$ is a basis for Col A. It follows that the rank of A is r.

We use this information to construct the diagonal matrix D. We choose the r positive eigenvalues $\lambda_1, \ldots, \lambda_r$ of $A^T A$ and form the $r \times r$ diagonal matrix

$$D_r = \text{diag}(\sigma_1, \ldots, \sigma_r)$$

whose diagonal entries σ_r are the square roots of the positive eigenvalues λ_i. We then extend D to the $m \times n$ generalized diagonal matrix D by adding zero rows and zero columns to D_r if necessary.

We conclude by constructing the orthogonal matrix U. We normalize the vectors in the orthogonal basis $\{A\mathbf{v}_1, \ldots, A\mathbf{v}_r\}$ by forming the vectors

$$\mathbf{u}_i = \frac{A\mathbf{v}_i}{\|A\mathbf{v}_i\|} = \frac{A\mathbf{v}_i}{\sigma_i} \qquad (1 \leq i \leq r)$$

and let $W_r = \{\mathbf{u}_1, \ldots, \mathbf{u}_r\}$. Then W_r is an orthonormal basis for Col A. Using Theorems 4.14 and 8.5, we extend W_r to an orthonormal basis $W_m = \{\mathbf{u}_1, \ldots, \mathbf{u}_r, \ldots, \mathbf{u}_m\}$ for $\mathbb{R}^m$. The vectors $\mathbf{u}_i$ are column vectors in $\mathbb{R}^m$, and we use them to form the $m \times m$ matrix

$$U = [\mathbf{u}_1 \ \cdots \ \mathbf{u}_m]$$

By Theorem 8.11, the matrix U is orthogonal. The columns of U are called the ***left singular vectors*** of A.

We conclude by showing that $A = UDV^T$. Theorem 2.1 tells us that

$$\begin{aligned} UD &= [\mathbf{u}_1 \ \cdots \ \mathbf{u}_m]\, D \\ &= [\sigma_1\mathbf{u}_1 \ \cdots \ \sigma_m\mathbf{u}_m] \\ &= [\sigma_1\mathbf{u}_1 \ \cdots \ \sigma_r\mathbf{u}_r \ \mathbf{0} \ \cdots \ \mathbf{0}] \\ &= [A\mathbf{v}_1 \ \cdots \ A\mathbf{v}_r \ \mathbf{0} \ \cdots \ \mathbf{0}] \end{aligned}$$

and that

$$\begin{aligned} AV &= A\,[\mathbf{v}_1 \ \cdots \ \mathbf{v}_n] \\ &= [A\mathbf{v}_1 \ \cdots \ A\mathbf{v}_n] \\ &= [A\mathbf{v}_1 \ \cdots \ A\mathbf{v}_r \ \mathbf{0} \ \cdots \ \mathbf{0}] \end{aligned}$$

Hence $AV = UD$. Since V is orthogonal, the matrix VV^T is the identity and we get $A = AI = AVV^T = UDV^T$. This proves the theorem. ■

We can use both the function **Svd** and the **SingularValues** function in the **LinearAlgebra** package to calculate singular value decompositions.

For any $m \times n$ real matrix A, the command `Svd(A,U,V)` returns the singular values of A and the left and right singular vectors of A in U and V, respectively. The singular vectors together with the singular values satisfy $A = UDV^T$, where U is an $n \times n$ orthogonal matrix whose columns are the left singular vectors of A, the matrix V is an $m \times m$ orthogonal matrix whose columns are the right singular vectors of A, and D is an mn generalized diagonal matrix whose diagonal entries are the singular values of A. To obtain the singular values of A from `Svd(A,U,V)`, we must apply the function **evalf** to `Svd(A,U,V)`, and to display the actual matrices U and V, we must apply the **evalm** function to `U` and `V`.

We can also use the **SingularValues** function in the **LinearAlgebra** package. For this function, we must specify explicitly that the entries of the matrix to be decomposed are real numbers. The command `S:=SingularValues(A)` produces the singular values of A in the form of a column vector, with the values listed in descending order of magnitude. The option **output=['U', 'Vt']** in the command `SingularValues(A,output=['U','Vt'])` produces the orthogonal matrices U and V^T. The output option **output=['S','U', 'Vt']**, on the other hand, returns both the vector of singular values and the matrices U and V^T. The option **output=['U','S', 'Vt']** also returns U, S, and V, but in a different order.

The more limited function **singularvals** in the **linalg** package returns only the singular values of square matrices.

EXAMPLE 9.3 A Singular Value Decomposition of a 2 x 2 Matrix

Use *Maple* to find the singular value decomposition of the matrix

$$A = \begin{bmatrix} 0 & 5 \\ 2 & 2 \end{bmatrix}$$

Solution. We use the **Svd** function.

```
[> A:=matrix([[0,5.0],[2,2]]):
[> S:=evalf(Svd(A,A2,A3)):
```

$$S := [5.442830582, 1.837279307]$$

```
[> U:=evalm(A2);
```

$$U := \begin{bmatrix} -.9076476244 & .4197329984 \\ -.4197329984 & -.9076476244 \end{bmatrix}$$

```
[> V:=evalm(A3);
```

$$V := \begin{bmatrix} -.1542333505 & -.9880344496 \\ -.9880344496 & .1542333505 \end{bmatrix}$$

```
[> Dm:=linalg[diag](S[1],S[2]);
```

$$Dm := \begin{bmatrix} 5.442830582 & 0 \\ 0 & 1.837279307 \end{bmatrix}$$

```
[> evalm(U&*Dm&*transpose(V));
```

$$\begin{bmatrix} .1 \times 10^{-9} & 5.000000000 \\ 2.000000000 & 2.000000000 \end{bmatrix}$$

Since $.1 \times 10^{-9}$ is close to 0, the matrix product UDV^T is approximately equal to A. ▲

Next we find a singular value decomposition of a rectangular real matrix.

EXAMPLE 9.4 A Singular Value Decomposition of a 2 × 3 Matrix

Use *Maple* to find the singular value decomposition of the matrix

$$A = \begin{bmatrix} 0 & 5 & 1. \\ 2 & 2 & 8 \end{bmatrix}$$

Solution. We use the **SingularValues** function.

```
[> with(LinearAlgebra):
[> A:=Matrix([[0,5,1.],[2,2,8]]):
[> S,U,Vt:=SingularValues(A,output=['S','U','Vt']):
[> Dm:=DiagonalMatrix([S[1],S[2]],2,3):
[> B:=evalf(U.Dm.Vt,5);
```

$$B := \begin{bmatrix} .00002 & 5.0000 & 1.0001 \\ 2.0000 & 2.0000 & 7.99999 \end{bmatrix}$$

As we can see, the matrix B is a close approximation of the given matrix A. ▲

We have already seen that the two-norm of a square matrix A can be described in terms of singular values of A. We now show that the singular value decomposition can also be used to describe the Frobenius norm of A.

THEOREM 9.4 *If $\sigma_1, \ldots, \sigma_n$ are the singular values of a real $n \times n$ matrix A, then the square of the Frobenius norm $\|A\|_F^2$ is the sum $\sigma_1^2 + \cdots + \sigma_n^2$ of the singular values of A.*

Proof. We know from Chapter 7, that $\|A\|_F^2 = \text{trace } A^T A$. It therefore suffices to prove that trace $A^T A = \sigma_1^2 + \cdots + \sigma_n^2$. Let $A = UDV^T$ be a singular value decomposition of A.

Then

$$\begin{aligned}\text{trace } A^T A &= \text{trace}(UDV^T)^T(UDV^T) \\ &= \text{trace } VDU^T UDV^T \\ &= \text{trace } VD^2V^T \\ &= \text{trace } V^T VD^2 \\ &= \text{trace } D^2\end{aligned}$$

Since $D = \text{diag}(\sigma_1, \ldots, \sigma_n)$, we have $D^2 = \text{diag}(\sigma_1^2, \ldots, \sigma_n^2)$, with trace $\sigma_1^2 + \cdots + \sigma_n^2$. Since the Frobenius norm of A is the trace of $A^T A$, the theorem follows. ■

EXAMPLE 9.5 A Frobenius Norm

Use *Maple* and Theorem 9.3 to approximate the Frobenius norm of the matrix

$$A = \begin{bmatrix} 9 & 2 & -2 & 4.0 \\ 5 & 1 & 1 & 2 \\ 6 & 1 & 2 & -5 \\ -2 & 3 & 2 & 7 \end{bmatrix}$$

and compare the result with the value trace $A^T A$.

Solution. We begin loading the **linalg** package and define the matrix A.

```
[>with(linalg):
[>A:=matrix([[9,2,-2,4.0],[5,1,1,2],[6,1,2,-5],
   [-2,3,2,7]]):
```

Next we calculate the trace of $A^T A$.

```
[>B:=evalm(transpose(A)&*A): trace(B);
                         268.00
```

We then calculate the singular value decomposition UDV^T of A and form the sum of the squares of the diagonal entries of D.

```
[> svd:=Svd(A);
```

$$svd := [12.28083683, 10.01776307, 3.990551816, .9491923852]$$

```
[> sum(svd[i]^ 2,i=1..4);
```

$$268.0000001$$

As we can see, the results are essentially the same. Both results yield $\|A\|_F = \sqrt{268.} = 16.3707$. ▲

Next we illustrate the calculation of a reduced singular value decomposition.

EXAMPLE 9.6 A Singular Value Decomposition of a 4 x 5 Matrix

Construct a 4×5 matrix of rank 3 and use *Maple* to find the singular value decomposition of the matrix.

Solution. The first three columns of the matrix

$$A = \begin{bmatrix} 1 & 0 & 0 & 0 & 1. \\ 0 & 1 & 0 & 1 & 0 \\ 0 & 0 & 1 & 1 & 1 \\ 0 & 0 & 0 & 0 & 0 \end{bmatrix}$$

are linearly independent and the last two columns are linear combinations of the first three. Therefore, A has rank 3. We use the **Svd** function to decompose A.

```
[> with(linalg):
[> A:=matrix([[1,0,0,0,1.],[0,1,0,1,0],[0,0,1,1,1],
    [0,0,0,0,0]]):
```

```
[> svd:=evalf(Svd(A,U,V),5);
```

$$svd := [2.0000, 1.4142, 1.0000, 0.]$$

```
[> Dm:=diag(svd[1],svd[2],svd[3],svd[4]):
[> W:=submatrix(transpose(V),1..4,1..5):
```

```
[> evalf(evalm(U&*Dm&*W),5);
```

$$\begin{bmatrix} .99999 & .00001 & .00001 & -.15563\ 10^{-15} & 1.0000 \\ -.00001 & .99999 & .00001 & 1.0000 & .22627\ 10^{-15} \\ 0. & 0. & 1.0000 & 1.0000 & 1.0000 \\ -0. & 0. & 0. & 0. & -0. \end{bmatrix}$$

As we can see, this matrix is approximately equal to the matrix A. ▲

EXAMPLE 9.7 A Singular Value Decomposition of a Matrix Without Eigenvalues

Use *Maple* to find the singular values decomposition of the rotation matrix

$$A = \begin{bmatrix} 0 & -1.0 \\ 1 & 0 \end{bmatrix}$$

Solution. Here are the calculations.

```
[> with(linalg):
[> A:=matrix([[0,-1],[1,0]]):
[> svd:=evalf(Svd(A,U,V));
```

```
[> Dm:=diag(svd[1],svd[2]);
```

$$\begin{bmatrix} 1. & 0 \\ 0 & 1. \end{bmatrix}$$

```
[> evalm(U);
```

$$\begin{bmatrix} 0. & -1. \\ -1. & 0. \end{bmatrix}$$

```
[> transpose(V);
```

$$\begin{bmatrix} -1. & 0. \\ 0. & 1. \end{bmatrix}$$

```
[> evalm(U&*Dm&*transpose(V));
```

$$\begin{bmatrix} -0 & -1. \\ 1. & 0. \end{bmatrix}$$

By multiplying the last three matrices, we get

$$\begin{bmatrix} 0 & -1 \\ -1 & 0 \end{bmatrix}\begin{bmatrix} 1 & 0 \\ 0 & 1 \end{bmatrix}\begin{bmatrix} -1 & 0 \\ 0 & 1 \end{bmatrix} = \begin{bmatrix} 0 & -1 \\ 1 & 0 \end{bmatrix}$$

The result is the expected singular value decomposition. ▲

EXERCISES 9.2

1. Show that if A is a positive definite matrix, then A has a singular value decomposition of the form QDQ^T.

2. Use Exercise 1 to conclude that if A is a positive definite matrix, then the eigenvalues and the singular values of A coincide.

3. Show that if $\mathbf{u}_i$ is a left singular vector and $\mathbf{v}_i$ is a right singular vector belonging to the singular value σ_i of a rectangular matrix A of rank r, then $A^T\mathbf{u}_i = \sigma_i\mathbf{v}_i$ for all $1 \le i \le r$.

4. Use Exercise 1 to explain why the following matrices are not positive definite.

$$\text{a. } \begin{bmatrix} 0 & 1 \\ 1 & 0 \end{bmatrix} \quad \text{b. } \begin{bmatrix} 2 & .5 & -1 \\ .5 & 3 & .5 \\ -1 & .5 & -1 \end{bmatrix} \quad \text{c. } \begin{bmatrix} 2 & 5 \\ 5 & 3 \end{bmatrix}$$

5. Show that if A is an $m \times n$ real matrix of rank r and $A = UDV^T$ is a singular value decomposition of A with $U = [\mathbf{u}_1 \ \cdots \ \mathbf{u}_m]$ and $V = [\mathbf{v}_1 \ \cdots \ \mathbf{v}_n]$ and positive singular values $\sigma_1, \ldots, \sigma_r$, then $A = \sigma_1\mathbf{u}_1\mathbf{v}_1^T + \cdots + \sigma_r\mathbf{u}_r\mathbf{v}_r^T$.

6. (*Maple V*) Show that the following matrices are positive definite and use the singular value decomposition to find their approximate eigenvalues.

$$\text{a. } \begin{bmatrix} 4 & 0 & 0 & 0 & 0 \\ 0 & 9 & 0 & -2 & 0 \\ 0 & 0 & 7 & 0 & 0 \\ 0 & -2 & 0 & 1 & 0 \\ 0 & 0 & 0 & 0 & 3. \end{bmatrix} \quad \text{b. } \begin{bmatrix} 2.7 & 1 & 0 & 0 & 0 \\ 1 & 2.5 & 1 & 0 & 0 \\ 0 & 1 & 3.8 & 1 & 0 \\ 0 & 0 & 1 & 1 & -.3 \\ 0 & 0 & 0 & -.3 & 2 \end{bmatrix}$$

7. (*Maple V*) Use singular values to calculate the Frobenius norm of the following matrices.

a. $\begin{bmatrix} 2 & -3 \\ -3 & -3 \end{bmatrix}$ b. $\begin{bmatrix} 9 & 0 & -2 \\ -2 & 0 & 1 \end{bmatrix}$ c. $\begin{bmatrix} 5 & 3 & 3 \\ -1 & 5 & -1 \\ -2 & -1 & 3 \end{bmatrix}$

8. (*Maple V*) Calculate the ratio σ_1/σ_n of the largest and smallest singular value for each of the matrices in Exercise 7.

9. (*Maple V*) Calculate the singular value decompositions of the following matrices and find their inverses in terms of V, D^{-1}, and U^T.

a. $\begin{bmatrix} 2 & -3 \\ -3 & -3 \end{bmatrix}$ b. $\begin{bmatrix} -1 & -1 & 2 \\ -1 & -3 & 3 \\ 2 & 3 & -1 \end{bmatrix}$ c. $\begin{bmatrix} 5 & 3 & 3 \\ -1 & 5 & -1 \\ -2 & -1 & 3 \end{bmatrix}$

10. (*Maple 6*) Generate three random 4×6 real matrices and show that the number of nonzero singular values of each matrix equals the rank of the matrix.

11. (*Maple 6*) Generate three random 4×4 invertible matrices and show that their singular values are nonzero.

APPLICATIONS

Fundamental Subspaces

An elegant application of the singular value decomposition is the calculation of orthonormal bases for the fundamental subspaces of real matrices.

THEOREM 9.5 (Singular vector bases theorem) *Suppose that* $\mathbf{u}_1, \ldots, \mathbf{u}_m$ *are the left singular vectors and* $\mathbf{v}_1, \ldots, \mathbf{v}_n$ *are the right singular vectors of a real* $m \times n$ *matrix* A *whose singular values are* $\sigma_1, \ldots, \sigma_n$ *and whose rank is* r. *Then*

1. $U_r = \{\mathbf{u}_1, \ldots, \mathbf{u}_r\}$ *is an orthonormal basis for* Col A
2. $U_{m-r} = \{\mathbf{u}_{r+1}, \ldots, \mathbf{u}_m\}$ *is an orthonormal basis for* Nul A^T
3. $V_r = \{\mathbf{v}_1, \ldots, \mathbf{v}_r\}$ *is an orthonormal basis for* Col A^T
4. $V_{n-r} = \{\mathbf{v}_{r+1}, \ldots, \mathbf{v}_n\}$ *is an orthonormal basis for* Nul A

Proof. We know from the proof of Theorem 9.3 that the set of left singular vectors

$$U = \{\mathbf{u}_1, \ldots, \mathbf{u}_r, \mathbf{u}_{r+1}, \ldots, \mathbf{u}_m\}$$

is an extension of an orthonormal basis W_r for Col A to a basis W for $\mathbb{R}^m$.

Since $\mathbb{R}^m = \text{Col } A \oplus \text{Nul } A^T$ and since Nul A^T is orthogonal to Col A, Theorem 4.40 implies that the complement

$$U_{m-r} = \{\mathbf{u}_{r+1}, \ldots, \mathbf{u}_m\}$$

of U_r in U is an orthonormal basis for Nul A^T.

Let us now consider the right singular case. Let

$$V = \{\mathbf{v}_1, \ldots, \mathbf{v}_r, \mathbf{v}_{r+1}, \ldots, \mathbf{v}_n\}$$

be a basis for $\mathbb{R}^n$ consisting of the right singular vectors of A. We show that the subset

$$V_{n-r} = \{\mathbf{v}_{r+1}, \ldots, \mathbf{v}_n\}$$

is an orthonormal basis for Nul A. Since $\mathbb{R}^n = \text{Nul } A \oplus \text{Col } A^T$ and since Nul A is orthogonal to Col A^T, Theorem 4.40 will then imply that

$$V_r = \{\mathbf{v}_1, \ldots, \mathbf{v}_r\}$$

is an orthonormal basis for Col A^T.

Since V is orthonormal, it suffices to show that $V_{n-r} = \{\mathbf{v}_{r+1}, \ldots, \mathbf{v}_n\}$ is a basis for Nul A. We note that $\lambda_i = 0$ for $i > r$ and that $A\mathbf{v}_i = \lambda\mathbf{u}_i$. Therefore,

$$A\mathbf{v}_{r+1} = \cdots = A\mathbf{v}_n = 0$$

Hence $\mathbf{v}_{r+1}, \ldots, \mathbf{v}_n \in \text{Nul } A$. Moreover, by Theorem 8.2, the vectors $\mathbf{v}_{r+1}, \ldots, \mathbf{v}_n$ are linearly independent. Since $\dim \text{Nul } A = n - \text{rank } A$, it follows that V_{n-r} is a basis for Nul A. ■

EXERCISES 9.3

1. Show that if $A = UDV^T$ is a singular value decomposition of A, then VDU^T is a singular value decomposition of A^T. Use this fact to show that A and A^T have the same nonzero singular values.

2. (*Maple V*) Use function **Svd** to find approximately orthonormal bases for the fundamental subspaces of the following matrices.

$$\text{a.} \begin{bmatrix} 0 & 0 & 0 & 0 \\ 1 & 0 & 2 & 3 \\ 0 & 0 & 0 & 0 \end{bmatrix} \quad \text{b.} \begin{bmatrix} 1 & 2 & 4 \\ 0 & 0 & 1 \\ 3 & 6 & 12 \end{bmatrix} \quad \text{c.} \begin{bmatrix} 1 & 2 & 4 & 0 & 1 \\ 0 & 0 & 1 & 9 & 0 \\ 0 & 2 & 1 & 0 & 0 \\ 0 & 0 & 0 & 3 & 0 \\ 0 & 0 & 0 & 0 & 3 \end{bmatrix}$$

3. (*Maple 6*) Let $U = [\mathbf{u}_1\ \mathbf{u}_2\ \mathbf{u}_3]$ and $V = [\mathbf{v}_1\ \mathbf{v}_2\ \mathbf{v}_3]$ be the orthogonal matrices in a singular value decomposition of the matrix

$$A = \begin{bmatrix} 1 & 2 & 4 \\ 0 & 0 & 1 \\ 3 & 6 & 12 \end{bmatrix}$$

Show that for all positive singular values σ of A, $A\mathbf{v}_i = \sigma_i \mathbf{u}_i$.

4. (*Maple 6*) Verify that the left singular vectors of the matrix

$$A = \begin{bmatrix} 0 & 1 & -1 & 1 \\ -1 & 2 & 0 & 1 \end{bmatrix}$$

form an orthonormal basis for Col A. What conclusion can you draw about Nul A^T?

5. (*Maple 6*) Verify that the first two right singular vectors of the matrix

$$A = \begin{bmatrix} 0 & 1 & -1 & 1 \\ -1 & 2 & 0 & 1 \end{bmatrix}$$

form an orthonormal basis for Col A^T. What conclusion can you draw about Nul A?

Pseudoinverses

The reduced singular value decompositions produced by *Maple* have several powerful applications. Recall that if A is an $m \times n$ real matrix of rank r and UDV^T is a singular value decomposition of A, then we can think of UDV^T as the product

$$A = \begin{bmatrix} U_r & U_{m-r} \end{bmatrix} \begin{bmatrix} D_r & 0 \\ 0 & 0 \end{bmatrix} \begin{bmatrix} V_r^T \\ V_{n-r}^T \end{bmatrix} = U_r D_r V_r^T$$

where U_r is the matrix $[\mathbf{u}_1 \cdots \mathbf{u}_r]$ and V_r is the matrix $[\mathbf{v}_1 \cdots \mathbf{v}_r]$. Suppose that $r > 0$. Then D_r is invertible since it is a diagonal matrix with nonzero diagonal entries. We can therefore define the matrix

$$A^+ = V_r D_r^{-1} U_r^T$$

by transposing the product $U_r D_r V_r^T$ and replacing D_r by its inverse, D_r^{-1}. The matrix A^+ is called the ***pseudoinverse*** of A. If A is invertible, then $A^+ = A^{-1}$ since

$$AA^+ = AV_r D_r^{-1} U_r^T = U_r D_r V_r^T V_r D_r^{-1} U_r^T = I$$

With the help of the functions in **LinearAlgebra** package, we can actually replicate the steps in the definition of the pseudoinverse and outline a *Maple* process for calculating pseudoinverses.

Once we have input the matrix A, we continue as follows:

```
[> rd:=A->RowDimension(A):
[> cd:=A->ColumnDimension(A):
[> U,S,Vt:=SingularValues(A,output=['U','S','Vt']):
[> V:=Transpose(Vt):
[> Vr:=SubMatrix(V,1..rd(V),1..Rank(A)):
[> Dr:=DiagonalMatrix(S[1..Rank(A)],Rank(A),Rank(A)):
[> Ur:=SubMatrix(U,1..rd(U),1..Rank(A)):
[> Aplus:=Vr.MatrixInverse(Dr).Transpose(Ur):
```

It is clear from the definition that the constructed matrix is the required pseudoinverse. We summarize the procedure for calculating an approximate pseudoinverse in Table 1.

EXAMPLE 9.8 A Pseudoinverse

Use *Maple* to find the pseudoinverse of the matrix

$$A = \begin{bmatrix} 1 & 0 & 0 \\ 0 & 2 & 0 \\ 0 & 0 & 3 \end{bmatrix}$$

Solution. First we use the **Svd** function.

```
[> with(linalg):
[> A:=matrix([[1,0,0],[0,2,0],[0,0,3]]):
[> svd:=evalf(Svd(A,U,V)):
[> Dm:=diag(svd[1],svd[2],svd[3]):
```

```
[> Aplus:=evalm(V&*inverse(Dm)&*transpose(U));
```

$$Aplus := \begin{bmatrix} 1.000000000 & 0. & 0. \\ 0. & .5000000000 & 0. \\ 0. & 0. & .3333333333 \end{bmatrix}$$

TABLE 1 FINDING APPROXIMATE PSEUDOINVERSES.

Step 1. For any real $m \times n$ matrix A of rank r, use *Maple* to calculate a singular value decomposition UDV^T of A, with $D = \text{diag}(\sigma_1, \ldots, \sigma_r, 0, \ldots, 0)$.

Step 2. Form the submatrix $U_r = [\mathbf{u}_1 \cdots \mathbf{u}_r]$ of $U = [\mathbf{u}_1 \cdots \mathbf{u}_m]$, the submatrix $V_r = [\mathbf{v}_1 \cdots \mathbf{v}_r]$ of $V = [\mathbf{v}_1 \cdots \mathbf{v}_n]$, and the diagonal submatrix $D_r = \text{diag}(\sigma_1, \ldots, \sigma_r)$ of D.

Step 3. The matrix $V_r D_r^{-1} U_r^T$ is an approximate pseudoinverse A^+ of the matrix A.

We conclude by showing that $AA^+ \approx I$.

```
[> evalm(A&*Aplus);
```

$$\begin{bmatrix} 1.000000000 & 0. & 0. \\ 0. & 1.000000000 & 0. \\ 0. & 0. & .9999999999 \end{bmatrix}$$

As we can see, the product AA^+ is equal to the 3×3 identity matrix. ▲

EXAMPLE 9.9 A Reduced Singular Value Decomposition of a 3 x 4 Matrix

Use *Maple* to find the reduced singular value decomposition of the matrix

$$A = \begin{bmatrix} 1 & 2 & 0 & 1 & 4 \\ 0 & 3 & 0 & 2 & 0 \\ 1 & 1 & 0 & 0 & 1 \end{bmatrix}$$

Solution. We use the **Svd** function to find the matrices U_r, D_r, and V_r.

```
[> with(linalg):
[> A:=matrix([[1,2,0,1,4],[0,3,0,2,0],[1,1,0,0,1]]):
[> rank(A);
```

$$3$$

Therefore $r = 3$, so $A^+ = V_3 D_3^{-1} U_3^T$.

```
[> svd:=evalf(Svd(A,U,V)):
[> Dm:=diag(svd[1],svd[2],svd[3]):
[> LinearAlgebra[Dimension](Matrix(U));
```

$$3, 3$$

```
[> LinearAlgebra[Dimension](Matrix(V));
```

$$5, 5$$

Therefore $U_3 = U$ and V_3 must be the submatrix W of V consisting of the first three column of V.

```
[> Vr:=evalf(submatrix(V,1..5,1..3),5);
```

$$Vr := \begin{bmatrix} -.20756 & -.19696 & .79039 \\ -.62713 & .55337 & .29141 \\ -.11120\ 10^{-16} & .14640\ 10^{-15} & .32507\ 10^{-15} \\ -.33150 & .44715 & -.45291 \\ -.67360 & -.67457 & -.29196 \end{bmatrix}$$

Next we construct D_r^{-1}.

```
[> invDr:=evalf(inverse(Dm),5);
```

$$invDr := \begin{bmatrix} .18606 & -0. & 0. \\ -0. & .34451 & -0. \\ 0. & -0. & 1.2072 \end{bmatrix}$$

It remains to calculate U_r^T.

```
[> UrT:=evalf(transpose(U),5);
```

$$UrT = \begin{bmatrix} -.83495 & -.47340 & -.28062 \\ -.46211 & .88002 & -.10961 \\ -.29884 & -.038158 & .95354 \end{bmatrix}$$

We can now form A^+ by multiplying V_r, the inverse of D_r^{-1}, and U_r^T.

```
[> Aplus:=evalf(evalm(Vr&*invDr&*UrT),5);
```

$$Aplus := \begin{bmatrix} -.22154 & -.77841e-1 & .92810 \\ -.095805 & .20959 & .34730 \\ -.13885\ 10^{-15} & .30390\ 10^{-16} & .36924\ 10^{-15} \\ .14370 & .18563 & -.52093 \\ .31736 & -.13174 & -.27544 \end{bmatrix}$$

We now multiply A and A^+.

```
[> evalm(A&*Aplus);
```

$$\begin{bmatrix} .999990 & .9\,10^{-5} & .00001 \\ -.000015 & 1.00003 & .00004 \\ .000015 & .9\,10^{-5} & .99996 \end{bmatrix}$$

As we can see, the product AA^+ is approximately equal to I_3. ▲

Pseudoinverses and least-squares solutions

Pseudoinverses provide a method for finding least-squares solutions for linear systems.

THEOREM 9.6 (Pseudoinverse theorem) *For any linear system* $A\mathbf{x} = \mathbf{b}$, *the vector* $\widehat{\mathbf{x}} = A^+\mathbf{b}$ *is a least-squares solution of the system.*

Proof. By definition, $A^+ = V_r D_r^{-1} U_r^T$. Therefore $A^+\mathbf{b} = V_r D_r^{-1} U_r^T \mathbf{b}$. If we multiply both sides of this equation by A, we get

$$\begin{aligned} AA^+\mathbf{b} &= AV_r D_r^{-1} U_r^T \mathbf{b} \\ &= (U_r D_r V_r^T)(V_r D_r^{-1} U_r^T)\mathbf{b} \\ &= U_r U_r^T \mathbf{b} \end{aligned}$$

since $\dot{V}_r^T V_r = I_r$. Moreover,

$$U_r U_r^T \mathbf{b} = \begin{bmatrix} \mathbf{u}_1 & \cdots & \mathbf{u}_r \end{bmatrix} \begin{bmatrix} \mathbf{u}_1^T \\ \vdots \\ \mathbf{u}_r^T \end{bmatrix} \mathbf{b} = \begin{bmatrix} \mathbf{u}_1 & \cdots & \mathbf{u}_r \end{bmatrix} \begin{bmatrix} \mathbf{u}_1^T \mathbf{b} \\ \vdots \\ \mathbf{u}_r^T \mathbf{b} \end{bmatrix}$$

$$= \begin{bmatrix} \mathbf{u}_1 & \cdots & \mathbf{u}_r \end{bmatrix} \begin{bmatrix} \mathbf{b} \cdot \mathbf{u}_1 \\ \vdots \\ \mathbf{b} \cdot \mathbf{u}_r \end{bmatrix} = (\mathbf{b} \cdot \mathbf{u}_1)\,\mathbf{u}_1 + \cdots + (\mathbf{b} \cdot \mathbf{u_r})\,\mathbf{u_r}$$

Since $(\mathbf{b} \cdot \mathbf{u}_1)\,\mathbf{u}_1 + \cdots + (\mathbf{b} \cdot \mathbf{u_r})\,\mathbf{u_r} = \text{proj}(\mathbf{b} \to \text{span}\,\{\mathbf{u}_1, \ldots, \mathbf{u}_r\})$, it follows from Theorem 8.38 that $\widehat{\mathbf{x}}$ is the least-squares solution of the system $A\mathbf{x} = \mathbf{b}$. ■

EXAMPLE 9.10 A Least-Squares Solution

Use the pseudoinverse to find a least-squares solution for the linear system

$$\begin{cases} 2x + 3y + z = 4 \\ x - y - z = 1 \\ x + y + 3z = 3 \\ 2x - 5y + z = 6 \end{cases}$$

Solution. The matrix equation $A\mathbf{x} = \mathbf{b}$ of the given linear system is

$$\begin{bmatrix} 2 & 3 & 1 \\ 1 & -1 & -1 \\ 1 & 1 & 3 \\ 2 & -5 & 1 \end{bmatrix} \begin{bmatrix} x \\ y \\ z \end{bmatrix} = \begin{bmatrix} 4 \\ 1 \\ 3 \\ 6 \end{bmatrix}$$

By Theorem 9.6, the least-squares solution of $A\mathbf{x} = \mathbf{b}$ is $A^{+}\mathbf{b}$.

```
[> with(linalg):
[> A:=matrix([[2,3,1],[1,-1,-1],[1,1,3],[2,-5,1]]):
[> b:=vector([4,1,3,6]):
[> svd:=evalf(Svd(A,U,V)):
[> Dm:=diag(svd[1],svd[2],svd[3]):
[> W:=submatrix(transpose(U),1..3,1..4);
```

```
[> Aplus:=evalm(V&*inverse(Dm)&*W):
```

$$Aplus := \begin{bmatrix} .3011869436 & .2195845697 & -.0682492582 & .1231454006 \\ .1216617211 & .00741839760 & .00445103858 & -.1275964392 \\ -.0875370919 & -.1943620177 & .2833827893 & .04302670622 \end{bmatrix}$$

Next we calculate $\mathbf{x} = A^{+}\mathbf{b}$.

```
[> x:=evalm(A&*b);
```

$$[1.958456973, -.2581602375, .5637982199]$$

Therefore, the vector $A\widehat{\mathbf{x}}$ is

$$\begin{bmatrix} 2 & 3 & 1 \\ 1 & -1 & -1 \\ 1 & 1 & 3 \\ 2 & -5 & 1 \end{bmatrix} \begin{bmatrix} 1.958456973 \\ -.2581602375 \\ .5637982199 \end{bmatrix} = \begin{bmatrix} 3.7062 \\ 1.6528 \\ 3.3917 \\ 5.7715 \end{bmatrix}$$

The distance between $A\widehat{\mathbf{x}}$ and $\mathbf{b}$ is

$$\begin{aligned} \|A\widehat{\mathbf{x}} - \mathbf{b}\|^2 &= (3.7062 - 4)^2 + (1.6528 - 1)^2 + (3.3917 - 3)^2 + (5.7715 - 6)^2 \\ &= 0.71811 \end{aligned}$$

Hence the distance between the vectors $A\widehat{\mathbf{x}}$ and $\mathbf{b}$ is $\sqrt{0.71811} = 0.84741$ units. ▲

EXAMPLE 9.11 Least-squares Solution Equals Exact Solution

Use *Maple* to show that the least-squares and exact solutions of the linear equation

$$\begin{bmatrix} 1 & 2 & 1 \\ 2 & 4 & 1 \\ 3 & -3 & 1 \end{bmatrix} \begin{bmatrix} x \\ y \\ z \end{bmatrix} = \begin{bmatrix} 4 \\ 5 \\ 6 \end{bmatrix}$$

coincide.

Solution. We will show that the exact solution produced by the **solve** function and the least-squares solution produced using the pseudoinverse are equal.

```
[> with(linalg):
[> A:=matrix([[1,2,1],[2,4,1],[3,-3,1]]):
[> evalm(A&*vector([x,y,z]));
```

$$[x + 2y + z, 2x + 4y + z, 3x - 3y + z]$$

We therefore solve the linear system

$$\begin{cases} x + 2y + z = 4 \\ 2x + 4y + z = 5 \\ 3x - 3y + z = 6 \end{cases}$$

```
[> solve({x+2*y+z=4,2*x+4*y+z=5,3*x-3*y+z=6},{x,y,z});
```

$$\{x = 1, z = 3, y = 0\}$$

This shows that $x = 1$, $y = 0$, $z = 3$ is the exact solution of the equation. We now use the pseudoinverse to approximate this solution.

```
[> svd:=evalf(Svd(A,U,V)):
[> Dm:=diag(svd[1],svd[2],svd[3]):
[> Aplus:=evalm(V &*inverse(Dm)&*transpose(U)):
[> evalm(Aplus&*vector([4,5,6]));
```

$$[1.000000001, 0., 2.999999999]$$

The singular value decomposition method thus tells us, as expected, that (x, y, z) is approximately equal to $(1, 0, 3)$. ▲

The real difference between the two methods shows up in examples that have no exact solution. If we ask *Maple* to solve the linear system

$$\begin{bmatrix} 1 & 2 & 1 \\ 2 & 4 & 2 \\ 3 & -3 & 1 \end{bmatrix} \begin{bmatrix} x \\ y \\ z \end{bmatrix} = \begin{bmatrix} 4 \\ 5 \\ 6 \end{bmatrix}$$

for example, then the command

```
solve({x+2*y+z=6,2*x+4*y+2*z=5,3*x-3*y+z=6},{x,y,z})
```

produces the empty output. We can conclude that the vector (4, 5, 6) does not belong to the column space of A. However, the command `evalm(Aplus&*vector([4,5,6]))` produces the output [1.727272727, .0509090910, .9709090908]. This means that of all the vectors in Col A, the vector

$$\begin{bmatrix} 1 & 2 & 1 \\ 2 & 4 & 2 \\ 3 & -3 & 1 \end{bmatrix} \begin{bmatrix} 1.727272727 \\ .0509090910 \\ .9709090908 \end{bmatrix}$$

is closest to the vector (4, 5, 6) in the Euclidean distance.

Pseudoinverses and orthogonal projections

In many practical cases, we have an easy formula for calculating pseudoinverses. This formula, in turn, provides us with yet another method of calculating an orthogonal projection of vector onto column spaces.

THEOREM 9.7 *If the real matrix A^TA is invertible, then the pseudoinverse of A is equal to the matrix $(A^TA)^{-1}A^T$.*

Proof. By Theorem 9.6, the vector $\widehat{\mathbf{x}} = A^+\mathbf{b}$ is the least-squares solution of the linear system $A\mathbf{x} = \mathbf{b}$. On the other hand, Theorem 8.32 tells us that $\widehat{\mathbf{x}}$ is a least-squares solution of $A\mathbf{x} = \mathbf{b}$ if and only if $\widehat{\mathbf{x}}$ is a solution of the normal equation $A^TA\mathbf{x} = A^T\mathbf{b}$. This means that if A^TA is invertible, then $\widehat{\mathbf{x}} = A^+\mathbf{b}$ if and only if $\widehat{\mathbf{x}} = (A^TA)^{-1}A^T\mathbf{b}$. Since this holds for all $\mathbf{b}$, we can conclude that $A^+ = (A^TA)^{-1}A^T$. ■

In Table 2, we summarize the process for calculating the pseudoinverse A^+ for a matrix A for which A^TA is invertible, described in Theorem 9.7.

TABLE 2 FINDING EXACT PSEUDOINVERSES.

Step 1. For any matrix A for which A^TA is invertible, form the matrix A^TA.
Step 2. Use the inverse $(A^TA)^{-1}$ of A^TA to form the matrix product $(A^TA)^{-1}A^T$.
Step 3. The matrix $(A^TA)^{-1}A^T$ is the pseudoinverse A^+ of the matrix A.

COROLLARY 9.8 *If the matrix $A^T A$ is invertible, then $AA^+\mathbf{b} = \text{proj}(\mathbf{b} \to \text{Col } A)$.*

Proof. Suppose A is an $m \times n$ real matrix, that $A^T A$ is invertible, and that $A\mathbf{x} = \mathbf{b}$ is a linear system. By Theorem 8.38, the vector $\text{proj}(\mathbf{b} \to \text{Col } A)$ is a best approximation of the vector $\mathbf{b} \in \mathbb{R}^m$ by vectors in Col A. Moreover, Theorem 8.32 tells us that $\text{proj}(\mathbf{b} \to \text{Col } A) = A\widehat{\mathbf{x}}$ if $\widehat{\mathbf{x}} = (A^T A)^{-1} A^T \mathbf{b}$, and Theorem 9.7 tells us that if $A^T A$ is invertible, then $(A^T A)^{-1} A^T = A^+$. Hence

$$\text{proj}(\mathbf{b} \to \text{Col } A) = A\widehat{\mathbf{x}} = A(A^T A)^{-1} A^T \mathbf{b} = AA^+\mathbf{b}$$

is the projection of $\mathbf{b}$ onto the columns space of A. ■

EXAMPLE 9.12 Pseudoinverses and Orthogonal Projections

Calculate the vector $AA^+\mathbf{b}$ for the linear system

$$A\mathbf{x} = \begin{bmatrix} 1 & 0 \\ 3 & 1 \\ 0 & 0 \end{bmatrix} \begin{bmatrix} x \\ y \end{bmatrix} = \begin{bmatrix} 1 \\ 2 \\ 3 \end{bmatrix}$$

and use Theorem 8.10 to show that $A\widehat{\mathbf{x}}$ is the projection of $\mathbf{b}$ onto the column space of A.

Solution. Since the determinant of the matrix

$$A^T A = \begin{bmatrix} 1 & 0 \\ 3 & 1 \\ 0 & 0 \end{bmatrix}^T \begin{bmatrix} 1 & 0 \\ 3 & 1 \\ 0 & 0 \end{bmatrix} = \begin{bmatrix} 10 & 3 \\ 3 & 1 \end{bmatrix}$$

is nonzero, the matrix $A^T A$ is invertible. By Theorem 9.7 that $A^+ = (A^T A)^{-1} A^T$. Therefore

$$AA^+\mathbf{b} = \begin{bmatrix} 1 & 0 \\ 3 & 1 \\ 0 & 0 \end{bmatrix} \left(\begin{bmatrix} 1 & 0 \\ 3 & 1 \\ 0 & 0 \end{bmatrix}^T \begin{bmatrix} 1 & 0 \\ 3 & 1 \\ 0 & 0 \end{bmatrix} \right)^{-1} \begin{bmatrix} 1 & 0 \\ 3 & 1 \\ 0 & 0 \end{bmatrix}^T \begin{bmatrix} 1 \\ 2 \\ 3 \end{bmatrix} = \begin{bmatrix} 1 \\ 2 \\ 0 \end{bmatrix}$$

Let us verify that if $A = QR$ is a QR decomposition of A, then

$$AA^+\mathbf{b} = QQ^T\mathbf{b}$$

```
[> with(LinearAlgebra):
[> A:=Matrix([[1,0],[3,1],[0,0]]):
[> B:=Transpose(A).A:
[> Determinant(B);
                                  1
```

```
[> Q,R:=QRDecomposition(A);
```

$$Q, R := \begin{bmatrix} \frac{1}{10}\sqrt{10} & -\frac{3}{10}\sqrt{10} \\ \frac{3}{10}\sqrt{10} & \frac{1}{10}\sqrt{10} \\ 0 & 0 \end{bmatrix}, \begin{bmatrix} \sqrt{10} & \frac{3}{10}\sqrt{10} \\ 0 & \frac{1}{10}\sqrt{10} \end{bmatrix}$$

```
[> b:=<1,2,3> :
[> proj:=Q.Transpose(Q).b;
```

$$\begin{bmatrix} 1 \\ 2 \\ 0 \end{bmatrix}$$

As we can see, the two methods produce the same vector. ▲

Pseudoinverses and large matrices

Pseudoinverses also provide an efficient method for approximating the inverses of large matrices. Here is an example. For the purpose of this example, we consider a 30×30 invertible matrix to be large. Since every computer has a smallest and a largest number it can store and process, the ability to manipulate large matrices on a particular computer is obviously hardware dependent.

EXAMPLE 9.13 The Inverse of a Large Matrix

Use the singular value decomposition to approximate the inverse of a random 30×30 invertible matrix.

Solution. We begin by generating a random matrix 30×30 matrix of rank 30.

```
[> with(linalg):
[> A:=randmatrix(30,30):
[> rank(A);
```

$$30$$

Next we construct a pseudoinverse for A.

```
[> svd:=evalf(Svd(A,U,V)):
[> Dm:=diag(seq(svd[i],i=1..30)):
[> Aplus:=evalm(V&*inverse(Dm)&*transpose(U)):
```

Since the matrices A and A^+ are too large to fit on small computer screens, we select some sample entries of the matrix AA^+ and convince ourselves that it is approximately equal to I_{100}.

```
[> P:=evalm(A&*Aplus):
[> P[1,1];
```

$$.999999996$$

```
[> P[25,25];
```

$$.999999999$$

```
[> P[30,29];
```

$$.2\ 10^{-9}$$

As we can see, the two chosen diagonal entries are approximately equal to 1, and the chosen nondiagonal entry is approximately equal to 0. If we were to work our way through the remaining 897 entries, we would find that these estimates hold for all entries. ▲

EXERCISES 9.4

1. (*Maple 6*) Verify that the pseudoinverse of the matrix

$$A = \begin{bmatrix} 1 & 2 \\ 3 & 4 \\ 5 & 6 \end{bmatrix}$$

is equal to the matrix $(A^T)^{-1}A^T$. Explain why an analogous formula fails for A^T.

2. (*Maple 6*) It can be proved that for any given matrix A, the pseudoinverse A^+ is the unique matrix B for which $ABA = A$, $BAB = B$, $(AB)^T = AB$, and $(BA)^T = BA$. Use these conditions to prove that $(A^T)^+ = (A^+)^T$ and that $A^{++} = A$. Show that these equations are approximately true for the matrix A in Exercise 1.

3. (*Maple 6*) Use the definition of A^+ to find a least-squares solution of the linear system $A\mathbf{x} = \mathbf{b}$ determined by

$$A = \begin{bmatrix} 1 & 2 & 0 & 1 & 4. \\ 0 & 3 & 0 & 2 & 0 \\ 1 & 1 & 0 & 0 & 1 \end{bmatrix} \quad \text{and} \quad \mathbf{b} = \begin{bmatrix} 1 \\ 2 \\ 3 \end{bmatrix}$$

4. (*Maple 6*) Generate a random 5×6 matrix and find a singular value decomposition for the matrix.

5. (*Maple 6*) Calculate the pseudoinverse A^+ for the following matrices and test whether A^+ is an actual inverse.

a. $\begin{bmatrix} -1 & 1 & 0 \\ 1 & -1 & 2 \\ -2 & 1 & 0 \end{bmatrix}$ b. $\begin{bmatrix} -1 & 1 & -2 \\ 1 & -1 & 2 \\ -2 & 1 & 0 \end{bmatrix}$ c. $\begin{bmatrix} 3 & 2 \\ 2 & 1 \\ 7 & 0 \end{bmatrix}$ d. $\begin{bmatrix} -2 & 2 & 0 & 0 \\ -2 & -1 & 1 & 1 \\ -1 & 0 & 1 & 1 \end{bmatrix}$

6. (*Maple 6*) Generate a large random invertible matrix A, and approximate the inverse of A with the pseudoinverse A^+. Experiment with your computer to assess how large a matrix it can process in this way in a reasonable time.

7. (*Maple 6*) Find a singular value decomposition UDV^T for the inverse of an invertible 3×3 matrix A and show that $A^{-1} = VD^{-1}U^T$.

Effective Rank

In many applications of linear algebra it is important to determine the rank of a matrix. One way of finding the rank is to use Gaussian elimination and reduce a matrix to its row echelon form. The number of nonzero rows of the echelon matrix is the rank of the matrix. If the matrix is large, the computer version of Gaussian elimination may fail to produce the rank of the matrix.

Accumulated round-off errors may introduce nonzero entries in what should be zero rows. A more reliable method is to find the singular value decomposition of the matrix and then discard small singular values. Let us refer to the cutoff point below which singular values are to be discarded as the rank ***tolerance***. The rank obtained in this way is known as an ***effective rank*** of a matrix.

EXAMPLE 9.14 An Effective Rank of a Matrix

Use *Maple* to calculate an effective rank of the matrix

$$A = \begin{bmatrix} 2.0 & 0 & 0 & 0 \\ 0 & 1.8 & 0 & 0 \\ 0 & 0 & 1.7 & 0 \end{bmatrix}$$

obtained by assigning to the matrix A a rank tolerance of 1.75.

Solution. Since A is in row echelon form and has three nonzero rows, its exact rank is 3. We can also tell that its singular values are the nonzero entries 2.0, 1.8, and 1.7. If we decide to discard all singular values smaller than 1.75, the effective rank of the matrix is 2. ▲

EXERCISES 9.5

1. Construct a 4×4 matrix with four nonzero singular values and specify a tolerance that eliminates all four singular values. Explain your reasoning.
2. Construct a 4×4 matrix with four nonzero singular values and specify a tolerance that retains all singular values. Explain your reasoning.
3. (*Maple 6*) Construct a 6×6 matrix whose two smallest singular values are $.8 \times 10^{-6}$ and $.9 \times 10^{-6}$, whose largest singular value is 5, and whose effective rank is 2.

Image Compression

In this section, we show how the singular value decomposition of a matrix can be used to compress images. We will use the result proved in Exercise 9.2.6, which tells us that if A is an $m \times n$ real matrix of rank r and

$$A = [\mathbf{u}_1 \ \cdots \ \mathbf{u}_m] \operatorname{diag}(\sigma_1, \ldots, \sigma_r, 0, \ldots, 0)[\mathbf{v}_1 \ \cdots \ \mathbf{v}_n]^T$$

is a singular value decomposition of A, with positive singular values $\sigma_1, \ldots, \sigma_r$, then

$$A = \sigma_1 \mathbf{u}_1 \mathbf{v}_1^T + \cdots + \sigma_r \mathbf{u}_r \mathbf{v}_r^T$$

We call this the ***Kronecker product expansion*** of A.

Let A be the real (floating-point) 560×560 matrix representing Image 1 in Table 3.

Each entry of the experimental image A was a real number between 0 and 255 and represented a particular shade of gray of one of the pixels of A. The matrix was sufficiently random to have full rank. Table 4 lists the first 70 singular values of A.

Based on an inspection of the reproductive quality of different compressions of Image 1 in Table 3, it turned out that it was reasonable to assign an effective rank of 70 to the matrix A. The rank tolerance was chosen to be 1.80. The processing of the images and the calculations of the singular values were carried out using the Image Processing toolbox of Matlab 5.

In terms of the Kronecker product expansion $A = \sigma_1 \mathbf{u}_1 \mathbf{v}_1^T + \cdots + \sigma_{70} \mathbf{u}_{70} \mathbf{v}_{70}^T$, we refer to the sum of the form

$$A_k = \sigma_1 \mathbf{u}_1 \mathbf{v}_1^T + \cdots + \sigma_k \mathbf{u}_k \mathbf{v}_k^T \qquad (k \leq 70)$$

as the kth compression of A. The images in Table 3 represent the compressions of A listed in Table 5.

Let us compare the number of values that need to be stored to represent the different images in Table 3. In general, an $m \times n$ matrix A requires mn values to be stored. The kth compression A_k of A, on the other hand, requires $k\,(m + n + 1)$ values to be stored. For each term of the sum A_k, we require m values for the left-singular vector $\mathbf{u}$, then n values for the right-singular vector $\mathbf{v}$, and one value for the associated singular value σ. Since there are k term, the total number of values to be stored is $k(m + n + 1)$. As long as

$$k(m + n + 1) < mn$$

TABLE 3 IMAGE COMPRESSIONS.

Image 1: $A = UDV^T$

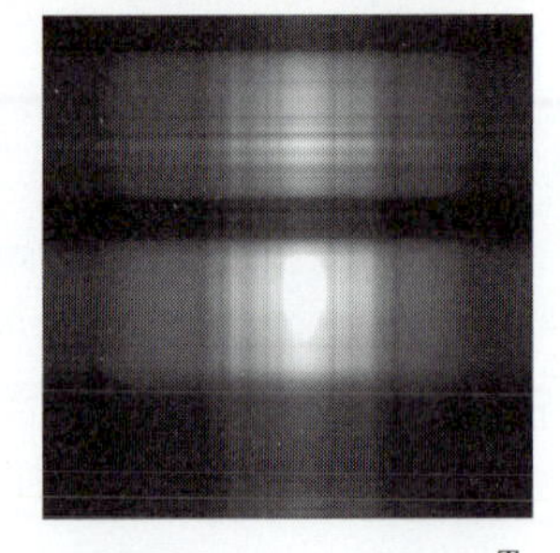

Image 2: $A_1 = UD_1V^T$

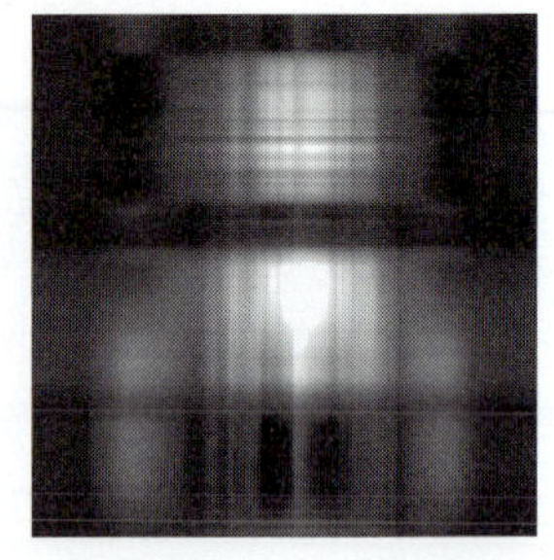

Image 3: $A_2 = UD_2V^T$

Image 4: $A_4 = UD_4V^T$

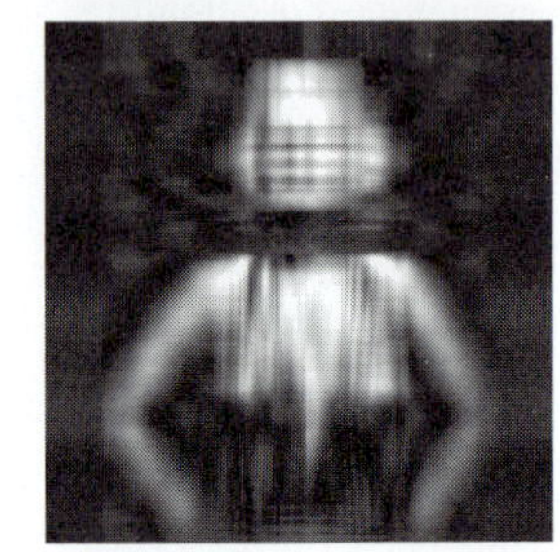

Image 5: $A_6 = UD_6V^T$

Image 6: $A_8 = UD_8V^T$

Image 7: $A_{50} = UD_{50}V^T$

Image 8: $A_{60} = UD_{60}V^T$

Image 9: $A_{70} = UD_{70}V^T$

TABLE 4 FIRST 70 SINGULAR VALUES OF A.

189.6168	48.2960	45.8549	36.8452	30.1904	20.2114	19.2047
15.7902	14.7003	14.3578	13.7295	12.6482	12.1814	10.8696
10.1879	9.9049	9.4828	8.6943	7.7952	7.4656	7.0791
6.8493	6.4498	6.1124	5.9161	5.6835	5.2712	4.9123
4.7572	4.7238	4.4918	4.3305	4.2279	4.0349	3.9554
3.8116	3.7402	3.6697	3.5925	3.4895	3.4430	3.2883
3.1089	3.0768	2.9760	2.9237	2.8936	2.8367	2.7465
2.7160	2.6643	2.6402	2.6144	2.4884	2.4598	2.4206
2.3615	2.3400	2.2890	2.2507	2.2162	2.1700	2.1543
2.1145	2.0697	2.0273	2.0006	1.9647	1.9366	1.8825

TABLE 5 KRONECKER PRODUCT EXPANSIONS.

$A = \sigma_1 \mathbf{u}_1 \mathbf{v}_1^T + \cdots + \sigma_{560} \mathbf{u}_{560} \mathbf{v}_{560}^T$
$A_1 = \sigma_1 \mathbf{u}_1 \mathbf{v}_1^T$
$A_2 = \sigma_1 \mathbf{u}_1 \mathbf{v}_1^T + \sigma_2 \mathbf{u}_2 \mathbf{v}_2^T$
$A_4 = \sigma_1 \mathbf{u}_1 \mathbf{v}_1^T + \sigma_2 \mathbf{u}_2 \mathbf{v}_2^T + \sigma_3 \mathbf{u}_3 \mathbf{v}_3^T + \sigma_4 \mathbf{u}_4 \mathbf{v}_4^T$
$A_6 = \sigma_1 \mathbf{u}_1 \mathbf{v}_1^T + \cdots + \sigma_6 \mathbf{u}_6 \mathbf{v}_6^T$
$A_8 = \sigma_1 \mathbf{u}_1 \mathbf{v}_1^T + \cdots + \sigma_8 \mathbf{u}_8 \mathbf{v}_8^T$
$A_{50} = \sigma_1 \mathbf{u}_1 \mathbf{v}_1^T + \cdots + \sigma_{50} \mathbf{u}_{50} \mathbf{v}_{50}^T$
$A_{60} = \sigma_1 \mathbf{u}_1 \mathbf{v}_1^T + \cdots + \sigma_{60} \mathbf{u}_{60} \mathbf{v}_{60}^T$
$A_{70} = \sigma_1 \mathbf{u}_1 \mathbf{v}_1^T + \cdots + \sigma_{70} \mathbf{u}_{70} \mathbf{v}_{70}^T$

the singular value decomposition provides us with a storage gain. If $m = n = 560$ and $k = 70$, then

$$k(m + n + 1) = 70(560 + 560 + 1) = 78\,470$$

and

$$mn = 560 \times 560 = 313\,600$$

As we can see, the storage gain is almost 75%, without a significant loss of image detail. The quantitative error involved in compressing an image A to its kth singular-value approximation can be measured in the matrix 2-norm using the remarkable fact that

$$\|A - A_k\|_2 = \sigma_{k+1}$$

EXERCISES 9.6

1. (*Maple 6*) Find the singular value decomposition UDV^T of the matrix

$$A = \begin{bmatrix} 255 & 0 & 255 \\ 0 & 255 & 0 \\ 255 & 0 & 255 \end{bmatrix}$$

Calculate the Kronecker product expansions $A_1 = \sigma_1 \mathbf{u}_1 \mathbf{v}_1^T$ and $A_2 = \sigma_1 \mathbf{u}_1 \mathbf{v}_1^T + \sigma_2 \mathbf{u}_2 \mathbf{v}_2^T$ and show that $\|A - A_1\|_2 = \sigma_2$. Suppose that an entry of 0 denotes a black square and an entry of 255 a white square, draw a graphic representation of the matrices A, A_1 and $\sigma_2 \mathbf{u}_2 \mathbf{v}_2^T$. Explain the geometric effect of superimposing the squares represented by $\sigma_1 \mathbf{u}_1 \mathbf{v}_1^T$ and $\sigma_2 \mathbf{u}_2 \mathbf{v}_2^T$.

2. (*Maple 6*) Repeat Exercise 1 using the matrix

$$A = \begin{bmatrix} 255 & 0 & 255 & 0 \\ 0 & 255 & 0 & 255 \\ 255 & 0 & 255 & 0 \\ 0 & 255 & 0 & 255 \end{bmatrix}$$

and calculate A_1, A_2, and A_3. Interpret the possible matrix sums geometrically.

3. (*Maple 6*) Calculate the singular values of the companion matrix A of the polynomial $p(t) = (t-2)^3 (t+1)^4$ and find the Kronecker product expansions A_1, A_2, and A_3 of A. Show in each case that

$$\|A - A_k\|_2 = \sigma_{k+1}$$

Orthogonal Matrices

Orthogonal matrices play an important role in computer graphics. They are the matrices that preserve length and angles relative to orthonormal bases. They are also the matrices whose inverses are easy to compute. Matrix transposition involves considerably fewer arithmetic operations than matrix inversion.

As we saw in Chapter 2, animation of graphic objects can be achieved through the multiplication of matrices. Usually the operations involved must be invertible. Often they also need to preserve length and angles. Since orthogonal matrices have both of these properties, they are at the center of computer-based graphic applications. Unfortunately, round-off errors often destroy the orthogonality of the matrices required in the process. For example, the numerical approximation

$$A = \begin{bmatrix} .40825 & .7071 & -.57735 \\ -.40825 & .7071 & .57735 \\ .8165 & 2.122 \times 10^{-6} & .57735 \end{bmatrix}$$

of the orthogonal matrix

$$Q = \begin{bmatrix} \frac{1}{\sqrt{6}} & \frac{1}{\sqrt{2}} & -\frac{1}{\sqrt{3}} \\ -\frac{1}{\sqrt{6}} & \frac{1}{\sqrt{2}} & \frac{1}{\sqrt{3}} \\ \frac{2}{\sqrt{6}} & 0 & \frac{1}{\sqrt{3}} \end{bmatrix}$$

is not orthogonal since the product

$$A^T A = \begin{bmatrix} .99999 & -1.0675 \times 10^{-5} & 4.603 \times 10^{-6} \\ -1.0675 \times 10^{-5} & .99999 & -1.602 \times 10^{-6} \\ 4.603 \times 10^{-6} & -1.602 \times 10^{-6} & 1.0 \end{bmatrix}$$

is not the identity matrix I_3. Therefore, it is often necessary to replace a matrix that should be orthogonal by the orthogonal matrix closet to it. The Frobenius norm is used to measure the distance between these matrices.

The calculation of orthogonal approximations is based on two facts: (1) the singular value decomposition UDV^T involves two orthogonal matrices U and V, and (2) the singular values of orthogonal matrices are 1. If A is an orthogonal matrix, then

$$A = UDV^T = UIV^T = UV^T$$

On the other hand, if A is a square matrix that is not orthogonal, then the matrix $B = UV^T$ is the orthogonal matrix closest to it. Thus if $\|A - B\|_F$ is sufficiently small, it may be possible to use B in place of A.

EXAMPLE 9.15 A Best Orthogonal Approximation

Find the best orthogonal approximation of the matrix

$$A = \begin{bmatrix} 0.0010 & -0.9990 \\ 1.0001 & 0.0003 \end{bmatrix}$$

Solution. First we find the singular value decomposition of A.

```
[> with(linalg):  Digits:=5:
[> A:=matrix([[0.001,-0.999],[1.0001,0.0003]]):
[> svd:=evalf(Svd(A,U,V)):
[> B:=evalm(matrix(U)&*transpose(V));
```

$$\begin{bmatrix} .00064 & -1.0000 \\ 1.0000 & .00064 \end{bmatrix}$$

We note that BB^T is approximately I_2. Hence B is approximately orthogonal. Moreover, the calculation

```
[> norm(evalm(A-B),frobenius);
```

$$.11204$$

shows that A and B are approximately equal in the frobenius norm. ▲

EXERCISES 9.7

1. (*Maple 6*) Find the orthogonal matrix that is closest to the matrix

$$A = \begin{bmatrix} .7075 & -.7071 \\ .7072 & .7079 \end{bmatrix}$$

in the Frobenius norm.

2. (*Maple 6*) Use orthogonal approximations of the matrices involved to move the triangle determined by the vertices

$$\begin{bmatrix} 0 \\ 0 \end{bmatrix}, \begin{bmatrix} 0 \\ 1 \end{bmatrix}, \begin{bmatrix} 2 \\ 0 \end{bmatrix}$$

through the following locations.

   a. Rotate the triangle counterclockwise through .3658 radians, using the best possible orthogonal matrix corresponding to this rotation.

   b. Reflect the resulting triangle about the line $y = x$, using the best possible orthogonal matrix corresponding to this reflection.

   c. Rotate the triangle counterclockwise through 1.7523 radians, using the best possible orthogonal matrix corresponding to this rotation.

   d. Reverse the three steps using the closest orthogonal matrices to the matrices representing the inverses of the matrices representing the operations in (a), (b), and (c).

   e. Discuss the difference between the triangles before and after the transformations.

Condition Numbers

A mathematical process is considered to be ill-conditioned if a relatively small change in inputs produces a relatively large change in outputs. Numerical measures are used to decide whether a process is or is not to be considered ill-conditioned. In linear algebra, this idea is applied mainly to the study of the effect of round-off errors on the properties of matrices.

If A is an invertible real $n \times n$ matrix, then there exists a unique $n \times n$ matrix A^{-1} such that $AA^{-1} = A^{-1}A = I_n$. If A is large, computer calculations of A^{-1} may produce a matrix that is actually singular due to round-off errors. The fact that this can happen leads to the following classification of invertible matrices using condition numbers. The ***condition number*** of a matrix is a number used to measure the extent to which an invertible matrix is sensitive to round-off errors.

DEFINITION 9.2 *Relative to a matrix norm $\|A\|$, the **condition number** $\operatorname{cond} A$ of an invertible matrix A is $\|A\| \, \|A^{-1}\|$.*

We use this definition to divide invertible matrices into three kinds. An invertible matrix is ***perfectly conditioned*** if cond $A = 1$, ***well-conditioned*** if cond A is slightly bigger than 1, and ***ill-conditioned*** if cond A is much larger than 1. The following property of condition number is as expected.

LEMMA 9.9 *The condition number of an identity matrix is* 1.

Proof. By definition, $\text{cond}\, I = \|I\| \left\|I^{-1}\right\| = \|I\| \, \|I\| = 1 \times 1 = 1.$ ■

The next theorem provides a lower bound on condition numbers.

THEOREM 9.10 *The condition number* cond A *of any invertible matrix is greater than or equal to* 1.

Proof. If follows from the definition of condition numbers and the properties of matrix norms that

$$\text{cond}\, A = \|A\| \left\|A^{-1}\right\| \geq \left\|AA^{-1}\right\| \geq \|I\| = 1$$

Hence 1 is a lower bound for condition numbers. ■

EXAMPLE 9.16 A Condition Number in the Infinity-Norm

Use the infinity-norm to calculate the condition number of the matrix

$$A = \begin{bmatrix} 2 & -1 & 0 \\ 2 & -4 & -1 \\ -1 & 0 & 2 \end{bmatrix}$$

Solution. First we calculate A^{-1}.

```
[> A:=matrix([[2,-1,0],[2,-4,-1],[-1,0,2]]):
[> B:=linalg[inverse](A);
```

$$B := \begin{bmatrix} \frac{8}{13} & \frac{-2}{13} & \frac{-1}{13} \\ \frac{3}{13} & \frac{-4}{13} & \frac{-2}{13} \\ \frac{4}{13} & \frac{-1}{13} & \frac{6}{13} \end{bmatrix}$$

Since the infinity-norm of a matrix is the maximum of the absolute row sums, we have

$$\begin{aligned} \|A\|_\infty &= \max\{|2| + |-1| + |0|, |2| + |-4| + |-1|, |-1| + |0| + |2|\} \\ &= 7 \end{aligned}$$

and

$$\begin{aligned}\left\|A^{-1}\right\|_{\infty} &= \max\left\{\tfrac{8}{13} + \left|\tfrac{-2}{13}\right| + \left|\tfrac{-1}{13}\right|, \tfrac{3}{13} + \left|\tfrac{-4}{13}\right| + \left|\tfrac{-2}{13}\right|, \tfrac{4}{13} + \left|\tfrac{-1}{13}\right| + \tfrac{6}{13}\right\} \\ &= \tfrac{11}{13}\end{aligned}$$

Therefore,

$$\operatorname{cond} A = \|A\|_{\infty} \left\|A^{-1}\right\|_{\infty} = 7 \times \tfrac{11}{13} \approx 5.9231$$

Depending on the application, we might consider this number to be reasonably small and conclude that A is reasonably well-conditioned. ▲

Some matrices are notoriously ill-conditioned. For example, consider the 3×3 Hilbert matrix

$$A = \begin{bmatrix} 1 & \frac{1}{2} & \frac{1}{3} \\ \frac{1}{2} & \frac{1}{3} & \frac{1}{4} \\ \frac{1}{3} & \frac{1}{4} & \frac{1}{5} \end{bmatrix}$$

whose entries are defined by $a_{ij} = 1/[i + j - 1]$. We show that the condition number of this matrix is quite large.

EXAMPLE 9.17 The Condition Number of a Hilbert Matrix

Find the condition number of the 3×3 Hilbert matrix relative to the one-norm.

Solution. We begin by finding the inverse of A.

```
[> A:=matrix([[1,1/2,1/3],[1/2,1/3,1/4],[1/3,1/4,1/5]]):
[> B:=linalg[inverse](A);
```

$$\begin{bmatrix} 9 & -36 & 30 \\ -36 & 192 & -180 \\ 30 & -180 & 180 \end{bmatrix}$$

The one-norm of a matrix is the maximum absolute column sum of the matrix. Therefore it is

$$\begin{aligned}\|A\|_1 &= \max\left\{1 + \tfrac{1}{2} + \tfrac{1}{3}, \tfrac{1}{2} + \tfrac{1}{3} + \tfrac{1}{4}, \tfrac{1}{3} + \tfrac{1}{4} + \tfrac{1}{5}\right\} \\ &= \tfrac{11}{6}\end{aligned}$$

and

$$\begin{aligned}\left\|A^{-1}\right\|_1 &= \max\{9 + |-36| + 30, |-36| + 192 + |-180|, 30 + |-180| + 180\} \\ &= 408\end{aligned}$$

Hence the condition number cond A of A is $\|A\|_1 \left\|A^{-1}\right\|_1 = \frac{11}{6} \times 408 = 748$. As we can see, this number is quite large. ▲

By adapting Example 9.17, we can easily confirm that the condition numbers of Hilbert matrices increase rapidly as the size of the matrices increases. Large Hilbert matrices are therefore considered to be extremely ill-conditioned.

We might think that if the determinant of a matrix is close to zero, then the matrix is ill-conditioned. However, this is false. Consider the matrix

$$A = \begin{bmatrix} 10^{-6} & 0 \\ 0 & 10^{-6} \end{bmatrix}$$

for which $\det A = 10^{-12} \approx 0$. In the one-norm,

$$\begin{aligned} \operatorname{cond} A &= \|A\|_1 \left\|A^{-1}\right\|_1 \\ &= \max\left\{10^{-6}, 10^{-6}\right\} \times \max\left\{10^{6}, 10^{6}\right\} = 10^{-6} \times 10^{6} = 1 \end{aligned}$$

The matrix A is therefore perfectly conditioned. Thus a small determinant is necessary but not sufficient for a matrix to be ill-conditioned.

The next theorem establishes a connection between singular values and condition numbers. It is often taken as the definition of condition numbers.

THEOREM 9.11 (Condition number theorem) *Let A be any invertible matrix. Then the two-norm condition number of A is* $\max \sigma_i / \min \sigma_i$.

Proof. Suppose that A is an $n \times n$ invertible matrix. Then its eigenvalues are nonzero and so are the eigenvalues of $A^T A$. Let

$$\lambda_1 \geq \lambda_2 \geq \cdots \geq \lambda_n > 0$$

be the eigenvalue of $A^T A$. Then it follows from Theorem 6.10 that the eigenvalues of $\left(A^T A\right)^{-1}$ are

$$0 < \frac{1}{\lambda_1} \leq \frac{1}{\lambda_2} \leq \cdots \leq \frac{1}{\lambda_n}$$

Therefore, $\sigma_1 = \sqrt{\lambda_1} \geq \cdots \geq \sigma_n = \sqrt{\lambda_n} > 0$ are the singular values of A and $0 < \frac{1}{\sigma_1} \leq \cdots \leq \frac{1}{\sigma_n}$ are the singular values of A^{-1}. We know from Chapter 7 that the two-norm $\|A\|_2$ of A is the largest singular value of A and that the two-norm of A^{-1} is the largest singular value of A^{-1}.

By the definition of condition numbers, we therefore have

$$\operatorname{cond} A = \|A\|_2 \left\|A^{-1}\right\|_2 = \frac{\sigma_1}{\sigma_n}$$

This proves the theorem. ■

Note that if a square matrix A has a zero singular value, then the smallest singular value σ_n is 0. In that case, we can define its condition number to be infinity. This means that A is singular. Hence we can tell whether A is invertible by simply looking at the singular value decomposition UDV^T of A. If a diagonal entry of D is 0, then A is not invertible.

EXAMPLE 9.18 The Singular Value Decomposition of a Singular Matrix

Use the singular value decomposition to show that the matrix

$$A = \begin{bmatrix} 2 & -1 & 0 \\ 2 & -1 & 0 \\ -1 & 0 & 2 \end{bmatrix}$$

is singular.

Solution. We use *Maple* to find the singular value decomposition of A.

```
[> with(linalg):
[> A:=matrix([[2,-1,0],[2,-1,0],[-1,0,2]]):
[> svd:=evalf(Svd(A,U,V));
```

$$svd := [3.357814351, 1.930047352, .2166933461\ 10^{-15}])$$

```
[> Dm:=diag(seq(svd[i],i=1..3)):
```

$$\begin{bmatrix} 3.357814351 & 0 & 0 \\ 0 & 1.930047352 & 0 \\ 0 & 0 & .2166933461\ 10^{-15} \end{bmatrix}$$

As we can see, the third row of the diagonal matrix Dm is approximately zero. It is therefore clear from Appendix E that from a computational point of view, the matrix A can be considered to be singular. Hence the condition numbers of A are infinite. This explains the fact that all *Maple* commands of the form `cond(A,#)`, where # is either `1`, `2`, `infinity`, or `frobenius`, yield the message

```
Error, (in linalg[inverse]) singular matrix.
```

Maple confirms that the matrix A is singular. ▲

EXAMPLE 9.19 The Condition Number of the 4 × 4 Identity Matrix

Use *Maple* to calculate the one-, two-, infinity-, and Frobenius condition numbers of the 4×4 identity matrix.

Solution.

```
[> with(LinearAlgebra):
[> A:=IdentityMatrix(4):
[> CondOne:=ConditionNumber(A,1):
[> CondTwo:=ConditionNumber(A,2):
[> CondInfinity:=ConditionNumber(A,infinity):
[> CondFrobenius:=ConditionNumber(A,Frobenius):
[> [CondOne,CondTwo,CondInfinity,CondFrobenius];
```

$$[1, 1, 1, 4]$$

As we can see, three of the four condition numbers yield the values 1, as expected. As these numbers indicate, the identity matrix I_4 is perfectly conditioned. The Frobenius condition number is used for a different purpose. ▲

Next we take the singular matrix

$$A = \begin{bmatrix} 1 & -1 \\ -1 & 1 \end{bmatrix}$$

and convert it to an invertible matrix by adding a small quantity to a_{22}. We add .00001.

EXAMPLE 9.20 An Ill-Conditioned Matrix

Use *Maple* to find the two-condition number of the matrix

$$A = \begin{bmatrix} 1 & -1 \\ -1 & 1.00001 \end{bmatrix}$$

Solution. We calculate the condition number of A as the ratio of the largest and smallest singular values of A.

```
[> with(linalg):
[> A:=matrix([[1,-1],[1,1.00001]]):
[> svd:=evalf(Svd(A));
```

$$svd := [2.000005000, .4999987500\ 10^{-5}]$$

```
[> Ratio:=evalf(svd[1]/svd[2]);
```

$$400002.0000$$

We confirm this fact by evaluating the **cond** function at A.

```
[> CondTwo:=cond(A,2);
```

$$CondTwo := 400002.0000$$

As we can see, the condition number is large. ▲

EXERCISES 9.8

1. (*Maple 6*) Calculate the infinity, Euclidean, and Frobenius condition numbers of the following matrices.

$$\text{a. } \begin{bmatrix} 4 & 0 & 0 \\ 0 & 5 & 0 \\ 0 & 0 & 4 \end{bmatrix} \quad \text{b. } \begin{bmatrix} 4 & 0 & 0 \\ 2 & 5 & 0 \\ 3 & 2 & 4 \end{bmatrix} \quad \text{c. } \begin{bmatrix} \frac{1}{\sqrt{2}} & \frac{1}{\sqrt{2}} & 0 \\ -\frac{1}{\sqrt{2}} & \frac{1}{\sqrt{2}} & 0 \\ 0 & 0 & 1 \end{bmatrix}$$

2. (*Maple 6*) Use the **HilbertMatrix** function in the **LinearAlgebra** package to generate the 2×2, 3×3, 4×4, and 5×5 Hilbert matrices, and use the function **LinearAlgebra[ConditionNumber]** to calculate their one-, two-, infinity-, and Frobenius condition numbers. What can you say about the rate at which these numbers increase?

3. (*Maple 6*) Use the **LinearAlgebra[SingularValues]** function to calculate an approximation B of the inverse of the matrix

$$A = \begin{bmatrix} 9 & 7 & 7 \\ 4 & 8 & 5 \\ 7 & 1 & 2 \end{bmatrix}$$

Compare the two-norm condition numbers of A and B.

4. (*Maple 6*) Let f be the generating function defined by the command

```
f:=(i,j)->1/factorial(i+j);
```

and put

```
An:=RandomMatrix(n,n,generator=f):
```

for $n = 3, \ldots, 10$. Discuss the rates of change of the Euclidean, Frobenius, and infinity condition numbers of An as n increases from 1 to 10.

5. (*Maple 6*) Show that the matrix

$$A = \begin{bmatrix} 1.0000000000000001 & 1 \\ 1 & 1 \end{bmatrix}$$

is ill-conditioned.

REVIEW

KEY CONCEPTS ▶ Define and discuss each of the following.

Invertible Matrices

Condition number of an invertible matrix, ill-conditioned invertible matrix, perfectly conditioned invertible matrix, well-conditioned invertible matrix.

Rectangular Matrices

Pseudoinverse of a rectangular matrix, reduced singular value decomposition of a matrix, singular value decomposition of a matrix, singular values of a rectangular matrix.

Singular Vectors

Left singular vector, right singular vector.

KEY FACTS ▶ Explain and illustrate each of the following.

1. For all real $m \times n$ matrices A, the eigenvalues of $A^T A$ are nonnegative.
2. The two-norm of any real square matrix A is the largest singular value of A.
3. For every $m \times n$ real matrix A, there exists an orthogonal $m \times m$ matrix U, an orthogonal $n \times n$ matrix V, and an $m \times n$ generalized diagonal matrix D such that $A = UDV^T$.
4. If $\sigma_1, \dots, \sigma_n$ are the singular values of a real $n \times n$ matrix A, then the square of the Frobenius norm $\|A\|_F^2$ is the sum $\sigma_1^2 + \cdots + \sigma_n^2$ of the singular values of A.
5. Suppose that $\mathbf{u}_1, \dots, \mathbf{u}_m$ are the left singular vectors and $\mathbf{v}_1, \dots, \mathbf{v}_n$ are the right singular vectors of a real $m \times n$ matrix A whose singular values are $\sigma_1, \dots, \sigma_n$ and whose rank is r. Then the following hold.

   a. $U_r = \{\mathbf{u}_1, \dots, \mathbf{u}_r\}$ is an orthonormal basis for Col A.

   b. $U_{m-r} = \{\mathbf{u}_{r+1}, \dots, \mathbf{u}_m\}$ is an orthonormal basis for Nul A^T.

   c. $V_r = \{\mathbf{v}_1, \dots, \mathbf{v}_r\}$ is an orthonormal basis for Col A^T.

   d. $V_{n-r} = \{\mathbf{v}_{r+1}, \dots, \mathbf{v}_n\}$ is an orthonormal basis for Nul A.

6. For any linear system $A\mathbf{x} = \mathbf{b}$, the vector $\widehat{\mathbf{x}} = A^+\mathbf{b}$ is a least-squares solution of the system.
7. The condition number of an identity matrix is 1.
8. The condition number cond A of any invertible matrix is ≥ 1.
9. Let A be any invertible matrix. Then the two-norm condition number cond A is $\max \sigma_i / \min \sigma_i$.

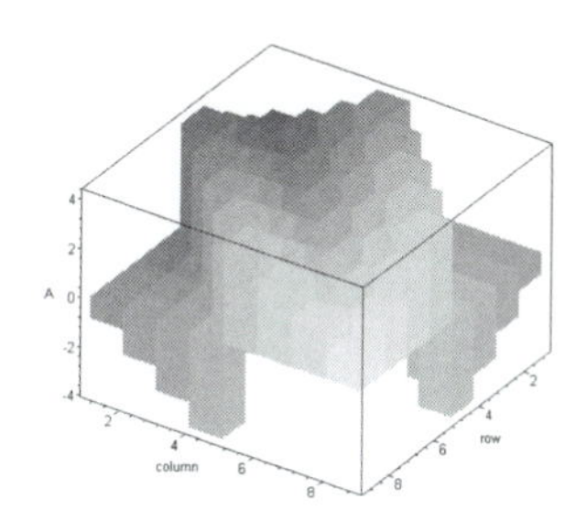

A

THE FUNDAMENTAL THEOREM OF ALGEBRA

In Chapter 8, we used the fundamental theorem of algebra to prove that real symmetric matrices have real eigenvalues. To do so, we had to work in complex vector spaces. We now provide a summary of the properties of complex numbers required for the definition of complex vector and complex inner product spaces.

The evolution of the usual number systems $\mathbb{N} \subset \mathbb{Z} \subset \mathbb{Q} \subset \mathbb{R} \subset \mathbb{C}$ can in many ways be explained as a step-by-step enrichment of mathematical systems in response to new needs. The names given to the systems in this chain also indicate some of the philosophical difficulties with which mathematicians struggled as they developed these systems.

The ***natural*** numbers $\mathbb{N} = \{1, 2, 3, \ldots\}$ were needed for counting and were considered to be altogether *natural*. The ***integers*** $\mathbb{Z} = \{0, 1, -1, 2, -2, 3, -3, \ldots\}$ are the whole numbers. The introduction of the number zero and of negative numbers was a major conceptual step forward in the history of mathematics. The ***rational*** numbers $\mathbb{Q}$ made up of ratios of integers p/q was probably an obvious extension of $\mathbb{N}$ required for measurements. The discovery that not all distances can be measured with rational numbers was an profound realization. The ruler-and-compass construction of $\sqrt{2}$ and the subsequent proof that $\sqrt{2} \notin \mathbb{Q}$ led to the introduction of ***irrational*** numbers, such as $\sqrt{2}$. Moreover, it was realized that other quantities occurring in nature, such as π—the ratio of the circumference of a circle to its diameter—are not rational numbers. Hence there was a need to extend $\mathbb{Q}$ to a more complete system. The result is the set $\mathbb{R}$ of ***real*** numbers. We have assumed throughout the text that the points on a straight line are in one–one correspondence with the real numbers. An actual construction of $\mathbb{R}$ starting from $\mathbb{Q}$ is usually discussed in courses in calculus. In linear algebra, the existence and the algebraic properties of $\mathbb{R}$ are taken for granted.

Unfortunately, the eigenvalue problem forces us to find roots of real polynomials, and some very simple polynomials have no real roots. We therefore sometimes need a more complete number system in which all polynomials have roots. The system $\mathbb{C}$ of ***complex*** numbers is that system. It is obtained from $\mathbb{R}$ by adding a single new object i, which has the property that $i^2 = -1$. We refer to i as the (positive) square root of -1 and write $i = \sqrt{-1}$. We note that $(-i)^2$ is also equal to -1. Hence there are in fact two square roots whose squares are -1.

Consider the polynomial $p(t) = t^2 + 1$ and suppose that $p(t) = (t - \lambda_1)(t - \lambda_2)$. Then

$$(t - \lambda_1)(t - \lambda_2) = t^2 - t\lambda_2 - \lambda_1 t + \lambda_1\lambda_2 = t^2 + 1$$

Therefore, $\lambda_1 + \lambda_2 = 0$ and $\lambda_1\lambda_2 = 1$. It follows that $\lambda_1(-\lambda_1) = \lambda_1^2 = -1$. Hence λ_1 must be $\pm i$.

DEFINITION A.1 *A **complex number** is any expression of the form $x + iy$, where x and y are real numbers and i is a symbol with the property that $i^2 = -1$.*

The sum and the product of two complex numbers are easily defined.

DEFINITION A.2 *If $z_1 = x_1 + iy_1$ and $z_2 = x_2 + iy_2$ are two complex numbers, then $z_1 + z_2 = (x_1 + x_2) + i(y_1 + y_2)$ and $z_1z_2 = (x_1x_2 - y_1y_2) + i(x_1y_2 + y_1x_2)$.*

It is easy to verify that these operations on the set $\mathbb{C}$ of complex numbers satisfy the usual axioms of arithmetic.

The main theorem about complex numbers required for linear algebra is the theorem that guarantees that $\mathbb{C}$ is rich enough to allow us to solve all polynomial equations.

THEOREM 1 (Fundamental theorem of algebra) *For every nonconstant polynomial $p(t) \in \mathbb{C}[t]$ there exists a $\lambda \in \mathbb{C}$ for which $p(\lambda) = 0$.* ■

EXAMPLE A.1 The Roots of a Complex Polynomial

Use *Maple* to find the roots of the polynomial $p(t) = t^4 + 4t^2 - 1$.

Solution. We use the **solve** function to find the roots of $p(t)$.

```
[> solve(t^4+4*t^2-1=0,t);
```

$$-\sqrt{-2+\sqrt{5}}, \sqrt{-2+\sqrt{5}}, -\sqrt{-2-\sqrt{5}}, \sqrt{-2-\sqrt{5}}$$

Thus $p(t)$ factors into the product

$$\left(t+\sqrt{-2+\sqrt{5}}\right)\left(t-\sqrt{-2+\sqrt{5}}\right)\left(t+i\sqrt{2+\sqrt{5}}\right)\left(t-i\sqrt{2+\sqrt{5}}\right)$$

over $\mathbb{C}$. Two of the roots of $p(t)$ are real and two are complex. ▲

If $p(t)$ is the characteristic polynomial of a real symmetric matrix, then Theorem 8.32 tells us that all roots of $p(t)$ are real.

EXAMPLE A.2 The Eigenvalues of a Symmetric Matrix

Use *Maple* to find the eigenvalues of the matrix

$$A = \begin{bmatrix} 8. & 4 & -5 \\ 4 & -5 & 3 \\ -5 & 3 & -5 \end{bmatrix}$$

Solution. We use *Maple* to find approximate eigenvalues of A.

```
[> A:=matrix([[8.,4,-5],[4,-5,3],[-5,3,5]]):
[> linalg[eigenvalues](A);
```

$$-10.22914670, -2.029738255, 10.25888495$$

Let us check that these real numbers are approximate roots of the characteristic polynomial of A.

```
[> p:=t->charpoly(A,t):
```

$$p := t^3 + 2.t^2 - 105.t - 213.$$

```
[> p(-10.22914670);
```

$$-.5\ 10^{-6}$$

```
[> p(-2.029738255);
```

$$.2\ 10^{-7}$$

```
[> p(10.25888495);
```

$$-.6\ 10^{-6}$$

As we can see, the approximate eigenvalues of A are approximate roots of $p(t)$. ▲

At various stages in this text, we showed that several concepts for real vector spaces had complex analogues. The relevant constructions required the idea of the ***conjugate*** of a complex number and of its ***modulus***.

DEFINITION A.3 *For every complex number $z = x + iy$, the complex number $\overline{z} = x - iy$ is the **conjugate** of z.*

It is easy to show that conjugates preserve addition and multiplication:

$$\overline{z_1 + z_2} = \overline{z_1} + \overline{z_2} \quad \text{and} \quad \overline{z_1 z_2} = \overline{z_1}\,\overline{z_2}$$

EXAMPLE A.3 The Conjugates of Complex Numbers

Calculate the complex conjugates of the complex numbers $z_1 = 3 - 4i$, $z_2 = 7$, and $z_3 = 9i$.

Solution. By definition,

$$\begin{aligned} \overline{z_1} &= \overline{3-4i} &&= 3+4i \\ \overline{z_2} &= \overline{7+0i} &&= 7 \\ \overline{z_3} &= \overline{9i} &&= -9i \end{aligned}$$

The equality $\overline{7} = 7$ is an example of the fact that a complex number z is real if and only if it equals its conjugate, $\overline{z}$. ▲

EXAMPLE A.4 The Conjugate Transpose of a Complex Matrix

Use *Maple* to calculate the conjugate transpose A^H of the matrix

$$A = \begin{bmatrix} 4+i & 7-2i \\ i & 5 \end{bmatrix}$$

Solution. We use the nested **transpose** and **conjugate** functions.

```
[> A:=matrix([[4+I,7-2*I],[I,5]]):
[> linalg[transpose](conjugate(A));
```

$$\begin{bmatrix} 4-I & -I \\ 7+2I & 5 \end{bmatrix}$$

As we can see, we can calculate the conjugate transpose of A in two stages. First we form the conjugate of each entry of A and then we transpose the resulting matrix. ▲

DEFINITION A.4 *The modulus $|z|$ of a complex number $z = x + iy$ is the real number $\sqrt{x^2 + y^2}$.*

As we can see, the modulus is the Euclidean distance function studied in Chapters 3 and 7. It is the length of the direction arrow determined by the point (x, y) in the xy-plane.

EXAMPLE A.5 Moduli of Complex Numbers

Calculate the modulus of the complex numbers $z_1 = 3 - 4i$, $z_2 = 7$, and $z_3 = 9i$.

Solution. By definition,

$$\begin{aligned}
|z_1| &= |3 - 4i| = \sqrt{3^2 + 4^2} = 5 \\
|z_2| &= |7 + 0i| = 7 \\
|z_3| &= |9i| = -9i = \sqrt{(-9)^2} = 9
\end{aligned}$$

are the required moduli. ▲

The modulus of a complex number can also be computed using the standard inner product on $\mathbb{C}$.

EXAMPLE A.6 The Modulus and the Standard Inner Product

Use the standard inner product on $\mathbb{C}$ to calculate the modulus of $z = 3 - 4i$.

Solution. Consider $\mathbb{C}$ as a complex vector space. Then the dimension of $\mathbb{C}$ is one. Hence every element of $\mathbb{C}$ is a 1×1 complex matrix with the property that $z^T = z$. According to the definition in Chapter 7, the square $\|z\|^2$ of the norm of z is therefore

$$z^T \overline{z} = (3 - 4i)(3 + 4i) = 25$$

Hence $\|z\| = \sqrt{25} = 5$, as expected. As we can see, the Euclidean norm of the vector z and the modulus of the complex number z coincide. ▲

Every complex number $z = x + iy$ determines an angle θ, with $-\pi < \theta \leq \pi$, called the ***principal argument*** $\arg(z)$ of z. The angle θ is the angle between the positive x-axis and the line segment terminating at the point (x, y). We know from geometry that if $r = |z|$, then $x = r\cos\theta$ and $y = r\sin\theta$. Therefore, $z = r(\cos\theta + i\sin\theta)$. If we assume that θ satisfies $-\pi < \theta \leq \pi$, then θ is uniquely determined by z.

EXAMPLE A.7 The Principal Argument of a Complex Number

Calculate the principal argument of $z = \sqrt{3} + i$.

Solution. The modulus r of z is $\sqrt{3+1} = 2$. Since $x = r\cos\theta$, we have $\sqrt{3} = 2\cos\theta$. Therefore $\theta = \cos^{-1}\frac{\sqrt{3}}{2} = \frac{1}{6}\pi$. ▲

The Algebraic Structure of the Set of Complex Numbers

In Example A.6, we used the fact that the set $\mathbb{C}$ of complex numbers is a complex vector space of dimension 1. The set $\mathbb{C}$ can also be made into a real vector space of dimension 2. Let $\mathcal{S} = \{\mathbf{x}_1 = 1, \mathbf{x}_2 = i\}$. Then $\mathcal{S}$ is linearly independent since no real multiple of 1—in other words, no real number $a = a \times 1$—is equal to the imaginary number i and no real multiple ai of i yields a nonzero real number. However, every complex number $z = a_1 + ia_2$ is a linear combination of the form $a_1\mathbf{x}_1 + a_2\mathbf{x}_2$. This example shows that the same set of vectors can give rise to very different vector spaces if the set of scalars is changed.

The set $\mathbb{C}$ of complex numbers is actually more than a vector space. Every nonzero complex number also has an inverse for multiplication.

DEFINITION A.5 *If $z = x + iy$ is a nonzero complex number, then the number $w = (x - iy) / \left(x^2 + y^2\right)$ is the multiplicative inverse of z.*

EXAMPLE A.8 The Multiplicative Inverse of a Complex Number

Show that the product of the complex numbers $z = 3 + 4i$ and $w = (3 - 4i) /25$ is 1.

Solution. By the definition of multiplication in $\mathbb{C}$,

$$zw = \tfrac{1}{25} (3 + 4i) (3 - 4i) = \tfrac{1}{25} (9 + 16 - 12i + 12i) = 1$$

Hence w is the multiplicative inverse of z. ▲

The laws of arithmetic satisfied by the complex numbers are the same laws that are satisfied by the system $\mathbb{R}$ of real numbers, the system $\mathbb{Q}$ of rational numbers, and other systems in which elements can be added, subtracted, multiplied, and divided, provided that any product ab is zero if and only if at least one of a or b is zero. We know from Chapter 2 that the algebras of matrices, for examples, satisfy many of these laws. However, we have also seen examples of nonzero matrices A and B of the same dimension whose product AB is zero, without A or B being zero. By contrasts, the complex number share with the rational and real numbers the property that the product of two nonzero elements is nonzero.

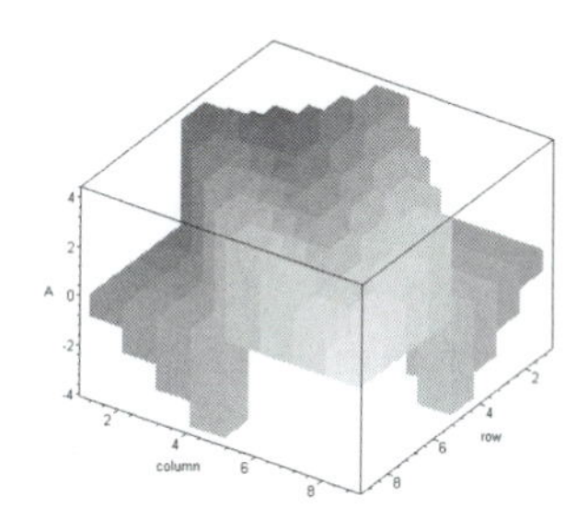

B

NUMERICAL CALCULATIONS

The scalars of the vector spaces studied in this text were assumed to be real or, sometimes, complex numbers. When calculating with such numbers, *Maple* provides us with two choices: calculating symbolically and calculating numerically. Thus $\pi + \sqrt{2}$ is a symbolic name for a specific real number, and in many situations we can use this symbol for that number. For example, we know that $\sin \pi = 0$. So any occurrence of $\sin \pi$ in a calculation can be replaced by 0. No approximation of π is necessary. We also know, for example, that $e^{\pi i} = -1$. So there is no need to approximate e and π in any calculation involving $e^{\pi i}$.

In real-world applications, however, we usually need numerical approximations for both rational and irrational numbers such as $1/3$ and $\pi + \sqrt{2}$. Moreover, no computer can calculate with more than a finite number of the digits of any decimal expansion of $\pi + \sqrt{2}$. Thus every numerical calculation with $\pi + \sqrt{2}$ involves round-off errors. In this section, we briefly discuss two examples of the impact of such errors on numerical calculations in linear algebra.

All computers store real numbers as ***floating point numbers*** using finite representations such as

$$a = \pm .d_1 d_2 \ldots d_m \times 10^n$$

where $d_i \in \{0, 1, 2, 3, 4, 5, 6, 7, 8, 9\}$ and $n = \pm 0, 1, 2, 3, \ldots$, with $d_1 \neq 0$ if $a \neq 0$. Table 1 gives some examples.

The actual floating-point numbers used as approximations of a given real number depend on the application and on the specific computer and computer software involved. According

TABLE 1 FLOATING-POINT NUMBERS.

Symbolic Names	Floating-Point Names
2.4^{45}	$.128678 \times 10^{18}$
$2.3 + 5.63$	$.793 \times 10^{1}$
π	$.314159 \times 10^{1}$
$\sin(45)$	$.850904 \times 10^{0}$
$2/333$	$.600601 \times 10^{-2}$

to the accuracy required, for example, we often use $.314159 \times 10^1$, $.31416 \times 10^1$, $.314 \times 10^1$, or other approximations of π in calculations involving this real number.

The following two examples illustrate how round-off errors can produce what appear to be erroneous results.

EXAMPLE B.1 $a + b = a$ **and** $b \neq 0$

Suppose we need to add the floating-point numbers $a = .451235 \times 10^4$ and $b = .123456 \times 10^{-2}$ and expect the answer to be accurate to within six decimal places. Then

$$\begin{aligned} a + b &= .451235 \times 10^4 + .000000123456 \times 10^4 \\ &= .451235123456 \times 10^4 \\ &= .451235 \times 10^4 \\ &= a \end{aligned}$$

Thus $b \neq 0$ and yet $a + b = a$. In *Maple,* this calculation takes the following form.

```
[> a:=.451235*10^4;
                    a := 4512.350000
[> b:=.123456*10^(-2);
                    b := .001234560000
[> evalf(a+b,6);
                    4512.35
```

Thus $a + b = a$ and $b \neq 0$. ▲

In this example, the round-off error is perfectly transparent. When multiple calculations are nested, however, the source and the nature of these errors may be more difficult to detect and their consequences may be more elusive.

The next example shows that there are times when a round-off error causes *Maple* to produce a singular approximation of a nonsingular matrix.

EXAMPLE B.2 An Approximately Singular Matrix

Use *Maple* to show that for the real numbers $r = .123456 \times 10^{-30}$, round-off errors cause the diagonal matrix

$$A = \begin{bmatrix} 10 & 10 \\ 10 & 10 - r \end{bmatrix}$$

to be approximately singular, relative to the default accuracy settings.

Solution. We first construct the floating-point real number r and then use it to create the matrix A.

```
[> with(linalg):
[> Digits:=5:
[> r:=.123456×10^(-10):
[> evalb(r=0);
```

$$false$$

This tells us that the number r is not 0. Hence $10 - r \neq 10$. We use this nonzero number to construct a seemingly singular matrix A.

```
[> A:=matrix([[10,10],[10,10-r]]);
```

$$A := \begin{bmatrix} 10 & 10 \\ 10 & 10. \end{bmatrix}$$

```
[> det(A);
```

$$0.$$

```
[> inverse(A);
Error, (in inverse) singular matrix
```

We can reverse this situation by changing the accuracy of floating-point arithmetic.

```
[> Digits:=20:
[> A:=matrix([[10,10],[10,10-r]]);
```

$$A := \begin{bmatrix} 10 & 10 \\ 10 & 9.999999999987654 \end{bmatrix}$$

```
[> det(A);
```

$$-.123460\ 10^{-9}$$

```
[> inverse(A);
```

$$\begin{bmatrix} -80997894054.654576381 & 80997894054.754576381 \\ 80997894054.754576381 & -80997894054.754576381 \end{bmatrix}$$

As we can see, from the point of view of *Maple*, the invertibility of the matrix A depends on the permitted round-off errors. ▲

Most linear algebra problems in science, engineering, the social sciences, and business are solved by computers using floating-point arithmetic. The impact of this arithmetic on the efficiency and reliability of algorithms used in these calculations is studied in *numerical linear algebra.*

Except for our work in Chapter 9, we have avoided any discussion of the difficulties inherent in using computers to solve linear algebra problems. In practice, however, this issue is particularly acute for eigenvalues since it has been proved that there are no algorithms for finding the roots of arbitrary polynomials of degree 5 or higher, even if their coefficients are integers.

The nonexistence of a suitable algorithm forces computer programs such as *Maple* to use numerical methods to calculate roots of polynomials. What is important, therefore, is that the computational processes used for this purpose are *stable*. This means that the approximations generated by the process must *converge*. If s is the exact solution of a problem, and s_n is an approximation of s, then s_n must approach s as n increases. In this text, we are assuming that the numerical methods implemented in *Maple* are stable.

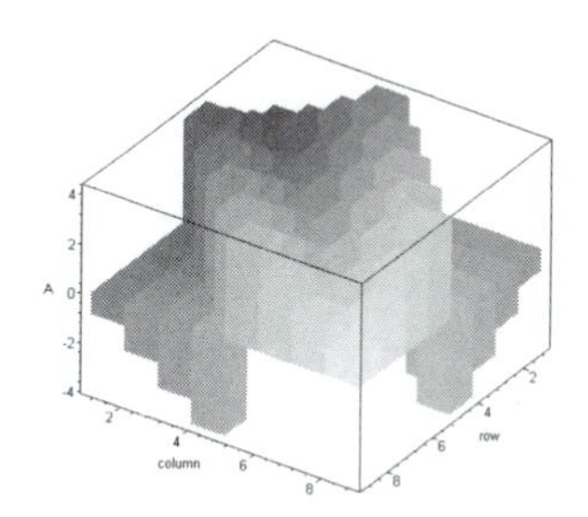

C

MATHEMATICAL INDUCTION

Mathematical induction is a powerful procedure for proving that certain statements involving natural numbers are true no matter which natural numbers are involved. For example, we might have a statement $\varphi(n)$ for calculating the sum $1 + 2 + \cdots + n$, and we want to prove that the formula works for all n. If $n = 15$, then the formula $\varphi(15)$ computes the sum $1+2+\cdots+15$; if $n = 1001$, then $\varphi(1001)$ computes the sum $1 + 2 + \cdots + 1001$; and so on. We would like to *prove* the fact that we can use the formula $\varphi(n)$ to find the sum $1 + \cdots + n$ for any natural number n. Mathematical induction allows us to do precisely that.

Behind mathematical induction is the assumption that the set $\mathbb{N}$ of natural numbers can be thought of as the set whose elements are either 1 or of the form $n+1 = (1+1+\cdots+1)+1$. The principle of mathematical induction uses this property to assert that if we can prove that the formula $\varphi(1)$ holds, and if we can deduce the validity of $\varphi(n + 1)$ from the validity of $\varphi(n)$, then the formula $\varphi(m)$ holds for all $m \in \mathbb{N}$.

A proof by mathematical induction consists of two parts. The first part, called the ***induction basis,*** consists of the proof that $\varphi(1)$ holds. Depending on the specific claim expressed by the formula $\varphi(n)$, the statement may actually only make sense for some $p > 1$. In that case, the induction basis is the proof that $\varphi(p)$ holds. We may also have to prove the validity of the first case or two by different means, depending on the nature of the formula $\varphi(n)$. If the induction basis involves more than one initial proof, the induction involved is known as *complete induction*.

The second part of the proof, called the ***induction step,*** consists of a calculation that shows that if $\varphi(q)$ holds for an arbitrary $q \in N$, then $\varphi(q+1)$ holds. The assumption that $\varphi(q)$ holds is called the ***induction hypothesis***.

The following example illustrates the two parts involved in a proof by mathematical induction. It is well known that the statement

$$\varphi(n) : 1 + 2 + \cdots + n = \frac{n(n+1)}{2}$$

holds for all $n \in \mathbb{N}$. The statement $\varphi(1)$, for example, holds because

$$1 = \frac{1\,(1+1)}{2}$$

The statement $\varphi(2)$ holds because

$$1+2=\frac{2(2+1)}{2}$$

Suppose now that $\varphi(q)$ holds. This means that we are assuming that

$$1+2+\cdots+q=\frac{q(q+1)}{2}$$

for an arbitrary natural number q. We would like to deduce the validity of $\varphi(q+1)$ from the assumed validity of $\varphi(q)$. We do this as follows.

$$\begin{aligned}
1+2+\cdots+q+(q+1) &= \frac{q(q+1)}{2}+(q+1)\\
&= \frac{q(q+1)}{2}+\frac{2(q+1)}{2}\\
&= \frac{(q+2)(q+1)}{2}\\
&= \frac{(q+1)(q+2)}{2}
\end{aligned}$$

The principle of mathematical induction now allows us to conclude that $\varphi(n)$ holds for all n.

In finite-dimensional linear algebra, statements of the form $\varphi(n)$ are often statements asserting that a certain fact φ holds for spaces of all dimensions. Other types of statements proved by mathematical induction are statements about linear combinations of arbitrary length or about polynomials of arbitrary degree. Good illustrations of proofs involving these kinds of statements are the proofs of the uniqueness of the reduced row echelon form, of the fact that determinants can be computed using elementary products, of the fact that eigenvectors belonging to distinct eigenvalues are linearly independent, of the fact that finite-dimensional real inner product spaces have orthonormal bases consisting of eigenvectors of self-adjoint linear transformations, and of the Gram–Schmidt process.

Using induction to define certain formulas is called a ***definition by induction***. Sometimes it is also called a ***recursive definition***. The factorial function $n!$, for example, is defined by induction. The statement $\varphi\ (1)$ is the statement that $1 = 0!$ and the statement $\varphi\ (n)$ is the statement that $n! \times (n+1) = (n+1)!$. The proof that this process defines the factorial function is by mathematical induction. The definition of the Laplace expansion of the determinant of a matrix is another example of a definition by induction.

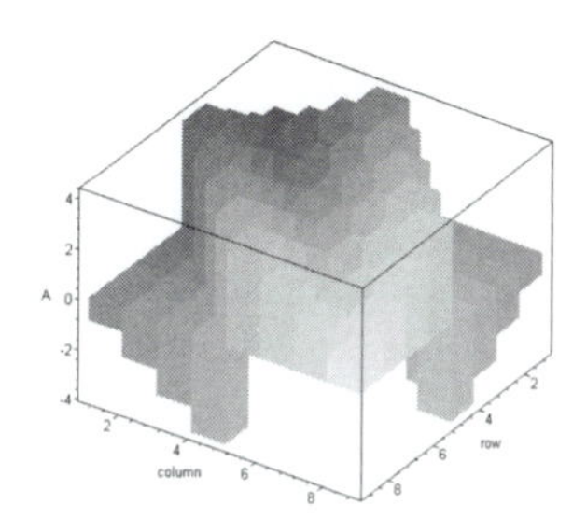

D

SIGMA NOTATION

An important application of mathematical induction is the definition of the ***sigma notation***. Arithmetic statements in linear algebra are often about sums made up of a fixed but arbitrary number of terms. There are two basic ways of dealing with this situation. We can either use an ***ellipsis*** (three dots) to represent missing elements, as in $\mathbf{x}_1 + \cdots + \mathbf{x}_n$, or we can use variable subscripts and the capital sigma and write $\sum_{i \in \mathbf{n}} \mathbf{x}_i$, where $\mathbf{n}$ denotes the indexing set $[1, \ldots, n]$. The sigma notation is defined by mathematical induction.

DEFINITION D.1 *If* $\mathbf{n} = \mathbf{1}$, *then* $\sum_{i \in \mathbf{1}} \mathbf{x}_i = \mathbf{x}_1$; *and if* $\mathbf{n} = \mathbf{k} + \mathbf{1}$, *then* $\sum_{i \in \mathbf{k+1}} \mathbf{x}_i = \left(\sum_{i \in \mathbf{k}} \mathbf{x}_i\right) + \mathbf{x}_{k+1}$.

However, we usually write the sum $\sum_{i \in \mathbf{n}} \mathbf{x}_i$ less formally as "$\mathbf{x}_1 + \cdots + \mathbf{x}_n$," using an ellipsis.

For any $m \times n$ matrix $A = \left[a_{ij}\right]$, the expression $\sum_{i \in \mathbf{m}} a_{ij}$ denotes the sum of entries in the jth column of A, and the expression $\sum_{j \in \mathbf{n}} a_{ij}$ denotes the sum of the entries in the ith row of A.

EXAMPLE D.1 A Sum of Vectors

The expressions $\mathbf{x}_1 + \mathbf{x}_2 + \mathbf{x}_3$ and $\sum_{i \in \mathbf{3}} \mathbf{x}_i$ denote the same sum of vectors. ▲

EXAMPLE D.2 A Linear Combination of Vectors

The expressions $a_1\mathbf{x}_1 + a_2\mathbf{x}_2 + a_3\mathbf{x}_3$ and $\sum_{i \in \mathbf{3}} a_i\mathbf{x}_i$ denote the same linear combination of the vectors $\mathbf{x}_1$, $\mathbf{x}_2$, and $\mathbf{x}_3$. ▲

EXAMPLE D.3 A Column Sum

If $A = \left[a_{ij}\right]$ is a $3 \times n$ matrix, then the two expressions $a_{1j} + a_{2j} + a_{3j}$ and $\sum_{i \in \mathbf{3}} a_{ij}$ both denote the sum of the entries in the jth column of A. ▲

EXAMPLE D.4 A Row Sum

If $A = \left[a_{ij}\right]$ is an $m \times 3$ matrix, then the two expressions $a_{i1} + a_{i2} + a_{i3}$ and $\sum_{j \in \mathbf{3}} a_{ij}$ both denote the sum of the entries in the ith row of A. ▲

The set of all permutations on an indexing set $\mathbf{n}$ is denoted by S_n, and $\sum_{\pi \in S_n} \varphi(\pi)$ denotes the sum of all terms $\varphi(\pi)$ determined by the permutations π.

EXAMPLE D.5 A Sum Indexed by a Set of Permutations

Let $\varphi(\pi)$ be a numerical expression indexed by permutations. For example, $\varphi(\pi)$ could be an elementary product of entries of a matrix. If $\mathbf{n} = \mathbf{2}$, then $S_2 = \{\pi_1 = [1, 2], \pi_2 = [2, 1]\}$, so

$$\varphi(\pi_1) + \varphi(\pi_2) = \sum_{\pi \in S_2} \varphi(\pi)$$

is the sum of instances of φ, indexed by the set of permutations S_2. ▲

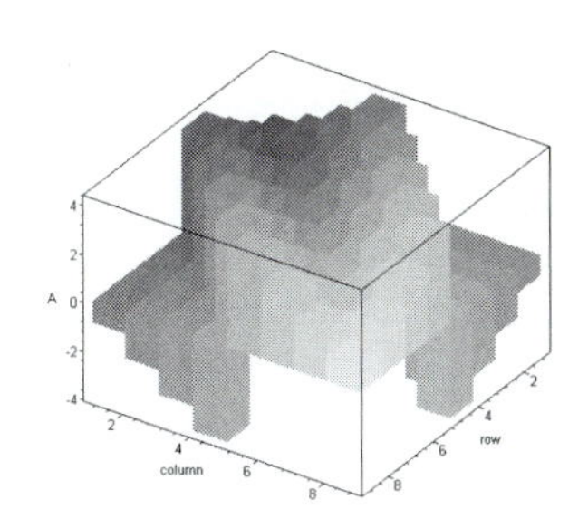

E

MAPLE PACKAGES

Many of the *Maple* functions used in this text are defined in special packages that must be loaded once *Maple* is running. We list the required packages and the functions defined in them.

PACKAGES

The combinat Package

```
[> with(combinat);
```

[Chi, bell, binomial, cartprod, character, choose, composition, conjpart, decodepart, encodepart, fibonacci, firstpart, graycode, inttovec, lastpart, multinomial, nextpart, numbcomb, numbcomp, numbpart, numbperm, partition, permute, powerset, prevpart, randcomb, randpart, randperm, stirling1, stirling2, subsets, vectoint]

The geometry Package

```
[> with(geometry);
```

[Appolonius, AreCollinear, AreConcurrent, AreConcyclic, AreConjugate, AreHarmonic, AreOrthogonal, AreParallel, ArePerpendicular, AreSimilar, AreTangent, CircleOfSimilitude, CrossProduct, CrossRatio, DefinedAs, Equation, EulerCircle, EulerLine, ExteriorAngle, ExternalBisector, FindAngle, GergonnePoint, GlideReflection, HorizontalCoord, HorizontalName, InteriorAngle, IsEquilateral, IsOnCircle, IsOnLine, IsRightTriangle, MajorAxis, MakeSquare, MinorAxis, NagelPoint, OnSegment, ParallelLine, PedalTriangle, PerpenBisector, PerpendicularLine, Polar, Pole, RadicalAxis, RadicalCenter, RegularPolygon, RegularStarPolygon, SensedMagnitude, SimsonLine, SpiralRotation, StretchReflection, StretchRotation, TangentLine, VerticalCoord, VerticalName, altitude, apothem, area, asymptotes, bisector, center, centroid, circle, circumcircle, conic, convexhull, coordinates, detail, diagonal, diameter, dilatation, directrix, distance, draw, dsegment, ellipse, excircle, expansion, foci, focus, form, homology, homothety, hyperbola, incircle, inradius, intersection, inversion, line, medial, median, method, midpoint, orthocenter, parabola, perimeter, point, powerpc, projection, radius, randpoint, reciprocation, reflection, rotation, segment, sides, similitude, slope, square, stretch, tangentpc, translation, triangle, vertex, vertices]

The linalg Package

```
[> with(linalg);
```

[BlockDiagonal, GramSchmidt, JordanBlock, LUdecomp, QRdecomp, Wronskian, addcol, addrow, adj, adjoint, angle, augment, backsub, band, basis, bezout, blockmatrix, charmat, charpoly, cholesky, col, coldim, colspace, colspan, companion, concat, cond, copyinto, crossprod, curl, definite, delcols, delrows, det, diag, diverge, dotprod, eigenvals, eigenvalues, eigenvectors, eigenvects, entermatrix, equal, exponential, extend, ffgausselim, fibonacci, forwardsub, frobenius, gausselim, gaussjord, geneqns, genmatrix, grad, hadamard, hermite, hessian, hilbert, htranspose, ihermite, indexfunc, innerprod, intbasis, inverse, ismith, issimilar, iszero, jacobian, jordan, kernel, laplacian, leastsqrs, linsolve, matadd, matrix, minor, minpoly, mulcol, mulrow, multiply, norm, normalize, nullspace, orthog, permanent, pivot, potential, randmatrix, randvector, rank, ratform, row, rowdim, rowspace, rowspan, rref, scalarmul, singularvals, smith, stackmatrix, submatrix, subvector, sumbasis, swapcol, swaprow, sylvester, toeplitz, trace, transpose, vandermonde, vecpotent, vectdim, vector, wronskian]

The LinearAlgebra Package

```
[> with(LinearAlgebra);
```

[Add, Adjoint, BackwardSubstitute, BandMatrix, Basis, BezoutMatrix, BidiagonalForm, BilinearForm, CharacteristicMatrix, CharacteristicPolynomial, Column, ColumnDimension, ColumnOperation, ColumnSpace, CompanionMatrix, ConditionNumber, ConstantMatrix, ConstantVector, CreatePermutation, CrossProduct, DeleteColumn, DeleteRow, Determinant, DiagonalMatrix, Dimension, Dimensions, DotProduct, Eigenvalues, Eigenvectors, Equal, ForwardSubstitute, FrobeniusForm, GenerateEquations, GenerateMatrix, GetResultDataType, GetResultShape, GivensRotationMatrix, GramSchmidt, HankelMatrix, HermiteForm, HermitianTranspose, HessenbergForm, HilbertMatrix, HouseholderMatrix, IdentityMatrix, IntersectionBasis, IsDefinite, IsOrthogonal, IsSimilar, IsUnitary, JordanBlockMatrix, JordanForm, LA_Main, LUDecomposition, LeastSquares, LinearSolve, Map, Map2, MatrixAdd, MatrixInverse, MatrixMatrixMultiply, MatrixNorm, MatrixScalarMultiply, MatrixVectorMultiply, MinimalPolynomial, Minor, Multiply, NoUserValue, Norm, Normalize, NullSpace, OuterProductMatrix, Permanent, Pivot,

QRDecomposition, RandomMatrix, RandomVector, Rank, Row, RowDimension, RowOperation, RowSpace, ScalarMatrix, ScalarMultiply, ScalarVector, SchurForm, SingularValues, SmithForm, SubMatrix, SubVector, SumBasis, SylvesterMatrix, ToeplitzMatrix, Trace, Transpose, TridiagonalForm, UnitVector, VandermondeMatrix, VectorAdd, VectorAngle, VectorMatrixMultiply, VectorNorm, VectorScalarMultiply, ZeroMatrix, ZeroVector, Zip]

The networks Package

```
[> with(networks);
```

[acycpoly, addedge, addvertex, adjacency, allpairs, ancestor, arrivals, bicomponents, charpoly, chrompoly, complement, complete, components, connect, connectivity, contract, countcuts, counttrees, cube, cycle, cyclebase, daughter, degreeseq, delete, departures, diameter, dinic, djspantree, dodecahedron, draw, duplicate, edges, ends, eweight, flow, flowpoly, fundcyc, getlabel, girth, graph, graphical, gsimp, gunion, head, icosahedron, incidence, incident, indegree, induce, isplanar, maxdegree, mincut, mindegree, neighbors, new, octahedron, outdegree, path, petersen, random, rank, rankpoly, shortpathtree, show, shrink, span, spanpoly, spantree, tail, tetrahedron, tuttepoly, vdegree, vertices, void, vweight]

The plots Package

```
[> with(plots);
```

[animate, animate3d, animatecurve, changecoords, complexplot, complexplot3d, conformal, contourplot, contourplot3d, coordplot, coordplot3d, cylinderplot, densityplot, display, display3d, fieldplot, fieldplot3d, gradplot, gradplot3d, implicitplot, implicitplot3d, inequal, listcontplot, listcontplot3d, listdensityplot, listplot, listplot3d, loglogplot, logplot, matrixplot, odeplot, pareto, pointplot, pointplot3d, polarplot, polygonplot, polygonplot3d, polyhedra_supported, polyhedraplot, replot, rootlocus, semilogplot, setoptions, setoptions3d, spacecurve, sparsematrixplot, sphereplot, surfdata, textplot, textplot3d, tubeplot]

The plottools Package

```
[> with(plottools);
```

[arc, arrow, circle, cone, cuboid, curve, cutin, cutout, cylinder, disk, dodecahedron, ellipse, ellipticArc, hemisphere, hexahedron, homothety, hyperbola, icosahedron, line, octahedron, pieslice, point, polygon, project, rectangle, reflect, rotate, scale, semitorus, sphere, stellate, tetrahedron, torus, transform, translate, vrml]

The stats Package

```
[> with(stats);
```

[anova, describe, fit, importdata, random, statevalf, statplots, transform]

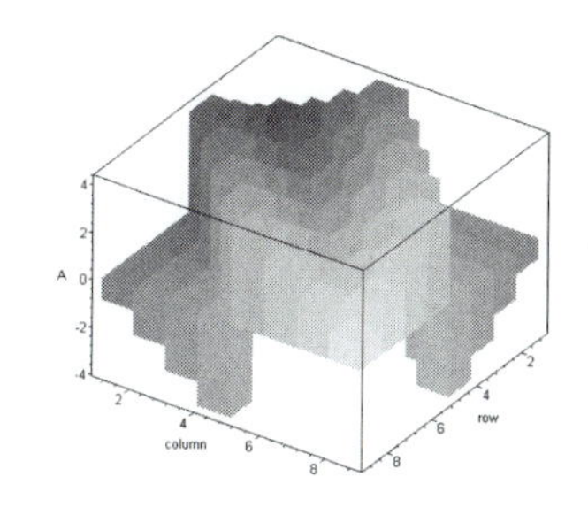

F

ANSWERS TO SELECTED EXERCISES

CHAPTER 1 - Linear Systems

EXERCISES 1.1

Exercise 1

$0x_1+0x_2=5$, $1x_1+0x_2=5$, $2x_1+0x_2=5$, $0x_1+1x_2=5$, $1x_1+1x_2=5$, $2x_1+1x_2=5$, $0x_1+2x_2=5$, $1x_1+2x_2=5$, $2x_1+2x_2=5$

Exercise 3

We need to construct a linear equation $ax+by=c$ for which $9a-7b=c$. If we let $a=1$, $b=1$, and $c=2$, then $9-7=2$. Hence $x=9$ and $y=-7$ is a solution of the equation $x+y=2$. We obtain another solution if we let $a=2$, $b=1$, and $c=11$, since the values $x=9$ and $y=-7$ are also a solution of the equation $2x+y=11$. It is clear from these two examples that there are infinitely many equations in x and y for which $x=9$ and $y=-7$ is a solution.

Exercise 5

All equations of the form $12x+by=12$.

Exercise 7

a. $\mathbf{x}\cdot\mathbf{y}=9$ b. $\mathbf{x}\cdot\mathbf{y}=29$ c. $\mathbf{x}\cdot\mathbf{y}=5a+c+6d$ d. $\mathbf{x}\cdot\mathbf{y}=6w+9x-y+4z$

Exercise 9 (*Maple V*)

A:=genmatrix({3*x[1]+7*x[2]-9*x[3]=12},{x[1],x[2],x[3]}): X:=[x[1],x[2],x[3]]: matrix(A)&*X=12:

Exercise 11 (*Maple V*)

v1:=vector([3,7,-9]): v2:=vector([x[1],x[2],x[3]]): solve(dotprod(v1,v2)=12, {x[1],x[2],x[3]}):

Exercise 13 (*Maple V*)

solve(3*x[1]+7*x[2]-9*x[3]=12,{x[2]}); returns $\left\{x_2=-\frac{3}{7}x_1+\frac{9}{7}x_3+\frac{12}{7}\right\}$.

Exercise 15 (*Maple V*)

with(linalg): solve(dotprod([3,7,-9],[x[1],x[2],x[3]])=12,x[1]); returns $3x_3+4-\frac{7}{3}x_2$.

Exercise 17 (*Maple 6*)

U:=Vector[row](7,[3,-8,9,0,11,12,1]): V:=Vector[column](7,[t,u,v,w,x,y,z]): DotProduct(U,V,conjugate=false)=-7; returns $3t-8u+9v+11x+12y+z=-7$.

EXERCISES 1.2

Exercise 1

Column form:

$$\text{a.} \begin{bmatrix} 2 \\ 3 \end{bmatrix} \quad \text{b.} \begin{bmatrix} -2 \\ 3 \end{bmatrix} \quad \text{c.} \begin{bmatrix} 2 \\ -3 \end{bmatrix} \quad \text{d.} \begin{bmatrix} -2 \\ -3 \end{bmatrix}$$

Point form: a. $(8, -2)$ b. $(7, 3)$ c. $(-15, -1)$ d. $(-8, 9)$

Exercise 3

$$\text{a.} \begin{bmatrix} 4 \\ 6 \end{bmatrix} \quad \text{b.} \begin{bmatrix} -2 \\ 6 \end{bmatrix} \quad \text{c.} \begin{bmatrix} 4 \\ -2 \end{bmatrix} \quad \text{d.} \begin{bmatrix} -2 \\ -2 \end{bmatrix}$$

Exercise 5

a. $2x + 3y = 18$ b. $2x - 9y = -30$ c. $6x + 3y = 30$ d. $4x - 3y = 0$

Exercise 7

$(x, y) = (4, -2) + t(7, 12)$

Exercise 9 (*Maple V*)

v1:=[2,3]: v2:=array([-2,3]): v3:=vector([2,3]): and v4:=vector(2,[-2,-3]): define vector-like objects that can be added and scaled using the **evalm** function.

Exercise 11 (*Maple V*)

vector(2,[1,2]): vector([1,2])+vector([4,5]):

Exercise 15 (*Maple V*)

evalb(-28*0+32*5+128=0); returns $false$.

evalb(-28*8+32*3+128=0); returns $true$.

evalb(-28*6+32*5+128=0); returns $false$.

Therefore only the point (8, 3) lies on the given line.

EXERCISES 1.3

Exercise 1

$\{x + y + z + w = 100, x + y + z = 50, y + z + w = 75\}$ is the required linear system. Its infinitely many solutions are of the form $(x, y, z, w) = (25, -z + 25, 50, z)$.

Exercise 3

The systems are consistent since $(x_1, \ldots, x_n) = (0, \ldots, 0)$ is a solution.

Exercise 5

a. The system is underdetermined. It is inconsistent since it is equivalent to the contradictory statement that $2 = 6$.

b. The system is determined. It is inconsistent since it is equivalent to the statement that $2 = 6$.

c. The system is overdetermined. It is inconsistent since the first two equations entail the contradictory statement that $6 = 9$.

Exercise 7 (*Maple V*)
a. plots[implicitplot]({x+2*y=5,x-2*y=1,2*x+y=6},x=-10..10,y=-10..10); shows that the three lines do not intersect. Hence the system has no solution.
b. The three lines intersect in a single point. Hence the system is consistent.
c. The three lines overlap. Hence the system has infinitely many solutions.

EXERCISES 1.4

Exercise 1
$15x - 2y - 22z + 9 = 0$
Exercise 3 (*Maple V*)
with(plots):
a. implicitplot({x+2*y=2,x-y=4},x=-10..10,y=-10..10); shows that the two lines intersect in a single point. The point of intersection is the unique solution.
b. implicitplot({x+y=2,-2*x-2*y=-4},x=-10..10,y=-10..10); shows that the two lines overlap. Hence every point on the line is a solution.
c. implicitplot({x+2*y=2,6*x+12*y=18},x=-10..10,y=-10..10); shows that the two lines are parallel but do not overlap. Hence the system has no solution.
Exercise 5
$\{x = 1 + 3t,\ y = 2 - t,\ z = 3 + 2t,\ u = 4 - 6t,\ v = 5 + 9t\}$
Exercise 7 (*Maple V*)
If the position vector $(x - a, y - b, z - c)$ is orthogonal to the position vector (p, q, r), then their dot product is zero.
dotprod([x-a,y-b,z-c],[p,q,r],'orthogonal')=0; returns $(x - a)p + (y - b)q + (z - c)r = 0$.
If $\mathbf{d} = (p, q, r) = (5, -2, 3)$ and $(a, b, c) = (1, 1, 1)$, then $5x - 6 - 2y + 3z = 0$.

EXERCISES 1.5

Exercise 1
a. $(x, y) = (1, 4)$ b. $(x, y) = (-31/13, 7/13)$ c. $(x, y) = (5, 2)$ d. $(x, y) = (9, 9)$ e. $(x, y) = (5, 21)$
Exercise 3 (*Maple V*)
a. solve({2*x+y-z=2,x-3*y+z=1,2*x+y-2*z=6},{x,y,z}); returns
$\{z = -4, y = \frac{-12}{7}, x = \frac{-1}{7}\}$.
b. solve({2*x+y-z=2,x-3*y+z=1,2*x+y-z=6},{x,y,z}); returns the empty output. Hence the system is inconsistent.
c. solve({2*x+y-z=2,x-3*y+z=1},{x,y,z}); returns $\{y = \frac{3}{7}z, x = 1 + \frac{2}{7}z, z = z\}$.

EXERCISES 1.6

Exercise 1 (*Maple V*)

```
f:=(i,j)->i+j:
array([[seq(f(1,j),j=1..5)],[seq(f(2,j),j=1..5)],[seq(f(3,j),j=1..5)],[seq(f(4,j),j=1..5)]]):
```

Exercise 3 (*Maple V*)

```
f(1,1):= 9: f(1,2):= -1: f(1,3):= 8: f(1,4):= -12: f(1,5):= -9: f(2,1):= 19: f(2,2):= 6:
f(2,3):= -8: f(2,4):= -2: f(2,5):=0: f(3,1):= 2: f(3,2):= -1: f(3,3):= 8: f(3,4):= -12:
f(3,5):= 55: f(4,1):= 4: f(4,2):= -1: f(4,3):= 8: f(4,4):= -12: f(4,5):=100
array([[seq(f(1,j),j=1..5)],[seq(f(2,j),j=1..5)],[seq(f(3,j),j=1..5)],[seq(f(4,j),j=1..5)]]):
```

Exercise 5 (*Maple V*)

```
r:=rand(-9..9):
Array([[seq(r(),i=1..5)],[seq(r(),i=1..5)],[seq(r(),i=1..5)],[seq(r(),i=1..5)]]):
```

EXERCISES 1.7

Exercise 1 (*Maple V*)

```
expr1:=(i-j)^2: expr2:=4*i: expr3:=i^2-3*j^4: unassign('i','j'): expr1:=(i,j)->(i-j)^2:
matrix(4,5,[seq(expr1(1,j),j=1..20)]): unassign('i','j'): expr2:=(i,j)->4*i:
matrix([
    [expr2(1,1),expr2(2,1),expr2(3,1),expr2(4,1)],
    [expr2(5,1),expr2(6,1),expr2(7,1),expr2(8,1)],
    [expr2(9,1),expr2(10,1),expr2(11,1),expr2(12,1)],
    expr2(13,1),expr2(14,1),expr2(15,1),expr2(16,1)],
    [expr2(17,1),expr2(18,1),expr2(19,1),expr2(20,1)]]):
unassign('i','j'): expr3:=(i,j)->  i^2-3*j^4:
matrix([
    [seq(expr3(1,j),j=1..5)],
    [seq(expr3(2,j),j=1..5)],
    [seq(expr3(3,j),j=1..5)],
    [seq(expr3(4,j),j=1..5)]]):
```

Exercise 3 (*Maple V*)

```
r:=rand(-2..2): s:=[seq(r(),j=1..20)]: matrix(4,5,s):
r:=rand(90..99): s1:=seq(r(),j=1..10): s2:=seq(r(),j=11..20): matrix(4,5,[s1,s2]):
r:=i->evalf(sin(rand()),2): s:=seq(r(i),i=1..20), matrix(4,5,[s]):
```

EXERCISES 1.8

Exercise 1

a. $\begin{bmatrix} 3 & -7 & \pi \\ 2.8 & 21 & -100 \end{bmatrix}$ b. $\begin{bmatrix} 3 & 2.8 \\ -7 & 21 \\ \pi & -100 \end{bmatrix}$ c. $\begin{bmatrix} 3 & 2.8 \end{bmatrix}$

d. $\begin{bmatrix} 3 \\ 2.8 \end{bmatrix}$ e. $\begin{bmatrix} 1 & 0 & 0 & 0 \\ 0 & 1 & 0 & 0 \\ 0 & 0 & 1 & 0 \\ 0 & 0 & 0 & 1 \end{bmatrix}$ f. $\begin{bmatrix} 1 & 0 & 0 & 0 \\ 0 & 2 & 0 & 0 \\ 0 & 0 & 3 & 0 \\ 0 & 0 & 0 & 4 \end{bmatrix}$

g. $\begin{bmatrix} 1 & 0 & 0 & 0 \\ 0 & \sqrt{2} & 0 & 0 \\ 0 & 0 & \sqrt{3} & 0 \\ 0 & 0 & 0 & \sqrt{4} \end{bmatrix}$ h. $\begin{bmatrix} 2 & 3 & 4 \\ 3 & 4 & 5 \\ 4 & 5 & 6 \end{bmatrix}$ i. $[99]$ j. $\begin{bmatrix} 4 & 0 & 16 \\ 0 & 16 & 0 \\ 16 & 0 & 36 \end{bmatrix}$

Exercise 3

The set of pairs $\{(1, 1), (2, 2), (2, 3)\}$ is not of the form $m \times n$.

Exercise 5 (*Maple V*)

a. matrix([[1,0],[0,1]]): b. matrix([[-85],[97],[49]]):

c. matrix([[-85,97,49]]): d. matrix([[0,1,0],[1,0,0],[0,0,1]]):

Exercise 7 (*Maple V*)

[seq(i,i=1)]; [seq(i,i=2)]; [seq(i,i=5)]; [seq(i,i=6)]; return **1**, **2**, **5**, and **6**.

EXERCISES 1.9

Exercise 1

a. $A : \mathbf{1} \times \mathbf{3} \to \mathbb{R}$, defined by $A(1, 1) = 1$, $A(1, 2) = 2$, and $A(1, 3) = -1$

b. $A : \mathbf{1} \times \mathbf{3} \to \mathbb{R}$, defined by $A(1, 1) = -8$, $A(1, 2) = 2$, and $A(1, 3) = 99$

c. $A : \mathbf{1} \times \mathbf{3} \to \mathbb{X} = \{x_1, x_2, x_3\}$

Exercise 3 (*Maple V*)

linalg[vector]([1,2,-1]): linalg[vector]([-8,2,99]): linalg[vector]([x[1],x[2],x[3]]):

Exercise 5 (*Maple V*)

3*[1,2,-1]-7*[-8,2,99]+[1,0,-100]; returns $[60, -8, -796]$.

Exercise 7 (*Maple 6*)

Vector[row]([1,2,-1]): Vector[row]([-8,2,99]): Vector[row]([x[1],x[2],x[3]]):

EXERCISES 1.10

Exercise 1

a. $\begin{bmatrix} 2 & 1 & -1 & 2 \\ 1 & -3 & 1 & 1 \\ 2 & 1 & -2 & 6 \end{bmatrix}$ b. $\begin{bmatrix} 2 & 1 & -1 & 2 \\ 1 & -3 & 1 & 1 \\ 2 & 1 & -1 & 6 \end{bmatrix}$ c. $\begin{bmatrix} 2 & 1 & -1 & 2 \\ 1 & -3 & 1 & 1 \end{bmatrix}$

d. $\begin{bmatrix} 1 & -1 & 1 & 2 \\ -1 & 1 & -7 & 4 \\ 2 & 1 & 0 & 5 \end{bmatrix}$ e. $\begin{bmatrix} 1 & 2 & -1 & 2 \\ 0 & -1 & 1 & 4 \\ 2 & 1 & 0 & 5 \end{bmatrix}$ f. $\begin{bmatrix} 1 & 0 & 2 \\ 0 & 1 & 3 \\ 2 & 1 & 5 \end{bmatrix}$

Exercise 3

a. $p(t) = 5 + \frac{2}{3}t - \frac{4}{3}t^2$

b. $p(t) = 5 + \frac{101}{420}t - \frac{433}{280}t^2 + \frac{179}{840}t^3$

c. $p(t) = 5 - \frac{529}{1890}t - \frac{1636}{945}t^2 + \frac{3859}{7560}t^3 - \frac{281}{7560}t^4$

Exercise 5 (*Maple V*)

a. A:=matrix([[1,2,-1,2],[1,-1,1,4],[2,1,-1,5]]): rref(A); shows that $x = 3$, $y = 0$, and $z = 1$.

b. A:=matrix([[1,-1,1,2],[1,-1,1,4],[2,1,-1,5]]): rref(A); shows that the third row of the row-reduced matrix is [0 0 0 1]. This tells us that the variable z has no value. Hence the linear system is inconsistent.

c. A:=matrix([[1,2,-1,1,2],[1,-1,1,0,4],[2,1,-1,0,5]]): rref(A);
The first row shows that $x = 3$, the second that $y + w = 0$, and the third that $z + w = 1$. The system has infinitely many solutions.

Exercise 7 (*Maple V*)

with(linalg): A:=randmatrix(15,15): X:=[seq(x[i],i=1..15)]: H:=geneqns(A,X):
The command H[3]; then displays the third equation.

Exercise 9 (*Maple 6*)

with(LinearAlgebra): eqns:=[x+2*y-z=2,x-y+z=4,2*x+y-z=5]: vars:=[x,y,z]:
GenerateMatrix(eqns,vars,augmented):

EXERCISES 1.11

Exercise 1

$\mathbf{r}_1(A) = [1\ 0\ 0\ 0\ 3]$, $\mathbf{r}_2(A) = [1\ 2\ 0\ 5\ 3]$, $\mathbf{r}_3(A) = [0\ 0\ 0\ 4\ -1]$

$\mathbf{r}_1(B) = [1\ 0\ 0\ 0\ 3]$, $\mathbf{r}_2(B) = [0\ 2\ 0\ 5\ 3]$, $\mathbf{r}_3(B) = [0\ 0\ 0\ 0\ 0]$

Exercise 3

$4\mathbf{r}_1(A) = [4\ 0\ 0\ 0\ 12]$, $4\mathbf{r}_2(A) = [4\ 8\ 0\ 20\ 12]$, $4\mathbf{r}_3(A) = [0\ 0\ 0\ 16\ -4]$

$4\mathbf{r}_1(B) = [4\ 0\ 0\ 0\ 12]$, $4\mathbf{r}_2(B) = [0\ 8\ 0\ 20\ 12]$, $4\mathbf{r}_3(B) = [0\ 0\ 0\ 0\ 0]$

$\mathbf{r}_2(A) = [1\ 2\ 0\ 5\ 3] \rightarrow \mathbf{r}_2(A) + 7\mathbf{r}_1(A) = [8\ 2\ 0\ 5\ 24]$

$\mathbf{r}_2(B) = [0\ 2\ 0\ 5\ 3] \rightarrow \mathbf{r}_2(B) + 7\mathbf{r}_1(B) = [7\ 2\ 0\ 5\ 24]$

$\mathbf{r}_2(A) = [1\ 2\ 0\ 5\ 3] \rightarrow \mathbf{r}_3(A) = [0\ 0\ 0\ 4\ -1]$

$\mathbf{r}_3(A) = [0\ 0\ 0\ 4\ -1] \rightarrow \mathbf{r}_2(A) = [1\ 2\ 0\ 5\ 3]$

$\mathbf{r}_2(B) = [0\ 2\ 0\ 5\ 3] \rightarrow \mathbf{r}_3(B) = [0\ 0\ 0\ 0\ 0]$

$\mathbf{r}_3(B) = [0\ 0\ 0\ 0\ 0] \rightarrow \mathbf{r}_2(B) = [0\ 2\ 0\ 5\ 3]$

Exercise 5 (a)

The following operations convert the matrix A to row echelon form:

$$A = \begin{bmatrix} -1 & 1 & 4 & 1 \\ -3 & 3 & 5 & 5 \\ -2 & -3 & 3 & -3 \end{bmatrix} \to \begin{bmatrix} -1 & 1 & 4 & 1 \\ 0 & 0 & -7 & 2 \\ -2 & -3 & 3 & -3 \end{bmatrix} \to \begin{bmatrix} -1 & 1 & 4 & 1 \\ 0 & 0 & -7 & 2 \\ 0 & -5 & -5 & -5 \end{bmatrix}$$

$$\to \begin{bmatrix} -1 & 1 & 4 & 1 \\ 0 & -5 & -5 & -5 \\ 0 & 0 & -7 & 2 \end{bmatrix} = B$$

The leading entries of the first, second, and third row of matrix B are -1, -5, and -7 and occur in the first, second, and third columns, respectively. All entries below -1 and -5 are 0. The matrix B has no zero row.

Exercise 7 (*Maple V*)

with(linalg):

A:=matrix([[-1,1,4,1],[-3,3,5,5],[-2,-3,3,-3]]):

A1:=addrow(A,1,2,-3): A2:=addrow(A1,1,3,-2): A3:=swaprow(A2,2,3):

Exercise 9 (*Maple V*)

A:=matrix([[4,-13,5,0,5],[4,4,2,5,-5],[1,0,-3,-5,-8],[-5,4,7,-9,5]]):

B:=matrix([[4,-13,5,0,5],[0,68,-12,20,-40],[0,0,-250,-405,-499],[0,0,0,5840,4522]]):

equal(rref(A),rref(B)); returns *true*.

Exercise 11 (*Maple V*)

with(LinearAlgebra): A:=Matrix([[-1,1,4,1],[-3,3,5,5],[-2,-3,3,-3]]):

A1:=RowOperation(A,[2,1],-3): A2:=RowOperation(A1,[3,1],-2):

A3:=RowOperation(A2,[2,3]):

EXERCISES 1.12

Exercise 1 (a, b)

a. $R_2 \to R_2 + (-1)\,R_1$, $R_3 \to R_3 + (-1/4)\,R_1$, $R_4 \to R_4 + (5/4)\,R_1$, $R_3 \to R_3 + \left(-\frac{13}{4}\frac{1}{17}\right) R_2$, $R_4 \to R_4 + \left(\frac{49}{4}\frac{1}{17}\right) R_2$, $R_4 \to R_4 + \left(\frac{377}{34}\frac{34}{125}\right) R_3$, $R_2 \to \left(\frac{8}{17}\right) R_2$, $R_3 \to \left(\frac{250}{1}\frac{34}{125}\right) R_3$

b. $R_2 \to \left(\frac{2}{17}\right) R_2$, $R_4 \to \left(-\frac{1}{5840}\frac{584}{25}\right) R_4$

Exercise 3

Any matrix A is row equivalent to itself since we reduce A to A by multiplying one of its rows by 1. Now suppose that A is row equivalent to B. This shows that row equivalence is reflexive. Then B results from A by a finite number of elementary row operations. If we reverse these operations, we recover A from B. Hence row equivalence is symmetric. If A is row equivalent to B and B is row equivalent to C, then we can combine the elementary row operations that convert A to B and B to C to convert A to C. Hence row equivalence is transitive.

Exercise 5 (*Maple V*)

```
with(linalg):
A:=matrix([[4,-13,5,0,5],[4,4,2,5,-5],[1,0,-3,-5,-8],[-5,4,7,-9,5]]):
B:=matrix([[4,-13,5,0,5],[0,68,-12,20,-40],[0,0,-250,-405,-499],[0,0,0,5840,4522]]):
equal(rref(A),rref(B)):
```

Exercise 7 (*Maple 6*)

```
with(LinearAlgebra):
A:=Matrix([[4,-13,5,0,5],[4,4,2,5,-5],[1,0,-3,-5,-8],[-5,4,7,-9,5]]):
A1:=RowOperation(A,[2,1],-1):
A2:=RowOperation(A1,[3,1],-1/4):
A3:=RowOperation(A2,[4,1],5/4):
A4:=RowOperation(A3,[3,2],(-13/4)*(1/17)):
A5:=RowOperation(A4,[4,2],(49/4)*(1/17)):
A6:=RowOperation(A5,[4,3],(377/34)*(34/125)):
A7:=RowOperation(A6,2,8/17):
B:=Matrix([[4,-13,5,0,5],[0,68,-12,20,-40],[0,0,-250,-405,-499],[0,0,0,5840,4522]]):
B1:=RowOperation(B,2,2/17):
B2:=RowOperation(B1,3,(1/250)*(125/34)):
B3:=RowOperation(B2,4,(1/5840)*(-584/25)):
```

EXERCISES 1.13

Exercise 1

The matrix (a) is not in reduced row echelon form since the first column contains two nonzero entries. The matrix (b) is not in reduced row echelon form since the fourth column contains two nonzero entries. The matrix (c) is not in reduced row echelon form since the leading entry of the first row is in a column to the right of the leading entries of the second row. The matrix (d) is not in reduced row echelon form since the second columns contains two nonzero entries. The matrices (e) and (f) are in reduced row echelon form.

Exercise 3 (a)

The reduced row echelon form of the matrix (a) is

$$\begin{bmatrix} 1 & 0 & 0 & 0 & 3 \\ 0 & 1 & 0 & 0 & \frac{5}{8} \\ 0 & 0 & 0 & 1 & -\frac{1}{4} \end{bmatrix}$$

This shows that the first, second, and fourth columns are pivot columns of the given matrix.

Exercise 5 (*Maple V*)

```
with(linalg): Id:=matrix([[1,0,0],[0,1,0],[0,0,1]]): A:=matrix([[1,2,0],[0,0,1],[0,0,0]]):
E1:=addrow(Id,1,2): E2:=addrow(Id,1,3):
E3:=addrow(Id,2,1): E4:=addrow(Id,2,3):
```

```
B:=evalm(E4&*E3&*E2&*E1&*A):
E5:=swaprow(E4,1,2): C:=evalm(E5&*B):
```
equal(evalm(A),rref(B)); returns *true*.

equal(evalm(A),rref(C)); returns *true*.

Exercise 7 (a) (*Maple V*)

with(linalg): A:=matrix([[-1,1,4,1],[-3,3,5,5],[-2,-3,3,-3]]): rref(A); returns

$$\begin{bmatrix} 1 & 0 & 0 & -\frac{6}{7} \\ 0 & 1 & 0 & \frac{9}{7} \\ 0 & 0 & 1 & -\frac{2}{7} \end{bmatrix}$$

EXERCISES 1.14

Exercise 1 (a, b, c)

a. $\begin{bmatrix} 2 & 1 & -1 \\ 1 & -3 & 1 \\ 2 & 1 & -2 \end{bmatrix}\begin{bmatrix} x \\ y \\ z \end{bmatrix} = \begin{bmatrix} 2 \\ 1 \\ 9 \end{bmatrix}$ b. $\begin{bmatrix} 2 & 1 & -1 & -5 \\ 1 & -3 & 1 & 0 \\ 2 & 1 & -2 & 1 \end{bmatrix}\begin{bmatrix} x \\ y \\ z \end{bmatrix} = \begin{bmatrix} 0 \\ 0 \\ 0 \end{bmatrix}$

c. $\begin{bmatrix} 2 & 1 & -1 \\ 1 & -3 & 1 \\ 6 & -3 & 7 \\ 1 & -3 & 1 \end{bmatrix}\begin{bmatrix} x \\ y \\ z \end{bmatrix} = \begin{bmatrix} 1 \\ 2 \\ 3 \\ 4 \end{bmatrix}$

Exercise 3

$$\begin{bmatrix} 1 & 0 & 0 \\ 0 & 1 & 0 \\ 0 & 0 & 1 \end{bmatrix}\begin{bmatrix} x \\ y \\ z \end{bmatrix} = \begin{bmatrix} 4+3t \\ 8+2t \\ 3+7t \end{bmatrix}$$

Exercise 5

$p(x) = 381 - \frac{4288}{5}x + \frac{10\,696}{15}x^2 - 288x^3 + \frac{182}{3}x^4 - \frac{32}{5}x^5 + \frac{4}{15}x^6$

EXERCISES 1.15

Exercise 1 (a, b, c)

a. $\begin{bmatrix} 1 & 3 & 3 & 1 \\ 0 & -8 & -6 & -3 \\ 0 & 0 & 30 & -29 \\ 0 & 0 & 0 & -38 \end{bmatrix}$

Each column of the original matrix is a pivot column.

b. $\begin{bmatrix} 2 & 3 & 3 & 2 \\ 0 & 2 & 1 & 1 \\ 0 & 0 & -9 & 5 \end{bmatrix}$

The first three columns of the original matrix are pivot columns.

c. $\begin{bmatrix} 4 & 0 & 4 & 1 & 4 \\ 0 & 16 & -8 & 7 & 0 \\ 0 & 0 & 8 & -15 & -32 \end{bmatrix}$

The first three columns of the original matrix are pivot columns.

Exercise 3

a. $\{(0, 0, 0)\}$ b. $\mathbb{R}^3$ c. $\{(0, 0, 0)\}$ d. $\{(0, 0, 0)\}$ e. The straight line $\{(x, -2x, x) : x \in \mathbb{R}\}$
f. The plane $\{(5y - 7z, y, z) : z, y \in \mathbb{R}\}$

Exercise 5 (*Maple V*)

solve({-lambda*x-6*y=0,x+(5-lambda)*y=0},{lambda,x,y}); returns
$\{\lambda = \lambda, x = 0, y = 0\}, \{y = y, \lambda = 2, x = -3y\}, \{y = y, \lambda = 3, x = -2y\}$.
For $\lambda = 2$ and 3, the system has nontrivial solutions.

Exercise 7 (*Maple V*)

A:=matrix([[1,0,0],[0,1,0],[0,0,1]]): b:=[0,0,0]: linalg[linsolve](A,b); returns [0, 0, 0].

Exercise 9 (*Maple V*)

a. solve({
x[1]+3*x[2]+3*x[3]+4*x[4]=0, 5*x[1]+3*x[4]=0,
3*x[1]+x[2]+3*x[3]=0, 3*x[1]+4*x[2]+2*x[3]+3*x[4]=0},
{x[1],x[2],x[3],x[4]}); returns $\{x_4 = 0, x_2 = 0, x_1 = 0, x_3 = 0\}$.

b. solve({
2*x[1]+3*x[2]+3*x[3] = 0, 5*x[1]+5*x[2]+4*x[3]= 0,
2*x[2]+x[3] = 0},{x[1],x[2],x[3]}); returns $\{x_2 = 0, x_1 = 0, x_3 = 0\}$.

c. solve({
x[1]+3*x[2]+3*x[3]+x4=0, 5*x[1]+3*x[4] = 0, 3*x[1]+x[2]+3*x[3] = 0,
3*x[1]+4*x[2]+2*x[3]+3*x[4] = 0, 2*x[1]+x[2]+3*x[3]+5*x[4] =0},
{x[1],x[2],x[3],x[4]}); returns $\{x_4 = 0, x_2 = 0, x_1 = 0, x_3 = 0\}$.

EXERCISES 1.16

Exercise 1

$T_A(1, 3, 3) = (14, 32, 46)$, $T_A(-3, 2, 1) = (4, 4, 0)$, $T_A(0, 0, 0) = (0, 0, 0)$, $T_A(1, 1, 1) = (6, 15, 22)$

Exercise 3

For every (a, b) in $\mathbb{R}^2$, the linear system $\{2x = a, 3y = b\}$ has precisely one solution.

Exercise 5

Since $A\mathbf{x} = \mathbf{0}$ for all $\mathbf{x} \in \mathbb{R}^3$, the range of T_A is $\{(0, 0)\}$.
Since $A\mathbf{x}$ is defined for all $\mathbf{x} \in \mathbb{R}^3$, the domain of T_A is $\mathbb{R}^3$.

Exercise 7

If $a\mathbf{x} + b\mathbf{y} = 3(3, 3, 3) + 2(0, 2, 4) = (9, 13, 17)$, then $A(a\mathbf{x} + b\mathbf{y}) = (-44, 38, 0)$.
Since $aA(\mathbf{x}) = (-36, 18, 0)$, and $bA(\mathbf{y}) = (-8, 20, 0)$, so $aA(\mathbf{x}) + bA(\mathbf{y}) = (-44, 38, 0)$.

Exercise 9 (*Maple V*)
We must find two distinct solutions of the equation $x_1 + 2x_2 + 3x_3 = y_1 + 2y_2 + 3y_3$.
Let $x_2 = x_3 = 2$, $y_2 = y_3 = 3$, $x_1 = 5$ and $y_1 = 0$.
The linear system associated with $T_A : \mathbb{R}^3 \to \mathbb{R}^3$ is underdetermined.

CHAPTER 2 - Matrix Algebra

EXERCISES 2.1

Exercise 5 (*Maple V*)
f:=rand(98..100): g:=x->evalf(f(x)):
Exercise 7 (*Maple 6*)
evalf(LinearAlgebra[RandomMatrix](3,4)):

EXERCISES 2.2

Exercise 1

$$A + B = B + A = \begin{bmatrix} 3 & 7 & 9 & 2 & 5 \\ 4 & 1 & 7 & 9 & 9 \\ -4 & 5 & 11 & -7 & 8 \end{bmatrix}$$

(Commutativity of the addition of real numbers)
Exercise 3
a. The rows of A are $\mathbf{r}_1 A = [1\ 2\ 3\ 4\ 5]$, $\mathbf{r}_2 A = [2\ 1\ 0\ 5\ 6]$, $\mathbf{r}_3 A = [6\ 4\ 3\ 1\ 2]$.
b. The columns of B are

$$\mathbf{c}_1 B = \begin{bmatrix} 2 \\ 2 \\ 2 \\ 3 \\ 3 \end{bmatrix}, \quad \mathbf{c}_2 B = \begin{bmatrix} 5 \\ 2 \\ 1 \\ 8 \\ 9 \end{bmatrix}, \quad \mathbf{c}_3 B = \begin{bmatrix} 6 \\ 7 \\ 8 \\ 2 \\ 4 \end{bmatrix}$$

c. The product AB of the matrices A and B is

$$\begin{bmatrix} 39 & 89 & 72 \\ 39 & 106 & 53 \\ 35 & 67 & 98 \end{bmatrix}$$

Exercise 5

$$AB = \begin{bmatrix} 39 & 89 & 72 \\ 39 & 106 & 53 \\ 35 & 67 & 98 \end{bmatrix}, BA = \begin{bmatrix} 48 & 33 & 24 & 39 & 52 \\ 48 & 34 & 27 & 25 & 36 \\ 52 & 37 & 30 & 21 & 32 \\ 31 & 22 & 15 & 54 & 67 \\ 45 & 31 & 21 & 61 & 77 \end{bmatrix}$$

The matrices AB and BA have different dimensions.

Exercise 7

$$B\mathbf{x} = \begin{bmatrix} 3x + 9y \\ 11y \\ 3x + 3y \end{bmatrix}, \quad A(B\mathbf{x}) = \begin{bmatrix} 12x + 40y \\ 6x + 29y \\ 27x + 107y \end{bmatrix}, \quad (AB)\mathbf{x} = \begin{bmatrix} 12x + 40y \\ 6x + 29y \\ 27x + 107y \end{bmatrix}$$

The vectors $A(B\mathbf{x})$ and $(AB)\mathbf{x}$ are equal because matrix multiplication is associative.

Exercise 9

$A + B = [a_{ij}] + [b_{ij}] = [a_{ij} + b_{ij}] = [b_{ij} + a_{ij}] = [b_{ij}] + [a_{ij}] = B + A$

Exercise 19

$0A = 0[a_{ij}] = [0a_{ij}] = [0] = \mathbf{0}$

Exercise 21

a. $\text{trace}\begin{bmatrix} 1 & 0 \\ 0 & 0 \end{bmatrix} \text{trace}\begin{bmatrix} 0 & 0 \\ 0 & 1 \end{bmatrix} = 1 \times 1 = 1$

b. $\text{trace}\begin{bmatrix} 1 & 0 \\ 0 & 0 \end{bmatrix}\begin{bmatrix} 0 & 0 \\ 0 & 1 \end{bmatrix} = \text{trace}\begin{bmatrix} 0 & 0 \\ 0 & 0 \end{bmatrix} = 0$

Exercise 23 (*Maple V*)

with(linalg): A:=randmatrix(3,5): B:=randmatrix(4,3): evalm(A&*B);

Error, (in linalg[multiply]) non matching dimensions for vector/matrix product

Since A has five rows and B has four columns, the product AB is not defined.

Exercise 25 (*Maple 6*)

with(LinearAlgebra):

A:=Matrix([[-91,-47,-61],[41,-58,-90],[53,-1,94]]):

B:=Matrix([[83,-86,23],[-84,19,-50],[88,-53,85]]):

Trace(A); returns −55. Trace(B); returns 187. Trace(A+B); returns 132. Trace(A.B); returns 428.

Trace(evalm(A&*B)); returns the message

Error, LinearAlgebra:- Trace expects its 1st argument, M, to be of type Matrix, but received array(1 .. 3, 1 .. 3, [(1, 2)=10166, (3, 1)=12755, (1, 3)=-4928, (2, 1)=355, (1, 1)=-8973, (2, 2)=142, (3, 3)=9259, (3, 2)=-9559, (2,3)=-3807]).

since the **Trace** function requires an object of type Matrix as its input.

Trace(Matrix(evalm(A&*B))); returns the correct answer 428.

EXERCISES 2.3

Exercise 1

a. The matrix is upper triangular since $a_{ij} = 0$ for $i > j$.
b. The matrix is upper triangular since $a_{ij} = 0$ for $i > j$, lower triangular since $a_{ij} = 0$ for $i < j$, diagonal since $a_{ij} = 0$ for $i \neq j$, and symmetric since $a_{ij} = a_{ji}$ for all i, j.
c. The matrix is neither upper nor lower triangular. It is not diagonal and not symmetric.
d. The matrix is upper, lower triangular, diagonal, and symmetric.
e. The matrix is upper and lower triangular, diagonal, and symmetric.
f. The matrix is symmetric, not triangular, and not diagonal.

Exercise 3

Right multiplication by the matrix (a) leaves A unchanged.
Right multiplication by the matrix (b) interchanges the first and third columns of A.
Right multiplication by the matrix (c) interchanges the first and last columns and the second and third columns of A.
Left multiplication by the matrix (a) leaves A unchanged.
Left multiplication by the matrix (b) interchanges the first and third rows of A.
Left multiplication by the matrix (c) interchanges the first and last rows and the second and third rows of A.

Exercise 5

A function $A : \mathbf{7} \times \mathbf{4} \to \mathbb{Z}$, with the property that $A(i, j) = 0$ if $i \neq j$.

Exercise 7

$$AB = \begin{bmatrix} 0 & 0 & 0 \\ 0 & 0 & 0 \\ 0 & 0 & 0 \end{bmatrix}, BA = \begin{bmatrix} 3 & 3 & 0 \\ -3 & -3 & 0 \\ 0 & 0 & 0 \end{bmatrix}$$

Exercise 9

Since $(EA)^T = A^T E^T$ for all 3×3 matrices A and E, it suffices to show that if E is an.elementary 3×3 permutation matrix, then $E = E^T$.

Exercise 11

The matrix A is row equivalent to B if and only if there exists a sequences of elementary row operations that convert A to B. The result therefore follows from Theorem 2.11.

Exercise 13

Let

$$A = \begin{bmatrix} a & b & c \\ d & e & f \\ g & h & i \end{bmatrix}$$

be a row echelon matrix. If $a = b = c = 0$, then it follows from the definition of row echelon matrices that A is a zero matrix. Hence it is upper triangular. Suppose that $a \neq 0$. Then a is a leading entry and all entries below a are zero. It therefore remains to show that $h = 0$.

Suppose that $e = f = 0$. Then $h = 0$ since all rows below a zero row must be zero. If $e = 0$ and $f \neq 0$, then f is the leading entry of the second row. If $h \neq 0$, then h is the leading entry of the third row. It must therefore occur in a column to the right of the column containing f. This is impossible. Hence $h = 0$. Hence A is upper triangular.

Exercise 15

$$\text{a. } \begin{bmatrix} 1 & 2 \\ 4 & 5 \end{bmatrix} \text{ b. } \begin{bmatrix} 5 & 6 \\ 8 & 9 \end{bmatrix} \text{ c. } \begin{bmatrix} 4 & 5 \\ 7 & 8 \end{bmatrix} \text{ d. } \begin{bmatrix} 2 & 3 \\ 5 & 6 \end{bmatrix}$$

Exercise 17 (*Maple V*)

```
with(linalg):
a. diag(rand(1..20)(),rand(4..9)(),rand(-7..-2)()):
b. f:=x-> evalf(sin(rand())): diag(f(1),f(2),f(3)):
c. randpolynomial := proc() Randpoly(5, x) mod 3 end proc:
diag(randpolynomial(),randpolynomial(),randpolynomial()):
```

Exercise 19 (*Maple V*)

```
with(linalg): A:=randmatrix(5,5): B:=evalm(transpose(A)&*A): B[18,18]: B[25,20]:
```

Exercise 21 (*Maple V*)

```
with(linalg): A:=randmatrix(2,2): B:=transpose(A):
equal(multiply(A,B),multiply(B,A)); returns false.
```

Exercise 23 (*Maple V*)

```
with(linalg): randompoly:=proc() Randpoly(3,x) mod 2 end proc:
A:=randmatrix(3,3,entries=randompoly):
B:=randmatrix(3,3,entries=randompoly):
equal(multiply(A,B),multiply(B,A)); returns false.
```

Exercise 25 (*Maple V*)

```
with(linalg): A:=evalf(randmatrix(3,7)):
B:=evalm(A&*transpose(A)): evalm(B-transpose(B)):
```

Exercise 27 (*Maple V*)

```
A:=randmatrix(6,6):
equal(multiply(A,transpose(A)),transpose(multiply(A,transpose(A)))); returns true.
equal(multiply(transpose(A),A),transpose(multiply(transpose(A),A))); returns true.
```

Exercise 29 (*Maple 6*)

```
with(LinearAlgebra):
KroneckerDelta:= proc(i,j) if i=j then 1 else 0 fi; end:
r:=i->[seq(KroneckerDelta(i,j),j=1..5)]: Id:=Matrix([r(1),r(2),r(3),r(4),r(5)]):
```

EXERCISES 2.4

Exercise 1

a. $\begin{bmatrix} 1 & 0 \\ 0 & \frac{1}{6} \end{bmatrix}\begin{bmatrix} \frac{1}{6} & 0 \\ 0 & 1 \end{bmatrix}\begin{bmatrix} 6 & 0 \\ 0 & 6 \end{bmatrix} = \begin{bmatrix} 1 & 0 \\ 0 & 1 \end{bmatrix}$

b. $\begin{bmatrix} 1 & 0 \\ 0 & \frac{1}{6} \end{bmatrix}\begin{bmatrix} \frac{1}{6} & 0 \\ 0 & 1 \end{bmatrix}\begin{bmatrix} 1 & 0 \\ -\frac{2}{3} & 1 \end{bmatrix}\begin{bmatrix} 6 & 0 \\ 4 & 6 \end{bmatrix} = \begin{bmatrix} 1 & 0 \\ 0 & 1 \end{bmatrix}$

Exercise 3

a. $\begin{bmatrix} 6 & 0 \\ 0 & 6 \end{bmatrix} = \begin{bmatrix} 6 & 0 \\ 0 & 1 \end{bmatrix}\begin{bmatrix} 1 & 0 \\ 0 & 6 \end{bmatrix}\begin{bmatrix} 1 & 0 \\ 0 & 1 \end{bmatrix}$

b. $\begin{bmatrix} 6 & 0 \\ 4 & 6 \end{bmatrix} = \begin{bmatrix} 1 & 0 \\ \frac{2}{3} & 1 \end{bmatrix}\begin{bmatrix} 6 & 0 \\ 0 & 1 \end{bmatrix}\begin{bmatrix} 1 & 0 \\ 0 & 6 \end{bmatrix}\begin{bmatrix} 1 & 0 \\ 0 & 1 \end{bmatrix}$

Exercise 5 (*Maple V*)

a. solve({6*x=0,6*y=0},{x,y}); returns $\{y = 0, x = 0\}$.

b. solve({6*x=0,4*x+6*y=0},{x,y}); returns $\{y = 0, x = 0\}$.

Exercise 7 (a)

$$\begin{bmatrix} 4 & 0 & 0 \\ 0 & 5 & 0 \\ 0 & 0 & 4 \end{bmatrix} = \begin{bmatrix} 4 & 0 & 0 \\ 0 & 1 & 0 \\ 0 & 0 & 1 \end{bmatrix}\begin{bmatrix} 1 & 0 & 0 \\ 0 & 5 & 0 \\ 0 & 0 & 1 \end{bmatrix}\begin{bmatrix} 1 & 0 & 0 \\ 0 & 1 & 0 \\ 0 & 0 & 4 \end{bmatrix}\begin{bmatrix} 1 & 0 & 0 \\ 0 & 1 & 0 \\ 0 & 0 & 1 \end{bmatrix}$$

Exercise 9 (b)

The matrix is equal to AB, where

$$A = \begin{bmatrix} 4 & 0 & 0 \\ 0 & 1 & 0 \\ 0 & 0 & 1 \end{bmatrix}\begin{bmatrix} 1 & 0 & 0 \\ 0 & 5 & 0 \\ 0 & 0 & 1 \end{bmatrix}\begin{bmatrix} 1 & 0 & 0 \\ 0 & 1 & 0 \\ 0 & 0 & 4 \end{bmatrix}$$

and

$$B = \begin{bmatrix} 1 & 0 & 0 \\ \frac{2}{5} & 1 & 0 \\ 0 & 0 & 1 \end{bmatrix}\begin{bmatrix} 1 & 0 & 0 \\ 0 & 1 & 0 \\ \frac{3}{4} & 0 & 1 \end{bmatrix}\begin{bmatrix} 1 & 0 & 0 \\ 0 & 1 & 0 \\ 0 & \frac{1}{2} & 1 \end{bmatrix}$$

Exercise 11 (c)

Let

$$\begin{bmatrix} \frac{1}{\sqrt{2}} & \frac{1}{\sqrt{2}} & 0 \\ -\frac{1}{\sqrt{2}} & \frac{1}{\sqrt{2}} & 0 \\ 0 & 0 & 1 \end{bmatrix}\begin{bmatrix} x \\ y \\ z \end{bmatrix} = \begin{bmatrix} \frac{1}{2}\sqrt{2}x + \frac{1}{2}\sqrt{2}y \\ -\frac{1}{2}\sqrt{2}x + \frac{1}{2}\sqrt{2}y \\ z \end{bmatrix} = \begin{bmatrix} 0 \\ 0 \\ 0 \end{bmatrix}$$

Then $z = 0$. Moreover, $\left(\frac{1}{2}\sqrt{2}x + \frac{1}{2}\sqrt{2}y\right) + \left(-\frac{1}{2}\sqrt{2}x + \frac{1}{2}\sqrt{2}y\right) = 0$ implies that $y = 0$.

Hence $x = 0$.

The result also follows directly from Theorem 2.21 since the coefficient matrix is invertible.

Exercise 13

If

$$A = \begin{bmatrix} 1 & 0 \\ 0 & 0 \\ 0 & 1 \end{bmatrix} \quad \text{and} \quad B = \begin{bmatrix} 1 & 0 & 0 \\ 0 & 0 & 1 \end{bmatrix}$$

then $BA = I_2$ and $AB = I_3$.

Theorem 2.24 holds only for square matrices.

Exercise 15 (*Maple V*)

with(linalg): A:=matrix([[4,-1,-4,-1],[-1,-4,-2,3],[4,4,-3,-3],[4,-1,0,5]]):
b:=vector(4,[b1,b2,b3,b4]): B:=augment(A,b): C:=rref(B):
evalm(A&*col(C,5)); returns [b1, b2, b3, b4].

Exercise 17 (*Maple 6*)

restart: with(LinearAlgebra): A:=Matrix(4,4,rand(-5..5)); returns

$$A := \begin{bmatrix} 4 & -1 & -4 & -1 \\ -1 & -4 & -2 & 3 \\ 4 & 4 & -3 & -3 \\ 4 & -1 & 0 & 5 \end{bmatrix}$$

B:=MatrixInverse(A): Equal(Matrix(evalm(A&*B)),IdentityMatrix(4)); returns *true*.

Exercise 19 (*Maple 6*)

A 3×3 real matrix A is either invertible or singular. We use the **RandomMatrix** function to generate a lower–triangular matrix with nonzero diagonal entries. This guarantees that the columns are linearly independent.

with(LinearAlgebra): A:=RandomMatrix(3,3,generator=1..9,
outputoptions=[shape=triangular[lower]]); returns

$$A := \begin{bmatrix} 1 & 0 & 0 \\ 2 & 7 & 0 \\ 8 & 2 & 7 \end{bmatrix}$$

The vector

$$\mathbf{x} = \begin{bmatrix} 1 & 0 & 0 \\ 2 & 7 & 0 \\ 8 & 2 & 7 \end{bmatrix}^{-1} \begin{bmatrix} x \\ y \\ z \end{bmatrix} = \begin{bmatrix} x \\ -\frac{2}{7}x + \frac{1}{7}y \\ -\frac{52}{49}x - \frac{2}{49}y + \frac{1}{7}z \end{bmatrix}$$

is a solution for the linear system $A\mathbf{x} = \mathbf{b}$.

If $\mathbf{b} = (0, 0, 0)$, then $\mathbf{x} = (0, 0, 0)$.

If A is singular, such as

$$\begin{bmatrix} 1 & 0 & 0 \\ 2 & 7 & 0 \\ 8 & 2 & 0 \end{bmatrix}$$

then

$$\begin{bmatrix} 1 & 0 & 0 \\ 2 & 7 & 0 \\ 8 & 2 & 0 \end{bmatrix}\begin{bmatrix} x \\ y \\ z \end{bmatrix} = \begin{bmatrix} x \\ 2x+7y \\ 8x+2y \end{bmatrix} = \begin{bmatrix} 0 \\ 0 \\ 0 \end{bmatrix}$$

if and only if $x = y = 0$. For any nonzero value of z, the vector $\mathbf{x} = (0, 0, z)$ is a nontrivial solution.

EXERCISES 2.5

Exercise 1 (a)

$$AB = \begin{bmatrix} 0 & 1 \\ -1 & 0 \end{bmatrix}\begin{bmatrix} 0 & -1 \\ 1 & 0 \end{bmatrix} = \begin{bmatrix} 1 & 0 \\ 0 & 1 \end{bmatrix}$$

Exercise 3
For any $y \neq 0$, the vector $(x, y) = \left(\frac{3}{2}y, y\right)$ is a nontrivial solution.
Exercise 5

$$AB = \begin{bmatrix} 1 & 1 \\ 0 & 0 \end{bmatrix}\begin{bmatrix} 1 & 2 \\ -1 & -2 \end{bmatrix} = CA = \begin{bmatrix} 0 & 3 \\ 0 & 4 \end{bmatrix}\begin{bmatrix} 1 & 1 \\ 0 & 0 \end{bmatrix} = \begin{bmatrix} 0 & 0 \\ 0 & 0 \end{bmatrix}$$

However, $B \neq C$ and A is not invertible.
Exercise 7 (a) (*Maple 6*)
with(LinearAlgebra): A:=Matrix([[2,-3],[-3,-3]]): A1:=MatrixInverse(A):
B:=Matrix([[0,-3],[-3,4]]): B1:=MatrixInverse(B):
Equal(MatrixInverse(A.B),B1.A1))); returns *true*.
Exercise 9 (*Maple 6*)
A:=Matrix([[1,a,a^2],[1,b,b^2],[1,c,c^2]]): B:=MatrixInverse(A):
Equal(Matrix(simplify(evalm(A&*B))),IdentityMatrix(3)); returns *true*.

EXERCISES 2.6

Exercise 1
Let A and B be two $n \times n$ orthogonal matrices. Since A and B are orthogonal, $(AB)^{-1} = B^{-1}A^{-1} = B^TA^T = (AB)^T$. Hence AB is orthogonal.

Exercise 3

$$AA^T = \begin{bmatrix} 0 & 1 \\ 1 & 0 \end{bmatrix}\begin{bmatrix} 0 & 1 \\ 1 & 0 \end{bmatrix}^T = \begin{bmatrix} 0 & 1 \\ 1 & 0 \end{bmatrix}\begin{bmatrix} 0 & 1 \\ 1 & 0 \end{bmatrix} = \begin{bmatrix} 1 & 0 \\ 0 & 1 \end{bmatrix}$$

Exercise 5 (c, d)

c. $\begin{bmatrix} \frac{15}{19} & -\frac{10}{19} & \frac{6}{19} \\ -\frac{10}{19} & -\frac{6}{19} & \frac{15}{19} \\ \frac{6}{19} & \frac{15}{19} & \frac{10}{19} \end{bmatrix}\begin{bmatrix} \frac{15}{19} & -\frac{10}{19} & \frac{6}{19} \\ -\frac{10}{19} & -\frac{6}{19} & \frac{15}{19} \\ \frac{6}{19} & \frac{15}{19} & \frac{10}{19} \end{bmatrix} = \begin{bmatrix} 1 & 0 & 0 \\ 0 & 1 & 0 \\ 0 & 0 & 1 \end{bmatrix}$

d. Let $\mathbf{u}$ be the vector $(4, 5)$. Then

$$A = I - \frac{2}{\mathbf{u}^T\mathbf{u}}\mathbf{u}\mathbf{u}^T = \begin{bmatrix} \frac{9}{41} & -\frac{40}{41} \\ -\frac{40}{41} & -\frac{9}{41} \end{bmatrix} \quad \text{and} \quad AA^T = \begin{bmatrix} 1 & 0 \\ 0 & 1 \end{bmatrix}$$

Exercise 7 (c) (*Maple V*)

with(linalg): A:=matrix([[15/19,-10/19,6/19],[-10/19,-6/19,15/19],[6/19,15/19,10/19]]): B:=matrix([[9/41,-40/41],[-40/41,-9/41]]): [orthog(A),orthog(B)]; returns $[true, true]$.

EXERCISES 2.7

Exercise 1

The matrices $A = [a_{ij}]$ in (a) and (c) are upper triangular since they are square and since in each case, $a_{ij} = 0$ for all $i > j$. The matrix (b) is not upper triangular since it is not square.

Exercise 3

$A = E_1^{-1}E_2^{-1}E_3^{-1}E_4^{-1}E_5^{-1}E_6^{-1}U$, where

$$E_1 = \begin{bmatrix} 1 & 0 & 0 & 0 \\ 0 & 1 & 2 & 0 \\ 0 & 0 & 1 & 0 \\ 0 & 0 & 0 & 1 \end{bmatrix}, E_2 = \begin{bmatrix} 1 & 0 & 0 & 0 \\ 0 & 1 & 0 & 0 \\ -2 & 0 & 1 & 0 \\ 0 & 0 & 0 & 1 \end{bmatrix}, E_3 = \begin{bmatrix} 1 & 0 & 0 & 0 \\ 0 & 1 & 0 & 0 \\ 0 & -2 & 1 & 0 \\ 0 & 0 & 0 & 1 \end{bmatrix}$$

and

$$E_4 = \begin{bmatrix} 1 & 0 & 0 & 0 \\ 0 & 1 & 0 & 0 \\ 0 & 0 & -\frac{1}{4} & 0 \\ 0 & 0 & 0 & 1 \end{bmatrix}, E_5 = \begin{bmatrix} 1 & 0 & 0 & 0 \\ 0 & 1 & 0 & 0 \\ 0 & 0 & 1 & 0 \\ 0 & 0 & -2 & 1 \end{bmatrix}, E_6 = \begin{bmatrix} 1 & 2 & 4 & 1 \\ 0 & -3 & -4 & 4 \\ 0 & 0 & 1 & \frac{3}{2} \\ 0 & 0 & 0 & 0 \end{bmatrix}$$

and where

$$U = \begin{bmatrix} 1 & 2 & 4 & 1 \\ 0 & -3 & -4 & 4 \\ 0 & 0 & 1 & \frac{3}{2} \\ 0 & 0 & 0 & 0 \end{bmatrix}$$

Exercise 5

$$\begin{bmatrix} 5 & 0 \\ -3 & 7 \end{bmatrix}\begin{bmatrix} x \\ y \end{bmatrix} = \begin{bmatrix} 1 & 0 \\ -\frac{3}{5} & 1 \end{bmatrix}\begin{bmatrix} 5 & 0 \\ 0 & 7 \end{bmatrix}\begin{bmatrix} x \\ y \end{bmatrix} = \begin{bmatrix} 1 & 0 \\ -\frac{3}{5} & 1 \end{bmatrix}\begin{bmatrix} y_1 \\ y_2 \end{bmatrix}$$

and

$$\begin{bmatrix} 1 & 0 \\ -\frac{3}{5} & 1 \end{bmatrix}\begin{bmatrix} y_1 \\ y_2 \end{bmatrix} = \begin{bmatrix} y_1 \\ -\frac{3}{5}y_1 + y_2 \end{bmatrix} = \begin{bmatrix} 3 \\ 4 \end{bmatrix}$$

Therefore $y_1 = 3$ and $-\frac{3}{5}(3) + y_2 = 4$, so $y_2 = 29/5$.

$$\begin{bmatrix} 5 & 0 \\ 0 & 7 \end{bmatrix}\begin{bmatrix} x \\ y \end{bmatrix} = \begin{bmatrix} 5x \\ 7y \end{bmatrix} = \begin{bmatrix} y_1 \\ y_2 \end{bmatrix} = \begin{bmatrix} 3 \\ \frac{29}{5} \end{bmatrix}$$

Hence $(x, y) = (3/5, 29/35)$.

Exercise 7

$$\begin{bmatrix} 1 & 0 & 0 \\ 4 & 1 & 0 \\ 9 & \frac{1}{2} & 1 \end{bmatrix}\begin{bmatrix} u \\ v \\ w \end{bmatrix} = \begin{bmatrix} u \\ 4u + v \\ 9u + \frac{1}{2}v + w \end{bmatrix} = \begin{bmatrix} 2 \\ 5 \\ 3 \end{bmatrix}$$

$u = 2$, $4u + v = 5$, $v = -3$, $9u + \frac{1}{2}v + w = 3$, $w = -\frac{27}{2}$

$$\begin{bmatrix} 1 & 0 & 2 \\ 0 & 2 & -3 \\ 0 & 0 & -\frac{27}{2} \end{bmatrix}\begin{bmatrix} x \\ y \\ z \end{bmatrix} = \begin{bmatrix} x + 2z \\ 2y - 3z \\ -\frac{27}{2}z \end{bmatrix} = \left[\begin{bmatrix} 2 \\ -3 \\ -\frac{27}{2} \end{bmatrix}\right]$$

$z = 1$; $2y - 3 = -3$, $y = 0$, $x = 0$

Exercise 9 (c)

$$\begin{bmatrix} 0 & 0 & 3 & 1 & 3 \\ 0 & 1 & 5 & 0 & 0 \\ 3 & 0 & 1 & 3 & 0 \\ 3 & 4 & 2 & 3 & 2 \end{bmatrix} = \begin{bmatrix} 0 & 0 & 1 & 0 \\ 0 & 1 & 0 & 0 \\ 1 & 0 & 0 & 0 \\ 0 & 0 & 0 & 1 \end{bmatrix}\begin{bmatrix} 1 & 0 & 0 & 0 \\ 0 & 1 & 0 & 0 \\ 0 & 0 & 1 & 0 \\ 1 & 4 & -\frac{19}{3} & 1 \end{bmatrix}\begin{bmatrix} 3 & 0 & 1 & 3 & 0 \\ 0 & 1 & 5 & 0 & 0 \\ 0 & 0 & 3 & 1 & 3 \\ 0 & 0 & 0 & \frac{19}{3} & 21 \end{bmatrix}$$

Exercise 11

$$L_1U_1 = \begin{bmatrix} 1 & 0 & 0 \\ 0 & 1 & 0 \\ 0 & 0 & 1 \end{bmatrix}\begin{bmatrix} 0 & 0 & 0 \\ 0 & 0 & 0 \\ 0 & 0 & 0 \end{bmatrix} = L_2U_2 = \begin{bmatrix} 1 & 0 & 0 \\ 1 & 1 & 0 \\ 1 & 1 & 1 \end{bmatrix}\begin{bmatrix} 0 & 0 & 0 \\ 0 & 0 & 0 \\ 0 & 0 & 0 \end{bmatrix}$$

Since the zero matrix is not invertible, the LU decomposition may not be unique.

Exercise 13

$$LU = \begin{bmatrix} 1 & 0 & 0 \\ 5 & 1 & 0 \\ 2 & 3 & 1 \end{bmatrix}\begin{bmatrix} 7 & 9 & 6 \\ 0 & 6 & 9 \\ 0 & 0 & 4 \end{bmatrix} = \begin{bmatrix} 7 & 9 & 6 \\ 35 & 51 & 39 \\ 14 & 36 & 43 \end{bmatrix}$$

Exercise 15 (a) (*Maple V*)

```
with(linalg): A:=matrix([[3,3,2],[5,5,4],[5,0,2]]):
LUdecomp(A,L='l',U='u'): evalm(evalm(l)&*evalm(u)); returns A.
```

Exercise 17 (*Maple 6*)

```
with(LinearAlgebra): A:=Matrix([[1,5,2,0],[3,5,3,2],[1,4,0,4],[3,3,1,5]]):
b1:=Vector[column]([1,2,0,2]):
b2:=Vector[column]([0,1,0,4]):
b3:=Vector[column]([2,8,4,-2]):
Decomp:=LUDecomposition(A):
V:=Vector[column]([x[1],x[2],x[3],x[4]]): L:=Decomp[2]: U:=Decomp[3]:
Y1:=ForwardSubstitute(L,b1): X1:=BackwardSubstitute(U,Y1); returns
```

$$X1 = (4, 7/9, -31/9, -16/9)$$

```
Y2:=ForwardSubstitute(L,b2): X2:=BackwardSubstitute(U,Y2); returns
```

$$X2 = (12, 22/9, -109/9, -49/9)$$

```
Y3:=ForwardSubstitute(L,b3): X3:=BackwardSubstitute(U,Y3); returns
```

$$X3 = (-44, -94/9, 442/9, 202/9)$$

EXERCISES 2.8

Exercise 1

$$A = \begin{bmatrix} 1 & 6 & i \\ i & -4i & 2 \end{bmatrix}, \quad B = \begin{bmatrix} i & 4 & -i \\ 5 & 21i & 6 \end{bmatrix} \quad A + B = \begin{bmatrix} 1+i & 10 & 0 \\ 5+i & 17i & 8 \end{bmatrix}$$

Exercise 3
Let

$$A = \begin{bmatrix} 5i & 2 & 3 & 4i \\ 3 & i & 4 & -i \\ i & 3 & -6i & 0 \end{bmatrix}, \quad B = \begin{bmatrix} -i & 3i & 3 \\ -i & -i & 0 \\ 5 & 1 & 6i \\ -i & 4 & 5 \end{bmatrix}$$

then

$$AB = \begin{bmatrix} 24-2i & -12+14i & 53i \\ 20-3i & 5+5i & 9+19i \\ 1-33i & -3-9i & 36+3i \end{bmatrix}$$

and

$$BA = \begin{bmatrix} 5+12i & 6-2i & -9i & 7 \\ 5-3i & 1-2i & -7i & 3 \\ -3+25i & 10+19i & 55 & 19i \\ 17+5i & 15+2i & 16-33i & 4-4i \end{bmatrix}$$

Exercise 5
Let A be the matrix

$$\begin{bmatrix} 4 & 1+it \\ 3i-t+7it^3 & -44i \end{bmatrix}$$

Then A^2 is the matrix

$$\begin{bmatrix} (16+3i)-4t-it^2+7it^3-7t^4 & (4-44i)+(44+4i)\,t \\ (132+12i)-(4-44i)\,t+(308+28i)\,t^3 & (-1936+3i)-4t-it^2+7it^3-7t^4 \end{bmatrix}$$

Exercise 7 (b) (*Maple V*)
with(linalg): P:=matrix([[0,0,0,-I],[0,0,I,0],[0,-I,0,0],[I,0,0,0]]):
equal(inverse(P),transpose(evalm(conjugate(P)))); returns *true*.
Exercise 9 (*Maple 6*)
a. with(linalg): A:=matrix([[1/sqrt(2),0,I/sqrt(2)],[0,1,0],[-1/sqrt(2),0,I/sqrt(2)]]):
equal(inverse(A),transpose(evalm(conjugate(A)))); returns *true*.
b. B:=matrix([[1/sqrt(2),I/sqrt(2)],[-1/sqrt(2),I/sqrt(2)]]):
equal(inverse(B),transpose(evalm(conjugate(B)))); returns *true*.

EXERCISES 2.9

Exercise 1 (*Maple V*)

with(plottools):

A:={arrow([0,0],[9,6],.02,.5,.05),arrow([0,0],[27,6],.02,.5,.05),arrow([0,0],[3,6],.02,.5,.05)}:

plots[display](A); returns a graph consisting of three arrows with tips on the line $y = 6$.

Exercise 3 (*Maple V*)

with(plottools),

A:={arrow([0,0],[4,-1],.02,.5,.05),arrow([0,0],[2,-1],.02,.5,.05),arrow([0,0],[4,7],.02,.5,.05)};

plots[display](A); returns a graph consisting of three arrows. The tips of the vectors $\mathbf{x}$ and $A\mathbf{x}$ determine the line $y = -1$, and the tips of the vectors $\mathbf{x}$ and $B\mathbf{x}$ determine the line $x = 4$.

Exercise 5 (*Maple V*)

with(plottools),

A:={arrow([0,0],[2,3],.02,.5,.05),arrow([0,0],[-9,-6],.02,.5,.05),
    arrow([0,0],[-9,-24],.02,.5,.05),arrow([0,0],[-3,-24],.02,.5,.05),
    arrow([0,0],[-24,-3],.02,.5,.05)}:

plots[display](A); returns a graph consisting of five arrows.

The last arrow corresponds to the vector $A\mathbf{x}$.

EXERCISES 2.10

Exercise 1

a. $\begin{bmatrix} 3 \\ 5 \\ 1 \end{bmatrix}$ b. $\begin{bmatrix} -8 \\ 0 \\ 1 \end{bmatrix}$ c. $\begin{bmatrix} 112 \\ 115 \\ 1 \end{bmatrix}$

Exercise 3

$$\begin{bmatrix} 1 & 0 & 2 \\ 0 & 1 & -5 \\ 0 & 0 & 1 \end{bmatrix}\begin{bmatrix} \cos\pi/4 & -\sin\pi/4 & 0 \\ \sin\pi/4 & \cos\pi/4 & 0 \\ 0 & 0 & 1 \end{bmatrix}\begin{bmatrix} 1 & \frac{1}{3} & 0 \\ 0 & 1 & 0 \\ 0 & 0 & 1 \end{bmatrix} = \begin{bmatrix} \frac{1}{2}\sqrt{2} & -\frac{1}{3}\sqrt{2} & 2 \\ \frac{1}{2}\sqrt{2} & \frac{2}{3}\sqrt{2} & -5 \\ 0 & 0 & 1 \end{bmatrix}$$

Exercise 5

a. $\begin{bmatrix} 4 \\ 5 \\ 2 \end{bmatrix}$ b. $\begin{bmatrix} \pi \\ -\pi \\ 8 \end{bmatrix}$ c. $\begin{bmatrix} -8 \\ -6 \\ -1 \end{bmatrix}$

Exercise 7

$$\begin{bmatrix} 1 & 0 & h \\ 0 & 1 & k \\ 0 & 0 & 1 \end{bmatrix}^{-1} = \begin{bmatrix} 1 & 0 & -h \\ 0 & 1 & -k \\ 0 & 0 & 1 \end{bmatrix} = \begin{bmatrix} 1 & 0 & h \\ 0 & 1 & k \\ 0 & 0 & 1 \end{bmatrix}^{T} = \begin{bmatrix} 1 & 0 & 0 \\ 0 & 1 & 0 \\ h & k & 1 \end{bmatrix}$$

Hence h and k must be zero.

EXERCISES 2.11

Exercise 1 (*Maple V*)

The required matrix is $[L|R]$, where

$$L = \begin{bmatrix} 16 & 174 & 9 & 30 & 98 & 49 & 9 & -8 \\ -14 & 12 & 5 & 52 & -24 & 49 & -3 & -14 \\ 22 & 294 & 22 & 86 & 150 & 118 & 14 & -18 \\ 159 & 333 & 116 & 109 & 313 & 216 & 48 & 99 \end{bmatrix}$$

and

$$R = \begin{bmatrix} 68 & 72 & 186 & -22 & 58 & 11 & 124 \\ -4 & 0 & 26 & -30 & -54 & -21 & -92 \\ 112 & 120 & 320 & -48 & 60 & 6 & 140 \\ 204 & 180 & 278 & 166 & 128 & 70 & 174 \end{bmatrix}$$

Exercise 3 (*Maple V*)

The decoded message reads "CONGRATULATIONS. YOU HAVE DONE IT."

CHAPTER 3 - Determinants

EXERCISES 3.1

Exercise 1 (a, b, c)

a. $\det(A) = (-85)(-35) - (-55)(-37) = 940$

b. $\det(A) = 97 \det \begin{bmatrix} 49 & 63 \\ -59 & 45 \end{bmatrix} - 50 \det \begin{bmatrix} 56 & 63 \\ 57 & 45 \end{bmatrix} + 79 \det \begin{bmatrix} 56 & 49 \\ 57 & -59 \end{bmatrix} = 146\,321$

c. $\det(A) = (0)(0) - (0)(0) = 0$

Exercise 3

If $A = \begin{bmatrix} t & 0 & 0 \\ 0 & t & 0 \\ 0 & 0 & t \end{bmatrix}$, then $\det(A) = t^3$. However, $a_{11}C_{11} + a_{22}C_{22} + a_{33}C_{33} = 3t^3$.

Exercise 5 (*Maple V*)

```
with(linalg): A:=matrix(3,3,x): B:=matrix(3,3,y):
evalb(det(evalm(A&*B))=det(evalm(B&*A))); returns true.
```

EXERCISES 3.2

Exercise 1

a. Since the matrix is upper triangular, the determinant is the product of the diagonal entries $(1)(2)(3) = 6$.

b. Since the matrix is upper triangular and since one of the diagonal entries is 0, we know immediately that the determinant is 0.

c. The determinant of the matrix is -32. We can use elementary row operations to simplify the calculation.

If we interchange the first and third rows, the determinant changes sign.

If we divide the entries of the second row of the matrix by 2, the determinant of the matrix is divided by 2.

$$(-1)\,(2)\det\begin{bmatrix} 2 & -1 & 0 \\ 0 & 1 & 2 \\ 0 & 0 & 8 \end{bmatrix} = \det\begin{bmatrix} 0 & -3 & 2 \\ 4 & 0 & 4 \\ 2 & -1 & 0 \end{bmatrix}$$

Hence $\det(A) = (-1)\,(2)\,(2)(1)(8) = -32$.

d. Since the matrix has a zero row, its determinant is zero.

e. It is not obvious how elementary row operations would help to simplify the manual calculation of the determinant of this matrix. Its determinant is $-91,708$.

f. The determinant of the matrix is -160. We can simplify its calculation somewhat by converting the given matrix to a new one with additional zero entries.

We subtract twice the fourth row from the third row.

We then add $3/2$ times the third row to the second row.

We then use the Laplace expansion along the first row and follow this up with a Laplace expansion along the first column. The determinant is -160.

Exercise 3

Every permutation matrix P is a product $E_1 \cdots E_p$ of elementary permutation matrices and $\det P = \det E_1 \cdots \det E_p$. Moreover, each matrix E_i is obtained from an identity matrix I by the interchange $(R_i \rightleftharpoons R_j)$ of two of its rows. Since $\det I = 1$, it follows that $\det E_i = 1$ if $i = j$ and $\det E_i = -1$ if $i \neq j$. Hence $\det(P)$ is a product of 1s and -1s and is therefore ± 1.

Exercise 5 (*Maple V*)

Generate a random 3×3 matrix $A = [a_{ij}]$ and calculate the signed elementary products $\operatorname{sgn}(\sigma)\, a_{1\sigma(1)} a_{2\sigma(2)} a_{3\sigma(3)}$ of A for the permutations $[3, 1, 2]$ and $[2, 1, 3]$.

```
e1:=6*1*7: e2:=6*4*6: e3:=7*5*7: e4:=7*6*2: e5:=3*1*2: e6:=3*4*5:
sigma:=x->x: tau:=x->-x:
```

$\operatorname{sgn}(\sigma)\,(e_i) = e_i$ and $\operatorname{sgn}(\tau)\,(e_i) = -e_i$.

Exercise 7 (*Maple V*)

```
with(linalg):
A:=randmatrix(3,3); returns the matrix
```

$$\begin{bmatrix} 10 & 57 & -82 \\ -48 & -11 & 38 \\ -7 & 58 & -94 \end{bmatrix}$$

cofactor:=(i,j)-> (-1)^(i+j)*det(minor(A,i,j)):
C11:=cofactor(1,1); returns -1170.
adjoint(A); returns the matrix of cofactors

$$\begin{bmatrix} -1170 & 602 & 1264 \\ -4778 & -1514 & 3556 \\ -2861 & -979 & 2626 \end{bmatrix}$$

Exercise 9 (*Maple V*)
a. with(linalg): A:=matrix([[1,1,2],[2,1,3],[3,1,4]]):
det(A); returns 0. Hence the matrix is not invertible.
b. The matrix is invertible since its determinant is -2.
c. The matrix is invertible since its determinant is 6.
Exercise 11 (*Maple V*)
with(linalg): A:=randmatrix(25,25): is(det(A)=det(transpose(A))); returns *true*.
Exercise 13 (*Maple 6*)
with(LinearAlgebra):
q:=x-> is(1/Determinant(x)=Determinant(MatrixInverse(x))):
A:=Matrix([[4,0,0],[0,5,0],[0,0,4]]):
B:=Matrix([[4,2,3],[0,5,2],[0,0,4]]):
C:=Matrix([[1/sqrt(2),1/sqrt(2),0],[-1/sqrt(2),1/sqrt(2),0],[0,0,1]]):
is(q(A)=true and q(B)=true and q(C)=true); returns *true*.

EXERCISES 3.3

Exercise 1
a. The matrix equation and solution are

$$\begin{bmatrix} 2 & 1 \\ 1 & -3 \end{bmatrix}\begin{bmatrix} x \\ y \end{bmatrix} = \begin{bmatrix} 2 \\ 1 \end{bmatrix}, x = \frac{\det\begin{bmatrix} 2 & 1 \\ 1 & -3 \end{bmatrix}}{\det\begin{bmatrix} 2 & 1 \\ 1 & -3 \end{bmatrix}} = 1, y = \frac{\det\begin{bmatrix} 2 & 2 \\ 1 & 1 \end{bmatrix}}{\det\begin{bmatrix} 2 & 1 \\ 1 & -3 \end{bmatrix}} = 0$$

b. The matrix equation is

$$\begin{bmatrix} 2 & 1 \\ -2 & -1 \end{bmatrix}\begin{bmatrix} x \\ y \end{bmatrix} = \begin{bmatrix} 2 \\ 1 \end{bmatrix}$$

Cramer's rule cannot be used since the determinant of the coefficient matrix is 0.

c. The matrix equation and solution are

$$\begin{bmatrix} 2 & 1 \\ 1 & -3 \end{bmatrix}\begin{bmatrix} x \\ y \end{bmatrix} = \begin{bmatrix} 0 \\ 0 \end{bmatrix}, x = \frac{\det\begin{bmatrix} 0 & 1 \\ 0 & -3 \end{bmatrix}}{\det\begin{bmatrix} 2 & 1 \\ 1 & -3 \end{bmatrix}} = 0, y = \frac{\det\begin{bmatrix} 2 & 0 \\ 1 & 0 \end{bmatrix}}{\det\begin{bmatrix} 2 & 1 \\ 1 & -3 \end{bmatrix}} = 0$$

EXERCISES 3.4

Exercise 1

$$\frac{1}{\det A}\begin{bmatrix} C_{11} & C_{21} \\ C_{12} & C_{22} \end{bmatrix} = \frac{1}{12}\begin{bmatrix} 4 & -3 \\ -4 & 6 \end{bmatrix} = \begin{bmatrix} \frac{1}{3} & -\frac{1}{4} \\ -\frac{1}{3} & \frac{1}{2} \end{bmatrix}$$

and

$$\begin{bmatrix} 6 & 3 \\ 4 & 4 \end{bmatrix}\begin{bmatrix} \frac{1}{3} & -\frac{1}{4} \\ -\frac{1}{3} & \frac{1}{2} \end{bmatrix} = \begin{bmatrix} 1 & 0 \\ 0 & 1 \end{bmatrix}$$

Exercise 3 (*Maple V*)

```
with(linalg): A:=matrix([[6,3],[4,4]]):
B:=matrix([[det(minor(A,1,1)),-det(minor(A,1,2))],
    [-det(minor(A,2,1)),det(minor(A,2,2))]]): scalarmul(B,1/det(A)); returns
```

$$\begin{bmatrix} \frac{1}{3} & -\frac{1}{4} \\ -\frac{1}{3} & \frac{1}{2} \end{bmatrix}$$

EXERCISES 3.5

Exercise 1

$$\begin{bmatrix} 3 & 2 \\ 1 & 4 \end{bmatrix} = \begin{bmatrix} 1 & 0 \\ \frac{1}{3} & 1 \end{bmatrix}\begin{bmatrix} 3 & 2 \\ 0 & \frac{10}{3} \end{bmatrix}$$

and $\det A = \det U = 10$.

Exercise 3 (*Maple V*)

```
with(linalg): A:=matrix([[8,4,-5,-5],[3,-5,8,5],[-1,0,3,-4],[-5,-2,-1,-9]]):
d:=LUdecomp(A,L='l',U='u'): det(evalm(u)); returns 2607.
```

Since $\det A = \det U$, the determinant of A is 2607.

EXERCISES 3.6

Exercise 1

Since the matrix

$$A = \begin{bmatrix} 1 & 0 & 0 \\ a & 1 & 0 \\ b & c & 1 \end{bmatrix}$$

is lower triangular, $\det A = \det A^T = 1$.

Exercise 3

The matrix A is upper triangular. Hence its determinant is the product of its diagonal entries, which is zero. The matrix B is also triangular. Hence its determinant equals $\left(4^2\right)(5)(2)(1) = 160$.

Exercise 5

Let a, b, c be nonzero diagonal entries of a 3×3 diagonal matrix. Then $a + b + c = abc$. If $a = 1$, $b = 2$, and $c = 3$, then $a + b + c = abc = 6$.

EXERCISES 3.7

Exercise 1

a. $3x + y - 13 = 0$ b. $11x - y - 29 = 0$ c. $-3x - 5y + 29 = 0$ d. $11x - 5y - 13 = 0$

Exercise 3

a. $-18x^2 - 18y^2 + 216x + 240y - 1158 = 0$ b. $-10x^2 - 10y^2 + 440x + 240y - 2030 = 0$
c. $36x^2 + 36y^2 + 216x - 264y - 492 = 0$ d. $44x^2 + 44y^2 + 440x - 264y - 1364 = 0$

Exercise 5

Let $\mathbf{u} = (a, b)$ and $\mathbf{v} = (c, d)$. Then $\mathbf{u}' = (a, b, 1)$ and $\mathbf{v}' = (c, d, 1)$.

$$\det \begin{bmatrix} a & c \\ b & d \end{bmatrix} = \det \begin{bmatrix} a & c & 0 \\ b & d & 0 \\ 1 & 1 & 1 \end{bmatrix}$$

Therefore the homogeneous coordinate embedding preserves the areas of parallelograms.

EXERCISES 3.8

Exercise 1

$$u_x = -\frac{6xv^2 - 6ux^2}{6v^3 + 6u^3}, \quad v_x = -\frac{6vx^2 + 6xu^2}{6v^3 + 6u^3}$$

Exercise 3

$$u_x = -\frac{-z - 2x}{z - v}, \quad u_y = -\frac{1}{343}\frac{343z - 768}{z - v}, \quad u_z = -1$$

$$v_x = -\frac{2x + v}{z - v}, \quad v_y = -\frac{-2y - v}{z - v}, \quad v_z = 0$$

Exercise 5

$f(x, y) = x^2 + y^2 + \frac{1}{x^2y^2}$, $f_x\left(x^2 + y^2 + \frac{1}{x^2y^2}\right) = 2\frac{x^4y^2 - 1}{x^3y^2} = 0$

Since x and y are nonzero, we get $x^4y^2 - 1 = x^2y^2(x^2 - 1) = 0$. Hence $x = \pm 1$. This shows that f has the four critical points $(x, y) = 1(\pm, \pm 1)$. We use the Hessian matrix to verify that f has local minima at these points. $D = 48 > 0$.

Exercise 7

a. $W(e^x, e^{2x}, e^{3x}) = 2e^x e^{2x} e^{3x}$ is nonzero for all $x \in \mathbb{R}$. Hence S_1 is linearly independent.

b. $W(x, x^2, \sin x) = -x^2 \sin x - 2x \cos x + 2 \sin x$ is nonzero for $x = \pi$. Hence S_2 is linearly independent.

c. $W(1, (x-2), (x-2)^3) = 6x^2 - 24x + 24$ is nonzero for $x = 0$. Hence S_3 is linearly independent.

Exercise 9 (*Maple V*)

a. with(linalg): f:=x^4+y^2-8*x-9*y:

Hf:=hessian(f,[x,y]): discriminant:=det(Hf); returns $24x^2$.

diff(f,x); returns $4x^3 - 8$. diff(diff(f,x),x); returns $12x^2$. diff(f,y); returns $2y - 9$.

$f(\sqrt[3]{2}, 9/2)$ is a local minimum since $12\left(\sqrt[3]{2}\right)^2 > 0$.

b. with(linalg): f:=x^3-6*x^2-7*x*y-4*y-5:

Hf:=Hessian(f,[x,y]): discriminant:=det(Hf); returns -49.

diff(x^3-6*x^2+7*x*y-4*y-5,y); returns $7x - 4$.

diff(x^3-6*x^2+7*x*y-4*y-5,x); returns $3x^2 - 12x + 7y$.

The critical point is $(4/7, 288/343)$, a saddle point.

c. with(linalg): f:=x^3-6*x^2-7*x-4*y-5:

Hf:=Hessian(f,[x,y]): discriminant:=det(Hf); returns 0. The test fails.

CHAPTER 4 - Vector Spaces

EXERCISES 4.1

Exercise 1

a. $a(-\mathbf{u}) = -(a\mathbf{u})$ since $\mathbf{0} = a\mathbf{0} = a(\mathbf{u} + (-\mathbf{u})) = a\mathbf{u} + a(-\mathbf{u})$

b. $(-a)\mathbf{u} = -(a\mathbf{u})$ since $\mathbf{0} = 0\mathbf{0} = (a + (-a))\mathbf{u} = a\mathbf{u} + (-a)\mathbf{u}$.

Exercise 3

a. $(x, y) + (x', y') = (x + x', y + y') = (x' + x, y' + y) = (x', y') + (x, y)$

b. $(a + b)(x, y) = ((a + b)x, (a + b)y) = (ax + bx, ay + by)$

and

$(ax + bx, ay + by) = (ax, ay) + (bx + by) = a(x, y) + b(x, y)$

Exercise 5

The origin $(0, 0)$ is not on the line $y = 3x + 5$.

Exercise 7

The set has no zero vector.

Exercise 9

The set is not closed under scalar multiplication by all real numbers.
For example, the matrix

$$\pi\begin{bmatrix} 1 & 2 \\ 3 & 4 \end{bmatrix} = \begin{bmatrix} \pi & 2\pi \\ 3\pi & 4\pi \end{bmatrix}$$

is not a 2×2 matrix with rational entries.

EXERCISES 4.2

Exercise 1

a. $\begin{bmatrix} 2 & 1 \\ -8 & 2 \end{bmatrix}\begin{bmatrix} 4 \\ 5 \end{bmatrix} = \begin{bmatrix} 13 \\ -22 \end{bmatrix}$ b. $\begin{bmatrix} 2 & 1 & 9 \\ -8 & 2 & 6 \end{bmatrix}\begin{bmatrix} 4 \\ 5 \\ 8 \end{bmatrix} = \begin{bmatrix} 85 \\ 26 \end{bmatrix}$

c. $\begin{bmatrix} 8 & 1 \\ -2 & 1 \\ 5 & 1 \end{bmatrix}\begin{bmatrix} 4 \\ -7 \end{bmatrix} = \begin{bmatrix} 25 \\ -15 \\ 13 \end{bmatrix}$ d. $\begin{bmatrix} 1 & 4 & 7 \\ 2 & 5 & 8 \\ 3 & 6 & 9 \end{bmatrix}\begin{bmatrix} 4 \\ 5 \\ -7 \end{bmatrix} = \begin{bmatrix} -25 \\ -23 \\ -21 \end{bmatrix}$

Exercise 3 (*Maple V*)

with(LinearAlgebra): u:=<u1,u2,u3>: v:=<v1,v2,v3>:
left:=ScalarMultiply(u+v,a):
Xleft:=<expand(left[1]),expand(left[2]),expand(left[3])> :
right:=ScalarMultiply(u,a)+ScalarMultiply(v,a):
Equal(Xleft,right); returns *true*.
left:=ScalarMultiply(u,a+b):
Xleft:=<expand(left[1]),expand(left[2]),expand(left[3])> :
right:=ScalarMultiply(u,a)+ScalarMultiply(u,b):
Equal(Xleft,right); returns *true*.
left:=ScalarMultiply(u,a*b):
right:=ScalarMultiply(ScalarMultiply(u,b),a):
Equal(left,right); returns *true*.
left:=ScalarMultiply(u,1): Equal(left,u); returns *true*.

EXERCISES 4.3

Exercise 1

a. The vectors in $\mathbb{R}$ are real numbers and those in $\mathbb{R}^{1\times 1}$ are 1×1 real matrices. In most applications, however, we treat 1×1 real matrices as real numbers. If ***expr*** is a *Maple* expression for a real number, then the corresponding 1×1 is denoted by [[***expr***]].

b. The vectors in $\mathbb{R}^3$ are column vectors of height 3, with real coordinates, the vectors in $\mathbb{R}^{3\times1}$ are 3×1 real matrices, and the vectors in $\mathbb{R}^{1\times3}$ are 1×3 real matrices. In *Maple,* vectors in $\mathbb{R}^3$ are denoted by [a,b,c], matrices in $\mathbb{R}^{1\times3}$ are denoted by [[a,b,c]], and matrices in $\mathbb{R}^{3\times1}$ are denoted by [[a],[b],[c]].
c. The vectors in $\mathbb{R}^{2\times3}$ are 2×3 matrices with real entries, and the vectors in $\mathbb{R}^{3\times2}$ are the transposes of the vectors in $\mathbb{R}^{2\times3}$.
d. The vectors in $\mathbb{R}^{2\times3}$ are 2×3 matrices and the vectors in $\mathbb{R}^6$ are column vectors of height 6.

Exercise 3

The set is not closed under the scalar multiplication by real numbers. For example, the matrix

$$(-1)\begin{bmatrix} 1 & 2 \\ 3 & 4 \\ 5 & \pi \end{bmatrix} = \begin{bmatrix} -1 & -2 \\ -3 & -4 \\ -5 & -\pi \end{bmatrix}$$

is not a matrix with nonnegative entries.

EXERCISES 4.4

Exercise 1

Let $p(t) = a_0 + a_1t + a_2t^2 + a_3t^3$ and $q(t) = b_0 + b_1t + b_2t^2 + b_3t^3$. Then

$$\begin{aligned} p(t) + q(t) &= \left(a_0 + a_1t + a_2t^2 + a_3t^3\right) + \left(b_0 + b_1t + b_2t^2 + b_3t^3\right) \\ &= (a_0 + b_0) + (a_1 + b_1)\,t + (a_2 + b_2)\,t^2 + (a_3 + b_3)\,t^3 \end{aligned}$$

and

$$a\,p(t) = a\left(a_0 + a_1t + a_2t^2 + a_3t^3\right) = (aa_0) + (aa_1)\,t + (aa_2)\,t^2 + (aa_3)\,t^3$$

Exercise 3

Let $p(t) = a_0 + a_1t + a_2t^2 + a_3t^3 + a_4t^4$. Then the polynomial $p(t) - p(t) = 0$ is not a polynomial of degree 4.

EXERCISES 4.5

Exercise 1

The vector space operations on the space $\mathbb{R}_2[t]$ are defined as in Exercise 4.4.1.
If $p(t) = a_0 + a_1t + a_2t^2$ and $q(t) = b_0 + b_1t + b_2t^2$ then

$$p(t) + q(t) = (a_0 + b_0) + (a_1 + b_1)\,t + (a_2 + b_2)\,t^2$$

and

$$a\,p(t) = (aa_0) + (aa_1)\,t + (aa_2)\,t^2$$

Since $a\, p(t) = r(t)$ is a real polynomial for real numbers a, the set $\mathbb{R}_2[t]$ is closed under the multiplication by real scalars. The verification of the vector space axioms is routine. The equation $1\, p(t) = p(t)$, for example, holds since $1\, p(t) = (1a_0) + (1a_1)\,t + (1a_2)\,t^2 = a_0 + a_1 t + a_2 t^2 = p(t)$.

Exercise 3

The vector space operations are defined in Example 4.12.

Suppose that f and g belong to $D^\infty(\mathbb{R}, \mathbb{R})$. Then

$$\frac{d}{dt}(af(t) + bg(t)) = a\frac{d}{dt}(f(t)) + b\frac{d}{dt}(g(t)).$$

Therefore $D^\infty(\mathbb{R}, \mathbb{R})$ is closed under vector addition and scalar multiplication. The verification of the vector space axioms is routine. For example, the following calculation shows that that every vector $\mathbf{x} \in D^\infty(\mathbb{R}, \mathbb{R})$ has the inverse $-\mathbf{x} \in D^\infty(\mathbb{R}, \mathbb{R})$.

$$\mathbf{x} + (-\mathbf{x}) = f(t) - f(t) = 0, \quad \tfrac{d}{dt}(-f(t)) = -\tfrac{d}{dt}(f(t))$$

Exercise 5

Let $V = \{y = mx + b : b \in \mathbb{R}\}$ be the set of all lines in the plane with a fixed slope m.

If $m = 0$, the V consists of the set of vertical lines in the plane.

Given any line L with equation $y = mx + b$, we let aL be the line with equation $y = mx + (ab)$.

Given two parallel lines L_1 and L_2 with the equations $y = mx + b_1$ and $y = mx + b_2$,we let $L_1 + L_2$ be the line determined by the equation $y = mx + (b_1 + b_2)$.

Exercise 7 (*Maple V*)

solve({3*x+7*y-z=0,x+y+z=0},{x,y,z}); returns $\{y = z, x = -2z, z = z\}$.

This tells us that the solutions are of the form $(-2z, z, z) = z(-2, 1, 1)$.

The equations

$$z(-2, 1, 1) + z'(-2, 1, 1) = \left(z + z'\right)(-2, 1, 1) \quad \text{and} \quad a\,z(-2, 1, 1) = (az)\,(-2, 1, 1)$$

show that the set is closed under the vector space operations of $\mathbb{R}^3$.

Exercise 9 (*Maple V*)

dsolve(diff(y(t),t)+3*y(t)=0,y(t)); returns $y(t) = \_C1e^{-3t}$.

This shows that the solutions are of the form $y(t) = a\,e^{-3t}$, where a is any real number.

Hence the set of solutions is closed under scalar multiplication.

Let $y_1(t) = a\,e^{-3t}$ and $y_2(t) = b\,e^{-3t}$ be two solutions. Then

$$y_1(t) + y_2(t) = a\,e^{-3t} + b\,e^{-3t} = (a + b)\,e^{-3t}$$

Hence $y_1(t) + y_2(t)$ is a solution.

For $a = 0$, the constant function $y(t) = 0\,e^{-3t} = 0$ is the zero vector of the space.

EXERCISES 4.6

Exercise 1

a. There there are nine possibilities:

$7\mathbf{x}+7\mathbf{y}$, $7\mathbf{x}+9\mathbf{y}$, $7\mathbf{x}+-4\mathbf{y}$, $9\mathbf{x}+7\mathbf{y}$, $9\mathbf{x}+9\mathbf{y}$, $9\mathbf{x}-4\mathbf{y}$, $-4\mathbf{x}+7\mathbf{y}$, $-4\mathbf{x}+9\mathbf{y}$, $-4\mathbf{x}-4\mathbf{y}$.

b. Since $\left(\pi+\pi^2\right)=\left(\pi^2+\pi\right)$, there are only three possibilities.

$$(\pi+\pi)\begin{bmatrix}1 & 2\\ 3 & 4\end{bmatrix},\quad \left(\pi+\pi^2\right)\begin{bmatrix}1 & 2\\ 3 & 4\end{bmatrix},\quad \left(\pi^2+\pi^2\right)\begin{bmatrix}1 & 2\\ 3 & 4\end{bmatrix}$$

Exercise 3

The span of this set of polynomials is the set of linear combinations

$$a_0\,(t-7)+a_1t^3+a_2t^5+a_3\left(t^6+1\right)$$

determined by all real numbers a_0, a_1, a_2, and a_3.

Exercise 5

a. The set is linearly independent since

$$a\begin{bmatrix}3\\ 5\end{bmatrix}+b\begin{bmatrix}-8\\ 2\end{bmatrix}=\begin{bmatrix}0\\ 0\end{bmatrix}$$

if and only if $a=b=0$.

b. The set is linearly dependent since the dimension of $\mathbb{R}^2$ is 2.

Any subset of $\mathbb{R}^2$ with more than two elements is linearly dependent.

c. The set is linearly independent since the associated homogeneous linear system

$$\{3a-8b=0, 5a+2b+c=0, 4a+3b+c=0\}$$

has only the trivial solution.

d. The set is linearly dependent since the dimension of $\mathbb{R}^3$ is 3.

Any subset of $\mathbb{R}^3$ with more than three elements is linearly dependent.

Exercise 7

a. The set of columns is linearly independent since the associated homogeneous linear system

$$\{3a-8b=0, 5a+c=0, 4a+3b+6c=0\}$$

has only the trivial solution.

b. The set of columns is linearly dependent since the associated homogeneous linear system

$$\{3a-8b+c=0, 0a+0b+0c=0, 4a+3b+c=0\}$$

has the nontrivial solution $a=-11b$, $c=41b$.

c. The set of columns is linearly dependent since the associated homogeneous linear system

$$\{3a - 8b - 5c = 0, 5a + 0b + 5c = 0, 4a + 3b + 7c = 0\}$$

has the nontrivial solution $(-c, -c, c)$.

EXERCISES 4.7

Exercise 1

For any nonzero real number a, the set $\mathcal{B} = \{a\}$ is a basis for $\mathbb{R}$.

Exercise 3

The set $\mathcal{B}$ is linearly independent since the homogeneous linear system

$$\{a + 2b + d = 0, 2a + 2c + d = 0, 3a + 5c + d = 0, 4a + 5b + d = 0\}$$

has only the trivial solution.

Since the space $\mathbb{R}^{2\times 2}$ has dimension 4, the set B is a basis.

Exercise 5

$\mathcal{B}_1 = \left\{1, t, t^2, t^3, t^4, t^5\right\}$ and $\mathcal{B}_2 = \left\{5, t, 1 + t^2, t^3, 4t^4, t^5\right\}$ are two bases.

Exercise 7 (*Maple V*)

dsolve(diff(y(t),t)-y(t)=0); returns $y(t) = \_C1e^t$.

This shows that all solutions of the given differential equation are of the form $y(t) = a\,e^t$.

Hence the set $\mathcal{B} = \left\{e^t\right\}$ is a basis for the solution space.

EXERCISES 4.8

Exercise 1

Let $\delta : \mathbf{4} \times \mathbf{4} \to \{0, 1\}$ be a Kronecker delta function.

Then $\mathcal{E} = \{\mathbf{e}_i = (\delta_{i1}, \delta_{i2}, \delta_{i3}, \delta_{i4}) : i = \mathbf{4}\}$ is the standard basis of $\mathbb{R}^4$.

Exercise 3

a. The standard basis of the space of real numbers is the set $\mathcal{B} = \{1\}$.

b. The standard basis of the space of real upper-triangular 3×3 matrices is the subset of upper-triangular matrices of the standard basis of $\mathbb{R}^{3\times 3}$.

c. The standard basis of the space of real diagonal 4×4 matrices is the subset of diagonal matrices of the standard basis of $\mathbb{R}^{4\times 4}$.

d. The standard basis of the space of real symmetric 3×3 matrices consists of the matrices

$$E_1 = \begin{bmatrix} 1 & 0 & 0 \\ 0 & 0 & 0 \\ 0 & 0 & 0 \end{bmatrix}, \quad E_2 = \begin{bmatrix} 0 & 0 & 0 \\ 0 & 1 & 0 \\ 0 & 0 & 0 \end{bmatrix}, \quad E_3 = \begin{bmatrix} 0 & 0 & 0 \\ 0 & 0 & 0 \\ 0 & 0 & 1 \end{bmatrix},$$
$$E_4 = \begin{bmatrix} 0 & 1 & 0 \\ 1 & 0 & 0 \\ 0 & 0 & 0 \end{bmatrix}, \quad E_5 = \begin{bmatrix} 0 & 0 & 1 \\ 0 & 0 & 0 \\ 1 & 0 & 0 \end{bmatrix}, \quad E_6 = \begin{bmatrix} 0 & 0 & 0 \\ 0 & 0 & 1 \\ 0 & 1 & 0 \end{bmatrix}$$

EXERCISES 4.9

Exercise 1 (*Maple V*)

with(linalg): A:=matrix([[4,4,2],[5,-5,9],[-4,9,0]]): B:=inverse(A):

C:=vector([3,-2,1]): x1:=evalm(B&*C); returns $x1 := \begin{bmatrix} \frac{233}{418} & \frac{75}{209} & \frac{-139}{418} \end{bmatrix}$.

C:=vector([7,5,3]): x2:=evalm(B&*C); returns $x2 := \begin{bmatrix} \frac{339}{418} & \frac{145}{209} & \frac{205}{418} \end{bmatrix}$.

C:=vector([1,1,1]): x3:=evalm(B&*C); returns $x3 := \begin{bmatrix} \frac{17}{418} & \frac{27}{209} & \frac{67}{418} \end{bmatrix}$.

The uniqueness of the solutions follows from the invertibility of A.

Exercise 3 (*Maple V*)

$$\begin{bmatrix} 1 \\ 2 \\ 3 \\ 4 \end{bmatrix} = \begin{bmatrix} 8a_1 + 4a_2 - 5a_3 - 5a_4 \\ 3a_1 - 5a_2 + 8a_3 + 5a_4 \\ -a_1 + 3a_3 - 4a_4 \\ -5a_1 - 2a_2 - a_3 - 9a_4 \end{bmatrix} = \begin{bmatrix} 8 & 4 & -5 & -5 \\ 3 & -5 & 8 & 4 \\ -1 & 0 & 3 & -4 \\ -5 & -2 & -1 & -9 \end{bmatrix} \begin{bmatrix} a_1 \\ a_2 \\ a_3 \\ a_4 \end{bmatrix}$$

with(linalg):

A:=matrix([[8,4,-5,-5],[3,-5,8,4],[-1,0,3,-4],[-5,-2,-1,9]]):

evalb(det(A)=0); returns $false$.

Hence the coefficient matrix A is invertible. This implies that a_1, a_2, a_3, a_4 are unique.

B:=inverse(A): C:=vector([1,2,3,4]):

evalm(B&*C); returns $\begin{bmatrix} \frac{25}{301} & \frac{4329}{301} & \frac{300}{43} & \frac{1343}{301} \end{bmatrix}$.

EXERCISES 4.10

Exercise 1

$\mathcal{B} = \{(1, 2), (2, 1)\}$ is linearly independent and spans $\mathbb{R}^2$.

It is therefore a basis.

Exercise 3

$\mathcal{B} = \{(1, 2, 3, 4), (5, 6, 7, 8), (8, 0, 0, 1), (0, 3, 0, 5)\}$ is linearly independent and spans $\mathbb{R}^4$.

It is therefore a basis.

Exercise 5

The third vector is linearly dependent on the first two vectors since $(-10, -11, -12) = 2(1, 2, 3) - 3(4, 5, 6)$. We delete the third vector and make a suitable replacement.

$\mathcal{B} = \{(1, 2, 3), (4, 5, 6), (-10, 11, -12)\}$ is linearly independent and spans $\mathbb{R}^3$.

Exercise 7

We know that the set S is linearly dependent since it has more than three elements. Since the first three vectors are linearly independent, we delete the fourth vector.

Exercise 9

For any nonzero constant a, the set $\mathcal{B} = \{a, t, t^2, t^3, t^4\}$ is a basis for $\mathbb{R}_4[t]$.

Exercise 11

The vector $t+t^2$ is linearly dependent on the remaining vectors since $t+t^2 = t + \left(\frac{1}{3}\right)(3t^2)$.

The vectors $\mathcal{B} = \{1, t, 3t^2, t^3, 8t^4, t^5\}$ form a basis for $\mathbb{R}_5[t]$.

Exercise 13

The sets $\mathcal{S}_1$ and $\mathcal{S}_2$ are linearly dependent.

However, the set $\mathcal{S}_1^{/} = \{(1, 2, 3), (2, 2, 2), (1, 0, 0)\}$ is linearly independent.

Since the dimension of $\mathbb{R}^3$ is 3, the set $\mathcal{S}_1^{/}$ is a basis.

Exercise 15

Since the space $\mathbb{R}^{2\times 2}$ has dimension 4, we require four matrices for a basis.

Consider the matrices

$$E_1 = \begin{bmatrix} 1 & 0 \\ 1 & 2 \end{bmatrix}, \quad E_2 = \begin{bmatrix} 1 & 0 \\ 0 & 0 \end{bmatrix}, \quad E_3 = \begin{bmatrix} 0 & 1 \\ 0 & 0 \end{bmatrix}, \quad E_4 = \begin{bmatrix} 0 & 0 \\ 0 & 1 \end{bmatrix}$$

Since

$$A = aE_1 + bE_2 + cE_3 + dE_4 = \begin{bmatrix} a+b & c \\ a & 2a+d \end{bmatrix} = \begin{bmatrix} 0 & 0 \\ 0 & 0 \end{bmatrix}$$

if and only if $a = b = c = d = 0$, the set $\{E_1, E_2, E_3, E_4\}$ is a basis for $\mathbb{R}^{2\times 2}$.

Exercise 17 (*Maple V*)

with(linalg): A:=matrix([[1,4,1,2,0],[2,4,0,4,1],[1,4,1,2,1]]):

rref(A); returns

$$\begin{bmatrix} 1 & 0 & -1 & 2 & 0 \\ 0 & 1 & \frac{1}{2} & 0 & 0 \\ 0 & 0 & 0 & 0 & 1 \end{bmatrix}$$

This shows that the first, second, and fifth columns are pivot columns. Since it takes three vectors to span $\mathbb{R}^3$, the set of pivot columns does exactly that.

Exercise 19 (*Maple V*)

A:=matrix([[1,4,5,6,10],[2,4,6,8,12],[1,4,5,6,10]]): linalg[rref](A); returns the matrix

$$\begin{bmatrix} 1 & 0 & 1 & 2 & 2 \\ 0 & 1 & 1 & 1 & 2 \\ 0 & 0 & 0 & 0 & 0 \end{bmatrix}$$

Since it has two pivot columns, the dimension of the span has dimension 2.

EXERCISES 4.11

Exercise 1 (*Maple V*)

with(linalg): A:=matrix([[-8,-5,7,7,-9],[-5,4,-9,-8,4],[7,-9,5,3,0],[7,-8,3,5,-1],[-9,4,0,-1,-5]]):

d:=det(A); returns $d := -6121$. Since $d \neq 0$, the matrix A is invertible.

Exercise 3 (*Maple 6*)

restart: with(LinearAlgebra): A:=RandomMatrix(4,4): z:=ZeroVector[column](4):
convert(LinearSolve(A,z),list); returns [0, 0, 0, 0].
Hence the columns of A are linearly independent and from a basis for $\mathbb{R}^4$.
convert(LinearSolve(MatrixInverse(A),z),list); returns [0, 0, 0, 0].
Hence the columns of A^{-1} are linearly independent and from a basis for $\mathbb{R}^4$.

EXERCISES 4.12

Exercise 1

On the standard basis for $\mathbb{R}_3[t]$, the coordinate-vector function $\mathbf{x} \to [\mathbf{x}]_\mathcal{E}$ from $\mathbb{R}_3[t]$ to $\mathbb{R}^4$ is defined by

$$[1]_\mathcal{E} = (1, 0, 0, 0), [t]_\mathcal{E} = (0, 1, 0, 0), [t^2]_\mathcal{E} = (0, 0, 1, 0), [t^3]_\mathcal{E} = (0, 0, 0, 1)$$

Therefore, $\left[3 + 7t - t^2\right]_\mathcal{E} = (3, 7, -1, 0)$, $\left[4 + t^3\right]_\mathcal{E} = (4, 0, 0, 1)$, $\left[21t^2\right]_\mathcal{E} = (0, 0, 21, 0)$, $\left[1 + t + t^2\right]_\mathcal{E} = (1, 1, 1, 0)$, $\left[3t^3 + 2t\right]_\mathcal{E} = (0, 2, 0, 3)$.

Exercise 3

Let $\mathcal{B} = \{\mathbf{x} = (x_1, x_2, x_3), \mathbf{y} = (y_1, y_2, y_3), \mathbf{z} = (z_1, z_2, z_3)\}$ be the required basis. Then

$$\begin{aligned}
(1, 0, 0) &= a_1\mathbf{x} + a_2\mathbf{y} + a_3\mathbf{z} = 2\mathbf{x} = (2x_1, 2x_2, 2x_3) \\
(0, 1, 0) &= b_1\mathbf{x} + b_2\mathbf{y} + b_3\mathbf{z} = 2\mathbf{z} = (2z_1, 2z_2, 2z_3) \\
(0, 0, 1) &= c_1\mathbf{x} + c_2\mathbf{y} + c_3\mathbf{z} = 2\mathbf{y} + 2\mathbf{z} = (2y_1 + 2z_1, 2y_2 + 2z_2, 2y_3 + 2z_3)
\end{aligned}$$

This implies that $\mathcal{B} = \{\mathbf{x} = (1/2, 0, 0), \mathbf{y} = (0, -1/2, 1/2), \mathbf{z} = (0, 1/2, 0)\}$.

Exercise 5

Since A is invertible, its columns form a basis $\mathcal{B} = \{\mathbf{x}, \mathbf{y}, \mathbf{z}\}$, where
$\mathbf{x} = (a_{11}, a_{21}, a_{31}) = (1)(a_{11}, a_{21}, a_{31}) + (0)(a_{12}, a_{22}, a_{32}) + (0)(a_{13}, a_{23}, a_{33})$
$\mathbf{y} = (a_{12}, a_{22}, a_{32}) = (0)(a_{11}, a_{21}, a_{31}) + (1)(a_{12}, a_{22}, a_{32}) + (0)(a_{13}, a_{23}, a_{33})$
$\mathbf{z} = (a_{13}, a_{23}, a_{33}) = (0)(a_{11}, a_{21}, a_{31}) + (0)(a_{12}, a_{22}, a_{32}) + (1)(a_{13}, a_{23}, a_{33})$
The coordinate vectors of $\mathbf{x}, \mathbf{y}, \mathbf{z}$ in the basis $\mathcal{B}$ are

$$\begin{bmatrix} a_{11} \\ a_{21} \\ a_{31} \end{bmatrix}_\mathcal{B} = \begin{bmatrix} 1 \\ 0 \\ 0 \end{bmatrix}, \begin{bmatrix} a_{12} \\ a_{22} \\ a_{32} \end{bmatrix}_\mathcal{B} = \begin{bmatrix} 0 \\ 1 \\ 0 \end{bmatrix}, \begin{bmatrix} a_{13} \\ a_{23} \\ a_{33} \end{bmatrix}_\mathcal{B} = \begin{bmatrix} 0 \\ 0 \\ 1 \end{bmatrix}$$

Moreover, the linearity of the coordinate–vector function implies that

$$\begin{bmatrix} a_{11} \\ a_{21} \\ a_{31} \end{bmatrix}_\mathcal{B} + \begin{bmatrix} a_{12} \\ a_{22} \\ a_{32} \end{bmatrix}_\mathcal{B} + \begin{bmatrix} a_{13} \\ a_{23} \\ a_{33} \end{bmatrix}_\mathcal{B} = \begin{bmatrix} a_{11} + a_{12} + a_{13} \\ a_{21} + a_{22} + a_{23} \\ a_{31} + a_{32} + a_{33} \end{bmatrix}_\mathcal{B} = \begin{bmatrix} 1 \\ 1 \\ 1 \end{bmatrix}$$

From the point of view of the basis $\mathcal{B}$, the given vector is the sum of three unit vectors, as illustrated in Figure 3.

Exercise 7 (*Maple 6*)
with(LinearAlgebra): Coord:=A -> <A[1,1],A[1,2],A[1,3],A[2,1],A[2,2],A[2,3]>
defines the required isomorphism.
A:=Matrix[[4,1,5],[4,3,-4]]): convert(Coord(A),list); returns $(4, 1, 5, 4, 3, -4)$.
B:=Matrix([[0,0,0],[0,1,0]]): convert(Coord(B),list); returns $(0, 0, 0, 0, 1, 0)$.
C:=Matrix([[1,-1,1],[-1,3/5,Pi]]): convert(Coord(C),list); returns $(1, -1, 1, -1, \frac{3}{5}, \pi)$.

EXERCISES 4.13

Exercise 1
The proper subspaces of $\mathbb{R}^2$ have dimension 0 or 1. The only subspace of dimension zero is the zero subspace $\{(0, 0)\}$. Moreover, every straight line of the form $S_a = \{(x, ax) : a \in \mathbb{R}\}$ is a one–dimensional subspace. As a basis vector, we choose the vector $(1, a)$. Since every subspace must contain the origin $(0, 0)$, there are no other possibilities.
Exercise 3
a. Since a plane is a two–dimensional subspace of $\mathbb{R}^3$, it suffices to find two linearly independent solutions of the equation $2x - y + 3z = 0$. One choice is $(3, 3, -1)$ and $(6, 12, 0)$.
b. The space W is a two dimensional subspace since it is determined by the two parameters a and b. The choice $a = 1$ and $b = 0$ determines the vector $(1, 1, 0)$, and the choice $a = 0$ and $b = 1$ determines the vector $(0, -1, 1)$. Since they are linearly independent, they form a basis for W.
c. A line is a one-dimensional subspace of $\mathbb{R}^3$. The assignment $t = 1$ determines the vector $(1, -2, 3)$. It forms a basis for the space.
d. The space is the solution space of the homogeneous linear system

$$\{2x - y + 3z = 0, x + y - z = 0\}$$

Geometrically, we can think of the solution space as the intersection of two planes. If we solve the given linear systems in terms of z, we get $\{y = -\frac{1}{3}z, x = \frac{4}{3}z\}$. For any nonzero value of z, the set $\mathcal{B} = \{(4/3z, -1/3z, z)\}$ is a basis for the solution space.
Exercise 5
If we consider 0 to be an even integer, the answer is yes. The set of real polynomials of even degrees is closed under addition and under the multiplication by real numbers.
Exercise 7
It suffices to show that the $n \times n$ zero matrix is symmetric and that the set of symmetric real $n \times n$ matrices is closed under linear combinations. The zero matrix is obviously symmetric. Suppose, therefore, that $A = [a_{ij}]$ and $B = [b_{ij}]$ are two symmetric $n \times n$ matrices. We must show that for all real scalars a and b, the matrix $aA + bB$ is symmetric.
Since $a\,a_{ij} + b\,b_{ij} = a\,a_{ji} + b\,b_{ij}$ for all i and j, the result follows.
Exercise 9
The set of invertible $n \times n$ matrices is not a vector space since the zero matrix is not invertible. If A is invertible, then $-A$ is invertible but $A - A = 0$ is not.

Exercise 11

Since the set of real polynomials of even degrees less than or equal to 6 is a subset of $\mathbb{R}_{10}[t]$ that is closed under addition and scalar multiplication, it is a subspace of $\mathbb{R}_{10}[t]$.

Exercise 13 (*Maple 6*)

$\mathbb{R}^4 = \text{Col } A \oplus \text{Col } B$, with A and B constructed as follows.

```
with(LinearAlgebra):
R:=RandomMatrix(4,4,generator=1..9,outputoptions=[shape=triangular[lower]]):
A:=[Column(R,1),Column(R,2)]: B:=[Column(R,3),Column(R,4)]:
IntersectionBasis([A,B]); returns the empty list [].
SumBasis([A,B]); returns the basis {(1, 3, 7, 3), (0, 1, 4, 9), (0, 0, 2, 6), (0, 0, 0, 8)} for R^4.
```

EXERCISES 4.14

Exercise 1

a. The column space of A is $\mathbb{R}^2$ since the first and third columns of A are linearly independent.

b. The row space of A is a two-dimensional subspace of $\mathbb{R}^3$ since the rows of A are linearly independent. Geometrically, it represents a plane through the origin of $\mathbb{R}^3$.

c. The null space of A consist of the solutions $(-2y, y, 0) = y\,(-2, 1, 0)$ of the homogeneous linear system

$$\begin{bmatrix} 1 & 2 & 4 \\ 0 & 0 & 1 \end{bmatrix} \begin{bmatrix} x \\ y \\ z \end{bmatrix} = \begin{bmatrix} 0 \\ 0 \end{bmatrix}$$

Geometrically, the space is a line through the origin of $\mathbb{R}^3$.

d. The left null space of A consists of the solutions of the homogeneous linear system

$$\begin{bmatrix} 1 & 0 \\ 2 & 0 \\ 4 & 1 \end{bmatrix} \begin{bmatrix} x \\ y \end{bmatrix} = \begin{bmatrix} 0 \\ 0 \\ 0 \end{bmatrix}$$

Since only the zero vector $(0, 0)$ satisfies this equation, the left null space of A is the zero subspace of $\mathbb{R}^2$.

Exercise 3

The following matrices have ranks 0, 1, 2, and 3, respectively.

$$\begin{bmatrix} 0 & 0 & 0 & 0 \\ 0 & 0 & 0 & 0 \\ 0 & 0 & 0 & 0 \end{bmatrix}, \begin{bmatrix} 1 & 0 & 0 & 0 \\ 0 & 0 & 0 & 0 \\ 0 & 0 & 0 & 0 \end{bmatrix}, \begin{bmatrix} 1 & 0 & 0 & 0 \\ 0 & 2 & 0 & 0 \\ 0 & 0 & 0 & 0 \end{bmatrix}, \begin{bmatrix} 1 & 0 & 0 & 0 \\ 0 & 2 & 0 & 0 \\ 0 & 0 & 3 & 0 \end{bmatrix}$$

Exercise 5

It is clear from the nature of elementary row operations that row reduction preserves the pivot columns of a matrix. By Theorem 4.22, these columns form a basis for the column space of the matrix. Since the rank of a matrix is the dimension of its column space, the result follows.

Exercise 7

By definition, the number of pivots of a matrix is equal to the number of pivot columns of the matrix. Moreover, by Theorem 4.22, the pivot columns of an $m \times n$ matrix A form a basis for Col A. By Definition 4.15, this number is equal to the rank of A.

Exercise 9

Corollary 2.19 tells us that a matrix is invertible if and only if it is a product of elementary matrices. Therefore $BA = E_p \cdots E_1 A$ and $AC = AE'_1 \cdots E'_q$ for appropriate elementary matrices $E_p, \ldots, E_1$ and $E'_1, \ldots, E'_q$. Moreover, we know from Chapter 2 that left-multiplication by an elementary matrix corresponds to an elementary row operation, and right-multiplication to an elementary column operation. Therefore left-multiplication of a matrix A by $E_p \cdots E_1$ preserves the row rank of A and right-multiplication preserves its column rank. However, since the row rank of A equals the column rank of A, its follows that the matrices A, BA, and AC have the same rank.

Exercise 11

By Corollary 4.31, a real $n \times n$ matrix A is invertible if and only if its rank is n, and by Theorem 4.29, the sum of the rank and nullity of A equals n. Hence the nullity of an invertible matrix A must be zero.

Exercise 13

$$\begin{aligned}\text{Col}\, A &= \{a_1\mathbf{c}_1 A + \cdots + a_n\mathbf{c}_n A : a_1, \ldots, a_n \in \mathbb{R}\} \\ &= \left\{ A \begin{bmatrix} a_1 \\ \vdots \\ a_n \end{bmatrix} : \begin{bmatrix} a_1 \\ \vdots \\ a_n \end{bmatrix} \in \mathbb{R}^n \right\}\end{aligned}$$

Since R is invertible,

$$\left\{ R \begin{bmatrix} a_1 \\ \vdots \\ a_n \end{bmatrix} : \begin{bmatrix} a_1 \\ \vdots \\ a_n \end{bmatrix} \in \mathbb{R}^n \right\} = \mathbb{R}^n$$

The fact that $A = QR$ therefore implies that

$$\text{Col}\, A = \left\{ QR \begin{bmatrix} a_1 \\ \vdots \\ a_n \end{bmatrix} : \begin{bmatrix} a_1 \\ \vdots \\ a_n \end{bmatrix} \in \mathbb{R}^n \right\}$$

$$= \left\{ Q \begin{bmatrix} a_1 \\ \vdots \\ a_n \end{bmatrix} : \begin{bmatrix} a_1 \\ \vdots \\ a_n \end{bmatrix} \in \mathbb{R}^n \right\} = \text{Col } Q$$

Exercise 15 (*Maple V*)

with(linalg), A:=matrix([[1,2,4],[0,0,1],[3,6,12]]):

colspace(A); returns {[1, 0, 3], [0, 1, 0]}, rowspace(A); returns {[1, 2, 0], [0, 0, 1]}, nullspace(A); returns {[−2, 1, 0]}, nullspace(transpose(A)); returns {[−3, 0, 1]}.

The column and row spaces are two planes, and the null and left null spaces are lines in $\mathbb{R}^3$.

Exercise 17 (*Maple V*)

If A is an $m \times n$ matrix of rank r, then the column space of A has dimension r.

nops(nullspace(A)); returns $(n - r)$, the dimension of the nullspace of A.

rowdim(transpose(A)); returns n, the number of columns of A.

rowdim(transpose(A))-nops(nullspace(A)); returns $n - (n - r) = r$, the column rank of A.

Exercise 19 (*Maple V*)

The row space of an $m \times n$ matrix A is determined by the nonzero rows of the reduced row echelon form of A. Moreover, the command

submatrix(rref(A),1..rank(A),1..coldim(A)); returns the $p \times n$ submatrix B of nonzero rows of the reduced row echelon form of A, and the rows of B are a basis for the row space of A.

Exercise 21 (*Maple 6*)

with(linalg): with(LinearAlgebra):

RowwSpace:=A-> submatrix(rref(A),1..rank(A),1..coldim(A)):

A:=RandomMatrix(3,4): C:=Matrix(ColumnSpace(Transpose(A))):

P:=transpose(augment(Column(C,1),Column(C,2),Column(C,3))):

Q:=RowwSpace(A): Equal(Matrix(P),Matrix(Q)); returns *true*.

EXERCISES 4.15

Exercise 1

The space U is spanned by the vector $(1, 3)$, and the space W is spanned by the vector $(1, -7)$. Since these vectors are linearly independent in $\mathbb{R}^2$, the spaces U and W are disjoint. Moreover, the span of $\{(1, 3), (1, -7)\}$ is $\mathbb{R}^2$. Therefore, $\mathbb{R}^2 = U \oplus V$.

Exercise 3

The set $\mathcal{B}_1 = \{(1, 0, 1)\}$ is a basis for the row space of A.

The null space of A is the set of all solutions of the homogeneous linear system

$$\begin{bmatrix} 1 & 0 & 1 \\ 1 & 0 & 1 \\ 1 & 0 & 1 \\ 0 & 0 & 0 \end{bmatrix} \begin{bmatrix} x \\ y \\ z \end{bmatrix} = \begin{bmatrix} x+z \\ x+z \\ x+z \\ 0 \end{bmatrix} = \begin{bmatrix} 0 \\ 0 \\ 0 \\ 0 \end{bmatrix}$$

Therefore, the set $\mathcal{B}_2 = \{((0,1,0),(1,0,-1)\}$ is a basis for the null space. Since the set $\mathcal{B}_1 \cup \mathcal{B}_2$ is linearly independent and spans $\mathbb{R}^3$, the result follows from Theorem 4.38.

Exercise 5

Since $\mathcal{B}$ is a basis for V, every vector $\mathbf{x} \in V$ is a unique sum of the form
$a_1\mathbf{v}_1 + \cdots + a_r\mathbf{v}_r + a_{r+1}\mathbf{v}_{r+1} + \cdots + a_n\mathbf{v}_n$.
Since $\mathbf{v}_{r+1}, \ldots, \mathbf{v}_n$ are linearly independent of $\mathbf{v}_1, \ldots, \mathbf{v}_r$, they do not belong to U. They therefore belong to W since $U \oplus W = V$. Moreover, since $\mathbf{v}_{r+1}, \ldots, \mathbf{v}_n$ are linearly independent in V, they are also linearly independent in W. Since the dimension of W is $n-r$, and since the vectors $\mathbf{v}_{r+1}, \ldots, \mathbf{v}_n$ are $n-r$ linearly independent in W, the span of the set of vectors $\mathcal{C} = \{\mathbf{v}_{r+1}, \ldots, \mathbf{v}_n\}$ is W. Hence $\mathcal{C}$ is a basis for W.

Exercise 7 (*Maple 6*)

with(linalg): with(LinearAlgebra): c1:=<1,1,1,1> : c2:=<1,0,1,0>:
A:=Matrix(augment(c1,c2,c1+c2,5*c1-7*c2)): Rank(A); returns 2.
nops(ColumnSpace(A)); returns 2, nops(RowSpace(A)); returns 2,
nops(NullSpace(Transpose(A))); returns 2, and nops(NullSpace(A)); returns 2.

Exercise 9 (*Maple 6*)

with(LinearAlgebra):
u1:=<3,2,2>: u2:=< 2,0,1>: v1:=<5,3,4>: v2:=< 0,1,0>: v3:=<5,3,1>:
U:={u1,u2}: V:={v1,v2,v3}: SumBasis([U,V]); returns $\{(0,1,0),(5,3,1),(5,3,4)\}$.

EXERCISES 4.16

Exercise 1

The set of all linear combinations $a\mathbf{x} + b\mathbf{y}$ of vectors $\mathbf{x}$ and $\mathbf{y}$ in $\mathcal{S}$ consists of all pairs

$$a(1,0) + b(0,1) = (a,b) = (a_0 + a_1 i, b_0 + b_1 i)$$

of complex number a, b, determined by the real numbers a_0, a_1, b_0, and b_1.

Exercise 3

Suppose that

$$a(1,0) + b(0,i) + c(0,1) + d(i,0) = (a + id, ib + c) = (0,0)$$

and a, b, c, d are real numbers. Then $a + id = ib + c = 0$ implies that $a = b = c = d = 0$. The vectors are therefore linearly independent over $\mathbb{R}$. On the other hand, $i(0,i) = (0,-1) = (-1)(0,1)$. Therefore, $i(0,i) + (1)(0,1) = (0,0)$. This tells us that the vectors $(0,i)$ and $(0,1)$ are linearly dependent over $\mathbb{C}$.

Exercise 5 (*Maple V*)

u:=(u0+u1*I): v:=(v0+v1*I): w:=(w0+w1*I): z:=0+0*I: a:=a0+a1*I: b:=b0+b1*I:
evalb(u+(v+w)=(u+v)+w); returns *true*, evalb(u+v=v+u); returns *true*,
evalb(z=z+z*I); returns *true*, evalb(u+z=u); returns *true*, evalb(u+(-u)=z); returns *true*.
evalb(expand(a*(u+v))=expand(a*u+a*v)); returns *true*.
evalb(expand((a+b)*u)=expand(a*u+b*u)); returns *true*.
evalb((a*b)*u=a*(b*u)); returns *true*, evalb(1*u=u); returns *true*.

Exercise 7 (*Maple V*)

The calculations are analogous to those in Exercise 4.16.5, with the scalars restricted to real numbers.

Exercise 9 (*Maple 6*)

Describe the vectors of $\mathbb{C}^2$ in three different ways as linear combinations of the vectors $\mathcal{S} = \{(1,0), (0,i), (0,1), (i,0)\}$ and real scalars a, b, c, d.

All vectors in $\mathbb{C}^2$ are of the form $(a+bi, c+di)$.

a. <a+b*I,c+d*I> ; returns a vector in $\mathbb{C}^2$.

b. Vector[column](evalm(a*<1,0>+b*< I,0>+c*<0,1>+d*<0,I>)); returns a vector in $\mathbb{C}^2$.

c. Vector[column](2,[a*Complex(1,0)+b*Complex(0,1)
+c*Complex(1,0)+d*Complex(0,1)]); returns a vector in $\mathbb{C}^2$.

EXERCISES 4.17

Exercise 1

If $\mathbf{u} = [0, x, y, z]$ is a unit quaternion, then $x^2 + y^2 + z^2 = 1$. Moreover, the square of the norm of the rotation quaternion $q = [\cos(\theta/2), \sin(\theta/2)[x, y, z]]$ is

$$\cos^2\frac{\theta}{2} + \sin^2\frac{\theta}{2}\left(x^2 + y^2 + z^2\right) = \cos^2\frac{\theta}{2} + \sin^2\frac{\theta}{2} = 1$$

Exercise 3 (*Maple 6*)

The following procedure for defining quaternions is provided in the *Maple 6* Programming Guide:

```
restart; interface(imaginaryunit=j);
‘type/Quaternion‘:={‘+‘,‘*‘,name, realcons, specfunc(anything,‘&^‘)}:
‘&^‘:=proc(x::Quaternion,y::Quaternion)
local Real, unReal, isReal;
isReal:=z-> evalb(is(z,real)=true);
if isReal(x) or isReal(y) then x*y;
elif type(x,‘+‘) then map(‘&^‘,x,y);
elif type(y,‘+‘) then map2(‘&^‘,x,y);
elif type (x,‘*‘) then Real, unReal:=selectremove(isReal,x);
if Real=1 then if type(y,‘*‘) then Real,unReal:=selectremove(isReal,x);
Real * ’‘&^‘’(x,unReal);
else ’‘&^‘’(x,y); end if;
else Real * ‘&^‘(unReal, y); end if;
if Real=1 then if type(y,‘*‘) then Real,unReal:=selectremove(isReal,x);
Real * ’‘&^‘’(x,unReal);
else ’‘&^‘’(x,y); end if;
else Real * ‘&^‘(unReal, y); end if;
elif type(y,‘*‘) then Real,unReal:=selectremove(isReal,y);
```

if Real=1 then ''&^''(x,y);
else Real * '&^'(x,unReal); end if;
else ''&^''(x,y); end if; end proc;
Using the operation &^, we can define the multiplication table for I, J, and K.
'&^'(I,I):=-1: '&^'(J,J):=-1: '&^'(K,K):=-1:'&^'(I,J):=K:
'&^'(J,I):=-K: '&^'(I,K):=-J: &^'(K,I):=J: '&^'(J,K):=I: '&^'(K,J):=-I:
Finally, we define the quaternion, conjugate, norm, and inverse functions.
quat:=(a,b,c,d)-> a+b*I+c*J+d*K:
qconj:=(a,b,c,d)-> a-b*I-c*J-d*K:
qnorm:=(a,b,c,d)-> sqrt(a^ 2+b^2+c^2+d^2):
qinv:=quart/qnorm:
Once we have completed our work with quaternions, we restore I as the imaginary unit.
interface(imaginaryunit=I);

Exercise 5 (*Maple 6*)

We use Exercise 3.
qnorm(0,1/3,1/4,a); returns $\frac{1}{12}\sqrt{25+144a^2}$.
solve(1/12*sqrt(25+144*a^2)=1,a); returns $\frac{1}{12}\sqrt{19}, -\frac{1}{12}\sqrt{19}$.

Exercise 7 (*Maple 6*)

x:=quat(0,1,1,1):
p:=quat(cos(Pi)/sqrt(3),sin(Pi/3)/sqrt(3),sin(Pi/3)/sqrt(3),sin(Pi/3)/sqrt(3)):
q:=quat(cos(Pi)/sqrt(3),-sin(Pi/3)/sqrt(3),-sin(Pi/3)/sqrt(3),-sin(Pi/3)/sqrt(3)):
is(simplify(p&^x&^q)=13/12*x); returns *true*.

Exercise 9 (*Maple 6*)

with(LinearAlgebra): If $\theta = \pi/3$, then $\cos\theta = 1/2$ and $\sin\theta = -\sqrt{3}/2$.
a. Rotation around the x-axis.
Rx:=Matrix([[1,0,0],[0,1/2,-1/2*sqrt(3)],[0,1/2*sqrt(3),1/2]]):
v:=<4,8,-1>: convert(Rx.v,list); returns $[4, \frac{1}{2}\sqrt{3}+4, -\frac{1}{2}+4\sqrt{3}]$.
w:=quat(0,4,8,-1): q:=quat(cos(Pi/6),sin(Pi/6),0,0):
qstar:=quat(cos(Pi/6),-sin(Pi/6),0,0):
simplify(q&^w&^qstar); returns $4I+4J-\frac{1}{2}K+4\sqrt{3}K+\frac{1}{2}\sqrt{3}J$.
b. Rotation around the y-axis.
Ry:=Matrix([[1/2,0,1/2*sqrt(3)],[0,1,0],[(-1/2)*sqrt(3),0,1/2]]):
convert(Ry.v,list); returns $\left[2-\frac{1}{2}\sqrt{3}, 8, -2\sqrt{3}-\frac{1}{2}\right]$.
q:=quat(cos(Pi/6),0,sin(Pi/6),0): qstar:=quat(cos(Pi/6),0,-sin(Pi/6),0):
simplify(q&^w&^qstar); returns $\left[2I+8J-\frac{1}{2}K-2\sqrt{3}K-\frac{1}{2}\sqrt{3}I\right]$.
c. Rotation around the z-axis.
Rz:=Matrix([[1/2,-1/2*sqrt(3),0],[1/2*sqrt(3),1/2,0],[0,0,1]]):
convert(Rz.v,list); returns $[2-4\sqrt{3}, 2\sqrt{3}+4, -1]$.
q:=quat(cos(Pi/6),0,0,sin(Pi/6)): qstar:=quat(cos(Pi/6),0,0,-sin(Pi/6)):
simplify(q&^w&^qstar); returns $2I+4J-K+2\sqrt{3}J-4\sqrt{3}I$.

CHAPTER 5 - Linear Transformations

EXERCISES 5.1

Exercise 1

The function T_θ is a matrix transformation, hence linear.

Exercise 3

a. T is not linear because $T(0) = 2 \neq 0$.

b. T is not linear since $T(\pi) = \cos\pi = -1$ and $T(\pi/2) = \cos\pi/2 = 0$, so

$$T(\pi/2 + \pi/2) \neq T(\pi/2) + T(\pi/2)$$

c. T is not linear since $[x + y]^2 \neq x^2 + y^2$ for most x and y.

d. T is not linear since $T(0) \neq 0$ and $T(x + y) = 3 \neq T(x) + T(y) = 3 + 3$.

e. T is not linear since $T(0) \neq 0$ since $\exp(x + y) \neq \exp x + \exp y$ for most x and y.

f. T is not linear since $T(x)$ is not defined if $x \leq 0$ and since $\log(x + y) \neq \log x + \log y$ for most x and y.

Exercise 5

a. yes b. yes c. yes d. no e. no f. yes

Exercise 7

a. no b. yes c. yes d. yes e. yes

Exercise 9

$T_{A^T}(\mathbf{e}_1) = (7, 17, 38, 27)$, $T_{A^T}(\mathbf{e}_2) = (0, 22, 7, 34)$, $T_{A^T}(\mathbf{e}_2) = (0, 0, 9, 25)$.

Exercise 11

The function F is not linear since $F(0) = 32 \neq 0$.

Exercise 13

$$\begin{aligned}(T_1 + T_2)(a\mathbf{x} + b\mathbf{y}) &= T_1(a\mathbf{x} + b\mathbf{y}) + T_2(a\mathbf{x} + b\mathbf{y}) \\ &= T_1(a\mathbf{x}) + T_1(b\mathbf{y}) + T_2(a\mathbf{x}) + T_2(b\mathbf{y}) \\ &= T_1(a\mathbf{x}) + T_2(a\mathbf{x}) + T_1(b\mathbf{y}) + T_2(b\mathbf{y}) \\ &= aT_1(\mathbf{x}) + aT_2(\mathbf{x}) + bT_1(\mathbf{y}) + bT_2(\mathbf{y}) \\ &= a(T_1 + T_2)(\mathbf{x}) + b(T_1 + T_2)(\mathbf{y})\end{aligned}$$

Exercise 15

$$\begin{aligned}f(ax + bx', ay + by') &= x_0(ax + bx') + y_0(ay + by') \\ &= ax_0x + bx_0x' + ay_0y + by_0y' \\ &= ax_0x + ay_0y + bx_0x' + by_0y' \\ &= a(x_0x + y_0y) + b(x_0x' + y_0y') \\ &= a\,f(x, y) + b\,f(x', y')\end{aligned}$$

Exercise 17 (*Maple V*)

```
f:=(x,y,z)-> 5*x+2*y-7*z:
left:=expand(a*f(u,v,w)+b*f(x,y,z)): right:=f(a*u+b*x,a*v+b*y,a*w+b*z):
evalb(left=right);
```
returns *true*.

Exercise 19 (*Maple V*)

Let $p(t) = a_0 + a_1 t + \cdots + a_n t^n$ and $q(t) = b_0 + b_1 t + \cdots + b_m t^m$.
Then $f(p(t)) = p(0) = a_0$ and $f(q(t)) = q(0) = b_0$.
Therefore, $f(0) = 0$ and $f(a\,p(t) + b\,q(t)) = a\,a_0 + b\,b_0 = a\,f(p(t)) + b\,f(q(t))$.
Hence f is linear. The following routine calculates the values of f in *Maple.*
p:=(a,n)->sum(a[i]*t^i,i=0..n); returns $p := (a, n) \to \sum_{i=0}^{n} a_i t^i$.
f:=(p,a,n)->p(a,0); returns $f := (p, a, n) \to p(a, 0)$. f(p,a,n); returns a_0.

Exercise 23 (*Maple V*)

Load the definition in Exercise 4.17.3.

```
q:=quat(0,1/sqrt(6),1/sqrt(6),2/sqrt(6)):
qstar:=qconj(0,1/sqrt(6),1/sqrt(6),2/sqrt(6)):
assume(a,real): assume(b,real): assume(c,real): assume(d,real):
assume(e,real): assume(f,real): assume(g,real): assume(h,real):
assume(r,real): assume(s,real): x:=quat(a,b,c,d): y:=quat(e,f,g,h):
T:=p->q&^p&^qstar:
first:=expand(collect(r*T(x),[I,J,K])): second:=expand(collect(s*T(y),[I,J,K])):
collect(T(r*x+s*y),[I,J,K]):
is(collect(T(r*x+s*y),[I,J,K])=collect(first+second,[I,J,K]));
```
returns *true*.

EXERCISES 5.2

Exercise 1

Let S and T be the matrix transformation defined by the matrices

$$S = \begin{bmatrix} 1 & 0 \\ 7 & 3 \end{bmatrix}, T = \begin{bmatrix} 1 & 0 \\ 7 & 3 \end{bmatrix}^T$$

Then

$$(ST)\begin{bmatrix} x \\ y \end{bmatrix} = \begin{bmatrix} 1 & 0 \\ 7 & 3 \end{bmatrix}\begin{bmatrix} 1 & 0 \\ 7 & 3 \end{bmatrix}^T \begin{bmatrix} x \\ y \end{bmatrix} = \begin{bmatrix} x + 7y \\ 7x + 58y \end{bmatrix}$$

and

$$(TS)\begin{bmatrix} x \\ y \end{bmatrix} = \begin{bmatrix} 1 & 0 \\ 7 & 3 \end{bmatrix}^T \begin{bmatrix} 1 & 0 \\ 7 & 3 \end{bmatrix}\begin{bmatrix} x \\ y \end{bmatrix} = \begin{bmatrix} 50x + 21y \\ 21x + 9y \end{bmatrix}$$

Exercise 3

Since S and T are isomorphisms, they have inverses S^{-1} and T^{-1}.
Since $S^{-1} \circ T^{-1} = (T \circ S)^{-1}$, the transformation $T \circ S$ is also invertible.

Exercise 5

Suppose that $T : V \to W$ is a linear transformation that is one–one and onto.
For every $\mathbf{y} \in W$ there exists an $\mathbf{x} \in V$ such that $T(\mathbf{x}) = \mathbf{y}$ since T is onto.
The vector $\mathbf{x}$ is unique since T is one–one.
We can therefore define a function $S : W \to V$ by $S(\mathbf{y}) = \mathbf{x}$.
It remains to show that S is linear.
Suppose that $S(\mathbf{u}) = \mathbf{x}$ and $S(\mathbf{v}) = \mathbf{y}$. Then $aS(\mathbf{u}) = a\mathbf{x}$ and $bS(\mathbf{v}) = b\mathbf{y}$.
Therefore, $aS(\mathbf{u}) + bS(\mathbf{v}) = a\mathbf{x}+b\mathbf{y}$.
By the definition of S, that $T(\mathbf{x}) = \mathbf{u}$ and $T(\mathbf{y}) = \mathbf{v}$.
Therefore $aT(\mathbf{x}) = a\mathbf{u}$ and $bT(\mathbf{y}) = b\mathbf{v}$.
Hence, $T(a\mathbf{x} + b\mathbf{y}) = aT(\mathbf{x}) + bT(\mathbf{y}) = a\mathbf{u} + b\mathbf{v}$.
By the definition of S, we therefore have $S(a\mathbf{u} + b\mathbf{v}) = a\mathbf{x} + b\mathbf{y}$.
Hence $S(a\mathbf{u} + b\mathbf{v}) = aS(\mathbf{u}) + bS(\mathbf{v})$.

Exercise 7

The claim is false. Let $n = 2$ and $m = 3$. The linear transformation $T(x, y) = (x, y, 0)$ is one–one, but $m \neq n$.

Exercise 9

Consider the matrix transformation $T_A : \mathbb{R}^2 \to \mathbb{R}^2$ defined by $T_A(x, y) = (x, 0)$.
Since $T_A(x, 1) = T_A(x, 2)$, the transformation T_A is not one–one.
If T_A had an inverse S, then we would have $ST_A(x, 1) = (x, 1) = ST_A(x, 2) = (x, 2)$.
Moreover T_A is not onto since (x, y) is not in the image of T_A if $y \neq 0$.
If T_A had an inverse S, then we would have $(0, y) = ST_A(0, y) = S(0, 0)$.
Since S has to be linear, we would have $S(0, 0) = (0, 0) = (0, y)$ for some nonzero y.

Exercise 11

If T is one–one, then T preserves linear independence.
Suppose that $\mathbf{x}_1, \ldots, \mathbf{x}_n$ are n linearly independent vectors in $\mathbb{R}^n$ and that $T(\mathbf{x}_1), \ldots, T(\mathbf{x}_n)$ are linearly dependent in $\mathbb{R}^m$.
Then there exist scalars $a_1, \ldots, a_n$, not all zero, such that $a_1T(\mathbf{x}_1) + \cdots + a_nT(\mathbf{x}_n) = 0$.
The fact that T is a linear transformation implies $0 = T(0)$ and that
$a_1T(\mathbf{x}_1) + \cdots + a_nT(\mathbf{x}_n) = T(a_1\mathbf{x}_1 + \cdots + a_n\mathbf{x}_n)$.
Since T is one–one, $a_1\mathbf{x}_1 + \cdots + a_n\mathbf{x}_n = \mathbf{0}$ for scalars $a_1, \ldots, a_n$, not all of which are zero.
This contradicts the fact that the vectors $\mathbf{x}_1, \ldots, \mathbf{x}_n$ are linearly independent.

Exercise 13

The inverse of the function T in Example 5.15 is the function

$$S(a_{11}, a_{12}, a_{21}, a_{22}) = \begin{bmatrix} a_{11} & a_{12} \\ a_{21} & a_{22} \end{bmatrix}$$

The inverse of the function T in Example 5.16 is the function

$$S\left(a_{11}+a_{12}t+a_{13}t^2+a_{21}t^3+a_{22}t^4+a_{23}t^5\right)=\begin{bmatrix} a_{11} & a_{12} & a_{13} \\ a_{21} & a_{22} & a_{23} \end{bmatrix}$$

Exercise 15
We know from Exercise 5.2.14, that Hom(V, V) is closed under sums and scalar multiples. Moreover, if $T, S : V \to V$ are two linear transformations, then the function $T \circ S : V \to V$ defined by $(T \circ S)(\mathbf{x}) = T(S(\mathbf{x}))$ is linear. Hence Hom(V, V) is closed under composition.
Exercise 17
Since

$$\varphi(ax+by)(f) = f(ax+by) = af(x)+bf(y) = a\varphi(x)(f)+b\varphi(y)(f)$$

and $\varphi(0)(f) = f(0) = 0$ for all f, it follows that φ is linear.
Exercise 19
The function $T(f) = (a_0, b_0, \ldots, b_n, c_0, \ldots, c_n)$ is the required isomorphism.
Exercise 21 (*Maple V*)
a. We show that with Hom$(\mathbb{R}, \mathbb{R})$ is a linear algebra. Every $f \in$ Hom$(\mathbb{R}, \mathbb{R})$ is a linear function of the form $f(x) = ax$, for some $a \in \mathbb{R}$. The laws of arithmetic therefore guarantee that Hom$(\mathbb{R}, \mathbb{R})$ is a linear algebra. In the following calculations, all variables range over the real numbers, with a, b, c playing the role of scalars and x, y, z playing the role of vectors. The multiplication of real numbers plays the role both of scalar multiplication and of the composition of functions.
evalb((a*x+b*y)+c*z=a*x+(b*y+c*z)); returns *true*.
evalb(a*x+b*y=b*y+a*x); returns *true*.
evalb(a*x+(-a*x)=0); returns *true*.
evalb(expand(a*(x+y))=a*x+a*y); returns *true*.
evalb(expand((a+b)*x)=a*x+b*x); returns *true*.
evalb((a*b)*x=a*(b*x)); returns *true*.
evalb(1*x=x); returns *true*.
evalb(s*(x*y)=(s*x)*y); returns *true*.
b. Consider Hom$(\mathbb{R}^2, \mathbb{R}^2)$. We use the fact, proved in the next section, that every linear transformation $T : \mathbb{R}^2 \to \mathbb{R}^2$ is represented by a 2×2 real matrix, with matrix multiplication playing the role of the composition of functions.
with(LinearAlgebra):
X:=Matrix([[x1,x2],[x3,x4]]), nX:=Matrix([[-x1,-x2],[-x3,-x4]]).
Y:=Matrix([[y1,y2],[y3,y4]]): Z:=Matrix([[z1,z2},{z3,z4]]).
Equal((X+Y)+Z = X+(Y+Z)); returns *true*.
Equal(X+Y = Y+X); returns *true*.
Equal(X+ZeroMatrix(2) = X); returns *true*.
Equal(X+nX=ZeroMatrix(2)); returns *true*.

```
L1:=evalm(a*(X+Y)): R1:=evalm(a*X+a*Y):
Matrix(simplify(evalm(L1-R1))):
Equal(Matrix(simplify(evalm(L1-R1))),ZeroMatrix(2)); returns true.
L2:=evalm((a+b)*X): R2:=evalm(a*X+b*X):
Equal(Matrix(simplify(evalm(L2-R2))),ZeroMatrix(2)); returns true.
L3:=Matrix(evalm((a*b)*X)): R3:=Matrix(evalm(a*(b*X))):
Equal(L3,R3); returns true.
Equal(1*X,X); returns true.
L4:=evalm(X&*(Y+Z)): R4:=evalm(evalm(X&*Y)+evalm(X&*Z)):
Equal(Matrix(simplify(evalm(L4-R4))),ZeroMatrix(2)); returns true.
L5:=evalm((X+Y)&*Z)): R5:=evalm(evalm(X&*Z)+evalm(Y&*Z)):
Equal(Matrix(simplify(evalm(L5-R5))),ZeroMatrix(2)); returns true.
L6:=Matrix(evalm(s*(X&*Y))): R6:=Matrix(evalm(X&*(s*Y))):
Equal(L6,R6); returns true.
L7:=Matrix(evalm(((s*X)&*Y))): R7:=Matrix(evalm(X&*(s*Y))):
Equal(L7,R7); returns true.
```

EXERCISES 5.3

Exercise 1

Since A is a singular matrix, the homogeneous linear system $A\mathbf{x} = \mathbf{0}$ has a nontrivial solution. Therefore $T(\mathbf{x}) = A\mathbf{x} = \mathbf{0}$ for some nonzero vector $\mathbf{x} \in \mathbb{R}^n$.

Exercise 3

The statement is true. Suppose that $T : U \to V$ is a singular linear transformation, that $T(\mathbf{x}) = 0$ for some nonzero vector $\mathbf{x} \in U$, and that $S : V \to W$ is any other linear transformation. Then $S(T(\mathbf{x})) = S(\mathbf{0}) = \mathbf{0}$. Hence $S \circ T$ is singular.

Exercise 5

Suppose that $T_{A^{n+1}}$ is singular for some n. Then there exists a nonzero vector $\mathbf{x}$ for which $T_{A^{n+1}}(\mathbf{x}) = A(A^n(\mathbf{x})) = \mathbf{0}$. But, by assumption, A and A^n are nonsingular. Since T_A is nonsingular, the vector $A^n(\mathbf{x})$ must be zero. On the other hand, the transformation T_{A^n} is also nonsingular, by assumption. Hence the vector $\mathbf{x}$ must be zero. This shows that $T_{A^{n+1}}$ is also nonsingular.

Exercise 7 (*Maple V*)

By Theorem 3.16, the determinant of a projection matrix is zero. Hence these matrices are singular. We now show that the projection $P(x, y, z) = (0, y, z)$ of $\mathbb{R}^3$ onto the yz-plane corresponds to a matrix transformation.

```
A:=submatrix(diag(1,1,1),1..3,2..3): z:=vector(3,[0,0,0]):
```

P:=concat(z,A); returns

$$P := \begin{bmatrix} 0 & 0 & 0 \\ 0 & 1 & 0 \\ 0 & 0 & 1 \end{bmatrix}$$

X:=vector(3,[x,y,z]); returns $X := [x, y, z]$. evalm(P&*X); returns $[0, y, z]$.

EXERCISES 5.4

Exercise 1

a. $A = \begin{bmatrix} 4 & 8 \\ 2 & 3 \end{bmatrix}$ b. $A = \begin{bmatrix} 4 & 1 & 8 \\ 2 & 1 & 3 \\ 1 & 1 & 4 \end{bmatrix}$ c. $A = \begin{bmatrix} 4 & 0 \\ \frac{5}{3} & \frac{1}{3} \end{bmatrix}$ d. $A = \begin{bmatrix} \frac{17}{3} & \frac{61}{9} & \frac{1}{3} \\ \frac{11}{3} & \frac{46}{9} & \frac{1}{3} \\ \frac{8}{3} & \frac{28}{9} & \frac{1}{3} \end{bmatrix}$

Exercise 3

$$A = \begin{bmatrix} 1 & 1 \\ 1 & 0 \end{bmatrix}$$

Exercise 5

$$[T]_{\mathcal{E}}^{\mathcal{B}} = \begin{bmatrix} -1 & -1 & 4 \\ 0 & 0 & 4 \\ 0 & 1 & 2 \end{bmatrix}$$

Exercise 7

$$[T]_{\mathcal{B}}^{\mathcal{B}} = \begin{bmatrix} -1 & \frac{5}{2} & -9 \\ 0 & -\frac{1}{2} & 3 \\ 0 & \frac{1}{4} & \frac{1}{2} \end{bmatrix}$$

Exercise 9

To be able to represent D_t and I_t by matrices, we must convert the spaces $\mathbb{R}_3[t]$ and $\mathbb{R}_2[t]$ to coordinate spaces, using appropriate isomorphisms.
We take $\varphi : \mathbb{R}_3[t] \to \mathbb{R}^4$ to be the isomorphism defined by
$\{\varphi(1) = (1, 0, 0, 0), \varphi(t) = (0, 1, 0, 0), \ \varphi(t^2) = (0, 0, 1, 0), \ \varphi(t^3) = (0, 0, 1, 0)\}$
and $\psi : \mathbb{R}_2[t] \to \mathbb{R}^3$ to be the isomorphism defined by
$\{\psi(1) = (1, 0, 0), \psi(t) = (0, 1, 0), \psi(t^2) = (0, 0, 1)\}$
Then the differentiation operation becomes the linear transformations $D_t' : \mathbb{R}^4 \to \mathbb{R}^3$, defined by $D_t'(a_0, a_1, a_2, a_3) = (a_1, 2a_2, 3a_3)$,
and the integration operation becomes the linear transformation $I_t' : \mathbb{R}^3 \to \mathbb{R}^4$, defined by $I_t'(a_0, a_1, a_2) = (0, a_0, 1/2a_1, 1/3a_2)$.

In the standard bases, we get

$$[D_t] = A = \begin{bmatrix} 0 & 1 & 0 & 0 \\ 0 & 0 & 2 & 0 \\ 0 & 0 & 0 & 3 \end{bmatrix} \quad \text{and} \quad [I_t] = B = \begin{bmatrix} 0 & 0 & 0 \\ 1 & 0 & 0 \\ 0 & 1/2 & 0 \\ 0 & 0 & 1/3 \end{bmatrix}$$

$$AB = \begin{bmatrix} 0 & 1 & 0 & 0 \\ 0 & 0 & 2 & 0 \\ 0 & 0 & 0 & 3 \end{bmatrix} \begin{bmatrix} 0 & 0 & 0 \\ 1 & 0 & 0 \\ 0 & 1/2 & 0 \\ 0 & 0 & 1/3 \end{bmatrix} = \begin{bmatrix} 1 & 0 & 0 \\ 0 & 1 & 0 \\ 0 & 0 & 1 \end{bmatrix}$$

$$BA = \begin{bmatrix} 0 & 0 & 0 \\ 1 & 0 & 0 \\ 0 & 1/2 & 0 \\ 0 & 0 & 1/3 \end{bmatrix} \begin{bmatrix} 0 & 1 & 0 & 0 \\ 0 & 0 & 2 & 0 \\ 0 & 0 & 0 & 3 \end{bmatrix} = \begin{bmatrix} 0 & 0 & 0 & 0 \\ 0 & 1 & 0 & 0 \\ 0 & 0 & 1 & 0 \\ 0 & 0 & 0 & 1 \end{bmatrix}$$

Exercise 11 (*Maple 6*)

```
with(LinearAlgebra):
A:=Matrix([[3,5,5,2], [5,2,-2,0], [5,-2,6,4], [2,0,4,-3]]):
B:=Matrix([[3,6,5,6], [4,6,7,7], [3,6,6,6], [7,7,2,7]]):
v:=[a,b,c,d]: b1:=B[1..-1,1]: b2:=B[1..-1,2]: b3:=B[1..-1,3]: b4:=B[1..-1,4]:
E1:=GenerateEquations(A,v,b1): E2:=GenerateEquations(A,v,b2):
E3:=GenerateEquations(A,v,b3): E4:=GenerateEquations(A,v,b4):
Eset1:=convert(E1,set): Eset2:=convert(E2,set):
Eset3:=convert(E3,set): Eset4:=convert(E4,set):
CB1:=solve(Eset1,{a,b,c,d}); returns
```

$$CB1 := \{a = \tfrac{343}{355}, b = -\tfrac{33}{1420}, d = -\tfrac{-392}{355}, c = -\tfrac{623}{1420}\}$$

CB2:=solve(Eset2,{a,b,c,d}); returns

$$CB2 := \{a = -\tfrac{466}{355}, c = -\tfrac{179}{355}, b = -\tfrac{79}{355}, d = -\tfrac{-279}{355}\}$$

CB3:=solve(Eset3,{a,b,c,d}); returns

$$CB3 := \{b = -\tfrac{349}{1420}, a = -\tfrac{454}{355}, c = -\tfrac{81}{1420}, d = -\tfrac{39}{355}\}$$

CB4:=solve(Eset4,{a,b,c,d}); returns

$$CB4 := \{d = -\tfrac{58}{71}, b = \tfrac{69}{284}, c = \tfrac{115}{284}, a = \tfrac{104}{71}\}$$

```
M:=Matrix([
    [343/355,33/1420,623/1420,-392/355],[466/355,79/355,179/355,-279/355],
    [454/355,349/1420,-81/1420,39/355],[104/71,69/284,115/284,-58/71]]):
BC:=Inverse(Transpose(M)); returns
```

$$\begin{bmatrix} \frac{37}{21} & \frac{-28}{3} & \frac{4}{7} & \frac{2}{3} \\ \frac{12}{7} & -9 & \frac{65}{7} & 5 \\ 2 & -7 & 1 & 2 \\ \frac{-71}{21} & \frac{61}{3} & \frac{-67}{7} & \frac{-20}{3} \end{bmatrix}$$

EXERCISES 5.5

Exercise 1

$$E_1E_2E_3E_4 = \begin{bmatrix} 1 & \frac{1}{4} \\ 0 & 1 \end{bmatrix}\begin{bmatrix} -\frac{3}{4} & 0 \\ 0 & 1 \end{bmatrix}\begin{bmatrix} 1 & 0 \\ 7 & 1 \end{bmatrix}\begin{bmatrix} 1 & 0 \\ 0 & 8 \end{bmatrix} = \begin{bmatrix} 1 & 2 \\ 7 & 8 \end{bmatrix}$$

T_{E_1} is a shear along the x-axis. T_{E_2} is an expansion along the x-axis. T_{E_2} is shear along the y-axis. T_{E_4} is expansion along the y-axis.

Exercise 3 (*Maple V*)

```
with(linalg): A:=matrix([[4,8],[2,3]]): B:=matrix([[6,2],[0,1]]): v:=vector([x,y]):
equal(evalm(B&*(A&*v)),evalm((B&*A)&*v)); returns true.
```

Exercise 5 (*Maple 6*)

```
with(LinearAlgebra): A:=Matrix([[0,2,5],[1,5,2],[3,2,1]]):
P,L,U:=LUDecomposition(A): X:=<x,y,z>: Equal(A.X,P.(L.(U.X))); returns true.
```

EXERCISES 5.6

Exercise 1

$$BA = \begin{bmatrix} \frac{-3}{4} & 2 \\ \frac{1}{2} & -1 \end{bmatrix}\begin{bmatrix} 4 & 8 \\ 2 & 3 \end{bmatrix} = \begin{bmatrix} 1 & 0 \\ 0 & 1 \end{bmatrix}$$

Hence $(T_A)^{-1} = T_B$.

Exercise 3

$$\left(T_{A^{-1}} \circ T_A\right)(\mathbf{x}) = T_{A^{-1}}\left(T_A(\mathbf{x})\right) = A^{-1}(Ax) = \left(A^{-1}A\right)\mathbf{x} = I\mathbf{x} = T_I(\mathbf{x})$$
$$\left(T_A \circ T_{A^{-1}}\right)(\mathbf{x}) = T_A\left(T_{A^{-1}}(\mathbf{x})\right) = A\left(A^{-1}x\right) = \left(AA^{-1}\right)\mathbf{x} = I\mathbf{x} = T_I(\mathbf{x})$$

EXERCISES 5.7

Exercise 1

$$\text{Since } T\begin{bmatrix} 1 \\ 0 \end{bmatrix} = \begin{bmatrix} -8 \\ 2 \end{bmatrix} \quad \text{and} \quad T\begin{bmatrix} 0 \\ 1 \end{bmatrix} = \begin{bmatrix} 3 \\ 77 \end{bmatrix}, \quad \text{we define}$$

$$T\begin{bmatrix} x \\ y \end{bmatrix} = \begin{bmatrix} -8 & 3 \\ 2 & 77 \end{bmatrix}\begin{bmatrix} x \\ y \end{bmatrix} = \begin{bmatrix} -8x+3y \\ 2x+77y \end{bmatrix}$$

Exercise 3

If $A\mathbf{x} = \begin{bmatrix} 1 & 0 \\ 0 & -1 \end{bmatrix}\begin{bmatrix} x \\ y \end{bmatrix} = \begin{bmatrix} x \\ -y \end{bmatrix} = \begin{bmatrix} 0 \\ 0 \end{bmatrix}$, then $x = y = 0$.

Therefore the kernel of T_A is the zero subspace of $\mathbb{R}^2$.

The basis for the kernel of T_A is the empty set.

Exercise 5

The kernel of T consists of the zero vector of $\mathbb{R}^{2\times 3}$. Hence the nullity of T is 0.

The image of T is $\mathbb{R}^6$. Hence the rank of T is 6.

T is one–one since its kernel contains only the zero vector.

T is onto since every (a, b, c, d, e, f) in $\mathbb{R}^6$ corresponds to the rows of a 2×3 matrix under T.

Exercise 7

The image of I_t is the set of all polynomials $a_0t + \frac{1}{2}a_1t^2 + \frac{1}{3}a_2t^3 + \frac{1}{4}a_3t^4$ of degree 4 or less without a constant term, together with the zero polynomial since $I_t(0) = 0t = 0$.

The kernel of I_t is the set $\{0\}$, where 0 is the zero polynomial since $a_0t + \frac{1}{2}a_1t^2 + \frac{1}{3}a_2t^3 + \frac{1}{4}a_3t^4 = 0$ if and only if $a_0 = a_1 = a_2 = a_3 = 0$.

Exercise 9

The kernel of D is the set of infinitely differentiable functions $f : \mathbb{R} \to \mathbb{R}$ satisfying the differential equation $f' = 0$. This means that $f(x) = a$ for some $a \in \mathbb{R}$.

Exercise 11 (*Maple V*)

with(linalg):

A:=matrix([[-8,3,0],[2,77,0],[3,5,0]]): T1:=<-8,2,3>: T2:=<3,77,5>: T3:=<0,0,0>:

T:=(x,y,z) -> x*T1+y*T2+z*T3; returns $T := (x, y, z) -> xT1 + yT2 + zT3$.

equal(evalm(T(x,y,z)),evalm(A&*[x,y,z])); returns *true*.

Exercise 13 (*Maple 6*)

a. with(LinearAlgebra): A1:=Matrix([[1,0,-2,6],[0,5,7,1]]):

ColumnSpace(A1); returns $\{(1, 0), (0, 1)\}$.

NullSpace(A1); returns $\{(44, 0, 1, -7), (30, 1, 0, -5)\}$.

Therefore, $\mathcal{K} = \{(-44, 0, 1, -7), (30, 1, 0, -5)\}$ is a basis for the kernel and $\mathcal{C} = \{(1, 0), (0, 1)\}$ is a basis for the image.

EXERCISES 5.8

Exercise 1

a. The rank of the matrix is 2, the nullity of the matrix is 2, and the dimension of the domain of the matrix transformation is 4.

b. The rank of the matrix is 2, the nullity of the matrix is 1, and the dimension of the domain of the matrix transformation is 3.

c. The rank of the matrix is 2, the nullity of the matrix is 1, and the dimension of the domain of the matrix transformation is 3.

Exercise 3

Suppose that T_A is an isomorphism. Then Theorem 5.27 guarantees that T_A is one–one. Theorem 5.22 implies that the nullity of T_A is 0. The converse follows from the assumption that the nullity of T_A is zero and from reversing the implications.

Exercise 5 (*Maple V, 6*)

with(linalg): A:=matrix([[1,0,0],[0,3,0],[0,0,7],[0,0,0]]):
rank(A); returns 3. nops(nullspace(A)); returns 0.
rank A + nullity $A = 3 + 0 = 3$ is the dimension of $\mathbb{R}^3$
B:=matrix([[1,0,4.5,0],[0,3,0,0],[0,0,0,0]]): rank(B); returns 2.
nops(nullspace(B)); returns 2.
rank A + nullity $A = 2 + 2 = 4$ is the dimension of $\mathbb{R}^4$.
C:=matrix([[1,0,0,9],[0,0,0,0],[0,0,0,0],[0,0,0,0]]): rank(C); returns 1.
nops(nullspace(C)); returns 3.
rank A + nullity $A = 1 + 3 = 4$ is the dimension of $\mathbb{R}^4$.
with(LinearAlgebra):
is(nops(nullspace(A)),nops(NullSpace(Matrix(A)))); returns *true*.
is(nops(colspace(B)),nops(ColumnSpace(Matrix(B)))); returns *true*.

Exercise 7 (*Maple 6*)

We know from Theorem 4.29 that for any $m \times n$ matrix A, $n =$ rank A + nullity A. Hence nullity $A = n -$ rank A.

EXERCISES 5.9

Exercise 1

The matrix

$$A = \begin{bmatrix} 1 & 1 \\ 0 & 1 \end{bmatrix}$$

represents T in the standard basis of $\mathbb{R}^2$.
The area is $(1/2)(1)(17) = 17/2$ units.

Exercise 3

Let

$$\text{a. } \mathbf{u} = \begin{bmatrix} -3\sqrt{2} + 2\sqrt{2} \\ 6\sqrt{2} - 5\sqrt{2} \end{bmatrix} = \begin{bmatrix} -\sqrt{2} \\ \sqrt{2} \end{bmatrix} \quad \text{b. } \mathbf{v} = \begin{bmatrix} 0 + 2\sqrt{2} \\ 7\sqrt{2} - 5\sqrt{2} \end{bmatrix} = \begin{bmatrix} 2\sqrt{2} \\ 2\sqrt{2} \end{bmatrix}$$

Then the area of the image of the transformed parallelogram is 8 units.

Exercise 5 (*Maple 6*)

A:=Matrix([[1,0,2],[0,1,0],[-6,0,1],[0,0,0]]):
C:=LinearAlgebra[ColumnSpace](A); returns $C := \{(1, 0, 0, 0), (0, 1, 0, 0), (0, 0, 1, 0)\}$.

EXERCISES 5.10

Exercise 1

$$\begin{bmatrix} 0 & 1 & 0 \\ 1 & 0 & 0 \\ 0 & 0 & 1 \end{bmatrix}\begin{bmatrix} 2 & 0 & 0 \\ 0 & 3 & 0 \\ 0 & 0 & 2 \end{bmatrix}\begin{bmatrix} 0 & 1 & 0 \\ 1 & 0 & 0 \\ 0 & 0 & 1 \end{bmatrix} = \begin{bmatrix} 3 & 0 & 0 \\ 0 & 2 & 0 \\ 0 & 0 & 2 \end{bmatrix}$$

Exercise 3

$$A = B = \begin{bmatrix} 1 & 3 \\ 3 & 2 \end{bmatrix} \quad \text{and} \quad C = D = \begin{bmatrix} 1 & 1 \\ 1 & 3 \end{bmatrix}$$

Then A is similar to B and C is similar to D.
The trace of $A + B$ is 6 and the trace of $C + D$ is 8. Hence $A + B$ and $C + D$ are not similar.
The trace of AB is 23 and the trace of CD is 12. Hence AB and CD are not similar.

Exercise 5

$$A = \begin{bmatrix} a & b \\ c & d \end{bmatrix} = P\begin{bmatrix} 0 & 0 \\ 0 & 0 \end{bmatrix}P^{-1} = \begin{bmatrix} 0 & 0 \\ 0 & 0 \end{bmatrix}$$

Exercise 7
If $A = PBP^{-1}$, then $sA = s\left(PBP^{-1}\right) = P\,(sB)\,P^{-1}$.

Exercise 9 (*Maple V*)
with(linalg): A:=matrix([[1,0,-2],[0,3,7],[0,0,2]]): B:=matrix([[6,0,0],[0,0,0],[0,0,0]]):
C:=matrix([[2,-8,3],[4,2,7],[4,-5,2]]): [det(A),det(B),det(C)]; returns [6, 0, -166].

Exercise 11 (*Maple V*)
with(linalg): a. A1:=matrix([[1,2,3],[4,5,6],[7,8,9]]): A2:=matrix([[0,3,1],[9,15,2],[-3,-3,0]]):
issimilar(A1,A2); returns *true*.
b. B1:=matrix([[1,1,0,0],[0,1,1,1],[0,0,1,0],[1,1,0,1]]):
B2:=matrix([[2,-2/7,4/7,-1],[2,1/2,1/2,-5/2],[1,1/2,3/2,1/2],[1,-1/7,2/7,0]]):
issimilar(B1,B2); returns *true*.

CHAPTER 6 - Eigenvalues and Eigenvectors

EXERCISES 6.1

Exercise 1
Consider the point $\mathbf{x} = (1, 1)$ with positive coordinates.
a.$T(x, y) = (3x, 4y)$ stretches $\mathbf{x}$ three units along the positive direction of the x-axis and four units along the positive direction of the y-axis. The result is the point $(3, 4)$.
b. $T(x, y) = (-3x, 4y)$ takes the point produced by a. and reflects it about the y-axis. The result is the point $(-3, 4)$.

c. $T(x, y) = (3x, -4y)$ takes the point produced by a. and reflects it about the x-axis. The result is the point $(3, -4)$.
d. $T(x, y) = (-3x, -4y)$ takes the point produced by a. and reflects it about the x- and y-axes. The result is the point $(-3, -4)$.

Exercise 3

a. The scalar 5 is an eigenvalue since it maps the nonzero vector $\mathbf{e}_1$ to the multiple $5\mathbf{e}_1$. The eigenvectors belonging to 5 are of the form $\mathbf{x} = a\mathbf{e}_1$, with $a \neq 0$.
b. The scalar -2 is an eigenvalue since it maps the nonzero vector $\mathbf{e}_2$ to the multiple $-2\mathbf{e}_2$. The eigenvectors belonging to -2 are of the form $\mathbf{x} = a\mathbf{e}_2$, with $a \neq 0$.
c. The scalar 7 is an eigenvalue since it maps the nonzero vector $\mathbf{e}_3$ to the multiple $7\mathbf{e}_3$. The eigenvectors belonging to 7 are of the form $\mathbf{x} = a\mathbf{e}_3$, with $a \neq 0$.

Exercise 5

Since A and PAP^{-1} have the same characteristic polynomial $-2 - 5t + t^2$, their eigenvalues are the same.

Exercise 7

The eigenvalues of the matrix

$$A = \begin{bmatrix} 0 & -1 \\ 1 & 0 \end{bmatrix}$$

are i and $-i$.
The eigenvalue of the matrix

$$A^2 = \begin{bmatrix} 0 & -1 \\ 1 & 0 \end{bmatrix}\begin{bmatrix} 0 & -1 \\ 1 & 0 \end{bmatrix} = \begin{bmatrix} -1 & 0 \\ 0 & -1 \end{bmatrix}$$

is -1. The matrix transformation T_A has no real eigenvalues. The matrix transformation T_{A^2} has the real eigenvalues -1.

Exercise 9

Since $I\mathbf{x} = \mathbf{x} = 1\mathbf{x}$ for all $\mathbf{x} \in V$, the scalar 1 is the only eigenvalue. By definition only nonzero vectors are eigenvectors.

Exercise 11

Suppose that λ is an eigenvalue of AA^T. The eigenvalue is either zero or not. If $\lambda = 0$, then $(AA^T)\mathbf{x} = 0\mathbf{x} = \mathbf{0}$. Hence the homogeneous linear system $(AA^T)\mathbf{x} = \mathbf{0}$ has a nontrivial solution. This means that the matrix $\det(AA^T) = 0$. By the property of determinants this, in turn, implies that $\det(A^T) = 0$. Hence the homogeneous linear system $(A^T)\mathbf{x} = \mathbf{0}$ has a nontrivial solution. Therefore the nonzero vector $\mathbf{x}$ is an eigenvector of both AA^T and A^TA associated with the eigenvalue zero.
Now suppose that $\lambda \neq 0$ that $(AA^T)\mathbf{x} = \lambda\mathbf{x}$ for some nonzero vector $\mathbf{x}$. Let $\mathbf{y} = A^T\mathbf{x}$. Then $A\mathbf{y} = AA^T\mathbf{x} = \lambda\mathbf{x}$. Since $\lambda \neq 0$ and $\mathbf{x} \neq \mathbf{0}$, it follows that $\mathbf{y} \neq \mathbf{0}$. Therefore $\mathbf{y}$ is an eigenvector of A^TA since $(A^T)\mathbf{y} = A^T(A\mathbf{y}) = A^T(AA^T\mathbf{x}) = A^T\lambda\mathbf{x} = \lambda A^T\mathbf{x} = \lambda\mathbf{y}$.

This result is a special case of the fact that for all $n \times n$ matrices A and B, the matrices AB and BA have the same eigenvalues.

Exercise 13

$(T_2 \circ T_1)\,\mathbf{x} = T_2\,(T_1\mathbf{x}) = T_2\,(\lambda\mathbf{x}) = \lambda T_2(\mathbf{x}) = \lambda\,(\mu\mathbf{x}) = (\lambda\mu)\,\mathbf{x}.$

Exercise 15 (*Maple V*)

with(linalg): A:=matrix([[0,0,60],[1,0,7],[0,1,-6]]):
eigenvaluesA:=eigenvalues(A); returns $eigenvaluesA := 3, -4, -5$.
eigenvectorsA:=eigenvects(A); returns
$eigenvectorsA := [3, 1, \{[20, 9, 1]\}], [-5, 1, \{[-12, 1, 1]\}], [-4, 1, \{[-15, 2, 1]\}]$.
The vector [20, 9, 1] is an eigenvector belonging to 3.
The vector [−15, 2, 1] is an eigenvector belonging to −4.
The vector [−12, 1, 1] is an eigenvector belonging to −5.

Exercise 17 (*Maple V*)

We use the **arrow** function in the **plottools** package to display the direction arrows

$$(1, 1, 1), (3, 4, 5), (-3, 4, 5), (-3, -4, 5), (3, 4, -5)$$

with(plottools):
a1:=arrow([0,0,0],[1,1,1],.3,.1,.1): a2:=arrow([0,0,0],[3,4,5],.3,.1,.1):
a3:=arrow([0,0,0],[-3,4,5],.3,.1,.1): a4:=arrow([0,0,0],[-3,-4,5],.3,.1,.1):
a5:=arrow([0,0,0],[3,4,-5],.3,.1,.1):
plots[display](a1,a2,a3,a4,a5,axes=normal,orientation=[66,48]); returns

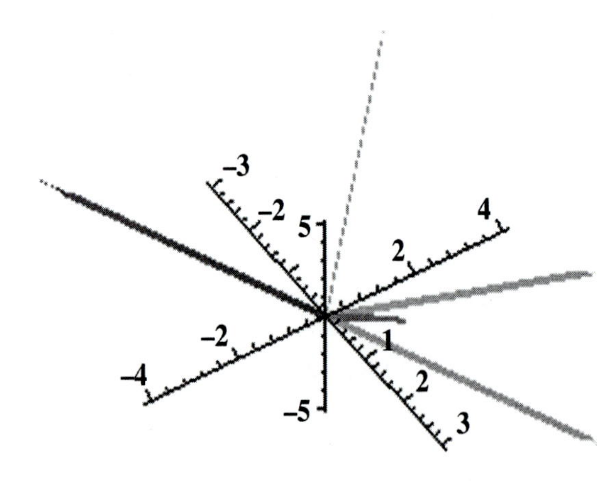

Exercise 19 (*Maple V*)

with(linalg): P:=matrix([[3,0,8],[0,1,3],[8,3,0]]): A:=matrix([[1,2,0],[3,4,0],[0,0,1]]):
is(eigenvalues(A)=eigenvalues(P&*A&*inverse(P))); returns *true*.

Exercise 21 (*Maple V*)

with(linalg): A:=matrix([[0,0,2],[1,0,1],[0,1,1]]):
eigenvalues(A); returns $2, \frac{-1}{2} + \frac{1}{2}I\sqrt{3}, -\frac{1}{2} - \frac{1}{2}I\sqrt{3}$.

EXERCISES 6.2

Exercise 1 (d, e, f)

d. $$\begin{bmatrix} 1 & 0 & -2 \\ 0 & 5 & 7 \\ 0 & 0 & 0 \end{bmatrix} - t \begin{bmatrix} 1 & 0 & 0 \\ 0 & 1 & 0 \\ 0 & 0 & 1 \end{bmatrix} = \begin{bmatrix} 1-t & 0 & -2 \\ 0 & 5-t & 7 \\ 0 & 0 & -t \end{bmatrix}$$

e. $$\begin{bmatrix} 0 & -1 & 0 \\ 1 & 0 & 0 \\ 0 & 0 & 1 \end{bmatrix} - t \begin{bmatrix} 1 & 0 & 0 \\ 0 & 1 & 0 \\ 0 & 0 & 1 \end{bmatrix} = \begin{bmatrix} -t & -1 & 0 \\ 1 & -t & 0 \\ 0 & 0 & 1-t \end{bmatrix}$$

f. $$\begin{bmatrix} 2 & 0 & 0 & 0 \\ 0 & 3 & 0 & 0 \\ 0 & 0 & 2 & 0 \\ 0 & 0 & 0 & 5 \end{bmatrix} - t \begin{bmatrix} 1 & 0 & 0 & 0 \\ 0 & 1 & 0 & 0 \\ 0 & 0 & 1 & 0 \\ 0 & 0 & 0 & 1 \end{bmatrix} = \begin{bmatrix} 2-t & 0 & 0 & 0 \\ 0 & 3-t & 0 & 0 \\ 0 & 0 & 2-t & 0 \\ 0 & 0 & 0 & 5-t \end{bmatrix}$$

Exercise 3

Since A has three distinct eigenvalues, it also has three linearly independent eigenvectors. Definition 6.3 therefore tells us that the matrix A is diagonalizable. In other words, $A = PDP^{-1}$. Since D is diagonal and since its diagonal entries are nonzero, it is invertible. Hence A is invertible since it is a product of invertible matrices. Therefore the system $A\mathbf{x} = \mathbf{0}$ can only have the trivial solution.

Exercise 5

This fact follows from the inductive definition of the Laplace expansion of the determinant of the matrix $A - tI$.

Exercise 7

Let A be any 3×3 real matrix. Then the characteristic polynomial $c_A(t)$ of A is a polynomial of the form $a_0 + a_1 t + a_2 t^2 + t^3$. By the intermediate value theorem of calculus, $c_A(t)$ has a real root. By Theorem 6.2, this root is a real eigenvalues of A.

Exercise 9 (*Maple V*)

We construct two matrices whose eigenvalues are 1 and 2 but whose associated eigenspaces have different dimensions.

with(linalg): A:=diag(1,2,2); returns

$$A := \begin{bmatrix} 1 & 0 & 0 \\ 0 & 2 & 0 \\ 0 & 0 & 2 \end{bmatrix}$$

B:=companion((t-1)*(t-1)*(t-2),t); returns

$$B := \begin{bmatrix} 0 & 0 & 2 \\ 1 & 0 & -5 \\ 0 & 1 & 3 \end{bmatrix}$$

eigenvectors(A); returns [2, 2, {[0, 0, 1], [0, 1, 0]}], [1, 1, {[1, 0, 0]}].
eigenvectors(B); returns [1, 2, {[2, −3, 1]}], [2, 1, {[1, −2, 1]}].

EXERCISES 6.3

Exercise 1

$$p_1(A) = \begin{bmatrix} -2 & 2 & 1 \\ 1 & -2 & 0 \\ -1 & 0 & 1 \end{bmatrix}, p_2(A) = \begin{bmatrix} -2 & -2 & 2 \\ -1 & -1 & 1 \\ -2 & -2 & 2 \end{bmatrix}, p_1(A)p_2(A) = \begin{bmatrix} 0 & 0 & 0 \\ 0 & 0 & 0 \\ 0 & 0 & 0 \end{bmatrix}$$

Exercise 3

Suppose that $c_A(t) = a_0 + a_1 t + \cdots + a_{n-1}t^{n-1} + t^n$ is the characteristic polynomial of A. By the Cayley–Hamilton theorem, $a_0 I + a_1 A + \cdots + a_{n-1}A^{n-1} + A^n = 0$. Therefore,

$$A\left(a_1 + a_2 A + \cdots + a_{n-1}A^{n-2} + A^{n-1}\right) = -a_0 I$$

It follows that $A^{-1} = \left(a_1 + a_2 A + \cdots + a_{n-1}A^{n-2} + A^{n-1}\right)/(-a_0)$.

Exercise 5 (*Maple 6*)

```
with(LinearAlgebra):
A:=Matrix([[-86,23,-84,19,-50],[88,-53,85,49,78],
     [17,72,-99,-85,-86],[30,80,72,66,-29],[-91,-53,-19,-47,68]]):
cA:=CharacteristicPolynomial(A,t); returns
```

$$cA := 4587563299 - 122673722t^2 - 539812t^3 - 17584t^4 + 104t + t^5$$

t:=A: Equal(cA,ZeroMatrix(5)); returns *true*.
B:=-(-122673722-539812*t-17584*t^2+104*t^ 3+t^4)/4587563299:
Equal(A.B,IdentityMatrix(5)); returns *true*.

EXERCISES 6.4

Exercise 1

$$\begin{bmatrix} 5 & 2 & 5 \\ 2 & 3 & 3 \\ 0 & 3 & 1 \end{bmatrix}, \begin{matrix} 5 & 2 & 5 & 0 \\ 2 & 3 & 3 & 0 \\ 0 & 3 & 1 & 0 \\ 0 & 0 & 0 & 0 \end{matrix}, \text{ and } \begin{bmatrix} 5 & 2 & 5 & 0 & 1 \\ 2 & 3 & 3 & 0 & 2 \\ 0 & 3 & 1 & 0 & 3 \\ 0 & 0 & 0 & 0 & 4 \\ 0 & 0 & 0 & 0 & 5 \end{bmatrix} \text{ are Hessenberg matrices.}$$

Exercise 3 (*Maple 6*)

with(LinearAlgebra): u:=<1,2,3>: A:=HouseholderMatrix(u,3):
Equal(A,Transpose(A)); returns *true*. IsOrthogonal(A); returns *true*.

v:=<0,1,0,1>: B:=HouseholderMatrix(v,4):
Equal(B,Transpose(B)); returns *true*. IsOrthogonal(B); returns *true*.
w:=<1,1,1,1,1>: C:=HouseholderMatrix(w,5):
Equal(C,Transpose(C)); returns *true*. IsOrthogonal(C); returns *true*.

Exercise 5 (*Maple 6*)

If $k = 2$, $r = \pm\sqrt{x_2^2 + x_3^2 + x_4^2}$ and $x_2 r \le 0$, then

$$\mathbf{x} = \begin{bmatrix} x_1 \\ x_2 \\ x_3 \\ x_4 \end{bmatrix}, \mathbf{u} = \begin{bmatrix} 0 \\ x_2 - r \\ x_3 \\ x_4 \end{bmatrix}, Q_\mathbf{u}\mathbf{x} = \begin{bmatrix} x_1 \\ r \\ 0 \\ 0 \end{bmatrix}$$

Therefore,

$$\mathbf{x} = \begin{bmatrix} 1 \\ 0 \\ 4 \\ 0 \end{bmatrix}, \mathbf{u} = \begin{bmatrix} 0 \\ 0 - \sqrt{4^2} \\ 4 \\ 0 \end{bmatrix} = \begin{bmatrix} 0 \\ -4 \\ 4 \\ 0 \end{bmatrix}, Q_\mathbf{u} = \begin{bmatrix} 1 & 0 & 0 & 0 \\ 0 & 0 & 1 & 0 \\ 0 & 1 & 0 & 0 \\ 0 & 0 & 0 & 1 \end{bmatrix}$$

and

$$Q_\mathbf{u} A Q_\mathbf{u} = \begin{bmatrix} 1 & 0 & 0 & 0 \\ 0 & 0 & 1 & 0 \\ 0 & 1 & 0 & 0 \\ 0 & 0 & 0 & 1 \end{bmatrix} \begin{bmatrix} 1 & 0 & 1 & 0 \\ 0 & 2 & 0 & 1 \\ 4 & 0 & 0 & 0 \\ 0 & -1 & 4 & 0 \end{bmatrix} \begin{bmatrix} 1 & 0 & 0 & 0 \\ 0 & 0 & 1 & 0 \\ 0 & 1 & 0 & 0 \\ 0 & 0 & 0 & 1 \end{bmatrix} = \begin{bmatrix} 1 & 1 & 0 & 0 \\ 4 & 0 & 0 & 0 \\ 0 & 0 & 2 & 1 \\ 0 & 4 & -1 & 0 \end{bmatrix}$$

If $k = 3$, $r = \pm\sqrt{x_3^2 + x_4^2}$ and $x_3 r \le 0$, then

$$\mathbf{x} = \begin{bmatrix} x_1 \\ x_2 \\ x_3 \\ x_4 \end{bmatrix}, \mathbf{v} = \begin{bmatrix} 0 \\ 0 \\ x_3 - r \\ x_4 \end{bmatrix}, Q_\mathbf{v}\mathbf{x} = \begin{bmatrix} x_1 \\ x_2 \\ r \\ 0 \end{bmatrix}$$

Therefore,

$$\mathbf{x} = \begin{bmatrix} 1 \\ 0 \\ 0 \\ 4 \end{bmatrix}, \mathbf{v} = \begin{bmatrix} 0 \\ 0 \\ 0 - \sqrt{4^2} \\ 0 \end{bmatrix} = \begin{bmatrix} 0 \\ 0 \\ -4 \\ 4 \end{bmatrix}, Q_\mathbf{v} = \begin{bmatrix} 1 & 0 & 0 & 0 \\ 0 & 1 & 0 & 0 \\ 0 & 0 & 0 & 1 \\ 0 & 0 & 1 & 0 \end{bmatrix}$$

and

$$Q_{\mathbf{v}}Q_{\mathbf{u}}AQ_{\mathbf{u}}Q_{\mathbf{v}} = \begin{bmatrix} 1 & 0 & 0 & 0 \\ 0 & 1 & 0 & 0 \\ 0 & 0 & 0 & 1 \\ 0 & 0 & 1 & 0 \end{bmatrix} \begin{bmatrix} 1 & 1 & 0 & 0 \\ 4 & 0 & 0 & 0 \\ 0 & 0 & 2 & 1 \\ 0 & 4 & -1 & 0 \end{bmatrix} \begin{bmatrix} 1 & 0 & 0 & 0 \\ 0 & 1 & 0 & 0 \\ 0 & 0 & 0 & 1 \\ 0 & 0 & 1 & 0 \end{bmatrix}$$

$$= \begin{bmatrix} 1 & 1 & 0 & 0 \\ 4 & 0 & 0 & 0 \\ 0 & 4 & 0 & -1 \\ 0 & 0 & 1 & 2 \end{bmatrix}$$

The HessenbergForm command, on the other hand, produces a slightly different matrix.
with(LinearAlgebra): A:=Matrix([[1,0,1,0],[0,2,0,1],[4,0,0,0],[0,-1,4,0]]):
HessenbergForm(A); returns

$$\begin{bmatrix} 1 & -1 & 0 & 0 \\ -4 & 0 & 0 & 0 \\ 0 & -4 & 0 & 1 \\ 0 & 0 & -1 & 2 \end{bmatrix}.$$

EXERCISES 6.5

Exercise 1

Consider the matrix transformation $T_{(A-\lambda I)} : \mathbb{R}^n \to \mathbb{R}^n$.
By Corollary 5.26, the sum of the rank and nullity of $T_{(A-\lambda I)}$ is n.
By Theorem 5.23, the rank of $T_{(A-\lambda I)}$ is equal to the rank r of $(A - \lambda I)$.
Hence the nullity of $T_{(A-\lambda I)}$ is $n - r$.
By Theorem 5.24, the nullity of $T_{(A-\lambda I)}$ is also equal to the dimension of the null space of $(A - \lambda I)$.
By Exercise 6.5.6, the null space of $(A - \lambda I)$ coincides with the eigenspace E_λ.
Hence the dimension of E_λ is $n - r$.

Exercise 3

Suppose that λ is an eigenvalue of a 3×3 real matrix A and that the dimension of E_λ is 3.
Then $E_\lambda = \mathbb{R}^3$. Hence every nonzero vector of $\mathbb{R}^3$ is an eigenvector.
This means that

$$A(1,0,0) = \lambda(1,0,0) = (\lambda,0,0) = (\lambda I)(1,0,0)$$
$$A(0,1,0) = \lambda(0,1,0) = (0,\lambda,0) = (\lambda I)(0,1,0)$$
$$A(0,0,1) = \lambda(0,0,1) = (0,0,\lambda) = (\lambda I)(0,0,1)$$

Hence the matrix transformation T_A agrees with the transformation $T_{\lambda I}$ on a basis for $\mathbb{R}^3$.
Therefore A is the diagonal matrix λI.

Exercise 5

By definition, the geometric multiplicity of an eigenvalue λ is the dimension of the eigenspace E_λ. If $\dim E_\lambda = n$, then $E_\lambda = \mathbb{R}^n$. By generalizing Exercise 6.5.4, we conclude that $A = \lambda I_n$.

Exercise 7 (*Maple 5*)

with(linalg): A:=diag(3,5,5): L:=matrix([[1,0,0],[2,3,0],[4,5,6]]):
B:=evalm(L&*A&*inverse(L)):
eigenvectors(B); returns [3, 1, {[1, 2, 4]}], [5, 2, {[0, 0, 1], [0, 1, 0]}].
The eigenspace E_3 has dimension 1 and the eigenspace E_5 has dimension 2.

EXERCISES 6.6

Exercise 1 (*Maple V*)

with(linalg): A:=matrix([[0,0,0,8],[1,0,0,-4],[0,1,0,-6],[0,0,1,5]]):
p:=factor(charpoly(A,t)); returns $p := (t+1)(t-2)^3$.
Since 0 is not an eigenvalue of A, it follows from Theorem 6.8 that A is invertible.

Exercise 3 (*Maple V*)

By Theorem 6.9, the eigenvalues of A^{-1} are the reciprocals of the eigenvalues of A.
with(linalg): A:=matrix([[-4,-3],[-3,-1]]):
evals:={simplify(1/eigenvals(A)[1]),simplify(1/eigenvals(A)[2])}; returns

$$evals := \{-2/5 + 3\sqrt{5}, 2/(-5 + 3\sqrt{5})\}$$

EXERCISES 6.7

Exercise 1

Suppose that λ is an eigenvalue of a matrix A and that $A\mathbf{x} = \lambda\mathbf{x}$. Then
$A(a\mathbf{x}) = a(A\mathbf{x}) = a\lambda\mathbf{x} = \lambda(a\mathbf{x})$. Hence $a\mathbf{x}$ is an eigenvector belonging to λ.

Exercise 3

The statement is false. The matrix

$$A = \begin{bmatrix} 3 & 4 \\ 0 & 5 \end{bmatrix}$$

for example, has the eigenvalues 3 and 5, with associated eigenvectors (1, 0) and (2, 1). However,

$$\begin{bmatrix} 1 \\ 0 \end{bmatrix}^T \begin{bmatrix} 2 \\ 1 \end{bmatrix} = 2 \neq 0$$

Exercise 5

Theorem 6.10 tells us that eigenvectors belonging to distinct eigenvalues are linearly independent. Let $\lambda_1, \ldots, \lambda_n$ be the n distinct eigenvalues of A and let $\mathbf{x}_1, \ldots, \mathbf{x}_n$ be n associated eigenvectors. Since the dimension of $\mathbb{R}^n$ is n, the vectors $\mathbf{x}_1, \ldots, \mathbf{x}_n$ span $\mathbb{R}^n$ and therefore form a basis.

Exercise 7 (d, e, f) (*Maple 6*)

d. with(LinearAlgebra): A:=Matrix([[1,0,-2,6],[0,5,7,1],[0,0,0,1],[0,0,0,1]]):
evA:=Eigenvectors(A); returns

$$evA := \begin{bmatrix} 0 \\ 1 \\ 1 \\ 5 \end{bmatrix}, \begin{bmatrix} 2 & 1 & 0 & 0 \\ \frac{-7}{5} & 0 & 0 & 1 \\ 1 & 0 & 0 & 0 \\ 0 & 0 & 0 & 0 \end{bmatrix}.$$

We show that the first, second, and fourth columns of the eigenvector matrix are linearly independent.

v:=ScalarMultiply(Column(evA[2],1),a) + ScalarMultiply(Column(evA[2],2),b)
  + ScalarMultiply(Column(evA[2],4),c):
LinearSolve(Matrix(v),ZeroVector(4)); returns [0]. This shows that $a = b = c = 0$.

A similar calculation shows that the first, third, and fourth columns of the eigenvector matrix are linearly independent.

e. Z:=ZeroMatrix(4,4):
evZ:=Eigenvectors(Z); returns

$$evA := \begin{bmatrix} 0 \\ 0 \\ 0 \\ 0 \end{bmatrix}, \begin{bmatrix} 0 & 0 & 1 & 0 \\ 0 & 0 & 0 & 1 \\ 1 & 0 & 0 & 0 \\ 0 & 1 & 0 & 0 \end{bmatrix}.$$

Since Z has only one eigenvalue, the result holds trivially because each set consisting of a single eigenvector is linearly independent.

f. A:=Matrix([[1,0,0,9],[0,1,7,0],[0,7,1,0],[9,0,0,1]]):
evA:=convert(Eigenvalues(A),set); returns $evA := \{-8, -6, 8, 10\}$.

Hence A has four distinct eigenvalues. The four sets containing the associated eigenvectors are linearly independent.

EXERCISES 6.8

All eigenvalues in the following exercises must be real numbers, and all eigenvectors must have real entries.

Exercise 1

Since eigenvectors belonging to distinct eigenvalues are linearly independent, the following eigenvector-eigenvalue pairs show that the given matrices have two linearly independent eigenvectors.

	$\mathbf{x}_\lambda \rightleftharpoons \lambda$	$\mathbf{x}_\lambda \rightleftharpoons \lambda$
a.	$(-1, 1) \rightleftharpoons 0$	$(1, 1) \rightleftharpoons 2$
b.	$(1, 2) \rightleftharpoons 3$	$(1, 0) \rightleftharpoons 1$
c.	$(1, \frac{1}{5} + \frac{1}{5}\sqrt{26}) \rightleftharpoons 2 + \sqrt{26}$	$(1, \frac{1}{5} - \frac{1}{5}\sqrt{26}) \rightleftharpoons 2 - \sqrt{26}$
d.	$(1, \frac{4}{5} + \frac{1}{5}\sqrt{41}) \rightleftharpoons -1 + \sqrt{41}$	$(1, \frac{4}{5} - \frac{1}{5}\sqrt{41}) \rightleftharpoons -1 - \sqrt{41}$

Exercise 3

The triangular 3×3 matrix (c) in Example 6.8.5 is not diagonalizable since it fails to have three linearly independent eigenvectors. On the other hand, we know from Theorem 3.20 that the determinant of this matrix is 8, hence nonzero. This tells us that the matrix is invertible. We know from Theorem 6.8 that a real square matrix is invertible if and only if 0 it not an eigenvalue of A. Since the eigenvalues of a diagonal matrix are its diagonal entries, any diagonal matrix with a zero on the diagonal is an example of a noninvertible diagonal matrix.

Exercise 5

The theorem tells us that $A = PDP^{-1} = P[\lambda I]P^{-1} = \lambda PP^{-1} = \lambda I$.

Exercise 7 (*Maple V, 6*)

We show that the first two matrices have three distinct eigenvalues. By Theorem 6.10, the associated eigenvectors are linearly independent.

A:=matrix([[2,0,-1],[0,1,0],[-1,1/2,1]]):

evA:=eigenvalues(A); returns $evA := 1, \frac{3}{2} + \frac{1}{2}\sqrt{5}, \frac{3}{2} - \frac{1}{2}\sqrt{5}$.

B:=matrix([[2,0,0],[5,0,5],[0,1/2,1]]):

evB:=eigenvalues(B); returns $evB := 2, \frac{1}{2} + \frac{1}{2}\sqrt{11}, \frac{1}{2} - \frac{1}{2}\sqrt{11}$.

Finally, consider matrix (c).

C:=matrix([[2,-1/2,0],[0,1,0],[-1,1/2,1]]):

evC:=eigenvalues(C); returns $evC := 2, 1, 1$.

Since C has a repeated eigenvalues, we cannot use the previous argument to show that C has three linearly independent eigenvectors. Instead, we show that the eigenvector matrix produced by *Maple* is invertible. Theorem 4.3 then guarantees that the vectors are linearly independent.

with(LinearAlgebra): evmatA:=Eigenvectors(Matrix(A)); returns

$$evmatA := \begin{bmatrix} 2 \\ 1 \\ 1 \end{bmatrix}, \begin{bmatrix} -1 & 1 & 0 \\ 0 & 2 & 0 \\ 1 & 0 & 1 \end{bmatrix}.$$

Determinant(evmatA[2]); returns -2.

Exercise 9 (*Maple 6*)
We solve the problem for matrix (c). The other cases are similar.
with(LinearAlgebra): A:=Matrix([[2,0,0,0],[-2,3,0,0],[0,0,3,0],[-1,0,0,3]]):
evecA:=Eigenvectors(A): is(Determinant(evecA[2])=0); returns $false$.
Exercise 11 (*Maple 6*) The expanded version of the polynomial $p(t)$ is

$$t^3 - 9t^2 + 26t - 24 = t^3 + a_2t^2 + a_1t + a_0$$

Therefore the companion matrix of $p(t)$ is

$$\begin{bmatrix} 0 & 0 & 24 \\ 1 & 0 & -26 \\ 0 & 1 & 9 \end{bmatrix}.$$

We use *Maple* to find matrices D and P for which $A = PDP^{-1}$.
with(LinearAlgebra): A:=Matrix([[0,0,24],[1,0,-26],[0,1,9]]):
convert(Eigenvalues(A),set); returns {2, 3, 4}.
evA:=Eigenvectors(A): diagA:=DiagonalMatrix([4,2,3]):
P:=Matrix([[seq(evA[2][1..3,i],i=1..3)]]): P.diagA.P^(-1); returns

$$\begin{bmatrix} 0 & 0 & 24 \\ 1 & 0 & -26 \\ 0 & 1 & 9 \end{bmatrix}.$$

EXERCISES 6.9

Exercise 1
We know from calculus that for any nonzero real number a, the function $f(x) = e^{ax}$ is a basis for the solution set of the this equation. Let ke^{ax} be any other solution. Then the equation $kae^{ax} = \lambda ke^{ax}$ implies that $\lambda = a$ is an eigenvalue, and the function ke^{ax} is an eigenvector associated with λ.
Exercise 3
Let $\mathcal{E} = \{1, t, t^2, t^3\}$ be the standard basis of $\mathbb{R}_3[t]$ and $p(t)\,(t/0)$ be the value of $p(t)$ at 0. Then $1\,(t/0) = 1$ and $t\,(t/0) = t^2\,(t/0) = t^3\,(t/0) = 0$. Hence

$$[T]_{\mathcal{E}} = \begin{bmatrix} 1 & 0 & 0 & 0 \\ 0 & 0 & 0 & 0 \\ 0 & 0 & 0 & 0 \\ 0 & 0 & 0 & 0 \end{bmatrix}$$

is a diagonal representation of T.

Exercise 5

Let $f(x) = \sin x$. Then $f'(x) = \cos x$ and $f''(x) = -\sin x$. Hence $f''(x) = (-1)\, f(x)$. Therefore $\sin x$ is an eigenvector associated with the eigenvalue -1. Hence f is an eigenvector of D^2. However, f is not an eigenvector of D since there is no λ for which $f'(x) \neq \lambda f(x)$.

EXERCISES 6.10

Exercise 1

If we think of the constants 1.08 as a 1×1 (diagonal) matrix A and P_n as a 1×1 vector $\mathbf{x}$, then the sequence

$$P_0,\, 1.08 P_0 = P_1,\, 1.08^2 P_0 = P_2, \ldots,\, 1.08^n P_0 = P_n, \ldots$$

is a discrete dynamical linear systems.

Exercise 5 (*Maple V*)

with(combinat): a:=rand(100..500):

f:=x->fibonacci(x+2)-fibonacci(x+1)-fibonacci(x): f(a()); returns 0.

Exercise 7 (*Maple 6*)

with(LinearAlgebra): A:=Matrix([[1,1],[1,0]]): x:=<1,1> : x101:=(A^100).x; returns

$$x101 := \begin{bmatrix} 927372692193078999176 \\ 573147844013817084101 \end{bmatrix}$$

with(combinat): [fibonacci(102),fibonacci(101)]; returns

$$[927372692193078999176, 573147844013817084101]$$

EXERCISES 6.11

Exercise 1

$$\begin{bmatrix} 1-p & q \\ p & 1-q \end{bmatrix} \begin{bmatrix} q \\ p \end{bmatrix} = \begin{bmatrix} (1-p)\,q + qp \\ qp + (1-q)\,p \end{bmatrix} = \begin{bmatrix} q \\ p \end{bmatrix}$$

Exercise 7 (*Maple V*)

A1:=matrix([[.5,.3],[.5,.7]]): evA1:=linalg[eigenvectors](A1); returns

$$\begin{aligned} evA1 := \quad & [1.000000000, 1, \{[-.5303300859, -.8838834765]\}], \\ & [.2000000000, 1, \{[-.7071067812, .7071067812]\}] \end{aligned}$$

A2:=matrix([[.4,.6],[.6,.4]]): evA2:=linalg[eigenvectors](A2); returns

$$\begin{aligned} evA2 := \quad & [-.200000000, 1, \{[.7071067812, -.7071067812]\}], \\ & [1.000000000, 1, \{[-.7071067812, -.7071067812]\}] \end{aligned}$$

A3:=matrix([[.1,.5,.2],[.1,.2,.5],[.8,.3,.3]]): evA3:=linalg[eigenvectors](A3):

evaA3[1]; returns [1.000000001, 1, {[−.4683989174, −.5369451003, −.7654323774]}].
evaA3[2]; returns

$$[-.2000000000 - .3316624790I, 1, \{[.1725098609 + .8344469337I,$$
$$.8650125099 - .4688658038I, -1.037522371 - .3655811293I]\}]$$

evaA3[3]; returns

$$[-.2000000000 + .3316624790I, 1, \{[.1725098609 - .8344469337I,$$
$$.8650125099 + .4688658038I, -1.037522371 + .3655811293I]\}]$$

Hence

$$\mathbf{x} = (-.5303300859, -.8838834765)$$
$$\mathbf{y} = (-.7071067812, -.7071067812)$$
$$\mathbf{z} = (-.4683989174, -.5369451003, -.7654323774)$$

are steady-state vectors of the matrices (a), (b), and (c), respectively.

Exercise 9 (*Maple V*)

A:=matrix([[1/2,1/2,1/3],[1/2,0,1/3],[0,1/2,1/3]]):
ev:=linalg[eigenvectors](A): ev[1]; returns [1, 1, [2, $\frac{4}{3}$, 1]].
Therefore 1 is an eigenvalues of A and that the vector (2, $\frac{4}{3}$, 1) is an associated eigenvector.

Exercise 11 (*Maple V*)

a. The matrix A^2 contains no zero entry. Therefore A is regular.
A:=matrix([[1/10,3/10,7/10,0],[1/10,2/10,0,8/10],[0,5/10,0,2/10],[8/10,0,3/10,0]]):
evalm(A1^2); returns

$$\begin{bmatrix} \frac{1}{25} & \frac{11}{25} & \frac{7}{100} & \frac{19}{50} \\ \frac{67}{100} & \frac{7}{100} & \frac{31}{100} & \frac{4}{25} \\ \frac{21}{100} & \frac{1}{10} & \frac{3}{50} & \frac{2}{5} \\ \frac{2}{25} & \frac{39}{100} & \frac{14}{25} & \frac{3}{50} \end{bmatrix}$$

b. The third power of the matrix contains no zero entry. Therefore the matrix is regular.
c. The third power of the matrix contains no zero entry. Therefore the matrix is regular.
d. The matrix is regular since it contains no zero entry.

CHAPTER 7 - Norms and Inner Products

EXERCISES 7.1

Exercise 1

a. $\|(3,4)\|_2 = 5$ b. $\|(0,0)\|_2 = 0$ c. $\|(-3,4)\|_2 = 5$ d. $\|(1,0)\|_2 = 1$ e. $\|(-9,11)\|_2 = \sqrt{202}$
f. $\|(-9,-11)\|_2 = \sqrt{202}$

Exercise 3

a. $d((4,5),(1,2)) = 3\sqrt{2}$ b. $d((4,5),(-1,2)) = \sqrt{34}$ c. $d((4,5),(0,2)) = \sqrt{65}$ d. $d((3,0),(4,5)) = \sqrt{26}$ e. $d((1,2),(0,0)) = \sqrt{5}$ f. $d((1,0),(0,1)) = \sqrt{2}$

Exercise 5 (*Maple V*)

with(linalg):

x1:=[3,4,5]: x2:=[0,0,0]: x3:=[-3,4,5]: x4:=[1,0,1]: x5:=[0,0,1]: x6:=[-9,-11,-5]:

vn:=x->norm(x,2):

[vn(x1),vn(x2),vn(x3),vn(x4),vn(x5),vn(x6)]; returns $\left[5\sqrt{2}, 0, 5\sqrt{2}, \sqrt{2}, 1, \sqrt{227}\right]$.

EXERCISES 7.2

Exercise 1

a. $\|A\|_F = \sqrt{30}$ b. $\|A\|_F = \sqrt{13}$ c. $\|A\|_F = \sqrt{2}$ d. $\|A\|_F = 2$ e. $\|A\|_F = \sqrt{42}$ f. $\|A\|_F = 2\sqrt{15}$

Exercise 3

The distance is 0.27 units.

Exercise 5

a. with(LinearAlgebra): A:=Matrix([[3,4],[-1,2]]): Norm(A,Frobenius); returns $\sqrt{30}$.

EXERCISES 7.3

Exercise 1

a. $\cos\theta = \frac{4}{205}\sqrt{5}\sqrt{41}$ b. $\cos\theta = \frac{4}{205}\sqrt{5}\sqrt{41}$ c. $\cos\theta = \frac{4}{41}\sqrt{41}$ d. $\cos\theta = -\frac{5}{41}\sqrt{41}$ e. undefined f. $\cos\theta = 0$

Exercise 3 (*Maple V*)

a. with(linalg): x:=<1,2,3>: y:=< 4,5,7>:

t:=dotprod(x,y): b1:=sqrt(dotprod(x,x)): b2:=sqrt(dotprod(y,y)):

cosine:=t/(b1*b2); returns *cosine* := $\frac{1}{12}\sqrt{14}\sqrt{10}$.

b. $\cos\theta = \frac{1}{25}\sqrt{10}\sqrt{5}$ c. $\cos\theta = \frac{2}{15}\sqrt{10}$ d. $\cos\theta = \frac{-31}{390}\sqrt{10}\sqrt{13}$ f. $\cos\theta = 0$

The angle between the two vectors in (e) is undefined.

EXERCISES 7.4

Exercise 1

	a.	b.	c.	d.	e.	f.
1-norm	7	0	7	1	20	20
∞-norm	4	0	4	1	11	11

Exercise 3 (*Maple V*)

with(linalg): [norm(<3,4>,1),norm(<0,0>,1),norm(<-3,4>,1)]; returns 7, 0, 7.

[norm(<3,4>,infinity),norm(< 0,0>,infinity),norm(<-3,4>,infinity)]; returns 4, 0, 4.

[norm(<1,0>,1),norm(<-9,11>,1),norm(<-9,-11>,1)]; returns 1, 20, 20.
[norm(<1,0>,infinity),norm(< -9,11>,infinity),norm(<-9,-11>,infinity)]; returns 1, 11, 11.

EXERCISES 7.5

Exercise 1
a. $\sqrt{6}$ b. $2\sqrt{2}$ c. $\sqrt{11}$ d. $\sqrt{6}$ e. $\sqrt{3}$ f. $\sqrt{2}$
Exercise 3
a. $\sqrt{10}$ b. $\sqrt{15}$ c. $\sqrt{21}$ d. $\sqrt{15}$ e. $\sqrt{6}$ f. $\sqrt{2}$
Exercise 5 (*Maple V*)
with(linalg):
sqrt(norm([1,2,3]-[4,5,7],infinity)); returns 2. sqrt(norm([4,5,7]-[0,-2,-3],infinity))]; returns $\sqrt{10}$.
sqrt(norm([1,2,3]-[0,0,0],infinity))]; returns $\sqrt{3}$. sqrt(norm([4,5,7]-[-1,2,0],infinity))]; returns $\sqrt{7}$.
sqrt(norm([1,0,0]-[4,5,7],infinity))]; returns $\sqrt{7}$. sqrt(norm([1,0,0]-[0,0,1],infinity))]; returns 1.

EXERCISES 7.6

All two-norms in these exercises are assumed to be matrix norms.
Exercise 1

$$\|A\mathbf{x}_1\|_1 = 22,\ \|A\mathbf{x}_2\|_1 = 26,\ \|A\mathbf{x}_3\|_1 = 8,\ \|A\mathbf{x}_4\|_1 = 0,\ \|A\mathbf{x}_5\|_1 = 32$$

$\max\{\|A\mathbf{x}\|_1 : \mathbf{x} \in S_1\} = \max\{22, 26, 8, 0, 32\} = 32$
Exercise 3

$$\|A\mathbf{x}_1\|_\infty = 18,\ \|A\mathbf{x}_2\|_\infty = 17,\ \|A\mathbf{x}_3\|_\infty = 7,\ \|A\mathbf{x}_4\|_\infty = 0,\ \|A\mathbf{x}_5\|_\infty = 19$$

$\max\{\|A\mathbf{x}\|_\infty : \mathbf{x} \in S_1\} = \max\{18, 17, 7, 0, 19\} = 19$
Exercise 5

	$d(A, B)$	$d(A, C)$	$d(B, C)$
1-norm distance	2	$\sqrt{5}$	$\sqrt{2}$
2-norm distance	2	$\left(11 + \sqrt{85}\right)^{1/4}$	$\sqrt{2}$
∞-norm distance	2	$\sqrt{6}$	$\sqrt{2}$

Exercise 7
with(LinearAlgebra): A:=Matrix([[3,4],[-1,2]]):
oneA:=MatrixNorm(A,1); returns $oneA := 6$.
twoA:=allvalues(MatrixNorm(A,2)); returns $twoA := \sqrt{15 + 5\sqrt{5}}$.

infinityA:=MatrixNorm(A,infinity); returns $infinityA := 7$.
A:=Matrix([[3,0],[0,2]]):
oneA:=MatrixNorm(A,1); returns $oneA := 3$.
twoA:=allvalues(MatrixNorm(A,2)); returns $twoA := 3$.
infinityA:=MatrixNorm(A,infinity); returns $infinityA := 3$.
A:=Matrix([[1,-1,0],[0,1,0],[0,0,1]]):
oneA:=MatrixNorm(A,1); returns $oneA := 2$.
twoA:=allvalues(MatrixNorm(A,2)); returns $twoA := \sqrt{\frac{3}{2} + \frac{1}{2}\sqrt{5}}$.
infinityA:=MatrixNorm(A,infinity); returns $infinityA := 2$.
A:=Matrix([[-3,2,0],[2,-3,0],[0,0,4]]):
oneA:=MatrixNorm(A,1); returns $oneA := 5$.
twoA:=allvalues(MatrixNorm(A,2)); returns $twoA := 5$.
infinityA:=MatrixNorm(A,infinity); returns $infinityA := 5$.
A:=Matrix([[-3,2,4],[2,-3,1],[1,0,4]]):
oneA:=MatrixNorm(A,1); returns $oneA := 9$.
twoA:=evalf(MatrixNorm(A,2)); returns $twoA := 6.156744090$.
infinityA:=MatrixNorm(A,infinity); returns $infinityA := 9$.

EXERCISES 7.7

Exercise 1
1-norm a. $\left(\frac{3}{7}, \frac{4}{7}\right)$ b. undefined c. $-\left(\frac{3}{7}, \frac{4}{7}\right)$ d. $(1, 0)$ e. $\left(-\frac{9}{20}, \frac{11}{20}\right)$ f. $\left(-\frac{9}{20}, -\frac{11}{20}\right)$
2-norm a. $\left(\frac{3}{5}, \frac{4}{5}\right)$ b. undefined c. $\left(-\frac{3}{5}, \frac{4}{5}\right)$ d. $(1, 0)$ e. $\left(-\frac{9}{\sqrt{202}}, \frac{11}{\sqrt{202}}\right)$ f. $\left(-\frac{9}{\sqrt{202}}, -\frac{11}{\sqrt{202}}\right)$
∞-norm a. $\left(\frac{3}{4}, \frac{4}{4}\right)$ b. undefined c. $\left(-\frac{3}{4}, \frac{4}{4}\right)$ d. $(1, 0)$ e. $\left(-\frac{9}{11}, 1\right)$ f. $\left(-\frac{9}{11}, -\frac{11}{11}\right)$
Exercise 3
a. 90 b. 50 c. -70 d. 0
Exercise 5 (*Maple V*)
with(linalg):
a. x:=<1,2>: y:=< 4,5>:
is(abs(dotprod(x,y))<=norm(x,2)*norm(y,2)); returns *true*.
b. x:=<-1,2>: y:=<4,5>: is(abs(dotprod(x,y))< =norm(x,2)*norm(y,2)); returns *true*.
c. x:=<0,-2>: y:=<4,5>: is(abs(dotprod(x,y))< =norm(x,2)*norm(y,2)); returns *true*.
d. x:=<1,2>: y:=<0,0>: is(abs(dotprod(x,y))< =norm(x,2)*norm(y,2)); returns *true*.
Exercise 7 (*Maple 6*)
with(LinearAlgebra):
a. x:=<3|4|5>: Normalize(x,1); returns $\left[\frac{1}{4}, \frac{1}{3}, \frac{5}{12}\right]$.
Normalize(x,2); returns $\left[\frac{3}{10}\sqrt{2}, \frac{2}{5}\sqrt{2}, \frac{1}{2}\sqrt{2}\right]$. Normalize(x,infinity); returns $\left[\frac{3}{5}, \frac{4}{5}, 1\right]$.
b. x:=<0|0|0>: Normalize(x,1); returns [0, 0, 0].
Normalize(x,2); returns [0, 0, 0]. Normalize(x,infinity); returns [0, 0, 0].

c. x:=<-3|4|5>: Normalize(x,1); returns $\left[\frac{-1}{4}, \frac{1}{3}, \frac{5}{12}\right]$.
Normalize(x,2); returns $\left[\frac{-3}{10}\sqrt{2}, \frac{2}{5}\sqrt{2}, \frac{1}{2}\sqrt{2}\right]$. Normalize(x,infinity); returns $\left[\frac{-3}{5}, \frac{4}{5}, 1\right]$.
d. x:=<1|0|1>: Normalize(x,1); returns $\left[\frac{1}{2}, 0, \frac{1}{2}\right]$.
Normalize(x,2); returns $[\frac{1}{2}\sqrt{2}, 0, \frac{1}{2}\sqrt{2}]$. Normalize(x,infinity); returns $[1, 0, 1]$.
e. x:=<0|0|1>: Normalize(x,1); returns $[0, 0, 1]$.
Normalize(x,2); returns $[0, 0, 1]$. Normalize(x,infinity); returns $[0, 0, 1]$.
f. x:=<-9|-11|-5>: Normalize(x,1); returns $\left[\frac{-9}{25}, \frac{-11}{25}, \frac{-1}{5}\right]$.
Normalize(x,2); returns $\left[\frac{-9}{227}\sqrt{227}, \frac{-11}{227}\sqrt{227}, \frac{-5}{227}\sqrt{227}\right]$.
Normalize(x,infinity); returns $\left[\frac{-9}{11}, -1, \frac{-5}{11}\right]$.
Exercise 9 (*Maple 6*)
with(LinearAlgebra):
a. A:=Matrix([[5,0,-3],[0,0,-3]]): B:=Matrix([[-1,4,5],[5,-2,-2]]):
Trace(Transpose(B).A); returns -14.
b. A:=Matrix([[5,0,1],[6,2,6],[1,4,-4],[4,2,6]]): B:=Matrix([[3,1,4],[-6,3,-2],[1,3,0],[1,-8,3]]):
Trace(Transpose(B).A); returns -4.
Exercise 11 (*Maple 6*)
with(LinearAlgebra): x:=<r*x1,x2>: y:=< s*y1,y2>: z:=<z1,z2>:
u:=ScalarMultiply(x,a): v:=ScalarMultiply(y,b):
left:=expand(DotProduct(u+v,z,conjugate=false)):
right1:=DotProduct(u,z,conjugate=false): right2:=DotProduct(v,z,conjugate=false):
is(left=right1+right2); returns *true*.
This shows that Axiom 1 holds for all r and s.
left:=DotProduct(x,y,conjugate=false): right:=DotProduct(y,x,conjugate=false):
is(left=right); returns *true*.
This show that Axiom 2 holds for all r and s.
DotProduct(x,x,conjugate=false); returns $r^2x_1^2 + x_2^2$.
DotProduct(y,y,conjugate=false); returns $s^2y_1^2 + y_2^2$.
This shows that Axiom 3 also holds.
Axiom 4 holds if and only if r and s are nonzero, for otherwise
$0^2x_1^2 + 0_2^2 = 0^2y_1^2 + 0_2^2$ for vectors $(x_1, 0)$ and $(y_1, 0)$ with x_1 and y_1 nonzero.

EXERCISES 7.8

Exercise 1

$$A = \begin{bmatrix} a & b \end{bmatrix}, A^TA = \begin{bmatrix} a^2 & ab \\ ab & b^2 \end{bmatrix}, \det(A^T) = t\left(-a^2 - b^2 + t\right)$$

Hence $t = 0$ is an eigenvalues of A^TA. Therefore A^TA is not positive definite.

Exercise 3
$A^T A$ is positive definite if and only if $xy < (a^2 + b^2 + c^2 + d^2)^2$,
where $x = (a - d)^2 + (b + c)^2$ and $y = (a + d)^2 + (b - c)^2$.

Exercise 5 (*Maple V*)
with(linalg): A:=matrix([[50,79],[56,49]]): B:=evalm(transpose(A)&*A):
evB:=eigenvalues(B): is(evB[1]> 0 and evB[2]> 0); returns *true*.
x:=<1,-1>: norm:=sqrt(evalm(evalm(x&*B)&*x)); returns $norm := \sqrt{890}$.
x:=<45,46>: norm:=sqrt(evalm(evalm(x&*B)&*x)); returns $norm := 2\sqrt{14353133}$.
x:=<10,0>: norm:=sqrt(evalm(evalm(x&*B)&*x)); returns $norm := 20\sqrt{1409}$.

Exercise 7 (*Maple 6*)
restart: with(LinearAlgebra):
A:=RandomMatrix(3,1,generator=rand(-2..2)); returns

$$A := \begin{bmatrix} -1 \\ -2 \\ 0 \end{bmatrix}$$

IsDefinite(A.Transpose(A)); returns *false*.
B:=RandomMatrix(2,2); returns $B := \begin{bmatrix} 62 & -71 \\ -79 & 28 \end{bmatrix}$.
IsDefinite(B.Transpose(B)); returns *true*.
C:=RandomMatrix(2,3); returns $C := \begin{bmatrix} -90 & -56 & -50 \\ -21 & -8 & 30 \end{bmatrix}$.
IsDefinite(C.Transpose(C)); returns *true*.

Exercise 9 (*Maple 6*)
with(LinearAlgebra): A:=RandomMatrix(3,3); returns

$$A := \begin{bmatrix} 23 & -75 & -82 \\ 25 & 38 & -66 \\ 5 & 97 & 55 \end{bmatrix}$$

IsDefinite(Transpose(A).A); returns *true*.
B:=RandomMatrix(3,3,generator=-rand(-1..1),
outputoptions=[shape=triangular[lower]]); returns

$$B := \begin{bmatrix} 1 & 0 & 0 \\ 1 & 0 & 0 \\ -1 & 1 & -1 \end{bmatrix}$$

IsDefinite(Transpose(B).B); returns *false*.

C:=RandomMatrix(3,3,density=.6,generator=-rand(-1..1)); returns

$$C := \begin{bmatrix} 0 & 0 & 0 \\ -1 & 0 & -1 \\ 0 & 0 & 0 \end{bmatrix}$$

IsDefinite(Transpose(C).C); returns $false$.

EXERCISES 7.9

Exercise 1

a. $\|(4,5)\| = \sqrt{41}$ b. $\|(-1,2)\| = \sqrt{5}$ c. $\|(0,0)\| = 0$ d. $\|(-4,5)\| = \sqrt{41}$ e. $\|(1,0)\| = 1$ f. $\|(-4,-5)\| = \sqrt{41}$

Exercise 3

$\|\mathbf{x}+\mathbf{y}\|^2 = \langle x+y, x+y\rangle = \langle x,x\rangle + 2\langle x,y\rangle + \langle y,y\rangle$

and

$\|\mathbf{x}-\mathbf{y}\|^2 = \langle x+y, x+y\rangle = \langle x,x\rangle - 2\langle x,y\rangle + \langle y,y\rangle$

Hence $\|\mathbf{x}+\mathbf{y}\|^2 - \|\mathbf{x}-\mathbf{y}\|^2 = 4\langle x,y\rangle$.

Exercise 5 (*Maple V*)

with(linalg): A:=matrix([[3,0,1],[0,5,0],[1,0,6]]):

eigenvalues(A); returns $5, \frac{9}{2}+\frac{1}{2}\sqrt{13}, \frac{9}{2}-\frac{1}{2}\sqrt{13}$.

is(eigenvalues(A)[3]> 0); returns $true$.

x:=<4,5,7>: normx:=sqrt(evalm(evalm(x&*A)&*x)); returns $normx := \sqrt{523}$.

x:=<4,-5,-7>: normx:=sqrt(evalm(evalm(x&*A)&*x)); returns $normx := \sqrt{411}$.

x:=<1,0,0>: normx:=sqrt(evalm(evalm(x&*A)&*x)); returns $normx := \sqrt{3}$.

x:=<-1,2,0>: normx:=sqrt(evalm(evalm(x&*A)&*x)); returns $normx := \sqrt{23}$.

x:=<0,1,1>: normx:=sqrt(evalm(evalm(x&*A)&*x)); returns $normx := \sqrt{11}$.

x:=<-1,-2,-3>: normx:=sqrt(evalm(evalm(x&*A)&*x)); returns $normx := \sqrt{83}$.

Exercise 7 (*Maple V*)

a. $\mathbf{x} = (1,2,-2)$ and $\mathbf{y} = (4,5,7)$ b. $\mathbf{x} = (1,0,0)$ and $\mathbf{y} = (0,5,7)$

c. $\mathbf{x} = (-1,2,0)$ and $\mathbf{y} = (4,5,7)$ d. $\mathbf{x} = (0,-2,-5)$ and $\mathbf{y} = (4,5,-2)$

e. $\mathbf{x} = (1,2,3)$ and $\mathbf{y} = (0,0,0)$ f. $\mathbf{x} = (4,5,7)$ and $\mathbf{y} = (-4,-5,-7)$

with(linalg): x:=<x1,x2,x3>: y:=< y1,y2,y3>:

SquaredNorm:=x-> linalg[dotprod](x,x):

PL:=(x,y)->SquaredNorm(x+y)+SquaredNorm(x-y)

=2*(SquaredNorm(x)+SquaredNorm(y)):

a. x:=<-1,2,-2>: y:=<4,5,7>: PL(x,y); returns $198 = 198$.

b. x:=<1,0,0>: y:=<0,5,7>: PL(x,y); returns $150 = 150$.

c. x:=<-1,2,0>: y:=<4,5,7>: PL(x,y); returns $190 = 190$.

d. x:=<0,-2,-5>: y:=<4,5,-2>: PL(x,y); returns $148 = 148$.

e. x:=<1,2,3>: y:=<0,0,0>: PL(x,y); returns $28 = 28$.
f. x:=<4,5,7>: y:=<-4,-5,-7>: PL(x,y); returns $360 = 360$.

EXERCISES 7.10

Exercise 1

a. $\cos\theta = \frac{14}{205}\sqrt{205}$ b. $\cos\theta = \frac{6}{205}\sqrt{205}$ c. $\cos\theta = \frac{4}{41}\sqrt{41}$
d. $\cos\theta = -\frac{5}{41}\sqrt{41}$ e. undefined f. $\cos\theta = 0$

Exercise 3 (*Maple V*)
a. evalf(arccos(14/sqrt(205)),4); returns .2111.
b. evalf(arccos(6/sqrt(205)),4); returns 1.138.
c. evalf(arccos(4/sqrt(41)),4); returns .8960.
d. evalf(arccos(-5/sqrt(41)),4); returns 2.468.
e. evalf(arccos(0),4); returns 1.571.
Exercise 5 (*Maple V*)
with(linalg):
a. theta:=evalf(angle(<1,2>,< 4,5>),4); returns $theta := .2111$.
b. theta:=evalf(angle(<4,5>,< -1,2>)); returns $theta := 1.138$.
c. theta:=evalf(angle(<1,0>,< 4,5>),4); returns $theta := .8961$.
d. theta:=evalf(angle(<4,5>,< 0,-2>)); returns $theta := 2.467$.
e. theta:=evalf(angle(<1,0>,< 0,1>)); returns $theta := 1.571$.
Exercise 7 (*Maple 6*)
with(LinearAlgebra):
a:=Trace(Transpose(y).x):
b:=sqrt(Trace(Transpose(x).x)):
c:=sqrt(Trace(Transpose(x).x)):
cosxy:=a/(b*c): theta:=evalf(arccos(a/(b*c)),5); returns $theta := 1.3634$.

EXERCISES 7.11

Exercise 1
a. $r = .82852$ b. $r = -.490\,56$ c. $r = 0$
Exercise 3 (*Maple V*)
a. with(stats): x:=[1,2,3,4,5]: y:=[-1,1,7,5,6]:
r:=evalf(describe[linearcorrelation](x,y),5); returns $r := .82853$.
b. x:=[1.5,-3,3,5,6.75]: y:=[2.7,4,4,0,3]:
r:=evalf(describe[linearcorrelation](x,y),5); returns $r := -.49056$.
c. x:=[1,2,3,4,5]: y:=[0,3,0,3,0]:
r:=evalf(describe[linearcorrelation](x,y),5); returns $r := 0$.

Exercise 5 (*Maple 6*)

We describe case (a) in detail.

a. with(LinearAlgebra): R:=(A,i)-> seq(A[i,j],j=1..ColumnDimension(A)):
Entries:=A-> [seq(R(A,i),i=1..RowDimension(A))]:
MatrixMean:=A-> stats[describe,mean](Entries(A)):
MeanDevForm:= A-> Matrix(evalm(A-ConstantMatrix(MatrixMean(A),
RowDimension(A),ColumnDimension(A)))):
A:=Matrix([[8,4,-5],[-5,3,-5]]): B:=Matrix([[8,5,-1],[0,3,-4]]):
mdA:=MeanDevForm(A): mdB:=MeanDevForm(B):
fnA:=MatrixNorm(mdA,Frobenius): fnB:=MatrixNorm(mdB,Frobenius):
ipAB:=Trace(Matrix(evalm(Transpose(mdA)&*mdB))):
PearsonAB:=evalf(ipAB/(fnA*fnB)); returns $PearsonAB := .9461926613$.

b. $r = 1$ c. $r = -1$ d. $r = -1$

EXERCISES 7.12

Exercise 1

a. $\begin{bmatrix} u \\ v \end{bmatrix}^T \begin{bmatrix} 6 & 5 \\ 5 & 8 \end{bmatrix} \begin{bmatrix} x \\ y \end{bmatrix}$ b. $\begin{bmatrix} u \\ v \end{bmatrix}^T \begin{bmatrix} 6 & 8 \\ 5 & 8 \end{bmatrix} \begin{bmatrix} x \\ y \end{bmatrix}$

c. $\begin{bmatrix} u \\ v \end{bmatrix}^T \begin{bmatrix} -6 & 8 \\ 5 & -8 \end{bmatrix} \begin{bmatrix} x \\ y \end{bmatrix}$

Exercise 3

a. $x^2a + 2xyb + y^2c = x + y$ (No solution)
b. $x^2a + 2xyb + y^2c = xy$ $(a = 0, b = 1/2, c = 0)$
c. $x^2a + 2xyb + y^2c = x^2 + x + y^2$ (No solution)
d. $x^2a + 2xyb + y^2c = 3x^2 + 2xy - y^2$ $(a = 3, b = 1, c = -1)$
e. $x^2a + 2xyb + y^2c = 5y^2$ $(a = 0, b = 0, c = 5)$
f. $x^2a + 2xyb + y^2c = 5y^2 + 7$ (No solution)
g. $x^2a + 2xyb + y^2c = 0$ $(a = b = c = 0)$
h. $x^2a + 2xyb + y^2c = x^2 + xy$ $(a = 1, b = 1/2, c = 0)$

Exercise 5

$$\begin{aligned} q(\mathbf{x}+\mathbf{y}) &= f(\mathbf{x}+\mathbf{y}, \mathbf{x}+\mathbf{y}) \\ &= f(\mathbf{x}, \mathbf{x}) + f(\mathbf{x}, \mathbf{y}) + f(\mathbf{y}, \mathbf{x}) + f(\mathbf{y}, \mathbf{y}) \\ &= q(\mathbf{x}) + f(\mathbf{x}, \mathbf{y}) + f(\mathbf{x}, \mathbf{y}) + q(\mathbf{y}) \\ &= q(\mathbf{x}) + f(\mathbf{x}, \mathbf{y}) + f(\mathbf{x}, \mathbf{y}) + q(\mathbf{y}) \\ &= q(\mathbf{x}) + 2f(\mathbf{x}, \mathbf{y}) + q(\mathbf{y}) \end{aligned}$$

Hence $f(\mathbf{x}, \mathbf{y}) = \frac{1}{2}(q(\mathbf{x}+\mathbf{y}) - q(\mathbf{x}) - q(\mathbf{y}))$.

Exercise 7 (*Maple 6*)

with(LinearAlgebra): A:=Matrix([[r,s],[t,u]]):

f:=(x,y)->Transpose(x).A.y:

[f(<1,0>,<1,0>),-f(< 1,0>,< 1,0>),f(<0,1>,<0,1>),

-f(<0,1>,<0,1>),f(< 0,1>,< 1,0>),-f(<1,0>,<0,1>)]; returns $[r, -r, u, -u, t, -s]$.

Since $r = -r$ and $u = -u$, we must have $r = u = 0$. Furthermore, $t = -s$. Hence

$$A := \begin{bmatrix} 0 & s \\ -s & 0 \end{bmatrix}$$

A:=Matrix([[0,s],[-s,0]]): f(<x,y>,< u,v>); returns $-uys + vxs$.

X:=Matrix([[x,u],[y,v]]): expand(s*det(X)); returns $-uys + vxs$.

EXERCISES 7.13

Exercise 1 (*Maple V*)

a. with(plots): implicitplot(4*x^2+7*y^2 = 1,x=-1/2..1/2,y=-1/sqrt(7)..1/sqrt(7))

b. implicitplot(4*x^2 -7*y^2 = 1,x=-10..10,y=-10..10)

c. implicitplot(-x^2+10*x*y+2*y^2 = 1,x=-10..10,y=-10..10)

d. implicitplot(x^2+10*x*y+2*y^2 = 1,x=-10..10,y=-10..10)

Exercise 3 (a) (*Maple 6*)

with(LinearAlgebra): A:=Matrix([[-5,-2],[-2,-1]]): evA:=Eigenvectors(A):

P:=evA[2]; returns $P := \begin{bmatrix} 1 & 1 \\ -1-\sqrt{2} & -1+\sqrt{2} \end{bmatrix}$.

diagA:=DiagonalMatrix([evA[1][1],evA[1][2]]):

simplify(evalm(P&*diagA&*MatrixInverse(P))); returns $\begin{bmatrix} -5 & -2 \\ -2 & -1 \end{bmatrix}$.

norm1:=VectorNorm(Column(P,1),2): norm2:=VectorNorm(Column(P,2),2):

Q:=Matrix([[1/norm1,1/norm2],[(-1-sqrt(2))/norm1,(-1+sqrt(2))/norm2]]); returns

$$Q := \begin{bmatrix} \frac{1}{\sqrt{1+\left(1+\sqrt{2}\right)^2}} & \frac{1}{\sqrt{1+\left(-1+\sqrt{2}\right)^2}} \\ \frac{-1-\sqrt{2}}{\sqrt{1+\left(1+\sqrt{2}\right)^2}} & \frac{-1+\sqrt{2}}{\sqrt{1+\left(-1+\sqrt{2}\right)^2}} \end{bmatrix}$$

norm1:=simplify(VectorNorm(Column(Q,1),2)); returns $norm1 := 1$.

norm2:=simplify(VectorNorm(Column(Q,2),2)); returns $norm2 := 1$.

simplify(DotProduct(Column(Q,1),Column(Q,2))); returns 0.

Hence the matrix Q is orthogonal. We now show that $QDQ^T = A$.

M:=simplify(evalm(Q&*diagA&*Transpose(Q))):

simplify(M[1,1]+5); returns 0. simplify(M[1,2]+2); returns 0.

simplify(M[2,1]+2); returns 0. simplify(M[2,2]+1); returns 0.

EXERCISES 7.14

Exercise 1

a. $\mathbf{x}^T\overline{\mathbf{y}} = -2 - 30i$ b. $\mathbf{x}^T\overline{\mathbf{y}} = -37 + 5i$

Exercise 3

a. $d\left(\begin{bmatrix} 3-5i \\ -i \end{bmatrix}, \begin{bmatrix} 4+i \\ 7+9i \end{bmatrix}\right) = \sqrt{186}$ b. $d\left(\begin{bmatrix} -i \\ -4+4i \end{bmatrix}, \begin{bmatrix} 7+9i \\ 5-2i \end{bmatrix}\right) = \sqrt{266}$

Exercise 5 (*Maple V*)

a. with(linalg): norm(<3,1,-4> -< 4,7,5>,2); returns $\sqrt{118}$.

b. norm(<3-5*I,-I,-4+4*I> -< 4+I,7+9*I,5-2*I>,2); returns $\sqrt{303}$.

CHAPTER 8 - Orthogonality

EXERCISES 8.1

Exercise 1

Theorem 8.2 tells us that orthogonal vectors are linearly independent.

a. $\mathbf{x}^T\mathbf{y} = \begin{bmatrix} 0 \\ 24 \\ 107 \\ 0 \end{bmatrix}^T \begin{bmatrix} 1 \\ 0 \\ 0 \\ 1 \end{bmatrix} = 0$, b. $\mathbf{x}^T\mathbf{y} = \begin{bmatrix} 1 \\ 24 \\ 0 \\ 1 \end{bmatrix}^T \begin{bmatrix} -1 \\ 0 \\ 7 \\ 1 \end{bmatrix} = 0$

c. $\mathbf{x}^T\mathbf{y} = \begin{bmatrix} 0 \\ 24 \\ 0 \\ 0 \end{bmatrix}^T \begin{bmatrix} 1 \\ 0 \\ 7 \\ 1 \end{bmatrix} = 0$ d. $\mathbf{x}^T\mathbf{y} = \begin{bmatrix} 1 \\ -1 \\ 1 \\ 1 \end{bmatrix}^T \begin{bmatrix} 1 \\ 1 \\ -1 \\ 1 \end{bmatrix} = 0$

Exercise 3 (*Maple V*)

with(linalg):

a. x:=<1,0,3>: y:=<1,2,3>: z:=<5,2,3>:

is(norm(z-x,2)^2=norm(z-y,2)^2+norm(y-x,2)^2); returns *true*.

b. x:=<1,0,-3>: y:=<1,2,-3>: z:=<5,2,-3> :

is(norm(z-x,2)^2=norm(z-y,2)^2+norm(y-x,2)^2); returns *true*.

c. x:=<-8,7,6>: y:=<3,7,6>: z:=<3,7,-12> :

is(norm(z-x,2)^2=norm(z-y,2)^2+norm(y-x,2)^2); returns *true*.

d. x:=<4,8,6>: y:=<4,3,6>: z:=<4,3,-12> :

is(norm(z-x,2)^2=norm(z-y,2)^2+norm(y-x,2)^2); returns *true*.

Exercise 5 (*Maple V*)

with(linalg): x:=vector(3,[1,0,3]): y:=vector(3,[1,2,3]): z:=vector(3,[5,2,3]):

A:=matrix([[9,0,-2],[0,7,0],[-2,0,1]]):

side1:=innerprod(z-y,A,z-y): side2:=innerprod(x-y,A,x-y): side3:=innerprod(z-x,A,z-x):

is(side1+side2=side3); returns *true*.

EXERCISES 8.2

Exercise 1

a. $\text{proj}\,((1,2) \to \text{span}\,\{\mathbf{e}_1\}) = (1,0)$

b. $\text{proj}\,((1,2) \to \text{span}\,\{\mathbf{e}_2\}) = (0,2)$

c. $\text{proj}\,((x,y) \to \text{span}\,\{\mathbf{e}_1,\mathbf{e}_2\}) = (x,y)$

Exercise 3

a. $\text{proj}\,((1,2,3) \to \text{span}\,\{\mathbf{e}_1\}) = (1,0,0)$

b. $\text{proj}\,((1,2,3) \to \text{span}\,\{\mathbf{e}_1,\mathbf{e}_2\}) = (1,2,0)$

c. $\text{proj}\,((1,2,3) \to \text{span}\,\{\mathbf{e}_1,\mathbf{e}_3\}) = (1,0,3)$

Exercise 5 (*Maple 6*)

with(LinearAlgebra):

proj:=(x,y)->(DotProduct(x,y)/DotProduct(y,y)).y:

perp:=(x,y)->x-proj(x,y):

a. x:=<1,3,1>: y:=<4,5,0>:

convert(proj(x,y),list); returns $\left[\frac{76}{41},\frac{95}{41},0\right]$.

convert(perp(x,y),list); returns $\left[\frac{-35}{41},\frac{28}{41},1\right]$.

b. z:=<-1,-9,12>:

convert(proj(x,z),list); returns $\left[\frac{8}{113},\frac{72}{113},\frac{-96}{113}\right]$.

convert(perp(x,z),list); returns $\left[\frac{105}{113},\frac{267}{113},\frac{209}{113}\right]$.

EXERCISES 8.3

Exercise 1

We use the basis $\mathcal{B} = \{\mathbf{x}_1 = (1,1), \mathbf{x}_2 = (1,-1)\}$ for $\mathbb{R}^2$ to construct the orthogonal basis

$$\mathcal{C} = \left\{\mathbf{y}_1 = \mathbf{x}_1, \mathbf{y}_2 = \mathbf{x}_2 - \frac{\langle \mathbf{x}_2, \mathbf{y}_1\rangle}{\langle \mathbf{y}_1, \mathbf{y}_1\rangle}\mathbf{y}_1\right\}$$

where $\langle \mathbf{x}, \mathbf{y}\rangle = \mathbf{x}^T\mathbf{y}$. Since $\mathcal{B}$ is already orthogonal, the Gram–Schmidt algorithm; returns $\mathbf{x}_2$ for $\mathbf{y}_2$.

$$\mathbf{y}_2 = \begin{bmatrix} 1 \\ -1 \end{bmatrix} - \frac{\begin{bmatrix} 1 \\ -1 \end{bmatrix}^T \begin{bmatrix} 1 \\ 1 \end{bmatrix}}{\begin{bmatrix} 1 \\ 1 \end{bmatrix}^T \begin{bmatrix} 1 \\ 1 \end{bmatrix}} \begin{bmatrix} 1 \\ 1 \end{bmatrix} = \begin{bmatrix} 1 \\ -1 \end{bmatrix} - 0 \begin{bmatrix} 1 \\ 1 \end{bmatrix} = \mathbf{x}_2$$

Exercise 3

We use the basis $\mathcal{B} = \{\mathbf{x}_1 = (1, 1, 0), \mathbf{x}_2 = (1, -1, 2), \mathbf{x}_3 = (2, 0, 0)\}$ for $\mathbb{R}^3$ to construct the orthogonal basis

$$\mathcal{C} = \left\{\mathbf{y}_1 = \mathbf{x}_1, \mathbf{y}_2 = \mathbf{x}_2 - \frac{\langle \mathbf{x}_2, \mathbf{y}_1\rangle}{\langle \mathbf{y}_1, \mathbf{y}_1\rangle}\mathbf{y}_1, \mathbf{y}_3 = \mathbf{x}_3 - \frac{\langle \mathbf{x}_3, \mathbf{y}_1\rangle}{\langle \mathbf{y}_1, \mathbf{y}_1\rangle}\mathbf{y}_1 - \frac{\langle \mathbf{x}_3, \mathbf{y}_2\rangle}{\langle \mathbf{y}_2, \mathbf{y}_2\rangle}\mathbf{y}_2\right\}$$

where $\langle \mathbf{x}, \mathbf{y}\rangle = \mathbf{x}^T\mathbf{y}$. We note that $\mathbf{y}_2 = \mathbf{x}_2$ since $\mathbf{x}_2$ is orthogonal to $\mathbf{x}_1$.

$$\mathbf{y}_3 = \begin{bmatrix} 2 \\ 0 \\ 0 \end{bmatrix} - (1)\begin{bmatrix} 1 \\ 1 \\ 0 \end{bmatrix} - \frac{1}{3}\begin{bmatrix} 1 \\ -1 \\ 2 \end{bmatrix} = \begin{bmatrix} \frac{2}{3} \\ -\frac{2}{3} \\ -\frac{2}{3} \end{bmatrix}$$

Exercise 5 (*Maple V*)

with(linalg): A:=randmatrix(5,5,unimodular); returns

$$A := \begin{bmatrix} 1 & 54 & -5 & 99 & -61 \\ 0 & 1 & -50 & -12 & -18 \\ 0 & 0 & 1 & 31 & -26 \\ 0 & 0 & 0 & 1 & -62 \\ 0 & 0 & 0 & 0 & 1 \end{bmatrix}$$

S:=[seq(col(A,i),i=1..5)]; returns

$$\begin{aligned} S := &[[1, 0, 0, 0, 0], [54, 1, 0, 0, 0], \\ &[-5, -50, 1, 0, 0], [99, -12, 31, 1, 0], \\ &[-61, -18, -26, -62, 1]] \end{aligned}$$

GramSchmidt(S); returns

$$[[1, 0, 0, 0, 0], [0, 1, 0, 0, 0], [0, 0, 1, 0, 0], [0, 0, 0, 1, 0], [0, 0, 0, 0, 1]]$$

EXERCISES 8.4

Exercise 1

$\mathbf{y}_1 = (1, 2), \mathbf{y}_2 = (\frac{6}{5}, -\frac{3}{5})$

Exercise 3

Consider the positive definite matrix $A = \begin{bmatrix} 9 & 1 \\ 1 & 7 \end{bmatrix}$.

We use f and the Gram–Schmidt process to convert the basis $\mathcal{S} = \{(1, 1), (3, 2)\}$ to the orthonormal basis $B = \left\{(\frac{1}{6}\sqrt{2}, \frac{1}{6}\sqrt{2}), (\frac{4}{93}\sqrt{31}, -\frac{5}{93}\sqrt{31})\right\}$.

Next we find the required coordinate vectors.

$$\begin{bmatrix} 3 \\ 4 \end{bmatrix}_{\mathcal{B}} = \begin{bmatrix} \frac{31}{3}\sqrt{2} \\ -\frac{1}{3}\sqrt{31} \end{bmatrix}, \begin{bmatrix} 2 \\ 5 \end{bmatrix}_{\mathcal{B}} = \begin{bmatrix} 10\sqrt{2} \\ -\sqrt{31} \end{bmatrix}$$

Hence,

$$\begin{aligned}[\mathbf{x}]_{\mathcal{B}}^T [\mathbf{y}]_{\mathcal{B}} &= \begin{bmatrix} \frac{31}{3}\sqrt{2} \\ -\frac{1}{3}\sqrt{31} \end{bmatrix}^T \begin{bmatrix} 10\sqrt{2} \\ -\sqrt{31} \end{bmatrix} = 217 \\ &= \begin{bmatrix} 3 \\ 4 \end{bmatrix}^T \begin{bmatrix} 9 & 1 \\ 1 & 7 \end{bmatrix} \begin{bmatrix} 2 \\ 5 \end{bmatrix} = \langle \mathbf{x}, \mathbf{y} \rangle\end{aligned}$$

Exercise 5 (*Maple V*)

We arbitrarily choose the positive definite matrix

$$A := \begin{bmatrix} 2 & 0 & 0 \\ 0 & 3 & 0 \\ 0 & 0 & 4 \end{bmatrix}$$

to define an inner product.

```
A:=matrix([[2,0,0],[0,3,0],[0,0,4]]): x:=vector(3,[3,4,-2]): y:=vector(3,[2,5,7]):
ipA:=(x,y)->evalm(x&*A&*y):
B:=diag(1,1,1): x1:=col(B,1): x2:=col(B,2): x3:=col(B,3):
is(ipA(x1,x2)=0 and ipA(x1,x3)=0 and ipA(x2,x3)=0); returns true.
```

Hence the vectors x1, x2, and x3 are orthogonal in the inner product ipA.

```
nx1:=sqrt(ipA(x1,x1)): u:=scalarmul(x1,1/nx1):
nx2:=sqrt(ipA(x2,x2)): v:=scalarmul(x2,1/nx2):
nx3:=sqrt(ipA(x3,x3)): w:=scalarmul(x3,1/nx3):
M:=Matrix(3,3,concat(u,v,w));
```

$$M := \begin{bmatrix} \frac{1}{2}\sqrt{2} & 0 & 0 \\ 0 & \frac{1}{3}\sqrt{3} & 0 \\ 0 & 0 & \frac{1}{2} \end{bmatrix}$$

xB:=linsolve(M,x); returns $xB := [3\sqrt{2}, 4\sqrt{3}, -4]$.

yB:=linsolve(M,y); returns $yB := [2\sqrt{2}, 5\sqrt{3}, 14]$).

is(dotprod(xB,yB)=ipA(x,y)); returns *true*.

Exercise 7 (*Maple 6*)

```
with(LinearAlgebra): x1:=<1,2,1>: x2:=< 1,-1,0>: x3:=<0,1,0>:
y1:=x1: sx2y1:=DotProduct(x2,y1)/DotProduct(y1,y1):
vsm1y1:=VectorScalarMultiply(y1,sx2y1):
y2:=x2-vsm1y1: sx3y1:=DotProduct(x3,y1)/DotProduct(y1,y1):
```

```
sx3y2:=DotProduct(x3,y2)/DotProduct(y2,y2):
vsm2y1:=VectorScalarMultiply(y1,sx3y1):
vsm2y2:=VectorScalarMultiply(y2,sx3y2):
y3:=x3-vsm2y1-vsm2y2:
A:=Matrix([[1,0,0],[0,2,0],[0,0,3]]):
ipA:=(x,y)-> evalm(x&*Transpose(A)&*A&*y):
z1:=VectorScalarMultiply(x1,1/sqrt(ipA(x1,x1))); returns
```

$$z1 := (\tfrac{1}{26}\sqrt{26}, \tfrac{1}{13}\sqrt{26}, \tfrac{1}{26}\sqrt{26})$$

```
sx2y1:=ipA(x2,y1)/ipA(y1,y1): vsm1y1:=VectorScalarMultiply(y1,sx2y1):
z2:=VectorScalarMultiply(x2-vsm1y1,1/sqrt(ipA(x2-vsm1y1,x2-vsm1y1))); returns
```

$$z2 := (\tfrac{11}{78}\sqrt{26}, \tfrac{-2}{39}\sqrt{26}, \tfrac{7}{234}\sqrt{26})$$

```
sx3y1:=ipA(x3,y1)/ipA(y1,y1): sx3y2:=ipA(x3,y2)/ipA(y2,y2):
vsm2y1:=VectorScalarMultiply(y1,sx3y1):
vsm2y2:=VectorScalarMultiply(y2,sx3y2):
z3:=VectorScalarMultiply(x3-vsm2y1-vsm2y2,
    1/sqrt(ipA(x3-vsm2y1-vsm2y2,x3-vsm2y1-vsm2y2))); returns
```

$$(\tfrac{2}{3}, \tfrac{1}{6}, \tfrac{-2}{9})$$

is(ipA(z1,z2)=0 and ipA(z1,z3)=0 and ipA(z2,z3)=0); returns *true*.
is(ipA(z1,z1)=1 and ipA(z2,z2)=1 and ipA(z3,z3)=1); returns *true*.

Exercise 9 (*Maple 6*)

```
with(LinearAlgebra): B:=[<8,-5,8>,< 4,3,5>,<-5,-5,-1>]:
GramSchmidt(B,normalized=false); returns
```

$$\begin{bmatrix} 8 \\ -5 \\ 8 \end{bmatrix}, \begin{bmatrix} \frac{52}{51} \\ \frac{248}{51} \\ \frac{103}{51} \end{bmatrix} \begin{bmatrix} \frac{-11809}{4401} \\ \frac{-1928}{4401} \\ \frac{10604}{4401} \end{bmatrix}$$

Exercise 11 (*Maple 6*)

```
with(LinearAlgebra): x:=<8,-5,8>: y:=< 4,3,5>: z:=<-5,-5,-1>:
GramSchmidt([x,y,z],normalized); returns
```

$$\left[\begin{bmatrix} \frac{8}{51}\sqrt{17} \\ -\frac{5}{51}\sqrt{17} \\ \frac{8}{51}\sqrt{17} \end{bmatrix}, \begin{bmatrix} \frac{52}{24939}\sqrt{8313} \\ \frac{248}{24939}\sqrt{8313} \\ \frac{103}{24939}\sqrt{8313} \end{bmatrix}, \begin{bmatrix} -\frac{49}{353547}\sqrt{28401609} \\ -\frac{8}{353547}\sqrt{28401609} \\ \frac{44}{353547}\sqrt{28401609} \end{bmatrix}\right]$$

EXERCISES 8.5

Exercise 1

Let $\mathcal{E} = \{\mathbf{e}_1, \mathbf{e}_2\}$ be the standard basis for $\mathbb{R}^2$. We convert $\mathcal{E}$ to a basis $\mathcal{B} = \{\mathbf{x}_1, \mathbf{x}_2\}$ that is orthonormal in the inner product defined by the matrix A.

$$\mathbf{x}_1 = \frac{1}{\|\mathbf{e}_1\|}\mathbf{e}_1 = \frac{1}{\sqrt{8}}\begin{bmatrix} 1 \\ 0 \end{bmatrix} = \begin{bmatrix} \frac{1}{\sqrt{8}} \\ 0 \end{bmatrix} \mathbf{x}_2 = \frac{1}{\|\mathbf{e}_2\|}\mathbf{e}_2 = \frac{1}{\sqrt{3}}\begin{bmatrix} 0 \\ 1 \end{bmatrix} = \begin{bmatrix} 0 \\ \frac{1}{\sqrt{3}} \end{bmatrix},$$

The coordinate conversion matrix from $\mathcal{E}$ to $\mathcal{B}$ is

$$C = \begin{bmatrix} \frac{1}{4}\sqrt{2} & 0 \\ 0 & \frac{1}{3}\sqrt{3} \end{bmatrix}$$

Let $\mathbf{p} = (p_1, p_2)$ and $\mathbf{q} = (q_1, q_2)$. Then

$$[\mathbf{p}]_{\mathcal{B}} = \begin{bmatrix} 2\sqrt{2}p_1 \\ \sqrt{3}p_2 \end{bmatrix}, [\mathbf{q}]_{\mathcal{B}} = \begin{bmatrix} 2\sqrt{2}q_1 \\ \sqrt{3}q_2 \end{bmatrix}$$

and

$$\begin{aligned} [\mathbf{p}]_{\mathcal{B}}^T [\mathbf{q}]_{\mathcal{B}} &= \begin{bmatrix} p_1 \\ p_2 \end{bmatrix}^T \begin{bmatrix} 8 & 0 \\ 0 & 3 \end{bmatrix} \begin{bmatrix} q_1 \\ q_2 \end{bmatrix} \\ &= 8p_1q_1 + 3p_2q_2 \end{aligned}$$

Exercise 3 (*Maple 6*)
We use the same method as that used in Exercise 1.

```
with(LinearAlgebra): A:=Matrix([[8,0,0],[0,3,0],[0,0,5]]):
Id3:=IdentityMatrix(3): E:=[Column(Id3,1),Column(Id3,2),Column(Id3,3)]:
a:=i-> 1/sqrt(Transpose(E[i]).A.E[i]):
x1:=VectorScalarMultiply(E[1],a(1)):
x2:=VectorScalarMultiply(E[2],a(2)):
x3:=VectorScalarMultiply(E[3],a(3)):
C:=Matrix([x1,x2,x3]): p:=<p1,p2,p3>: q:=< q1,q2,q3>:
DotProduct(MatrixInverse(C).p,MatrixInverse(C).q,conjugate=false);
```
returns

$$8p1q1 + 3p2q2 + 5p3q3$$

```
ipA:=(x,y)->evalm(x.A.y): ipA(Transpose(p),q);
```
returns

$$8p1q1 + 3p2q2 + 5p3q3$$

EXERCISES 8.6

Exercise 1
The determinants of matrices (a) and (e) are zero. Hence their columns are linearly dependent and Theorem 8.8 does not apply. The determinant of matrix (c) is -37 and the theorem applies. The matrices (b) and (f) have more columns than rows. Hence the columns must be linearly dependent and Theorem 8.8 does not apply. The two columns of matrix (d), finally, are linearly independent and the theorem applies.

Exercise 3

$$A_1 = \begin{bmatrix} 0 & 1 \\ -1 & 0 \end{bmatrix}, Q_1R_1 = \begin{bmatrix} 0 & 1 \\ -1 & 0 \end{bmatrix}, A_2 = \begin{bmatrix} 0 & 1 \\ -1 & 0 \end{bmatrix}, Q_2R_2 = \begin{bmatrix} 0 & 1 \\ -1 & 0 \end{bmatrix}$$

Hence

$$A_3 = \begin{bmatrix} 1 & 0 \\ 0 & 1 \end{bmatrix}\begin{bmatrix} 0 & 1 \\ -1 & 0 \end{bmatrix} = \begin{bmatrix} 0 & 1 \\ -1 & 0 \end{bmatrix}$$

so $A = A_1 = A_2 = A_3$.

Exercise 5

One of the requirements for the existence of a QR decomposition of a matrix A is that the columns of A are linearly independent. Hence the number of columns of A cannot exceed the number of rows. Suppose A is an $m \times n$ matrix and $n < m$. Then it is possible for the n columns to be linearly independent. On the other hand, A^T is an $n \times m$ matrix and $m > n$. Hence A^T has more columns than rows. Hence the columns of A^T cannot be linearly independent.

Exercise 7 (*Maple 6*)

with(LinearAlgebra): A:=Matrix([[2,1,0],[0,2,3],[3,0,2]]): b:=< 5,4,9>:
QR:=QRDecomposition(A): y:=Vector(Transpose(QR[1]).b):
x:=BackwardSubstitute(QR[2],y); returns $x := (\frac{39}{17}, \frac{7}{17}, \frac{18}{17})$.
A:=Matrix([[2,5,0],[2,0,3],[4,1,3]]): b:=<1,1,1> :
QR:=QRDecomposition(A): y:=Vector(Transpose(QR[1]).b):
x:=BackwardSubstitute(QR[2],y); returns $x := (\frac{-1}{8}, \frac{1}{4}, \frac{5}{12})$.

Exercise 9 (a) (*Maple 6*)

with(LinearAlgebra): A:=Matrix([[2,1],[0,2],[3,0]]): b:=< 5,4,9>:
QR:=QRDecomposition(Transpose(A).A)):
y:=Vector[column](Transpose(QR[1]).Transpose(A).b):
BackwardSubstitute(QR[2],y); returns $(\frac{159}{61}, \frac{95}{61})$.
This solution coincides with the least-squares solution of the original equation.
LeastSquares(A,b); returns $(\frac{159}{61}, \frac{95}{61})$.
The determinants of the matrix $A^T A$ in (b) is 108. Hence the matrix $A^T A$ is invertible.
We can therefore use the previous method to solve the normal equations.
The solution is $(x, y) = (1/3, 0)$.

EXERCISES 8.7

Exercise 1

Let A be an orthogonal matrix. Then $A^T = A^{-1}$. Moreover, $\det A \neq 0$ since A is invertible.
We also know from the properties of determinants that $\det A = \det A^T$. Hence
$\det I = 1 = \det A \det A^{-1} = \det A \det A^T = \det A \det A$.
This implies that $\det A = \pm\sqrt{1} = \pm 1$.

Exercise 3

$$\begin{bmatrix} \frac{20}{29} & -\frac{21}{29} \\ -\frac{21}{29} & -\frac{20}{29} \end{bmatrix}$$

Exercise 5 (*Maple V*)

$$[I]_{\mathcal{B}}^{\mathcal{E}} = \begin{bmatrix} \frac{2}{3} & \frac{2}{3} & -\frac{1}{3} \\ -\frac{1}{3} & \frac{2}{3} & \frac{2}{3} \\ \frac{2}{3} & -\frac{1}{3} & \frac{2}{3} \end{bmatrix}$$

with(linalg): A:=matrix([[2/3,2/3,-1/3],[-1/3,2/3,2/3],[2/3,-1/3,2/3]]):
equal(inverse(A),transpose(A)); returns *true*.

Exercise 7 (*Maple V*)

with(linalg): H:=A->transpose(conjugate(A)):
A:=matrix([[1,-I],[0,3]]): H(A); returns

$$\begin{bmatrix} 1 & 0 \\ I & 3 \end{bmatrix}$$

A:=matrix([[1,5],[0,6-14*I],[1+2*I,4*I]]): H(A); returns

$$\begin{bmatrix} 1 & 0 & 1-2I \\ 5 & 6+14I & -4I \end{bmatrix}$$

A:=matrix([[1,-I,5],[0,3,6-14*I],[1+2*I,23,4*I]]): H(A); returns

$$\begin{bmatrix} 1 & 0 & 1-2I \\ I & 3 & 23 \\ 5 & 6+14I & -4I \end{bmatrix}$$

Exercise 9 (*Maple 6*)

A:=Matrix([[1,-I],[I,3]]):
IsUnitary(A); returns *false*.
A:=Matrix([[1,0,0],[0,(1/2)*sqrt(2),(1/2)*sqrt(2)],[0,(1/2)+(1/2)*I,-(1/2)-(1/2)*I]]):
IsUnitary(A); returns *true*.
A:=Matrix([[(1/2)*sqrt(2),(1/2)*sqrt(2)],[(1/2)+(1/2)*I,-(1/2)-(1/2)*I]]):
IsUnitary(A); returns *true*.

Exercise 11 (*Maple 6*)

with(LinearAlgebra): f:=(i,j)-> if i< j then I else -I fi:
A:=RandomMatrix(4,4,generator=f):; returns

$$A := \begin{bmatrix} -I & I & I & I \\ -I & -I & I & I \\ -I & -I & -I & I \\ -I & -I & -I & -I \end{bmatrix}$$

and det(A); returns 8. Hence A is invertible. This tells us that its columns are linear independent. We can therefore use the **GramSchmidt** function to convert its columns to an orthonormal basis for $\mathbb{C}^4$.

S:=[seq(Column(A,i),i=1..4)]; returns

$$S := \left[\begin{bmatrix} -I \\ -I \\ -I \\ -I \end{bmatrix}, \begin{bmatrix} I \\ -I \\ -I \\ -I \end{bmatrix}, \begin{bmatrix} I \\ I \\ -I \\ -I \end{bmatrix}, \begin{bmatrix} I \\ I \\ I \\ -I \end{bmatrix} \right]$$

G:=GramSchmidt(S,normalized); returns

$$G := \left[\begin{bmatrix} \frac{-1}{2}I \\ \frac{-1}{2}I \\ \frac{-1}{2}I \\ \frac{-1}{2}I \end{bmatrix}, \begin{bmatrix} \frac{1}{2}I\sqrt{3} \\ \frac{-1}{6}I\sqrt{3} \\ \frac{-1}{6}I\sqrt{3} \\ \frac{-1}{6}I\sqrt{3} \end{bmatrix}, \begin{bmatrix} 0 \\ \frac{1}{2}I\sqrt{6} \\ \frac{-1}{6}I\sqrt{6} \\ \frac{-1}{6}I\sqrt{6} \end{bmatrix}, \begin{bmatrix} 0 \\ 0 \\ \frac{1}{2}I\sqrt{2} \\ \frac{-1}{2}I\sqrt{2} \end{bmatrix} \right]$$

U:=Matrix(G): IsUnitary(U); returns $true$.

EXERCISES 8.8

Exercise 1

We show that the vectors (1, 2, 0) and (0, 4, 3) are both orthogonal to the vector (6, −3, 4).

$$\begin{bmatrix} 1 \\ 2 \\ 0 \end{bmatrix}^T \begin{bmatrix} 6 \\ -3 \\ 4 \end{bmatrix} = 0, \begin{bmatrix} 0 \\ 4 \\ 3 \end{bmatrix}^T \begin{bmatrix} 6 \\ -3 \\ 4 \end{bmatrix} = 0$$

Exercise 3

We find two vectors orthogonal to the vectors (1, 2, 1, 2) and (1, −1, 1, 1) by solving the homogeneous linear system

$$\begin{bmatrix} 1 & 2 & 1 & 2 \\ 1 & -1 & 1 & 1 \end{bmatrix} \begin{bmatrix} w \\ x \\ y \\ z \end{bmatrix} = \begin{bmatrix} 0 \\ 0 \end{bmatrix}$$

The solutions satisfy the equations $x = -\frac{1}{3}z$, $w = -\frac{4}{3}z - y$.

Let $z = 3$ and $y = 0$. Then $(w, x, y, z) = (-4, -1, 0, 3)$.
If we let $z = 0$ and $y = 1$, then $(w, x, y, z) = (-1, 0, 1, 0)$.
These vectors are linearly independent. By construction, they are orthogonal to $(1, 2, 1, 2)$ and $(1, -1, 1, 1)$.
The space $W = \text{span}\{(-4, -1, 0, 3), (-1, 0, 1, 0)\}$ is one of the two-dimensional subspace of $\mathbb{R}^4$ that is orthogonal to the space U.

Exercise 5

Let $V = \text{span}\{(-1, -10, 0, 1, 0)\}$ and $W = \text{span}\{(-1, -10, 0, 1, 0), (-1, -19, 1, 1, 1)\}$.
Then V is a subspace of W, and both V and W are orthogonal to U.

Exercise 7 (*Maple 6*)

with(LinearAlgebra): A:=Matrix([[9,-9,4,0],[-8,-5,4,7]]),
cspA:=ColumnSpace(A); returns

$$\left[\begin{bmatrix} 1 \\ 0 \end{bmatrix}, \begin{bmatrix} 0 \\ 1 \end{bmatrix}\right]$$

rspA:=RowSpace(A); returns $rspA := \left[\left[1, 0, \frac{-16}{117}, \frac{-7}{13}\right], \left[0, 1, \frac{-68}{117}, \frac{-7}{13}\right]\right]$.
nspA:=NullSpace(A); returns $nspA := \left\{(0, \frac{-7}{4}, \frac{-63}{16}, 1), (1, \frac{17}{4}, \frac{117}{16}, 0)0\right\}$.
lnspA:=NullSpace(Transpose(A)); returns $lnspA := \{\}$.
Since the left null space of A is empty, it is orthogonal to the column space of A by default.
It therefore remains to verify the orthogonality of the row space of A and the null space of A.
d11:=DotProduct(rspA[1],nspA[1]): d12:=DotProduct(rspA[1],nspA[2]):
d21:=DotProduct(rspA[2],nspA[1]): d22:=DotProduct(rspA[2],nspA[2]):
is(d11=0 and d12=0 and d21=0 and d22=0); returns *true*.

EXERCISES 8.9

Exercise 1

The orthogonal complement $S^\perp = \{(x, y) : x, y \in \mathbb{R}\}$ is the set of all vectors in $\mathbb{R}^2$ that are orthogonal to the vector $(1, 3)$.

$$\begin{bmatrix} x \\ y \end{bmatrix}^T \begin{bmatrix} 1 \\ 3 \end{bmatrix} = x + 3y = 0$$

This shows us that $S^\perp$ is the straight line determined by the equation $y = -1/3x$.

Exercise 3

The orthogonal complement $S^\perp = \{(x, y, z) : x, y, z \in \mathbb{R}\}$ is the set of all vectors in $\mathbb{R}^3$ that are orthogonal to both the vector $(1, 3, 2)$ and the vector $(2, 1, 0)$.

$$\begin{bmatrix} x \\ y \\ z \end{bmatrix}^T \begin{bmatrix} 1 \\ 3 \\ 2 \end{bmatrix} = x + 3y + 2z = 0, \quad \begin{bmatrix} x \\ y \\ z \end{bmatrix}^T \begin{bmatrix} 2 \\ 1 \\ 0 \end{bmatrix} = 2x + y = 0$$

Hence $S^{\perp}$ is the intersection of the planes given by the equation $x + 3y + 2z = 0$ and $2x + y = 0$.

Exercise 5 (*Maple 6*)

with(LinearAlgebra):

A:=Matrix([[3,1,0,2,4],[1,1,0,0,2],[5,2,0,3,7]]):

ColumnSpace(Transpose(A)); returns $\{(1, 0, 0, 1, 1), (0, 1, 0, -1, 1)]\}$

This is the same basis for $\mathbb{R}^4$ found in Example 4.44.

Exercise 7 (*Maple 6*)

We generate a 4×3 real matrix A of rank 3 and use it to construct a 4×4 invertible matrix B using bases for Col A and Nul A^T. The theorem then follows from the fact that the columns of B form a basis for $\mathbb{R}^4$ and that Col A and Nul A^T are orthogonal complements. If $\mathbf{b} \in$ Col A, then $\mathbf{b}$ is orthogonal to every vector in the nonzero subspace Nul A^T of $\mathbb{R}^4$. Hence $\langle \mathbf{b}, \mathbf{y} \rangle = 0$ for $\mathbf{y} \in$ Nul A^T. If $\mathbf{b} \notin$ Col A then $\mathbf{b} \neq \mathbf{0}$ since Col A is a subspace of $\mathbb{R}^4$ and therefore contains $\mathbf{0}$. Therefore $\mathbf{b}$ is a nonzero vector in Nul A^T. Hence it follows from the property of the standard inner product that $\mathbf{b}^T\mathbf{b} \neq 0$. Fredholm's alternative theorem therefore follows.

with(LinearAlgebra): A:=RandomMatrix(4,3,generator=1..9,
outputoptions=[shape=triangular[lower]]); returns

$$A := \begin{bmatrix} 6 & 0 & 0 \\ 8 & 5 & 0 \\ 9 & 9 & 8 \\ 1 & 8 & 4 \end{bmatrix}$$

and Rank(A); returns 3.

C:=ColumnSpace(A); returns

$$C := \left[\begin{bmatrix} 1 \\ 0 \\ 0 \\ \frac{-91}{60} \end{bmatrix}, \begin{bmatrix} 0 \\ 1 \\ 0 \\ \frac{7}{10} \end{bmatrix}, \begin{bmatrix} 0 \\ 0 \\ 1 \\ \frac{1}{2} \end{bmatrix}\right]$$

Nt:=convert(NullSpace(Transpose(A))[1],list); returns $Nt := \left[\frac{-91}{30}, \frac{7}{5}, -1, -2\right]$.

B:=Matrix([Matrix(C),Nt[1]]); returns

$$B := \begin{bmatrix} 1 & 0 & 0 & -\frac{91}{30} \\ 0 & 1 & 0 & \frac{7}{5} \\ 0 & 0 & 1 & -1 \\ -\frac{91}{60} & \frac{7}{10} & \frac{1}{2} & -2 \end{bmatrix}$$

EXERCISES 8.10

Exercise 1

$T^*(1,0) = (2,-3)$ and $T^*(0,1) = (1,1)$.

Exercise 3 (*Maple V*)

```
X:=vector(3,[x,y,z]): Y:=vector(3,[u,v,w]):
A:=matrix([[1,0,2],[4,3,0],[0,0,3]]): B:=evalm(A&*X):
evalm(B&*Y); returns (x + 2z)u + (4x + 3y)v + 3zw.
C:=evalm(X&*transpose(A)):
evalm(C&*Y) also; returns (x + 2z)u + (4x + 3y)v + 3zw.
```

EXERCISES 8.11

Exercise 1

For any orthogonal matrix, $\det A = \det A^T = \det A^{-1}$. This tells us that λ is an eigenvalue of A if and only if λ is an eigenvalue of A^{-1}. We also know that λ is an eigenvalue of A, then the corresponding eigenvalue of A^{-1} is $1/\lambda$. Therefore $\lambda = 1/\lambda$, so $\lambda^2 = 1$. Hence $\lambda = \pm 1$.

Exercise 3

Since $\mathbb{R}^n$ is finite-dimensional, $T^{**} = T$.

$\langle (T+T^*)(\mathbf{x}), \mathbf{y}\rangle = \langle T(\mathbf{x}) + T^*(\mathbf{x}), \mathbf{y}\rangle = \langle T(\mathbf{x}), \mathbf{y}\rangle + \langle T^*(\mathbf{x}), \mathbf{y}\rangle$

$\langle T(\mathbf{x}), \mathbf{y}\rangle + \langle T^*(\mathbf{x}), \mathbf{y}\rangle = \langle \mathbf{x}, T^*(\mathbf{y})\rangle + \langle \mathbf{x}, T^{**}(\mathbf{y})\rangle = \langle \mathbf{x}, T^*(\mathbf{y})\rangle + \langle \mathbf{x}, T(\mathbf{y})\rangle$

$\langle \mathbf{x}, T^*(\mathbf{y})\rangle + \langle \mathbf{x}, T(\mathbf{y})\rangle = \langle \mathbf{x}, (T^*+T)(\mathbf{y})\rangle = \langle \mathbf{x}, (T+T^*)(\mathbf{y})\rangle$

Hence $\langle (T+T^*)(\mathbf{x}), \mathbf{y}\rangle = \langle \mathbf{x}, (T+T^*)(\mathbf{y})\rangle$ for all $\mathbf{x}, \mathbf{y} \in \mathbb{R}^n$.

Exercise 5 (*Maple 6*)

Load the quaternion routine defined in Exercise 4.17.3.

```
v:=quat(0,1,-1,-1):
q:=quat(cos(Pi/4),sin(Pi/4)/sqrt(6),2*sin(Pi/4)/sqrt(6),-sin(Pi/4)/sqrt(6)):
qconj:=quat(cos(Pi/4),-sin(Pi/4)/sqrt(6),-2*sin(Pi/4)/sqrt(6),sin(Pi/4)/sqrt(6)):
w:=simplify(q&^v&^qconj); returns (-√6/2, 0, -√6/2).
```

Exercise 7 (*Maple 6*)

Load the quaternion routine defined in Exercise 4.17.3.

```
q:=quat(0,1/sqrt(6),1/sqrt(6),2/sqrt(6)):
qstar:=qconj(0,1/sqrt(6),1/sqrt(6),2/sqrt(6)):
assume(a,real): assume(b,real): assume(c,real): assume(d,real):
assume(e,real): assume(f,real): assume(g,real): assume(h,real):
assume(r,real): assume(s,real):
x:=quat(a,b,c,d): y:=quat(e,f,g,h):
T:=p-> q&^p&^ qstar:
first:=expand(collect(r*T(x),[I,J,K])):
second:=expand(collect(s*T(y),[I,J,K])):
collect(T(r*x+s*y),[I,J,K]):
is(collect(T(r*x+s*y),[I,J,K])=collect(first+second,[I,J,K])); returns true.
```

EXERCISES 8.12

Exercise 1

	Eigenvalues	Eigenvectors
a.	$-1, 3$	$(-1, 1), (1, 1)$
b.	$0, 2$	$(-1, 1), (1, 1)$
c.	$-3, 3$	$(-1, 1), (1, 1)$

In all three cases, the set $\mathcal{B} = \{(-1, 1), (1, 1)\}$ is a basis for $\mathbb{R}^2$.

Exercise 3

Eigenvalues	$\lambda_1 = 2 + 2\sqrt{2}$	$\lambda_2 = 2 - 2\sqrt{2}$
Eigenvectors	$\begin{bmatrix} 1 \\ 1+\sqrt{2} \end{bmatrix}$	$\begin{bmatrix} 1 \\ 1-\sqrt{2} \end{bmatrix}$

$$Q = \begin{bmatrix} \mathbf{u}_1 & \mathbf{u}_2 \end{bmatrix} = \begin{bmatrix} \frac{1}{\sqrt{\left(1+\left(1+\sqrt{2}\right)^2\right)}} & \frac{1}{\sqrt{\left(1+\left(-1+\sqrt{2}\right)^2\right)}} \\ \frac{\left(1+\sqrt{2}\right)}{\sqrt{\left(1+\left(1+\sqrt{2}\right)^2\right)}} & \frac{\left(1-\sqrt{2}\right)}{\sqrt{\left(1+\left(-1+\sqrt{2}\right)^2\right)}} \end{bmatrix} \quad A = A_1 + A_2$$

where

$$A_1 = \left(2+2\sqrt{2}\right) \begin{bmatrix} \frac{1}{\sqrt{\left(1+\left(1+\sqrt{2}\right)^2\right)}} \\ \frac{\left(1+\sqrt{2}\right)}{\sqrt{\left(1+\left(1+\sqrt{2}\right)^2\right)}} \end{bmatrix} \begin{bmatrix} \frac{1}{\sqrt{\left(1+\left(1+\sqrt{2}\right)^2\right)}} \\ \frac{\left(1+\sqrt{2}\right)}{\sqrt{\left(1+\left(1+\sqrt{2}\right)^2\right)}} \end{bmatrix}^T$$

and

$$A_2 = \left(2-2\sqrt{2}\right) \begin{bmatrix} \frac{1}{\sqrt{\left(1+\left(-1+\sqrt{2}\right)^2\right)}} \\ \frac{\left(1-\sqrt{2}\right)}{\sqrt{\left(1+\left(-1+\sqrt{2}\right)^2\right)}} \end{bmatrix} \begin{bmatrix} \frac{1}{\sqrt{\left(1+\left(-1+\sqrt{2}\right)^2\right)}} \\ \frac{\left(1-\sqrt{2}\right)}{\sqrt{\left(1+\left(-1+\sqrt{2}\right)^2\right)}} \end{bmatrix}^T$$

Exercise 5

Let A be an $n \times n$ real symmetric matrix. We know from the spectral theorem (Corollary 8.35) that $A = QDQ^T$, for some diagonal matrix D and an associated orthogonal matrix Q. The diagonal entries of D are the eigenvalues of A and the columns of Q are the eigenvectors associated with the eigenvalues in the corresponding columns in D. It is also clear from the construction of characteristic polynomials that the algebraic multiplicity of an eigenvalue λ of D correspond to the number of repetitions of λ on the diagonal of D. We also know from Theorem 6.7 that the dimension of the eigenspace E_λ is less than or equal to the number of repetitions of λ on the diagonal of D. This number is the same as the number of occurrences of eigenvectors in Q that belong to λ. Theorem 4.3 tells us that since Q is invertible, its columns are linearly independent. Hence the eigenvectors in Q corresponding to the same eigenvalue λ form a linearly independent set of vectors that is larger than or equal to the dimension of E_λ. They therefore constitute a basis for E_λ. Hence the algebraic multiplicity of λ equals its geometric multiplicity.

Exercise 7 (*Maple 6*)

with(LinearAlgebra):

a. A:=Matrix([[4,3],[3,2]]): evecsA:=Eigenvectors(A):

evecsA[1]; returns

$$\begin{bmatrix} 3+\sqrt{10} \\ 3-\sqrt{10} \end{bmatrix}$$

diagA:=DiagonalMatrix([evecsA[1][1],evecsA[1][2]]):

c1:=Column(evecsA[2],1): c2:=Column(evecsA[2],2):

u:=Normalize(c1,Euclidean): v:=Normalize(c2,Euclidean):

P:=Matrix([u,v]): IsOrthogonal(P); returns *true*

simplify(evalm(P&*diagA&*MatrixInverse(P))); returns

$$\begin{bmatrix} 4 & 3 \\ 3 & 2 \end{bmatrix}$$

b. Similar to a.

c. C:=Matrix([[-1,2,2],[2,-1,2],[2,2,-1]]):

evecsC:=Eigenvectors(C); returns

$$evecsC := \left[\begin{bmatrix} -3 \\ -3 \\ 3 \end{bmatrix} \begin{bmatrix} -1 & -1 & 1 \\ 1 & 0 & 1 \\ 0 & 1 & 1 \end{bmatrix}\right]$$

diagC:=DiagonalMatrix([-3,-3,3]):

gsC:=GramSchmidt([Column(evecsC[2],1),Column(evecsC[2],2)]); returns

$$gsC := \left[\begin{bmatrix} -1 \\ 1 \\ 0 \end{bmatrix}, \begin{bmatrix} \frac{-1}{2} \\ \frac{-1}{2} \\ 1 \end{bmatrix}\right]$$

ngsC1:=VectorNorm(gsC[1],Euclidean): ngsC2:=VectorNorm(gsC[2],Euclidean):
u:=VectorScalarMultiply(gsC[1],1/ngsC1); returns $u := (-\frac{1}{2}\sqrt{2}, \frac{1}{2}\sqrt{2},)0$.
v:=VectorScalarMultiply(gsC[2],1/ngsC2); returns $V := (-\frac{1}{6}\sqrt{6}, -\frac{1}{6}\sqrt{6}, \frac{1}{3}\sqrt{6})$.
c3:=Column(evecsC[2],3):
w:=VectorScalarMultiply(c3,1/Norm(c3,Euclidean)); returns $(\frac{1}{3}\sqrt{3}, \frac{1}{3}\sqrt{3}, \frac{1}{3}\sqrt{3})$.
P:=Matrix([u,v,w]): IsOrthogonal(P); returns *true*.
Equal(B,P.(diagB.MatrixInverse(P))); returns *true*.

Exercise 9 (*Maple 6*)

We show that A fails to have four linearly independent eigenvectors.
with(LinearAlgebra): A:=Matrix([[0,0,0,4],[1,0,0,11],[0,1,0,9],[0,0,1,1]]):
Eigenvectors(A); returns

$$\begin{bmatrix} -1 \\ -1 \\ -1 \\ 4 \end{bmatrix}, \begin{bmatrix} -4 & 0 & 0 & 1 \\ -7 & 0 & 0 & 3 \\ -2 & 0 & 0 & 3 \\ 1 & 0 & 0 & 1 \end{bmatrix}$$

EXERCISES 8.13

Exercise 1

$$\begin{bmatrix} 1 \\ 1 \end{bmatrix}^T \begin{bmatrix} 1 \\ 1 \end{bmatrix} [x] = \begin{bmatrix} 1 \\ 1 \end{bmatrix}^T \begin{bmatrix} 1 \\ 2 \end{bmatrix}$$

This equation reduces to the numerical equation $2x = 3$. Hence $x = 3/2$.
The least-squares solution of the normal equation is the arithmetic mean of 1 and 2.

Exercise 3 (*Maple V*)

A:=matrix(5,1,[1,1,1,1,1]): b:=vector(5,[1,2,5,6,7]):
linalg[leastsqrs](A,b); returns $\frac{21}{5}$.

Exercise 5 (*Maple V*)

with(linalg): A:=matrix(3,2,[1,1,1,-1,2,1]): b:=vector(3,[1,2,7]):
linsolve(evalm(transpose(A)&*A),evalm(transpose(A)&*b)); returns $\left[\frac{39}{14}, \frac{1}{7}\right]$.
We confirm the result using the **leastsqrs** function.
leastsqrs(A,b); returns $\left[\frac{39}{14}, \frac{1}{7}\right]$.

Exercise 7 (*Maple V*)

with(linalg): A:=matrix([[1,-2],[2,1]]): b:=vector(2,[-7,5]),
linsolve(A,b); returns $\left[\frac{3}{5}, \frac{19}{5}\right]$. leastsqrs(A,b); returns $\left[\frac{3}{5}, \frac{19}{5}\right]$.

Exercise 9 (*Maple V*)

A:=matrix([[1,3,0],[0,-1,8],[1,2,4]]): b:=vector(3,[1,2,3]):

evalm((inverse(transpose(A)&*A)&*(transpose(A)&*b))); returns $[7, -2, 0]$.

linsolve(A,b); returns $[7, -2, 0]$.

Exercise 11 (*Maple V*)

with(stats): Xvalues:=[1,2,3,4,5]: Yvalues:=[6,9,10,21,30]:

fit[leastsquare[[x,y],y=a+b*x+c*x^2+d*x^3+e*x^4,{a,b,c,d,e}]]([Xvalues, Yvalues]); returns

$$y = -35 + 78x - 48x^2 + 12x^3 - x^4$$

CHAPTER 9 - Singular Values and Singular Vectors

EXERCISES 9.1

Exercise 1

$$A^T A = \begin{bmatrix} a & b \\ c & d \end{bmatrix}^T \begin{bmatrix} a & b \\ c & d \end{bmatrix} = \begin{bmatrix} a^2 + c^2 & ab + cd \\ ab + cd & b^2 + d^2 \end{bmatrix}$$

and

$$AA^T = \begin{bmatrix} a & b \\ c & d \end{bmatrix} \begin{bmatrix} a & b \\ c & d \end{bmatrix}^T = \begin{bmatrix} a^2 + b^2 & ac + bd \\ ac + bd & c^2 + d^2 \end{bmatrix}$$

If these matrices are equal, then $a^2 + c^2 = a^2 + b^2$.

This implies that $c = \pm b$. Moreover, $ab + cd = ac + bd$.

We have two cases.

If $c = b$, then $ab + bd = ab + bd$ holds for all a, b, d.

If $c = -b$, then $ab - bd = -ab + bd$, so $2ab = 2bd$. This implies that $a = d$.

Hence $A^T A = AA^T$ provided that $a = d$ and $c = \pm b$.

Exercise 3

Let A be an orthogonal matrix. Then $A^T = A^{-1}$. Moreover, the singular values of A are the eigenvalues of the matrix $A^T A = A^{-1} A = I$. Since the only eigenvalue of an identity matrix is 1, the result follows.

Exercise 5 (*Maple V*)

a. with(linalg): A:=matrix([[0,1],[-1,0]]):

eigenvalues(transpose(A)&*A); returns 1, 1. eigenvalues(A&*transpose(A)); returns 1, 1.

b. A:=matrix([[0,1,0],[-1,0,0]]):

eigenvalues(transpose(A)&*A); returns 1, 1, 0. eigenvalues(A&*transpose(A)); returns 1, 1.

c. A:=matrix([[0],[-1],[0]]):

eigenvalues(transpose(A)&*A); returns 1. eigenvalues(A&*transpose(A)); returns 0, 0, 1.

Exercise 7 (*Maple 6*)
with(LinearAlgebra):
a. A:=Matrix([[-1,-2,0],[1,-1,1],[-2,1.,-1]]):
mnA:=MatrixNorm(A,Euclidean); returns $mnA := 2.976212710.$
svA:=SingularValues(A); returns

$$svA := \begin{bmatrix} 2.97621271010417888 \\ 2.24784029025438858 \\ .298951390234940216 \end{bmatrix}$$

mnA-svA[1,1]; returns 0.
b. A:=Matrix([[-2,2,-2,1],[0,-1,0,-2],[0,0,-1,2.],[0,0,0,0]]):
mnA:=MatrixNorm(A,Euclidean); returns $mnA := 4.123105626.$
svA:=SingularValues(A); returns

$$svA := \begin{bmatrix} 4.12310562561766059 \\ 2.23606797749978981 \\ .99999999999999978 \\ 0. \end{bmatrix}$$

mnA-svA[1,1]; returns 0.
c. A:=Matrix([[0,2,9.],[-2,0,0],[-9,0,0]]):
mnA:=MatrixNorm(A,Euclidean); returns $mnA := 9.219544457.$
svA11:=evalf(SingularValues(A),11); returns

$$svA11 := \begin{bmatrix} 9.2195444574 \\ 9.2195444569 \\ .10846522892\ 10^{-9} \end{bmatrix}$$

svA10:=evalf(SingularValues(A),10); returns

$$svA10 := \begin{bmatrix} 9.219544461 \\ 9.219544461 \\ .2169304578\ 10^{-9} \end{bmatrix}$$

We compare how *Maple* treats the equality of these absolute values.
a:=evalf(svA11[1],9); returns $a := 9.21954446.$
b:=evalf(svA10[1],9); returns $b := 9.21954446.$
is(a=b); returns $true.$
c:=evalf(svA11[1],10); returns $c := 9.219544457.$
d:=evalf(svA10[1],10); returns $d := 9.219544461.$
is(c=d); returns $false.$

Maple treats the approximate singular values of A calculated to nine decimal places as equal, but as unequal if calculated to ten decimal places.

EXERCISES 9.2

Exercise 1

Let A be a positive definite matrix. Then A is real symmetric, and by Theorem 8.35 there therefore exists a diagonal matrix D and an orthogonal matrix Q such that $A = QDQ^T$.
Let $U = Q = V$. Then $A = UDV^T$. Moreover,
$A^TA = (QDQ^T)^T QDQ^T = QDQ^TQDQ^T = QD^2Q^T$.
Therefore the singular values of A are the square roots of the diagonal entries of D^2.
Since A is positive definite, these roots are the diagonal entries of D.

Exercise 3

Let $A = UDV^T$ be a singular value decomposition of A. Then $A^T = VDU^T$ and $A^TU = VD$.
By Theorem 2.1, for $1 \le i \le r$, the ith column of A^TU is $A^T\mathbf{u}_i$ and the ith column of VD is $\sigma_i\mathbf{v}_i$ since D is diagonal. Therefore $A^T\mathbf{u}_i = \sigma_i\mathbf{v}_i$ for all $1 \le i \le r$.

Exercise 5

Suppose that A is a 3×4 matrix whose singular value decomposition is UDV^T, with

$$UD = \begin{bmatrix} a & c & 0 \\ b & d & 0 \\ 0 & 0 & 1.0 \end{bmatrix} \begin{bmatrix} \sigma_1 & 0 & 0 & 0 \\ 0 & \sigma_2 & 0 & 0 \\ 0 & 0 & 0 & 0 \end{bmatrix} = \begin{bmatrix} a\sigma_1 & c\sigma_2 & 0 & 0 \\ b\sigma_1 & d\sigma_2 & 0 & 0 \\ 0 & 0 & 0 & 0 \end{bmatrix}$$

and

$$UDV^T = \begin{bmatrix} a\sigma_1 & c\sigma_2 & 0 & 0 \\ b\sigma_1 & d\sigma_2 & 0 & 0 \\ 0 & 0 & 0 & 0 \end{bmatrix} \begin{bmatrix} v_{11} & v_{12} & v_{13} & v_{14} \\ v_{21} & v_{22} & v_{23} & v_{24} \\ v_{31} & v_{32} & v_{33} & v_{34} \\ v_{41} & v_{42} & v_{43} & v_{44} \end{bmatrix}^T$$

$$= \begin{bmatrix} a\sigma_1 v_{11} & a\sigma_1 v_{21} & a\sigma_1 v_{31} & a\sigma_1 v_{41} \\ b\sigma_1 v_{11} & b\sigma_1 v_{21} & b\sigma_1 v_{31} & b\sigma_1 v_{41} \\ 0 & 0 & 0 & 0 \end{bmatrix} + \begin{bmatrix} c\sigma_2 v_{12} & c\sigma_2 v_{22} & c\sigma_2 v_{32} & c\sigma_2 v_{42} \\ d\sigma_2 v_{12} & d\sigma_2 v_{22} & d\sigma_2 v_{32} & d\sigma_2 v_{42} \\ 0 & 0 & 0 & 0 \end{bmatrix}$$

Then

$$UDV^T = \sigma_1 \begin{bmatrix} av_{11} & av_{21} & av_{31} & av_{41} \\ bv_{11} & bv_{21} & bv_{31} & bv_{41} \\ 0 & 0 & 0 & 0 \end{bmatrix} + \sigma_2 \begin{bmatrix} cv_{12} & cv_{22} & cv_{32} & cv_{42} \\ dv_{12} & dv_{22} & dv_{32} & dv_{42} \\ 0 & 0 & 0 & 0 \end{bmatrix}$$

$$= \sigma_1 \begin{bmatrix} a \\ b \\ 0 \end{bmatrix} \begin{bmatrix} v_{11} & v_{21} & v_{31} & v_{41} \end{bmatrix} + \sigma_2 \begin{bmatrix} c \\ d \\ 0 \end{bmatrix} \begin{bmatrix} v_{12} & v_{22} & v_{32} & v_{42} \end{bmatrix}$$

$$= \sigma_1 \mathbf{u}_1 \mathbf{v}_1^T + \sigma_2 \mathbf{u}_2 \mathbf{v}_2^T$$

The general case is analogous.

Exercise 7 (*Maple V*)

with(linalg):

a. A:=matrix([[2,-3],[-3,-3]]): svA:=evalf(Svd(A)):
sqrt(svA[1]^2+svA[2]^2); returns 5.567764363.

b. B:=matrix([[9,0,-2],[-2,0,1]]): svB:=evalf(Svd(B)):
sqrt(svB[1]^2+svB[2]^2); returns 9.486832980

c. C:=matrix([[5,3,3],[-1,5,-1],[-2,-1,3]]): svC:=evalf(Svd(C)):
sqrt(svC[1]^2+svC[2]^2+svC[3]^2); returns 9.165151390

Exercise 9 (*Maple V*)

a. with(linalg):
A:=matrix([[2,-3],[-3,-3]]):
svA:=evalf(Svd(A,U,V)); returns $svA := [4.405124838, 3.405124838]$.
diagA:=diag(svA[1],svA[2]); returns

$$diagA := \begin{bmatrix} 4.405124838 & 0 \\ 0 & 3.405124838 \end{bmatrix}$$

B:=evalf(evalm(V&*inverse(diagA)&*transpose(U)),4); returns

$$B := \begin{bmatrix} .2000 & -.2000 \\ -.2000 & -.1333 \end{bmatrix}$$

evalm(A&*B); returns

$$\begin{bmatrix} 1.0000 & -.0001 \\ 0. & .9999 \end{bmatrix}$$

As we can see, the matrix AB is approximately equal to the 2×2 identity matrix. The inverses of the matrices (b) and (c) can be approximated in the same way.

Exercise 11 (*Maple 6*)

with(LinearAlgebra):
A:=evalf(RandomMatrix(4,density=0.75,generator=0..0.5),2); returns

$$A := \begin{bmatrix} .28 & .40 & .14 & 0. \\ .25 & .48 & .12 & .22 \\ .29 & .25 & .49 & .088 \\ .34 & .18 & .18 & .15 \end{bmatrix}$$

Determinant(A); returns 8.1489×10^{-3}. Hence A is invertible.
svA:=SingularValues(A); returns

$$svA := \begin{bmatrix} 1.02374528801233611 \\ .339188141121371632 \\ .169867419272460046 \\ .138152271298308704 \end{bmatrix}$$

As we can see, the singular values are nonzero.
B:=evalf(RandomMatrix(4,4),3); returns

$$B := \begin{bmatrix} -84. & 50. & -13. & 51. \\ -22. & 90. & 51. & 15. \\ 72. & 22. & -60. & 44. \\ -61. & 63. & 74. & 12. \end{bmatrix}$$

Determinant(B); returns 1.6646×10^{7}. Hence B is invertible.
svB:=SingularValues(B); returns

$$svB := \begin{bmatrix} 177.774222240681894 \\ 102.523541065364768 \\ 77.4347449912877011 \\ 11.7944776176540582 \end{bmatrix}$$

As we can see, the singular values are nonzero.
g:=x->evalf(rand(),3): C:=RandomMatrix(4,generator=g); returns

$$C := \begin{bmatrix} .321\ 10^{11} & .723\ 10^{12} & .604\ 10^{12} & .746\ 10^{12} \\ .260\ 10^{12} & .310\ 10^{12} & .797\ 10^{12} & .392\ 10^{11} \\ .884\ 10^{11} & .960\ 10^{12} & .813\ 10^{12} & .454\ 10^{12} \\ .644\ 10^{12} & .921\ 10^{12} & .951\ 10^{12} & .146\ 10^{12} \end{bmatrix}$$

Determinant(C); returns -1.8229×10^{-4}. Hence C is invertible.
svC:=SingularValues(C); returns

$$svC := \begin{bmatrix} 2364098559649.50390 \\ 718506491388.516846 \\ 299823938346.867310 \\ 188739603925.293916 \end{bmatrix}$$

The singular values are nonzero.

EXERCISES 9.3

Exercise 1

Suppose that A is an $m \times n$ real matrix of rank $r > 0$ and that $A = UDV^T$ is a singular value decomposition of A. Then D is an $m \times n$ generalized diagonal matrix and with r nonzero diagonal entries and

$$A^T = \left(UDV^T\right)^T = VD^TU^T$$

The product VD^TU^T is a singular values decomposition of A^T since V and U are orthogonal and since D^T is a generalized diagonal matrix. Since D and D^T have the same r nonzero diagonal entries, the matrices A and A^T have the same singular values.

Exercise 3 (*Maple 6*)

with(LinearAlgebra):

A:=Matrix([[1,2,4],[0,0,1],[3,6,12.]]):

S := evalf(SingularValues(A,output=list),5); returns $S := [14.518, .48705, .00013418]$

U,Vt:=evalf(SingularValues(A,output=['U','Vt']),5):

evalf(A.Column(Transpose(Vt),1),5); returns $(-4.5827, -.87389, -13.748)$.

evalf(S[1]*Column(U,1),5); returns $(-4.582, -.87394, -13.748)$.

evalf(A.Column(Transpose(Vt),2),5); returns $(.0095, -.48616, .0284)$.

evalf(S[2]*Column(U,2),5); returns $(.0090479, -.48617, .027888)$.

evalf(A.Column(Transpose(Vt),3),5); returns $(.00014, 0., .0004)$.

evalf(S[3]*Column(U,3),5); returns $(-.00012729, .64414\ 10^{-7},).000042430$.

The values $A\mathbf{v}_i$ and $\sigma_i\mathbf{u}_i$ are relatively close but are not equal, due to round-off errors.

Exercise 5 (*Maple 6*)

with(LinearAlgebra):

A:=Matrix([[0,1,-1,1],[-1,2,0,1.]]):

Rank(A); returns 2.

Vt:=evalf(SingularValues(A,output=['Vt']),5):

RowSpace(A); returns $[[1., 0., -2., 1.], [0., 1., -1., 1.]]$.

V:=Transpose(Vt); returns

$$\begin{bmatrix} -0.30359 & -0.49113 & -0.57737 & 0.57737 \\ 0.79472 & 0.18753 & -0.5773 & -0.00002 \\ -0.18757 & 0.79466 & 0 & 0.57738 \\ 0.49114 & -0.30357 & 0.57738 & 0.57732 \end{bmatrix}$$

evalf(Transpose(Column(V,1)).Column(V,2),3); returns 0.

evalf(Transpose(Column(V,1)).Column(V,1),3); returns 1.00.

evalf(Transpose(Column(V,2)).Column(V,2),3); returns .999.

The first two columns of V are an approximately orthonormal basis for the columns space of A^T.

The last two columns of V form an approximately orthonormal basis for the null space of A.

EXERCISES 9.4

Exercise 1 (*Maple 6*)

with(LinearAlgebra):
A:=Matrix([[1,2],[3,4],[5,6]]):
B:=MatrixInverse(Transpose(A).A).Transpose(A):
C:=evalf(B,4); returns

$$C := \begin{bmatrix} -1.333 & -.3333 & .6667 \\ 1.083 & .3333 & -.4167 \end{bmatrix}$$

with(linalg):
Am:=matrix([[1,2],[3,4],[5,6]]):
svd:=evalf(Svd(Am,U,V)); returns svd := [9.525518092, .5143005807].
Vr:=evalm(V); returns

$$Vr := \begin{bmatrix} -.6196294838 & -.7848944533 \\ -.7848944533 & .6196294838 \end{bmatrix}$$

Dr:=diag(1/svd[1],1/svd[2]); returns

$$Dr := \begin{bmatrix} .1049811664 & 0 \\ 0 & 1.944388238 \end{bmatrix}$$

Urt:=submatrix(evalm(transpose(U)),1..2,1..3); returns

$$Urt := \begin{bmatrix} -.2298476964 & -.5247448188 & -.8196419411 \\ .8834610177 & .2407824921 & -.4018960334 \end{bmatrix}$$

Aplus:=evalf(evalm(Vr&*Dr&*Urt),4); returns

$$Aplus := \begin{bmatrix} -1.333 & -.3334 & .6666 \\ 1.084 & .3334 & -.4168 \end{bmatrix}$$

The matrices $Aplus$ and C are approximately equal. The formula fails for A^T since AA^T is not invertible.

Exercise 3 (*Maple 6*)

By Theorem 9.6, the vector $A^+\mathbf{b}$ is a least-squares solution of the system $A\mathbf{x} = \mathbf{b}$.

with(LinearAlgebra):
A:=Matrix([[1,2,0,1,4.],[0,3,0,2,0],[1,1,0,0,1]]): b:=< 1,2,3>:
rd:=A-> RowDimension(A):
cd:=A-> ColumnDimension(A):
U,S,Vt:=SingularValues(A,output=['U','S','Vt']):
V:=Transpose(Vt):
U.DiagonalMatrix(S[1..Rank(A)],cd(U),rd(V)).Vt:

```
Vr:=SubMatrix(V,1..rd(V),1..Rank(A)):
dsvA:=DiagonalMatrix(S[1..Rank(A)],Rank(A),Rank(A)):
Ur:=SubMatrix(U,1..rd(U),1..Rank(A)):
Aplus:=Vr.MatrixInverse(dsvA).Transpose(Ur):
convert(evalf(Aplus.b,5),list); returns
```

$$[2.4072, 1.3652, 0., -1.0479, -.77245]$$

Exercise 5 (*Maple 6*)

```
with(LinearAlgebra):
a. A:=Matrix([[-1,1,0],[1,-1,2],[-2,1,0]]):
rd:=A-> RowDimension(A): cd:=A-> ColumnDimension(A):
U,S,Vt:=SingularValues(A,output=['U','S','Vt']):
V:=Transpose(Vt):
U.DiagonalMatrix(S[1..Rank(A)],cd(U),rd(V)).Vt:
Vr:=SubMatrix(V,1..rd(V),1..Rank(A)):
dsvA:=DiagonalMatrix(S[1..Rank(A)],Rank(A),Rank(A)):
Ur:=SubMatrix(U,1..rd(U),1..Rank(A)):
Aplus:=evalf(Vr.MatrixInverse(dsvA).Transpose(Ur)); returns
```

$$\begin{bmatrix} 1.0000 & .7\ 10^{-5} & -1.0000 \\ 2.0000 & -.2\ 10^{-5} & -.99997 \\ .50000 & .50000 & -.00001 \end{bmatrix}$$

```
b. A:=Matrix([[-1,1,-2],[1,-1,2],[-2,1,0]]):
U,S,Vt:=SingularValues(A,output=['U','S','Vt']):
V:=Transpose(Vt):
U.DiagonalMatrix(S[1..Rank(A)],cd(U),rd(V)).Vt:
Vr:=SubMatrix(V,1..rd(V),1..Rank(A)):
dsvA:=DiagonalMatrix(S[1..Rank(A)],Rank(A),Rank(A)):
Ur:=SubMatrix(U,1..rd(U),1..Rank(A)):
Aplus:=evalf(Vr.MatrixInverse(dsvA).Transpose(Ur),5); returns
```

$$Aplus := \begin{bmatrix} .00001 & -1.0000 & -1.0000 \\ 1.0000 & -.99998 & -1.0000 \\ .50000 & .49999 & 0. \end{bmatrix}$$

```
c. A:=Matrix([[3,2],[2,1],[7,0]]):
U,S,Vt:=SingularValues(A,output=['U','S','Vt']):
V:=Transpose(Vt):
U.DiagonalMatrix(S[1..Rank(A)],cd(U),rd(V)).Vt:
Vr:=SubMatrix(V,1..rd(V),1..Rank(A)):
dsvA:=DiagonalMatrix(S[1..Rank(A)],Rank(A),Rank(A)):
```

Ur:=SubMatrix(U,1..rd(U),1..Rank(A)):
Aplus:=evalf(Vr.MatrixInverse(dsvA).Transpose(Ur),5); returns

$$Aplus := \begin{bmatrix} -.004066 & .008129 & .14228 \\ .40651 & .18700 & -.22765 \end{bmatrix}$$

d. A:=Matrix([[-2,2,0,0],[-2,-1,1,1],[-1,0,1,1]]):
U,S,Vt:=SingularValues(A,output=['U','S','Vt']):
V:=Transpose(Vt):
U.DiagonalMatrix(S[1..Rank(A)],cd(U),rd(V)).Vt:
Vr:=SubMatrix(V,1..rd(V),1..Rank(A)):
dsvA:=DiagonalMatrix(S[1..Rank(A)],Rank(A),Rank(A)):
Ur:=SubMatrix(U,1..rd(U),1..Rank(A)):
Aplus:=evalf(Vr.MatrixInverse(dsvA).Transpose(Ur),5); returns

$$Aplus := \begin{bmatrix} -.25000 & -.49999 & .49999 \\ .25000 & -.49999 & .49999 \\ -.12500 & -.24999 & .74999 \\ -.12500 & -.24999 & .74999 \end{bmatrix}$$

Exercise 7 (*Maple 6*)
with(LinearAlgebra):
A:=RandomMatrix(3,3,outputoptions=[datatype=float]); returns

$$A := \begin{bmatrix} -21. & -50. & -79. \\ -56. & 30. & -71. \\ -8. & 62. & 28. \end{bmatrix}$$

invA:=evalf(MatrixInverse(A),5); returns

$$invA := \begin{bmatrix} .13633 & -.090976 & .15396 \\ .055554 & -.031731 & .076283 \\ -.084059 & .044266 & -.089206 \end{bmatrix}$$

U,S,Vt:=SingularValues(A,output=['U','S','Vt']):
svD:=DiagonalMatrix(S[1..3],3,3):
evalf(Transpose(Vt).MatrixInverse(svD).Transpose(U),5); returns

$$\begin{bmatrix} .13635 & -.090983 & .15398 \\ .055559 & -.031733 & .076289 \\ -.084069 & .044272 & -.089219 \end{bmatrix}$$

EXERCISES 9.5

Exercise 1

Consider the matrix

$$A = \begin{bmatrix} .4 & 0 & 0 & 0 \\ 0 & .3 & 0 & 0 \\ 0 & 0 & .2 & 0 \\ 0 & 0 & 0 & .1 \end{bmatrix}$$

Since A is a positive definite matrix, its eigenvalues coincide with its singular values. Hence its largest singular value is 0.4. If we specify a tolerance of 0.5, then all singular values are eliminated from A and the effective rank of the resulting matrix is 0.

Exercise 3 (*Maple 6*)

We know from Exercise 9.2.2 that the singular values and the eigenvalues of a positive definite matrix coincide. We also know from Theorem 8.36 that a real symmetric matrix is positive definite if its eigenvalues are positive. Hence it suffices to use the given values as the diagonal entries of a diagonal matrix. This means that the matrix

$$A = \begin{bmatrix} 5. & 0 & 0 \\ 0 & .9 \times 10^{-6} & 0 \\ 0 & 0 & .8 \times 10^{-6} \end{bmatrix}$$

satisfies the required conditions.

with(LinearAlgebra): A:=Matrix([[5.,0,0],[0,.9*10^(-6),0],[0,0,.8*10^(-6)]]); returns

$$A := \begin{bmatrix} 5. & 0 & 0 \\ 0 & .9000000000\ 10^{-6} & 0 \\ 0 & 0 & .8000000000\ 10^{-6} \end{bmatrix}$$

U,S,Vt:=SingularValues(A,output=['U','S','Vt']):
B:=U.DiagonalMatrix(S[1..2],3,3).Vt; returns

$$B := \begin{bmatrix} 5. & 0. & 0. \\ 0. & 899999999999999960.\ 10^{-6} & 0. \\ 0. & 0. & 0. \end{bmatrix}$$

Rank(B); returns 2.

EXERCISES 9.6

Exercise 1 (*Maple 6*)

with(LinearAlgebra): A:=Matrix([[255,0,255],[0,255,0],[255,0,255.]]):
U,S,Vt:=SingularValues(A,output=['U','S','Vt']):

s1:=S[1]; returns $s1 := 510.$
u1:=Column(U,1): V:=Transpose(Vt): v1:=Column(V,1):
A1:=evalf(s1.u1.Transpose(v1),5); returns

$$A1 := \begin{bmatrix} 255.00 & 0. & 255.00 \\ 0. & 0. & 0. \\ 255.00 & 0. & 255.00 \end{bmatrix}$$

evalf(A-A1,5); returns

$$\begin{bmatrix} 0. & 0. & 0. \\ 0. & 255. & 0. \\ 0. & 0. & 0. \end{bmatrix}$$

evalf(MatrixNorm(A-A1,2),3); returns 255.
s2:=S[2]; returns $s2 := 255.$
u2:=Column(U,2): v2:=Column(V,2):
A2:=evalf(s1.u1.Transpose(v1)+s2.u2.Transpose(v2),5); returns

$$A2 := \begin{bmatrix} 255.00 & 0. & 255.00 \\ 0. & 255. & 0. \\ 255.00 & 0. & 255.00 \end{bmatrix}$$

Exercise 3 (*Maple 6*)
with(LinearAlgebra): p:=(t-2)^3*(t+1)^4: A:=Matrix(evalf(CompanionMatrix(p,t))):
Next we find a singular value decomposition of A and calculate the terms $\sigma_1\mathbf{u}_1\mathbf{v}_1^T$, $\sigma_2\mathbf{u}_2\mathbf{v}_2^T$ and $\sigma_3\mathbf{u}_3\mathbf{v}_2^T$.
U,S,Vt:=SingularValues(A,output=['U','S','Vt']):
evalf(U.DiagonalMatrix(S[1..7],7,7).Vt,3):
s1:=S[1]: u1:=Column(U,1): V:=Transpose(Vt): v1:=Column(V,1):
K1:=evalf(s1.u1.Transpose(v1),3):
s2:=S[2]: u2:=Column(U,2): v2:=Column(V,2):
K2:=evalf(s2.u2.Transpose(v2),3):
s3:=S[3]: u3:=Column(U,3): v3:=Column(V,3):
K3:=evalf(s3.u3.Transpose(v3),3):
Next we calculate the matrix norms $\|A - A_1\|_2$, $\|A - A_2\|_2$, and $\|A - A_3\|_2$ and compare them with σ_2, σ_3, and σ_4.
A1:=K1: A2:=evalf(K1+K2,3): A3:=evalf(K1+K2+K3,3):
evalf(s2,4)-evalf(MatrixNorm(A-A1,2),4); returns 0.
evalf(s3,4)-evalf(MatrixNorm(A-A2,2),4); returns 0.
evalf(s4,3)-evalf(MatrixNorm(A-A3,2),3); returns 0.
This shows that for $k = 1, 2, 3$, $\|A - A_k\|_2 \approx \sigma_{k+1}$.

EXERCISES 9.7

Exercise 1 (*Maple 6*)

We use the fact that if A is a square matrix and UDV^T is a singular value decomposition of A, then UV^T is the orthogonal matrix closest to A in the Frobenius norm.

with(LinearAlgebra): A:=Matrix([[.7075,-.7071],[.7072,.7079]]):
U,Vt:=SingularValues(A,output=['U','Vt']): Q:=evalf(U.Vt,5); returns

$$Q := \begin{bmatrix} .70739 & -.70683 \\ .70683 & .70739 \end{bmatrix}$$

evalf(Norm(A-Q,Frobenius),5); returns .00069426.

EXERCISES 9.8

Exercise 1 (*Maple 6*)

with(LinearAlgebra): cn:=(A,type)-> ConditionNumber(A,type):
a. A:=Matrix([[4,0,0],[0,5,0],[0,0,4]]):
[cn(A,infinity),cn(A,Euclidean),cn(A,Frobenius)]; returns [5/4, 5/4, $1/20\sqrt{66}\sqrt{57}$].
b. B:=Matrix([[4,0,0],[2,5,0.],[3,2,4]]):
[cn(B,infinity),cn(B,Euclidean),cn(B,Frobenius)]; returns

$$[2.250000000, 2.422243829, 3.884464249]$$

c. C:=Matrix([[1/sqrt(2),1/sqrt(2),0],[-1/sqrt(2),1/sqrt(2),0],[0,0,1]]):
[cn(C,infinity),cn(C,Euclidean),cn(C,Frobenius)]; returns [2, 1, 3].

Exercise 3 (*Maple 6*)

with(LinearAlgebra):
A:=<<9,4,7>|< 7,8,1>|<7,5,2.> >:
U,S,Vt:=SingularValues(A,output=['U','S','Vt']):
Rank(A); returns 3.
B:=evalf(Transpose(Vt).MatrixInverse(DiagonalMatrix(S[1..3],3,3)).Transpose(U),20):
evalf(A.B,4); returns

$$\begin{bmatrix} 1.000 & 0. & .001 \\ 0. & .999 & .001 \\ 0. & 0. & 1.000 \end{bmatrix}$$

ConditionNumber(A,Euclidean); returns 21.47230457.
ConditionNumber(B,Euclidean); returns 21.47230457.

Exercise 5 (*Maple 6*)
We attempt to compute the determinant of A.
We know from Chapter 3, that $\det A = 0.0000000000000001 \neq 0$.
However, *Maple* produces the value 0 and thinks that A is a singular matrix.
A:=Matrix([[1.0000000000000001,1],[1,1]]):
LinearAlgebra[Determinant](A); returns 0.

Index